Coordination Chemistry

Volume 2

Coordination Chemistry Volume 2

ARTHUR E. MARTELL, Editor

Texas A&M University
College Station, Texas

ACS Monograph **174**

AMERICAN CHEMICAL SOCIETY

WASHINGTON, D. C. 1978

Library of Congress CIP Data

Coordination Chemistry.
(ACS monograph 174; ISSN 0065-7719)

Includes bibliographies and indexes.

1. Coordination compounds.
I. Martell, Arthur Earl, 1916– , ed. II. Series: American Chemical Society. ACS monograph 174
QD471.C63 541'.2242 74-151255

ISBN 0-8412-0292-3 (v. 2) ACMOAG 174 1-636 (1978)

PRINTED IN THE UNITED STATES OF AMERICA

GENERAL INTRODUCTION

American Chemical Society's Series of Chemical Monographs

By arrangement with the interallied Conference of Pure and Applied Chemistry, which met in London and Brussels in July 1919, the American Chemical Society undertook the production and publication of Scientific and Technologic Monographs on chemical subjects. At the same time it was agreed that the National Research Council, in cooperation with the American Chemical Society and the American Physical Society, should undertake the production and publication of Critical Tables of Chemical and Physical Constants. The American Chemical Society and the National Research Council mutually agreed to care for these two fields of chemical progress.

The Council of the American Chemical Society, acting through its Committee on National Policy, appointed editors and associates (the present list of whom appears at the close of this sketch) to select authors of competent authority in their respective fields and to consider critically the manuscripts submitted. Since 1944 the Scientific and Technologic Monographs have been combined in the Series. The first Monograph appeared in 1921, and up to 1972, 168 treatises have enriched the Series.

These Monographs are intended to serve two principal purposes: first to make available to chemists a thorough treatment of a selected area in form usable by persons working in more or less unrelated fields to the end that they may correlate their own work with a larger area of physical science; secondly, to stimulate further research in the specific field treated. To implement this purpose the authors of Monographs give extended references to the literature.

American Chemical Society
F. Marshall Beringer
Editor of Monographs

Contents

Preface

In accordance with the objectives presented in the preface of Volume 1, the purpose of this book is to provide an authoritative treatment on a unified area of coordination chemistry, with all subjects presented in accordance with their importance to the field. In other words, this volume is not intended to be a review of recent developments but rather a well-rounded presentation in depth of the subjects being treated. No aspect of the field has been omitted merely because of its classical flavor; similarly, new developments are included only if they seem to present basic information of lasting importance, rather than mere novelty.

The evolution of the contents of the present volume, as well as the fate of the material projected for future volumes, is an interesting reflection of the exponential growth of scientific literature and the changing practices of publishers who must deal with the expansion of information, the corresponding increases in publication expenses, and the declining sales of works dealing with advanced scientific literature. As time for publication of the second volume approached, it became obvious that the proposed content described in the preface of Volume 1 had become too extensive for a single volume. Therefore a logical division into three parts was made, with the material described for the second section on reaction kinetics and mechanisms to be published first. Presentation of the three-chapter manuscript to Van Nostrand Reinhold was first made four years ago, but was held up by a change in policy of the parent Litton Industrial Corporation which eventually led to termination of ACS Monograph publication and subsequent transfer of the manuscript to the American Chemical Society. After further delays required for review and up-dating of material, the present volume finally evolved.

Recently the ACS Publication Board has made policy changes for future monographs that virtually eliminate multi-authored monographs. The authors of the remaining chapters originally intended for our ACS Monograph Series on Coordination Chemistry will, however, be invited to offer their work for publication, either as individual ACS monographs under the new plan, or as contributions to appropriate review journals.

Finally, the editor wishes to thank the authors of the present volume for their patience with the extensive delays that have occurred as a result of the change of publishers and the consequent requirement to initiate a second review procedure. Thanks are also expressed to the reviewers of the chapters in this monograph for their careful analysis of subject matter content, for their valuable aid in the selection of the most significant material, and for their help in guiding the emphasis given to each subject in proportion to its importance to the field of inorganic reaction mechanisms.

ARTHUR E. MARTELL

Texas A&M University
College Station, Texas
January 1978

1

Kinetics and Mechanisms of Complex Formation and Ligand Exchange

Dale W. Margerum

Department of Chemistry
Purdue University
West Lafayette, Indiana 47907

George R. Cayley

Chemistry Department
Rothamsted Agricultural Station
Hertfordshire, AL5 2JQ England

David C. Weatherburn

Chemistry Department
Victoria University of New Zealand
Wellington, New Zealand

Gordon K. Pagenkopf

Department of Chemistry
Montana State University
Bozeman, Montana 59715

I. INTRODUCTION

This chapter concerns the dynamics of substitution reactions of coordination complexes, which is an extensive area when the possible variations of metal ions, monodentate ligands, multidentate ligands, types of complexes and solvents are considered. In fact, the area is too broad to permit a comprehensive survey even in a review of moderate length. Therefore, we are forced to be selective and have

0-8412-0292-3/78/36-174-001$30.00/1

chosen to deal primarily with reactions of the more labile metal ions and their complexes in aqueous solution. The kinetics and mechanisms of the sluggish metal complexes, particularly those of Co(III), Cr(III), and Pt(II), have been reviewed in considerable depth elsewhere.[1,2,3,4] With the slower-reacting metal ions it is often possible to isolate reaction intermediates or products and prove their configuration, a task which is more difficult with labile metal complexes. Therefore, the mechanisms proposed with the slower-reacting metal ions serve as a foundation for understanding the faster substitution reactions. However, the types of reactions studied are often quite different. With the sluggish metal ions the emphasis has been on monodentate substitutions and the replacement of weakly coordinated ligands. With labile metal ions many important reactions are with chelates or multidentate ligands, and often the replacements of strongly coordinated ligands (such as amines with transition metal ions) are studied. The reactions of nickel(II) have been the most extensively studied of the labile metal ions, and much of this work has been reviewed.[5] Chapters on the kinetics of labile metal complexes have been included in the series of Specialist Periodical Reports on Inorganic Reaction Mechanisms.[6]

This review places considerable weight on the reactions of multidentate ligands, a topic of special interest to the authors. In the process it is natural that a number of examples are taken from our own studies in order to illustrate general effects. We have attempted to give a critical review of the existing literature, and this has involved incorporating several of our own ideas in re-evaluating some reaction mechanisms. Reactions in nonaqueous solvents have been a subject of increasing interest in the past few years, and some aspects of these studies which seem to be of particular significance to aqueous chemistry are included, but the topic as a whole is not considered.

The kinetic behavior of the aquometal ion serves as a reference point for other substitution reactions of the metal complexes. Therefore, our first topic concerns water exchange rate constants and rate constants for monodentate ligand replacement of water from aquometal ions. There have been a number of reviews of this topic (Refs. 7–13) and of the experimental techniques used to measure fast reactions (Refs. 14–20). We have tried to bring the information up to date and to indicate the extent to which the *characteristic water exchange rate constants,* spanning values of 18 orders of magnitude, can be used to predict monodentate substitution reaction rates.

The majority of complexation reactions of metal ions involve chelation or multidentate coordination rather than monodentate substitution. The question which then arises is, "are the characteristic water exchange rate constants applicable?" The answer is yes if the first coordinate bond formation of the ligand is the rate-determining step, but even this statement must be further qualified because such rates can be accelerated by interactions in the outer-sphere or inhibited by steric effects. In cases where the first coordinate bond formation is not the rate-determining step, the reaction rates may be many orders of magnitude less than for the corresponding monodentate substitutions. Therefore, the position of the rate-determining step is of practical as well as mechanistic interest. Steric effects, ligand protonation, and electrostatic factors can help to determine which step in a multistep reaction is limiting. Furthermore, in a given reaction the concentration of reactants can shift the rate-determining step.

What effect do coordinated ligands have upon subsequent replacement reactions of coordinated water? Obviously, if there is a large effect, this must be considered in multidentate ligand formation reactions. Also it is important to know if a coordinated metal, for example, in a metalloenzyme, has a significantly altered water replacement rate constant. The answer is that coordinated ligands can change the water exchange rate constants by several orders of magnitude but that such large effects occur in relatively few cases. Coordinated ligands can influence substitution rates in other ways, however. Thus, large rate enhancements in complex formation can occur when there are stacking interactions between the incoming ligand and the coordinated ligand.[21]

The direct replacement of one monodentate ligand by another (other than solvent) is not a commonly studied reaction of labile metal complexes. On the other hand, there are many examples of multidentate ligand exchange reactions in which one ligand directly replaces another. The solvent path in the complete dissociation of a multidentate ligand may be an extraordinarily slow process, and hence steps which assist the process may provide important reaction paths. Thus, ligands and metal ions both may assist the removal of ligands. In ligand–ligand exchange reactions and in metal–metal exchange reactions, steric factors and the relative stabilities of intermediate mixed complexes become important. The kinetics can appear to be quite complicated. For instance, in the reaction of nickel(II)–EDTA with copper(II) the reaction order in copper shifts from first to zero, back to first, and finally back to zero order as the copper(II) concentration increases.[22] However, a relatively simple unwrapping mechanism explains the results, and, in general, a consideration of possible mixed complexes often indicates the reaction pathway.

Multidentate ligand exchange reactions are frequently subject to catalysis because of multiple reaction pathways in which hydrogen ion, metal ions, or ligands may play a role. Thus, traces of multidentate ligands can catalyze the exchange of other ligands in metal complexes. Such catalytic processes have been shown to cause coordination chain reactions[23,24,25] in which as little as 10^{-8} to $10^{-9} M$ free ligand controls the reaction rate. Steric effects and ligand catalysis are particularly noticeable in the reactions of oligopeptide complexes.

Eigen[7] has characterized the rates of complex formation in aqueous solution into three groups:

(*1*) Substitution rate constants for metal ions where the water exchange rate constant is greater than 10^7 sec^{-1} and diffusion-controlled processes of ion pairing may influence or limit the rate.

(*2*) Substitution rate constants for metal ions where the water exchange rate constant is less than 10^7 sec^{-1} and the rate-limiting step is H_2O substitution. The rates are characteristic of the metal ion only.

(*3*) Substitution reactions, usually quite slow, where internal hydrolysis of the metal ion–ligand outer-sphere complex occurs so that the rate depends on the ligand basicity and aquometal acidity.

In the present case we have tried to give separate consideration to monodentate substitution *vs.* chelate or multidentate substitution. Therefore, the discussion of the rates of complex formation is grouped somewhat differently from the arrangement used by Eigen, but the major factors considered are similar. Most of the detailed kinetic knowledge concerns group (*2*) above. There are still only limited examples in group (*3*), at least for monodentate ligand reactions. There are also relatively few instances where diffusion is clearly rate-controlling for metal ion complexation although the diffusion process may affect the rate.

II. REPLACEMENT OF WATER MOLECULES FROM THE FIRST COORDINATE SPHERE OF METAL IONS

A. Water Exchange

The rate of exchange of water between the inner coordination sphere of a metal ion and bulk water [Eq. (1)] has been measured by a number of techniques.

$$M(H_2O)_x{}^{n+} + H_2O^* \rightleftarrows M(H_2O)_{x-1}(H_2O^*)^{n+} + H_2O \qquad (1)$$

If the exchange reaction is slow, the rate of incorporation of labelled solvent into the coordination sphere of the metal ion can be determined directly using the isotope dilution technique.[26] Nuclear magnetic resonance methods have been used to measure most of the water exchange rates, and the results are given in Table 1-1. The various NMR methods are adaptable to a wide range of rate constants covering ~10^8-fold variation and are suitable for very rapid reactions. The upper limit for exchange rate constants determined by NMR techniques is about 10^8 sec^{-1}. The use of NMR techniques has been reviewed elsewhere.[10,27,28,29] Isotopic dilution and NMR techniques also have been used to determine the hydration number of cations. This has been reviewed,[10,30,31,32] and the results of such measurements have been tabulated and discussed.

TABLE 1-1
RATE CONSTANTS FOR WATER EXCHANGE OF AQUATED METAL IONS MEASURED BY NMR AT 25° C

Metal Ion	$k_{M\text{-}H_2O}$ (sec^{-1})	$\Delta H^\ddagger$ (kcal mol^{-1})	$\Delta S^\ddagger$ (cal °K^{-1} mol^{-1})	Ref.
Be^{2+}	2.1×10^3	8.3	−15	36,37
Mg^{2+}	5.3×10^5	10.2	2	38

TABLE 1-1 CONTINUED
RATE CONSTANTS FOR WATER EXCHANGE OF AQUATED METAL IONS MEASURED BY NMR AT 25° C

Metal Ion	$k_{M\text{-}H_2O}$ (sec^{-1})	$\Delta H^{\ddagger}$ ($kcal\ mol^{-1}$)	$\Delta S^{\ddagger}$ ($cal\ {}^{\circ}K^{-1}\ mol^{-1}$)	Ref.
V^{2+}	90	16.4	5.5	39
Cr^{2+}	$>10^8$			34
Mn^{2+}	3.1×10^7	8.1	2.9	33
	2.3×10^7	7.8	1.4	40
Fe^{2+}	3.2×10^6	7.7	−3.0	33
Co^{2+}	2.4×10^6	10.4	5.0	41
	2.2×10^6	10.3	5.1	42
	1.1×10^6	8.0	−4.1	33
Ni^{2+}	2.7×10^4	11.6	0.6	33
	3.0×10^4	10.8		43
	3.6×10^4	12.3	3.6	44
Cu^{2+}	$>10^8$			34,45
	$\sim 5 \times 10^9$			46,35
Al^{3+}	16 (0.17)	15.6 (27)	0 (28)	37(47)
Ga^{3+}	$7.6 \times 10^2 (1.8 \times 10^3)$	16.5 (6.3)	+10 (−22)	37 (47)
In^{3+}	4.0×10^4	4.6	−23	48
Ti^{3+}	1×10^5	6.2	−15	49
V^{3+}	1×10^3	9	−15	50
	1.6×10^3	6.2	−23	51
Cr^{3+}[a]	5.8×10^{-7}	27	+10	39
	4.3×10^{-7}	26.2	+0.3	52
Fe^{3+}	$(0.8–2.0) \times 10^4$			53,10
	1.5×10^2			54
Rh^{3+}[a]	3×10^{-8}	32.6	14	55
Gd^{3+}	6.3×10^7	12	18	56
	9×10^8	3.2	−7	57
Tb^{3+}	$(2.1–7.8) \times 10^7$			58
Dy^{3+}	$(1.4–3.2) \times 10^7$			58
	6.3×10^7			59
Ho^{3+}	$(0.9–6.1) \times 10^7$			58
Er^{3+}	$(0.54–13.5) \times 10^7$			58
Tm^{3+}	$(0.33–0.64) \times 10^7$			58
VO^{2+}	5×10^2	13.7	−0.6	60
	5.9×10^2 (27° C)	15	−1	61
	8×10^2	13.3	−2	62
	$\sim 10^{11}$[b]			63,62

[a]Measured by isotopic dilution.
[b]One axial water.

The rate constants in Table 1-1 include the isotopic dilution measurements for Cr(III) and Rh(III). The rates of exchange for Cr(II) and Cu(II) were investigated,[33,34] and the rate constants are greater than 10^8 sec^{-1}. There is some evidence that because of Jahn-Teller distortion the axial water molecules in Cu(II) exchange much more rapidly than the equatorial water molecules. However, interconversion of axial and equatorial molecules is believed to be fast, and only the rapid exchange rate is observable.[35] The results for Fe(III) should be regarded as approximate. Because of the presence of $Fe(OH)^{2+}$ and $Fe_2(OH)_2{}^{4+}$ in the solutions studied, exact interpretation was difficult.[53] Some of the uncertainties in the lanthanide-exchange rate constants are a result of uncertainties concerning the hydration number of the metal ion (which is believed to be 8 or 9) as well as of interpretation of the data.[58]

There has been some confusion in regard to the manner in which the water exchange rate constant is defined. The rate constants reported by Connick and co-workers[33,34,36,37,38] refer to the exchange of a particular water molecule in the metal first coordination sphere rather than to the possible exchange of all the coordinated water molecules. Some workers[48,64] in reporting water exchange rate constants have multiplied the exchange rate of an individual water molecule by the coordination number of the metal. This gives a different reference value, and in Table 1-1 we have corrected the rate constant reported for In^{3+} to be consistent with the other $k^{\mathrm{M\text{-}H_2O}}$ values. When these rate constants ($k_1 = \tau^{-1} = k^{\mathrm{M\text{-}H_2O}}$) are applied to ligand–solvent exchange of octahedral complexes, they are multiplied by a factor of $3/4$ on the basis that the exchanging ligand on the face of the octahedron has the chance to replace any of three adjacent coordinated water molecules and that any open site could be replaced by solvent or ligand at the four adjacent faces. The factor of $3/4$ tends to be obscured by other effects and is generally neglected.

All the exchange rate constants in Table 1-1 are first order because the exchanging water is in large excess over the aquated metal ion. In nonaqueous solutions, where it is sometimes possible to alter the concentration of the exchanging solvent molecule by the addition of an inert diluent, it has been shown that solvent exchange of trimethyl phosphate with Be^{2+} [65] and DMF and DMSO with Ni^{2+} [66] is independent of the solvent concentration. This does not mean, however, that the mechanisms of all exchange reactions are necessarily dissociative processes. In general at least three types of processes for substitution reactions could occur:[1,2] a dissociative process [S_N1(lim) or D mechanism]; an associative process (S_N2(lim) or A mechanism); or a concerted interchange process (I_a or I_d). It should be possible to draw some conclusions regarding the possible mechanisms of the solvent exchange process from a consideration of the activation parameters given in Table 1-1. The use of activation parameters as a criterion of mechanism in octahedral substitution reactions has been reviewed recently.[67] All other considerations being equal, the formation of an activated complex with an expanded coordination shell (A or I_a process) should lead to a negative value of $\Delta S^\ddagger$ while an activated complex with a reduced coordination shell (D process) should lead to a positive value of $\Delta S^{\ddagger}$. The values of $\Delta S^{\ddagger}$ in Table 1-1 divide the metal ions roughly into two classes, those with $\Delta S^{\ddagger}$ values less than -10 eu and those with slightly negative or positive values of $\Delta S^{\ddagger}$. Only Be^{2+} and the trivalent ions In^{3+}, Ti^{3+}, and V^{3+} have large negative $\Delta S^{\ddagger}$ values. It is tempting to assign an associative type mechanism (A or I_a) for the solvent exchange reactions of these ions and a dissociative type mechanism to the exchange reactions of the other metal ions. However, the accuracy and therefore the significance of the $\Delta S^{\ddagger}$ values are questionable. Thus, Connick and co-workers first reported[47] $\Delta S^{\ddagger}$ values of $+28$ and -22 eu for Al^{3+} and Ga^{3+}, respectively but have more recently revised[37] these values to 0 and $+10$ eu. Realizing that the reported $\Delta S^{\ddagger}$ values could be subject to shifts of ± 30 eu after reevaluation, we are hesitant to draw conclusions. The other ions in Table 1-1 may react via a dissociative type mechanism, but the large variation in the positive $\Delta S^{\ddagger}$ values suggests significant differences in the degree of the dissociative process.

B. Monodentate Ligand Replacement

The reaction of a monodentate ligand L with an aquated metal ion may be regarded as a generalization of Eq. (1) where the ligand L is the solvent molecule. The study of Eq. (2) for the majority of metal ions has become feasible since the development of techniques to study fast reactions.

$$M(H_2O)_x{}^{n+} + L^{m-} \rightarrow M(H_2O)_{x-1}L^{n-m} + H_2O \qquad (2)$$

In particular, the relaxation techniques developed by Eigen and DeMaeyer,[14] the stopped-flow techniques by Roughton and Chance,[68] and the NMR techniques used by Connick and others[28] have been very important. Relaxation and flow techniques have been reviewed in several books.[15,16,17,20]

1. Sound Absorption. Ultrasonic methods can be used over a wide frequency range and are relaxation techniques which in theory can give considerable information about the detailed mechanism of ligand substitution processes.[69,70] A wide range of relaxation times (10^{-4}–10^{-9} sec) may be measured, but because of technical difficulties, sound absorption is not used as much as other relaxation methods.[16] Some of the data obtained by different workers (or the interpretation of the data) has led to a controversy concerning the number of relaxations which are observed for a given system.[71,72,73,91]

A three-step mechanism was proposed by Eigen and Tamm to explain the relaxations observed in $MgSO_4$ and other electrolyte solutions.[69] The first step [Eq. (3)] is the diffusion-controlled association of the ions to form an outer-sphere complex. The rate constant for this process is about $4 \times 10^{10} M^{-1}$ sec^{-1} from diffusion data. The next step [Eq. (4)] involves the loss of one water molecule not bound in the inner coordination sphere of the metal ion; this results in a second outer-sphere complex. The third step [Eq. (5)] is the replacement of a water molecule in the inner coordination sphere of the cation by the uni-

$$M_{aq}^{2+} + L_{aq}^{2-} \underset{k_{21}}{\overset{k_{12}}{\rightleftharpoons}} [M(H_2O)_2L]_{aq} \quad (3)$$

$$[M(H_2O)_2L]_{aq} \underset{k_{32}}{\overset{k_{23}}{\rightleftharpoons}} [M(H_2O)L]_{aq} + H_2O \quad (4)$$

$$[M(H_2O)L]_{aq} \underset{k_{43}}{\overset{k_{34}}{\rightleftharpoons}} (ML)_{aq} + H_2O \quad (5)$$

dentate ligand L. Some sound absorption measurements have been interpreted as indicating the presence of three relaxations.[72,74,75] However, this work has been critized by Jackopin and Yeager,[71,73] who interpreted the measurements as indicating only two relaxations. The latter authors feel that Eq. (4) should not be treated as a separate process from the overall diffusional approach of the ions and prefer a two-step mechanism. Despite these differences in interpretation, the calculated values of the rate constants corresponding to k_{34} in Eq. (5) (which are of greatest interest in the present case) are very similar for the two interpretations as is evident for a number of the rate constants given in Table 1-2, where both two- and three-step mechanisms have been used. The ligand most widely used in sound absorption measurements has been sulfate ion, and the reasonable assumption is made that sulfate acts as a monodentate ligand in aqueous solution. The values of the rate constants for the step corresponding to k_{34} are given in Table 1-2 and in many instances are in good agreement with the rate constants of water exchange of the cations measured by NMR (Table 1-1). The equality of the ligand substitution rates and the water exchange rates was the basis for the suggestion by Eigen et al.[7,12] that the rate-determining step in the substitution of most metal ions is the rate of loss of water from the first coordination sphere of the metal ion. Thus, many reactions seem insensitive to the ligands in the second coordintion sphere—i.e., whether SO_4^{2-} or H_2O. However, the rates of SO_4^{2-} and H_2O exchange for the lanthanide ions mea-

TABLE 1-2
RATES OF WATER EXCHANGE FROM METAL IONS MEASURED BY SOUND ABSORPTION AT 25° C

Metal	Ligand	$k_{M\text{-}H_2O}$ (sec^{-1})	Ref.	Metal	Ligand	$k_{M\text{-}H_2O}$ (sec^{-1})	Ref
Be^{2+}	SO_4^{2-}	5×10^{2a}	72	Cr^{2+}	SO_4^{2-}	5×10^{8a}	72
		1×10^{2}	69	Mn^{2+}	SO_4^{2-}	2.5×10^{7a}	72
Mg^{2+}	SO_4^{2-}	3×10^{5a}	72			4×10^{6}	69
	$S_2O_3^{2-}$					2×10^{7d}	71
	CrO_4^{2-}					4.8×10^{7}	77, 74,75,78
		1×10^{5}	69				
Ca^{2+}	CrO_4^{2-}	$\geqslant 5 \times 10^{7}$	69,7,11	Fe^{2+}	SO_4^{2-}	2×10^{6a}	72
	CH_3COO^-	$6\text{–}9 \times 10^{8a}$	76			1×10^{6}	69

TABLE 1-2 CONTINUED
RATES OF WATER EXCHANGE FROM METAL IONS MEASURED BY SOUND ABSORPTION AT 25° C

Metal	Ligand	kM-H_2O(sec^{-1})	Ref.
Co^{2+}	SO_4^{2-}	1×10^{6}[a]	72
		2×10^{5}	69
		1.8×10^{6}	77
Ni^{2+}	SO_4^{2-}	2×10^{4}[a]	72
		1.5×10^{4}	69,12
		1×10^{4}	76
Cu^{2+}	SO_4^{2-}	5×10^{8}[a]	72
		2×10^{8}	79
Zn^{2+}	SO_4^{2-}	5×10^{8}[a]	72
	CH_3COO^-	3×10^{7}	79
	SO_4^{2-}	3×10^{7}	80
		7.2×10^{7}	79
Cd^{2+}	SO_4^{2-}	$>10^{8}$	79
	CH_3COO^-	2.5×10^{8}	79
	Cl^-	5×10^{8}	12
Hg^{2+}	Cl^-	2×10^{9}	12
Pb^{2+}	CH_3COO^-	7.5×10^{9}	81
La^{3+}	SO_4^{2-}	2.1×10^{8}[c]	83
		8.4×10^{7}	82
		2.4×10^{7}[e]	82
	CH_3COO^-	4.8×10^{6}[b]	84
Ce^{3+}	SO_4^{2-}	2.7×10^{8}[c]	85
		2.0×10^{8}[d]	85
	CH_3COO^-	7.4×10^{6}[b]	84
Pr^{3+}	SO_4^{2-}	3.1×10^{8}[c]	85
		2.5×10^{8}[d]	85
		2.1×10^{8}	86
		1.5×10^{8}[e]	86
Nd^{3+}	SO_4^{2-}	3.9×10^{8}[c]	85
		3.1×10^{8}[d]	85
		1.9×10^{8}	87, 82
		8.0×10^{7}[e]	82
	NO_3^-	6×10^{8}	88
Nd^{3+}	NO_3^-	1.8×10^{8}	89
Sm^{3+}	SO_4^{2-}	5.9×10^{8}[c]	85
		4.7×10^{8}[d]	85
		2.1×10^{8}	90
		3.3×10^{8}	86
		2.9×10^{8}[e]	86
	CH_3COO^-	8.9×10^{6}[b]	84
	NO_3^-	1×10^{8}	91
Eu^{3+}	SO_4^{2-}	6.5×10^{8}[c]	85
		4.9×10^{8}[d]	85
Gd^{3+}	SO_4^{2-}	6.4×10^{8}[c]	85
Dy^{3+}	SO_4^{2-}	4.0×10^{8}[c]	85
		2.9×10^{8}[d]	85
		1.3×10^{8}[e]	86
		9.1×10^{7}	82
		1.1×10^{8}[e]	82
	CH_3COO^-	8.5×10^{6}[b]	84
Ho^{3+}	SO_4^{2-}	2.5×10^{8}[c]	85
		1.6×10^{8}[d]	85
		1.9×10^{8}[c]	86
Gd^{3+}	SO_4^{2-}	4.9×10^{8}[d]	85
		2.8×10^{8}	86
		2.9×10^{8}[e]	86
		8.2×10^{8}	82
		6.9×10^{8}[e]	82
	NO_3^-	5×10^{8}	88
Tb^{3+}	SO_4^{2-}	5.2×10^{8}[c]	85
		3.8×10^{8}[d]	85
		1.9×10^{8}	86
Er^{3+}	SO_4^{2+}	1.9×10^{8}[c]	83
	NO_3^-	6.0×10^{7}	89
Tm^{3+}	SO_4^{2-}	1.4×10^{8}[c]	83
Yb^{3+}	SO_4^{2-}	$\sim 8 \times 10^{7}$[c]	83
Lu^{3+}	SO_4^{2-}	$\sim 6 \times 10^{7}$[c]	83

[a] Assumed values used as constants in the calculations.
[b] The acetate values are thought to indicate that the acetate acts as a bidentate ligand.
[c] Three-step mechanism.
[d] Two-step mechanism.
[e] In D_2O.

sured by sound absorption[83,84,85] are an order of magnitude higher than the water exchange rates measured by the NMR technique.[58] The differences are sufficiently consistent to make it difficult to brush them off as experimental uncertainty. The rates of exchange of SO_4^{2-} with lanthanide ions in D_2O and H_2O have been investigated by two groups. Silber[82,87] claims a solvent isotope effect which is attributed to differences in hydrogen bonding in the second coordination sphere. Purdie,[86] however, found no solvent isotope effects.

Other ligands have been examined by sound absorption techniques with lanthanide ions. Studies with acetate have been interpreted as indicating that it acts as a bidentate ligand,[84] and studies with nitrate indicate that more than one reaction occurs.[83] Wang and Hemmes[91] have interpreted their data from yttrium nitrate solutions as indicating a one-step process ($k_f = 1.0 \times 10^8 M^{-1}\text{sec}^{-1}$). These authors feel that the multiple relaxations seen in previous work with nitrate ion are a result of multiligand complexation reactions. Other substitution reactions with bi- and tridentate ligands have been used to test or to estimate water substitution rate constants for the lanthanides. The ligands used include murexide,[93,94] oxalate,[95] anthranilic acid,[96,97] malonate,[90] and xylenol orange.[98] All but the xylenol orange are in reasonable agreement with the NMR studies. Nevertheless it has been suggested that the loss of the second water molecule is the rate-determining step in the complexation of malonate, murexide, oxalate, and anthranilate.[70,83,90] This theory would be consistent with some of the sound absorption studies but not with the NMR results. Further studies of substitution reactions of monodentate ligands with the lanthanides are needed to solve this question.

In a recent study using the sound absorption technique[72] the value of k_{34} was set equal to the NMR rate of water exchange and was used as a constant in calculating the other rate constants in the three-step mechanism. Naturally, this gives no new information about the k^{M-H_2O} value, and we have footnoted these cases in Table 1-2.

2. Outer-Sphere Association Constants. From the viewpoint of estimating the rates of most substitution reactions, the three-step mechanism proposed by Eigen and Tamm[69] may be simplified to a two-step process. The outer-sphere reactions are generally faster than rate-controlling substitution within the inner sphere. The reaction in Eq. (6) may be considered as a rapid pre-equilibrium with K_{os}

$$M(H_2O)_x{}^{n+} + L^{m-} \underset{}{\overset{K_{os}}{\rightleftharpoons}} [M(H_2O)_x{\cdot}L]^{n-m} \tag{6}$$

$$[M(H_2O)_x{\cdot}L] \xrightarrow{k^{M-H_2O}} M(H_2O)_{x-1}L^{n-m} + H_2O \tag{7}$$

known as the outer-sphere association constant. The forward rate constant, k_f, is given by Eq. (8). The value of K_{os} may be determined experimentally and has been with a number of slow-to-substitute complexes. Experimentally determined

$$\frac{d[ML]}{dt} = k_f[M_{aq}{}^{n+}][L_{aq}{}^{m-}] \qquad k_f = K_{os}k^{M-H_2O} \tag{8}$$

values of K_{os} vary from $\sim 10M^{-1}$ (for 3+, 1− ions) to $\sim 3 \times 10^3 M^{-1}$ (for 3+, 2− ions). The nature and properties of outer-sphere complexes have been reviewed.[99]

For labile metal ions K_{os} values are more difficult to determine, and theoretical expressions derived from a consideration of statistical arguments[100] and the theory of diffusion[101] are used to estimate values. The theoretical expression is given in Eq. (9) where $b = Z_M Z_L e_0^2/a\epsilon kT$,

$$K_{os} = \frac{4\pi N a^3}{3000} e^{-b} \cdot \exp\{b\kappa a/(1 + \kappa a)\} \tag{9}$$

a is the center to center distance (cm) between the species M and L, N is Avogadro's number, Z_M and Z_L are the formal charges on the reacting species, e_0 is the electronic charge (esu), ϵ is the dielectric constant of the solvent, k is the Boltzmann constant (ergs), κ is the Debye-Hückel ion atmosphere parameter,[1] and T is the absolute temperature. The exponential expression results since K_{os} is a concentration constant and can be corrected by the Debye-Hückel theory where $\kappa^2 = (8\pi N e^2/1000\epsilon kT)\,\mu$, where μ is the ionic strength of the solution. The value of K_{os} is sensitive to the choice of a, which has been varied judiciously to obtain desired K_{os} values although values of a of 4–5Å are frequently used. Using approximations for a, the simpler expression, $K_{os} = (\frac{4}{3}\pi N a^3 e^{-b})10^{-3}M^{-1}$ is often sufficient. Rorabacher et al.[102] have modified the expression for cases where a bidentate ligand such as enH^+ reacts [Eq. (10)]. In this equation, a is defined as before for the unprotonated nitrogen donor of en, and a' is the center-to-center distance of the metal ion and the protonated nitrogen. This behavior has been further investigated for the reaction of $H_2NCH_2CH_2N(CH_3)_3{}^+$ with Ni^{2+} as a function of the dielectric constant of the solvent.[103]

$$K_{os} = \frac{4}{3}\pi a^3 N \exp\left(\frac{-Z_M Z_L e_0^2}{a'\epsilon kT}\right) \times 10^{-3}M^{-1} \tag{10}$$

3. Diffusion-Controlled Reaction Rates. The maximum possible rate constant for a reaction can be calculated if it is assumed that species A and B will react upon encounter. An abbreviated treatment is summarized by Caldin;[15] using Fick's law of diffusion and the Stokes-Einstein equation, for solute molecules A and B, the rate constant for a diffusion-controlled reaction k_D, is given by Eq. (11):

$$k_D = \frac{2RT}{3000\eta}(2 + r_A/r_B + r_B/r_A)M^{-1}\mathrm{sec}^{-1} \tag{11}$$

where η is the viscosity of the solvent and r_A, r_B are the radii of the molecules. As Eq. (11) indicates, k_D is not very sensitive to the size of the molecules, and for $r_A \simeq r_B$ in water at room temperature, $k_D = 7 \times 10^9 M^{-1}\text{sec}^{-1}$. However, when A and B are ions, their ionic attraction or repulsion makes a difference, and Eq. (12) gives an approximate value for k_D where b is the same as in Eq.

$$k_D = \left(\frac{8RT}{3000\eta}\right)\left(\frac{b}{e^b - 1}\right) M^{-1}\text{sec}^{-1} \tag{12}$$

(9). This expression holds for dilute solutions and can be multiplied by $\exp(b\kappa a/1 + \kappa a)$ to correct for the screening effects of an ionic atmosphere. Substituting into Eq. (12) for $Z_A Z_B = -2$ with $a = 5$Å gives $k_D = 2 \times 10^{10} M^{-1}\text{sec}^{-1}$. In general, diffusion-controlled rate constants for metal ions reacting with neutral or negatively charged ligands at low ionic strength can be expected to fall between 10^9 and $10^{11} M^{-1}\text{sec}^{-1}$. Therefore, rate constants below $10^7 M^{-1}\text{sec}^{-1}$ will not be affected by the rate of diffusion, those in the vicinity of $10^8 M^{-1}\text{sec}^{-1}$ may be slightly influenced, and those significantly larger than $10^9 M^{-1}\text{sec}^{-1}$ may be limited by diffusion.

Many proton-transfer rates are diffusion controlled,[14] and in sound-absorption studies, the outer-sphere sulfate species appear to form at diffusion-controlled rates. However, relatively few complexation reactions are known which are as fast. The k_f values for Pb^{2+} with NTA^{3-}, $H(EDTA)^{3-}$, and other aminocarboxylate ligands have been calculated to be between 2×10^{10} and $2.4 \times 10^{11} M^{-1}\text{sec}^{-1}$.[104] The alkali metal ion reactions with some polyphosphates are reported to have rate constants larger than $5 \times 10^9 M^{-1}\text{sec}^{-1}$ for K^+, Rb^+, and Cs^+,[7] but the reactions of the alkali metal ions with $EDTA^{4-}$, NTA^{3-}, and IDA^{2-} are one to two orders of magnitude slower and are characteristic of the ligand. A few monodentate ligand reactions with metal ions begin to approach diffusion limits—e.g., the reaction of Cu^{2+} and acetate ion has a $k_f = 1.5 \times 10^9 M^{-1}\text{sec}^{-1}$—but they are not actually diffusion limited.

4. Bivalent Metal Ions. Table 1-3 presents available data for the formation kinetics of complexes with monodentate ligands and bivalent metal ions. Where comparisons are possible, the ratios of k_f (second-order rate constant for the reaction of M^{2+} and L) to $k^{M\text{-}H_2O}$ are given. The first-row transition metal ions would be expected to have $k_f/k^{M\text{-}H_2O}$ (= K_{os}) values of 0.1 to 0.2 for neutral ligands and greater than unity for uninegative ligands. Many of the neutral ligands have $k_f/k^{M\text{-}H_2O}$ ratios consistent with this and, in general, support the theory of outer-sphere formation followed by a ligand-independent water-loss step in the substitution process. Certainly, the Ni(II) reactions follow this mechanism in most instances, but the data for the alkylamines give lower ratios as discussed below.

The $k_f/k^{M\text{-}H_2O}$ ratios for F^- appear to be smaller than expected, but this is probably a result of the high ionic strength used (0.5M) which invalidates the Debye-Hückel estimates in Eq. (9). The $k_f/k^{M\text{-}H_2O}$ ratios found at this ionic strength with F^- vary from 0.1–0.4M, and the formation rate constants are in general agreement with the SO_4^{2-} data.[105] However, there are some unexplained

TABLE 1-3
RATE CONSTANTS MEASURED FOR MONODENTATE LIGANDS REACTING WITH BIVALENT METAL IONS AT 25° C

Metal	Ligand	k_f (M^{-1} sec^{-1})	$k^{M\text{-}H_2O}$ (sec^{-1})	$k_f/k^{M\text{-}H_2O}$ (M^{-1})	Ref.
Be^{2+}	HF	73	$\geqslant 3 \times 10^3$	0.02	107
	F^-	7.2×10^2		0.24	107
Mg^{2+}	$P_2O_7^{4-}$	7.1×10^7	5.3×10^5	134	108
	$HP_2O_7^{3-}$	5.1×10^6		9.6	108
	$H_2P_2O_7^{2-}$	5.4×10^5		1.01	108
V^{2+}	F^-	28,15,9	90	~0.02	109,110,111
	NCS^-	28,15		~0.02	109,110
Mn^{2+}	HF	2.2×10^{6} c	3.1×10^7	0.07	105
	F^-	2.7×10^{6} c		0.09	105
	Cl^-	1.5×10^7		0.5	112
Fe^{2+}	NO	6.2×10^5	3.2×10^6	0.2	113
	HF	9.3×10^{5} c		0.3	105
	F^-	1.4×10^{6} c		0.4	105
Co^{2+}	NH_3	1.1×10^5	2×10^6	0.05	114
	imidazole	1.3×10^5			105
	HF	5.5×10^{5} c		0.3	105
	F^-	1.8×10^{5} c		0.09	105
Ni^{2+}	H^+-imidazole	4×10^2 (23.7° C)	3×10^4	0.016	116
		2×10^2		0.007	117
	H^+-NMe-imid	2.3×10^2		0.007	118
	$H_2N(CH_2)N^+Me_3$	4×10^2		0.01	119
		4.8×10^2		0.016	103
	NH_3	4.5×10^3		0.15	114,106
	N_2H_4	2.5×10^3		0.08	119
	Py	4.0×10^3		0.09–0.13	119
		3.9×10^3		0.21	120
		2.4×10^3		0.17	121
	imidazole	6.4×10^3		0.11	117
		5×10^3			115
		3.2×10^3			116
	NMe-imid	4.5×10^3		0.15	118
	$MeNH_2$	1.3×10^{3} b		0.04	106
	$EtNH_2$	8.7×10^{2} b		0.03	106
	i-$PrNH_2$	6.1×10^{2} b		0.02	106
	Me_2NH	3.3×10^{2} b		0.01	106
	theophylline	1×10^2		0.003	122
	HF	3.1×10^{3} c		0.1	105
	F^-	8.4×10^{3} c		0.3	105
	NCS^-	5×10^3		0.13	123
		3×10^4 (20° C)		1	124

TABLE 1-3 CONTINUED
RATE CONSTANTS MEASURED FOR MONODENTATE LIGANDS REACTING WITH BIVALENT METAL IONS AT 25° C

Metal	Ligand	k_f (M^{-1} sec^{-1})	$k^{M\text{-}H_2O}$ (sec^{-1})	$k_f/k^{M\text{-}H_2O}$ (M^{-1})	Ref.
Ni^{2+}	$MePO_4^{2-}$	2.9×10^5 [a] (7×10^3)[a]		41[a]	125
	OAc^-	1.5×10^5		5	126
	$Ru(NH_3)_5pyr$	1.1×10^3		0.04	92
Cu^{2+}	NH_3	2.0×10^8			127
	imidazole	5.7×10^8			127
	HF	3.1×10^7 [c]			127
	F^-	2.2×10^8 [c]			128
	OAc^-	1.5×10^9			79
	$ClCH_2COO^-$	8.5×10^8			129
Zn^{2+}	NH_3	3.6×10^6 (11° C)	7×10^7	0.05	114
	Br^-I^-	$\sim 5 \times 10^5$			130
Cd^{2+}	Br^-	1.4×10^9			130
Ru^{2+}	Cl^-	8.5×10^{-3}			131
	Br^-	9.7×10^{-3}			131
	I^-	9.0×10^{-3}			131
	ClO_4^-	3.2×10^{-3}			131

[a] Relaxation measurements gave K_{os} and the $k^{M\text{-}H_2O}$ values as in Eq. (6) and (7), and k_f is the calculated product.
[b] $1M$ ionic strength.
[c] $0.5M$ ionic strength.

variations, for instance, with Co^{2+}, the HF molecule reacts more rapidly than the F^- ion.

5. Steric Effects. Relatively little attention has been paid to the importance of steric effects in monodentate ligand formation rates with aquometal ions. However, the study by Rorabacher and Melendez-Cepeda[106] shows that alkylamines react more slowly than ammonia with nickel(II) (Table 1-3) so that k_f for $(CH_3)_2NH$ is only 7% of that for NH_3. The effect can be attributed in part to the diminished K_{os} value because the ligand molecule must be properly oriented in the outer-sphere complex for bonding to occur at the time the metal–water bond breaks.[106] However, there are also likely to be steric effects in the rate process of the ligand entering the inner coordination sphere.

6. Trivalent Metal Ions. Substitution rate constants for the reaction of monodentate ligands with the labile trivalent metal ions are given in Table 1-4. A recent survey[4] includes the reactions of the more sluggish trivalent metal ions. A general comparison of $k_f/k^{M\text{-}H_2O}$ values is difficult because of the uncertainty of many of the water exchange rate constants. There have not been direct de-

TABLE 1-4
RATE CONSTANTS FOR SUBSTITUTION REACTIONS OF TRIVALENT METAL IONS WITH MONODENTATE LIGANDS AT 25° C

Metal	Ligand	k_f (M^{-1} sec^{-1})	$k^{M\text{-}H_2O}$ (sec^{-1})	Ref.
Ti^{3+}	SCN^-	8×10^3 (8–9° C)		145
		1×10^4	1×10^5 (Table 1-1)	146
	$CH_3CO_2^-$	2×10^6 (15° C)		145
	CH_3CO_2H	1.1×10^3 (15° C)		145
	$ClCH_2CO_2^-$	2.1×10^5 (15° C)		145
	$Cl_2CHCO_2^-$	$\sim 6 \times 10^4$ (15° C)		145
	$HC_2O_4^-$	4×10^5 (10° C)		145
	NH_2OH	42.1		147
V^{3+}	SCN^-		1.6×10^3	110
		1.1×10^2	1×10^3 (Table 1-1)	148
	Cl^-	$\leqslant 3$		149
	HN_3[a]	0.39		150
	Br^-	$\leqslant 10$		149
	NH_2OH	1.8×10^3		151
	$HC_2O_4^-$	1.3×10^3		149
Cr^{3+}	SCN^-	3.5×10^{-6}	4.3×10^{-7} (Table 1-1)	132
	Cl^-	3.0×10^{-7}		152
	Br^-	3.0×10^{-7}		153
	I^-	8×10^{-10}		154
	F^-	1.5×10^{-5}		155
Mo^{3+}	SCN^-	10		132
		0.28		156
	Cl^-	4.6×10^{-3}		156
Mn^{3+}[b]	F^-[c]	$5–10 \times 10^3$		146
	quinol[d]	$> 5 \times 10^5$		157
	H_2O_2	7.3×10^4		158
	HNO_2	2.2×10^4		159
	p-$C_6H_4(OH)_2$	4.8×10^3		160
Co^3[b]	Cl^-	3	~ 0.1	161
	Br^-	$\leqslant 5$		162
	SCN^-	86.5		162
	I^-	8000		162
	ClO_2	$\leqslant 1$		163
	HN_3	17.5 (2M $HClO_4$)		164
	HNO_3	18		162
	H_2O_2	$\leqslant 2$		162
Fe^{3+}	HF	11.4		138
	Cl^-	9.4		165
		19		166
		8.4		167
	Br^-	20		168
		50		166

TABLE 1-4 CONTINUED
RATE CONSTANTS FOR SUBSTITUTION REACTIONS OF TRIVALENT METAL IONS WITH MONODENTATE LIGANDS AT 25° C

Metal	Ligand	k_f (M^{-1} sec^{-1})	$k^{M\text{-}H_2O}$ (sec^{-1})	Ref.
Fe^{3+}	SCN^-	150		169
		130		170
		127		133
		132		171
		90		172
	SO_4^{2-}	3.5—4.6 × 10^3		173
		4.35 × 10^3		174
		3.2 × 10^3		175
	HSO_4^-	51		173
	HOAc	4.8		176
		27		177
	HOPr	5.7		176
	$HOCOCH_2Cl$	2.2		176
	HN_3	4		178
		2.6		179
		4		139
	$HC_2O_4^-$	14.4 × 10^2		180
		8.6 × 10^2		181
	$Fe(CN)_6^{3-}$	1.7 × 10^3		182
		8 × 10^2		183
	H_3PO_2	35		184
	phenol	25		143
In^{3+}	SO_4^{2-}	2.6 × 10^5	4.0 × 10^4 (Table 1-1)	64
			~3 × 10^2 [f]	64
Ga^{3+}	SO_4^{2-}	2.1 × 10^4	7.6 × 10^2 (Table 1-1)	64
			~20 [g]	64
		6.5 × 10^3 [e]	5 [f]	144
Al^{3+}	SO_4^{2-}	1.2 × 10^3	16 (Table 1-1)	64
			~1 [g]	64
		7.8 × 10^2 [e]	0.5 [f]	144
	$Co(CN)_6^{3-}$	1.62 × 10^3	8.5 × 10^{-2}	185

[a] If N_3^- is assumed to be the reactant, then $k_f = 9 \times 10^2 M^{-1}$ sec^{-1}.[150]

[b] These values are obtained from a study of the oxidation reactions of these ions. The rates are believed to be controlled by the water substitution rates of these ions.

[c] If HF is assumed to be the reactant, then $k_f = 10^8 – 10^9 M^{-1}$ sec^{-1}.

[d] This value is the minimum possible value calculated on the basis that the reaction was too fast to observe under the conditions given in Ref. 157.

[e] Calculated from theoretical K_{os} values and the measured relaxation exchange rate constant.

[f] Measured from sulfate exchange for water by pressure-jump relaxation assuming theoretical K_{os} values.

[g] SO_4^{2-}—H_2O exchange calculated from an assumed value of K_{os}.

terminations of the $k^{M\text{-}H_2O}$ values for $Co_{aq}{}^{3+}$ or $Mn_{aq}{}^{3+}$. The $k^{M\text{-}H_2O}$ value for $Fe_{aq}{}^{3+}$ is uncertain as a result in part of the considerable contribution of $FeOH^{2+}$ to the exchange rate. Many of the numerical values for Fe(III) in the literature seem to have been passed along from preliminary reports. Similarly, the value for $V_{aq}{}^{3+}$ is quoted only from a private communication. Hydrolysis and the difficulties inherent in ^{17}O-NMR determinations of $k^{M\text{-}H_2O}$ have affected the values for $Ga_{aq}{}^{3+}$ and $Al_{aq}{}^{3+}$. These constants have very recently been redetermined[37] with large changes for $k^{M\text{-}H_2O}$ and ought to be reliable now. The $k^{M\text{-}H_2O}$ values for $Ti_{aq}{}^{3+}$ and $In_{aq}{}^{3+}$ have been determined by NMR (Table 1-1), but in giving the value for In^{3+} it was noted that the exchange rate could be faster.[48] The exchange constant for Cr^{3+} is accurately known. With Cr(III), the contribution of $CrOH^{2+}$ and other hydrolyzed species are also important.[28]

The calculated K_{os} values ($\mu \rightarrow 0$, a = 5Å) are $23M^{-1}$ for 3+, 1− ions and $1.7 \times 10^3 M^{-1}$ for 3±, 2− ions. The outer-sphere association constant has been measured spectrophotometrically for $Cr(H_2O)_6{}^{3+}{\cdot}NCS^-$ and has a value of $7M^{-1}$ when extrapolated to zero ionic strength.[132] Postmus and King also measured the k_f values for these species as a function of ionic strength, and the value in Table 1-4 is for $\mu = 0.16M$. If extrapolated to zero ionic strength, $k_f = 1.12 \times 10^{-5} M^{-1}\text{sec}^{-1}$, which corresponds to a K_{os} value of $26M^{-1}$.

The ratios of $k_f/k^{M\text{-}H_2O}$ for $In_{aq}{}^{3+}$, $Ga_{aq}{}^{3+}$, and $Al_{aq}{}^{3+}$ with $SO_4{}^{2-}$ (12, 37, and $81M^{-1}$, respectively) are all smaller than would be expected for these ions at zero ionic strength. [Miceli and Stuehr[133] calculated the $k_f/k^{M\text{-}H_2O}$ values to be six times greater, but as discussed earlier they used a different reference value for $k^{M\text{-}H_2O}$ which is not valid for the needed comparison.] The reason the $k_f/k^{M\text{-}H_2O}$ values are relatively small is not clear. Outer-sphere complexes are known with $Co(NH_3)_6{}^{3+}$ and $SO_4{}^{2-}$, and the association constant is equal to $2.2 \times 10^3 M^{-1}$,[134] in agreement with the range expected from calculations. [Constants evaluated[135,136] for the corresponding complexes with halide ions are, however, smaller than the theoretical values when the possibility of ion triplets is considered.] It may be that the substitution reactions of these metal ions have more associative character than some of the other ions and that the outer-sphere to inner-sphere exchange is slower for $SO_4{}^{2-}$–H_2O than for H_2O–H_2O exchange. The $\Delta S^\ddagger$ value for $In_{aq}{}^{3+}$ and H_2O exchange tends to support an associative mechanism, but this is less so for the revised $\Delta S^\ddagger$ values for the H_2O exchange of $Al_{aq}{}^{3+}$ and $Ga_{aq}{}^{3+}$.

The reactions of $Ti_{aq}{}^{3+}$ and $V_{aq}{}^{3+}$ with NCS^- appear to be slower than would be expected from the water exchange data (after temperature corrections). Here again, an associative mechanism may be involved as indicated by the negative $\Delta S^\ddagger$ values reported for the water exchange rate constants of these metal ions. Chmelnick and Fiat[49] suggest an S_N2 path. If this is the case, then smaller k_f values would occur with NCS^- only if it were a poor nucleophile for these metal ions. This point is not established.

The rate constants given in Table 1-4 for Mn_{aq}^{3+} and Co_{aq}^{3+} were obtained for the most part from studies of redox reactions with these ions. The constants given are thought to be for oxidation processes controlled by the rate of substitution reaction. These reactions are discussed fully in a review.[137]

The only trivalent ion which has been extensively studied with a large variety of ligands is Fe_{aq}^{3+}. The observed kinetics are invariably of the form $(k + k'[H^+]^{-1})[L]$ because of the greater substitution reactivity of $FeOH^{2+}$. The rate constants in Table 1-4 are resolved from studies in which the H^+ concentration is varied, and the constants given are from the acid-independent path. Although these are the acid-independent rate constants, it is still possible for internal hydrolysis to occur between Fe(III) and ligand—i.e., Eigen's third mechanism.[7] Thus, the reactants may be $Fe(H_2O)_6^{3+}\cdot L$ or $Fe(H_2O)_5OH^{2+}\cdot HL$. Larger k_f values are found for ligands which are weaker acids—i.e., SO_4^{2-}, $HC_2O_4^-$, and possibly NCS^-. The reaction of F^- with Fe^{3+} has been reported[138] ($k_f = 5.4 \times 10^3 M^{-1}$ sec^{-1}), but this has been reevaluated[139] as the reaction between $FeOH^{2+}$ and HF($k_f = 3 \times 10^3 M^{-1}$ sec^{-1}). All the other neutral or uninegative ligands in Table 1-4 are not capable of removing a proton from $Fe(H_2O)_6^{3+}$ and give k_f values between 2 and $50 M^{-1}$ sec^{-1}.

Relaxation studies of the reaction of a number of substituted phenols[140,141,142] show all to react via $FeOH^{2+}\cdot HL$ although with phenol itself, the $Fe^{3+}\cdot HL$ pathway can be observed.[143]

Other metal ions which show accelerated substitution reactions because of hydroxy complexes include Al(III), Ga(III), and In(III),[64] but as already pointed out, their substitution rate constants with SO_4^{2-} are somewhat smaller, not larger, than expected, so there is no reason to suspect the internal hydrolysis pathway in this case. Knoche and co-workers[144] also have investigated the reactions of Al^{3+} and Ga^{3+} with SO_4^{2-} and have suggested that the values of $k^{M\text{-}H_2O}$ are smaller than those quoted by Miceli and Stuehr.[64] They use smaller values of ion-contact distance to calculate the outer-sphere association constant so that the concentrations of outer-sphere complexes are not negligible under the experimental conditions used. This is taken into account in their relaxation expressions and rate constants. They also suggest that some gallium hydroxide could have been present in the pH range used by Miceli and Stuehr.

Table 1-4 compares the SO_4^{2-}–H_2O and H_2O–H_2O exchange rate constants for In^{3+}, Ga^{3+}, and Al^{3+} in the fourth column (under the heading $k^{M\text{-}H_2O}$). The H_2O–H_2O exchange rate is much greater than the SO_4^{2-}–H_2O exchange rate for all of the three metals. In the case of Al^{3+}, the older NMR value[47] for H_2O–H_2O exchange is in better agreement with the sulfate exchange rate constants than is the newer value from the same laboratory.[37] However, with Ga^{3+} the older NMR value[47] is even larger than the newer one. It seems safe to conclude that the exchange rate for these ions is ligand dependent and is therefore not a simple dissociative mechanism. As more kinetic studies of the substitution reactions of trivalent metal ions (Ti^{3+}, V^{3+}, Cr^{3+}, Fe^{3+}, and Mo^{3+})

TABLE 1-5
RATE CONSTANTS FOR SUBSTITUTION REACTIONS OF OXOCATIONS

Cation	Ligand	k_f (M^{-1} sec^{-1})	Ref.
VO^{2+}	SO_4^{2-}	1.5×10^4 (25°C, 0.01μ)	186
	SCN^-	1.15×10^4 (25°C, 0μ)	187
	SCN^-	1.6×10^2 (20°C, 0.2μ)	188
UO_2^{2+}	OAc^-	1.05×10^3 (20°C, 0.15μ)	188
	NO_3^-	$\sim 1.5 \times 10^5$ (3°C, 0.2μ)	179, 189
	SCN^-	$\sim 2 \times 10^5$ (3°C, 0.2μ)	179, 189
	SO_4^{2-}	1.8×10^2 (20°C, 0.15μ)	188
	$ClCH_2CO_2^-$	1.1×10^2 (20°C, 0.15μ)	188
	SCN^-	2.9×10^2 (20°C, 1.2μ)	188
$Mo_2O_4^{2+}$	SCN^-	2.9×10^4 (25°C, 2.0μ)	190

are reported, the results are easier to understand in terms of mechanisms with some associative character.[145,67,156]

7. Oxocations. The substitution reactions of VO^{2+} and UO_2^{2+} have been studied, and the results are given in Table 1-5. The water exchange rate has not been measured for UO_2^{2+} although it has been estimated to be as high as 10^5 sec^{-1} at 3°C.[191] There is serious disagreement concerning the substitution reactions of UO_2^{2+}, with values differing by three orders of magnitude. The detailed results for the faster exchange reactions have not been published but appear to be consistent with the formation kinetics of $(UO_2OH)_2^{2+}$.[191]

With VO^{2+}, two rates of water exchange have been measured by Wüthrich and Connick.[60] The four equatorial waters exchange with a rate constant of 500 sec^{-1} (25°C), while the rate of exchange in the axial position (opposite the V=O group) was too fast to measure accurately, and a lower limit of 10^6 sec^{-1} was given. The substitution reactions require replacement in the equatorial position for stable coordination. The earlier work of Strehlow and Wendt[186] with SO_4^{2-} exchange estimated a K_{os} value of 25M^{-1} (0.01μ) and an inner-sphere–outer-sphere exchange rate constant for H_2O and SO_4^{2-} equal to 6.0×10^2 sec^{-1}. This is in good agreement with the water exchange rate constant. The VO^{2+} reaction has a reciprocal [H^+] dependence which is taken into account in these values.

C. General Conclusions

In the preceding sections we have tried to point out some of the inconsistencies and areas that need more study concerning water and monodentate ligand exchange reactions. This effort does not contradict the general picture that emerged around 1960[192,193] in which the rates of ligand substitution reactions in aqueous solution were described as being controlled by the rate of exchange of water molecules from the inner coordination sphere of the metal ion. This model has

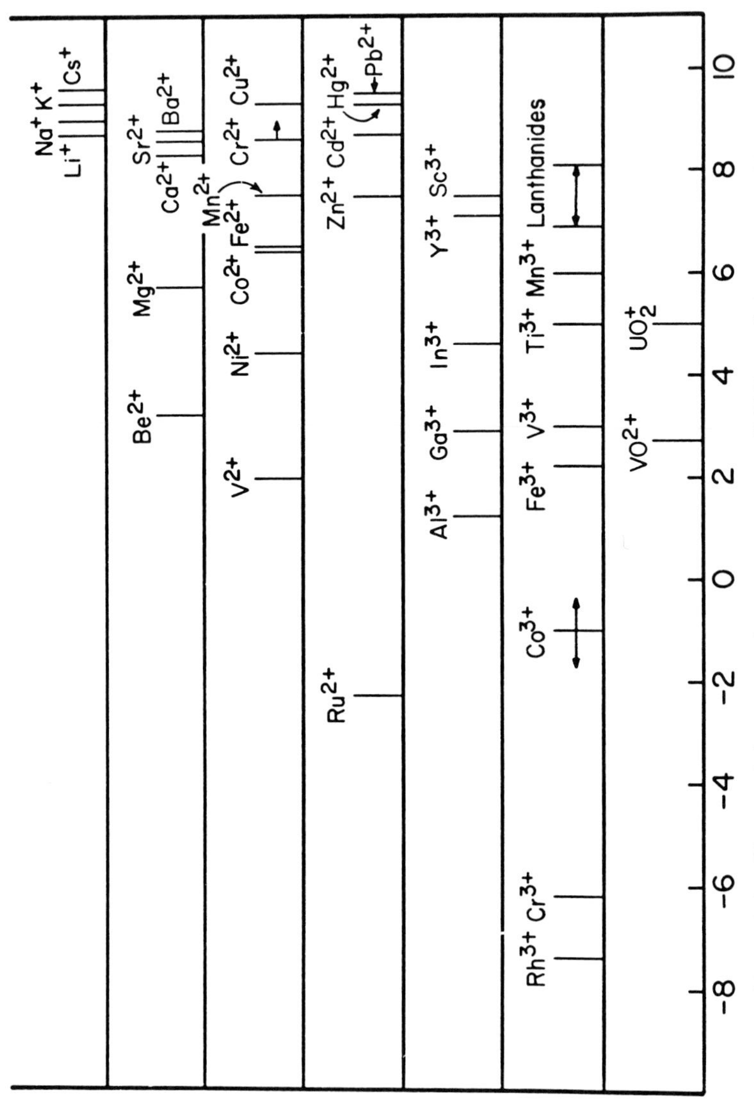

Figure 1-1 Characteristic water exchange rate constants (25°C, sec^{-1}) for aquo metal ions

proved to be consistent with most of the data collected since then. Even for the reactions of Ti_{aq}^{3+}, V_{aq}^{3+}, Cr^{3+}, Fe^{3+}, Mo^{3+}, and In_{aq}^{3+}, where there is some evidence for associative mechanism, it is still convenient to use Eq. (8) for predictive purposes. The characteristic water exchange rate constants are very useful for estimating and comparing reactivities. These constants are presented in Figure 1-1. Inspection will show that for some of these aquometal ions there are no data via isotopic dilution, NMR, or the reactions of monodentate ligands. Where then do the $k^{M\text{-}H_2O}$ values originate for the alkali metal ions for Ba^{2+}, Sr^{2+}, Pb^{2+}, Sc^{3+}, and Y^{3+}? They originate from the reactions of chelates or multidentate ligands, and this, of course, necessitates some additional assumptions. These reactions and the assumptions are discussed in the next section.

It has previously been suggested that an electrostatic model can account for the variation of $k^{M\text{-}H_2O}$, and thus $k^{M\text{-}H_2O}$ is related to the charge/radius ratio of the metal ion. This treatment has been extended by Neely,[37] and he suggests that

$$-\log \boldsymbol{k}^{M\text{-}H_2O} + \log\left(\frac{\boldsymbol{k}T}{h}\right) = \frac{2Ze\mu\Delta r}{2.303\boldsymbol{k}Td^3D} \qquad (13)$$

where $\boldsymbol{k}$ and h are Boltzmann's and Planck's constants, respectively, Z is the charge of the metal ion, e is the basic unit of electric charge, μ is the dipole moment of water, D is the effective dielectric constant, and T is the absolute temperature. The terms d and Δr are related to r_1 and r_2, the separation of the metal ion and the oxygen in the hydrated ion (taken to be the sum of the ionic radii of the metal and oxygen) and in the activated complex, respectively, by the following equations

$$d = \frac{r_1 + r_2}{2} \quad \text{and} \quad r_2 - r_1 = \Delta r \qquad (14)$$

The value of Δr is not critical to this evaluation, but a physically reasonable value of 0.5Å has been used. If the approximations in this treatment are valid, then a plot of $\log \boldsymbol{k}^{M\text{-}H_2O}$ vs. Z/d^3 should give a straight line of an intercept $\log(\boldsymbol{k}T/h)$ (= 12.8 at 25°C) and a slope equal to -1×10^{-25} cm^3.

Figure 1-2 shows a plot of Z/d^3 for a range of metal ions and gives a reasonably straight line for many 2+ and 3+ metal ions where ligand field or other effects do not enter. The slope of this plot is in reasonable agreement with the value calculated from Eq. (13) using values of the bulk dielectric constant and dipole moment for water. The relatively slow rates for V^{2+}, Ni^{2+}, etc. can be attributed to crystal field effects,[1] and the rapid exchange rates for Cu(II), Cr(II), and Mn(III) are attributed to Jahn-Teller distortion and the rapid interconversion of the elongated and compressed axes as discussed earlier. That the observed rate constants fall off for the alkali and alkaline earth metals is not surprising since the $k^{M\text{-}H_2O}$ values for these ions are obtained from k_f rate constants which are close to the diffusion limit. Furthermore, the values for the alkali metal ions,

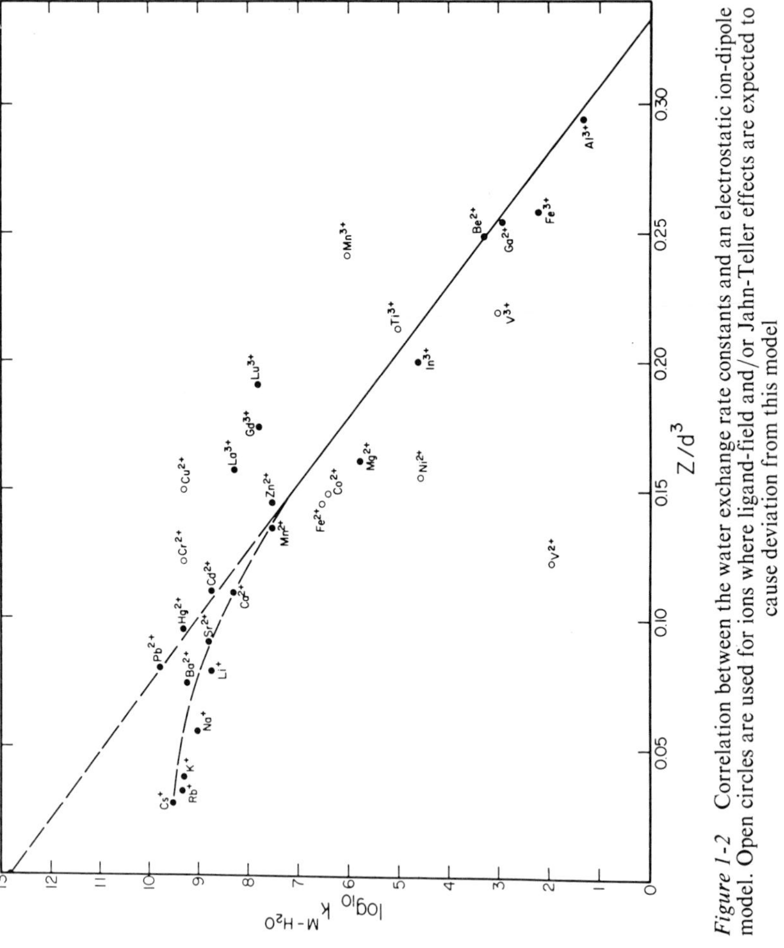

Figure 1-2 Correlation between the water exchange rate constants and an electrostatic ion-dipole model. Open circles are used for ions where ligand-field and/or Jahn-Teller effects are expected to cause deviation from this model

Ba^{2+}, Sr^{2+}, and Ca^{2+}, were obtained from sound absorption measurements using aminopolycarboxylates (cf. Table 1-19) where the rate-determining step may involve chelation effects as well. The large $k^{M\text{-}H_2O}$ values for La^{3+}, Gd^{3+}, and Lu^{3+} may be a result of non-equivalent water molecules around the metal ions (*see* section III.H.), but there also are sizeable variations in the reported rate constants obtained by different techniques.

The pressure dependence of reaction rates gives the volume of activation, which represents the difference in volume between the reactants and the transition state. In principle it should be possible to distinguish between different reaction mechanisms since the volume requirements of the transition states can often be easily visualized. Where the values of $\Delta V^{\ddagger}$ are not complicated by charge effects, negative values of $\Delta V^{\ddagger}$ can be taken to indicate an associative-type mechanism because the volume of a water molecule in the bulk phase is larger than in a complex. Conversely, positive values indicate a dissociative mechanism or charge dispersal in the transition state. This subject has been reviewed recently.[194]

In some recent work on the reactions of neutral molecules with aquo-Ni(II) and aquo-Co(II),[195,196,197] the positive values of $\Delta V^{\ddagger}$ (Table 1-6) are interpreted as representing a stretching of a metal–water bond in the transition state. The similarity between the activation volumes implies a common rate-determining step which does not involve the ligands, while the positive sign suggests either a dissociative mechanism or a mechanism involving charge dispersal in the transition state. These results when taken in conjunction with all the other kinetic

TABLE 1-6
ACTIVATION PARAMETERS FOR THE COMPLEXATION OF NEUTRAL LIGANDS

System	$\Delta H^{\ddagger}$ (kcal mol^{-1})	$\Delta S^{\ddagger}$ (cal mol^{-1} deg^{-1})	$\Delta V^{\ddagger}$ (cm^3 mol^{-1})	Ref.
$Ni^{2+} + NH_3$	10	−6	6.0	196
Ni^{2+} + PADA	13.6	1	7.7	196
$Co^{2+} + NH_3$	6	−15	4.8	196
Co^{2+} + PADA	10.3	−2	7.2	196
Ni^{2+} + imidazole	—	—	11	197
$H_2{}^{18}O$ exchange with:				
$Co(NH_3)_5H_2O^{3+}$	26.6	+6.7	+1.2	198
$Co(en)_2(H_2O)_2{}^{3+}$		+22	+14.3	199
$Rh(NH_3)_5H_2O^{3+}$	24.6	+0.8	−4.1	200
$Cr(NH_3)_5H_2O^{3+}$	23.2	0.0	−5.8	200
$Cr(H_2O)_5H_2O^{3+}$	26.2	+0.3	−9.3	52
$Ir(NH_3)_5H_2O^{3+}$	28.1	+2.7	−3.2	201

TABLE 1-7
VOLUME CHANGES FOR THE AQUATION REACTIONS OF $Co^{III}(NH_3)_5X$

X	$\Delta V^\ddagger$ ($cm^3\ mol^{-1}$)	ΔV [a] ($cm^3\ mol^{-1}$)
NO_3^-	−5.9 [b]	−7.2
Br^-	−8.6 [b]	−10.8
Cl^-	−9.7 [b]	−11.6
N_3^- (75°C)	+16.0 [c]	
NCS^- (88°C)	−4.0 [c]	
SO_4^{2-}	−16.9 [b]	−19.2
H_2O	+1.2 [d]	0.0

[a] Estimated for $\mu = 0.1$ from data $\mu = 1.0$ (30°C), Ref. 207.
[b] Ref. 202.
[c] Ref. 203.
[d] Ref. 198.

evidence provide support for the reactions of Ni(II) and Co(II) aquo proceeding via the Eigen mechanism.

Early work on the tripositive transition metal ions has not produced such conclusive results[202,203] (Table 1-7). The interpretation of the volume changes for the aquation reactions of $Co^{III}(NH_3)_5X$ is complicated by the electrostrictive effects arising from changes in the formal charge. However, the fact that $\Delta V^\ddagger$ is a linear function of ΔV, and the slope of the line is 0.94—i.e. near unity—provides some evidence for a dissociative mechanism since it implies that the

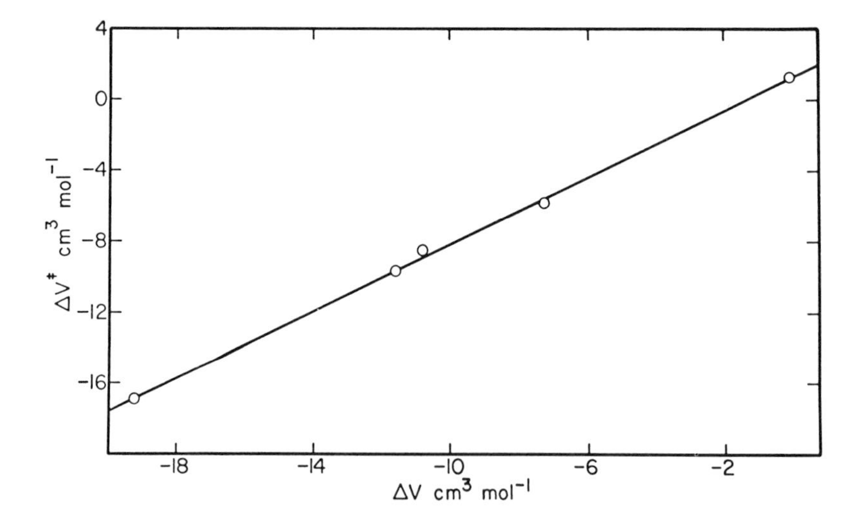

Figure 1-3 Correlation of the activation volume and the change of reaction volume for the aquation reactions of $Co^{III}(NH_3)_5X$

departing ion X^{n-} must be essentially as fully formed in the transition state as in the final products (Figure 1-3).

For $H_2{}^{18}O$ reactions, where electrostrictive effects should not be important, Stranks and Swaddle[200] suggest that the positive value of $\Delta V^\ddagger$ for the Co(III) reaction indicates a dissociative mechanism. In contrast to the Co(III) work, studies with $M(NH_3)_5OH_2$ where M is Rh(III), Ir(III), or Cr(III) and with $Cr^{III}(H_2O)_6$, the pressure dependence of $H_2{}^{18}O$ exchange (cf. negative $\Delta V^\ddagger$ values in Table 1-6) has given rise to the postulation of an associative mechanism for these species.[200,201] This theory supports the conclusions of Poë[204] and Monacelli[205,206] based on anation studies of substituted Rh(III) and Ir(III) species, and it leads to the further possibility that associative activation may be the norm in substitution reactions of trivalent metal ions, with Co(III) complexes being anomalous.

In theory it should be possible to distinguish further between a dissociative and an associative mechanism by investigating the pressure dependence of $\Delta V^\ddagger$, but so far it has not been possible to measure $(\partial \Delta V^\ddagger/\partial P)T$ (the compressibility coefficient of activation) with a sufficient degree of accuracy.

Although it is worthwhile to consider the pros and cons of dissociative vs. associative mechanisms, in either case, the characteristic water exchange rate constants are useful in predicting the rates of formation of monodentate ligand complexes. In other words, most reaction rates are not highly dependent on the nucleophilic properties of the entering ligand.

D. Other Solvents

Substitution reactions in nonaqueous solvents have attracted considerable attention recently. The early work on solvent exchange reactions has been reviewed.[208] Although the dissociative mechanism provides a working model for complex formation of divalent metal ions in water, this mechanism need not apply to other solvents. The substitution rate constants for Ni(II) and Cu(II) in methanol are in agreement with a dissociative mechanism.[209,210,211,212] The interpretation of data for DMSO is not so obvious as in earlier studies. It was suggested that a distinct five-coordinate intermediate exists in this solvent,[213,214] but more recently some doubt has been cast on these results, and evidence for a dissociative mechanism has been provided.[215,216]

Bennetto and Caldin have made an extensive study of solvent effects on the kinetics and activation parameters of Ni(II) and Co(II) reactions with bipyridyl.[217–221] Their results were interpreted as indicating that factors other than the solvent exchange rate must be taken into account to explain the kinetics. The structural properties of the bulk solvent were used to account for the ratios of $k_f/k^{\text{M-solvent}}$ and the activation parameters. Subsequent work on these systems suggests that the nature of the bipyridyl ligand is important in that ring closure may become rate determining,[222,223,224,225] and the outer-sphere complex may be stabilized.[226] Recently, Buck and Moore[227] concluded that steric hindrance in the formation of the first Ni—N bond is significant in DMSO for bipyridyl and terpyridyl. The solvent exchange rates have been shown to be independent of the solvent structure.[228] The nature of the solvent is important in solvents such as glycerol where the rate of diffusion of the reacting species is greatly reduced.[229,230]

MacKellar and Rorabacher[231] interpret the behavior of the Ni(II)–ammonia system in water–methanol mixtures in terms of a dissociative model in which the differing labilities of the varying solvated species are related primarily to the inner-sphere composition. In a further study[232] of Ni(HEEDTA) with ammonia in water–methanol mixtures, the observed trend in the formation rate constant could be attributed to a labilized species with methanol in the exchanging site, and the reverse reaction provides evidence for an interaction with the outer coordination sphere.

III. FORMATION REACTIONS OF MULTIDENTATE LIGAND COMPLEXES

A. Introduction

The reaction rates of multidentate ligands with aquometal ions vary more than the rates of the corresponding reactions of monodentate ligands. Depending on the nature of the multidentate ligand and the metal ion, it is possible to shift the rate-determining step from the first coordination reaction to some later chelation step. Steric effects may become important in ring closure reactions as well as in the initial coordination step. Electrostatic repulsion is important in determining the K_{os} value of a partially protonated multidentate ligand. In some cases the proton-transfer reaction from a partially coordinated ligand will limit the overall reaction rate. All these factors tend to decrease the rate of formation reactions, but there are other factors which can make multidentate ligand reactions faster than monodentate ligand complexation reactions. Electrostatic attraction between metal ions and highly negatively charged multidentate ligands is an obvious factor leading to enhanced rates. In addition, significant enhancement can result from outer-sphere interaction of the multidentate ligand with coordinated water molecules.

It is convenient to consider first a general mechanism for a flexible bidentate ligand (L—L) coordination reaction [Eq. (15), (16), and (17)], where species II is an outer-sphere complex, species III is mono-coordinated, and species IV is chelated.

$$\underset{\text{I}}{M(H_2O)_x{}^{n+}} + (L\text{—}L)^{m-} \underset{k_{21}}{\overset{k_{12}}{\rightleftharpoons}} \underset{\text{II}}{M(H_2O)_x \cdot (L\text{—}L)^{n-m}} \tag{15}$$

$$\underset{\text{II}}{M(H_2O)_x \cdot (L\text{—}L)^{n-m}} \underset{k_{32}}{\overset{k_{23}}{\rightleftharpoons}} \underset{\text{III}}{[M(H_2O)_{x-1}\text{—}L\text{—}L]^{n-m}} + H_2O \tag{16}$$

$$[M(H_2O)_{x-1}{-}L{-}L]^{n-m} \underset{k_{43}}{\overset{k_{34}}{\rightleftarrows}} \left[(H_2O)_{x-2}M\begin{matrix} L \\ | \\ L \end{matrix}\right]^{n-m} + H_2O \quad (17)$$

III IV

Assuming that the outer-sphere reaction is rapid ($K_{os} = k_{12}/k_{21}$), that species III is a steady-state intermediate, and neglecting the reverse reaction from species IV leads to the rate expression in Eq. (18). For ligands with a higher denticity

$$\frac{d[\text{IV}]}{dt} = \frac{K_{os}k_{23}k_{34}}{k_{32} + k_{34}}[M(H_2O)_x{}^{n+}][(L{-}L)^{m-}] \quad (18)$$

than two, similar, expanded expressions can be derived. Eq. (18) is not valid when the concentrations of species II and III become appreciable. Although this is usually not the case for low concentrations of reactants, it is possible for electrostatically favorable outer-sphere complexes and for certain partially chelated species to exist in appreciable concentrations, and this must be kept in mind.

One example of the latter behavior is found in the reaction of La^{3+} with H_2CyDTA^{2-}($CyDTA^{4-}$ = cyclohexanediaminetetraacetate ion) where the observed reaction rate is first order although equal concentrations of reactants are used.[233] The reaction sequence was shown to be

$$La^{3+} + H_2Cy^{2-} \overset{\text{fast}}{\rightleftarrows} \underset{\mathbf{1}}{La{*}HCy} + H^+ \quad (19)$$

$$La{*}HCy \xrightarrow[\text{rds}]{k^*} LaCy^- + H^+ \quad (20)$$

where La*HCy is a complex in which La^{3+} is coordinated to acetate groups, but the metal ion is outside the coordination cage formed by the nitrogen and oxygen donors of CyDTA. One of the nitrogens remains protonated in La*HCy as suggested in structure **1**. The La*HCy complex forms rapidly, and it becomes

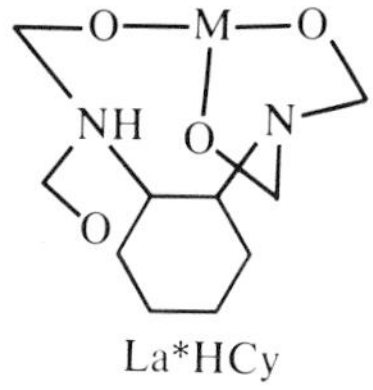

O—M—O
NH O N
O

La*HCy
1

the main reactant giving a first-order conversion to the final product. The value of k^* is only 9 sec^{-1} compared with the $k^{La\text{-}H_2O}$ value of 2×10^8 sec^{-1}.[83] To form the final complex, $LaCy^-$, a proton must be transferred, and additional chelate rings must be formed. In this example the rate of formation of the final complex bears no resemblance to the characteristic water exchange rate constant. Nevertheless, there are many instances in which such a correlation is excellent.

B. Formation Rate Constants Predicted from Characteristic Water Exchange Rate Constants

The second-order formation rate constant, k_f, for a chelate complex is the combination of constants from Eqs. (15)–(18) as expressed in Eq. (21). If chelate

$$k_f = \frac{K_{os}k_{23}k_{34}}{k_{32} + k_{34}} \qquad (21)$$

ring closure is much faster than loss of the singly bound intermediate, then $k_{34} \gg k_{32}$, and the k_f value reduces to that in Eq. (22), which is the same as the rate

$$k_f = K_{os}k_{23} = K_{os}k^{M\text{-}H_2O} \qquad (22)$$

constant expected for a monodentate ligand substitution reaction. This condition often appears to be satisfied (even if $k_{34} = k_{32}$, then $k_f = \frac{1}{2}K_{os}k^{M\text{-}H_2O}$, and the predicted rate constant may be within experimental error of the observed rate constant). Only when k_{34} is significantly less than k_{32} does k_f become obviously less than expected for a monodentate ligand.

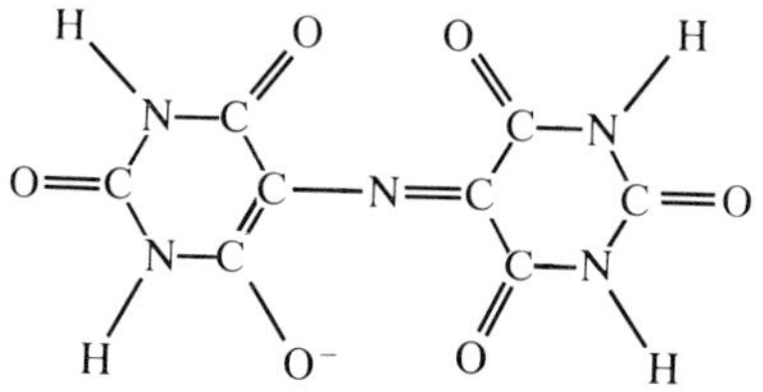

Murexide$^-$

2

Geier[93,94,234] has measured the k_f values for murexide (structure **2**) with a considerable number of metal ions (Table 1-8). Although the ligand is uninegatively charged, K_{os} values of $\sim 0.1 M^{-1}$ for its reaction with 2+ metal ions and $\sim 1 M^{-1}$ for its reaction with 3+ metal ions have been estimated under the ionic strength conditions used.[94] The values for $k_f/k^{M\text{-}H_2O}$ in Table 1-8 are calculated where $k^{M\text{-}H_2O}$ values are obtained from NMR data and where the ratios fit the assigned magnitude of the K_{os} values for Co^{2+}, Ni^{2+}, and Gd^{3+}. Where no values for $k^{M\text{-}H_2O}$ are available, the murexide data can be used to predict this characteristic rate constant. The predicted $k^{M\text{-}H_2O}$ values for Tb^{3+}, Dy^{3+}, Ho^{3+}, and Er^{3+} (using $K_{os} = 1 M^{-1}$) in Table 1-8 are within the range established by NMR studies. The predicted values also are in general agreement with the $k^{M\text{-}H_2O}$ values measured by sound absorption (Table 1-2). The murexide values are more precise, in general, than the sound absorption or NMR values, and they are useful for comparing relative reactivities of metal ions. However, Lin and Bear[235] have estimated K_{os} to be $2M^{-1}$ for the divalent metal ions (25°C, $\mu = 0.1M$) using the Fuoss equation and have suggested that although the first bond formation

TABLE 1-8
RATE CONSTANTS FOR THE REACTION OF MUREXIDE WITH METAL IONS IN 0.1M KNO_3 AT pH 4.0[a]

Metal Ion	Temperature (°C)	k_f (M^{-1} sec^{-1})	$k_f/k^{M\text{-}H_2O}$ [b]	Predicted[c] $k^{M\text{-}H_2O}$ (sec^{-1})
Mn^{2+}	10	$\sim 9 \times 10^6$	$\sim .6$	
Co^{2+}	10	1.5×10^5	0.17	
Ni^{2+}	10	1.0×10^3	0.09	
	25	3.5×10^3	0.12	
Cu^{2+}	10	1.2×10^8		1.2×10^9
Zn^{2+}	10	2.0×10^7		2.0×10^8
Cd^{2+}	10	1.1×10^8		1.1×10^9
Pb^{2+}	10	$>5 \times 10^8$		$>5 \times 10^9$
In^{3+}	12[d]	2×10^6[e]	74[e]	
Sc^{3+}	12[d]	4.8×10^7		4.8×10^7
Y^{3+}	12	1.3×10^7		1.3×10^7
La^{3+}	12	8.6×10^7		8.6×10^7
Ce^{3+}	12	9.5×10^7		9.5×10^7
Pr^{3+}	12	8.6×10^7		8.6×10^7
Nd^{3+}	12	9.3×10^7		9.3×10^7
Sm^{3+}	12	9.6×10^7		9.6×10^7
Eu^{3+}	12	8.2×10^7		8.2×10^7
Gd^{3+}	12	5.2×10^7	2.2	5.2×10^7
Tb^{3+}	12	3.0×10^7	0.5–4	3.0×10^7
Dy^{3+}	12	1.7×10^7	1.4–3.2	1.7×10^7
Ho^{3+}	12	1.4×10^7	0.6–4.1	1.4×10^7
Er^{3+}	12	1.0×10^7	1.9–4.9	1.0×10^7
Tm^{3+}	12	1.1×10^7	4.5–8.8	1.1×10^7
Yb^{3+}	12	1.1×10^7		1.1×10^7
Lu^{3+}	12	1.3×10^7		1.3×10^7

[a] Refs. 93, 94, amd 234.
[b] Values of $k^{M\text{-}H_2O}$ from NMR which are corrected to the indicated temperature.
[c] For the indicated temperature.
[d] At pH 3.0.
[e] Hydrolysis of indium may be causing enhanced reactivity.

is rate-determining for Co_{aq}^{2+}, this is not the case for Ni_{aq}^{2+}. We discuss the general problem of the location of the rate-determining step later.

The $k_f/k^{M\text{-}H_2O}$ ratio for In^{3+} is larger than expected by a factor of 74. The most likely explanation is that hydrolysis of indium (even at pH3) significantly increases the substitution rate. The rate constant for $InOH^{2+}$ with SO_4^{2-} measured by Miceli and Stuehr[64] is $2.5 \times 10^7 M^{-1}$ sec^{-1}, which is 100 times faster than the reaction of In^{3+} with SO_4^{2-}. This value was obtained using a hydrolysis constant of $10^{-3.42} M^{-1}$ and suggests that at pH3 the murexide rate constant could be 30 times too large because of the reaction of $InOH^{2+}$. Indium(III) hydrolysis constants as low as $10^{-2.1}M^{-1}$ have been reported.[236] All the murexide data are in KNO_3, and Geier showed no effect on changing to KCl with Sc^{3+}, Y^{3+}, and La^{3+}. However, comparable reactions with Sm^{3+} in $NaClO_4$ can have rate constants 30% lower.[97]

C. Effect of Ionic Charge on k_f

It is well known that electrostatic attraction or repulsion of reactants can strongly influence the K_{os} value and hence affect the formation rate constant, but the magnitude of the effect is sometimes forgotten. The experimental rate constants in Table 1-9 show the electrostatic effect for the reaction of nickel ion with

TABLE 1-9
FORMATION RATE CONSTANTS FOR THE REACTION OF NICKEL(II) WITH LIGANDS OF VARIOUS CHARGE AT 25°C, μ = 0.1 M

Ligand	$-Z_M Z_L$	k_f (M^{-1} sec^{-1})	Ref.
$H_3\,tetren^{3+}$	−6	3.5	239
$H_2\,trien^{2+}$	−4	9.7×10^1	239
Hen^+	−2	1.7×10^2	240
NH_3	0	4.5×10^3	114, 106
NCS^-	+2	5×10^3[a]	123
$C_2O_4^{2-}$	+4	7.5×10^4	241
IDA^{2-}	+4	8.8×10^4[b]	242
NTA^{3-}	+6	2.0×10^6[c]	243
$HP_3O_{10}^{4-}$	+8	6.8×10^6	244

[a] 0.5 μ.
[b] 0.3 μ.
[c] 0.4 μ.

various ligands. Although the charges are dispersed over moderate distances on many of the multidentate ligands, the total electrostatic effects are sufficient to change rate constants by factors of as great as 10^6. Ionic strength variations have pronounced effects in these reactions and should be kept in mind if detailed comparisons are to be made. An example of ionic strength effects is the variation of the rate constant for Ni^{2+} with malonate ion, $CH_2(COO)_2^{2-}$. In 0.1*M* $NaClO_4$ the rate constant was found to be $7.0 \times 10^4 M^{-1}$ sec^{-1} by temperature-jump relaxation,[237] while a value of $4.2 \times 10^5 M^{-1}$ sec^{-1} was found at ionic strength extrapolated to zero using pressure jump.[238] The factor of six can be accounted for completely using ionic strength corrections for the rate constant.[238]

D. Internal Conjugate Base Mechanism

The discrepancy in the relative formation rate constants of ammonia vs. the polyamines with nickel(II), wherein the polyamine rates were 50 times faster, was pointed out a number of years ago.[239] Rorabacher[114] proposed an explanation of this effect in terms of an internal conjugate base (ICB) mechanism. This mechanism might be considered a partial combination of a conjugate base effect as elucidated by Basolo and Pearson[1] and the third category of substitution reactions proposed by Eigen[7] for internal hydrolysis of the metal ion–ligand outer-sphere complex as applied to chelating ligands. However, enhanced water exchange rates appear to be less important than increased K_{os} values.[240] The ICB mechanism is outlined in Figure 1-4 for ethylenediamine. The essential concept is that one strongly basic donor atom of the ligand is presumed to hydrogen bond to a coordinated water molecule to give a stronger outer-sphere

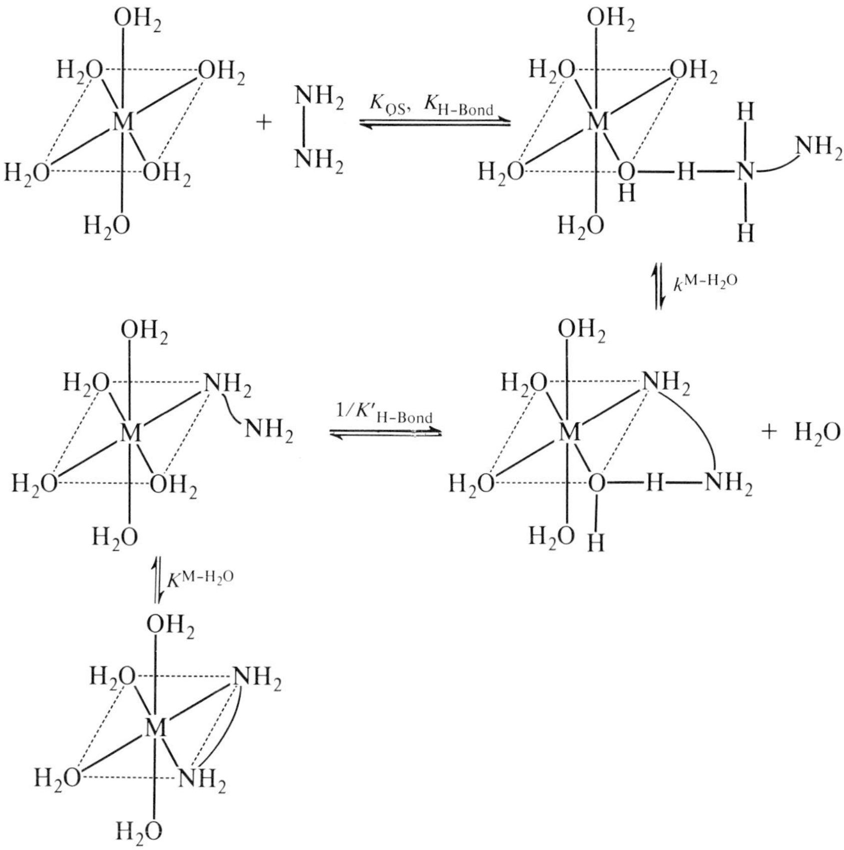

K_{OS}, $K_{H\text{-}Bond}$

$k^{M\text{-}H_2O}$

$1/K'_{H\text{-}Bond}$

$+ H_2O$

$k^{M\text{-}H_2O}$

Figure 1-4 Schematic of the internal conjugate base (ICB) effect in the coordination of ethylenediamine

complex (larger K_{os}) and to labilize subsequent water exchange (larger $k^{M\text{-}H_2O}$) from an adjacent position on the metal ion so that the other donor atom from the ligand can form an inner-sphere complex. At least two donor atoms in the same ligand are needed for the ICB mechanism to permit one to hydrogen bond while the other rotates freely enough to readily occupy an inner-sphere coordination position.

Table 1-10 summarizes the type of data which supports the ICB postulate. Ethylenediamine reacts 80 and 20 times faster than NH_3 with Ni^{2+} and Cu^{2+}, respectively. The enhanced reactivity is even more striking when the 500-fold ratio of rate constants for $C_2H_5NH_2$ and $H_2NCH_2CH_2NH_2$ is considered. The rate constant for Cu(II) with en is so large that it is in part diffusion limited.[245,246] Rorabacher and co-workers have also suggested evidence for ICB effects in Co(II) reactions with poly(amino)alcohols,[102,247] but the argument is not direct and requires a consideration of relative steric effects. Turan[248] has observed a linear relationship between the pK_a of the basic nitrogen atom and the degree of ICB enhancement.

The internal conjugate base mechanism requires at least one of the donor groups of the multidentate ligand to be a moderately strong base to facilitate specific hydrogen bonding to a metal-coordinated water molecule. This hydrogen bonding produces some degree of metal–hydroxide ion character, and hydroxide complexes of Cr(III), Fe(III), Co(III), and Cu(II) are known to give larger k_f values than the aquometal ion for many substitution reactions. However, substitution of a monodentate ligand into the inner sphere of Ni(II) results in a much smaller enhancement of the water exchange rate. For example, $Ni(H_2O)_5OAc^+$ [249] and $Ni(H_2O)_5NH_3^{2+}$ [44] show a three- and seven-fold increase in $k^{Ni\text{-}H_2O}$, respectively. The reaction of $Ni(H_2O)_5OH^+$ with PAR^- [249] shows an enhancement of seven-fold compared with the hexaquo Ni(II) rate constant. Therefore, it is felt that the major contribution to the 200-fold increase of the formation rate constant of Ni(II) with ethylenediamine is from an increase in the outer-sphere association constant, K_{os}.[240] Hence, ICB effects might be expected for ligands with one amine group and one less basic group such as the amino acids. However, the primary criterion used to detect the ICB effect has

TABLE 1-10
COMPARATIVE LOGARITHMIC RATE CONSTANTS FOR FORMATION REACTIONS SHOWING THE ICB EFFECT FOR ETHYLENEDIAMINE[a]

	Ni(II)	Cu(II)
NH_3	3.7	8.3
CH_3NH_2	3.1	—
$C_2H_5NH_2$	2.9	—
en	5.5	9.6

[a] Refs. 106, 114, 127, 246, and 240.

been an enhanced rate. In a way, this turns out to be a self-limiting criterion because the ICB effect can, as we see in the next section, help contribute to a shift in the location of the rate-determining step in amino acids and in other similar ligands. This shift in turn often leads to smaller chelation rate constants and therefore hides any effect which ICB formation may have in the initial coordination step. Perhaps additional evidence to support the ICB mechanism with other ligands is unnecessary because the data in Table 1-10 indicate a very significant effect for which no other explanation has been given. Of course, it is important to see which metal ions exhibit the effect.

E. Location of the Rate-Determining Step

Some multidentate ligands do not react with metal ions as rapidly as would be expected from the $K_{os}k^{M\text{-}H_2O}$ product. This can be readily understood in terms of the reaction scheme in Eqs. (15), (16), and (17). For the alkali metal ions, where it has been necessary to use multidentate ligands to obtain complexes of suitable stability for kinetic measurements, all three formation rate steps in Eqs. (15), (16), and (17) (diffusion, first-bond coordination, and subsequent chelation) can be of comparable magnitude. This gives rise to rates of formation which depend on the number of binding groups of the ligand or the radius of the metal ion.[7] Thus, rate constants for Li^+ with NTA^{3-}, IDA^{2-}, and $P_3O_{10}{}^{5-}$ are 4.7×10^7, 2.5×10^8, and 9×10^8 sec^{-1}, respectively, and for NTA^{3-}, the rate constants for Li^+ to Cs^+ increase by a factor of 7.5.[13]

With divalent metal ions of the first-row transition elements, the rate of ligand–water exchange at the metal ion is usually much slower than the rate of formation of the outer-sphere complex. Therefore, when a multidentate ligand exhibits a slower rate than that for similar monodentate ligand, that chelate ring closure may be limiting the rate. The k_f value for a bidentate ligand is given by Eq. (21), and the reverse rate constant is

$$k_r = \frac{k_{43}k_{32}}{k_{32} + k_{34}}$$

As the rate-limiting step shifts to ring closure, the observed k_f value decreases by the ratio $k_{34}/k_{32} + k_{34}$, and k_r tends toward the limiting value of k_{43}. In fact, it is often the case that a smaller k_f value for comparable reactions is the only evidence for a shift in the rate-determining step from first bond formation to chelate ring closure. This can be a sufficient criterion provided there are no steric effects in first bond formation, but it is not a necessary one because other factors such as the ICB mechanism can offset the effect.

Jones and Margerum[250] studied the kinetics of the equilibrium between $Nien_2(H_2O)_2{}^{2+}$ and $Nien_3{}^{2+}$ where the pH dependence provides more direct evidence for the contribution of ring closure to the limiting rate in the chelation of the third en molecule. The data permitted an evaluation of several of the individual rate constants for the reaction scheme shown in Figure 1-5. The two fully coordinated en molecules increase the lability at the remaining coordination

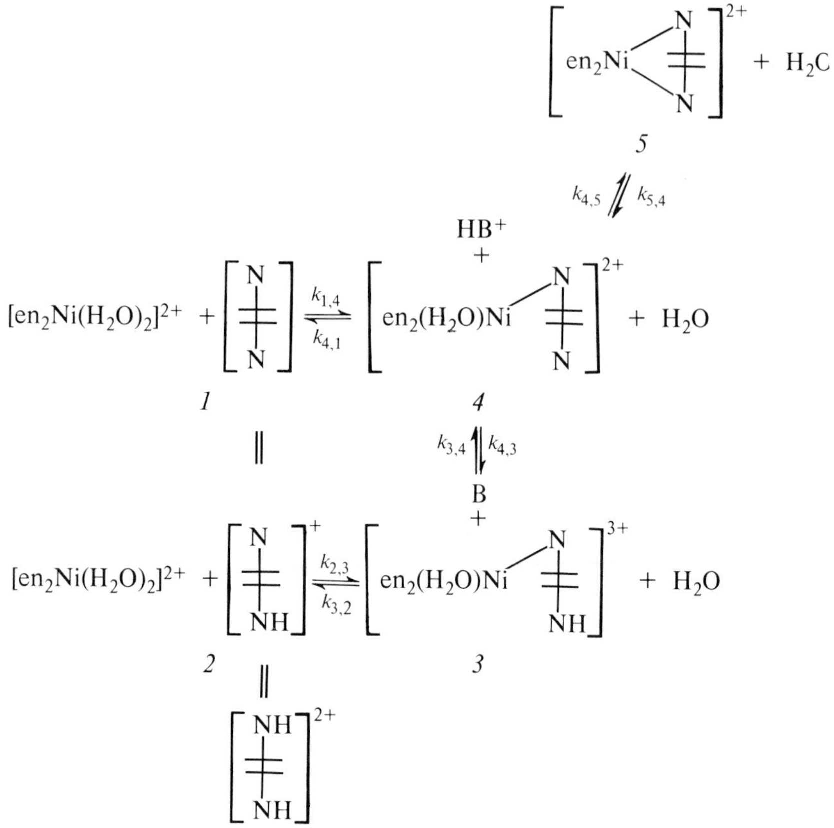

Monodentate en formation: $k_{1,4} = 5.5 \times 10^6\ M^{-1}\ sec^{-1}$
Monodentate en dissociation: $k_{4,1} = 1.2 \times 10^6\ sec^{-1}$
en ring closure: $k_{4,5} = 2.2 \times 10^6\ sec^{-1}$
en ring opening: $k_{5,4} = 87\ sec^{-1}$
$H(en)^+$ formation: $k_{2,3} = 1.1 \times 10^5\ M^{-1}\ sec^{-1}$
Ratio en/$H(en)^+$ monodentate formation: $k_{1,4}/k_{2,3} = 50$
Stability constant of monodentate species: $k_{1,4}/k_{4,1} = 44\ M^{-1}$
Stability constant for ring closure: $k_{4,5}/k_{5,4} = 2.5 \times 10^3$

Figure 1-5 Proposed mechanism for the reaction of en and $H(en)^+$ with $Ni(en)_2(H_2O)_2^{2+}$, where B and HB^+ are a base and its conjugate acid in the rapid proton transfer reaction between *3* and *4*

sites and cause the monodentate en dissociation rate constant ($= 1.2 \times 10^5\ sec^{-1}$) to approach the value of the en ring closure rate constant ($= 2.2 \times 10^5\ sec^{-1}$). By comparison, the reaction of $Ni(H_2O)_6^{2+}$ with ethylenediamine to form $Nien(H_2O)_4^{2+}$ does not have such a complex pH behavior. The ring closure rate is faster than the monodentate en dissociation rate with the result that the first bond formation is the rate-determining step.[12] Other polyamines (triethylenetetramine and tetraethylenepentamine) also react with aquonickel ion with the first bond formation as the rate-determining step.[239]

In the reactions of polyamines and amino acids with transition metal ions, the location of the rate-determining step cannot be evaluated by comparing the chelation rate with corresponding rates of formation of monodentate ligand complexes. As seen in the previous section, the k_f values may be larger for the chelates because of the ICB effect. This accelerating effect is not observed with dicarboxylate ligands because of their weaker basicity. Hoffmann and co-workers[126,251,252] have compared the rate constants of malonate and tartrate complexes with those of other dicarboxylate ligands which do not form chelates—e.g., adipate. This permits estimates of the individual rate constants corresponding to Eqs. (16) and (17), and these are given in Table 1-11. Closure of the six-membered chelate rings with these two ligands is believed to contribute to the limiting rate with both Ni(II) and Mg(II). The $k_{34}/(k_{32} + k_{34})$ ratio varies from 0.6 for nickel–malonate to 0.2 for magnesium–tartrate.

Recently, Cavasino and co-workers[256,257,262] have studied the formation of the nickel complexes of methyl-, ethyl-, *n*-butyl-, and benzyl-malonic acids and phthalic acid. From the pH variation it was possible to calculate the rate constants for a number of chelation steps. From this analysis it appears that there is only a small contribution (6–7%) of ring closure in limiting the rate of these reactions and that the same would be expected for the nickel–malonate system. They suggest that Hoffmann[126] has assigned too large a value to the K_{os} term in the malonate reaction.

The reaction of phthalate ion with nickel has a larger contribution (12%) because of ring closure, while Harada and co-workers[261] suggest that phthalate ion forms a monodentate complex. The reactions of glycolic[254] and lactic acid[253] with nickel ion are believed to have ring closure as the rate-determining step. The results of reactions of nickel ion with various mono- and dicarboxylate ligands are summarized in Table 1-11.

Kustin, Pasternack, and co-workers[263–268] have examined the substitution kinetics of a number of amino acids with transition metal ions and have noted the importance of *sterically controlled substitution* for chelates with six-membered rings as opposed to those with five-membered rings. The most recent rate constants for nickel(II) and cobalt(II), revised after a number of studies, are summarized in Table 1-12 together with copper(II) rate constants by Stuehr and co-workers.[269] The k_f values are consistently smaller for β-alanine compared with α-alanine and for β-aminobutyrate compared with α-aminobutyrate. This has been interpreted as being a result of a shift in the location of the rate-determining step to ring closure for the six-membered chelates. It was suggested initially[263–267] that the smaller k_f values (and the shift to sterically controlled substitution in the ring closure step) were directly related to the lability of the metal ion and that therefore the effect was more important as the $k^{M\text{-}H_2O}$ value increased for the sequence Ni(II), Co(II), Mn(II), and Cu(II). However, as already discussed, the location of the rate-determining step depends on the ratio

TABLE 1-11
STEPWISE RATE CONSTANTS FOR THE FORMATION AND CHELATION OF Ni(II) COMPOUNDS AT 25.0°C, $\mu = 0$

Ligand	$k_f \times 10^{-4}$ (M^{-1} sec^{-1})	k_d (sec^{-1})	$K_0 k_{23} \times 10^{-4}$ (M^{-1} sec^{-1})	k_{43} (sec^{-1})	$\frac{k_{34}}{k_{34} + k_{32}}$	Ref.
Oxalate^{2-}[a]	7.4	3.6				241
	52.0	3.6	85.0[b]	9	0.59	126
Lactate$^-$	2.6	170	15.0[c]	210	0.17	253
	2.5	900	15.0[c]	1200	0.17	126
Glycolate$^-$	2.6	260	15.0[c]	320	0.17	254
	3.3	200	15.0[c]	260	0.22	126
Malonate^{2-}[a]	7.0	44				237
	42.0	35	85.0[b]	70	0.49	255
	45.0	45	85.0[b]	100	0.56	126, 251
Methyl malonate^{2-}[a]			7.3	2200	0.93	256
Ethyl malonate^{2-}[a]			7.6	2700	0.92	257
n-Butyl malonate^{2-}[a]			7.5	3900	0.94	256
Benzyl malonate^{2-}[a]			9.4	3400	0.93	257
Malate^{2-}	56.0	17.0				258
	11.0	50.0				259
Tartrate^{2-}	25.0	200.0	85.0[b]	250	0.30	252
	15.0	160.0	110.0[d]	180	0.11	259
Maleate^{2-}	84.0	2100				260
Succinate^{2-}	58.0	4000				255
Phthalate^{2-}[a]			26.0	13000	0.88	262
	63.0	2600	85.0[b]	(10000)	0.73	261
Adipate^{2-}	85.0	4000				126
Acetate$^-$	15.0	5000				126

[a] $\mu = 0.1M$.
[b] Assumed like adipate.
[c] Assumed like acetate.
[d] Assumed like succinate.

of k_{34}/k_{32}, and because both of these rate constants increase as the lability of the metal ion increases, it is not possible to assign a smaller value of k_{34}/k_{32} as a function of lability. The actual process of bringing the free dentate group around into a position suitable for substitution—i.e., adjacent to a coordinated water—cannot be rate-limiting in itself because there is no reason for this process to be slow. Our proposed transition state involves the loss of water and its replacement by the adjacent free dentate group. On the other hand, the sterically controlled substitution (SCS) mechanism implies a five-coordinate species as a reaction intermediate rather than as a transition state. This is a necessary assumption to have the ligand rotation and ring closure become a rate-limiting step in the SCS mechanism. However, if there were such a limiting rate of rotation, then it would be impossible to explain the fact that the effect is seen for all the metal ions when the k_f values for a given ligand vary by more than four orders of magnitude. The revised data in Table 1-12 show the k_5/k_6 ratio to be important for Ni(II) as well as Co(II) and Cu(II). Hence, alternate explanations are needed as to why the k_{34}/k_{32} ratio is smaller for the six-membered chelate rings than for the five-membered ones.

The rate constant for the ring closure step, k_{34}, should be no greater than the product of the water exchange rate constant of the monodentate-bonded metal ion and the fraction of the free dentate group which is in an exchange position in the outer sphere. As the distance between the two dentate groups increases, this fraction becomes smaller, in accord with the proximity explanation of the chelate effect,[270] and the value of k_{34} becomes smaller. It is unlikely that k_{34} could be further reduced by rotational barriers in moving the second dentate group into the inner sphere from its exchange position. The rotational barriers would be expected to be large for dentates in a ring system such as *trans*-1,2-diaminocyclohexane (chx). Nevertheless, the formation rate constants for Ni^{2+} with chx and en (calculated from the dissociation rate constants[271] and the stability constants[272]) are virtually the same. The term *sterically controlled substitution* does not sufficiently emphasize the importance of the effect of proximity and has sometimes been used to imply that the rotation of the ligand itself is a limiting step independent of the metal ion. We prefer the term *chelation-controlled substitution* for reactions where the ring-closure step is important in limiting the formation rate.

The magnitude of the *proximity effect* can be estimated from the relative distribution of the free end of bidentate ligands of varying chain length when one end is coordinated. Cotton and Harris[273] have calculated a distribution of

TABLE 1-12
FORMATION RATE CONSTANTS (M^{-1} sec^{-1}) FOR THE 1:1 COMPLEX WITH FIVE-MEMBERED (k_5) AND SIX-MEMBERED (k_6) CHELATE RINGS AT 25.0°C

	Ni(II)	Co(II)	Cu(II)
α-alaninate, k_5	3.1×10^4 [a]	1.9×10^6 [a]	1.3×10^9 [c]
β-alaninate, k_6	5×10^3 [a]	1.3×10^5 [a]	2.0×10^8 [c]
ratio k_5/k_6 =	6.2	14.6	6.5
α-aminobutyrate, k_5	1.0×10^4 [b]	2.5×10^5 [b]	
β-aminobutyrate, k_6	4.0×10^3 [b]	2.0×10^4 [b]	
ratio k_5/k_6 =	2.5	12.5	

[a] Ref. 268.
[b] Ref. 264.
[c] Ref. 269.

this type for bidentates with four to seven links. They calculate a difference of 5.4 eu for five- and six-membered chelate rings after correcting for the relative population of ring conformations. However, in the present case, we need the probability ratio for the free end of a four- and five-membered chain to be in a position adjacent to a leaving water molecule. A ratio of 2 was obtained from the Cotton and Harris distribution diagrams (for end-to-end distances) by assuming that the free dentate group has to be in an outer-sphere position but that it cannot be separated from the metal ion by two water molecules. This factor becomes larger when the hydrogen–hydrogen repulsions within the ligand are considered—i.e., conformational population—but should be smaller than the factor of four estimated by Cotton and Harris for five- vs. six-membered rings because of the intervening water molecule. Hence, the proximity effect should cause k_5/k_6 to be between 2 and 8 and thus can account for a major portion of the experimental k_5/k_6 ratios. The end-to-end distribution diagrams indicate that any additional increase in ring size will cause a much greater decrease in the ring-closure rate.

Chelation-controlled substitution depends not only on the number of atoms which separate the two donor groups but on the nature of the two donor groups. The type of the donor group and metal ion are very important in affecting the k_{32} value (and k_{34}), and this has a much bigger influence on the location of the rate-determining step of chelate formation than has been realized.

The mechanisms given by Kustin and Pasternack for the chelation of amino acids have ignored the question of whether the amino group or the carboxylate group coordinates first. Initial coordination of the amino group has been suggested by Wilkins;[5] however, Rorabacher[103] has suggested that the carboxylate group binds first. Arguments can be made from the experimental data to show that the amino group cannot bond first if ring closure is the rate-determining step (rds).

Why the amino group cannot coordinate before the carboxylate group if ring closure is the rds with Ni(II)–β-alanine:

(*1*) If the nitrogen coordinates first, then the ring-closure rate constant should be at least as large as that found for the ring closure of the six-membered ring of monodentate Ni(II)–malonate or the malonate derivatives, and, therefore, $k_{34} \geqslant 5 \times 10^3\ sec^{-1}$. There are two reasons why the ring-closure rate would be faster for $Ni^{2+}—NH_2CH_2CH_2COO^-$ than for $Ni^+—OOCCHRCOO^-$: (a) greater electrostatic attraction of the free COO^- group to the metal site and (b) Ni–amine bond formation tends to labilize coordinated water more than Ni–carboxylate does. Therefore, $k_{34}^{Ni\text{-}\beta\text{-}Ala} > 5 \times 10^3\ sec^{-1}$.

(*2*) The value of the forward rate constant for β-alanine compared with α-alanine requires that $k_{34}/k_{32} \leqslant 1/5$ for β-alanine, provided that the $K_{os}k_{23}$ value for α-alanine is equal to or greater than the value for β-alanine.

(*3*) From (*1*) and (*2*) it follows that $k_{32}^{Ni\text{-}\beta\text{-}Ala} > 2.5 \times 10^4\ sec^{-1}$.

(*4*) However, it is known that $k_{32}^{NiNH_3} = 6\ sec^{-1}$, and there is no reason to expect a value more than 4×10^3 greater for β-alanine. Hence, the initial postulate that the nickel–amino bond forms first is not valid if ring closure is rate determining.

Why the carboxylate group bonds first in metal–amino acid complex formation.

(*1*) Electrostatic attraction of the COO^- group for the positive metal ion.

(*2*) The internal conjugate base effect where outer-sphere association of a coordinated water molecule with the amino group speeds the replacement of the adjacent water by the carboxylate group.

(*3*) In order for ring closure rates to become limiting, the k_{34}/k_{32} ratio must be less than unity. This is favored if the carboxylate group bonds first: (a) k_{32} is much larger for metal–carboxylate than for metal–amino because of weaker bonding and possible ICB effects and (b) k_{34} should be smaller for amino group ring closure because of less electrostatic attraction and because the coordinated carboxylate group does not labilize other coordinated water molecules as much as a coordinated amine.

When the (III) to (II) step corresponds to the loss of a monodendate carboxylate group, the value of k_{32} is quite large ($5 \times 10^3\ sec^{-1}$ for nickel–acetate and $9 \times 10^3\ sec^{-1}$ for nickel-glycollate[126]) and increases as the pK_a decreases for the carboxylic acid. Thus, k_{32} for the α-amino acids should be larger than for β-amino acids. To explain the six-fold decrease in k_f values in going from α-alanine to β-alanine with Ni(II) (when carboxylate bonds first), the value of k_{34} must be reduced for β-alanine more than k_{32} is increased for α-alanine, and it is necessary to have ring closure as the rate-determining step for β-alanine. It follows that $k_r = 0.12\ sec^{-1}$ [268] for nickel–β-alanine and is approximately equal to k_{43}. This is similar to the k_{43} value for an amine nitrogen ring opening step ($0.14\ sec^{-1}$ for $Ni(en)(H_2O)_4^{2+}$ [274]) and quite different from a carboxylate group ring opening step. According to Hoffmann,[126] the value of k_{43} is $100\ sec^{-1}$ for nickel–malonate, and according to Cavasino,[256] it is $2.2 \times 10^3\ sec^{-1}$ and $3.9 \times 10^3\ sec^{-1}$ for substituted malonates.

The possible pathways for the chelation of an aminocarboxylate ligand are shown in more detail in Figure 1-6. The arguments given thus far show that if ring closure is the rate-determining step, then the pathway is through $c \rightarrow f$ and not through $e \rightarrow f$. It can be shown that even for an aminocarboxylate ligand which forms a five-membered ring, the *abb′cf* pathway is the most favorable one. We have qualified the arguments concerning the reactions of β-alanine because its formation rates have become sufficiently slow for the alternate pathway *adef* to contribute. The proximity effect for β-alanine can account for its slower (than α-alanine) reaction via the *abb′cf* pathway, and it also makes the $b \rightarrow e$ pathway unfavorable. However, if the amine group coordinates first,

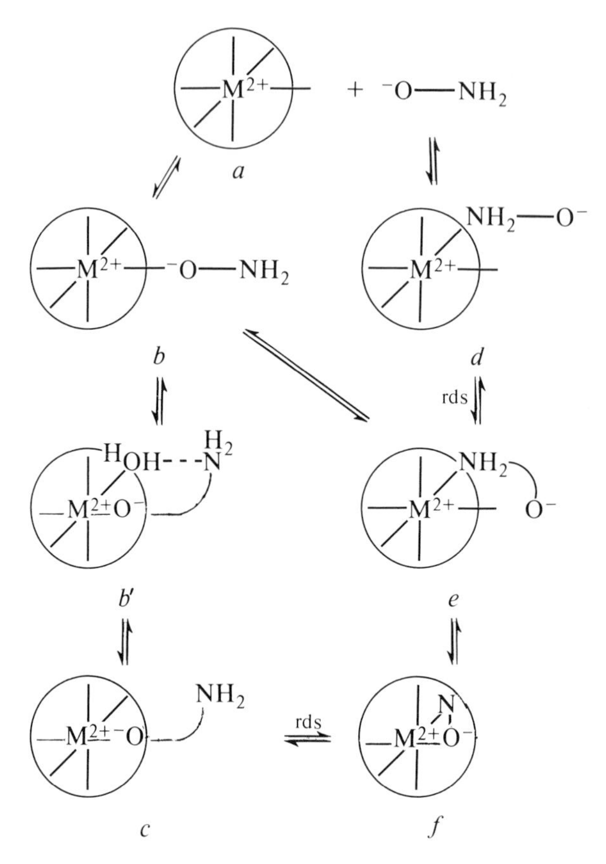

Figure 1-6 Possible pathways for the reaction of aminocarboxylate ligands with aquo metal ions. The circle represents the division between inner-sphere coordination and outer-sphere association. The ICB effect assists initial coordination of the carboxylate end as shown in *b′*. The position of the rate-determining step (rds) depends on which end of the ligand coordinates first.

then the rate-determining step is $d \rightarrow e$ and is not ring closure. The rate constant for the *ade* steps should be greater than that for $C_2H_5NH_2$ reacting with Ni^{2+} ($k_f = 8.7 \times 10^2 M^{-1}$ sec^{-1}) because of the electrostatic attraction of the carboxylate group; therefore, a significant portion of the β-alanine reaction may take place via the initial coordination of the amine group. In both the *abb′cf* and *adef* pathways, it is the amine coordination which is the rate-determining step.

The fact that the k_f values for α-alaninate ion and for glycinate ion reacting with Ni(II) are only $3 \times 10^4 M^{-1}$ sec^{-1} and $4.1 \times 10^4 M^{-1}$ sec^{-1} [268,275] whereas the value for ethylenediamine is $3.6 \times 10^5 M^{-1}$ sec^{-1} suggests that the rate-determining step has shifted towards ring closure for these amino acids. The ethylenediamine reaction is 13–18 times faster despite the greater electrostatic attraction of Gly^- and Ala^-, and an explanation is needed. The ICB effect is possible for the amino acids, and although a statistical disadvantage of two must be considered, this is more than balanced by the increased electrostatic attraction. The Gly^- and α-Ala^- reactions appear to proceed via carboxylate first-bond formation, and their k_{34}/k_{32} ratio must be much less than unity to explain the substantial decrease in rate compared with ethylenediamine.

Therefore, our analysis of these mechanisms suggests that the amino-carboxylate ligands, which form five-membered rings, react by chelation-controlled substitution with Ni^{2+} and probably with other transition metal ions. The proximity effect slows the rate of the aminocarboxylate ligands, which form six-membered rings, to such an extent that the initial coordination of the amine end becomes a competitive pathway. This is a very different explanation from that proposed by Kustin, Pasternack, and co-workers.[263–268]

The experimental observation[242,276,264] that the zwitterion form of amino acids is kinetically much less reactive[277] does not mean that the carboxylate group fails to bond first, but rather it means that such a reaction with the zwitterion is not as favorable a pathway to chelation. Jordon[277] presents evidence that the initial complexation is with the carboxylate group of the zwitterion followed by proton loss from the amino group and chelate ring closure. If the proton-transfer step were limiting, it might be possible to have general base catalysis of the reaction of a metal ion with an amino acid zwitterion [Eq. (23)], but there would be a narrow range of acidity and base concentration where this pathway would compete favorably with the reaction of the free $^-O{=}N$ ligand. A complete mechanism involving the protonated species is discussed in section III, H.

$$\mathrm{M{-}O{-}NH^+} + \mathrm{B} \rightleftarrows \mathrm{BH^+} + \mathrm{M{-}O{-}N} \rightleftarrows \mathrm{M}\langle^{\mathrm{O}}_{\mathrm{N}} \tag{23}$$

The observation that there is an acid dependence for the dissociation rate at low pH has been given as an argument for the Ni—N bond's being the last to break in the acid dissociation of nickel-glycine.[5] However, direct proton attack on the coordinated carboxylate group may shift the acid reaction to this pathway, and, in any case, the pathway in acid need not be the same as that for the glycinate anion.

Initial coordination of the carboxylate group(s) is observed with CyDTA reactions.[233,278] For the Cr(III)–EDTA reaction, the sluggishness of it permits the identification of three individual rate steps.[279,280] The first step is the formation of a chromium–carboxylate bond followed by the slow formation of a chromium–nitrogen bond[279] and a second slow addition of the remaining nitrogen group.[280] First coordination of a carboxylate group also fits other EDTA[281,282] and NTA kinetics.[283] These multidentate ligand reactions have ring closure, rate-determining steps, and indeed this appears to be true for even the bidentate, five-membered ring aminocarboxylates.

Aminocarboxylates are not the only ligands which show evidence of rates' being limited to some extent by ring closure. Another example is the reaction of bipyridyl with nickel ion. The rate constants with pyridine are known[12,119] and can be used to assign $K_{os}k_{23}$ ($= 5 \times 10^3 M^{-1}$ sec^{-1}) and k_{32} ($= 38.5$ sec^{-1}) values for bipyridyl ($k_f = 1.6 \times 10^3 M^{-1}$ sec^{-1}). This results in a k_{34}/k_{32} ratio of 0.5 and indicates that the ring closure contributes to the limiting rate. Other metal ion reactions with bipyridyl also give smaller k_f values than are expected from the water exchange rate constants. For Cu(II), k_f is $4 \times 10^7 M^{-1}$ sec^{-1} and for Cr(II), k_f is $3.5 \times 10^7 M^{-1}$ sec^{-1}.[284] By comparison, the NH_3 and imidazole rate constants with Cu(II) are 2×10^8 and $5.7 \times 10^8 M^{-1}$ sec^{-1}.[127] Hence, ring closure appears to be rate-limiting for bipyridyl with both of these metal ions.[284]

Other ligands with aromatic nitrogen donors also appear to give k_f values which are smaller than expected from the water exchange rate constants. The reactions of the 8-quinolinol (oxine) molecule with Mg(II), Mn(II), Ni(II), and Cu(II)[285-288] are all slower than expected, and the values for Ni(II), Cu(II), and Mn(II) are similar to the bipyridyl and terpyridyl rate constants.[289,284] However, for the reactions of the oxine anion with Mg(II) and Mn(II), the k_f values are in good agreement with the predicted values, suggesting that water exchange is the rate-determining step. The slow rates for the oxine molecule could be a result either of rate-limiting proton transfer or of ring closure, but the former has been discounted by Wilkins[288] and Hague.[285] Hence, the most likely explanation for the slow rates is that the ring closure step, involving bonding of the phenolic oxygen, is rate limiting. For the reaction of the oxine anion, the highest electron density is on the oxygen atom rather than the nitrogen in the oxine molecule,[290] and so, presumably, the metal–oxygen bond is the first to form, giving a smaller value of k_{32}, which prevents a ring-closure limiting rate. Thus, the formation rate constants for the anion are characteristic of the water exchange rate constant of the metal ion.

The reactions of 2-(2-aminoethyl)pyridine (AEP) and 2-aminomethylpyridine (AMP) with Ni(II) are interesting in that it is possible for either the pyridine nitrogen or the amine nitrogen to be involved in the first bond formation. If the metal-amine nitrogen bond forms first, the formation rate constants should be similar to those for $MeNH_2$ and $EtNH_2$ with Ni(II). If the pyridine group is bound first, the rate constant should be enhanced by the ICB effect and be similar to that for ethylenediamine. Naturally, the faster reaction path would be expected, but the rate constants for AEP and AMP are much less than for en. However, we have already suggested that when a pyridine nitrogen is involved in first-bond formation, the ring-closure step can contribute to the formation rate constant. The formation rate constant for $Ni(AEP)^{2+}$ is $9.5 \times 10^3 M^{-1}$ sec^{-1} [291] which is slightly larger than that for $Ni(AMP)^{2+}$, $8.6 \times 10^3 M^{-1}$ sec^{-1}.[242] This is surprising since chelation-controlled substitution would be expected to cause a slower formation of the six-membered ring in $Ni(AEP)^{2+}$ than the formation of the five-membered ring in $Ni(AMP)^{2+}$. However, the pK_a values[292] of AMP (8.62) and AEP (9.52) are considerably different, and this may affect their reactivity. When the amino group in AEP is protonated, the first bond formed must be the metal–pyridine bond, and with no possibility of an ICB effect, the formation rate constant is very small ($7.3M^{-1}$ sec^{-1}).[291]

There are several factors which contribute to chelation-controlled substitution.

(*1*) A weak donor group bonds first causing k_{32} to be large. There can be several reasons why the weaker donor group is the first to bond: (a) electrostatic attraction, (b) ICB contribution from the stronger donor, and (c) steric hindrance of the strong donor—i.e., secondary or tertiary nitrogen atoms.

(*2*) The k_{34} value is reduced. This can be a result of larger ring size and conformational preference (proximity effect) in going from the outer sphere to the inner sphere.

(*3*) A ligand with one strong donor and one weak donor can have a large ICB effect which will help to shift the rate-determining step to ring closure. (Examples include aminocarboxylates and aminoalkylpyridines.)

(*4*) A ligand with two weak donors can partially or completely shift the rate-determining step to ring closure. The weaker the donor, the larger the value of k_{32}, while k_{34} is less dependent on the strength of the donor.

Perhaps the appropriate question is which multidentate ligands really do have first-coordinate bond formation as the rate step? One answer is that those which have two or more strong donor groups, are not sterically hindered, are relatively flexible, and have the donor groups in good proximity—e.g., polyamines. On this basis, murexide is not a very good candidate, which may help to explain the somewhat low K_{os} values assigned to it. In many instances, the ratio $k_{34}/(k_{32} + k_{34})$ may be between 0.1 and 1.0, but the effect on k_f is lost in the uncertain assignment of the K_{os} value.

Thus far, we have emphasized the evidence for the rate-limiting contribution of the second-coordinate bond formation. In a multidentate ligand, can the third, fourth, or fifth coordination step be rate-determining? For the general case, the answer is no because once the first chelate ring is formed the reverse ring-opening step (k_{43}) becomes much smaller in comparison to k_{32} whereas the rate constant for the coordination of the third group (k_{45}) is frequently equal to k_{34} or larger if the first two groups labilize coordinated water. As a result, $k_{45}/k_{43} \gg 1$ and the subsequent ring-closure reactions are fast by comparison to the first or second coordination steps. However, as in most chemical systems, there are exceptions resulting from steric hindrance or the need for proton-transfer reactions. One exception is the reaction of $HCyDTA^{3-}$ with La^{3+} (Eqs. (19) and (20) and structure **1**), and a possible exception is Cu^{2+} with HTKED and $HTHPED^{+}$, in which proton loss following chelation is apparently rate determining.[245]

F. Steric Effects

The importance of steric hindrance in chelation kinetics has been shown in studies by Turan[248] and Rorabacher[293] of the reactions between Ni(II) and *N*-alkyl derivatives of ethylenediamine. The formation rate constant (Table 1-13) for chelation of ethylenediamine is reduced by a factor of 36 with the *N,N'*-diethyl derivative, of 11 with the *N,N*-dimethyl derivative, and of 10^3 with the *N,N,N',N'*-tetramethyl derivative. The tetraethyl derivative is so sterically hindered that it does not form a complex. The triethyl derivative appears to react quite slowly, but its reactions with Ni(II) are not well characterized. Thus, there is a very large steric effect related to the crowding of coordinated water molecules by the alkyl groups. The location of the rate-determining step is believed to shift from first-bond formation with ethylenediamine to second-bond formation with the highly substituted diamines, with TeMeen at the crossover point. Several important conclusions are drawn in this work in regard to the mechanism of multidentate ligand reactions: (a) in linear multidentate ligands such as polyamines, the terminal primary nitrogens are much more reactive than the internal secondary nitrogens; (b) for branched multidentate ligands such as EDTA, the first coordination step must be at terminal donor atoms and not at tertiary nitrogen donors; and (c) for molecules with a high degree of steric hindrance, such as the macrocyclic ligands, the reactions will be slow, and the rate-determining step will shift to other than the first-bond formation. Another interesting aspect of this study is the evidence that *N,N*-diethylenediamine forms a monodentate complex with nickel. Its k_f value is comparable with the value for en rather than the value for $EtNH_2$ and much faster than *N,N*-dimethylethylenediamine. The conclusion is that a sterically hindered nitrogen donor atom can promote the ICB mechanism even if it is too sterically hindered to coordinate significantly

TABLE 1-13

STERIC EFFECTS OF N-ALKYL-SUBSTITUTED ETHYLENEDIAMINES ON THE RATE CONSTANTS FOR THEIR REACTIONS WITH NICKEL(II) AT 25.0°C, μ = 0.1M

Ligand	$k_{Ni}{}^{L}$ (k_f) (M^{-1} sec^{-1})	k^{NiL} (sec^{-1})
en	3.5×10^5	0.008
N,N-DiEten	3.4×10^5	5.4
N,N-DiMeen	3.2×10^4	0.05
N,N'-DiMeen	4.3×10^4	0.01
N,N'-DiEten	9.7×10^3	0.015
TriMeen	2.3×10^3	0.02
TriEten	< 10	—
TeEten	N.R.	N.R.
TeMeen	3.6×10^2	0.1

[a] Refs. 240, 248, and 293.

TABLE 1-14
RATE CONSTANTS FOR THE REACTIONS OF METAL IONS WITH POLY-(AMINOALCOHOL) LIGANDS AT 25.0°C, $\mu = 0.1M$

Metal Ion	Ligand	$k_M{}^L$ (M^{-1} sec^{-1})	$k_M{}^{HL}$ (M^{-1} sec^{-1})	Ref.
Ni^{2+}	TEA[a]	3×10^2		294
	TKED[b]	2.4×10^2	8.6	102
	THPED[c]	1.7×10^2	2.8	102
Co^{2+}	TKED	6.2×10^4	7.5×10^2	247
	THPED	2.3×10^4	2.7×10^2	247
Cu^{2+}	TEA	3×10^7[d]		245
	TKED	3.5×10^7	5.4×10^4	245
	THPED		4.7×10^4	245

[a] TEA = $N(CH_2CH_2OH)_3$
[b] TKED, R = CH_2CH_2OH } for $(R_2NCH_2CH_2NR_2)$
[c] THPED, R = $CH_2CH(OH)CH_3$ }
[d] $\mu = 0.3M$.

to the metal ion. With *N,N*-dimethylethylenediamine, ring closure presumably contributes to the rate.

This type of steric effect is not limited to the reactions of nickel. The rate constant for TeMeen with Cu(II) is $1.7 \times 10^8 M^{-1}$ sec^{-1} [245] compared with a value of $3.8 \times 10^9 M^{-1}$ sec^{-1} for en.[246] It is interesting that the rate constant is reduced by only a factor of 22 with Cu(II) while the ratio is 10^3 for Ni(II). One explanation[245] is that the en rate with Cu(II) is so fast that it is diffusion-limited and that, therefore, the ratios are for different controlling mechanisms for the two metal ions.

Rorabacher and co-workers[102,247,245,294] also have demonstrated the importance of steric hindrance in the reactions of poly(amino alcohol) ligands with Ni(II), Co(II), and Cu(II) (Table 1-14). Here the consequence of the steric interaction is not only to give smaller formation rate constants but to force the second-bond formation—i.e., metal-nitrogen bond formation to be the rate-determining step. Compared with ethylenediamine, the TKED and THPED, ligands react slower by factors of 100–2500 with Cu(II) and Ni(II). The suggested mechanism is first-bond formation involving an alcoholic oxygen donor followed by second-bond formation to a nitrogen donor atom as the rate-determining step. The rate constant for the rupture of the cobalt(II)–alcoholic oxygen bond was estimated as 6×10^6 sec^{-1}, which is about eight times faster than the $k^{Co\text{-}H_2O}$ value; for nickel, the alcoholic oxygen bond dissociation rate constant is about 25 times greater than the $k^{Ni\text{-}H_2O}$ value. The rate constants for TEA are very similar to those for TKED for both Ni(II) and Cu(II), indicating (despite earlier statements to the contrary) that there is no appreciable ICB effect on the overall formation rate constant. This is logical for the postulated mechanism because the ICB effect would influence only the metal–alcohol bond formation, which is not the rate-determining step. There is a ten-fold decrease in the formation rate constant for Ni(II) with TEA compared with NH_3 as a result of chelation-controlled substitution, which in this instance is directly related to steric effects which prevent the amine bond from forming first. Note that there is a significant difference in this mechanism and the mechanism which has been termed sterically controlled substitution.[263,264]

Other structural features of the ligand may lead to small rate constants, as in the case of cis,cis-tach (structure **3**). The equatorial NH_2 groups in the free

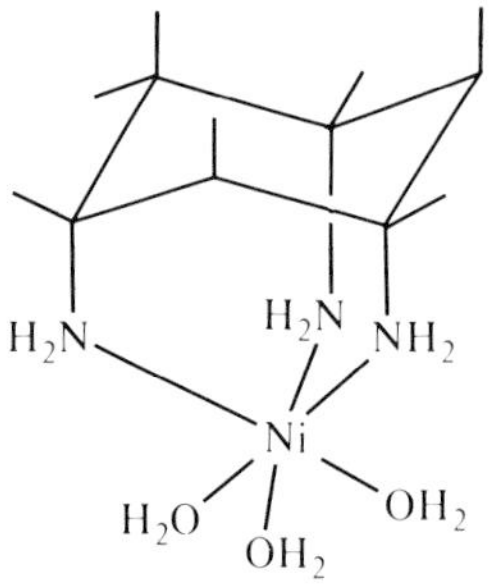

H_2N H_2N NH_2 Ni H_2O OH_2 OH_2

Ni(*cis, cis*-tach)$(H_2O)_3^{2+}$

3

ligand have to be rotated into axial positions so as to occupy the trigonal face of an octahedral complex (structure **3**). As a result, the k_f value is only $9M^{-1}$ sec^{-1} at 45°C for the formation of the nickel complex.[5]

G. Macrocyclic Ligands

The reactions of metal ions with macrocyclic ligands provide additional examples of the effect of steric hindrance on the formation rates of complexes.[295] In the overall reaction, the insertion of a metal ion into the center of a macrocyclic ligand is accompanied by extensive desolvation of the metal ion. The energy required for this multiple desolvation is much too large for it to take place before insertion, and therefore a stepwise coordination pathway is sought in which the macrocyclic donors successively replace the coordinated solvent, as is the case for flexible open-chain multidentate ligands.[239,296,297] However, it is sometimes difficult for macrocyclic ligands to twist and fold in order to maintain a stepwise desolvation path. Some degree of multiple desolvation may be necessary as the metal becomes more highly coordinated by the macrocyclic ligand.

The rate constants for the reactions of Cu(II) with open-chain and macrocyclic polyamines have been measured in acidic and in basic solution as summarized in Table 1-15. The formation rate constants vary by eight orders of magnitude in comparing the open-chain ligand. 2,3,2-tet, and hematoporphyrin IX (*see* Figure 1-7 for ligand structures), In acid solution the aliphatic macrocyclic ligands react much more slowly with $Cu(aquo)^{2+}$ than do the porphyrins. This is because of the low reactivity of the protonated macrocyclic tetramines, which

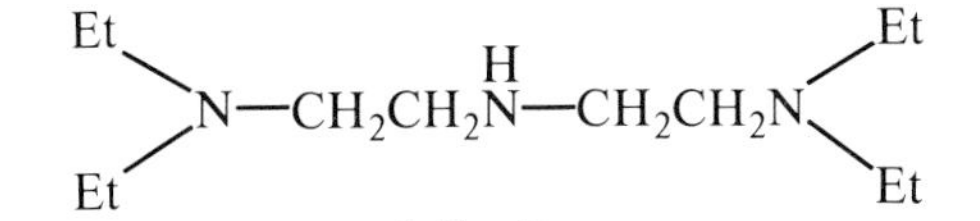

6, Et_4dien

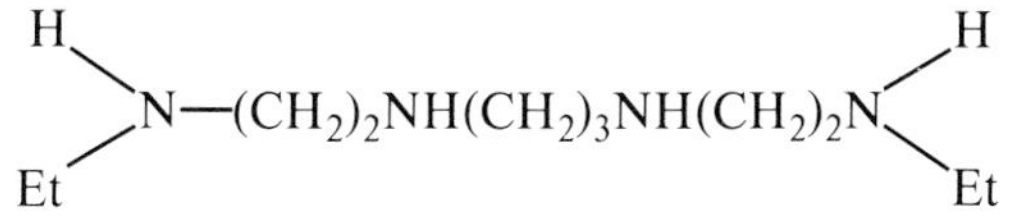

5, Et_2-2,3,2-tet

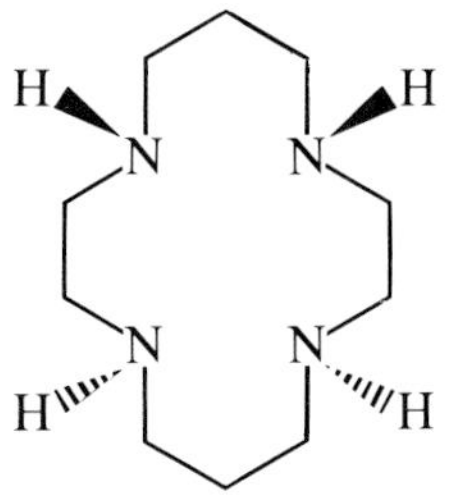

7, Cyclam

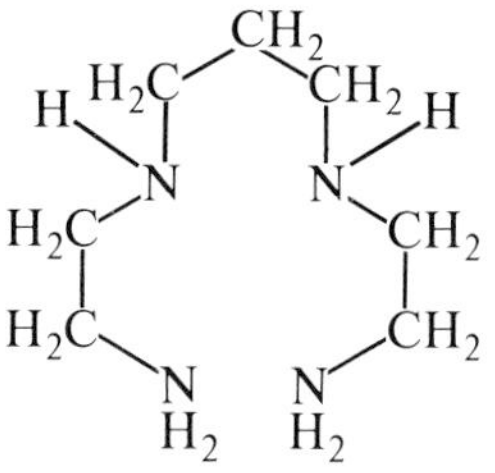

4, 2,3,2-tet

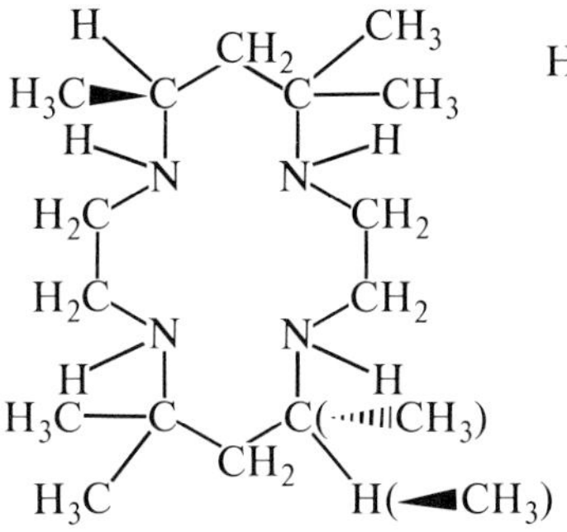

11, tet *a* or *meso*-1,7-CTH
10, tet *b* or *rac*-1,7-CTH

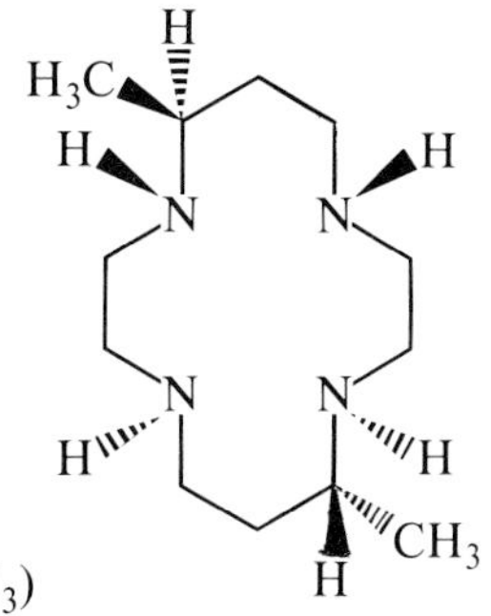

8, Me_2cyclam

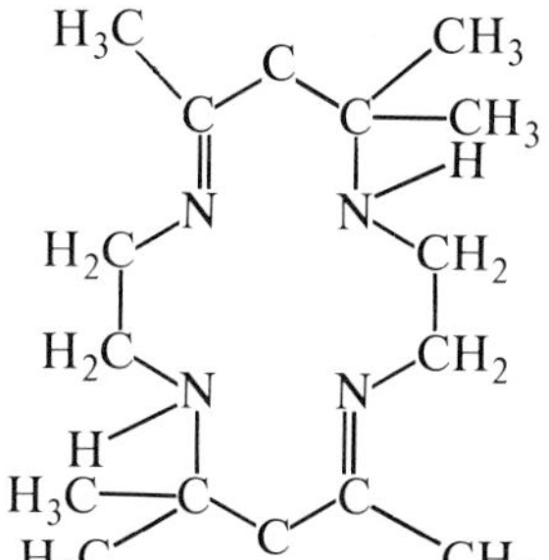

9, trans-(14)-diene or 1,7-CTH

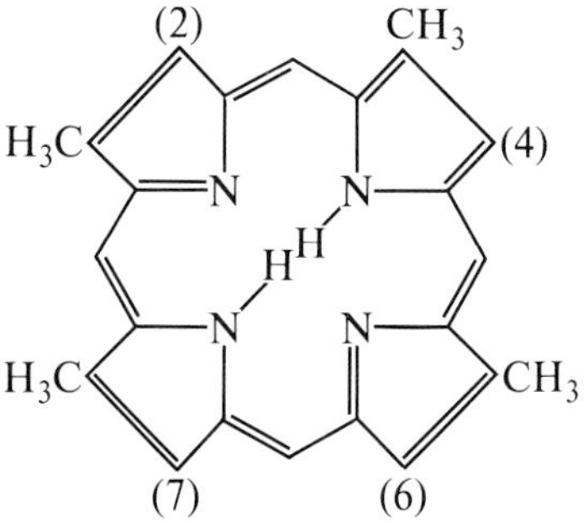

12, hematoporphyrin IX
(2),(4) = $CHOHCH_3$
(6),(7) = CH_2CH_2COOH

13, deuteroporphyrin-2,4-disulfonic acid dimethylester
(2),(4) = SO_3H
(6),(7) = $CH_2CH_2COOCH_3$

Figure 1-7 Structures of ligands in Table 1-15

TABLE 1-15
SECOND-ORDER RATE CONSTANTS FOR THE FORMATION OF Cu(II) COMPLEXES WITH OPEN-CHAIN AND MACROCYCLIC LIGANDS AT 25.0°C

Ligand	Structure	Formation Rate Constant (M^{-1} sec^{-1})	
		In 0.5*M* NaOH	At pH 4.7
2,3,2-tet	4	4.7×10^{6}	
tetren			8.9×10^{4} [b]
Et_2–2,3,2-tet	5	1.7×10^{6}	
Et_4dien	6	4.0×10^{4}	
cyclam	7	2.5×10^{5}	
meso-Me_2cyclam	8	6.9×10^{4}	0.12 [b]
trans-[14]-diene	9	5.6×10^{3}	0.28 [c]
tet *b*	10	3.6×10^{3}	2.7×10^{-2} [c]
tet *a*	11	1.6×10^{3}	2.0×10^{-2} [b] 5.8×10^{-2} [c]
Hematoporphyrin	12	2.0×10^{-2}	
Deuterporphyrin	13		3.2 [d]

[a] Refs. 295, 298, and 299.
[b] $\mu = 0.1M$.
[c] $\mu = 1.0M$.
[d] 0.1*M* HAc–NaAc, $\mu = 0.5M$, Ref. 300.

is in marked contrast to the high reactivity of the protonated open-chain polyamines. (This is discussed in section III, H.) Because acid has such a large effect on the kinetics, the formation rate constants are compared in 0.5*M* NaOH where copper is present as $Cu(OH)_3^-$ and $Cu(OH)_4^{2-}$. The hydroxide complexes are not ideal substitutes for the aquo complex, but strong base is necessary to deprotonate the ligands completely. The ligands are neutral, so it is assumed that the general effects observed would carry over to the behavior of aquo metal ions as well. The rate constants given in Table 1-15 for the reactions in 0.5*M* NaOH show a regular progression. The open chain tetramine 2,3,2-tet, which has two primary amine groups, is the most reactive, and *N,N'''*-diethyl derivative with only secondary amines is less reactive by a factor of four. The tridentate ligand Et_4dien has terminal tertiary nitrogens and is much less reactive than either of the 2,3,2-tet ligands because of the steric hindrance encountered in its coordination.

Rate constants have been resolved for the reactions of $Cu(OH)_3^-$ and $Cu(OH)_4^{2-}$ with these polyamines.[298] The comparative kinetic behavior of the most closely correlated open-chain and cyclic ligands (Et_2-2,3,2-tet and cyclam) indicates that ligand cyclization itself has only a relatively small influence upon the complex formation rate constants. Instead, the more significant kinetic effects arise from substitution at the nitrogen donor atoms of the open-chain ligands or on the alkyl backbone of the cyclic compounds. Jahn-Teller (tetragonal) inversion following the first-bond formation is proposed as the rate-determining step with $Cu(OH)_3^-$. Second-bond formation is proposed as the rate-determining step for the $Cu(OH)_4^{2-}$ reaction with the macrocyclic ligands and the more sterically hindered open-chain polyamines. Since in both cases the ligands are coordinated at the time of the rate-determining step, the reactions exhibit associative character.

As methyl groups are substituted for hydrogens on the carbon atoms of the 14-membered aliphatic macrocyclic tetramines, the formation rate constants decrease, falling by a factor of 3.6 with two methyl groups and by a factor of 150 with six methyl groups in tet *a* (*see* Appendix for full names of tet *a* and tet *b* and other abbreviated substances) (Table 1-15). This is a consequence of the relative ease of folding and twisting the ligands. There is direct evidence that the tet *a* and tet *b* ligands fold in their reaction with Cu(II) because intermediate blue complexes have been isolated for both (the final square-planar product is red), and, for tet *b*, the crystal structure shows copper to be five-coordinate and the ligand to be folded.[301] The blue → red conversion is relatively fast in base but is quite slow in acid.[302] The rate constant given for tet *a* at pH 4.7 is for the formation of the $Cu(tet\ a)(blue)^{2+}$ complex. Although Kaden[303] initially reported a blue form of $Cu(tet\ c)^{2+}$, this apparently was a tet *a* complex.[304] The suggestion by Kaden[303] that the blue form has only two nitrogens coordinated to copper can be ruled out on the basis of the stability constant, the reactivity, and the spectrum of the blue complex. Assuming an initial coordination number of six for Cu(II), the crystal structure of the five-coordinate $Cu(tet\ b)(blue)^{2+}$ salt indicates that stable intermediates with reduced coordination number are possible. This means that at some point in the reaction the addition of one nitrogen donor has caused the loss of two coordinated water molecules. In general it is energetically more favorable if the multiple desolvation step occurs after the coordination of several strong donor groups. Reaction intermediates were not observed with the less sterically hindered macrocyclic ligands such as cyclam and Me_2cyclam.[299]

Recently Kaden[305,306,307] has studied the reactions of cyclam with various degrees of *N*-methyl substitution. Table 1-15A gives the resolved rate constants for the monoprotonated ligands (HL^+) for Ni^{2+}, Cu^{2+}, Co^{2+}, and Zn^{2+}. The mono-*N*-methyl derivative appears to be more reactive than cyclam. This could be the result of forcing the macrocyclic ligand into a more reactive conformation. Further substitution of *N*-methyl groups causes a decrease in rate as observed with the straight chain amines.

Kaden has found that the geometry of the final product depends on the degree of substitution of the nitrogen atoms of the macrocycles. Where one or two of the amine groups are methylated, essentially square-planar products are formed although five-coordinate intermediates can be observed. However, when all four nitrogens are methylated, the final product appears to possess five-coordinate geometry. Kaden states that this is the consequence of the inability of the tet-

TABLE 1–15A
REACTIONS OF METAL(II) IONS WITH VARIOUS METHYL-SUBSTITUTED DERIVATIVES OF MONOPROTONATED MACROCYCLIC TETRAMMINES[a]

Metal Ion	Ligand	k^{M}_{HL} (M^{-1} sec^{-1})	Product Geometry
Ni(II)	cyclam	145,7.4	sq. planar
	1-Mecyclam	55	sq. planar
	2-Mecyclam	10	sq. planar
	4-Mecyclam	1.4	five-coord.
	A	110	sq. planar
	B	55	sq. planar
	C	14	sq. planar
	D	4.5×10^{-2}	five-coord.
Cu(II)	cyclam	8.1×10^{5}	sq. planar
		1.1×10^{7}	sq. planar
	1-Mecyclam	1.2×10^{7}	sq. planar
	2-Mecyclam	2.8×10^{6}	sq. planar
	4-Mecyclam	2.9×10^{5}	five-coord.
	A	1.2×10^{7}	sq. planar
	B	6.5×10^{6} ($k_3 = 1.1 \times 10^{3}$)	sq. planar
	C	3.4×10^{6} ($k_3 = 2.5 \times 10^{3}$)	sq. planar
	D	2.2×10^{3}	five-coord.

TABLE 1–15A CONTINUED

REACTIONS OF METAL(II) IONS WITH VARIOUS METHYL-SUBSTITUTED DERIVATIVES OF MONOPROTONATED MACROCYCLIC TETRAMMINES[a]

Metal Ion	Ligand	k^{M}_{HL} (M^{-1} sec^{-1})	Product Geometry
Co(II)	4-Mecyclam	160	five-coord.
	A	180	octahedral
	D	1.1×10^{-1}	five-coord.
Zn(II)	4-Mecyclam	4.5×10^{3}	five-coord.
	A	8.0×10^{4}	—
	D	88	—

Ligand	n	R_1	R_2	R_3	R_4	R_5	R_6
1-Mecyclam	3	CH_3	H	H	H	H	H
2-Mecyclam	3	CH_3	CH_3	H	H	H	H
4-Mecyclam	3	CH_3	CH_3	CH_3	H	H	H
A	2	H	H	H	H	H	CH_3
B	2	H	H	H	H	CH_3	H
C	2	H	H	H	CH_3	CH_3	H
D	2	CH_3	CH_3	CH_3	H	H	CH_3

[a] Refs. 305, 306, 307, and 308.

ramethyl derivatives to form a conjugate base by deprotonation of one of the coordinated amine groups. The same trends also appear with the 13-membered macrocyclic ligands. In several cases, the rate constant, k_3, of the base-catalyzed pathway for the reaction of Cu(II) with cyclen derivatives has been resolved.[306]

The *trans*-[14]-diene macrocyclic ligand is somewhat more reactive than either of its saturated products, tet *a* or tet *b*. The location of the double bonds does not significantly reduce the ligand flexibility in this case but gives less steric hindrance for some ligand conformations. On the other hand, hematoporphyrin IX with a relatively rigid 16-membered conjugated ring system is much slower to react than any of the other macrocycles. The incorporation of metal ions into porphyrin rings has been the subject of much interest and has been reviewed recently.[309-313] Care must be taken not to overemphasize the rigidity of the porphyrin ring system because there is evidence which indicates a moderate amount of flexibility. Crystal structures of porphyrin diacids[314] and the existence of *N*-methyl-, *N,N'*-dimethyl-, and *N,N',N''*-trimethylporphyrins[315,316,317] indicate the flexibility of the porphyrin nucleus. The acid-catalyzed solvolysis reactions of zinc porphyrins and zinc *N*-methylporphyrins indicate successive porphyrin protonations and metal ion ligations which deform the porphyrin and reduce the coordination number of the porphyrin to the zinc ion.[318] Nevertheless, the porphyrins are more difficult to force into a stepwise reaction pathway than the other macrocyclic ligands in Table 1-15.

H. Protonated Ligands as Reactants and Rate-Limiting Proton-Transfer Reactions

The previous discussions have emphasized the rate behavior of the unprotonated ligand with the aquometal ion, but often the partially protonated multidentate ligand is present in solution in much higher concentration than the unprotonated species. In general, fully protonated ligands can be considered to be unreactive, with the notable exceptions of HF (cf. Table 1-3) and HCN.[319,320] However, the relative reactivity of partially protonated compared with unprotonated (or less protonated) multidentate ligands can vary enormously depending on the nature of the ligand and the metal ion. Two rough classifications can be made based on the ratios of k_f values for L compared with HL (or of $H_{n-1}L$ compared with H_nL): (a) protonation causes a significant decrease in the formation rate resulting in a k_f ratio for $H_{n-1}L$ compared with H_nL which is between 10 and 10^3 and (b) protonation has a greater effect, causing the k_L/k_{HL} ratio to be 10^3 to 10^7.

There are a number of examples of the first case with flexible multidentate ligands where the protonation site is well removed from the initial coordination site of the ligand. Thus, Mg^{2+} reacts with $HADP^{2-}$ ($k_f = 1 \times 10^6 M^{-1}\ sec^{-1}$)[321] and with ADP^{3-} ($k_f = 3 \times 10^6 M^{-1}\ sec^{-1}$) at comparable rates,[322] and Ni^{2+} is reported to react with $H(TPTZ)^+$ ($k = 1.7 \times 10^3 M^{-1}\ sec^{-1}$) and with TPTZ ($k = 2.0 \times 10^3 M^{-1}\ sec^{-1}$)[323] at nearly the same rate. A more common situation for the first case is for the k_f value to decrease by a factor of 10 to 100 with the addition of one proton to the ligand as seen in Table 1-16 for Ni^{2+} reacting with trien vs. $Htrien^+$, mal^{2-} vs. $Hmal^-$, bipy vs. $Hbipy^+$, and terpy vs. $Hterpy^+$. For some ligands where there are many dentate groups a similar effect is seen in comparing multiprotonated species—e.g., Ni^{2+} with $H_2tetren^{2+}$ vs. $H_3tetren^{3+}$ and with $HEDTA^{3-}$ vs. H_2EDTA^{2-}, and Cu^{2+} with $H_3tetren^{3+}$ vs. $H_4tetren^{4+}$. With $Htrien^+$ and with $Hterpy^+$ the protonation site can be fairly distant from

TABLE 1-16
EFFECT OF LIGAND PROTONATION ON THE RATIO OF FORMATION RATE CONSTANTS AT 25°C

Ligand L	Metal	k_L/k_{HL}	Ref.
	Class (a): Moderate Effect		
ADP^{3-} (str. **14**)	Mg^{2+}	3	321,322
TPTZ	Ni^{2+}	1.2	323
$H(TPTZ)^+$	Ni^{2+}	~170	324
trien	Ni^{2+}	~40	239
$Htrien^+$	Ni^{2+}	96	239
$H_2tetren^{2+}$	Ni^{2+}	90	239
$H_2tetren^{2+}$	Cu^{2+}	270	325
$H_3tetren^{3+}$	Cu^{2+}	11	325
$HEDTA^{3-}$	Ni^{2+}	92,63	326,242
$HEDTA^{3-}$	Cu^{2+}	233	327
mal^{2-}	Ni^{2+}	23	237
bipy	Ni^{2+}	64	117,242
	Cu^{2+}	192	284
terpy	Ni^{2+}	14	117,242
$oxine^-$	Mg^{2+}	46	285,287
	Class (b): Large Effect		
en	Ni^{2+}	1.9×10^3	240
en	Cu^{2+}	2.7×10^4	246
histamine	Ni^{2+}	3×10^3	117
histamine	Cu^{2+}	3.7×10^4	328,329
$serine^-$ (37°)	Cu^{2+}	1.5×10^6	328
$H(TPTZ)^+$	Cu^{2+}	$>2.4 \times 10^4$	324
$Hdien^+$	Cu^{2+}	$\sim 3.7 \times 10^5$	324
$Hcyclam^+$	Ni^{2+}	4.2×10^3	330
IDA^{2-}	Ni^{2+}	5.8×10^5	331
NTA^{3-}	Ni^{2+}	6.4×10^4	283
gly^-	Cu^{2+}	$\sim 10^7$	332
$CyDTA^{4-}$	Sr^{2+}	1.0×10^4	333
$CyDTA^{4-}$	Ba^{2+}	4.8×10^3	333

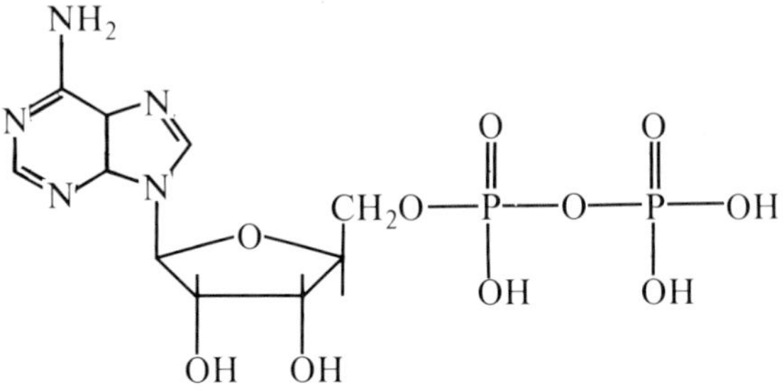

Adenosine-5′-diphosphoric acid (H_3ADP)

14

the initial coordination site and therefore does not have too large an effect. In other multidentate ligands such as H_2EDTA^{2-}, H_2trien^{2+}, and $H_2tetren^{2+}$, the two protons can be distributed on the ligands in such a way as to minimize the effect on the metal coordination reaction—e.g., a possible *kinetically* reactive species for H_2EDTA^{2-} is

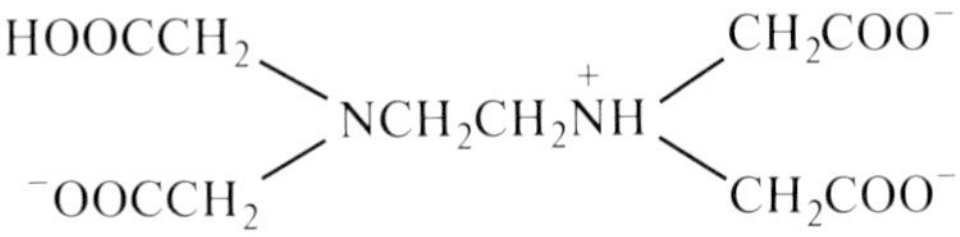

$$(HOOCCH_2)(^{-}OOCCH_2)NCH_2CH_2\overset{+}{N}H(CH_2COO^{-})_2$$

although the thermodynamically stable species is

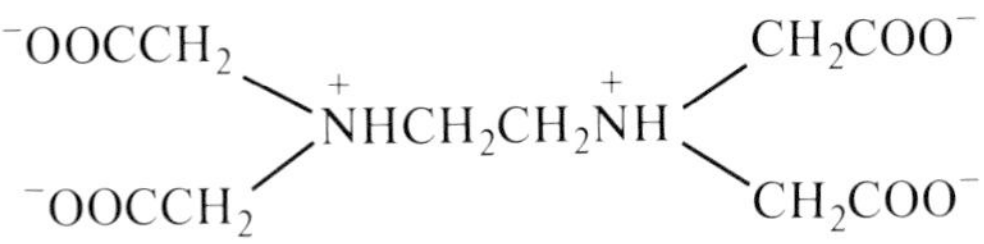

$$(^{-}OOCCH_2)_2\overset{+}{N}HCH_2CH_2\overset{+}{N}H(CH_2COO^{-})_2$$

Protonation will have less effect if the rate-determining step occurs early in the coordination of the multidentate ligand. The bidentate ligands which fall into class (a) exhibit no ICB effects, and ring closure may limit the rate for the unprotonated species. Therefore, the decrease in the rate constant is not too great when these bidentate ligands add a proton.

Several factors can cause a large decrease in the formation rate as a result of protonation [class (b)]. The Hen^+ species loses the ICB rate enhancement of en, and its k_f value is greatly decreased because of this as well as electrostatic repulsion, a statistical effect, and internal hydrogen bonding.[240] The loss of ICB enhancement is also important with H(histamine)$^+$.

In the reactions of IDA^{2-} and NTA^{3-} the rate-determining step is believed to be second-bond formation as discussed previously. This pathway is blocked for the protonated ligands, and in order for them to react, the proton must first be shifted from the nitrogen to one of the acetate groups. The k_L/k_{HL} ratio corresponds to a log K_{HL} difference of about five for this change in protonation site. (The similar shift for H_2EDTA^{2-} is not so unfavorable; the Δlog K_{HL} is less than two.) In the reactions of Sr^{2+} and Ba^{2+} with $CyDTA^{4-}$ the metal ions must coordinate to both nitrogens in the CyDTA coordination cage during the rate-determining step.[333,278,281] In order for these metal ions to react with $HCyDTA^{3-}$ the proton apparently must be shifted away from the nitrogens and the Δlog K_{HL} for this shift is very large, causing k_L/k_{HL} to be large. The $HCyDTA^{3-}$ reactions are interesting in that some metal ions do not require this proton shift until after the rate-determining step and therefore have much faster formation rates relative to the $k^{M\text{-}H_2O}$ value.[278] The $k_M^{HCy}/k^{M\text{-}H_2O}$ ratio is 13M, 86M, 60M, 24M, and 7M for Ni^{2+}, Cu^{2+}, Zn^{2+}, Co^{2+}, and Cd^{2+}, respectively, and $1.1 \times 10^{-5}M$, $2 \times 10^{-4}M$, 5.8×10^{-3} M, $2.9 \times 10^{-2}M$, and $7.2 \times 10^{-3}M$ for Ba^{2+}, Sr^{2+}, Ca^{2+}, Mg^{2+}, and Pb^{2+}, respectively.[333,334] The rate constants for k_{Ni}^{HCy}, k_{Cu}^{HCy}, k_{Zn}^{HCy}, k_{Co}^{HCy}, and k_{Cd}^{HCy} can be considered normal because they correspond to the values expected if an acetate group of $HCyDTA^{3-}$ reacts first with the metal ion and if the ring closure reaction to the first nitrogen bond is the rate-determining step. The ratios of $k_M^{HCy}/k^{M\text{-}H_2O}$ are abnormally small for the alkaline earth ions and Pb^{2+} in accord with the suggested necessity of shifting the proton to another part of the molecule.

The doubly protonated macrocyclic ligand, $H_2cyclam^{2+}$, is very unreactive because the macrocycle keeps the two protons close to the coordination site, thus increasing the electrostatic repulsion much more than in an open chain tetramine. A similar effect is observed with the reactions of tet *a* with copper(II) where the doubly protonated species does not contribute to the rate, even at pH 1.[295]

In general, the zwitterion forms of amino acids are very unreactive because the normal chelation-controlled substitution pathway is blocked, and the alternative path [eq. (23)] involves a proton transfer from $M\text{—}OOCCHRNH_3^+$ to H_2O. This proton transfer has an unfavorably large ΔpK_a change and is therefore relatively slow,[335,336] with a first-order rate constant of about 10^2 sec^{-1}. The stability constant for most $M\text{—}OOCCHRNH_3^+$ species also would be small so that the overall formation rate constant for this path is relatively small. Therefore, the k_L/k_{HL} values are very large for metal–amino acid formation reactions. With Ni^{2+} the zwitterion rate constant is negligible or is very small,[277] and with Cu^{2+} the k_L/k_{HL} ratios are 10^6–10^7 for serine[328] and glycine.[332] A more comprehensive treatment is presented later.

Another comparison of interest in Table 1-16 is the increase by an order of magnitude of the k_L/k_{HL} ratio in the reactions of Cu^{2+} vs. Ni^{2+} with ethylenediamine and with histamine. The Cu^{2+} reactions have been studied recently,[246,328,329] and the postulated rate-limiting step is the release of a proton to the solvent from $CuLH^{3+}$. The lability of copper is sufficient to cause the proton-transfer step to limit its rate, while this is not the case for the slower nickel reactions with these ligands. The mechanism which must be considered is essentially that given in Figure 1-5 replacing $en_2Ni(H_2O)_2^{2+}$ with $Cu(H_2O)_6^{2+}$. The k_f value in terms of all the rate constants is complex and is given by Jones and Margerum.[250] However, the ring-closure step does not limit the rate in the formation of $(H_2O)_4Cuen^{2+}$, and this permits the rate to be expressed in terms of the rate constants which lead up to the formation of species 4, the monodentate $(Cu\text{-}NH_2CH_2CH_2NH_2)^{2+}$. The resulting expression is given in Eq. (24) using the rate constant designations in Figure 1-5 with the stipulation that $k_{45} \gg k_{41}$ and $k_{45} \gg k_{43}[HB^+]$.

$$\text{rate} = k_f[Cu^{2+}][en_T] \text{ where } k_f = \frac{(k_{32} + k_{34}[B])k_{14} + k_{23}k_{34}K_{H(en)}[H^+][B]}{(1 + K_{H(en)}[H^+] + \beta_{2H}[H^+]^2)(k_{32} + k_{34}[B])} \tag{24}$$

This expression can be simplified by writing the rate in terms of the two individual reactive ethylenediamine species [Eqs. (25) and (26)]. It is only permissible to do this when the two paths have their rate-determining steps prior

$$\text{rate} = k_{en}^{Cu}[en][Cu^{2+}] + k_{Hen}^{Cu}[Hen^+][Cu^{2+}] \tag{25}$$

$$k_{en}^{Cu} = k_{14}, \quad k_{Hen}^{Cu} = \frac{k_{23}k_{34}[B]}{k_{32} + k_{34}[B]} \tag{26}$$

to their point of junction—i.e., before or during the formation of species 4. If H_2O is the only form of base B present to accept the proton in step 3 → 4, then $k_{32} \gg k_{34}$ (k_{34} can be estimated to be ~250 sec^{-1} for the transfer to H_2O) and $k_{Hen}^{Cu} = (k_{23}/k_{32})k_{34}$. The values measured by Kirschenbaum and Kustin[246] at 25°C are for k_{en}^{Cu}, $3.8 \times 10^9 M^{-1}\ sec^{-1}$ and for k_{Hen}^{Cu}, $1.4 \times 10^5 M^{-1}\ sec^{-1}$. The k_{34} value for proton transfer to water is obtained from the estimated acidity constant[250] of $10^{7.6} M$ for $(Cu{-}NH_2CH_2CH_2NH_3)^{3+}$ and from the reverse proton-transfer rate for H_3O^+.[336] A value of 500 sec^{-1} is estimated by Sharma and Leussing[328] for reactions at 37°C. The k_{Hen}^{Cu} rate constant will be highly sensitive to catalysis by even low concentrations of base B. The relative proton-transfer rate constants as a function of the ΔpK_a values of MLH and BH^+ can be calculated,[336] giving $k_{34} \simeq 10^8$ (for B = Hen^+) and $k_{34} \simeq 10^6$ (if B is an indicator with a pK_a of 4). Therefore, as little as $2.5 \times 10^{-6} M$ Hen^+ or $2.5 \times 10^{-4} M$ indicator or buffer could double the rate. Experimental work[246] using relaxation techniques avoided this problem since no buffer was used; the indicator concentration was ~$10^{-5} M$, and the free Hen^+ concentration was at least a factor of 10 below the level needed for catalysis to be important. Nevertheless, it is important to keep in mind that the k_{Hen}^{Cu} value in the presence of even low concentrations of buffer is predicted from Eq. (26) to become greater, approaching a value of $2 \times 10^6 M^{-1}\ sec^{-1}$.

The ligand-exchange kinetics of nitrilotriacetate complexes with Cd^{2+}, Zn^{2+}, and Pb^{2+} have been studied by nuclear magnetic resonance techniques.[337,339] The k_f values for the reaction in Eq. (27) are summarized in Table 1-17 along

$$M^{2+} + HNTA^{2-} \underset{}{\overset{k_f}{\rightleftarrows}} M(NTA)^- + H^+ \tag{27}$$

TABLE 1-17
EXPERIMENTAL AND PREDICTED FORMATION RATE CONSTANTS FOR THE REACTIONS OF METAL IONS WITH $HNTA^{2-}$ AT 25°C

Metal	Ionic Strength (M)	Experimental k_f (M^{-1} sec^{-1})	Predicted[a] k_f (M^{-1} sec^{-1})	Ratio of Exp./Pred.	Ref.
Zn^{2+}	1–2	5.1×10^5	1.5×10^9	3.4×10^{-4}	337
Cd^{2+}	1–2	2.1×10^5	1.2×10^{10}	1.7×10^{-5}	337
Pb^{2+}	0.1	4.4×10^6	2.6×10^{10}	1.7×10^{-4}	339
Ni^{2+}	1.25	7.5	9×10^5	8.3×10^{-6}	283
Cu^{2+}	0.5	1.1×10^5	6×10^{10}	1.9×10^{-6}	338

[a] Based on $k^{M\text{-}H_2O}$ (sec^{-1}) = 5×10^7 (Zn^{2+}), 4×10^8 (Cd^{2+}), 2×10^9 (Pb^{2+}), 2×10^9 (Cu^{2+}), 3×10^4 (Ni^{2+}), and $K_{os} = 30M^{-1}$ except for Pb^{2+} where a K_{os} value of 13 was used, Ref. 339.

with that for the nickel[283] and copper[338] reactions studied spectrophotometrically. The experimental k_f values are all 10^{-6} to 10^{-4} smaller than expected from the predicted $K_{os}k^{M\text{-}H_2O}$ product. These are chelation-controlled substitution reactions, and the proton must be shifted from the nitrogen-protonated $HNTA^{2-}$ to the carboxylate-protonated $HNTA^{2-}$ to give a reactive species. The estimated stability of the protonated-carboxylate form relative to the protonated-nitrogen form is 3.3×10^{-5} [337] which approximately agrees with the ratio of experimental to predicted k_f values. Rabenstein and Kula[337] proposed that proton transfer from the nitrogen atom of $HNTA^{2-}$ governs the rate of its reaction with M^{2+} in an intermediate metal–carboxylate bonded M(HNTA) species. However, an alternate path in which the proton is shifted before reaction with the metal ion would not be limited by the proton transfer rate. (Also, the value of 175 sec^{-1} for migration of the proton from the amino group to the carboxylate group in glycine[340] seems unusually small considering the possible H-bonded pathways.) In any case, the fact that the very much slower reaction of Ni^{2+} with $HNTA^{2-}$ has an even smaller k_f ratio of exp./pred. rules out proton-transfer limiting steps as a cause for the lower ratio for Cd^{2+} compared with Zn^{2+}. When the variations in ionic strength and the uncertainty of the $k^{M\text{-}H_2O}$ values for Zn^{2+}, Cd^{2+}, Cu^{2+}, and Pb^{2+} are considered, the fit of the data to the carboxylate-protonated $HNTA^{2-}$ model seems rather good. One obvious refinement would be to use K_{os} values based on the metal–acetate stability constants and the electrostatic attraction of a nearby carboxylate group in a manner similar to the treatment used for $HCyDTA^{3-}$ kinetics.[278] This would predict a K_{os} ratio for Pb^{2+}/Ni^{2+} of 13 and would give nearly the same exp./pred. k_f ratio for these ions.

Maguire[338] has made the interesting observation that H_2NTA^- reacts faster with Cu^{2+} than does $HNTA^{2-}$. This is an unusual situation where the more protonated ligand is more reactive. An explanation for the observed kinetic behavior may be that in the case of the monoprotonated species, there is a rate-limiting transfer of a proton from nitrogen before complex formation takes place and that this transfer is facilitated with increasing acidity.[338]

Kinetic studies of metal ion reactions with amino acids have recognized the much greater reactivity of L^- compared with the zwitterion $HL^{\pm}$,[242,264,268,269,276,341] and usually rate contributions for the zwitterion have not been observed even in quite acidic solutions. A recent study by Sharma and Leussing[328] of the complexation of Cu^{2+} with serine at pH 2.5–4.0 necessitated the introduction of rate constants for the reaction of $SerH^{\pm}$ to form $Cu(SerH)^{2+}$ and its loss of a proton to the solvent to give $Cu(Ser)^+$.

$$Cu^{2+} + SerH^{\pm} \underset{k_{-27}}{\overset{k_{27}}{\rightleftharpoons}} Cu(SerH)^{2+} \tag{28a}$$

$$Cu(SerH)^{2+} \underset{k_{-28}}{\overset{k_{28}}{\rightleftharpoons}} Cu(Ser)^{+} + H^{+} \tag{28b}$$

Although these authors discuss two possible forms of $Cu(SerH)^{2+}$ and suggest two pathways, it appears that the rate constants calculated from the experimental data did not consider both pathways simultaneously. In any case, the reported constant, $k_{27} = 1.2 \times 10^3 M^{-1}\ sec^{-1}$ (37°C), fits the expected formation rate constant for $[CuNH_2CH(CHOH)CO_2H]^{2+}$ but does not fit the expected value for $[CuO_2CCH(CH_2OH)NH_3]^{2+}$, which should be much larger. However, k_{28}

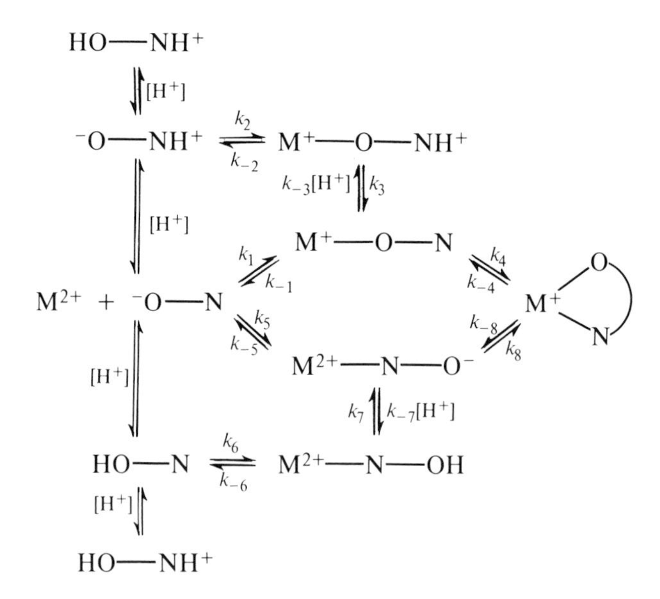

Figure 1-8 General mechanism for the reaction of a metal ion with an amino acid considering the reactions of the various protonated forms

$= 1.9 \times 10^2\ \mathrm{sec}^{-1}$ (37°C) is in the predicted range for the proton transfer from $[CuO_2CCH(CH_2OH)NH_3]^{2+}$ but not from $[CuNH_2CH(CH_2OH)CO_2H]^{2+}$, which would give a much larger k_{28} value.

The full mechanism for the formation of a metal–aminocarboxylate complex has not been considered as a whole. We give such a mechanism in Figure 1-8. There is sufficient information about metal ion coordination rates and proton-transfer rates to predict which pathways will contribute to the formation of the chelate as a function of pH and the metal ion. The ligand may be present in the L^-, $HL^\pm$, HL^0, and H_2L^+ forms. The mono-protonated ligand is predominantly in the zwitterion form, $HL^\pm(^-O{-}NH^+)$, but the HL^0 form (HO—N) also is considered as a possible reactive species. The rate of formation of the chelated ligand, v_T, is the sum of the rates via the L^-, $HL^\pm$ and the L^-, HL^0 paths [Eq. (29)], and L_T is the sum of all the amino acid forms. Using steady-state ap-

$$v_T = v_{L,HL^\pm} + v_{L,HL^0} = k_T[L_T][M^{2+}] \tag{29}$$

proximations for all the unidentate-coordinated species, assuming a pre-equilibration of L^-, $HL^\pm$, HL^0, and H_2L^+, and defining $K_{HL^\pm} = [HL^\pm]/[H^+][L^-]$, $K_{HL^0} = [HL^0]/[H^+][L^-]$, and $\beta_{2H} = [H_2L^+]/[H^+]^2[L^-]$ leads to the general expression for the rate constant given in Eq. (30).

$$\begin{aligned} k_T(1 + K_{HL^\pm}[H^+] + K_{HL^0}[H^+] + \beta_{2H}[H^+]^2) \\ = \frac{k_1(k_3 + k_{-2})k_4 + k_2k_3k_4K_{HL^\pm}[H^+]}{k_{-1}(k_3 + k_{-2}) + k_4(k_3 + k_{-2}) + k_{-2}k_{-3}[H^+]} \\ + \frac{k_5(k_7 + k_{-6})k_8 + k_6k_7k_8K_{HL^0}[H^+]}{k_{-5}(k_7 + k_{-6}) + k_8(k_7 + k_{-6}) + k_{-6}k_{-7}[H^+]} \end{aligned} \tag{30}$$

This general expression is rather formidable, but it is readily simplified. Since $K_{HL^\pm} \gg K_{HL^0}$, the $K_{HL^0}[H^+]$ term on the left side of the equation can be dropped. For Ni(II), Co(II), or Cu(II) as the metal ions, it can be shown that $k_7 \gg k_{-6}$ but that $k_{-2} \gg k_3$, based on proton-transfer rates to the solvent vs. metal–ligand dissociation rate constants. This gives Eq. (31).

$$k_T(1 + K_{HL^\pm}[H^+] + \beta_{2H}[H^+]^2) = \frac{k_1k_4 + \left(\frac{k_2}{k_{-2}}\right)k_3k_4K_{HL^\pm}[H^+]}{k_{-1} + k_4 + k_{-3}[H^+]} + \frac{k_5k_8 + k_6k_8K_{HL^0}[H^+]}{k_{-5} + k_8 + k_{-6}\left(\frac{k_{-7}}{k_7}\right)[H^+]} \tag{31}$$

To resolve the rate constant, k_T, into individual rate constants for the unprotonated and protonated forms of the amino acid, it is necessary to drop the

$[H^+]$ terms in the denominators. It is easily shown for Ni^{2+}, Co^{2+}, and Cu^{2+} that $k_8 \gg k_{-5} + k_{-6}(k_{-7}/k_7)[H^+]$; at least this is the case below $1M$ $[H^+]$. However, the $k_{-3}[H^+]$ term is approximately $10^{10}[H^+]$, and it is important even at pH 6–7 for nickel ion. Hence it is not accurate to refer to an L^- rate constant and an $HL^{\pm}$ rate constant for data taken below pH 7 for the reaction of Ni^{2+} and an amino acid. The same restriction holds for Cu^{2+} reactions below pH 4. The $HL^{\pm}$ pathway may contribute to the rate as well as the L^- pathway, but the two pathways become indistinguishable if $k_4/k_{-1} \ll 1$. This can be shown by using the ratio of rate constants required by the cyclic pathway, namely, $k_2k_3/k_{-2}k_{-3}\, K_{HL^{\pm}} = k_1/k_{-1}$, and the earlier inequality involving k_8, to give Eq. (32). If $k_4/k_{-1} \ll 1$, then the first term on the right side of Eq. (32) becomes

$$k_T(1 + K_{HL^{\pm}}[H^+] + \beta_{2H}[H^+]^2) = \frac{k_1k_4\left(1 + \frac{k_{-3}}{k_{-1}}[H^+]\right)}{k_{-1}\left(1 + \frac{k_4}{k_{-1}} + \frac{k_{-3}}{k_{-1}}[H^+]\right)} + (k_5 + k_6K_{HL^0}[H^+]) \quad (32)$$

k_1k_4/k_{-1} for all pH values, yet $k_{-3}/k_{-1}[H^+]$ may be much greater than unity so that the reaction can be considered to proceed via the $HL^{\pm}$ species as well as the L^- species. The pathways are indistinguishable when $k_4/k_{-1} \ll 1$ because the rate-determining step is k_4 and because the other species are all in pre-equilibrium. On the other hand, if k_4 is not the rate-determining step, then the k_4/k_{-1} term cannot be dropped. At high pH $(k_{-3}/k_{-1})[H^+]$ is small, and the first term on the right side of Eq. (32) becomes $k_1k_4/(k_{-1} + k_4)$, while at low pH $(k_{-3}/k_{-1})[H^+]$ is larger, and the term again becomes equal to k_1k_4/k_{-1}. Thus, the pH dependence of the rate constant indicates the location of the rate-determining step. The reactions of glycine and lysine with nickel were studied by stopped-flow methods[242] from pH 5.8–7.0 and did not give evidence of a shift in rate constant from $k_1k_4/(k_{-1} + k_4)$ to k_1k_4/k_{-1}. This agrees with our earlier postulate *that for five-membered chelation of aminocarboxylates to nickel, the location of the rate-determining step is ring closure of the amino group.* We have discussed previously the relative contribution of k_5, which for nickel we estimate would carry less than 3% of the reaction because of the preference for initial carboxylate coordination as opposed to initial amine group coordination. As the acidity increases, the relative contribution from the $k_6K_{HL^0}[H^+]$ term increases, but it should not become appreciable until pH < 3 for nickel ion. Also, if $k_4 \ll k_{-1}$, then proton-transfer steps are not rate-limiting, and the rates will not depend on buffer concentrations. Another estimate of the k_5 rate constant for nickel(II) can be obtained from the k_f values[242,275] of diglycine ($GlyGly^-$) which are approximately one-tenth of those for Gly^-. The $Ni(GlyGly)^+$ complex is coordinated by the amino group and the peptide oxygen (the peptide nitrogen is not deprotonated and will not be coordinated) so this complex must form by the initial coordination of the $—NH_2$ group.

The approximations which gave Eq. (32) are valid for the reactions of Cu^{2+} as well as Ni^{2+}. Table 1-18 gives a summary of the effect of pH on the relative contributions of L^-, $HL^{\pm}$, and HL^0. One of the main differences in the expected behavior of Cu^{2+} and amino acids is that k_{-3}/k_{-1} is only about $10^2 M$ compared with about $10^6 M$ for the Ni^{2+} reactions. As a result the $HL^{\pm}$ path for Cu^{2+} does not begin to contribute until pH 3, and at this low a pH the HL^0 path also contributes. The second difference is that k_1 becomes diffusion limited in the copper reactions (the ICB effect cannot be fully realized), and k_5 is therefore more important. For the same reason, the contribution from the HL^0 pathway (k_6 term) becomes significant about one pH unit higher than in the case with nickel. Finally, the k_{-1} and k_4 terms should be on the same order of magnitude for the copper reactions because of the stronger carboxylate complexes. Therefore, at pH 4 or higher the right side of Eq. (32) should equal $k_1 k_4/(k_{-1} + k_4) + k_5$ for the reactions of copper and amino acids. At pH 2 and below, the right side of Eq. (32) should equal $k_6 K_{HL^0}[H^+]$. In the pH range 2.5–4.0, where the Cu^{2+}

TABLE 1-18

METAL-ION AND pH DEPENDENCE PREDICTED FOR THE RELATIVE CONTRIBUTIONS OF REACTION PATHWAYS WITH α-AMINO ACIDS

Mechanism	Ni^{2+}	Cu^{2+}
only L^- path	>pH 7	>pH 3
L^- and $HL^{\pm}$ paths	pH 2–7	~pH 3
HL^0 path favored over $HL^{\pm}$ path	<pH 2	<pH 3

Estimated Rate Constants for the Mechanism in Figure 1–8[a]
($k_{-3} = k_{-7} = 10^{10}$, $k_3 = 10^{2.5}$, $k_7 = 10^{7.7}$)

	Ni^{2+}	Cu^{2+}
k_1	$>10^{5.6}$	10^{10}
k_1/k_{-1}	$10^{1.3}$	$10^{2.3}$
k_2	10^4	10^9
k_{-2}	10^4	10^8
k_4	10^3	10^8
k_5	$10^{3.8}$	$10^{8.8}$
k_{-5}	$10^{0.4}$	10^4
k_6	$10^{2.6}$	$10^{7.6}$
k_{-6}	10^1	$10^{4.6}$
k_8	$>10^{4.6}$	$>10^8$

[a] Ref. 342.

and serine reactions were studied,[328] the full expression in Eq. (32) is needed, and a complex pH dependence can be expected. The complete mechanism must be considered to draw valid conclusions about relative rate constants and reaction pathways.

Another interesting point about the copper reactions at pH 4–6 is that the k_T rate constant could be affected by general acid–base catalysis in strongly buffered solutions. If the k_3 and $k_{-3}[H^+]$ terms are supplemented by larger terms from $k'_3[B]$ and $k'_{-3}[HB^+]$, where B and HB^+ are buffers, then it would be possible to shift the value of the right side of Eq. (31) to $(k_1k_4/k_{-1}) + k_5$.

The formation of metal–peptide complexes involves reactions in which proton-transfer steps are important and in some cases may be rate limiting. These are not the reactions of protonated ligands but rather the deprotonation of ligands already in their basic form. The deprotonated peptide nitrogen cannot exist as a free entity in aqueous solution, but it forms during the process of coordination to metal ions such as Pd^{2+}, Ni^{2+}, and Cu^{2+}. The kinetics and mechanisms of metal-ion and proton-transfer reactions of oligopeptide complexes have been reviewed recently by Margerum and Dukes.[343] Examples of the types of reaction are the formation of $Cu(H_{-2}GGG)^-$, which is general base catalyzed, and the formation of $Ni(H_{-3}GGGG)^{2-}$, which depends on the square of the hydroxide ion concentration [$GGG^- = H_2NCH_2CONHCH_2CONHCH_2COO^-$ and $GGGG^- = H_2NCH_2CONHCH_2CONHCH_2CONHCH_2COO^-$]. The rate expression for the formation of the nickel–tetraglycine complex is given in Eq. (33) where the reversibility of the formation reaction is neglected. The significant

$$\frac{d[Ni(H_{-3}GGGG)^{2-}]}{dt} = \frac{4.5 \times 10^{-20}}{[H^+]^2}[NiGGGG^+] \qquad (33)$$

point is that even at pH 9 the forward rate constant is only 0.045 sec^{-1}, and the reaction is seen to be very slow compared with typical multidentate ligand reactions with Ni^{2+}.

Sutin and co-workers[344] have studied the reactions of monothenoyltrifluoroacetone with Ni^{2+}, Co^{2+}, Cu^{2+}, and Fe^{3+}. This ligand can exist in the hydrated keto, enol, or enolate forms. The keto form is inert to attack by metal ions, but the enol form reacts via parallel acid-independent and inverse acid paths. The rate constant for the reaction of Ni^{2+} with the enolate ion is within a factor of 2 of the value predicted by the water loss mechanism, but the rate constant for the reaction of the neutral species is much smaller than would be expected. The low formation rate of the monodentate intermediate may be a result of internal hydrogen bonding or of steric hindrance, but the authors are doubtful whether either effect could reduce the formation rate constant by such a large amount. The rate-determining step is thus likely to be the ring-closure step either because of the large dissociation rate constant for the monodentate intermediate or because of this influence of the proton transfer (it is unlikely that steric factors slow ring closure because the enolate reaction should also be abnormally slow).

From the available data it is not possible to distinguish between the two possible alternatives. However, it is possible to estimate the proton-transfer rate from the enol form to be about 10^6 sec^{-1}, and thus it is most likely that the ring-closure step is rate limiting because of weak first-bond formation.

I. Summary

The data presented in Table 1-19 are a comprehensive collection of published rate constants. In general, values of $k^{M\text{-}H_2O}$ have been recorded when these were reported in the original communication. In some cases, particularly for the alkali and alkaline earth metals, the rate constants have not been published other than in review articles, and we have simply recorded these values. Where the rate

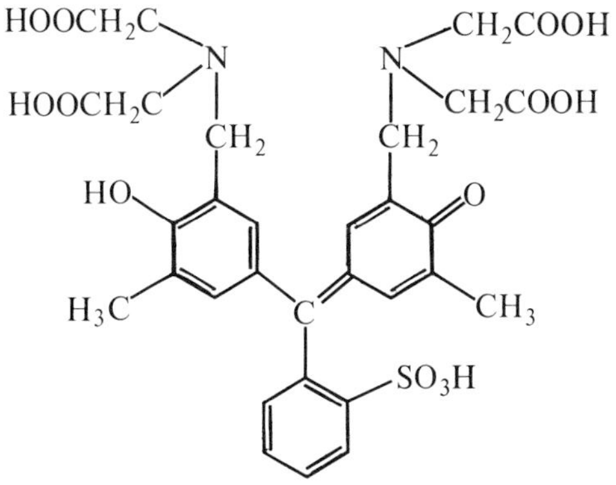

Xylenol Orange

15

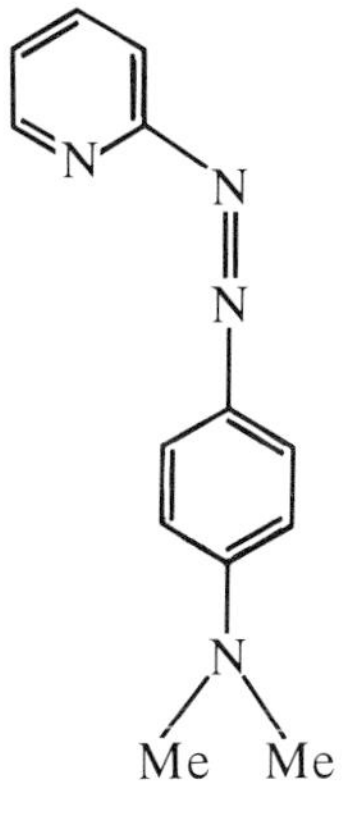

PADA

16

TABLE 1–19
FORMATION RATE CONSTANTS OF AQUOMETAL IONS WITH MULTIDENTATE LIGANDS AT 25.0°C

Metal	Ligand	k_f (M^{-1} sec^{-1})	k^{M-H_2O} (sec^{-1})	Ref.
Li^+	IDA^{2-}(20°C)	2.5×10^{8}[a]		7
	NTA^{3-}(20°C)	4.7×10^{7}[a]		7
	$EDTA^{4-}$(20°C)	1×10^8	4.8×10^7	7,11
	$DGITA^{4-}$	1.7×10^8		11
	TP^{5-}(20°C)	9×10^8		7
Na^+	IDA^{2-}(20°C)	2.5×10^{8}[a]		7
	NTA^{3-}(20°C)	1.9×10^8	8.8×10^7	7
	$EDTA^{4-}$(20°C)	1.7×10^8	4.7×10^7	7,11
	$DGITA^{4-}$	2.0×10^8		11
	TP^{5-}(20°C)	$>2 \times 10^9$		7
K^+	NTA^{3-}(20°C)	1.5×10^{8}[a]		7
	$EDTA^{4-}$(20°C)	8.5×10^7	7.5×10^7	7,11
	$DGITA^{4-}$	2.0×10^8		11
	TP^{5-}(20°C)	$>5 \times 10^9$		7
Rb^+	NTA^{3-}(20°C)	2.3×10^{8}[a]		7
	$EDTA^{4-}$(20°C)	1.4×10^{8}[a]		7
	TP^{5-}(20°C)	$>5 \times 10^9$		7
Cs^+	NTA^{3-}(20°C)	3.5×10^{8}[a]		7
	$EDTA^{4-}$(20°C)	2.1×10^{8}[a]		7
	TP^{5-}(20°C)	$>5 \times 10^9$		7
Mg^{2+}	HOX	1.3×10^4	1.3×10^5	285,287
	ATP	1.3×10^7		352
	ADP	5.6×10^6		352
	CDP	5.6×10^6		352
	HOXS	1.2×10^4	1.2×10^5	350
	OX^-	6.0×10^5	6×10^5	285,287
	OXS^{2-}	3.8×10^5	3.8×10^5	350
	NSA^{2-}	7.0×10^5	7.0×10^5	290
	$Oxalate^{2-}$(20°C)	6×10^6	1×10^5	351
	$Malonate^{2-}$(20°C)	4.2×10^6	1.8×10^{7}[b]	252
	$Tartrate^{2-}$(20°C)	2.5×10^6	1.8×10^{7}[b]	252
	IDA^{3-}	9×10^5		11
	$HADP^{2-}$	1×10^6		321,322
	ADP^{3-}	3×10^6		321,322
	(15°C)	3.8×10^6	1.1×10^5	353
	CDP^{3-}(15°C)	3.8×10^6	1.1×10^5	353
	$HP_2O_7^{3-}$(15°C)	3.85×10^6	1.1×10^5	353
	HTi^{3+}	3.5×10^6		354
	$HATP^{3-}$	3×10^6		355
	$CyDTA^{4-}$(trans)	2.5×10^6		358
	ATP^{4-}	1.2×10^7		355

TABLE 1-19 CONTINUED
FORMATION RATE CONSTANTS OF AQUOMETAL IONS WITH MULTIDENTATE LIGANDS AT 25.0°C

Metal	Ligand	k_f (M^{-1} sec^{-1})	$k^{M\text{-}H_2O}$ (sec^{-1})	Ref.
Mg^{2+}	ATP^{4-} (15° C)	8.7×10^6	1.1×10^5	353
	CTP^{4-}(15°C)	8.7×10^6	1.1×10^5	353
	$HP_3O_{10}^{4-}$(15°C)	8.5×10^6	1.1×10^5	353
	H(metalphth)$^{5-}$	$\sim 2 \times 10^6$		356
Ca^{2+}	Glycine$^-$	4×10^{8a}		11
	Murexide$^-$(10°C)	$\geq 6 \times 10^7$		93
	OX^-	$\geq 2 \times 10^8$		286
	IDA^{2-}	2.5×10^{8a}		11,357
	ADP^{3-}	$> 2.5 \times 10^8$		321,322
	$HEDTA^{3-}$	1×10^7		359
	$HEDTA^{3-}$	1.7×10^7		334
	$HPDTA^{3-}$	1.7×10^7		334
	$HBDTA^{3-}$	7.5×10^6		334
	$HCyDTA^{3-}$	1.7×10^6		334
	$HEEDTA^{3-}$	1.3×10^8		359
	ATP^{4-}	$> 1 \times 10^9$		363
	$HEDTA^{3-}$(30°C)	3.7×10^7		364
	$EDTA^{4-}$(30°C)	6.1×10^9		364
	$EDTA^{4-}$	$\leq 2.5 \times 10^9$		359
	$EDTA^{4-}$	2.4×10^{10}		334
	$PDTA^{4-}$	1.7×10^{10}		334
	$BDTA^{4-}$	1.2×10^{10}		334
	$CyDTA^{4-}$	1.4×10^{10}		334
	H(metalphth)$^{5-}$	$\sim 7 \times 10^8$		356
Sr^{2+}	Murexide$^-$	$> 6 \times 10^7$		93
	IDA^{2-}	3.5×10^{8a}		11
	$HEDTA^{3-}$	5×10^5		359
	$EDTA^{4-}$	8.1×10^8		359
	$CyDTA^{4-}$(trans)	1.3×10^8		358
Ba^{2+}	Murexide$^-$	$> 6 \times 10^7$		93
	IDA^{2-}	7×10^{8a}		11
Sc^{3+}	HCy^3	6×10^9		233
	Murexide$^-$(12°C) (pH = 3.0)	4.8×10^7		94
Y^{3+}	HCy^{3-}	2.4×10^7		233
	Murexide$^-$(12°C) (pH = 4.0)	1.3×10^7		94
La^{3+}	$HPAR^+$	38		360
	PAR	430		360
	Murexide$^-$(12°C) (pH = 4.0)	8.6×10^7		94

TABLE 1-19 CONTINUED
FORMATION RATE CONSTANTS OF AQUOMETAL IONS WITH MULTIDENTATE LIGANDS AT 25.0°C

Metal	Ligand	k_f (M^{-1} sec^{-1})	M^{M-H_2O} (sec^{-1})	Ref.
La^{3+}	An^-(12.5°C)	5.5×10^7		97
	$Oxalate^{2-}$	8.0×10^7		95
	HCy^{3-}	8.0×10^6		233
	XO(str. **15**)	8.5×10^6		98
Ce^{3+}	HCy^{3-}	1.0×10^7		233
	$Murexide^-$(12°C) (pH = 4.0)	9.5×10^7		94
	XO	3.7×10^6		98
Ce^{4+} (0.5*M* H^+)	Isobutyl alcohol (10°C)	8.3×10^4		361
	2,2-Dimethyl-1-propanol (13°C)	2×10^4		361
	sec-Butyl alcohol (10°C)	9.2×10^3		361
	tert-Butyl alcohol (10°C)	3.7×10^2		361
Pr^{3+}	An^-(12.5°C)	4.6×10^7		97
	HCy^{3-}	1.6×10^7		233
	$Murexide^-$(12°C) (pH = 4.0)	8.6×10^7		94
	XO	1.2×10^7		98
Nd^{3+}	HCy^{3-}	2.1×10^7		233
	$Murexide^-$(12°C)	9.3×10^7		94
	XO	9.5×10^6		98
	$Oxalate^{2-}$	8.6×10^7		95
Sm^{3+}	An^-(12.5°C)	6.3×10^7		97
	$Oxalate^{2-}$	8.2×10^7		95
	HCy^{3-}	5.6×10^7		233
	XO	4.0×10^6		98
	$Murexide^-$	8.2×10^8		362
	$Murexide^-$(12°C) (pH = 4.0)	9.6×10^7		94
Eu^{3+}	An^-(12.5°C)	1.05×10^8		97
	$Murexide^-$(12°C) (pH = 4.0)	8.2×10^7		94
	$Oxalate^{2-}$	7.7×10^7		95
	HCy^{3-}	3.4×10^7		233
Gd^{3+}	An^-(12.5°C)	5.9×10^7		97
	$Murexide^-$ (12°C) (pH = 4.0)	5.2×10^7		94
	$Oxalate^{2-}$	4.6×10^7		95

TABLE 1-19 CONTINUED
FORMATION RATE CONSTANTS OF AQUOMETAL IONS WITH MULTIDENTATE LIGANDS AT 25.0°C

Metal	Ligand	k_f (M^{-1} sec^{-1})	$k^{M\text{-}H_2O}$ (sec^{-1})	Ref.
Gd^{3+}	HCy^{3-}	2.2×10^7		233
	XO	5.4×10^6		98
Tb^{3+}	An^-(12.5°C)	3.5×10^7		97
	$Murexide^-$ (12°C) (pH = 4.0)	3.0×10^7		94
	$Oxalate^{2-}$	2.4×10^7		95
	HCy^{3-}	3.0×10^7		233
Dy^{3+}	An^-(12.5°C)	1.4×10^7		96,97
	$Murexide^-$(12°C) (pH = 4.0)	1.7×10^7		94
	$Oxalate^{2-}$	1.3×10^7		95
	HCy^{3-}	3.4×10^7		233
Ho^{3+}	$Murexide^-$(12°C)	1.4×10^7		94
	$Oxalate^{2-}$	1.0×10^7		95
	HCy^{3-}	2.7×10^7		233
Er^{3+}	An^-(12.5°C)	5.8×10^7		97
	$Murexide^-$(12°C) (pH = 4.0)	1.0×10^7		94
	$Oxalate^{2-}$	6.3×10^6		95
	HCy^{3-}	2.2×10^7		233
Tm^{3+}	$Murexide^-$(12.5°C)	1.1×10^7		94
	$Oxalate^{2-}$	6.3×10^6		95
	HCy^{3-}	2.8×10^7		233
Yb^{3+}	An^-(12.5°C)	6.9×10^7		97
	$Murexide^-$(12°C) (pH = 4.0)	1.1×10^7		94
	HCy^{3-}	3.9×10^7		233
Lu^{3+}	An^-(12.5°C)	9.5×10^7		97
	$Murexide^-$(12°C)	1.3×10^7		94
	HCy^{3-}	3.7×10^7		233
V^{2+}	Phen	3.0		365
	Bipy	30		366
	Terpy	0.8		367
V^{3+}	$HC_2O_4^-$	1.3×10^3		149
Cr^{2+}	$Hbipy^+$ (24°C)	6.5×10^3		284
	Bipy (24°C)	3.5×10^7		284
Mn^{2+}	ACA° (pH = 6.5)	2.4		368
	TMTPyP (22°C) (M^{-2} sec^{-1})	2.5×10^{-3}		313
	BCA(B) (pH = 6.25)	9.0		369

TABLE 1-19 CONTINUED
FORMATION RATE CONSTANTS OF AQUOMETAL IONS WITH MULTIDENTATE LIGANDS AT 25.0°C

Metal	Ligand	k_f (M^{-1} sec^{-1})	k^{M-H_2O} (sec^{-1})	Ref.
Mn^{2+}	Phen (11°C)	1.1×10^5		289
	Bipy	2.8×10^5		370
	Terpy	1×10^5		289
	HOX	4.9×10^5		286,287
	Murexide$^-$(10°C)	9×10^6		93,94
	β-Alanine$^-$(20°C)	5×10^4		263
	OX^-	3.3×10^8		286,287
	$HNTA^{2-}$	2.5×10^5		371
	NTA^{3-}	5×10^8		371
	ATP^{3-}	6×10^8 or 3×10^9		372
	ATP^{4-}	$>10^9$		363
	$HCyDTA^{3-}$j	1.2×10^8		373
Fe^{2+}	Phen	5.6×10^4		289
	Bipy	1.6×10^5		289
	Terpy	8×10^4		289
	TPTZ	1.3×10^5		297
Fe^{3+}	CH_3COOH	27		177
	TTF^-	2.4×10^4		344
	TTF (enol)	1.4		344
	HRSH	~1		374
	SaliH	24.6		375
	$HC_2O_4^-$	1.4×10^3		180
	H_2SSA^-	1.5		376
	H_2Sal	3.0		377
	NADH	5.0×10^3		378
	Acac (enol)	5.2		379
	Acac (keto)	0.29		379
Co^{2+}	$HTKED^+$	7.5×10^2		247
	$HTHPED^+$	2.7×10^2		247
	$HDPA^+$	4.8×10^3		380
	4-*N*-MethylcyclamH$^+$	16		305
	β-DimethylcyclamH$^+$	1.8×10^2		306
	β-Dimethyl-4-*N*-methylcyclamH$^+$	1.1×10^{-1}		306
	β-Dimethylcyclam	1.0×10^6		306
	β-Dimethyl-4-*N*-methylcyclam	1.1×10^2		306
	DPA	2.5×10^5		380
	Imidazole	1.3×10^5		115
	PADA (15°C)(str. **16**)	4×10^4		381
		7.6×10^4		382,383

TABLE 1-19 CONTINUED
FORMATION RATE CONSTANTS OF AQUOMETAL IONS WITH MULTIDENTATE LIGANDS AT 25.0°C

Metal	Ligand	k_f (M^{-1} sec^{-1})	k^{M-H_2O} (sec^{-1})	Ref.
Co^{2+}	Phen	3×10^5		289
	NH_3 (20°C)	9.5×10^4		114
		1.1×10^5		384
	$5NO_2$-phen	1.6×10^5		289
	Terpy	2.4×10^4		289
	Bipy	6.3×10^4		289
	4,4′DiMebipy	1.5×10^5		385
	4,4′-Dichlorobipy	4×10^5		386
	TKED	6.2×10^4		247
	THPED	2.3×10^4		247
	ACA°	4.0		368
		2.0		387
	TMTPyP (M^{-2}, sec^{-1}) (22°C)	2.2×10^{-3}		313
	L-Carnosine	4.2×10^5		266
	L-Arginine	1.5×10^5		388
	Salicylate$^-$(20°C)	1.3×10^5		389
	Sarcosine$^-$	9.2×10^5		390
	Serine$^-$	2.0×10^6		341
	Leucine$^-$	1.3×10^6		267
	α-Alanine$^-$(20°C)	6×10^5		263
		1.9×10^6		268
	α-Aminobutyric acid	2.5×10^5		264
	β-Aminobutyric acid	2.0×10^4		264
	L-Dopa$^-$	4.3×10^5		391
	L-Tyrosine	1.3×10^6		392
	Hydroxyproline$^-$	7.0×10^4		393
	Proline$^-$	3.5×10^5		393
	β-Alanine$^-$(20°C)	7.5×10^4		263
		1.3×10^5		268
	Glycine$^-$	1.5×10^6		275
		4.6×10^{5c}	2.6×10^5	115
		3.0×10^6		394
	DiGly$^-$	4.6×10^5	2.6×10^5	115
		2×10^5		275
	TriGly$^-$	3.1×10^5		275
	TetraGly$^-$	2.6×10^5		275
	Pyc$^-$(20°C)	1×10^7		395
	Murexide$^-$(10°C)	1.5×10^5		93,94
	L-HCystein$^-$(20°C)	5.6×10^5		396
	TTF$^-$	$\geqslant 3 \times 10^4$		344
	GlyLeu$^-$	3.5×10^5		397

TABLE 1–19 CONTINUED

FORMATION RATE CONSTANTS OF AQUOMETAL IONS WITH MULTIDENTATE LIGANDS AT 25.0°C

Metal	Ligand	k_f (M^{-1} sec^{-1})	k^{M-H_2O} (sec^{-1})	Ref.
Co^{2+}	$LeuGly^-$	1.4×10^5		397
	$Glysa^-$	4.6×10^5		398
	IDA^{2-}(30°C)	1.1×10^7		268
	H_2tiron^{2-}(20°C)	1.04×10^6		354
	5-Sulfosalicylate^{2-} (20°C)	2.7×10^5		389
	$Malonate^{2-}$	9×10^6		399
	$Tartrate^{2-}$	5×10^6		400
	$Cystein^{2-}$(20°C)	5.6×10^6		396
	ASP^{2-}(30°C)	4.3×10^6		268
	IDP^{2-}(30°C)	2.7×10^5		268
	$HP_2O_7^{3-}$	9.3×10^7		244
	$Htiron^{3-}$	4.2×10^6		354
	$HEDTA^{3-}$	4×10^6		401
		10^7		402,403
	$EDTA^{4-}$	$\simeq 10^9$		402
	$HP_3O_{10}^{4-}$	4.1×10^8		244
	ATP^{4-}	9.2×10^7		363
	$HCyDTA^{3-j}$	2.5×10^7		404
		2.8×10^7		373
Ni^{2+}	$H_3tetren^{3+}$	3.5		239
	$H_2tetren^{2+}$	3.2×10^2		239
	H_2trien^{2+}	97		239
	H_2terpy^{2+}	0.5		117
	$H_2cyclam^{2+}$	3.3×10^{-3}		330
	H_2TPTZ^{2+}	~ 10		323
	H_2BPEDA^{2+}	1.2×10^2		405
	$Htrien^+$	9.3×10^3		239
	$HENAO^+$(24.3°C)	3.8		406
	$HAMP^+$	35		242
	Hen^+	5.9×10^2		242
		1.7×10^2		240
	$Tmen^+$	5.0×10^2		103
	$Hphen^+$	3.0		117
	$Hterpy^+$	90		117
	$Hpald^+$	2.0×10^2		117
	$Himid^+$(23.7°C)	4.0×10^2		116
		2.0×10^2		117
	N-MethylimidH$^+$ (23.7°C)	2.3×10^2		118
	$Himac^+$	2×10^2		117
	$HisH_2^+$	~ 0		117

TABLE 1-19 CONTINUED
FORMATION RATE CONSTANTS OF AQUOMETAL IONS WITH MULTIDENTATE LIGANDS AT 25.0°C

Metal	Ligand	k_f (M^{-1} sec^{-1})	k^{M-H_2O} (sec^{-1})	Ref.
Ni^{2+}	1-MethylHisH_2	~4.8×10^2		118
	$Hcyclam^+$	14		330
	$(Hcyclam\text{-}14)^+$	7.4		307
	$(1\text{-}HMecyclam\text{-}14)^+$	55		307
	$(2\text{-}HMecyclam\text{-}14)^+$	10		307
	$(4\text{-}HMecyclam\text{-}14)^+$	1.4		307
	$(Hcyclam\text{-}14)^+$	145		306
	H-β-Dimethylcyclam$^+$ (I)	110		306
	H-α,α'-Dimethyl-cyclam$^+$ (II)	55		306
	H-$\alpha,\alpha,\alpha',\alpha'$-Tetra-methylcyclam$^+$ (III)	14		306
	H-β-Dimethyl-4-N-methylcyclam$^+$ (IV)	4.5×10^{-2}		306
	$Hbipy^+$	25		117
	$HDPA^+$	3.4×10^2		380
	$Htach^+$	2.0		407
	Tach	45		407
	H-N,N-DiEten$^+$	123		293
	H-N,N'-DiEten$^+$	8.0		293
	$HTeMeen^+$	0.2		293
	H-N,N-DiMeen	69		248
	H-N,N'-DiMeen	29		248
	H-N,N,N'-TriMeen	9.6		248
	$HTKED^+$	8.6		102
	$HTHPED^+$	2.8		102
	$HMeHis^+$(23.7°C)	6.0×10^2		116
	Hhm^+	2.0×10^2		117
	$AEPH^+$	7.3		291
	$HBPEDA^+$	2.6×10^2		405
	$HDBEDA^+$	22.5		405
	$HTPTZ^+$	1.7×10^3		323
	$Htach^+$	9		407
	NH_3	4.6×10^3		324
		4.3×10^3		114
	En	3.6×10^5		240
	Bipy	2.26×10^3	2.26×10^4	216
		2.0×10^3	2.0×10^4	117
		1.6×10^3	1.6×10^4	289
		1.61×10^3		408
		1.63×10^3	1.6×10^4	217

TABLE 1-19 CONTINUED
FORMATION RATE CONSTANTS OF AQUOMETAL IONS WITH MULTIDENTATE LIGANDS AT 25.0°C

Metal	Ligand	k_f (M^{-1} sec^{-1})	kM-H_2O (sec^{-1})	Ref.
Ni^{2+}	4,4′-DiMebipy	3.0×10^3	3×10^4	217
		2.7×10^3	2.7×10^4	385
	4,4′-Cl_2bipy	9.7×10^2	9.7×10^3	217
	4,4′-Cl_2bipy	9.0×10^2		386
	4,4′-$(EtO)_2$bipy	1.3×10^3		386
	Phen	4.1×10^3	4.1×10^4	216
		3.9×10^3	3.9×10^4	289
		2.95×10^3		408
		4×10^{3i}	4×10^4	409,410
		2.7×10^4		411
		4×10^4		412
		3.5×10^3		117
	5-NO_2phen	1.9×10^3	1.9×10^4	289
	5-CH_3phen	3.2×10^3	3.2×10^4	289
	2-CH_3phen	5×10^{2i}	5×10^3	409,413
	5-Cl-Phen	2.8×10^2	2.8×10^3	289
	2-Cl-Phen	1×10^3	1×10^4	289
	ENAO (24.5°C)	2.5×10^4		406
	1-Arg	2.3×10^3	2.3×10^4	388
	AMP	8.6×10^3	8.6×10^4	242
	Ga	2.2×10^3	2.2×10^4	242
	DPA	5.8×10^3		380
	ACA°	5.0		368
	Theophylline	$\sim 1 \times 10^2$		122
	PAD	1.3×10^3	1.3×10^4	242
	PADA	1.35×10^3	1.35×10^4	414
		1.32×10^3		378
	(15°C)	4×10^3		381
	PADA(D_2O)	1.01×10^3		378
	PAR	3.2×10^3	3.2×10^4	289
	Terpy	1.4×10^3	1.3×10^4	289
		1.52×10^3	1.52×10^4	408
		4.9×10^2	4.9×10^3	217
		2.1×10^3	2.1×10^4	117
		1.49×10^3	1.49×10^4	216
	DBEPA	3.1×10^2		405
	BPEDA	1.8×10^3		405
	$NH_3^+CH_2CH$-$(NH_2)COO^-$	1.9×10^3	1.9×10^4	242
	$NH_3^+(CH_2)_2CH$-$(NH_2)COO^-$	2.8×10^3	2.8×10^4	242

TABLE 1-19 CONTINUED
FORMATION RATE CONSTANTS OF AQUOMETAL IONS WITH MULTIDENTATE LIGANDS AT 25.0°C

Metal	Ligand	k_f (M^{-1} sec^{-1})	$k^{M\text{-}H_2O}$ (sec^{-1})	Ref
Ni^{2+}	$NH_3^+(CH_2)_3CH(NH_2)COO^-$	2.0×10^3	2.0×10^4	242
	$NH_3^+(CH_2)_4CH(NH_2)COO^-$	4.4×10^3	4.4×10^4	242
	HOX^-	1.3×10^3	1.3×10^4	288
	$HOXS^-$	1.4×10^3	1.4×10^4	288
	HHis	1.2×10^3		117
	(23.7°C)	2.2×10^3	2.2×10^4	116
	MeHis(23.7°C)	2.6×10^3	2.6×10^4	116
	3-MethylHis(23.7°C)	2.1×10^3		118
	1-MethylHis(23.7°C)	1.8×10^3		118
	Imidazole	3.2×10^3	3.2×10^4	116
		5.0×10^3	5.0×10^4	115
		6.4×10^3	6.4×10^4	117
	N-Methylimid(23.7°C)	4.5×10^3		118
	Imac	7.3×10^3	7.3×10^4	117
	$PenH_2$	3.5×10^2	3.5×10^3	117
	L-Carnosine	6.9×10^3	6.9×10^4	388
	$CysH_2$	3.5×10^2	3.5×10^3	117
	PhencarbH	70		117
	4,7-Dihydroxyphen	0.5		117
	Hm	6.0×10^5		117
	Pald	1.3×10^3	1.3×10^4	117
	9-Methylpurine	6.7×10^3		415
	Adenine	3×10^2		415
	Adenine(19°C)	1.3×10^2		416
	Adenosine(19°C)	6.3×10^2		416
	AEP	9.5×10^3		291
	N,N-DiMeen	3.2×10^4		248
	N,N'-DiMeen	4.3×10^4		248
	N,N,N'-TriMeen	2.9×10^3		248
	N,N'-DiEten	9.7×10^3		293
	N,N-DiEten	3.4×10^5		293
	TriEten	<10		293
	TeMeen	3.6×10^2		293
	TMTPyP (22°C) (M^{-2} sec^{-1})	5×10^{-5}		313
	N,N'-DiEten	9×10^3		293
	TPTZ	2.0×10^3		323
	TTF	2.3		344
	Hacac	2.7		417
	TKED	2.4×10^2		102

TABLE 1-19 CONTINUED
FORMATION RATE CONSTANTS OF AQUOMETAL IONS WITH MULTIDENTATE LIGANDS AT 25.0°C

Metal	Ligand	k_f (M^{-1} sec^{-1})	$k^{M\text{-}H_2O}$ (sec^{-1})	Ref.
Ni^{2+}	THPED	1.7×10^2		102
	Proline$^-$	3.4×10^4		393
	Hydroxyproline$^-$	1.2×10^4		393
	NSA$^-$	1.02×10^5		418
	Salicylate$^-$	5.3×10^3		389
	Anthranilate$^-$	2.3×10^3		389
	HC_2O_4	5×10^3	3×10^3	241
	Hmal$^-$	3.1×10^3	3.1×10^3	237
	Glycine$^-$	4.1×10^4	4.1×10^4	275
		1.5×10^{4c}	9×10^3	115
		2.2×10^4	2.2×10^4	242
		2.1×10^4		394
	L-dopa$^-$	2.2×10^3		391
	L-tyrosine	1.4×10^4		392
	DiGly$^-$	2.1×10^4		115
		3.6×10^3		242
		3.2×10^3		275
	Glysa$^-$	2.0×10^3		398
	TriGly$^-$	8.0×10^3		419
		3.7×10^3		242
		1.7×10^3		275
	TetraGly$^-$	4.2×10^3		242
		1.8×10^3		275
	α-Alanine$^-$	3.1×10^4	3.1×10^4	268
	(20°C)	2.0×10^4	2.0×10^4	264
	β-Alanine$^-$	5.0×10^3		268
	(20°C)	1.0×10^4		264
	Sarcosine$^-$	1.3×10^4		390
	α-Aminobutyrate (20°C)	1×10^4		264
	β-Aminobutyrate (20°C)	4×10^3		264
	Serine$^-$	2.9×10^4		341
	$NH_2COCH_2CH(NH_2)CO_2^-$	8.7×10^3		242
	HIDA$^-$	7.7×10^{-2}		331
	Murexide$^-$	1.8×10^3		235
	(10°C)	1.0×10^3		58,59
	(25°C)	3.5×10^3		58,59
	His	1.5×10^5		396
	Pyc$^-$	2.6×10^4	2.6×10^4	242
	Pya$^-$	1.0×10^4	1.0×10^4	242

TABLE 1-19 CONTINUED
FORMATION RATE CONSTANTS OF AQUOMETAL IONS WITH MULTIDENTATE LIGANDS AT 25.0°C

Metal	Ligand	k_f (M^{-1} sec^{-1})	$k^{M\text{-}H_2O}$ (sec^{-1})	Ref.
Ni^{2+}	$PydicH^-$	5.0×10^3		242
	$Leucine^-$	1.7×10^4		267
	$GlyLeu^-$	3.5×10^3		397
	$LeuGly^-$	2.0×10^3		397
	L-His^-	3.8×10^5		117
	L-$Hcys^-$	1.5×10^4		396
		4×10^4		117
	$Hpen^-$	2.2×10^4		117
	$Phencarb^-$	2.5×10^4		117
	4,7-$(OH)_2$phen	~20		117
	$ChelH^-$	60		117
	TTF^-	1.0×10^4		344
	DTZ^-	1.3×10^3		420
		1.1×10^3		421,422
	DTZ^--*p*F	2.7×10^3		421,422
	DTZ^--*p*Cl	3.3×10^3		421,422
	DTZ^--*p*Br	2.0×10^4		421,422
	DTZ^--*p*I	1.3×10^5		421,422
	DTZ^--*o*CH_3	5.2×10^3		420
	DTZ^--*p*CH_3	2.4×10^4		421,422
	DTZ^--*p*OCH_3	7.0×10^4		421,422
	DTZ^--*p*CF_3	2.7×10^4		421,422
	DTZ^--α-naphthyl	1.3×10^5		420
	α-$Hydroxyacetate^-$	3.3×10^4	1.5×10^{5} [g,e]	126
			2.5×10^{3} [f,e]	126
	α-$Hydroxyacetate^-$	2.6×10^4 ($\mu \to 0$)		254
	β-$Hydroxypropionate^-$	2.5×10^4	1.5×10^{5} [g,e]	126
			1.0×10^{3} [f,e]	126
	β-$Hydroxypropionate^-$	2.6×10^4 ($\mu \to 0$)		253
	H-$Methylmalonate^-$	5.1×10^{3} [d]		256
	H-*n*-$Butylmalonate^-$	4.0×10^{3} [d]		256
	$Oxalate^{2-}$	7.5×10^4		241
	L-Leucyl-L-tyrosine	2.2×10^3		423
	D-Leucyl-L-tyrosine	2.4×10^3		423
	$CH_2(COO^-)_2$	7.0×10^4		237
	(20°C)	2.7×10^5		255
		4.5×10^5	8.5×10^{5} [h,e]	126
			3.5×10^{3} [f]	126
	$(CH_2COO^-)_2$ (20°C)	4.3×10^5		255
	$NH_2CH(COO^-)_2$	3.4×10^5		242
	IDA^{2-}	8.8×10^4		242
		1.2×10^6		424

TABLE 1–19 CONTINUED
FORMATION RATE CONSTANTS OF AQUOMETAL IONS WITH MULTIDENTATE LIGANDS AT 25.0°C

Metal	Ligand	k_f (M^{-1} sec^{-1})	k^{M-H_2O} (sec^{-1})	Ref.
Ni^{2+}	IDA^{2+}	4.5×10^4		283
	5-Sulfosalicylate^{2-}	4.7×10^3		389
	$^{-}OOCCH_2CH$-$(NH_2)COO^{-}$	3.9×10^4		242
		2×10^5		268
	Pydic^{2-}	6.3×10^4		242
	Pidic^{2-}	5.1×10^4		242
	H_2ti^{2-} (20°C)	8.2×10^4		354
	L-Cystein^{2-}	1.5×10^5		396
	MNT^{2-}	6.6×10^4	4.7×10^3	211
	DTO^{2-}	5.8×10^4	4.1×10^3	211
	Mephosphate^{2-}		7×10^3	125
	Malate^{2-}	5.6×10^5		258
	Ama^{2-}	3.4×10^5		242
	ChelH^{2-}	1.7×10^2		117
	Methylmalonate^{2-}	7.3×10^4[e]	4.2×10^4[f]	256
	Ethylmalonate^{2-}	7.6×10^4[e] ($\mu = .1$)		257
	n-Butylmalonate^{2-}	7.5×10^4[e]	5.1×10^4[f]	256
	Benzylmalonate^{2-}	9.4×10^5[e] ($\mu = .1$)		257
	Tartrate^{2-}	1.8×10^5	4.8×10^5[e]	252
	L-Tartrate^{2-}	2.1×10^5		425
	Phthalate^{2-}	6.3×10^5 ($\mu \to 0$)		261
		2.6×10^5[e] ($\mu = .1$)		262
	Adipate^{2-}	8.5×10^5	8.5×10^5[e]	252
	meso-Tartrate^{2-}	2.5×10^5	8.5×10^5[h]	126
			2.8×10^3[f]	126
		1.5×10^5		259
	Glutarate^{2-}		1.3×10^3	426
	2-Hydroxyglutarate^{2-}	1.2×10^4	1.3×10^3	426
			3×10^3	426
			300[f]	426
	3-Hydroxyglutarate^{2-}	4×10^4	1.2×10^3	426
			450[f]	426
	Maleate^{2-}	2.1×10^5		426
		8.4×10^5		260
	HNTA^{2-}	4		427
	(1.25μ)	7.5		283
	H_2EDTA^{2-}	3×10^3		242
		2×10^3		326
		1×10^3[k]		428
	H_2CyDTA^{2-}	8×10^3		242
	$H_2P_2O_7^{2-}$	5.7×10^4		429

TABLE 1-19 CONTINUED
FORMATION RATE CONSTANTS OF AQUOMETAL IONS WITH MULTIDENTATE LIGANDS AT 25.0°C

Metal	Ligand	k_f (M^{-1} sec^{-1})	kM-H_2O (sec^{-1})	Ref.
Ni^{2+}	Hti^{3-}	2.85×10^5		354
	NTA^{3-}	4.8×10^5		283
		5.6×10^4		427
		8.7×10^4		427
	$HP_2O_7^{3-}$	2.1×10^6		244
	$HP_2O_7^{3-}$	1.6×10^6		429
	TPP^{2-}		4.2×10^3	430
	Phenol Red	2.1×10^3		430
	$HEDTA^{3-}$	1×10^6		431
		1.9×10^5		242
		1.7×10^6		432
		1.8×10^5		326
		3×10^{5k}		428
	$HCyDTA^{3-}$	3.6×10^5		278
		1.9×10^5		242
	$HGEDTA^{3-}$	8.1×10^4		433
	$GEDTA^{4-}$	1.5×10^5		424
	$EDTA^{4-}$	6×10^6		431
	$HP_3O_{10}^{4-}$	6.8×10^6		244
	ATP^{4-}	4.1×10^6		363
Cu^+	$Maleate^{2-}$	2.0×10^9		434
	$Fumarate^{2-}$	1.7×10^9		434
Cu^{2+}	$H_4tetren^{4+}$	1.4×10^4		435
		3.3×10^4		245
	H_3tren^{3+}	$\sim 1 \times 10^2$		324
	$H_3tetren^{3+}$	1.55×10^5		435
		9.0×10^4		245
	H_3trien^{3+}	$\sim 2 \times 10^3$		324
		1.6×10^4		245
	H_2trien^{2+}	7.0×10^6		324
		7.5×10^6		245
	H_2tren^{2+}	3.8×10^6		324
	H_2dien^{2+}	$\sim 1.0 \times 10^3$		324
	$H_2tetren^{2+}$	4.2×10^7		435
	$H_2cyclam^{2+}$ (50°C)	17.4		303
	$Hcyclam^+$	2.6×10^5		307
	$H(1\text{-Mecyclam})^+$	1.2×10^7		307
	$H(2\text{-Mecyclam})^+$	2.8×10^6		307
	$H(4\text{-Mecyclam})^+$	2.9×10^5		307
	$Hcyclam^+$	1.1×10^7		306
	H-β-Dimethyl-cyclam$^+$	1.2×10^7		306

TABLE 1-19 CONTINUED
FORMATION RATE CONSTANTS OF AQUOMETAL IONS WITH MULTIDENTATE LIGANDS AT 25.0°C

Metal	Ligand	k_f (M^{-1} sec^{-1})	k^{M-H_2O} (sec^{-1})	Ref.
Cu^{2+}	H-α,α′ Dimethyl-cyclam$^+$	6.5×10^6		306
	H-α,α,α′,α′-tetra-methylcyclam$^+$	3.4×10^6		306
	H-β-Dimethyl-4-*N*-methylcyclam$^+$	2.2×10^3		306
	H_2TPTZ^{2+}	$<1 \times 10^3$		324
	H_2terpy^{2+}	$<5 \times 10^2$		436
	$Hterpy^+$	8.0×10^7		436
	$HTPTZ^+$	2.4×10^7		324
	Hbipy	2.9×10^5		284
		2.6×10^5		436
	$Hphen^+$	$<5 \times 10^2$		436
	Hhm^+ (37°C)	7.1×10^4		437,328
	$Hdien^+$	3.7×10^8		324
	Hen^+	1.4×10^5		246
	(37°C)	1.7×10^5		329,328
	$HTeMeen^+$	1.0×10^3		245
	$HTKED^+$	5.4×10^4		245
	$HTHPED^+$	5.3×10^4		245
	(Htet a^+)	7.6×10^3		295
	$Hcarnosine^+$	8.9×10^2		265
	$Hamp^+$	8.6×10^3		324
	SerH (37°C)	1.2×10^3		328
	PADA (15°C)	1×10^8		381
	Bipy	$\simeq 4 \times 10^7$		284
		5×10^7		436
	Adenine	1.8×10^3		438
	En	3.8×10^9		246
	TeMeen	1.7×10^8		245
	TEA	3.0×10^7		245
	TKED	3.5×10^7		245
	Terpy (6.5°C)	2×10^7		289
	Phen	6.4×10^7		436
	ACA°	6.0×10^4		368,387
	Apo-polyphenoloxidase	$>5 \times 10^6$		439
	Acac (enol)	2×10^4		417
	Acac (keto)	15		417
	Deuteroporpyhrin	4.3		295
	TMTPyP (22°C) (M^{-2} sec^{-1})	2.3		313
	(26°C) (M^{-1} sec^{-1})	3.1		311

TABLE 1-19 CONTINUED
FORMATION RATE CONSTANTS OF AQUOMETAL IONS WITH MULTIDENTATE LIGANDS AT 25.0°C

Metal	Ligand	k_f (M^{-1} sec^{-1})	kM-H_2O (sec^{-1})	Ref.
Cu^{2+}	L-Carnosine	3.5×10^6		265
	Hm(37°C)	2.6×10^9		437
	Proline$^-$	2.5×10^9		393
	Proline$^-$	2.6×10^9		440
	Hydroxyproline$^-$	7.4×10^8		393
	Bicine$^-$	9.5×10^8		441
	L-Dopa$^-$	1.1×10^8		442
	Histidine$^-$	1.3×10^7		269
	GlyLeu$^-$	3.1×10^8		397
	GlyGly$^-$	3.5×10^8		443
	Glysa$^-$	4.0×10^8		443
	LeuGly$^-$	1.0×10^8		397
	Leucine$^-$	1.6×10^9		267
	Sarcosine$^-$	2.8×10^9		390
	L-Phenylalanine$^-$	1.2×10^9		442
	Glycine$^-$	4.3×10^9		441
		4.0×10^9	2×10^9	274
		3.4×10^9		394
	N,N-Dimethylglycine$^-$	1.4×10^9		440
	α-Alanine$^-$	1.3×10^9		269
		1.1×10^9		441
	β-Alanine$^-$	2.0×10^8		269
	Valine$^-$	1.1×10^9		441
	Serine$^-$(37°C)	1.8×10^9		328
		2.5×10^9		341
	Murexide$^-$(11.5°C)	7×10^7		444
	(10°C)	1.2×10^8		93
	Acac$^-$	$\leq 2 \times 10^{8}$[l]		344
	L-Tyrosine$^-$	1.1×10^9		392
	Asparagine	4×10^9		445
	Glutamine	3.3×10^9		445
	HIDA$^-$	1.2×10^4		324
	HMIDA$^-$	1.1×10^4		324
	TTF$^-$	$\geq 3 \times 10^6$		344
	HNTA^{2-}	5×10^3		446
		1.1×10^5		338
	ASP^{2-}(30°C)	5×10^9		268
	IDP^{2-}(30°C)	$\sim 2 \times 10^8$		268
	IDA^{2-}	3.0×10^9		324
	MIDA^{2-}	7.4×10^9		324
	H_2ATP^{2-}	8.8×10^8		447
	H_2EDTA^{2-}(0°C)	3.6×10^5		327
	$HEDTA^{3-}$(0°C)	8.4×10^7		327

TABLE 1-19 CONTINUED
FORMATION RATE CONSTANTS OF AQUOMETAL IONS WITH MULTIDENTATE LIGANDS AT 25.0°C

Metal	Ligand	k_f (M^{-1} sec^{-1})	kM-H$_2$O (sec^{-1})	Ref.
Cu^{2+}	HEDTA^{3-} (0°C)	3.0×10^9		448
	(15°C)	6.2×10^8		449
	NTA^{3-}	2.0×10^8		446
	HCyDTA^{3-}j	8.3×10^9		450
Ag^+	Phen(12.5°C)	$\geqslant 5 \times 10^6$		289
Zn^{2+}	H(4-*N*-Methylcyclam)$^+$	4.5×10^3		305
	H(β-Dimethylcyclam)$^+$	8.0×10^4		306
	H-β-Dimethyl(4-*N*-methylcyclan)$^+$	88		306
	β-Dimethylcyclam	1.2×10^{11}		306
	β-Dimethyl-4-*N*-methylcyclam	3.0×10^7		306
	Terpy	1.1×10^6		289
	Bipy	1×10^6		289
	Phen(15°C)	1.1×10^6		289
	PADA(15°C)	4×10^6		381
		6.8×10^6		452
	Porphyrin	0.2		451
		0.237		452
	TMT PyP (22°C) (M^{-2} sec^{-1})	5		313
	ACA°	$\sim 10^4$		453
	TPPS	0.476		454
	8-SQ	4×10^9		420,421
	2M8SQ	1.1×10^9		420,421
	Glycine$^-$	1.5×10^8	7×10^7	455
	(10°C)	7×10^7		394
	DTZ$^-$	6.2×10^6		420
		6.9×10^6		422
	DTZ$^-$*p*F	3.7×10^7		422
	DTZ$^-$*p*Cl	4.6×10^7		422
	DTZ$^-$*p*Br	1.9×10^8		422
	DTZ$^-$*p*I	6.4×10^8		422
	DTZ$^-$*o*CH_3	7.5×10^6		420
	DTZ$^-$*p*CH_3	6.9×10^8		422
	DTZ$^-$*p*OCH_3	2.7×10^8		422
	DTZ$^-$*p*CF_3	1.7×10^8		422
	DTZ$^-$-α-naphthyl	9.7×10^8		420
	Murexide$^-$(11.5°C)	8×10^6		444
	(10°C)	2.0×10^7		93
	Serine$^-$	$\sim 1 \times 10^8$		341
	$H_2(GEDTA)^{2-}$	2.5		433

TABLE 1-19 CONTINUED
FORMATION RATE CONSTANTS OF AQUOMETAL IONS WITH MULTIDENTATE LIGANDS AT 25.0°C

Metal	Ligand	k_f (M^{-1} sec^{-1})	kM-H_2O (sec^{-1})	Ref.
Zn^{2+}	H_2EDTA^{2-}	3×10^6		456
	$HHEEDTA^{2-}$	9×10^8		456
	$HNTA^{2-}$	2.7×10^5		457
		5.1×10^5		337
	NTA^{3-}	$<1 \times 10^{10}$		337
		2.1×10^9		104
	$HEDTA^{3-}$	$\sim 10^9$		458
		2×10^9		456
	$H(1,3\text{-}PDTA)^{3-}$	4.0×10^9		459
	$HEDDDA^{3-}$	8.3×10^8		459
Cd^{2+}	Phen (12°C)	$\sim 5 \times 10^6$		289
	Terpy	3.2×10^6		289
	PADA (15°C)	$\sim 1.3 \times 10^7$		381
	Murexide$^-$ (10°C)	1.2×10^8		93,94
	$HNTA^{2-}$	2.1×10^5		337
		4×10^6		12
		8×10^5		371
	H_2EGTA^{2-}	1.5×10^6		456
	NTA^{3-}	2×10^{10}		371
		4×10^9		104
		$<2.5 \times 10^9$		337
	$HEDTA^{3-}$	4×10^9		460
		2.6×10^8		461
		8×10^8		462
		6.4×10^9		463
	$HEGTA^{3-}$	$\leq 1 \times 10^9$		456
	$HCyDTA^{3-}$	1.1×10^{9} [j]		373
		1.4×10^{9} [j]		464
	$H(1,3\text{-}PDTA)^{3-}$	4.1×10^9		459
	$HEDDDA^{3-}$	4.0×10^8		459
Hg^{2+}	Kojic Acid	2.8×10^5		465
	3,5-DiNSA	3.1×10^5		465
Ga^{3+}	Methylthymol Blue (15°C)	77.1		466
Al^{3+}	Methylthymol Blue	0.58		467
	Salicylate$^-$	9.1×10^{-1}		468
	$Co(CN)_6$	1.62×10^3	8.5×10^{-2}	185
	$Fe(CN)_6^{3-}$	0.66 [a]		469
In^{3+}	H_2SXO^-	6.9×10^2		470
		2.8×10^5		470
	Murexide$^-$ (12°C) [m]	2×10^{6} [d]		94

TABLE 1-19 CONTINUED
FORMATION RATE CONSTANTS OF AQUOMETAL IONS WITH MULTIDENTATE LIGANDS AT 25.0°C

Metal	Ligand	k_f (M^{-1} sec^{-1})	$k^{M\text{-}H_2O}$ (sec^{-1})	Ref.
Tl^{3+}	H_3SXO	7×10^5		470
Pb^{2+}	$Murexide^-$ (10°C)	$>5 \times 10^8$		93
	$HIDA^-$	2.0×10^{10}		104
	$HNTA^{2-}$	4.4×10^6		339
		1.5×10^8		371
	NTA^{3-}	$<1.5 \times 10^{11}$		339
		1.5×10^{11}		104
	$HEDTA^{3-}$	2×10^9		458,471
		6.1×10^{11}		104
	HEEDTA	2.4×10^{11}		104
	$HGEDTA^{3-}$	7.5×10^{11}		104
	$HEDTA^{3-}$	3×10^9		472
	$HCyDTA^{3-}$	8×10^8		472
	$HEDDDA^{3-}$	2.5×10^{10}		459
	$HDPTA^{3-}$	6.8×10^{10}		459

[a] First-order rate constant (sec^{-1}).
[b] Calculated using $K_{os} = 60M^{-1}$.
[c] Estimated value, *see* Ref. 115.
[d] Assuming proton transfer is not rate limiting.
[e] $K_{os}\, k^{M\text{-}H_2O}$.
[f] Ring-closure rate constant (sec^{-1}).
[g] Assumed to be like $acetate^-$.
[h] Assumed to be like $adipate^{2-}$.
[i] Calculated from the dissociation rate and stability constants.
[j] Calculated from $k_f/k_b = K_{MY}\, K_{HY}$.
[k] Values calculated in Ref. 428.
[l] Estimated from data in Ref. 417.
[m] pH = 3.0.

constants were obtained by sound absorption techniques, they are quoted as first-order rate constants.

On examining Table 1-19, it is immediately obvious that the formation rate constant often depends on the incoming ligand. This is not to say that we must abandon the dissociative mechanism but that we must take into account other effects which are superimposed on this mechanism. If these other effects are recognized, many of the reactions which appear to be contrary to a dissociative mechanism can now be explained in terms of one. The reactions of many polyamines with metal ions are much faster than expected because of the internal conjugate base effect.[114] This enhances the outer-sphere association between the metal and ligand by hydrogen bonding while simultaneously imparting some hydroxide character to one of the inner-sphere water molecules which may have a labilizing effect on the remaining water molecules.

There are many examples of reactions of ligands of the same charge type with the metal ion which have rate constants within a factor of two of each other. These values have often been compared with the water exchange rate constants determined by NMR using values of K_{os} calculated from the Fuoss equation, and agreement within a factor of two has been attributed to uncertainties in the value of K_{os}. However, it seems likely, particularly for the pyridine-containing ligands, that ring closure can contribute to the rate-determining step and that this may explain the differences in $k^{M\text{-}H_2O}$ (a 50% contribution from ring closure only reduces the formation rate constant by a factor of two). Steric effects can influence the formation rate constants for monodentate ligands by changing the magnitude of K_{os} and by affecting the penetration of the inner sphere. For multidentate ligands, steric hindrance can be large enough to change the rate-determining step from first- to second-bond formation.

The nature of the ligand donor groups and the strength of the first metal–ligand bond are very important in determining the rate step. If the first bond formed is weak, then ring closure can make a significant contribution to the formation rate constant. This leads to chelation-controlled substitution where the formation rate constant depends on the stability of the monodentate intermediate and the chelate ring size. This behavior is found for the reactions of several amino acids with Ni^{2+}, Co^{2+}, and Cu^{2+}.

When the incoming ligand is protonated and if the proton is remote from the first binding site, there is seldom a large effect on the formation rate unless the proton blocks the normal reaction path. When the proton is adjacent to the first binding site, the rate-determining step is sometimes shifted to proton transfer, and this can lower the formation rate constant.

It might seem from the above summary that the general mechanisms of inorganic reactions in aqueous solution are well understood, but for some reactions there are still no adequate explanations, particularly the lanthanide reactions. Because of the unavailability of the 4f electrons of the lanthanides to form hybrid orbitals with those of the donor ligands, the reactions of the lanthanides might be expected to be similar to those of the d^0 or d^{10} systems. From this it would be expected that the water exchange rate for the lanthanide would simply depend on the radius of the metal ion. However, the results plotted in Figure 1–9 suggest that not only does the water exchange rate constant depend on some other function of the metal ion but also that the formation rate constants depend on the nature of the incoming ligand. For the reaction of $HCyDTA^{3-}$ with the lanthanides, the rate-determining step is not the loss of a water molecule from the inner coordination sphere of the metal ion but is the rearrangement of a rapidly formed intermediate which then incorporates the metal ion into the coordination cage of the ligand.[233] Earlier we suggested that there might be a contribution to the formation rate constant as a result of ring closure for the reactions of murexide with the lanthanides. If this is so, then it is also the case

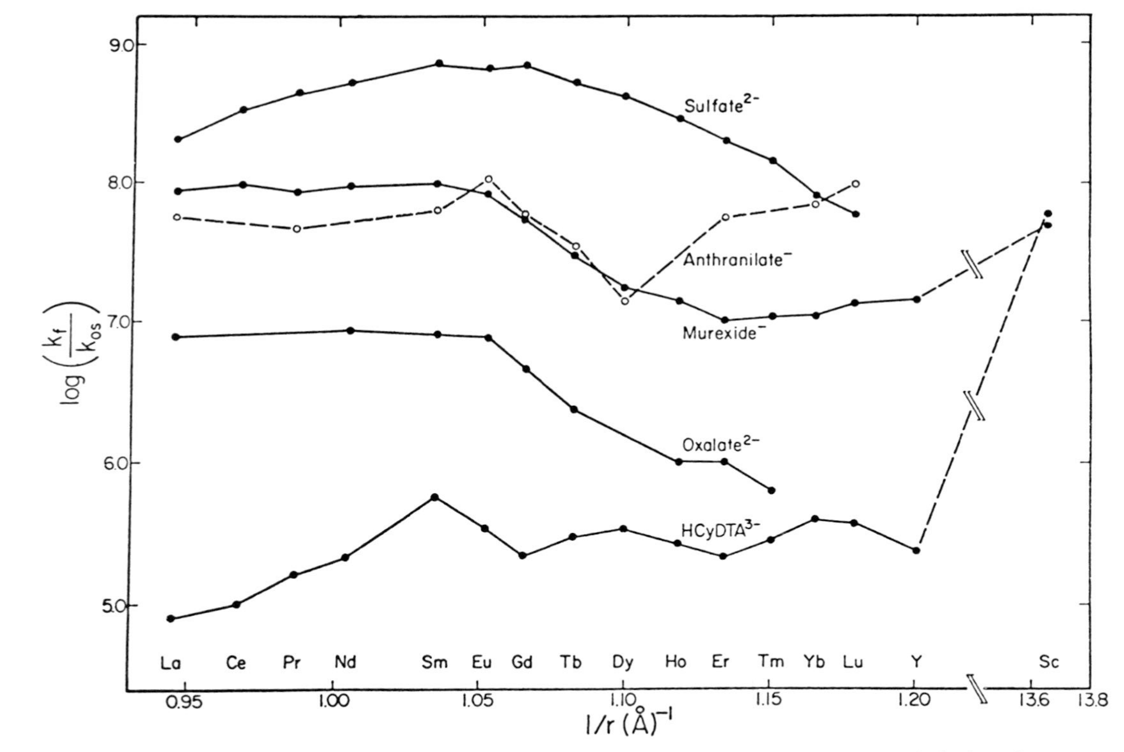

Figure 1-9 The dependence of the formation rate constant (corrected for outer-sphere association) on the nature of the ligand for the complexation of the lanthanide ions

for the oxalate reactions because they have rate constants similar to those for murexide although electrostatics would predict that they should be much larger. It has often been suggested that the properties of the lanthanides can be explained in terms of a change of coordination number for the aquo ions.[345,346] However, this proposal has been discounted by Geier[347] who suggests that all the lanthanides have nine-coordinated water molecules, but that as the size of the metal ion decreases it becomes more difficult to pack nine equivalent water molecules around the metal ion. This eventually gives rise to nonequivalent water molecules and could explain the observed trends in the water exchange rate constants. Geier has also found that the substitution of a ligand into the inner coordination sphere of the metal ion can change the coordination number of some of the lanthanides.[347,348,349]

For the discussions presented in this section it does not matter whether the reactions of some labile metals follow an associative or dissociative pathway. Although there seems to be evidence that the latter pathway is more commonly found, an associative mechanism was found for the reactions of $Cu(OH)_3^-$ and $Cu(OH)_4^{2-}$ with linear and macrocyclic polyamines.[298]

IV. DISSOCIATION REACTIONS OF METAL COMPLEXES

For the reaction in Eq. (34) the dissociation rate constant, k_d, can be

$$M + L \underset{k_d}{\overset{k_f}{\rightleftarrows}} ML \tag{34}$$

written in terms of the stability constant of the complex (K_{ML}).

$$k_d = \frac{k_f}{K_{ML}} \tag{34a}$$

Therefore, for reactions of monodentate ligands or for multidentate ligands where the first-bond formation is the rate-determining step, k_d can be calculated from the K_{ML}, $k^{M\text{-}H_2O}$, and K_{os} values. Eq. (35) is useful for cases

$$k_d = \frac{K_{os} k^{M\text{-}H_2O}}{K_{ML}} \tag{35}$$

where chelation-controlled substitution or ICB effects do not influence the rate, and of course Eq. (34a) is valid in general if the correct equilibrium constant is used.

A. Polyamines

The simple relationship in Eq. (34a) has some interesting consequences for multidentate ligands such as the polyamines because the predicted k_d values (without the assistance of acid) are extraordinarily small for metal ion complexes that usually are considered to be labile. Thus, for nitrien^{2+} the k_f value is estimated to be the same as that for Nien^{2+} ($3.5 \times 10^5 M^{-1}sec^{-1}$), and the K_{ml} value is $10^{14.4} M^{-1}$ with the result that k_d is 1.4×10^{-9} sec^{-1}. This corresponds to a dissociation half life of 16 years! It is not too surprising that the pure-solvent dissociation rate of Nitrien^{2+} has not been measured. Individual steps in the unwrapping of trien from nickel (Figure 1-10) are not so slow, but unless the reaction is assisted by protonation of the released amine groups (or by other means), the successive equilibria leading up to the last bond cleavage of trien are very unfavorable and a slow dissociation rate results. At pH 6.2, where protonation helps to speed the reaction, a dissociation rate constant of $10^{-5.5}$ sec^{-1} has been determined for Nitrien^{2+} by measuring the induction period in a coordination chain reaction.[473] The rate constant for the reaction in Eq. (36) has been measured,[239] $k_H{}^{Nitrien} = 0.27 M^{-1}sec^{-1}$, and also can be calculated from Eq. (37).

$$\text{Nitrien}^{2+} + \text{H}^+ \rightleftarrows \text{Ni}^{2+} + \text{Htrien}^+ \qquad (36)$$

$$k_H{}^{NiL} = \frac{k_{Ni}{}^{HL}\, K_{HL}}{K_{NiL}} \qquad (37)$$

At pH 6.2 the $k_H{}^{NiL}[H^+]$ term does not contribute as much to the observed k_d value of Nitrien^{2+} as the term with two protons, $k_{2H}{}^{NiL}[H^+]^2$. Even in neutral solutions the hydrogen ion dissociation pathway is important. Despite these slow rates in neutral solution, when Nitrien^{2+} reacted with $0.5M$ H^+, Melson and Wilkins[474] found three relatively fast reactions with rate constants of 15 sec^{-1}, 4 sec^{-1}, and 2 sec^{-1}. The high acid concentration permits the very first step of the dissociation reaction to be seen because it stabilizes the intermediate protonated species relative to Nitrien^{2+} (but *not* relative to the products). The three rate constants correspond to single nickel–nitrogen bond cleavage steps for Nitrien^{2+}, Ni(Htrien)$^{3+}$, and Ni(H$_2$trien)$^{4+}$, respectively. The dissociation of the fourth (last-remaining) nitrogen does not have to break a chelate ring and therefore is faster than that for the third nitrogen, with the result that the last bond-breaking step is not seen. Other electrophilic agents such as metal ions also interrupt the slow dissociation process pictured along the top of Figure 1-10. Thus, the transfer of trien from Ni^{2+} to Cu^{2+} is fairly rapid with a second-order rate constant $k_{Cu}{}^{Nitrien} = 2.7 M^{-1}sec^{-1}$ at 25°C, 0.1μ.[475] The mechanism of this reaction is discussed later, and it is proposed that Cu^{2+} reacts after the first chelate ring of Nitrien^{2+} opens.

Nucleophilic attack will also speed the dissociation of Nitrien^{2+} or other complexes in which the incoming nucleophile has little difficulty in coordinating to the metal ion. Thus, the reactions of Nitrien^{2+} with CN$^-$ [476] and with EDTA^{4-} [282] are relatively fast. In fact, it is difficult to find a reactant which will force the dissociation of Nitrien^{2+} without accelerating the dissociation rate via electrophilic or nucleophilic interactions with the complex. Hence, most of

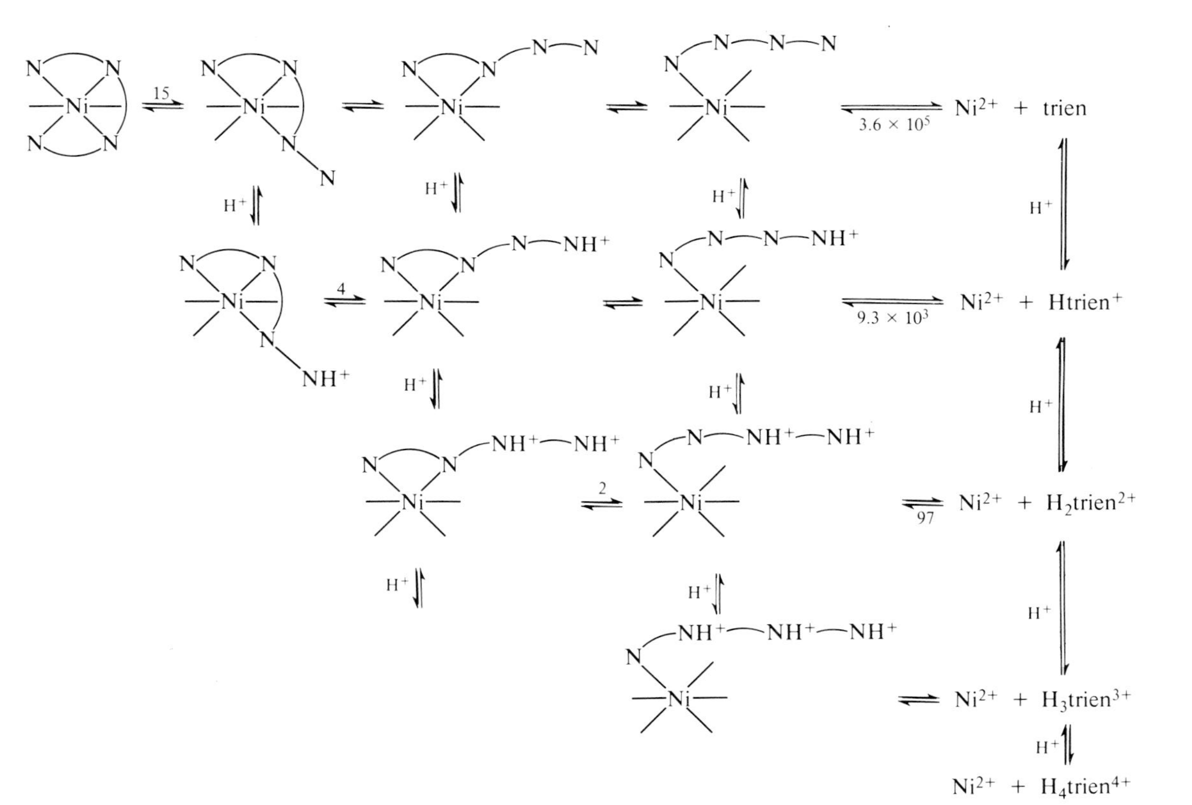

Figure 1-10 Proposed stepwise mechanism for the formation and dissociation of triethylenetetraminenickel(II). Resolved rate constants for bond-breaking steps (sec^{-1}) and for bond-formation steps ($M^{-1}\ sec^{-1}$) are given where they have been evaluated.

the reactions of $Nitrien^{2+}$ and similar complexes are truly labile despite the possibility of extremely slow dissociation rates if only the solvent is displacing the ligand. Measurement of the small dissociation rate constant at pH 6.2 was possible only because the coordination chain reaction is capable of detecting minute traces of free trien.[23] The actual extent of dissociation at this pH is very small.

It is interesting that in the reaction of $Ni(2,3,2\text{-tet})^{2+}$ with strong acid ($0.5M$) it is possible to observe only one rate constant, 0.38 sec^{-1}.[477] The 2,3,2-tet complex is 100 times more stable than the trien complex,[475] and the decrease by a factor of 40 in the first bond-dissociation rate reflects much of this change in stability. The additional methylene group in the center ring of $Ni(2,3,2\text{-tet})^{2+}$ decreases the strain caused by the linked consecutive rings in the polyamine complex and thus stabilizes the complex. Only one rate step is seen in the acid reaction because the subsequent steps are faster (their rates should be about the same as those for Nitrien). With $Ni(2,3,2\text{-tet})^{2+}$ the high acid concentration has forced the rate-determining step back to the first bond cleavage rather than the next to last bond cleavage.

A shift in the rate-determining step to earlier stages in the dissociation process on mixing complexes with strong acid is typical. It permits chelate ring-opening steps to be measured for a number of complexes. Wilkins and co-workers[474,479–482,271,272] have studied the rates of chelate ring-opening reactions of nickel and copper complexes by this technique (Table 1-20). Sequential steps are observed for many of the nickel–polyamine complexes in which nickel–nitrogen bond rupture is measured and the proton scavenges the released amine group.[474] Interestingly, despite the electrostatic repulsion in the highly charged protonated complexes the relative rates of the first ring opening are nevertheless:

$$Nidien^{2+} > Ni(Hdien)^{3+};\; Nitrien^{2+} > Ni(Htrien)^{3+} > Ni(H_2trien)^{4+}$$

This is related to two factors: (1) the ring strain in linked consecutive rings and (2) the enhanced lability of one amine group caused by metal coordination to other amine groups. The latter effect is very noticeable in the relative reactivity of $Nien_3^{2+} > Nien_2^{2+} > Nien^{2+}$ where the initial ring-opening rate changes by a factor of 580 from the tris to the mono complex. There is an accelerating effect of charge placed near the reacting donor center as seen by comparing[474] the dissociation rate constant of 0.32 sec^{-1} for $Ni(N\text{-pren})^{2+}$ with the value of 2.8 sec^{-1} for $Ni(Hdien)^{3+}$. A similar effect is seen for $Ni(Htren)^{3+}$ with a dissociation rate constant of 0.22 sec^{-1} compared with 4.0 sec^{-1} for $Ni(Htrien)^{3+}$ This is attributed to the difference in the distance between the protonated site and the reacting donor center. The reverse situation holds for the addition of one more proton to the trien complex; the dissociation of $Ni(H_2trien)^{4+}$ is not greatly accelerated over that of $Ni(Htrien)^{3+}$ because the charges are now well separated.

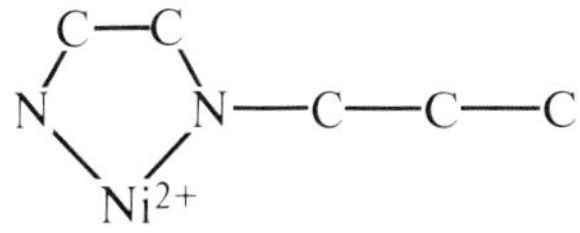

$Ni(N\text{-}pren)^{2+}$

17

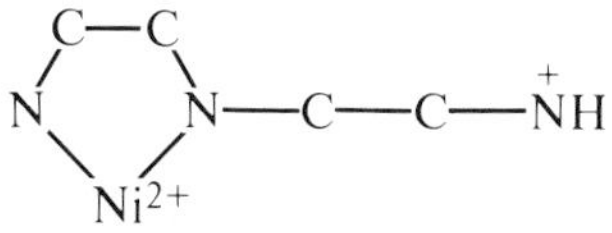

$Ni(Hdien)^{3+}$

18

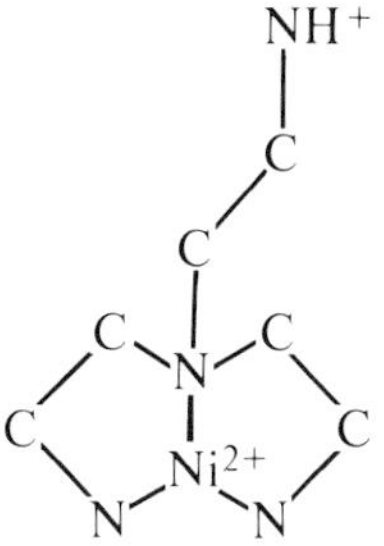

$Ni(Htren)^{3+}$

19

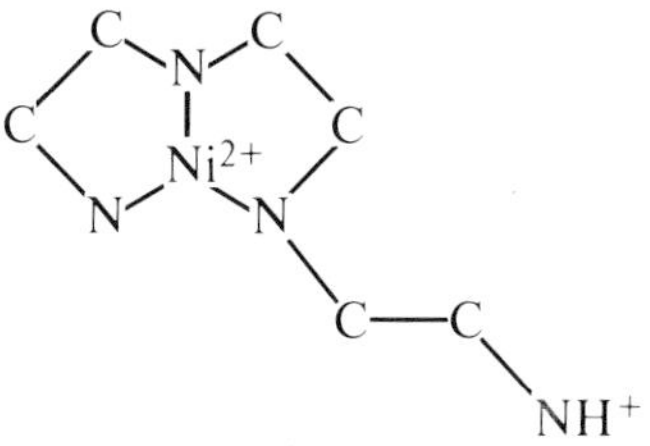

$Ni(Htrien)^{3+}$

20

However, for $Ni(H_2tren)^{4+}$ both protonated groups are now near the reactive center, and k_3 must be larger than k_2 (Table 1-20) so that only two rates are observable in the reaction of acid with $Nitren^{2+}$. The reaction of the five-coordinate complex, $[Ni(Me_6tren)H_2O)]^{2+}$ or $Ni(tda)^{2+}$, with acid gives only one observable rate step, which is the first bond cleavage.

The small effect of ring size on the decomposition rate is shown by the behavior of $Ni(dien)^{2+}$ and $Ni(3,3'\text{-}dapa)^{2+}$ which in the first stage of their protonation reactions have rate constants of 14 and 11 sec^{-1}, respectively.[474] Similarly, $Nien^{2+}$ and $Nitn^{2+}$ at 0.3°–0.6°C have rate constants of 6.6×10^{-3} sec^{-1} and 9.6×10^{-3} sec^{-1}, respectively.[482] Thus, it is hard to justify a steric effect—i.e., sterically controlled substitution—in regard to relative rates of ring opening of six- vs. five-membered chelates. There is no question that opening a chelate ring is a much slower process than the loss of a monodentate ligand. The factor is 40 to 100, depending upon the comparison taken in Table 1-20.

Why is chelate ring opening slower? Since it is not because of stronger metal–donor bonds in the chelate, it must be related to the restrictive properties of the ring itself. Margerum[486] proposed that the reason a chelate ring is more difficult to break is that the minimum energy path for the dissociation of a donor group from a metal ion is a linear displacement rather than an angular displacement forced by the chelate. For a chelate ring to open with a minimum arc to avoid repulsions of nearby donor groups, the various possible internal rotational barriers of the chelate molecule will be encountered. Accordingly, C-substituted chelates with higher rotational barriers should give smaller dissociation rate constants. This certainly seemed to be supported by the decrease in acid dissociation rate constants by factors of 7.5, 20, and 11 for the mono, bis,

TABLE 1-20
METAL AMINE DISSOCIATION RATE CONSTANTS IN ACID

Ni Complex	k_1 (sec^{-1})	k_2 (sec^{-1})	k_3 (sec^{-1})	pH	Temp. °C	Ref.
N-Pren	0.32			0.5*M* [H$^+$]	25	474
Ptn	63	0.34		0.5*M* [H$^+$]	25	474
Dien	14	2.8		0.2*M* [H$^+$]	25	474
3,3′-Dapa	11	1.7		0.2*M* [H$^+$]	25	474
Tren	66	0.22		0.5*M* [H$^+$]	25	474
Trien	15	4.0	2.1	0.5*M* [H$^+$]	25	474
Penten	70	49	3	0.5*M* [H$^+$]	25	474

Ni Complex	k_1 sec^{-1}	pH	Temp. °C	Ref.
cis,cis-tach	$k = 10^{-4}(1.3 + 1.4\,[H^+])$	0.5–5*M* [H$^+$]	25	483
En	$k = 10^{-3}(6.3 + 4.3\,[H^+])$	0.5–3*M* [H$^+$]	0.6	484
En	0.15	0.2*M* [H$^+$]	25	480
	0.173	1*M* [H$^+$]	21.8	274
	6.6×10^{-3}	4.0	0.3	271
$(En)_2$	5.2	0.2*M* [H$^+$]	25	480
$(En)_3$	86.6	0.2*M* [H$^+$]	25	480
Tn	8.8×10^{-3}	4.8	0.3	271
	9.9×10^{-3}	2*M* [H$^+$]	0.6	271
N,N′-DiMeen	1.7×10^{-3}	4.0	0.3	271
N,N-DiMeen	3.85×10^{-2}	4	0.3	271
N,N′-DiEten	1×10^{-2}	4	0.3	271
	.015	5.8–7.0	25	293
N,N-DiEten	5.4	5.9–7.0	25	293
N,N′-Di-*n*-Pren	> 200	4	0.3	271
1,3 Ddmp	5.2×10^{-3}	1.5	0.6	271
	6.5×10^{-3}	2*M* [H$^+$]	0.6	271
cis-Chx	5.6×10^{-3}	3.6	0.3	271
trans-Chx	1.3×10^{-4}	6.8	0.3	271
	1.4×10^{-3}	4.0	0.3	271
	2.3×10^{-3}	1*M* [H$^+$]	0.6	271
N,N,N-Tri-methylen	40	0.5*M* [H$^+$]	25	119
rac-Bn	1.6×10^{-4}	6.8	0.3	271
	7×10^{-4}	4.3	0.3	271
	1.02×10^{-3}	1.08*M* [H$^+$]	0.6	271
Bn	5.8×10^{-3}	4.8	0.3	271
	1.42×10^{-2}	2*M* [H$^+$]	0.6	271
$(rac\text{-Bn})_2$	0.257	0.2*M* [H$^+$]	25	274
$(rac\text{-Bn})_3$	8.25	0.2*M* [H$^+$]	25	274
$(C\text{-Tetrameen})_2$	3.3×10^{-3}	0.5*M* [H$^+$]	25	482
(*N*-Tetrameen)	0.14	0.5*M* [H$^+$]	25	482
NH_3	3.4[a]	6.5–7.0	25	114

TABLE 1-20 (CONTINUED)
METAL AMINE DISSOCIATION RATE CONSTANTS IN ACID

Ni Complex	k_1 sec^{-1}	pH	Temp. °C	Ref.
NH_3	7.1[b]	6.5–7.0	25	106
	5.7	0.2M [H^+]	25	119
$MeNH_2$	7.7	6.5–7.5	25	106
$EtNH_2$	13.3	6.5–7.5	25	106
iPrNH$_2$	16.8	6.5–7.5	25	106
Me_2NH	11.3	6.5–7.5	25	106
Pyridine	38.5	0.5M [H^+]	25	119
Aziridine	5.0	0.2–0.5M [H^+]	25	119
Hydrazine	3.6	0.1–0.5M [H^+]	25	119
Tda	7.87	1M [H^+]	20	485
Cu Complex	**k_1 sec^{-1}**	**pH**	**Temp. °C**	**Ref.**
$(En)_2$	43.3	0.15M [H^+]	0.8	482
En	~115	0.15M [H^+]	25	482
$(Bn)_2$	15.1	0.2M [H^+]	25	482
(Bn)	4.56	0.2M [H^+]	25	482
(C-Tetrameen)$_2$	0.80	0.15M [H^+]	25	482
C-Tetrameen	0.42	0.15M [H^+]	25	482
N-Tetrameen	38.5	0.5M [H^+]	25	482
NH_3	1.3×10^4	3.6–3.8	25	127
Tda	28.0	1M [H^+]	20	485
Cu(tet a)$^{2+}$ (red)	3.6×10^{-7}	6.1M [H^+]	25	295
Cu(tet a)$^{2+}$ (unstable red)	4.3×10^{-6}	6.1M [H^+]	25	295
Cu(tet a)$^{2+}$ (blue)	3.8×10^{-3}	6.1M [H^+]	25	295
Cu($trans$-[14]-diene)$^{2+}$	1.2×10^{-3}	6.1M [H^+]	25	295
Cu(2,3,2-tet)$^{2+}$	4.1	6.1M [H^+]	25	295
Co complex				
Co(tda)$^{2+}$	0.97	1M [H^+]	20	485

[a] $\mu = 0.1$.
[b] $\mu = 1.0$.

and tris bn [$H_2NCH(CH_3)CH(CH_3)NH_2$] complexes of nickel compared with the constants for the corresponding en complexes. For the Ni(C-tetraMeen)$_2^{2+}$ complex the k_d value in HNO_3 is only 0.0033 sec^{-1}, which is 1/1600 of the corresponding value for Nien$_2^{2+}$. The copper(II) complexes of en, bn, and C-tetraMeen also show large effects of the same type (Table 1-20). There are some serious problems with this explanation, however, because Ahmed and Wilkins[271] found that at 0.3°C the k_d (= k_1) values (at low pH) of the mono nickel com-

TABLE 1-21
STABILITY CONSTANTS AND ARRHENIUS PARAMETERS FOR THE ACID DISSOCIATION OF NiL^{2+} IN $HClO_4$ AT 0.3°C (RING-OPENING STEPS FOR THE DIAMINES)

L	log K_1	k (sec^{-1})	E_a (kcal mol^{-1})	log A	Ref.
en	7.92	6.6×10^{-3}	20.5	14.2	484
N,N'-diEten	7.42	1.0×10^{-2}	17.1	11.8	271
rac-2,3-bn	8.30	7×10^{-4}	19.4	12.4	271
trans-chx	8.22	1.4×10^{-3}	19.5	12.9	271
NH_3 (∼pH 7)	2.8	0.43[a]	14	11	114

[a]Corrected to 0.3°C using the experimental value of E_a.

plexes of en, *meso*-2,3-diaminobutane, and *cis*-1,2-diaminocyclohexane are all similar, as shown in Table 1-20. The *rac*-2,3-diaminobutane and *trans*-1,2-diaminocyclohexane are slower to react, but the values indicate less effect as a result of strongly hindered rotation than expected for the mechanism suggested. Furthermore, the activation energies for the acid dissociation are nearly the same for 2,3-diaminobutane and *trans*-1,2-diaminocyclohexane as for ethylenediamine (Table 1-21), yet the activation energy for the loss of NH_3 from Ni^{2+} is 5–6 kcal/mol less than the diamines with primary amine groups, a difference too great to be assigned to difference in basicity. We suggest two explanations. First, the initial phase of the chelate ring-opening step involves some angular expansion of the bond angles in the chelate ring and depends less on the hindered internal rotations. This would cause much larger activation energies (or larger $\Delta H^{\ddagger}$) for bidentate chelates compared with monodentates in accord with experiment. Second, an intermediate state with one donor displaced from the normal chelation position is proposed prior to the solvation of the metal ion and the solvation of the leaving donor group—i.e., prior to complete solvent separation of the donor and acceptor. The breakup of this intermediate would be influenced by steric effects on the permissible conformations of the one-bonded bidentate ligand and by the chances of rotating one group out of the coordination sphere of the metal ion. The proposed mechanism is pictured in Figure 1-11, where the k_{21} rate constant is very large because it does not involve solvent displacement from the metal ion, and hence the rate-determining step in dilute acid is from *2* → *3*. The k_{23} term affects the Arrhenius frequency factor (or the $\Delta S^{\ddagger}$ term). The rate constant for the overall ring-opening step is $k_{12}k_{23}/k_{21}$.

There is an additional reason to propose the detailed reaction pathway shown in Figure 1-11. The papers by Ahmed and Wilkins[271,484] reported an accelerated rate of decomposition of nickel–diamine complexes in strong acid. The dissociation rate constants increase by a factor of about 20 in changing from neutral pH to pH 4. The rate constants do not change from pH 4 to 1, but they increase again below pH 1 with a first-order, hydrogen-ion dependence. To our knowledge these data have not been refuted. The initial explanation was that H_3O^+ aids the bond rupture directly by competing for the nickel–nitrogen bonding electrons.[484] The mechanism in Figure 1-11 may be considered as a modification of this explanation. If we use Ahmed and Wilkins rate constants (0.6°C) for the H^+ effect in strong acid on the dissociation of $Nien^{2+}$, then according to our interpretation, $k_{24}[H^+] = k_{23}$ at $[H^+] = 1.4M$. This means that the amine group which has moved partially away from the metal ion (species *2* of Figure 1-11) is a poor protonation site. Its proximity to the Ni^{2+} ion reduces its basicity so that only at very high H^+ concentration does protonation occur prior to or during the solvation of the metal ion and donor group. More detailed studies of the acid dependence could give information about the relative preference of the *2* → *3* path compared with the *2* → *4* path. Unfortunately, many of the comparisons of dissociation rate constants are based on data at 0.5*M* or 0.2*M* H^+ where presumably both the *2* → *3* and the *2* → *4* paths make contributions.

The acid dissociation rate of $Ni(phen)^{2+}$ also has a complex $[H^+]$ dependence in the range of 0.01–1*M* $[H^+]$. Protonated intermediates of one-bonded complexes were again suggested[412] except in this case the ligand structure makes impossible the rotation of a single donor group from the vicinity of the metal ion.

There are other ligands where the structure tends to prevent the stepwise unwrapping of the ligand from around the metal ion. One example is the nickel complex of the ligand *cis,cis*-1,3,5-triaminocyclohexane (*cis,cis*-tach) which is attacked by acid very slowly.[483] In 0.5–5.0*M* HNO_3 the observed first-order dissociation rate constant was $k_{obsd} = k + k_H[H^+]$, where $k = 1.30 \times 10^{-4} sec^{-1}$

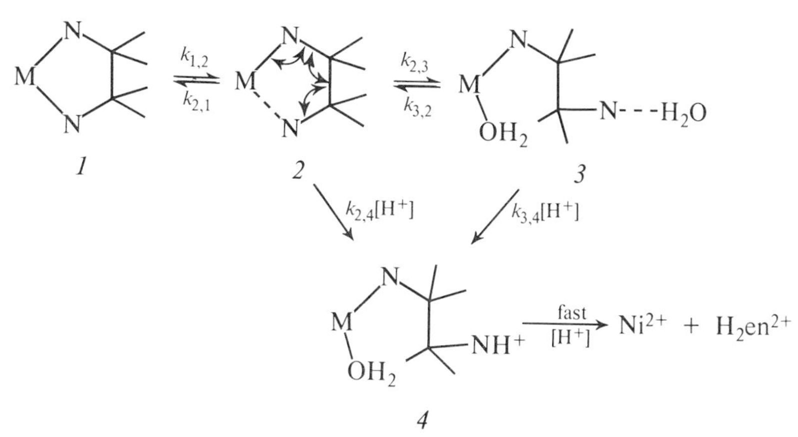

Figure 1-11 Proposed reaction pathway for the reaction of metal–ethylenediamine chelates with strong acid. Species *2* is a reaction intermediate where one metal–amine bond is broken (or weakened) by angular expansion, but the metal ion and amine group are not far enough apart to solvate.

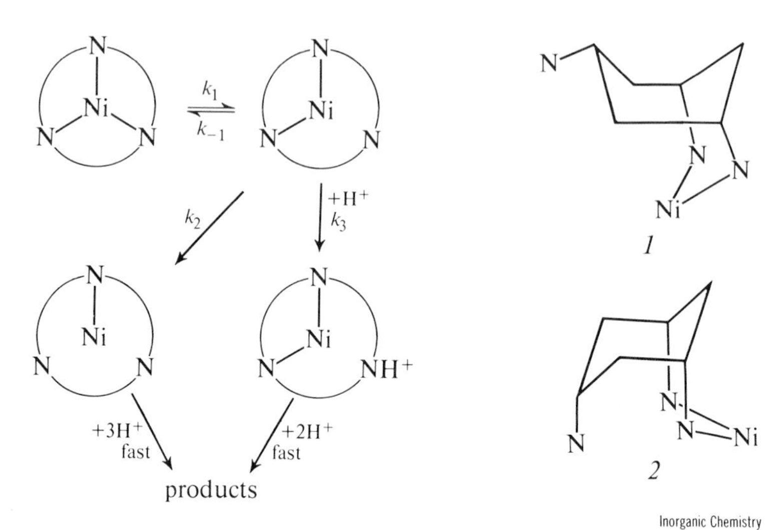

Figure 1-12 Proposed mechanism and intermediates for the acid dissociation of Ni(*cis,cis*-tach)$^{2+}$ [483]

and $k_H = 1.39 \times 10^{-4} M^{-1} sec^{-1}$ at 25.0°C. The proposed mechanism is given in Figure 1-12 where the rate-determining steps were assigned to the second nickel–nitrogen bond cleavage so that $k = (k_1/k_{-1})\ k_2$ and $k_H = (k_1/k_{-1})\ k_3[H^+]$. Childers and Wentworth[483] assume that the formation of either structure *1* or *2* in Figure 1-12 is difficult and that the repulsive nonbonded contacts are best relieved by returning to the tridentate form of the ligand. However, the activation energy for the non-proton rate constant (k) is only 19.3 kcal mol^{-1}, which is less than that for Nien^{2+} in strong acid and does not correspond to a pathway with an additional Ni—N bond broken in the rate-determining step. It is very hard to understand how the proposed mechanism could be correct. If the breaking of the first Ni—N bond forces one of the remaining six-membered rings into an unstable boat conformation as the authors suggest, then surely this nitrogen must immediately add a proton, and the extent of protonation would be great. There are several alternative explanations for the observed data. Perhaps the rate-determining step is the first-bond dissociation in accord with the mechanism given in Figure 1-11. The ratio of k/k_H is about unity, in agreement with that proposed for k_{23}/k_{24} in Figure 1-11. The smaller rate constant for the reaction of Ni(*cis,cis*-tach)$^{2+}$ compared with Nien^{2+} in strong acid is a result primarily of the negative $\Delta S^‡$ value (-13.5 cal deg^{-1} mol^{-1}) for k.

B. Macrocyclic Ligand Complexes

Other examples in which the ligand structure makes a stepwise unwrapping pathway difficult are found with the macrocyclic ligand complexes. Thus, Ni-(cyclam)$^{2+}$, if thrown into very strong hydrochloric acid, simply shows no evidence of dissociation over prolonged periods of time.[487,488] The Cu(tet *a*)(red)$^{2+}$ complexes dissociates in 6.1M HCl at 25°C with a half-life of 22 days.[295] Additional dissociation rate constants for copper–macrocyclic ligands are given in Table 1-20. The rate of dissociation of the Cu(14-ane-N_4)$^{2+}$ complex—i.e., Cu(cyclam)$^{2+}$—has not been measured because it is so slow, but the rates of dissociation of Cu(12-ane-N_4)$^{2+}$ and Cu(13-ane-N_4)$^{2+}$ have been determined in aqueous acetic acid solution.[489,490] The values for k_{-1} ($M^{-1}sec^{-1}$) when the dissociation rate is $k_{-1}[CuZ^{2+}][HOAc]$ are $10^{-10.9}$ and $10^{-14.2}$ for Z = 12-ane-N_4 and 13-ane-N_4, respectively. There also is a combined H^+ and HOAc path where the dissociation rate is $k_{-2}[CuZ^{2+}][HOAc][H^+]$ and the k_{-2}($M^{-2}sec^{-1}$) values are $10^{-8.2}$ and $10^{-9.9}$, respectively, for the 12- and 13-membered macrocyclic ligand complexes.

Macrocyclic ligands including polythiaethers and polyethers as well as polyamines are slow to dissociate from otherwise labile metal ions. The dissociation rate constants in 80% CH_3OH–H_2O for the 12-ane-S_4 and 14-ane-S_4 complexes of Cu(II) are only 4.4 and 9 sec^{-1}, respectively.[491] The 13-ane-S_4, 15-ane-S_4, and 16-ane-S_4 ligands dissociate more rapidly, but all are much slower than an open-chain polythiaether ligand. The suggested rate-determining step in these macrocyclic ligand reactions is the breaking of the last remaining chelate ring rather than the removal of the final thiaether group from the copper ion. Recently,[492] the dissociation rate constant for Ca(cryptate)$^{2+}$ has been reported to be 0.16 sec^{-1} at 25°C and high pH. The cryptate ligand is $N(CH_2CH_2O$-$CH_2CH_2OCH_2CH_2)_3N$, and its calcium complex has a stability constant of $1.8 \times 10^4 M^{-1}$.

C. Aminocarboxylates

Studies of the dissociation reactions of aminocarboxylate complexes in acid once again bring up the question of whether the amino group or the carboxylate group dissociates first. NMR studies on diamagnetic polyaminocarboxylates indicate that the carboxylate bonds are much more labile than the nitrogen bonds.[493] The dissociation rate constant for $NiGly^+$ is reported to be independent of acidity from pH 7 to 3.5 and is significantly enhanced by acid only below pH 3.[5] At pH 1 the $NiGly^+$ complex rapidly picks up one proton and decomposes at a measureable rate to form Ni^{2+} and H_2Gly^+.[479] Wilkins[5] interpreted these results as evidence that the carboxylate group is detached first in the dissociation reaction. A consideration of the full mechanism (see section III, G. and Figure 1-8) and the behavior of protonated ligands indicates only partial agreement with this conclusion. At pH 1 the HL° pathway will predominate, and therefore the rate-determining dissociation step will be the breakup of (Ni—NH_2CH_2COOH)$^{2+}$ ($k_d = k_{-6}$ in Figure 1-8), in agreement with Wilkins' interpretation. However, at pH 3.5–7, if the amine group detaches first, then this ring opening is the rate-determining step, and there should be no H^+ dependence. Hence, the lack of H^+ dependence does not require carboxylate ring opening but supports our suggestion that $k_{-1} \gg k_4$ so that above pH 3 $k_d = k_{-4}$ (Figure

1-8). Unfortunately, the actual k_d value from pH 3.5–7 has not been published. Hammes and Steinfeld[115] estimated $k_d = 0.024\ sec^{-1}$ at high pH by comparison with the behavior of the bis and tris glycine complexes. Our estimated value for $k_d = k_{-4}$ from Table 1-18 and log $K_1 = 5.7$[481] is 0.04 sec^{-1}. An approximate value of 0.53 sec^{-1} (20°C) for the dissociation rate constant in strong acid (presumably in 0.2*M* $[H^+]$) has been reported.[480]

The dissociation rate constants of several nickel–aminopolycarboxylates have been studied in conjunction with metal ion-exchange reactions[281,283,331,428] and are summarized in Table 1-22. Bydalek[283,331] has shown that the best fit for the $NiNTA^-$ and NiIDA data is a mechanism in which the rate-determining step is the cleavage of the Ni—N bond of the last glycine segment of these complexes. The proposed intermediate structures for $NINTA^-$ are:

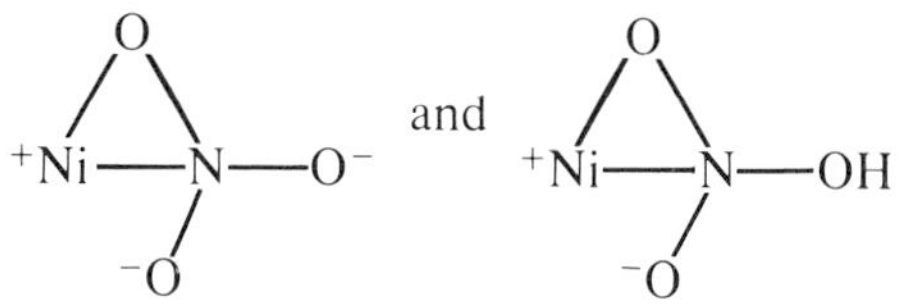

```
      O                        O
     / \          and         / \
+Ni-----N----O-          +Ni-----N----OH
        /                        /
      -O                       -O
```

This is consistent with the behavior of the EDTA and CyDTA complexes except that the second nitrogen is protonated in the acid-attack path. The proposed reaction intermediate for $NiEDTA^{2-}$ reaction with acid is:

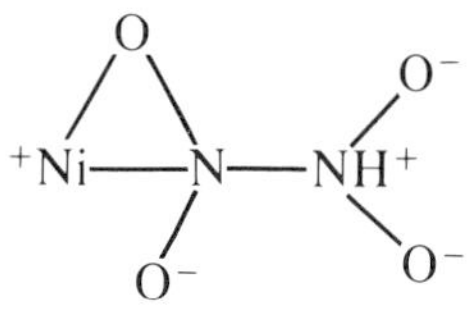

```
      O
     / \          O-
    /   \        /
+Ni------N----NH+
        /        \
      O-          O-
```

TABLE 1-22
DISSOCIATION RATE CONSTANTS FOR NICKEL–AMINOCARBOXYLATE COMPLEXES (25°C)

NiL		NiL k_d (sec^{-1})	NiL k_H ($M^{-1}\ sec^{-1}$)	Ref.
$NiGly^+$	($\mu = 0.15$)	~0.024[a]	—	115
NiIDA	($\mu = 0.1$)	2.8×10^{-4}	2.8	331
	($\mu = 1.25$)	1.7×10^{-4}	1.2	331
$NiNTA^-$	($\mu = 0.4$)	5.8×10^{-6}	0.77	283
	($\mu = 1.25$)	3.5×10^{-6}	0.43	283
$NiEDTA^{2-}$	($\mu = 1.25$)	—	8×10^{-4}	428
$NiCyDTA^{2-}$	($\mu = 1.25$)	—	2.5×10^{-4}	281

[a] Estimated value.

There are a number of reactions of nickel–aminopolycarboxylate complexes with metal ions[494-497,22,278,281] where the reaction proceeds to the point at which only a nickel–glycinate chelate remains, and then the rate-determining step is the nickel–nitrogen bond cleavage.[496] This is consistent with the mechanism we have proposed for the amino acid complexes.

D. CyDTA

In the exchange reactions of metal–CyDTA complexes (where $CyDTA^{4-}$ is *trans*-1,2-diaminocyclohexane-*N,N,N′,N′*-tetraacetate ion) the direct displacement of one metal ion by another is not a favorable pathway.[281,498,278,464] This is the result of the ligand structure in which it is sterically impossible to have one metal ion associated with each nitrogen donor, as is the case for EDTA exchange reactions.[22,494] As a result, the rate of a reaction such as that in Eq. (38) does not increase with the Pb^{2+} concentration but depends strongly on the H^+ concentration. The general mechanism for the transfer of CyDTA between metal ions M and M′ is given by Eq. (39) and (40).

$$[\text{Cd(CyDTA)}]^{2-} + Pb^{2+} \longrightarrow [\text{Pb(CyDTA)}]^{2-} + Cd^{2+} \quad (38)$$

$$\text{MCyDTA} \overset{k_d}{\rightleftarrows} \text{CyDTA} + \text{M} \quad (39)$$

$$\text{CyDTA} + \text{M}' \rightleftarrows \text{M}'\text{CyDTA} \quad (40)$$

If M′ is a labile metal ion which forms a strong complex (such as Pb^{2+}, Cu^{2+}, or Sc^{3+}), then the rate expression is given by Eq. (41), and M′ acts only as a

$$\frac{-d[\text{MCyDTA}]}{dt} = k_d^{\text{MCy}}[\text{MCyDTA}] \quad (41)$$

scavenger for the released CyDTA. The CyDTA complexes of most metal ions are very stable so the dissociation rate constants are small. Hydrogen ion is able to assist the dissociation reaction because it can penetrate into the cage-like cavity formed by the ligand donor groups without completely displacing the metal ion, M. The proposed reaction intermediate structure[278] for one class of metal ions is:

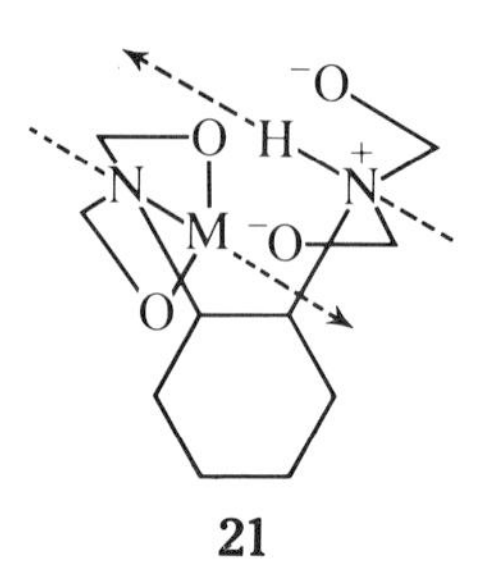

21

As a result of this type of protonated intermediate and the protonation of the acetate groups at higher acidity, the general expression for k_d is given by Eq. (42). For many metal ions, $k_H{}^{HCy}$ is the easiest term to characterize, and

$$k_d = k^{MCy} + k_H{}^{MCy}[H^+] + k_H{}^{MHCy}[H^+] \qquad (42)$$

this constant has been measured for 33 metal ions[499,308] as given in Table 1-23. Its value varies by 10 orders of magnitude for $BaCyDTA^{2-}$ compared with $NiCyDTA^{2-}$. In most cases, the $k_M{}^{HCy}$ constant can be calculated quite accurately from the K_{HCy}, and K_{MCy} constants and Eq. (43). This permits a com-

$$MCyDTA + H^+ \underset{k_M{}^{HCy}}{\overset{k_H{}^{MCy}}{\rightleftarrows}} M + HCyDTA^{3-} \qquad (43)$$

parison of the $k_M{}^{HCy}/k^{M\text{-}H_2O}$ ratio for a great many metal ions. For a number of metals, including Co(II), Ni(II), Cu(II), and Zn(II), this ratio is proportional to the metal–acetate stability constant. A stepwise mechanism is proposed in accord with the diagram in Eq. (44) where the metal–acetate bond formation precedes the rate-determining step. The latter is the coordination of the metal to one nitrogen atom of CyDTA without displacing the proton from the other nitrogen atom.[278] On the other hand, there are many metal ions where the $k_M{}^{HCy}/k^{M\text{-}H_2O}$ ratio is very much smaller than would be expected by the above mechanism. This is true, for example, with Ba^{2+}, Sr^{2+}, Ca^{2+}, Mg^{2+}, and the lanthanide ions. In some instances, the $k_M{}^{HCy}$ values are one thousandth of that predicted from the water exchange rate constants. The larger ions which do not coordinate strongly to nitrogen donors can coordinate with one or more of the acetate groups of CyDTA outside the coordination cage but must wait for the proton to be transferred from nitrogen to an acetate group of CyDTA before the reaction can proceed toward full coordination—i.e., before the metal ion enters the coordination cage. This accounts in part for the relatively small values calculated for $k_M{}^{HCy}$, but even without the protonated species the $k_M{}^{Cy}$ values are smaller than expected for Sr^{2+} and Ba^{2+} [428] by factors of 20–50. It can be

TABLE 1-23
DISSOCIATION RATE CONSTANTS FOR METAL–CYDTA COMPLEXES AT 25.0°C[a]

Metal	k_H^{MCy} (M^{-1} sec^{-1})
Ba(II)	1.1×10^8
Sr(II)	6.1×10^6, 3.5×10^6[b]
Ca(II)	7.1×10^5, 5.4×10^5[b], 4.4×10^5[c]
Mg(II)	5.6×10^4, 7.1×10^4[b]
Mn(II)	3.2×10^2
Zn(II)	1.7×10^2
La(III)	1.3×10^2, 2.8×10^2[b]
Ce(III)	53
Cd(II)	42
Pr(III)	35
Pb(II)	23
Nd(III)	16
Sm(III)	5.1
Cu(II)	3.9
Co(II)	3.2
Hg(II)	3.1
Eu(III)	2.2
Gd(III)	1.3
Tb(III)	0.58
Y(III)	0.36
Dy(III)	0.26
Th(IV)	0.23
Ho(III)	0.13
Er(III)	0.053
Tm(III)	0.037
Yb(III)	0.023
Sc(III)	0.019
Lu(III)	0.017
Al(III)	8.5×10^{-3}[d]
Ni(II)	3.4×10^{-4}
In(III)	1.3×10^{-6}[e]
	k_{2H}^{MCy}, (M^{-2} sec^{-1})[f]
Fe(III)	2.0×10^{-4}[e]
Ga(III)	6.4×10^{-5}[e]

[a] Ref. 499.
[b] Ref. 358.
[c] Ref. 334.
[d] Ref. 308.
[e] MH_nCy where n is uncertain but is in the range of 0–2, Ref. 308.
[f] These reactions are second-order in [H^+] and first-order in complex.

concluded that even for the unprotonated $CyDTA^{4-}$ the rate-determining step for these metal ions occurs after the first-bond formation and presumably involves the movement of the metal ion into the coordination cage of the ligand.

$$M^{2+} + HCy^{3-} \rightleftarrows [\text{structure}] \xrightleftharpoons{\text{rds}} MCy^{2-} + H^+ \quad (44)$$

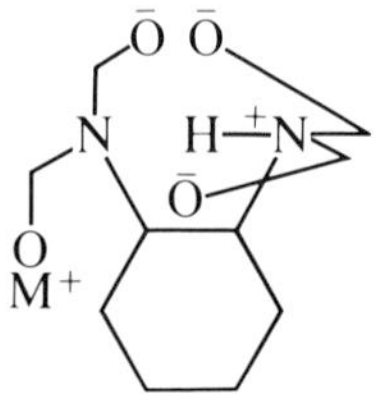

The behavior of La^{3+} with H_2CyDTA^{2-} was described earlier [Eq. (19) and (20)] in which a La*HCy complex forms rapidly. The metal ion in La*HCy is believed to be outside the coordination cage, and the proton is on one of the nitrogen atoms[233]:

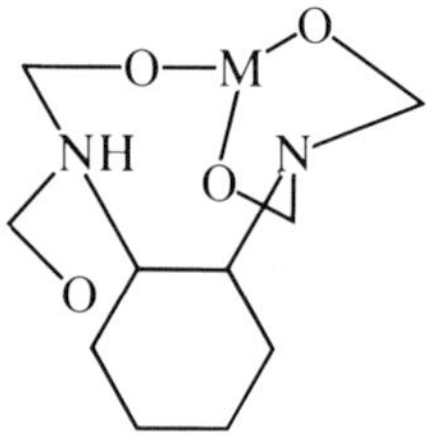

The La^{3+} moves into position for nitrogen coordination, and the proton is shifted away in the observed reaction to give the stable $LaCy^-$ complex.

The kinetic behavior of the Pb^{2+} exchange for Cd^{2+} in CyDTA is somewhat more complicated that indicated earlier because Pb^{2+} can actually suppress the rate of its own exchange reaction. The mechanism is:

$$CdCyDTA^{2-} + H^+ \xrightleftharpoons{k_H^{CdCy}(\text{rds})} HCyDTA^{3-} + Cd^{2+}$$

$$CdCyDTA^{2-} \underset{}{\overset{Pb^{2+}}{\rightleftarrows}} Cd(CyDTA)Pb \overset{Pb^{2+}}{\rightleftarrows} Cd(CyDTA)Pb_2^{2+} \quad \left.\right\} \text{no reaction with } H^+$$

$$HCyDTA^{3-} \xrightarrow[\text{rapid}]{+Pb^{2+},\ -H^+} PbCyDTA^{2-} \quad (45)$$

where Cd(CyDTA)Pb is:

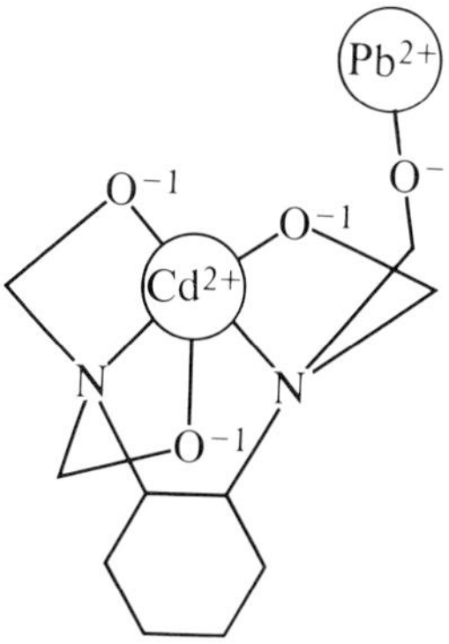

22

This mixed complex is not readily attacked by H^+, and as a result its formation suppresses the rate at which Pb^{2+} replaces Cd^{2+} in the inner-coordination sphere. Blocking the H^+ reaction in this way is not limited to Pb^{2+} but also is found for Cu^{2+} and can be expected for any metal ion which forms reasonably strong acetate complexes.

Although it is sterically impossible for two metal ions to each bind to a nitrogen of CyDTA, it is possible for each to bind to two of the acetate groups without binding to nitrogen. For class *a* metals which prefer oxygen donors this provides a metal–metal exchange pathway, but it is a relatively unfavorable one.[233]

A novel method for removing a metal ion from the cage of the CyDTA which does not involve a proton or solvent-assisted dissociation has been studied by Glentworth *et al.*[500,501] The β-decay of ^{144}Ce and ^{143}Ce results in the breakup of the complex; the different isotopes give daughter atoms which cause different percentage breakups of the metal complex.

The kinetic behavior of CyDTA has been studied with more metal ions than any other ligand. The reactions are easily monitored, and most of them can be adjusted to speeds convenient for stopped-flow or conventional spectrophotometric methods. As a result, the rate constants are known with relatively good accuracy. Although the general trend of the rate constants parallels the behavior expected from the $k^{M\text{-}H_2O}$ values and Eq. (35), the specific behavior of individual ions indicates acetate group bonding prior to the rate-determining step in the formation of the complex. Furthermore, some rate constants are significantly smaller (by factors of 10^3) than expected from the $k^{M\text{-}H_2O}$ values because of slower chelation steps. The relative formation and dissociation reaction rates of CyDTA illustrate steric effects and the behavior of protonated ligands. For the most part, the H^+ dissociation rates of the metal–CyDTA complexes are not affected by other metal ions (except for the suppression effect already noted). Hence, a mixture of metal–CyDTA complexes will dissociate with different first-order rate constants, characteristic of the metal ion and of the acidity but without synergistic effects. Therefore, the reactions are excellent for qualitative and quantitative analysis. If the acidity is adjusted to make the half-life of the $BaCyDTA^{2-}$ reaction 10 msec, then the half-life of the $NiCyDTA^{2-}$ dissociation is 100 yr. However, in $1M$ acid the barium dissociation reaction is over upon mixing while the nickel reaction now has a half-life of about 300 sec.[281] By varying the acidity and examining the resulting reactions as a function of time, one can identify and determine over 30 metal ions.[499]

E. Partial Dissociation of Multidentate Ligand Complexes

Some complexes undergo a partial dissociation before they react with other nucleophiles or with other metal ions. This is reflected in a first-order dissociative rate process which is itself independent of the concentration of the species driving the reaction. One example is the partial dissociation of $NiEDTA^{2-}$ before it reacts with Cu^{2+}. The complete mechanism of the transfer of EDTA from Ni^{2+} to Cu^{2+} is discussed later, but suffice it to state now that the dissociation rate constant for the half-unwrapping of EDTA from Ni^{2+} has been measured[22] and equals $k_d^{NiY-1/2} + k_H^{NiY-1/2}[H^+]$, where these rate constants are many orders of magnitude greater than those calculated for the complete dissociation reactions.

It is common for a complex in which the metal ion is coordinatively saturated by strong donors to replace one of its multidentate ligands by solvent before it reacts with another ligand. Thus, the reaction [Eq. (46)] does not occur by

$$Ni(dien)_2^{2+} + EDTA \rightarrow NiEDTA^{2-} + 2dien \tag{46}$$

direct attack of EDTA on $Ni(dien)_2^{2+}$ but by the following sequence:[296]

$$Ni(dien)_2^{2+} \underset{k_{-1}}{\overset{k_1}{\rightleftarrows}} Ni(dien)^{2+} + dien \tag{47}$$

$$Ni(dien)^{2+} + EDTA \xrightarrow{k_2} NiEDTA^{2-} + dien \tag{48}$$

In this case, it was possible to determine all the rate constants and to show that $k_1 \simeq k_H^{Ni(dien)_2}[H^+]$, where $k_H^{Ni(dien)_2} = 1 \times 10^6 M^{-1}sec^{-1}$. This rate constant is several orders of magnitude greater than the corresponding rate constant for the acid attack on the mono complex, $Nidien^{2+}$. Section V considers the effect of one or more coordinated ligands on the speed of substitution reactions of other ligands.

The $Fe(TPTZ)_2^{2+}$ complex is another example in which one tridentate ligand in a bis complex completely dissociates before EDTA can react.[297] On the other hand, the reaction of $Ni(IDA)_2^{2-}$ [502] as well as $Fe(terpy)_2^{3+}$ [222] with cyanide ion occurs by both solvent- and cyanide ion-assisted dissociation.

$$\text{rate} = (k_d + k_{CN}[CN^-])[Ni(IDA)_2{}^{2-}] \quad (49)$$

The $Ni(phen)_3{}^{2+}$ complex must lose one phenanthroline molecule before it can react with either $EDTA^{4-}$ or with CN^-.[503] Before jumping to an unfortunate generalization, however, note that by contrast there are direct reactions between CN^-, $N_3{}^-$, and OH^- with $Fe(phen)_3{}^{2+}$.[504]

The metal–N(peptide) complexes offer several examples in which partial dissociation of the peptide ligand occurs before reaction with other multidentate ligands. Thus, the deprotonated tripeptide–copper(II) complexes are not very reactive with EDTA as a nucleophile, and one of their main reaction pathways is a partial dissociation of the tripeptide by breaking the copper–carboxylate bond and the adjacent copper–N(peptide) bond (between sites *2* and *3* in the diagram). The partially dissociated complex then reacts very rapidly with EDTA. The rate expression for this partial dissociation step is given in Eq. (50)[505,506,343]

$$\frac{-d[CuH_{-2}L^-]}{dt} = (k_d + k_H[H^+] + k_{HX}[HX])[CuH_{-2}L^-] \quad (50)$$

where HX refers to any general acid present in solution—e.g., buffer acids or H_2EDTA^{2-}. In this case the actual proton-transfer reactions from H_3O^+ or from HX are shown to limit the reaction rate.[343] There are other metal–peptide complexes where it appears that protonation occurs prior to the metal–N(peptide) bond cleavage. This is the case for nickel–tetraglycine, $Ni(H_{-3}GGGG)^{2-}$, and nickel–triglycinamide, $Ni(H_{-3}GGGa)^-$,[507] and copper–glycylglycyl-L-histidine, $Cu(H_{-2}GGHis)^-$.[508] The proton goes either to the peptide oxygen or to the peptide nitrogen and then the M—N(peptide) rearranges to a M—O(peptide) as shown in Eq. (51).

$$\text{M–N}{=}\text{C}(\text{O})^{(-)} + \text{HX} \underset{k_{-1}}{\overset{k_1}{\rightleftharpoons}} \text{X} + \left[\begin{array}{c}\text{M–N}{=}\text{C–O–H} \\ \text{or} \\ \text{M–N(H)–C}{=}\text{O}\end{array}\right] \xrightarrow{k_2} \text{M–O}{=}\text{C–N–H} \quad (51)$$

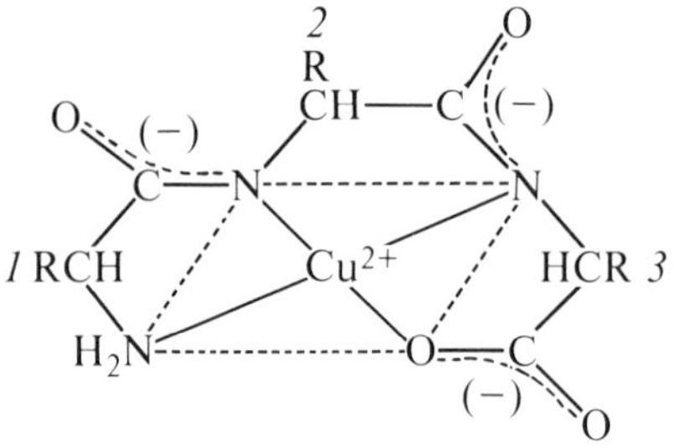

$Cu(H_{-2}tripeptide)^-$

23

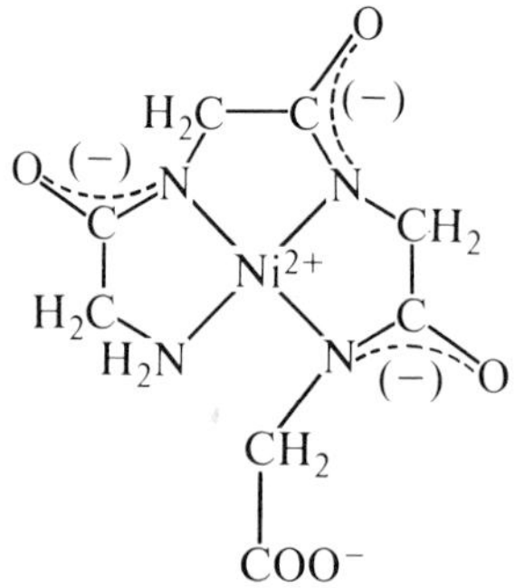

$Ni(H_{-3}GGGG)^{2-}$

24

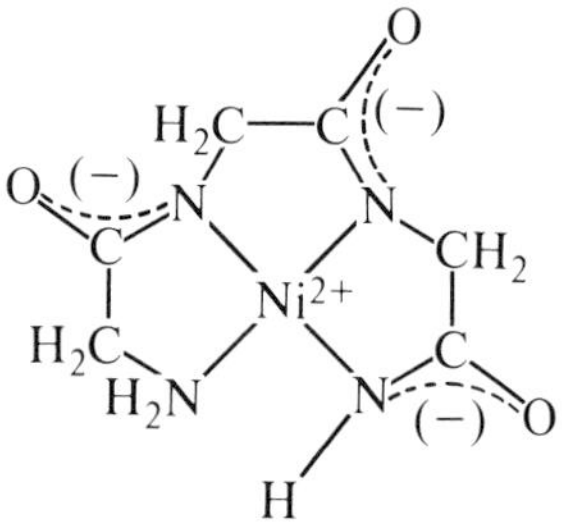

$Ni(H_{-3}GGGa)^-$

25

Ligand dissociation after protonation is initiated at a terminal end. The controlling factor as to whether the reaction is proton-transfer limited or whether protonation occurs prior to bond cleavage is determined by the overall ligand field stabilization of each complex. Thus, even though Cu^{II} and Ni^{II}-triglycine dissociation reactions are general-acid catalyzed, Pd^{II}-triglycine is not.[509] The effect of increase in ligand-field stabilization is reflected in the dissociation reactions for CN^--mixed complexes of Ni^{II}-peptides. The dissociation rate for $[Ni^{II}(H_{-2}Gly{\cdot}Gly{\cdot}amide)CN]^-$ is 1/38,000 that of $[Ni^{II}(H_{-2}Gly{\cdot}Gly{\cdot}amide)]$.[510] The acid dissociation for both the $[Ni^{II}(H_{-2}GlyGlyGly)CN]^{2-}$ and $[Ni^{II}(H_{-2}Gly{\cdot}Gly{\cdot}amide)CN]^-$ [511] is not general-acid catalyzed. Because of the increase in ligand-field stabilization, the rate step shifts from protonation to bond breaking.

Cu(II) complexes with ligands such as GlyGlyHis (structure **25b**), GlyGlyHisGly, etc., where histidine is the third amino acid residue, can have either protonation or bond breaking as the limiting step.[512] In strong acid, the peptide complexes protonates, taking up two protons. The limiting step is the bond breaking and dissociation of the ligand (no general-acid catalysis is observed). In the presence of a nucleophile (trien) above pH 7, the rate step can be shifted so that the reaction is proton limited; the reaction is then general-acid catalyzed. These reactions are different from the typical peptide dissociation reactions in that the peptide ligand dissociation is initiated at a nonterminal position. Reactions for copper(II)–protein complexes, the protein being human serum albumin (HSA) and bovine serum albumin (BSA), also follow this mechanism.[512]

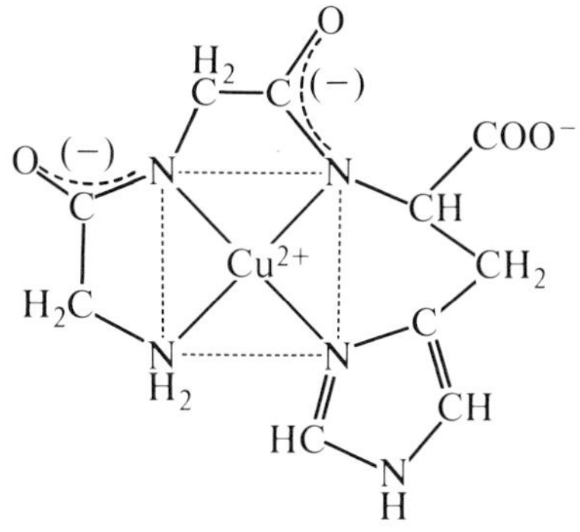

25b

Protonation of the donor atom or group before it dissociates from the metal ion is a reasonable mechanism for the reaction of acid with a number of ligands (those with extra electron-pair donors). It might be expected for carboxylate, phenolate, or sulfhydryl donors, for example, but there is relatively little experimental evidence available. One case where protonation of a monodentate ligand is known to accelerate its dissociation is the acid attack on $Ni(CN)_4^{2-}$.[320] An example for a bidentate ligand is the dissociation of $Ni(cysteine)_2^{2-}$ where protonation of a sulfhydryl group may explain the acid dependence of the reaction, and the following reaction intermediate was proposed as one of two possible mechanisms.[513] (H_2O is believed coordinated above and below the planar structure.)

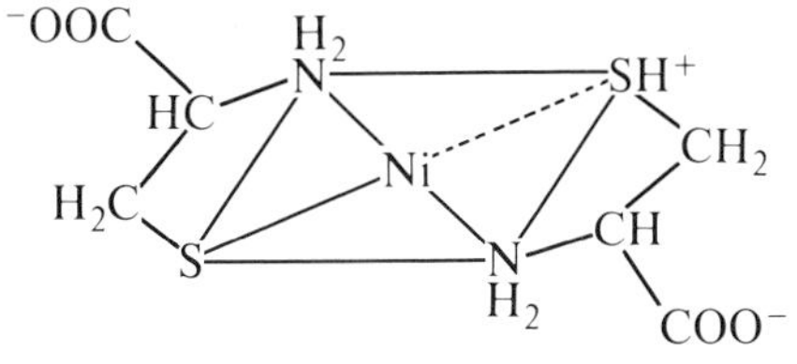

26

V. THE EFFECT OF COORDINATED LIGANDS ON THE RATE OF ADDITIONAL COORDINATION IN WATER-REPLACEMENT REACTIONS

The effect of ligands coordinated to a metal ion on the rate of exchange of the remaining water molecules is a subject that has attracted considerable attention over the past few years. Until recently, the available data were restricted almost entirely to Ni(II) complexes, but evidence of ternary complex formation in the reactions of several metalloenzymes (cf. Refs. 514–516) has led to considerable interest in the reactions of other metal ions.

Most of the data discussed in this section were obtained by studying substitution reactions of the metal complexes. Wherever possible, for comparison, the water exchange rate constants have been calculated from the substitution rate constant and a suitable value of K_{os}. Calculating K_{os} values for metal complexes is an even more uncertain procedure than for the aquo metal ions. The approximations are less likely to hold because the ions are not spherical and because the charge density is nonuniform in the complexes. Where the incoming ligand is a neutral molecule, these approximations appear to have little effect, as shown by the excellent agreement between the values of $k^{\text{M-H}_2\text{O}}$ calculated from substitution rate constants and those obtained by NMR methods. However, when the incoming ligand is charged and the formal charge on the metal ion differs from the aquo metal ion, it is usually not possible to calculate a reliable value of K_{os}. This may be because most of the available data have been collected under ionic strength conditions of 0.1 or greater. The ionic atmosphere considerably modifies the value of K_{os} because of the unsymmetrical nature of the charge distribution. It has been observed for reactions of NSA^{2-} [290] and OX^- [285,287] with a variety of MgL complexes of differing charge types that the most reactive form of the ligand is always the deprotonated form whatever the formal charge on the metal ion. Ammonia reacts at a similar rate with $Ni(H_2O)_6^{2+}$, $2.8 \times 10^3 M^{-1}sec^{-1}$ [114] and $NiNTA^-$, $5.4 \times 10^3 M^{-1}sec^{-1}$.[517,518] If the overall charge on the complex is used to calculate the value of K_{os}, it would be expected that Gly^- would react much more rapidly with $Ni(H_2O)_6^{2+}$ than with $Ni(NTA)^-$, but the experimental rate constants are $2.2 \times 10^4 M^{-1}sec^{-1}$ [242] and $1.4 \times 10^4 M^{-1}sec^{-1}$ respectively.[5] Previously we have discussed the possible mechanisms for the reaction of Gly^- with aquo metal ions. It might have been expected that the negative charge on $NiNTA^-$ would change the mechanism of the glycine reaction so that the amine group bonds first. However, the formation rate appears to be too large for this pathway to contribute significantly to the formation rate constant.

To compare the effects of different bound ligands, a statistical correction is made for the observed rate constant to allow for the varying number of water molecules available for substitution in the different complexes. Thus, $K_{os}k^{\text{M-H}_2\text{O}} = k_f \times 6/(6\text{-}n)$ where the bound ligand is n-dentate.

Other than changing the value of the outer-sphere association constant, the presence of bound ligands on the metal ion might be expected to have several effects.

(*1*) The rate of exchange of the remaining water molecules may be affected. These molecules may not be equivalent and may exchange at different rates.

(*2*) If it is necessary to rearrange the bound ligand, the mechanism may change.

(*3*) The rate-determining step for an incoming chelate may shift because of electrostatic or steric factors.

In a recent paper[519] it has been shown that a change of the coordinated ligands about Rh(III) changes the mechanism from I_a to D.

A. Effect of Monodentate Ligands on the Rate of Water Exchange of Metal Ions

The results presented in Table 1-24 indicate that, in general, the replacement of water molecules by monodentate ligands increases the rate of exchange of the remaining water molecules. For nickel(II) the order of this labilizing effect is $F^- < OAc^- < OH^- \sim N_3^- \sim NH_3$, and the substitution of more than one water molecule increases still further the rate of water exchange. The substitution of two or more chloride or thiocyanate groups on cobalt(II)[520,521] gives rise to a considerable increase in the water exchange rate constants. This increase has been attributed to a change in the geometry of the metal from an octahedral to a tetrahedral configuration, and therefore the mechanisms for water exchange may be different.

The labilizing effect of the hydroxide ion in $M(H_2O)_5OH^{X+}$ complexes is observed with Fe(III), Rh(III), Al(III), Ga(III), In(III), and Ni(II) (Table 1-24). For the hydroxygallium(III) and -indium(III) species it has been suggested that the reaction with SO_4^{2-} may be S_N2 in character,[64] but more data are needed to give conclusive evidence. The methyl group as a ligand has a very large labilizing effect on the reactions of Pt(IV) and Au(III).[48,522] The effect on Sn(IV) is difficult to evaluate because of a lack of comparative data, but $(CH_3)_2Sn^{2+}$ exchanges H_2O rapidly.[48] The $(CH_3)_2Ga^+$ ion is about as labile in its H_2O exchange as $GaOH^{2+}$. The CH_3HgX reactions are definitely associative,[523] but the interchange of Cl^- and OH^- is slower than the reaction between $HgCl^+$ and Cl^-. It is interesting to compare the effect of replacing five water molecules by ammonia molecules on the rate of water exchange.

The results for a series of metals are shown in Table 1-25. In all cases except for Co(III), substituting ammonia for water molecules causes an increase in the rate of water exchange of the remaining water molecules. For Co(III) it seems likely that ammonia substitution decreases the rate of water exchange, although there are experimental difficulties in measuring the water exchange rate of the $Co(H_2O)_6^{3+}$ ion because it is reduced to Co^{2+} by water. $k^{M\text{-}H_2O}$ can be estimated

TABLE 1-24
EFFECT OF COORDINATED MONODENTATE LIGANDS ON THE VALUE OF FORMATION RATE CONSTANTS

Complex	Ligand	k_f (M^{-1} sec^{-1})	kM-H_2O (sec^{-1})	Ref.
$DyOAc^{2+}$	OAc^-	$>1 \times 10^7$		524
$VO(H_2O)_3Cl^+$	H_2O		2.3×10^4	61
$VO(H_2O)_2Cl_2$	H_2O		8.5×10^5	61
VO^{2+}	HGly	1.3×10^3		525
	$Oxal^{2-}$	$\sim 4 \times 10^4$		526
	Mal^{2-}	$\sim 2 \times 10^4$		526
	L-$Htart^-$	3×10^4		526
	D-$Htart^-$	1.70×10^2		527
	BCA	$\sim 10^8$		528
$VO_2(OH)_2^-$	$VO_2(OH)_3^{2-}$	3.1×10^4		529
	H_2EDTA^{2-}	2.34×10^4		530
	$Halizarin^-$	2.28×10^4		530
$VO_2(OH)_3^{2-}$	H_2EDTA^{2-}	2.4×10^3		530
	$Halizarin^-$	1.2×10^3		530
VO_2^{2+}	catechol	1.84×10^4		533
	pyrogallol	4.36×10^4		533
	1,2,4-benzenetriol	9.91×10^4		533
	L-dopa	1.12×10^4		533
	epinephrine	1.70×10^4		533
MoO_4^{2-}	H_2OX^+	5.0×10^5		531
	H_2OXS	5.0×10^6		532
	HOX	4.1×10^2		531
	HOXS	2.5×10^3		532
	catechol	2.9×10^2		534
$MoO_3(OH)^-$	HOX	$\sim 4.5 \times 10^6$		531
	HOXS	$\sim 3.9 \times 10^6$		532
	OX^-	$\sim 1.5 \times 10^8$		531
	OXS^-	$\sim 4 \times 10^7$		532
	catechol	1.9×10^8		534
	H_2EDTA^{2-}	2.26×10^5		535
$MoO_2(OH)_2$catechol	catechol	5.4×10^2		534
WO_4^{2-}	H_2OX^+	2.0×10^5		532
	H_2OXS	$<1.5 \times 10^{6}$[a]		532
	HOX	2.4×10^3		532
	HOXS	1.28×10^4		532
	catechol	1.2×10^2		536
	gallic acid	6.05×10^2		536
	pyrogallol	6.92×10^2		536
	3,4-dihydroxybenzene	2.4×10^2		536
	L-dopa	2.41×10^2		536

TABLE 1-24 (CONTINUED)
EFFECT OF COORDINATED MONODENATATE LIGANDS ON THE VALUE OF FORMATION RATE CONSTANTS

Complex	Ligand	k_f (M^{-1} sec^{-1})	$k^{M\text{-}H_2O}$ (sec^{-1})	Ref.
$WO_3(OH)^-$	HOX	$\sim 5.4 \times 10^6$		532
	HOXS	$< 3 \times 10^{6a}$		532
	OX^-	$\sim 2.6 \times 10^9$		532
	OXS^-	$\sim 6.2 \times 10^8$		532
	catechol	8.4×10^7		536
	gallic acid	7×10^7		536
	pyrogallol	2.0×10^8		536
	3,4-dihydroxybenzene	5.0×10^7		536
	L-dopa	5.6×10^8		536
$Mn(H_2O)_5(NH_3)^{2+}$	H_2O		3.6×10^7	654
$Mn(OH)^{2+}$	H_2O_2	3.2×10^4		158
	HF (10°C)	10^4–10^5		146
$Fe(H_2O)_5(OH)^{2+}$	H_2O		$\sim 9 \times 10^3$	177
			$\sim 3 \times 10^4$	177
			$\sim 1 \times 10^4$	166
	HN_3	6.8×10^3		139
		6.1×10^3		179
		7.4×10^3		178
	HF	3.1×10^3		138
	H_3PO_2	2.1×10^4		184
	phenol	7.2×10^2		143
		1.1×10^3		142
		1.5×10^3		140
	o-NH_2phenol	1.1×10^5		143
	o-Clphenol	1.0×10^3		142
		3.5×10^2		140
	o-Brphenol	1.9×10^3		142
	m-Mephenol	9×10^2		142
		1.6×10^3		140
	m-NO_2phenol	7.0×10^2		142
		5.9×10^2		144
		8.0×10^2		140
	m-Clphenol	1.3×10^3		140
	p-NO_2phenol	8.0×10^2		142
		8.2×10^2		140
	p-Clphenol	6.6×10^2		140
		1.2×10^3		142
	p-Brphenol	7.0×10^2		142
	catechol	3.1×10^3		537

TABLE I-24 (CONTINUED)
EFFECT OF COORDINATED MONODENATATE LIGANDS ON THE VALUE OF FORMATION RATE CONSTANTS

Complex	Ligand	k_f (M^{-1} sec^{-1})	k^{M-H_2O} (sec^{-1})	Ref.
$Fe(H_2O)_5(OH)^{2+}$	H_2sal	2.9×10^3		375
		5.5×10^3		377
	Hsali	6.5×10^2		375
	mandelic acid	2.6×10^3		538
	benzylmalonic acid	5.4×10^3		539
	n-butylmalonic acid	2.6×10^3		539
	methylmalonic acid	4.3×10^3		539
	malonic acid	6.2×10^3		539
	cyclobutane-1,1-dicarboxylic acid	3.3×10^3		539
	CH_3COOH	5.3×10^3		176
		2.8×10^3		177
	$CH_2ClCOOH$	6.8×10^3		176
	CH_3CH_2COOH	5.1×10^3		176
	H_2IDA	2.5×10^3		540
	H_3NTA	1.5×10^4		540
	H_4EDTA	3.0×10^4		541
	H_5DTPA	5.3×10^4		541
	acac(enol)	2.5×10^3		542
		4.4×10^3		379
	acac(keto)	5.4		379
	TTF(enol)	1.3×10^5		344
	H_2SSA^-	5.7×10^3		375
		5.5×10^3		376
	$HSal^-$	1.4×10^4		377
	CH_2ClCOO^-	2.8×10^4		176
	$Hmalonate^-$	1.3×10^5		539
	MeHmalonate	1.2×10^5		539
	n-butyl-$Hmalonate^-$	1.0×10^5		539
	benzyl-$Hmalonate^-$	1.0×10^5		539
	cyclobutane-1,1-Hdicarboxylic acid	1.1×10^5		539
	$HIDA^-$	8.8×10^3		540
	H_2NTA^-	5.6×10^4		540
		1.1×10^5		172

TABLE 1-24 (CONTINUED)
EFFECT OF COORDINATED MONODENTATE LIGANDS ON THE VALUE OF FORMATION RATE CONSTANTS

Complex	Ligand	k_f (M^{-1} sec^{-1})	$k^{M\text{-}H_2O}$ (sec^{-1})	Ref.
$Fe(H_2O)_5(OH)^{2+}$	H_3EDTA^-	1.1×10^5		541
	H_4DTPA^-	1.6×10^5		541
	Hoxalate$^-$	2×10^4		181
		4.8×10^4		180
	SCN^-	4.2×10^4		171
		1×10^4		133
		2.4×10^4		169
		1.3×10^4		170
		5.1×10^4		172
	N_3^-	$\sim 1 \times 10^4$		178
	Cl^-	1.1×10^4		165
		1.15×10^4		166
		1.81×10^4		167
	Br^-	2.6×10^4		543
		4.1×10^4		166
	HSO_4^-	2.4×10^4		168
		1.45×10^5		169
	$HCrO_4^-$	2.12×10^4		544
	HSO_3^-	2.7×10^2		545
	SSA^{2-}	1.2×10^4		376
	SO_4^{2-}	2.4×10^5		169
		1.14×10^5		173
		1.1×10^5		174
	8-hydroxyquinoline	6.5×10^2		375
$Fe(CN)_5(H_2O)^{3-}$	4-picoline	3.54×10^2		546
	pyridine	3.65×10^2		546
	isonicotinamide	2.95×10^2		546
	pyrazine	3.80×10^2		546
	N-methylpyrazinium	5.50×10^2		546
	DMSO	2.40×10^2		546
$Fe(CN)_5OH^{3-}$	SCN^-	1.3×10^6		547
$Fe(CN)_5(H_2O)^{2-}$	SCN^-	1.96×10^{-2}		547
$Co(H_2O)_5(NH_3)^{2+}$	H_2O		1.6×10^7	548
$Co(H_2O)_4(NH_3)_2^{2+}$			6.5×10^7	548
$Co(H_2O)_5Cl^+$			1.7×10^7	521
$Co(H_2O)_2Cl_2$			$>10^8$	521
$Co(H_2O)_5(NCS)^+$			9.5×10^6	520
$Co(H_2O)_2(NCS)_2$			2.4×10^8	520
$Co(H_2O)(NCS)_3^-$			$>5 \times 10^8$	520

TABLE 1-24 (CONTINUED)
EFFECT OF COORDINATED MONODENTATE LIGANDS ON THE VALUE OF FORMATION RATE CONSTANTS

Complex	Ligand	k_f M^{-1} sec^{-1}	$k^{M\text{-}H_2O}$ sec^{-1}	Ref.
$Rh(H_2O)_6^{3+}$			2×10^{-5}	55
$Rh(H_2O)_5(OH)^{2+}$			3×10^{-3}	55
$Ni(H_2O)_5(OH)^+$	PAR^-	1.0×10^4	2×10^5	249
$Ni(H_2O)_5(N_3)^+$		1.1×10^4	2.2×10^5	249
$Ni(H_2O)_5(F)^+$		1.1×10^3	2.2×10^6	249
$Ni(H_2O)_5(OAc)^+$		4.5×10^3	9.0×10^4	249
	$HNTA^{2-}$	24		427
		8		427
$Ni(CN)_4^{2-}$	CN^-(0°C)	$>5 \times 10^3$		550
$Ni(H_2O)_5$-(Imidazole)$^{2+}$	imid	4.3×10^3	1.6×10^4	115
$Ni(H_2O)_5$-(Theophylline)$^{2+}$	theophylline	1×10^2		122
$Ni(H_2O)_4(imid)_2^{2+}$	imid	2.4×10^3	1.1×10^4	115
$Ni(H_2O)_5(NH_3)^{2+}$	H_2O		2.5×10^5	44
$Ni(H_2O)_4(NH_3)_2^{2+}$			6.1×10^5	44
$Ni(H_2O)_3(NH_3)_3^{2+}$			2.5×10^6	44
$Ni(H_2O)(NH_3)_5^{2+}$			4.3×10^6	518
$Ni(H_2O)_5(Cl)^+$			6.9×10^5	551
$Ni(H_2O)_4(Py)_2^{2+}$	malonate$^-$	4.1×10^5		251
$Ni(H_2O)_2(NCS)_4^{2-}$	H_2O		1.1×10^6	552
$Pt(NH_3)_2(H_2O)_2^{2+}$			~1	48
$Me_3Pt(H_2O)_3^+$			1.3×10^4	48
$Cu(acetate)^+$	$HNTA^{2-}$	1×10^3		446
$AgS_2O_3^-$	$S_2O_3^{2-}$	3.5×10^9		553
$Ag(OH)_4^{2-}$	$H_2IO_6^{3-}$	4.5×10^4		554
	$H_4TeO_4^{2-}$	2.72×10^3		554
$Ag(OH)_2(H_2IO_6)^{2-}$	$H_2IO_6^{3-}$	9.3×10^3		554
$Ag(OH)_2(H_4TeO_4)$	$H_4TeO_4^{2-}$	1.00×10^4		554
$(CH_3)_2Au^+$	H_2O		4.5×10^4	48
$AuCl_4^-$	Hen^+	4.60×10^2		555
	H_2dien^{2+}	7.3		556
$AuCl_3(OH)^-$	Hen^+	6.76×10^2		555
	H_2dien^{2+}	1.2×10^3		556
ZnX^+	Br^-, I^-	$\sim 4 \times 10^5$		130
ZnX_2^-	Br^-, I^-	$\sim 4 \times 10^5$		130
$ZnBr_3^-$	Br^-	8×10^7		130
ZnI_3^-	I^-	6×10^6		130
$CdBr^+$	Br^-	1.4×10^8		130
$CdCl_2$	Cl^-	1×10^8		557

TABLE 1-24 (CONTINUED)
EFFECT OF COORDINATED MONODENTATE LIGANDS ON THE VALUE OF FORMATION RATE CONSTANTS

Complex	Ligand	k_f M^{-1} sec^{-1}	k^{M-H_2O} sec^{-1}	Ref.
$CdBr_2$	Br^-	1.4×10^7		130
$Cd(CN)_2$	CN^-	4×10^{10}		558
$CdCl_3^-$	Cl^-	4×10^9		557
$CdBr_3^-$	Br^-	1.2×10^7		130
CdI_3^-	I^-	$\sim 4 \times 10^8$		130
$Cd(CN)_3^-$	CN^-	1.5×10^8		559
		8×10^7		560
$CH_3Hg(OH_2)$	pada	1.6×10^9		561
	Br^-	5.0×10^9		561
	I^-	5.0×10^9		561
CH_3HgOH	pyridine	2.1×10^4		561
	Cl^-	2.5×10^4		561
	pada	2.5×10^5		561
	SCN^-	6.3×10^5		561
	Br^-	3.7×10^5		561
	I^-	1.4×10^7		561
	$(C_6H_5)_2C_6H_4SO_3^-$	1.2×10^7		561
	1-methylchinaldin-thion-4	5.0×10^7		561
	p-nitrothiophenolate	5.0×10^8		561
	CN^-	8.0×10^8		562
	SO_3^{2-}	2.5×10^5		523
	Cl^-	1.1×10^4		523
	Br^-	2.2×10^5		523
	I^-	7.0×10^6		523
	SCN^-	2.0×10^5		523
CH_3HgCl	OH^-	1.5×10^8		523
CH_3HgBr		1.2×10^8		523
CH_3HgI		4.1×10^7		523
CH_3HgSCN		5.0×10^8		523
CH_3HgSO_3		5.0×10^6		523
$HgCl^+$	Cl^-	7×10^9		563
	poly(U)	1.4×10^6 (15°C)		564
	poly(A)	1.2×10^5 (15°C)		564
	poly(A)·poly(U)	2.7×10^3 (15°C)		564
$HgBr_2$	Br^-	5×10^9		130
$HgBr_3^-$		1×10^9		130

TABLE 1-24 (CONTINUED)
EFFECT OF COORDINATED MONODENTATE LIGANDS ON THE VALUE OF FORMATION RATE CONSTANTS

Complex	Ligand	k_f M^{-1} sec^{-1}	k_{M-H_2O} sec^{-1}	Ref.
HgI_3^-	I^-	7.1×10^8		565
$Al(H_2O)_5OH^{2+}$	SO_4^{2-}	8.5×10^5		64
	$[Fe(CN)_6]^{3-}$		1.9×10^4	469
$AlCl_3$[b]	H_2O		5×10^6	566, 567
$GaCl_4^-$	Cl^-		1.8×10^6	568
$Ga(H_2O)_5OH^{2+}$	SO_4^{2-}	9.8×10^4		64
	H_2SXO^-	1.2×10^4		470
	H_3SXO^-	1.1×10^3		470
$(CH_3)_2Ga^+$	H_2O		9.1×10^4	48
$In(H_2O)_5OH^{2+}$	SO_4^{2-}	6.8×10^6		64
	H_2SXO^-	3.1×10^6		470
		5.0×10^6		470
$Tl(H_2O)_5OH^{2+}$	H_3SXO	5.3×10^5		470
$(CH_3)_2Sn^{2+}$	H_2O		$>8.3 \times 10^4$	48

[a] Calculated upper limit.
[b] Value calculated in Ref. 9 from data in Refs. 566 and 567.

TABLE 1-25
WATER EXCHANGE RATE CONSTANTS FOR $M(H_2O)_6^{n+}$ AND $M(NH_3)_5(H_2O)^{n+}$

Metal	k_{M-H_2O}(sec^{-1})		Ref.
	$M(H_2O)_6^{n+}$[a]	$M(NH_3)_5(H_2O)^{n+}$	
Ni(II)	3×10^4	4.3×10^6	518
Ru(II)	$\sim 5 \times 10^{-3}$	7.0×10^{-1}[c]	570,571
Cr(III)	3×10^{-6}	6.0×10^{-5}	572
Co(III)	10^{-1}[b]	5.9×10^{-6}	283,573
Rh(III)	3×10^{-8}	1.07×10^{-5}	574
Ir(III)		2.2×10^{-4}[d]	575

[a] Results taken from Table 1-1.
[b] Value calculated from oxidation rates assuming substitution is rate controlling, Ref. 161. For a review *see* Ref. 576.
[c] These values calculated from substitution rate constants assuming a dissociative mechanism.
[d] 88°C.

only by considering its redox reactions. The minimum rate constant observed in these reactions is $\sim 1 M^{-1}sec^{-1}$, and assuming that the oxidation–reduction reaction is a substitution-controlled inner-sphere process, a value for $k^{M\text{-}H_2O}$ of approximately 0.1 sec^{-1} is obtained. The water exchange rate constant for $Co(H_2O)_6^{3+}$ is abnormally high for a d^6 ion, and two mechanisms have been proposed to account for this. It has been suggested that the water exchange rate of $Co(H_2O)_6^{3+}$ is catalyzed by $Co(H_2O)_6^{2+}$, but this has also been criticized. Taube et al.[569] indicate that there is no reason why the electron transfer in the pair $Co^{2+}{}_{aq}/Co^{3+}{}_{aq}$ should be rapid when this is not the case for the electronically similar pair $Co(NH_3)_6^{2+}/Co(NH_3)_6^{3+}$. Taube has therefore suggested that the high-spin form of $Co(H_2O)_6^{3+}$ may be only a few kilocalories above the ground state in energy and that this additional activation energy might still result in a system which is fairly labile with respect to substitution. The fact that fluoride ion forms a high-spin complex with Co(III) while ammonia forms a low-spin complex supports Taube's argument, and so it might be expected that water, which is intermediate in polarizability compared with these two, forms a complex in which the two-spin states have nearly the same energy. However, Johnson and Sharpe[577] have estimated that the energy barrier for spin interconversion is in the range 15–20 kcal mol^{-1} and suggest that the water exchange occurs through the electron-transfer reaction. The problem seems to be unresolved.

For the substitution reactions of a series of nucleophiles with $Co(CN)_5H_2O^{2-}$ the evidence is that the reaction proceeds via an S_N1 limiting pathway with the five-coordinate $Co(CN)_5^{2-}$ species as a reactive intermediate.[578,579]

$$Co(CN)_5H_2O^{2-} + X^- \rightarrow Co(CN)_5^{2-} + X^- \rightarrow Co(CN)_5X^{3-} \quad (52)$$

The anation reactions can be interpreted in terms of the rate of water loss from the complex and an anion-dependent rate. The value of $k^{M\text{-}H_2O}$ obtained from the anation experiments ($\sim 1.6 \times 10^{-3}$ sec^{-1}) is in reasonable agreement with the direct determination ($1\text{–}1.3 \times 10^{-3}$ sec^{-1})[580] and is considerably larger than the value for $Co(NH_3)_5(H_2O)^{3+}$, but it is still smaller than for the aquo metal ion.

The corresponding complex with iron(III) ($Fe(CN)_5H_2O^{2-}$) has also been studied.[547] In its reaction with thiocyanate, $Fe(CN)_5H_2O^{2-}$ exhibits kinetic behavior more consistent with an associative mechanism involving a hepta-coordinate intermediate rather than an S_N1-limiting pathway with $Fe(CN)_5^{2-}$ as intermediate. The complex $Fe(CN)_5H_2O^{3-}$, however, is thought[546] to react by a predominantly dissociative mechanism.

B. Effect of Multidentate Ligands on Additional Complex Formation

A reasonably comprehensive summary of the effect of multidentate ligands on the rate of water exchange in metal complexes is presented in Table 1-26. Whenever possible the values of $k^{M\text{-}H_2O}$ have been calculated (using the values of K_{os} adopted in the references quoted) and compared with the values obtained

TABLE 1-26
THE EFFECT OF MULTIDENTATE LIGANDS ON THE RATES OF ADDITIONAL COMPLEX FORMATION AT 25°C

Complex (Coordinated Waters Not Shown)	Ligand	k_f (M^{-1} sec^{-1})	kM-H_2O (sec^{-1})	Ref.
$MgNTA^-$	HOX	2.5×10^4		285,287
	OX^-	1.5×10^5		285,287
	NSA^{2-}	1.2×10^5		290
$MgUDA^-$	HOX	9.3×10^4		285,287
	OX^-	5.5×10^4		285,287
$MgADP^-$	HOX	2.5×10^4		285,287
	OX^-	2.5×10^5		285,287
$MgATP^{2-}$	HOX	2.4×10^4		285,287
	OX^-	1.2×10^5		285,287
	NSA^{2-}	1.2×10^5		290
$MgTP^{3-}$	HOX	4.5×10^4		285,287
	OX^-	5.6×10^4		285,287
	NSA^{2-}	2.3×10^5		290
$LaEDTA^-$	OXS^{2-} (10°C)	3×10^6		349,583
$CeEDTA^-$	(10°C)	2×10^6		349,583
$PrEDTA^-$	(10°C)	9×10^5		349,583
$NdEDTA^-$	(10°C)	5×10^5		349,583
$SmEDTA^-$	(10°C)	5×10^5		349,583
$EuEDTA^-$	(10°C)	5×10^5		349,583
$GdEDTA^-$	(10°C)	5.6×10^5		349,583
$TbEDTA^-$	(10°C)	3×10^5		349,583
$DyEDTA^-$	(10°C)	1.2×10^4		349,583
$HoEDTA^-$	(10°C)	5×10^4		349,583
$ErEDTA^-$	(10°C)	2.5×10^4		349,583
$TmEDTA^-$	(10°C)	1.6×10^4		349,583
$YbEDTA^-$	(10°C)	6×10^3		349,583
$LuEDTA^-$	(10°C)	2.5×10^3		349,583
$YEDTA^-$	(10°C)	5.6×10^4		349,583
VO(IDA)	H_2O		1.2×10^5	60,63
VO(SSA)			1.5×10^5	60,63
$VO(tiron)^{2-}$			5.3×10^5	60,63
VO(tart)	$Htart^-$	277		527
$(VOOH)_2^{2+}$	Mal^{2-}	1.2×10^7		526
	$Hmal^-$	8×10^5		526
VO(tart)	VO(tart)	4×10^4		526
VO(oxal)	$Hoxal^-$	8×10^{2} [b]		526
	$oxal^{2-}$	9.5×10^{4} [a]		526
		1×10^{5} [b]		526
VO(mal)	$Hmal^-$	6×10^{4} [b]		526
	mal^{2-}	2.2×10^{2} [b]		526

TABLE 1-26(CONTINUED)
THE EFFECT OF MULTIDENTATE LIGANDS ON THE RATES OF ADDITIONAL COMPLEX FORMATION AT 25°C

Complex (Coordinated Waters Not Shown)	Ligand	k_f (M^{-1} sec^{-1})	kM-H_2O (sec^{-1})	Ref.
$VO(mal)_2^{2-}$	mal^{2-}	1.1×10^3		587
VO_2IDA^-	H_2O_2	9.2×10^2		584
$Mn(phen)^{2+}$	H_2O		5.3×10^7	585
$Mn(phen)^{2+}$			1.2×10^8	585
MnBCA	sulfacetamide	9×10^6		586
PC(avidin)Mn	H_2O		1.3×10^6	588
	oxalacetate	8×10^6		589
Enolase Mn^{2+}	H_2O		7.7×10^6	590
	CH_2PEP	1.38×10^{10}		590
	L-phospholacetate	$\geq 2.77 \times 10^8$		590
	D-phospholacetate	$\geq 2.05 \times 10^8$		590
	phosphoglycolate	$\geq 2.02 \times 10^{10}$		590
$MnNTA^-$	H_2O		1.5×10^9	40
	HOX	1.7×10^6		286,287
	OX^-	2.3×10^7		286,287
	bipy	1.4×10^6		370
$MnUDA^-$	HOX	1.4×10^6		286,287
	OX^-	2.2×10^7		286,287
$MnADP^-$	HOX	1.4×10^6		286,287
	OX^-	8.5×10^7		286,287
$MnATP^{2-}$	H_2O		5.0×10^7	591
	HOX	9.3×10^5		286,287
	OX^-	2.5×10^6		286,287
	ATP	1.8×10^6		372
	bipy	2.7×10^5		370
$MnEDTA^{2-}$	H_2O		4.4×10^8	40
$MnTP^{3-}$	HOX	4.0×10^5		286,287
	OX^-	9.1×10^5		286,287
	bipy	9.4×10^4		370
PCMn	H_2O		1.2×10^6	588
	oxalacetate	3.1×10^7		588
	pyruvate	4.5×10^6		589
$Mntren^{2+}$	bipy	2.1×10^5		370
$Mntrien^{2+}$		2.2×10^5		370
$Mndien^{2+}$		5.3×10^5		370
MnEDDA		2.6×10^5		370
$MnNDA^-$		1.8×10^6		370
Mn-2,3,2-tet		1.9×10^5		370
$MnMIDA^-$		6.8×10^5		370

TABLE 1-26(CONTINUED)
THE EFFECT OF MULTIDENTATE LIGANDS ON THE RATES OF ADDITIONAL COMPLEX FORMATION AT 25°C

Complex (Coordinated Waters Not Shown)	Ligand	k_f (M^{-1} sec^{-1})	k^{M-H_2O} (sec^{-1})	Ref.
MnHCA	Cl^-	5×10^6 (pH 6.46)		592
		3.7×10^6 (pH 8.59)		592
Mn(ribulose-1,5-diphosphate)	HCO_3^-	3.9×10^6		593
Mnglutamine synthetase			1.3×10^{-7}	594
Mnglutamine synthetase L-glutamate			1.0×10^{-7}	594
Mocitrate	ClO_3^-	1.25×10^5		595
$MoO_3(EDTA)^{4-}$	$HMoO_4^-$	$\leqslant 3.3 \times 10^4$		533
$MoO_3(HEDTA)^{3-}$	MoO_4^{2-}	$\leqslant 1.0 \times 10$		533
$Febipy^{2+}$	bipy	1.3×10^5	2×10^6	289
$Feterpy^{2+}$	terpy (5°C)	$\sim 10^7$		289
$FePen^{1+k}$	imidazole	1.0×10^7		596
		8.5×10^6		596
$FePen(OH)^k$	imidazole	6.0×10^6		596
		8.9×10^6		596
ferricytochrome	N_3^-	20		597
cytochrome C oxidase	HF	5.1×10^7		598
	HCN	1.1×10^5		598
$Coterpy^{2+}$	terpy (5°C)	$\sim 5 \times 10^6$		289
	PAR^-	8×10^6		289
	bipy	2.48×10^6		599
	phen	2.3×10^5		261
$Cophen^{2+}$	phen	4×10^5	8×10^6	289
$CoArg^{2+}$	Arg	8.7×10^5	1.7×10^7	388
$Co(Arg)_2^{2+}$	Arg	2.0×10^5	6×10^6	388
$Cobipy^{2+}$	Gly^-	1.6×10^6	$\sim 2 \times 10^6$	601
$Copyc^+$	pyc^- (20°C)	2.4×10^7		395
$CoGly^+$	Gly^-	2×10^6		275
		2.2×10^6	4.4×10^6	115
Co(L-histidine$^+$)	L-histidine$^-$	1.4×10^6		602
	D-histidine$^-$	1.2×10^6		602
$CodiGly^+$	$diGly^-$	9×10^4	1.8×10^5	115
		1.6×10^5		275
$CotriGly^+$	$triGly^-$	1.0×10^5		275

TABLE 1-26(CONTINUED)
THE EFFECT OF MULTIDENTATE LIGANDS ON THE RATES OF ADDITIONAL COMPLEX FORMATION AT 25°C

Complex (Coordinated Waters Not Shown)	Ligand	k_f (M^{-1} sec^{-1})	k^{M-H_2O} (sec^{-1})	Ref.
$CotetraGly^+$	$tetraGly^-$	2.3×10^5		275
CoGlySar	$Glysa^-$	8.0×10^5		398
$CoGlyLeu^+$	$GlyLeu^-$	3.3×10^5		397
$CoLeuGly^+$	$LeuGly^-$	1.4×10^5		397
$Co(\alpha\text{-Ala})^+$	$\alpha\text{-Ala}^-$	1×10^6		268
	(20°C)	8×10^5		263
$CoLeu^+$	Leu^-	1.7×10^6		267
$CoSar^+$	Sar^-	1.5×10^6		390
$CoSer^+$	Ser^-	2.0×10^6		341
Co(L-tyrosine)	L-tyrosine	1.5×10^6		392
$Coproline^+$	$proline^-$	9.6×10^5		393
$Cohydroxyproline^+$	$hydroxyproline^-$	9×10^5		393
$Co(\beta\text{-Ala})^+$	$\beta\text{-Ala}^-$	1.2×10^5		268
	(20°C)	9×10^4		263
$Co(\alpha\text{-abu})^+$	$\alpha\text{-abu}^-$ (20°C)	1.9×10^6		264
$Co(\beta\text{-abu})^+$	$\beta\text{-abu}^-$ (20°C)	7×10^4		264
CoCys	$cystein^{2-}$ (20°C)	1.5×10^6		396
$Co(Gly)_2$	Gly^-	8×10^5		275
		9×10^5	8.6×10^6	115
$Co(pyc)_2$	pyc^-	1.9×10^6		395
$Co(\alpha\text{-Ala})_2$	$\alpha\text{-Ala}^-$ (20°C)	9.0×10^4		263
Comal	H_2O		2.2×10^7	401
CoIDA	PADA	6.5×10^5		382,383
	NH_3	7.5×10^4		603
	IDA^{2-}	2.4×10^6		268
	apoBCA (pH 7.5)	8.8		607
CoASP	ASP^{2-}	2×10^6		268
$CoNTA^-$	PADA	3.7×10^6		382,383
	NH_3	7.2×10^4		603
	apoBCA (pH 7.5)	$\ll 1$		607
$Co(mal)_2^{2-}$	H_2O		$> 10^8$	548
$CoCyDTA^{2-}$	CN^-	1.8×10^6		604
$CoMe_4py(OH)H_2O^+$	NCS^-	1.1×10^2		606
$CoMe_4py(H_2O)^{2+}$		2.11		606
$CoMe_4py(H_2O)NCS^+$		2.8×10^4		606
$CoTP^{3-}$	PADA	8.0×10^4		382,383
$Coterpy^{2+}$	apoBCA (pH 7.5)	2.9		607
$Cophen^{2+}$	(pH 7.5)	30.0		607
Copyridine-2,6-dicarboxylate	(pH 7.5)	70		607

TABLE 1-26 (CONTINUED)
THE EFFECT OF MULTIDENTATE LIGANDS ON THE RATES OF ADDITIONAL COMPLEX FORMATION AT 25°C

Complex (Coordinated Waters Not Shown)	Ligand	k_f (M^{-1} sec^{-1})	$k^{M\text{-}H_2O}$ (sec^{-1})	Ref.
CoCAC[c]	formate	3.9×10^8		605
	monofluoro-acetate	2.2×10^8		605
	difluoroacetate	1.9×10^8		605
	trifluoroacetate	2.0×10^8		605
CoHCA	CNO^-	8.0×10^6		368,608
CoHCAC[e]	*p*-nitrobenzene-sulfonamide	1.77×10^6		609
Cobalamin	SCN^-	2.3×10^3		610
	I^-	1.4×10^3		610
		2.6×10^3		611
	Br^-	1.0×10^3		610
	N_3^-	1.2×10^3		610
	NCO^-	4.7×10^2		610
	$S_2O_3^{2-}$	2.0×10^2		610
	HSO_3^-	1.7×10^2		610
	SO_3^{2-}	$\lesssim 2 \times 10^2$		610
CoP(SCN)	SCN^-	2.8×10^4		612
CoP		2.1		612
CoP(OH)		1.1×10^2		612
Co^{III}hemato-porphyrin	OOH^-	5.01×10^6		613
	CN^-	65		614
	SCN^-	1850		614
$Ni(2,3,2\text{-tet})^{2+}$	H_2O	5.6×10^4		615
			$\sim 1 \times 10^6$	616,617
	NH_3	$\sim 3 \times 10^5$		616,617
Ni(*cis,cis*-tach)$^{2+}$		1.9×10^5	3.8×10^6	518
	H-*cis,cis*-tach$^+$	2.0		618
	cis-cis-tach	45.0		618
$Ni(pn)^{2+}$	NH_3	1.5×10^4	2.3×10^5	517,518
	malonate^{2-}	3.3×10^6		251
$Ni(pn)_2^{2+}$		$>10^7$		251
$Ni(tn)^{2+}$	NH_3	2.9×10^4	4.4×10^5	517,518
$Ni(dap)^{2+}$	malonate^{2-}	4×10^6		251
Ni(*N*-Meen)$^{2+}$	NH_3	1.3×10^4	2.0×10^5	517,518
	meso-tartrate^{2-}	6×10^5	1.8×10^6[i]	619
			1.8×10^4[j]	619
	hydroxyglutarate$^-$	1.35×10^5	9×10^5[i]	619
			9×10^3[j]	619

TABLE 1-26 (CONTINUED)
THE EFFECT OF MULTIDENTATE LIGANDS ON THE RATES OF ADDITIONAL COMPLEX FORMATION AT 25°C

Complex (Coordinated Waters Not Shown)	Ligand	k_f (M^{-1} sec^{-1})	k^{M-H_2O} (sec^{-1})	Ref.
Ni(*N,N'*-diMeen)$^{2+}$	NH_3	8×10^3	1.2×10^5	517,518
	meso-tartrate^{2-}	2.9×10^5		619
	hydroxyglutarate$^-$	3.8×10^4		619
Ni(*N*-Eten)$^{2+}$	NH_3	3.1×10^4	4.7×10^5	517,518
	meso-tartrate^{2-}	6.8×10^5		619
Ni(*N,N*-diMeen)$^{2+}$	NH_3	3.6×10^4	5.4×10^5	517,518
	meso-tartrate^{2-}	12×10^6		619
Ni(triEten)$^{2+}$	*meso*-tartrate^{2-}	2.6×10^5		619
Ni(bipy)$^{2+}$	H_2O		4.9×10^4	620
	malonate^{2-}	2.7×10^5		251
	terpy	5.6×10^4		21
	PADA	5.25×10^4		21
Ni(bipy)$_2{}^{2+}$	H_2O		6.6×10^4	620
	bipy	4.7×10^3		289
	PAR$^-$	8.3×10^3		289
	terpy	$\sim 5 \times 10^3$		289
Ni(en)$^{2+}$	H_2O		4.4×10^5	621
	NH_3	1.2×10^4	1.8×10^5	517,518
	phen	9.5×10^3		408
	bipy	5.1×10^3		408
	malonate^{2-}	1.5×10^6		251
Ni(en)$_2{}^{2+}$	H_2O		5.4×10^6	621
	phen	9.4×10^3		408
	bipy	8.2×10^3		408
	malonate^{2-}	$>10^7$		251
	en	7×10^6		250
Nidien^{2+}	H_2O		1.2×10^6	622
	NH_3	4.3×10^4	8.6×10^5	517,518
	terpy	7.7×10^3		408
	phen	2.3×10^4		408
	bipy	1.1×10^4		408
	PADA	4.7×10^4	9.4×10^5	414
	malonate^{2-}	8.5×10^6		251
	5-nitrosalicylate$^-$	1.53×10^6	3.1×10^6	418
Nitrien^{2+}	H_2O		2.9×10^6	622
	NH_3 (8°C)	1.2×10^5	3.6×10^6	518
	PADA	2.5×10^4		414
	malonate^{2-}	$>10^7$		251
	5-nitrosalicylate$^-$	4.6×10^6	1.4×10^7	418

TABLE 1-26 (CONTINUED)
THE EFFECT OF MULTIDENTATE LIGANDS ON THE RATES OF ADDITIONAL COMPLEX FORMATION AT 25°C

Complex (Coordinated Waters Not Shown)	Ligand	k_f (M^{-1} sec^{-1})	k^{M-H_2O} (sec^{-1})	Ref.
$Nitren^{2+}$	H_2O		6.0×10^5	581
			1.0×10^7	581
	ornithine	1.2×10^4		5,623
	NH_3 (6°C)	2.6×10^5	7.8×10^6	518
	PADA	1.8×10^4		414
	phen	1.3×10^4		5,623
	bipy	1.0×10^4		5,623
	Gly^-	9.0×10^4		5,623
	5-nitrosalicylate$^-$	5.2×10^6	1.6×10^2	418
$NiCR^{2+}$	H_2O		4.5×10^4	624
$NiCRCH_3^{2+}$			5.2×10^4	624
$Nitheophylline^{2+}$	theophylline	$\sim 1 \times 10^2$		122
$Ni(pya_2tn)^{2+}$	NH_3	7.0×10^3		625
	$EtNH_2$	1.5×10^3		625
	Gly^-	5.7×10^3		625
	α-Ala^-	8.9×10^3		625
$Ni(terpy)^{2+}$	H_2O		5.2×10^4	626
	PAR^-	6.3×10^4		289
	phen	6.3×10^3		5
	bipy	2×10^4		5
	terpy	2.2×10^5		289
	ornithine	7×10^2	1.4×10^4	5
	NH_3	2.0×10^4		627
	PADA	4.45×10^5		21
	terpy	2.76×10^5		21
$Ni(phen)^{2+}$	NH_3	1.5×10^3	2.2×10^4	517
	H_2trien^{2+}	43		628
	$Htrien^+$	5.7×10^5		628
	$Hdien^+$	3.59×10^2		243,629
	NTA^{3-}	2.82×10^6		243,629
	benzylamine	1.53×10^3		21
	bipy	3.70×10^4		21
	Me_2bipy	1.58×10^5		21
	PADA	9.64×10^4		21
	terpy	1.00×10^5		21
	TPTZ	1.29×10^5		21
$Ni(5\text{-}Clphen)^{2+}$	$Hdien^+$	3.27×10^2		243,629
	NTA^{3-}	2.34×10^6		243,629

TABLE 1-26 (CONTINUED)
THE EFFECT OF MULTIDENTATE LIGANDS ON THE RATES OF ADDITIONAL COMPLEX FORMATION AT 25°C

Complex (Coordinated Waters Not Shown)	Ligand	k_f (M^{-1} sec^{-1})	kM-H_2O (sec^{-1})	Ref.
$Ni(5\text{-}NO_2phen)^{2+}$	$Hdien^+$	2.81×10^2		243,629
	NTA^{3-}	1.15×10^6		243,629
$Ni(5\text{-}SO_3phen)^+$	$Hdien^+$	5.9×10^2		243
	NTA^{3-}	2.64×10^6		243
$Ni(5\text{-}Mephen)^{2+}$	$Hdien^+$	4.08×10^2		243,629
	NTA^{3-}	3.78×10^6		243,629
	terpy	1.75×10^5		21
$Ni(5,6\text{-}diMephen)^{2+}$	$Hdien^+$	4.26×10^2		243,629
	NTA^{3-}	4.15×10^6		243,629
$Ni(3,4,7,8\text{-}Me_4phen)^{2+}$	terpy	1.12×10^6		21
	PADA	1.38×10^6		21
	TPTZ	5.8×10^6		21
$Ni(2AMP)^{2+}$	$malonate^{2-}$	1.9×10^6		251
$Ni(2AEP)^{2+}$		1.9×10^6		251
$Ni(Arg)^{2+}$	Arg	2.4×10^4		388
$Ni(Arg)_2{}^{2+}$		3.5×10^4		388
$Ni(proline)^+$	$proline^-$	8.7×10^3		393
$Ni(hydroxyproline)^+$	$hydroxyproline^-$	1.8×10^4		393
$Ni(Gly)^+$	NH_3	1.4×10^4		517,518
	Gly^-	5.6×10^4	1.2×10^5	115
		4.0×10^4		275
$Ni(\alpha\text{-}Ala)^+$	$\alpha\text{-}Ala^-$	3×10^4		268
	(20°C)	4×10^4		263
$Ni(\beta\text{-}Ala)^+$	$\beta\text{-}Ala^-$	6×10^3		268
	(20°C)	7×10^3		263
$Ni(Leu)^+$	Leu^-	4.1×10^4		267
$Ni(Ser)^+$	Ser^-	3.4×10^4		341
$Ni(Sar)^+$	Sar^-	1.2×10^4		390
$Ni(diGly)^+$	$diGly^-$	9.2×10^3		275
		4.0×10^3	8×10^3	115
$Ni(triGly)^+$	$triGly^-$	5.5×10^3		275
$Ni(tetraGly)^+$	$tetraGly^-$	4.9×10^3		275
$Ni(\alpha\text{-}abu)^+$	$\alpha\text{-}abu^-$ (20°C)	1.5×10^4		264
$Ni(\beta\text{-}abu)^+$	$\beta\text{-}abu^-$ (20°C)	8.0×10^3		264
$Ni(pyc)^+$	pyc^- (20°C)	1.2×10^5	1.8×10^5	395
$Ni(GlyLeu)^+$	$GlyLeu^-$	8.6×10^3		397

TABLE 1-26 (CONTINUED)
THE EFFECT OF MULTIDENTATE LIGANDS ON THE RATES OF ADDITIONAL COMPLEX FORMATION AT 25°C

Complex (Coordinated Waters Not Shown)	Ligand	k_f (M^{-1} sec^{-1})	$k_{M\text{-}H_2O}$ (sec^{-1})	Ref.
Ni(LeuGly)$^+$	LeuGly$^-$	2.4×10^3		397
Ni(Glysa)$^+$	Glysa$^-$	8.0×10^3		398
Ni(L-histidine)$^+$	L-histidine$^-$	4.1×10^5		602
	D-histidine$^-$	4.3×10^5		602
Ni(acac)$^+$	malonate^{2-}	1.5×10^5		251
Ni(α-β-DPA)$^+$	NH_3	2.7×10^3	5.4×10^4	517,518
Ni(L-tyrosine)$^+$	L-tyrosine$^-$	2.4×10^4		391
Ni(Cys)	cysteine^{2-} (20°C)	4.4×10^4		396
Ni(IDA)	NH_3	3.2×10^3	6.4×10^4	517,518
	PADA	6.6×10^3		414
	IDA^{2-} (30°C)	2.5×10^4		268
		1×10^4		502
	5-nitrosalicylate$^-$	6.3×10^4	1.3×10^4	418
Ni(pyc)$_2$	pyc$^-$ (20°C)	1.6×10^5		395
Ni(ASP)	ASP^{2-} (30°C)	3.6×10^4		268
Ni(MIDA)	NH_3	5.1×10^3	1×10^5	517,518
	PADA	2.7×10^3		414
Ni(α-abu)$_2$	α-abu$^-$ (20°C)	3.0×10^4		264
Ni(β-abu)$_2$	β-abu$^-$ (20°C)	3.0×10^3		264
Ni(Gly)$_2$	Gly$^-$	4.2×10^4	4×10^5	115
Ni(Ser)$_2$	Ser$^-$	3.0×10^4		341
Ni(diGly)$_2$	diGly$^-$	3.3×10^3	2.9×10^4	115
Ni(EDDA)	NH_3	6.1×10^3	1.8×10^5	517,518
	5-nitrosalicylate$^-$	2.6×10^4	7.8×10^4	418
Ni(HEIDA)	NH_3	3.0×10^3	6×10^4	517,518
Ni(NTA)$^-$	Hbipy$^+$	1.0×10^2		117
	Hen$^+$	3.5×10^3		117
	en	7.0×10^3		117
	NH_3	4.6×10^3	1.4×10^5	517,518
	PADA	4.3×10^4		414
	phen	3.6×10^3		117
	bipy	2.4×10^3		117
	terpy	2.1×10^3		5
	ornithine	2.0×10^3		5
	Gly$^-$	1.4×10^4		5
	5-nitrosalicylate$^-$	1.48×10^4	4.4×10^4	418
Ni(HEDTA)$^-$	H_2O		2×10^5	630
Ni(HEEDTA)$^-$	NH_3	2.9×10^2	8.7×10^3	517,518
	HCN	4.1×10^2		319
	CN$^-$	2.0×10^2		319

TABLE 1-26 (CONTINUED)
THE EFFECT OF MULTIDENTATE LIGANDS ON THE RATES OF ADDITIONAL COMPLEX FORMATION AT 25°C

Complex (Coordinated Waters Not Shown)	Ligand	k_f (M^{-1} sec^{-1})	k^{M-H_2O} (sec^{-1})	Ref.
$Ni(EDTA)^{2-}$	H_2O		7.0×10^5	630
	NH_3	4.4×10^2	2.6×10^4	517,518
	HCN	6.0×10^2		319
	CN^-	36		319
$Ni(TP)^{3-}$	PADA	2.0×10^3		414
	5-nitrosalicylate$^-$	1.48×10^4	3×10^4	418
$Pt(dien)^{2+}$	H_2O		0.5	631
$Cu(bipy)^{2+}$	Hen^+	2.2×10^4		632
	en	2.0×10^9		632
	Gly^-	1.6×10^9		632
	α-Ala^-	1.0×10^9		632
	β-Ala^-	3.4×10^8		632
	PADA	$>10^8$		633
$Cu(bipy)_2{}^{2+}$		$\sim 3 \times 10^7$		633
$Cu(en)_2{}^{2+}$		3.9×10^5		633
$Cu(en)^{2+}$		$\sim 5 \times 10^8$		633
	Hen^+	3.1×10^4		246
	(37°C)	6.3×10^4		329
	en	1.9×10^9		246
	(37°C)	3.6×10^9		329
	hm (37°C)	3.6×10^8		329
	Ser^- (37°C)	6.2×10^8		328
$Cu(asparagine)^+$	asparagine$^-$	2.5×10^8		445
$Cu(glutamine)^+$	glutamine$^-$	1.4×10^8		445
$Cu(hm)^{2+}$	hm (37°C)	5.6×10^8		328
	en (37°C)	1.2×10^{10}		329
	Ser^- (37°C)	1.1×10^8		328
$Cu(te)^{2+}$	te (0°C)	$\sim 2 \times 10^8$		634
$Cu(dien)^{2+}$	PADA	2.0×10^8		633
$Cu(tren)^{2+}$	H_2O		2.5×10^5	581
$Cu(\alpha\text{-}Ala)^+$	α-Ala^-	1.5×10^8		269
$Cu(\beta\text{-}Ala)^+$	β-Ala^-	8×10^6		269
$Cu(histidine)^+$	histidine$^-$	3×10^6		269
$Cu(Ser)^+$	Ser^- (37°C)	2.8×10^8		328
		5×10^8		341
$Cu(Gly)^+$	Gly^-	4×10^8	5×10^8	276
$Cu(Leu)^+$	Leu^-	8×10^8		267
$Cu(Sar)^+$	Sar^-	2.8×10^9		390
$Cu(L\text{-}tyrosine)^+$	L-tyrosine	3.1×10^8		391
$Cu(L\text{-}dopa)^+$	L-dopa$^-$	4.2×10^7		442

TABLE 1-26 (CONTINUED)
THE EFFECT OF MULTIDENTATE LIGANDS ON THE RATES OF ADDITIONAL COMPLEX FORMATION AT 25°C

Complex (Coordinated Waters Not Shown)	Ligand	k_f (M^{-1} sec^{-1})	kM-H_2O (sec^{-1})	Ref.
Cu(hydroxy-proline)$^+$	hydroxyproline$^-$	2.8×10^8		393
Cu(proline)$^+$	proline$^-$	2.7×10^8		393
Cu(L-phenyl-alanine)$^+$	L-phenylalanine$^-$	3×10^8		442
Cu(Gly)$_2$	Gly$^-$	$\sim 10^7$		635
	PADA	1.5×10^6		633
Cu(ASP)	ASP^{2-} (30°C)	5×10^8		268
Cu(IDP)	IDP^{2-} (30°C)	8×10^5		268
Cu(adenine)$^+$	adenine$^-$ (26°C)	4.2×10^4		636
Cu(valine)$^+$	valinate$^-$	2.3×10^8		441
Cu(*N*-*N*-dihydroxy-ethylglycine)	*N*,*N*-dihydroxy-ethylglycine	3.2×10^7		441
Zndien^{2+}	PADA	2.3×10^6		452
Zntrien^{2+}		1.1×10^6		452
Zn(L-histidine)$^+$	L-histidine$^-$	6.0×10^7		602
	D-histidine$^-$	5.7×10^7		602
ZnIDA	PADA	1.5×10^7		452
ZnEDDA		9.1×10^5		452
ZnCys		5.4×10^6		452
ZnNTA$^-$		$>2 \times 10^6$		452
	HNTA^{2-}	5.3×10^2		337
	NTA^{3-}	2.0×10^6		337
	DTO	1.73		457
	MNT	27.9		457
ZnTP^{3-}	PADA	1.4×10^6		452
Zn(OXS)$^{2-}$	HEDTA	33.3		457
Zn(MNT)$_2$	EDTA	5		457
Zn(DTO)$_2$	EDTP	45		457
	PDTA	20.9		457
Zn(HCB)	Cl$^-$	5×10^6 (pH 6.46)		592
ZnHCAC[c]				
$^-$OOC, HO–C_6H_3–N=N–C_6H_4–SO_2NH_2		1.13×10^7		637

TABLE 1-26 (CONTINUED)
THE EFFECT OF MULTIDENTATE LIGANDS ON THE RATES OF ADDITIONAL COMPLEX FORMATION AT 25°C

Complex (Coordinated Waters Not Shown)	Ligand	k_f (M^{-1} sec^{-1})	kM-H_2O (sec^{-1})	Ref.
ZnHCAC[c]	4-[(2-hydroxy-6,8-disulfonato-1-naphthyl)azo]benzenesulfonamide (^-O_3S, SO_3^-, OH, $-N{=}N-C_6H_4-SO_2NH_2$)	5.81×10^5		637
	O_2N-thiophene-SO_2NH_2	8.60×10^6		637
	CH_3CONH-thiadiazole-SO_2NH_2	4.83×10^6		637
	$C_6H_5SO_2NH_2$	1.06×10^6		637
	$CH_3C_6H_4SO_2$-NH_2-p	1.17×10^5		637
	$^-OOCC_6H_4SO_2$-NH_2-p	2.71×10^5		637
	$O_2NC_6H_4SO_2$-NH_2-p	7.37×10^5		637
		1.49×10^{6g}		609
	$CH_3CONHC_6H_4$-SO_2NH_2-p	1.12×10^5		637
	$ClC_6H_4SO_2NH_2$-p	5.13×10^5		637
	CN^-	$>2.8 \times 10^9$		637
	$(CH_3)_2N$-naphthalene-SO_2NH_2	2.40×10^5		637
	3,4-$Cl_2C_6H_3SO_2$-NH_2	6.08×10^5		637
	$^-OOC(3,5\text{-}(NO_2)_2)$-$C_6H_2SO_2$-$NH_2$-$p$	1.26×10^6		637
	$^-OOC(3\text{-}NO_2)C_6$-$H_3SO_2NH_2$-p	5.75×10^5		637
	$HO(3\text{-}NO_2)C_6H_3$-SO_2NH_2-p	5.48×10^4		637
	$Cl(3\text{-}NH_2)C_6H_3$-SO_2NH_2-p	2.04×10^5		637

TABLE 1-26 (CONTINUED)
THE EFFECT OF MULTIDENTATE LIGANDS ON THE RATES OF ADDITIONAL COMPLEX FORMATION AT 25°C

Complex (Coordinated Waters Not Shown)	Ligand	k_f (M^{-1} sec^{-1})	kM-H_2O (sec^{-1})	Ref.
ZnHCAC[c]	$^{-}OOC(3\text{-}NH_2)C_6H_3SO_2NH_2\text{-}p$	7.56×10^4		637
	$CH_3(2\text{-}Cl,5\text{-}NH_2)C_6H_2SO_2NH_2\text{-}p$	1.78×10^5		637
	$2,4,5\text{-}Cl_3C_6H_2SO_2NH_2\text{-}p$	1.75×10^6		637
	$2,4,6\text{-}Cl_3C_6H_2SO_2NH_2\text{-}p$	6.58×10^6		637
	$NC(2\text{-}NO_2)C_6H_3SO_2NH_2\text{-}p$	2.50×10^6		637
	$Cl(2\text{-}NO_2)C_6H_3SO_2NH_2\text{-}p$	1.96×10^6		637
carboxymethyl ZnHCAB	*p*-nitrobenzenesulfonamide[g]	9.0×10^5		609
		3×10^6[f]		609
	p-carboxymethylsulfonamide[f]	1.08×10^4		638
ZnBCA	5-dimethylaminonaphthalene-1-sulfonamide[d]	1.43×10^5		639
	4-nitrothiophenol[f]	7×10^8– 3×10^9		640
	2,4-dinitrothiophenol[f]	2–6×10^9		640
ZnHCAB[c]	$^{-}OOC(3\text{-}NO_2)C_6H_3SO_2NH_2\text{-}p$	5.12×10^5		637
	^{-}OOC, HO–, –N=N–, –SO_2NH_2	6.8×10^6		637
	OH, –N=N–, –SO_2NH_2, $^{-}O_3S$–, SO_3^{-}	9.7×10^5		637
	O_2N, S, SO_2NH_2	2.52×10^6		637

TABLE 1-26 (CONTINUED)
THE EFFECT OF MULTIDENTATE LIGANDS ON THE RATES OF ADDITIONAL COMPLEX FORMATION AT 25°C

Complex (Coordinated Waters Not Shown)	Ligand	k_f (M^{-1} sec^{-1})	$k^{M\text{-}H_2O}$ (sec^{-1})	Ref.
ZnHCAB[c]	5-(dimethylamino)naphthalene-1-sulfonamide [structure: $(CH_3)_2N$–naphthalene–SO_2NH_2]	1.3×10^5		637
	$Cl(2\text{-}NO_2)C_6H_3\text{-}SO_2NH_2\text{-}p$	7.7×10^5		637
	$NC(2\text{-}NO_2)C_6H_3\text{-}SO_2NH_2\text{-}p$	1.8×10^6		637
	p-nitrobenzene-sulfonamide	2.86×10^5		637
		3.5×10^6[f]		638
		2.7×10^6[g]		609
	p-carboxybenzenesulfonamide[f]	2.7×10^6		638
	Cl^- (pH 6.46)	5×10^6		592
	Cl^- (pH 8.59)	3.7×10^6		592
Cd(L-histidine)$^+$	L-histidine$^-$	$\sim 7 \times 10^7$		602
$CdNTA^-$	$HNTA^{2-}$	2.9×10^2		337
	NTA^{3-}	1.8×10^7		337
$PbNTA^-$	$HNTA^{2-}$	3.1×10^3		339
	NTA^{3-}	6.6×10^7		339

[a] Axial substitution.
[b] Equatorial substitution.
[c] pH 6.5 unless otherwise stated.
[d] pH 7.4.
[e] pH 7.6.
[f] pH 7.6–10.2.
[g] pH 8.
[h] Rate calculated assuming ionized form reacting with aquo form of enzyme.
[i] $K_{os}\, k^{M\text{-}H_2O}$ (M^{-1} sec^{-1}).
[j] Ring-closure rate (sec^{-1}).
[k] Rate constant depends on reaction scheme used.

by NMR techniques. Care must be taken in the comparison of $k^{M\text{-}H_2O}$ values for different ligands reacting with the same metal complex since it is possible that two or more nonequivalent water molecules exist in a complex. The reactions of bidentate ligands—e.g., Gly^-, bipy, and PADA—with $Nitren(H_2O)_2^{2+}$ have appeared to be anomalous since the formation rate constants are much smaller than those for the reaction with ammonia.[5] However, it has now been shown that the two water molecules in $Nitren^{2+}$ exchange at different rates, $k^{M\text{-}H_2O} \simeq 9$

$\times 10^6$ sec^{-1} and 8.2×10^5 sec^{-1} (at 25°C).[581] Thus, the results can be explained using a similar mechanism for both monodentate and bidentate ligands. In the reaction with ammonia, $k_f = 2.6 \times 10^5 M^{-1}$sec^{-1} (at 6°C), the rate-determining step can be assigned to the exchange of the more labile water molecule. For the bidentate ligands the rate-limiting step must involve the exchange of the other water molecule—e.g., $k_f = 9 \times 10^4 M^{-1}$sec^{-1} for Gly$^-$ at 25°C and $k_f = 1 \times 10^4 M^{-1}$sec^{-1} for bipy at 25°C. For several reactions[251,264,268] it has been suggested that ring closure contributes appreciably to the observed rate constant, and for these reactions $k^{M\text{-}H_2O}$ values have not been used in Table 1-25 unless quoted by the authors.

In complexes of magnesium, the coordination number is usually six,[582] and as it has a d^0 electronic configuration, it might be expected that bound ligands would affect the rate of complex formation by changing the electron density at the metal and by reducing the number of sites at which substitution can take place. However, in temperature-jump studies with oxine[285,286,287] and NSA^{2-} [290] the rate constants and activation parameters for the substitution process are almost independent of the different bound ligands. There is a marked similarity between the reactions of NSA^{2-} and the reactions of oxine$^-$. At first one would expect considerably different K_{os} values for the NSA complexes and the oxine complexes. However, the local charge densities on the metal ion and the coordinating atom of the incoming ligand need to be considered rather than the overall charges.[290] Hückel calculations indicate that the sums of the charges on the two atoms in NSA^{2-} and oxine$^-$ which are involved in coordination to the metal differ by less than half of an electronic charge (Figure 1-13), and thus it is unreasonable to treat these two ligands as if they were acting as -1- and -2-charged ligands.

For the reactions of a series of manganese(II) complexes with oxine and bipy[286,287,370] the bound ligand, including NTA, does not markedly affect the formation rate constants. Although there is usually a decrease in the activation energy as compared with the aquo metal ion, this tends to be compensated by a corresponding change in the entropy of activation. Hague et al. suggest that the variation in rate constants and activation parameters originates in the water exchange process at the metal. In a recent NMR study on the water exchange of $MnNTA^-$ and $MnEDTA^{2-}$ [40] it was found that for both, the bound ligand labilizes the remaining water molecules with respect to $Mn(H_2O)_6^{2+}$. Zetter et al.[40] suggest that "some kind of concerted process may be implicated in the manganese systems."

The ternary complexes of Ni(II) have received considerably more attention than the other divalent metal ions, and the bound ligands can have a large influence on the value of $k^{M\text{-}H_2O}$. The values of k_f and $k^{M\text{-}H_2O}$ increase regularly with increasing numbers of aliphatic nitrogens coordinated to the nickel as shown in Figure 1-14. The values for $k^{M\text{-}H_2O}$ increase from 2.5×10^5 sec^{-1} for $Ni(H_2O)_5NH_3^{2+}$ to 4.3×10^6 sec^{-1} for $Ni(NH_3)_5(H_2O)^{2+}$ regularly. The originally reported results for ammonia substitution with Nitrien^{2+} and Nite-

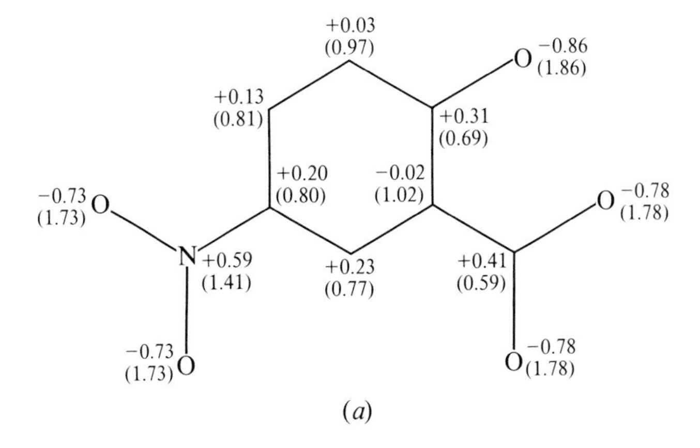

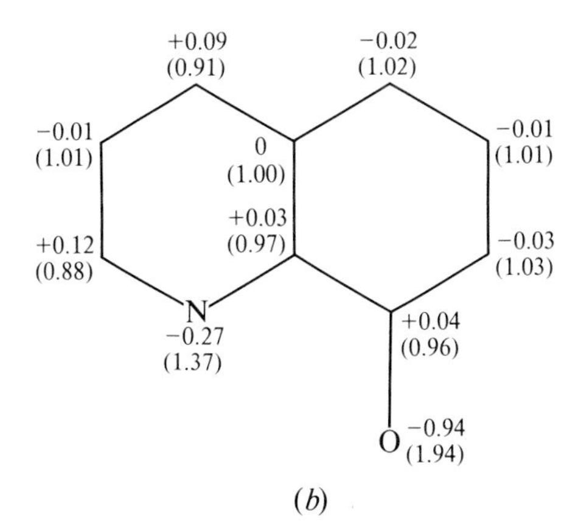

Figure 1-13 Net atom charges for (*a*) NSA^{2-} (total charge −2) and (*b*) OX^- (total charge −1). The numbers of π-electrons localized on the individual atoms are shown in brackets (the total being 16 for NSA^{2-} and 12 for OX^-). The net atom charges for HOX are the same as those for OX^- except that the total charge on the O atom is equal to the number of π-electrons (+0.06)

$tren^{2+}$ [517] have been shown to be incorrect. The corrected results for Ni-$trien^{2+}$ [518] are in excellent agreement with the water exchange rates measured by NMR. The calculated $k^{M\text{-}H_2O}$ values for other polyamines also agree well with the results for water exchange rates as determined by NMR.[581,621,622]

Using the results in Table 1-27, it is possible to compare ligands with the same number of nitrogen donors but with different stereochemistry of the complex. The nickel-*cis,cis*-tach complex, where the ligand is tridentate and coordinates on the faces of an octahedron, has a k_f value for ammonia which is larger than the corresponding value for nickel–dien by a factor of 4.4. The k_r values for the same reactions are larger by a factor of 2 for the *cis,cis*-tach complex. Further comparisons can be made for the quadridentate ligands tren, trien, 2,3,2-tet, and the bis(en)complexes. Tren is restricted to one structure about the metal ion, and the two water molecules are necessarily cis to each other.[641,642] The reactions of $Nitrien^{2+}$ and $Nitren^{2+}$ with ammonia were too fast to measure on a conventional T-jump apparatus at 25°C but could be observed at 6°–8°. The values of k_f given in Table 1-27 are for 8°C, but the value of $k^{M\text{-}H_2O}$ has been corrected to 25°C using the temperature dependence of the $Nidien^{2+}$ and NH_3 reaction. The reaction of $Ni(2,3,2\text{-tet})^{2+}$ with NH_3 was also too fast to measure on the conventional T-jump apparatus even at 8°C, but it has been possible to

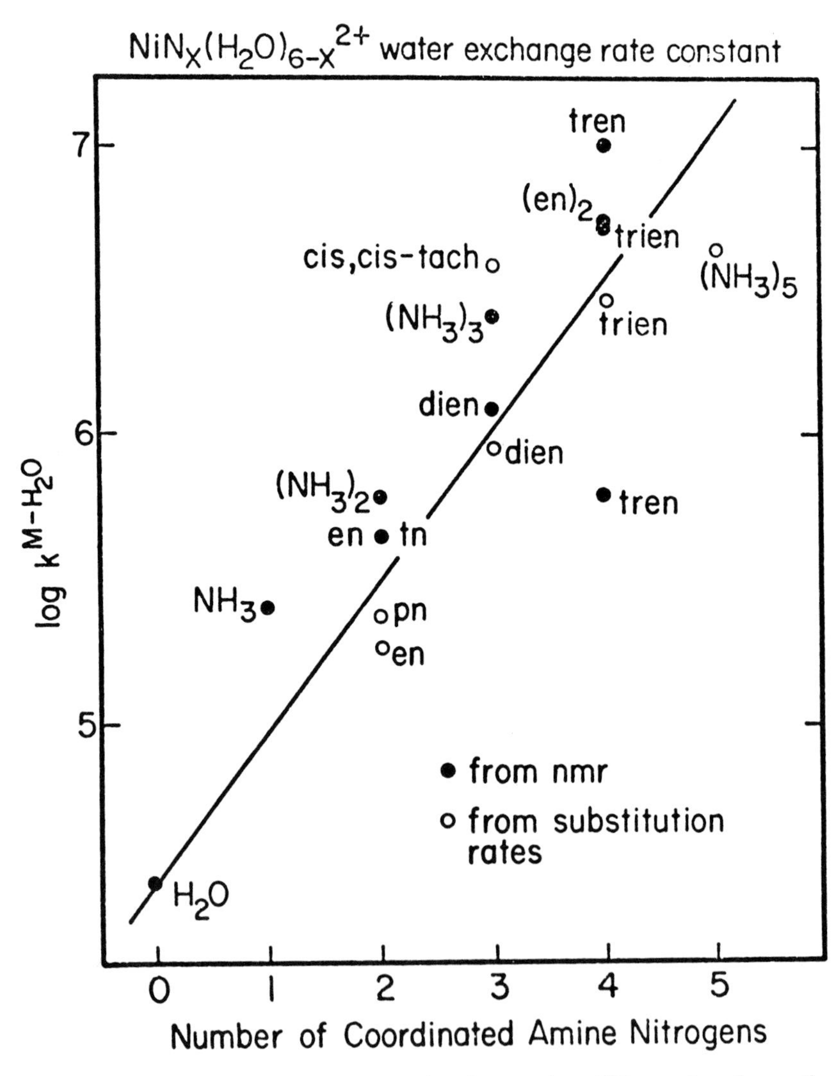

Figure 1-14 The effect of coordinated amine nitrogens from NH_3 or polyamines on the value of the water exchange rate constant for Ni(II)

TABLE 1-27
THE EFFECT OF AMINE LIGANDS ON $k^{Ni\text{-}H_2O}$ AT 25°C

Complex	$k^{Ni\text{-}H_2O}$ (sec^{-1})[a]	Ref.
$Ni(H_2O)_6^{2+}$	2.8×10^4	518
	3.0×10^4	43
$Ni(NH_3)(H_2O)_5^{2+}$	2.5×10^5	44
$Ni(NH_3)_2(H_2O)_4^{2+}$	6.1×10^5	44
$Ni(en)(H_2O)_4^{2+}$	4.4×10^5	621
	1.8×10^5	518
$Ni(tn)(H_2O)_4^{2+}$	4.4×10^5	518
$Ni(pn)(H_2O)_4^{2+}$	2.8×10^5	518
$Ni(NH_3)_3(H_2O)_3^{2+}$	2.5×10^6	44
$Ni(dien)(H_2O)_3^{2+}$	1.2×10^6	622
	8.6×10^5	517,518
Ni(*cis,cis*-tach)$(H_2O)_3^{2+}$	3.8×10^6	518
$Ni(trien)(H_2O)_2^{2+}$	5×10^6[b]	518
	2.9×10^6	622
Ni(2,3,2-tet)$(H_2O)_2^{2+}$	$\sim 8 \times 10^6$	616,617
$Ni(tren)(H_2O)_2^{2+}$	8.2×10^5	581
	$\sim 9 \times 10^6$	581
$Ni(en)_2(H_2O)_2^{2+}$	5.4×10^6	621
$Ni(NH_3)_5(H_2O)^{2+}$	4.3×10^6	518

[a] $k^{Ni\text{-}H_2O} = \dfrac{k_f}{K_{os}} \dfrac{(6)}{(6\text{-}n)}$ where n is the bound ligand(s) coordination number.
[b] Corrected to 25°C using $\Delta H^{\ddagger} = 7$ kcal mol^{-1}.

measure this rate constant and the rate of interconversion of the blue six-coordinate (presumably *trans*-diaquo) to the yellow square-planar form using a nanosecond temperature-jump apparatus.[616,617] This also has been measured using 1.06-μm radiation from a neodymium laser to initiate a photochemical displacement of the equilibrium.[615]

For the reactions of pyridine-2-azo-*p*-dimethylaniline (PADA)[414,418] and 5-nitrosalicilyate[418] with a series of nickel complexes, the rate constants generally agree with those previously obtained from ammonia.[517,518] There is not such a large rate enhancement with the polyamine complexes presumably because of a contribution from ring closure of PADA in the rate-determining step.[414] However, ring closure cannot have become completely rate limiting since the activation enthalpies for the reverse step do not differ significantly from the values obtained for other nickel–ligand complexes with PADA. No evidence of complex formation between $Ni(2,3,2\text{-}tet)^{2+}$ and PADA was found under conditions similar to those where $Ni(trien)(PADA)^{2+}$ forms, supporting the suggestion that the two remaining water molecules in the nickel–2,3,2-tet complex occupy trans positions.

The substitution of groups onto the nitrogens of ethylenediamine greatly affects the rates of ammonia substitution.[517,518] *N*-alkyl substitution generally increases the rate of complex formation in the order

$$N,N\text{-diMeen} > N\text{-Eten} > N\text{-Meen} > \text{en} > N,N'\text{-diMeen}$$

while the substitution of carboxylate groups decreases the rate of water exchange (compare en and EDDA). Additional carboxyl groups in EDTA and HEEDTA make the substitution reaction with ammonia even slower. However, there is a considerable difference between the water exchange rate constant for $Ni(EDTA)^{2-}$ determined by NMR and that determined by ammonia substitution, suggesting that different processes are being monitored in these two reactions since it is possible that ammonia replaces a bound carboxylate. The replacement of an acetate group in EDTA by a hydroxyethyl group reduces the ammonia formation-rate constant. A similar effect may be present for IDA and HEIDA.

The effect of coordinated aminocarboxylates on the values of $k^{M\text{-}H_2O}$ for other metal ions directly contrasts the results obtained for nickel. Thus, as indicated previously, the water exchange rate constants for $MnEDTA^{2-}$ and $MnNTA^{-}$ [40] are greater than the exchange rate constant for the aquo metal. For the reactions of cobalt(II) aminocarboxylate complexes with PADA[382] and ammonia[603] the value of $k^{M\text{-}H_2O}$ increases considerably in progressing from $Co(H_2O)_6^{2+}$ to CoIDA to $CoNTA^{-}$.[382] Cobb and Hague point out that paralleling this increase in the value of $k^{M\text{-}H_2O}$ is an increase in the extinction coefficient of the d–d bands in the range of 450–550 nm, which indicates either an increasing degree of distortion of the ligand field of the metal as a result of the particular stereochemistry of the bound ligand or an increasing covalent nature of the metal–ligand bonding.[382] In an earlier communication[383] on ternary complex formation Co(II) complexes it was found that, in contrast to the corresponding nickel complexes, there is no marked rate enhancement as the number of aliphatic nitrogens bound to the metal is increased.

For the reaction of $CoCyDTA^{2-}$ with cyanide ion Jones and Margerum[604] found an unusually low activation energy, 0.8 kcal/mol, which is much less than the generally accepted value of 3.5 ± 0.4 kcal/mol for diffusion-controlled reactions in aqueous solution.[335,643] The formation rate constant equals $1.8 \times 10^6 M^{-1}sec^{-1}$ at 25°C and is much larger than would be expected from a calculated value of K_{os} and the water exchange rate constant of Co(II), but it is still much smaller than the diffusion-controlled limit. On the other hand, the rate constant for the formation of $Ni(CyDTA)CN^{3-}$ is only $27 M^{-1}sec^{-1}$ [319] which is a factor of 5 smaller than would be predicted using a K_{os} value of $0.005 M^{-1}$ and the aquonickel water exchange rate constant. The only way to account for such a low activation energy in the cobalt reaction is to conclude that the energy required for diffusion is offset by a large degree of cyanide bonding before the transition state, leading to the formation of a seven-coordinate complex.

Most of the other work on Co(II) complexes has been on the formation of bis-amino acid complexes by Kustin, Pasternack, and co-workers. The addition of a second amino acid is often faster than the first despite electrostatic and statistical effects. This may be a result of a labilizing effect of the bound ligand on the remaining water molecules.

Because the rate of water exchange of $Cu(H_2O)_6{}^{2+}$ is close to the diffusion-controlled limit, it is unlikely that bound ligands will have large rate-enhancing effects. The fast rate of water exchange of $Cu(II)_{aq}$ can be understood in terms of Jahn-Teller distortion followed by a rapid inversion such that the axial and equatorial molecules interchange. Therefore, it might be expected that a bound ligand could prevent the rapid inversion from occurring and cause a slow reaction to be observed in subsequent substitution. Similarly, if the geometry of the copper complex is such that the replaceable water molecule is in an equatorial position, the rate for $k^{M\text{-}H_2O}$ might be expected to be slow. NMR water exchange measurements by Hunt et al.[581] have shown that the water exchange rate for copper–tren, 2×10^5 sec^{-1}, is extremely slow compared with other copper(II) reactions. X-ray measurements show that in the solid state the isothiocyanate complex of Cu–tren^{2+} exists in the trigonal bipyramidal form.[644] The copper–trien thiocyanate complex has a square pyramidal structure[645] in the solid state, and it would be interesting to compare its water exchange rate with that of Cu–tren^{2+}. For the reaction of some copper–bis-amino acid complexes it has been suggested that the rate of inversion is the rate-determining step.[269] This conclusion was based on the extrapolation from the Co(II)- and Ni(II)-amino acid results whre the formation of the bis complex is faster than the mono complex.

Substitution kinetics of zinc complexes might be expected to be very complicated since zinc(II) complexes can exist in several geometries.[646,647,648] From work with the neutral molecule PADA it has been found that the charge type of the bound ligand has little effect on the rate of ternary complex formation. It is not easy to vary the number of bound aliphatic nitrogen groups systematically without changing the coordination number of the zinc, but it does appear that nitrogen groups do not markedly increase the rate of substitution. Coordination number changes seem to have little effect on the rate of complex formation as the tetrahedral Zndien$(H_2O)^{2+}$ complex adds PADA at a rate very similar to aquozinc.[452] This casts doubt on the very slow rates for tetrahedral–octahedral interconversion previously observed.[647] If this interconversion is slow, two different rate constants for the reaction of ligands with aquozinc, a result of reactions with the octahedral and with the tetrahedral species, should be seen, but such reactions have not been reported. Recently in a study of zinc glycine[455] it has been suggested that a slow relaxation effect observed at high concentrations and high pH values results from the tetrahedral–octahedral interconversion of

$Zn(Gly)_2$. However, this could also be explained by an internal rearrangement of the ligands in the bis and tris complexes.

In a study on the reactivity of the lanthanides Geier and co-workers[349,583] have found that the binding of EDTA to the aquo metal ion has a large effect on the subsequent substitution reactions as seen in Figure 1-15. There is a much larger variation in k_f for the reaction of OXS^{2-} with $LnEDTA^-$ than with the aquo metal ions. Previously it has been shown spectroscopically and thermodynami-

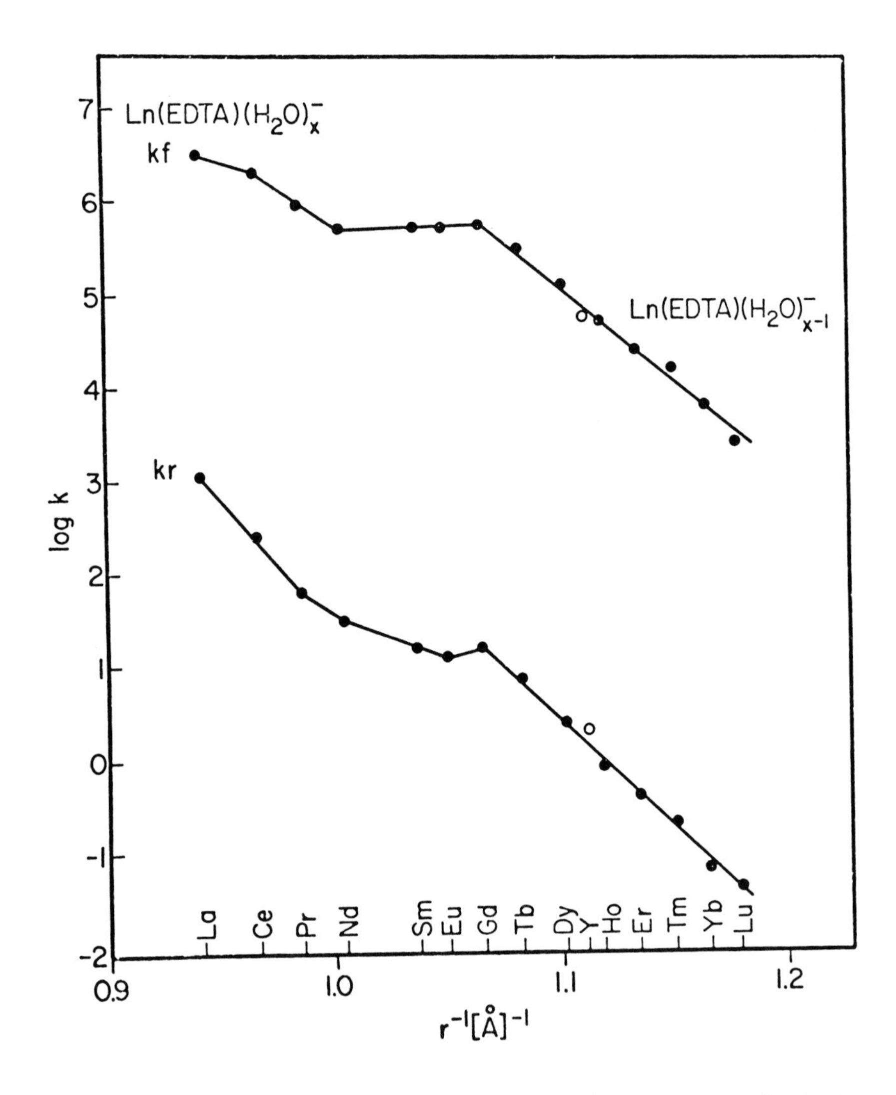

Figure 1-15 Formation rate constants (k_f) and dissociation rate constants (k_r) for the reaction of 8-hydroxyquinoline-5-sulfonate(OXS^{2-}) with lanthanide–EDTA complexes

cally[347,348] that the formation of EDTA complexes induces a change in coordination numbers along the lanthanide series. The variation in the ternary complex formation rate constants parallels the coordination number change in the $LnEDTA^-$ complexes. Thus, between La and Nd the coordination number of the metal ion is probably nine while from Tb to Lu it is probably eight. For the metal ions between these two regions there is an equilibrium where both eight- and nine-coordinate complexes are present. The monotonic change in radius of the lanthanides across the series would be expected to produce a steady decrease in k_f, and this is observed with the effect of the changing coordination number superimposed. Similar formation rate constants for other ternary complexes of the lanthanides suggest that the values of k_f for OXS^{2-} are characteristic of the $k^{M\text{-}H_2O}$ values for $LnEDTA^-$.[583]

In nonaqueous solvents coordinated ligands strongly affect the rate of solvent exchange,[231,232] but few systematic studies have been performed. The rate of methanol exchange in the complex $Mn(MeOH)_5Cl^+$ is about twice the exchange rate of $Mn(MeOH)_6^{2+}$,[649] and the rate of methanol exchange with the nickel and cobalt complexes is accelerated by the addition of water.[650,651,652] For the species $Co(MeOH)_5Cl^+$, the four equatorial MeOH molecules exchange faster than the axial molecule,[653] with $Co(MeOH)_5py^{2+}$ the axial molecule exchanges faster,[654] and with $Co(MeOH)_5NCS^+$ [655] and $Co(MeOH)_5(H_2O)$[652] the exchange rates are equal. Substitution reactions of NiS^{2+} (S = solvent) by NCS^- have been investigated in the solvents methanol, dimethylformamide, acetonitrile, and dimethyl sulfoxide,[124,656] and the observed rates have been correlated with the solvent donicity scale. Large labilizing effects ($>10^5$) have been observed with porphyrin rings and Fe^{3+} in DMF and MeOH.[657] Lincoln and West[658-662] have reported the exchange rates of acetonitrile with a series of cobalt(II), nickel(II), and copper(II) complexes. Large effects are observed, particularly with copper(II) where the rates are slow[661] and for the five-coordinate complexes formed by Me_6tren where the exchange rates are too slow to observe using NMR.[661,658] A number of papers have discussed kinetics of nickel substitutions in mixed solvents.[103,211,231,232,663,664] The solvent composition affects the nature of the metal ion in the solution, and the different species exchange their coordinated solvent molecules at different rates. In most of the studies little quantitative information about the distribution of the solvated metal species is available so that only qualitative interpretations regarding the rates of solvent exchange are possible. These results led Rorabacher and MacKellar[231] to suggest a general theory which explains the kinetic solvent effects. In mixtures of two coordinating solvents, the solvated species with the greatest lability will always be the one which contains one molecule of the weakly bonding solvent, the remaining sites being occupied by the strongly bonding solvent. As more weakly bonding solvent molecules enter the coordination sphere, the dissociation rate of the other coordinated solvent molecules decreases, and the least labile species will be the metal ion species saturated with the weakly bonding solvent.

In a study of the substitution reaction of ammonia with $NiHEEDTA^-$ in water–methanol mixtures, an increase in the formation rate constant has been attributed to a labilized species with methanol in the exchanging position.[232] The trends in the reverse rate constant are consistent with an outer-sphere interaction between the solvent and the exchanging ligand.

The effect of substituents on a phenanthroline ring on the rates of substitution of $Hdien^+$ and NTA^{3-} at a nickel center have been investigated by Steinhaus and Margerum.[243,629] The substituents were quite remote from the reaction site, and so the differences observed in the rates of complex formation are not a result of steric effects but of changes in the metal–ligand electron density. Excellent Hammett free-energy relationships were obtained for the reaction of both $Hdien^+$ and NTA^{3-} (Figure 1-16), and negative ρ values suggest that the

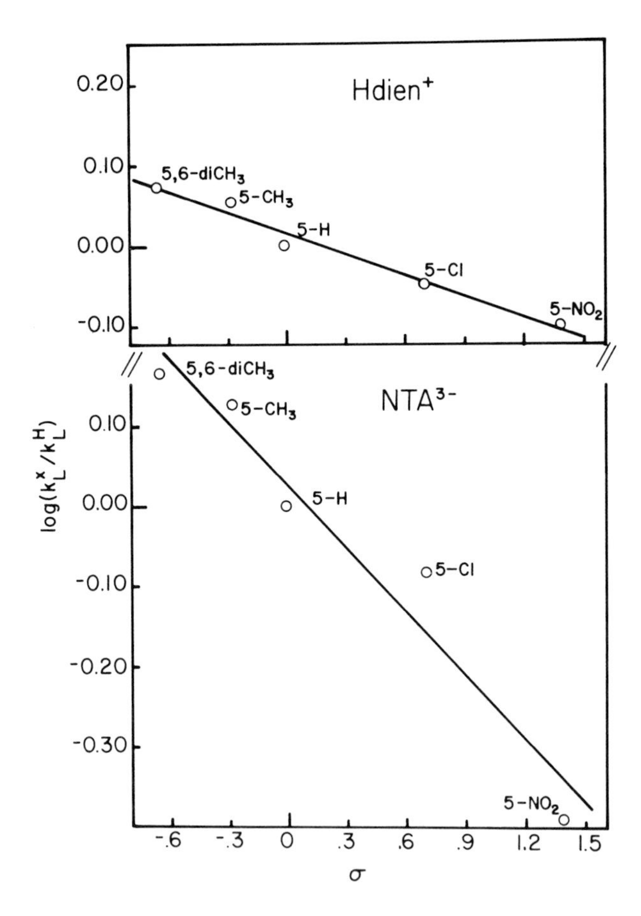

Journal of the American Chemical Society

Figure 1-16 Hammet plots showing substitutent effects for the reactions of $Hdien^+$ and NTA^{3-} with substituted nickel phenanthrolines[629]

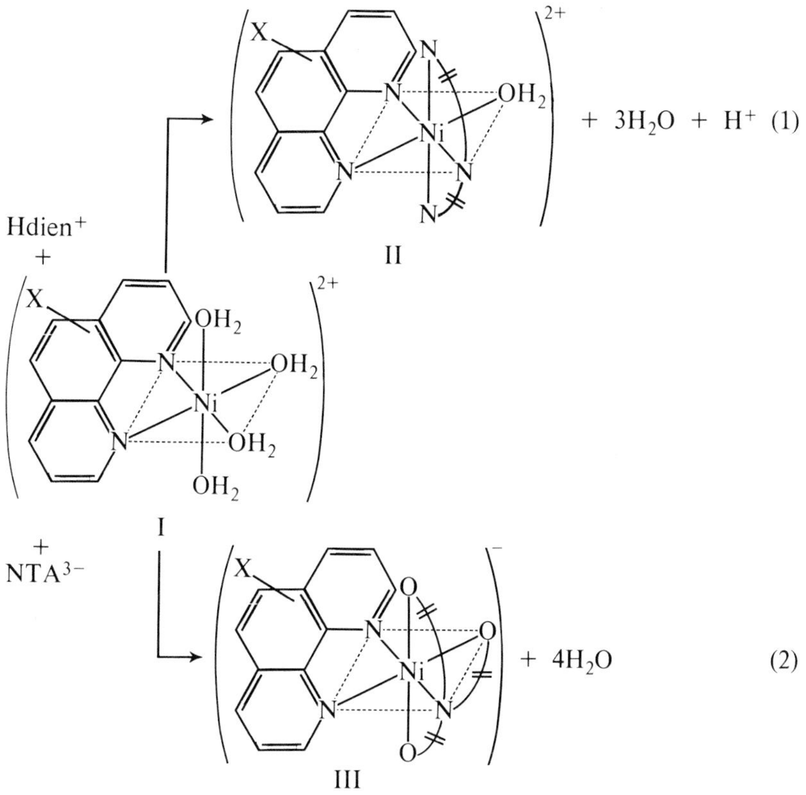

$Hdien^+$ + I → II + $3H_2O$ + H^+ (1)

NTA^{3-} + I → III + $4H_2O$ (2)

Rate Constants (M^{-1} sec.$^{-1}$) at 25.0°C

Substituent	pK_a	k_{Hdien}[a]	k_{NTA}[b]
5,6-Di-CH_2	5.6	426	4.15×10^6
5-CH_2	5.23	408	3.78×10^6
5-H	4.96	359	2.82×10^6
5-Cl	4.26	327	2.34×10^6
5-NO_2	3.57	281	1.15×10^6

[a] Ionic strength = 0.13. [b] Ionic strength = 0.16.

Figure 1–16 Continued

mechanism is a dissociative one. The differences in ρ for the two reacting species are attributed to a difference in the axial and equatorial substitution rates. A similar explanation for the reaction of nickel pyridinealdoxime with malonate has been suggested by Hoffmann and Yeager.[251] In this system two observed relaxation effects were attributed to different water exchange rates for the axial and equatorial positions.

A series of papers on the stability of ternary complexes involving bipyridyl showed that the values of ΔK (= $\log K_{M(bipy)L}^{M(bipy)} - \log K_{ML}^{M}$) are often positive where M is Co^{2+}, Ni^{2+}, Cu^{2+} and L has donor oxygen atoms. This has been attributed to the π-acceptor properties of the pyridyl group.[478,549,600] A similar type of system, $Co(bipy)^{2+} + Gly^- \rightleftharpoons Co(bipy)(Gly)^+$, has been investigated kinetically by Pasternack et al.,[601] and the increase in the stability of the $Co(bipy)(Gly)^+$ complex compared with the $Co(Gly)_2$ system results not from an increase in the formation rate constant but from the smaller value of the dissociation rate constant.

C. Stacking Interactions

Comparison of the rates of complex formation of nickel with some aromatic nitrogen donors introduces another effect which must be considered in ternary complex kinetics. The rates of formation of the 1:1, 1:2, and 1:3 complexes of bipy and phen with nickel in water are close to the values predicted by the water exchange rate constant. However, the formation of the bis-terpyridyl complex of nickel proceeds much faster[289] than predicted from the water exchange rate constant of $Niterpy^{2+}$ as determined by NMR.[626] For the rate of complexation and the water exchange rate constant to be compatible, it is necessary for the outer-sphere association constant for $Niterpy^{2+}$ with terpy to be $4M^{-1}$, in contrast to the "normal" value of $0.1\ M^{-1}$, for the association of a neutral molecule with a metal ion. As shown in Table 1-28A the rate enhancement for the formation of a ternary complex involving nickel–terpyridyl depends on the incoming ligand, suggesting that the terpy does not have a pronounced effect on $k^{M\text{-}H_2O}$ but rather changes the value of K_{os}.

TABLE 1-28A
STACKING INTERACTIONS: FORMATION RATE CONSTANTS FOR THE REACTIONS OF NiTerpy^{2+} WITH VARIOUS LIGANDS

Ligand	k_f (M^{-1} sec^{-1})	$\frac{k_f}{k^{M\text{-}H_2O}}$ (M^{-1})	Ref.
H_2O	5×10^{4}[a]		626
PAR^-	6.3×10^4	1.26	289
phen	6.3×10^3	0.126	289
bipy	2×10^4	0.40	289
terpy	2.2×10^5	4.4	289
	2.76×10^5	5.52	21
ornithine	7×10^2	0.014	5
NH_3	2×10^4	0.40	665
PADA	4.45×10^5	8.9	21

[a] $k^{M\text{-}H_2O}$ rate constant in sec^{-1}.

TABLE 1–28B
STACKING INTERACTIONS: FORMATION RATE CONSTANTS AND RATE ENHANCEMENTS

Complex	Ligand	k_f (M^{-1} sec^{-1})	Rate Enhancement
$[Ni(phen)]^{2+}$	NH_3	1.5×10^3 [b]	1.0
	benzylamine	1.53×10^3	1.7
	phen	3×10^3 [c]	0.8
	bipy	3.70×10^4	23
	Me_2bipy	1.58×10^5	52
	PADA	9.64×10^4	70
	terpy	1.00×10^5	65
	TPTZ	1.29×10^5	63
$[Ni(terpy)]^{2+}$	PADA	4.45×10^5	190
	bipy	2×10^4 [d]	7
	terpy	2.76×10^5	106
$[Ni(bipy)]^{2+}$	terpy	5.60×10^4	23
	PADA	5.25×10^4	24
$[Ni(5\text{-}Mephen)]^{2+}$	terpy	1.75×10^5	108
$[Ni(3,4,7,8\text{-}Me_4phen)]^{2+}$	terpy	1.12×10^6	543
	PADA	1.38×10^6	744
	TPTZ	5.8×10^6	2110
$[Ni(phen)]^{2+}$	NH_3 [e]	1.01×10^3	—
	terpy[e]	4.83×10^3	~4.8
$[Ni(3,4,7,8\text{-}Me_4phen)]^{2+}$	terpy[e]	2.53×10^4	~19

[a] 25.0°C in 0.1M $NaClO_4$ solution.
[b] Ref. 517.
[c] Ref. 666.
[d] Ref. 5.
[e] In $MeOH-H_2O$ (65:35 w/w).

This effect is not specific to complexes of nickel, having also been observed for Co(II), Fe(II), and Zn(II). For the reaction of TPTZ with Fe(II)[297] the 1:1 complex forms at a rate consistent with the $k^{Fe\text{-}H_2O}$ value, yet the bis complex is formed at least 100 times faster. The $FeTPTZ^{2+}$ complex shows a strong kinetic preference to react with a second neutral TPTZ molecule rather than with the negatively charged $EDTA^{4-}$ ion; thus, it appears that the rate enhancement results from an outer-sphere association between the complex and the ligand rather than from an increase in $k^{M\text{-}H_2O}$. Similar effects are found for the reactions of substituted diphenylthiocarbazones with Ni(II) and Zn(II) in which the bis complex is formed much more rapidly than the mono complex.[420,421,422]

The nature of the outer-sphere association was investigated by Cayley and Margerum[21] who systematically varied incoming and bound ligands of Ni(II) complexes seen in Table 1-28B. It is proposed that the rate enhancements are a result of stacking interactions which will assist the substitution rate only if the incoming ligand is sufficiently flexible to permit one donor group to be oriented in a position suitable to replace a coordinated water molecule while the rest of the molecule is in a stacked arrangement. Thus, bipy addition to $Ni(phen)^{2+}$ is 25 times faster than NH_3 addition while phen addition is only twice as fast. The more flexible bipy can interact as shown in Figure 1-17 while phen cannot.

An increase in the number of aromatic groups in the bound ligand also increases the formation rate constants as seen for the reactions of $[Ni(bipy)(H_2O)_4]^{2+}$, $[Ni(phen)(H_2O)_4]^{2+}$, and $[Ni(terpy)(H_2O)_3]^{2+}$ with terpy. The enhancement factor is only seven for $[Ni(terpy)(H_2O)_3]^{2+}$ + bipy because only one pair of rings can stack; the enhancement factor is 23 for $[Ni(bipy)(H_2O)_4]^{2+}$ + terpy where two pairs of rings can stack.

The magnitude of the formation rate constants also depends on the number of methyl substituent groups on the phenanthroline rings. As seen in Table 1-28B, the rate enhancements are large even after correcting for the increased lability of the coordinated water molecules.[243] Methyl groups on the incoming ligands also increase the rate constants (cf. Me_2bipy with bipy). The effectiveness of PADA may result in part from its methyl groups. The general mechanism proposed for these reactions is given in Figure 1-18. The rate-determining step must occur early in the formation reaction in order to maintain any advantage from stacking interactions. A consequence of this mechanism is that the dissociation rate constants also increase because of stacking interactions. The stacking phenomena can be attributed to hydrophobic bonding and polarization effects. Such effects may be important in biological systems in attracting and positioning substrates in enzymatic reactions.

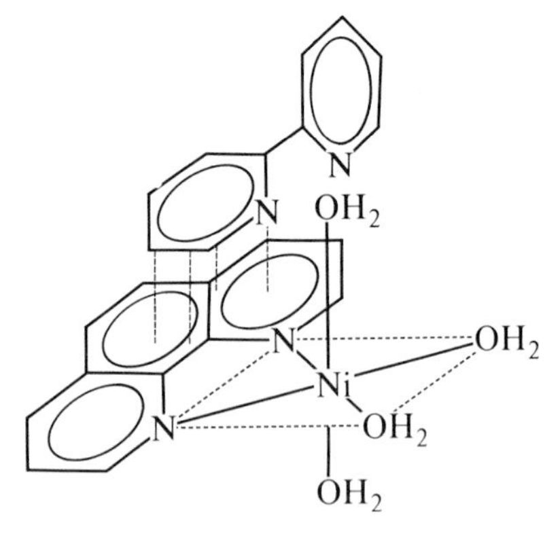

Journal of the Chemical Society, Chemical Communications

Figure 1-17 A possible orientation of the stacking interaction between bipy and $[Ni(phen)(H_2O)_4]^{2+}$ immediately prior to first-bond co-ordination of bipy.

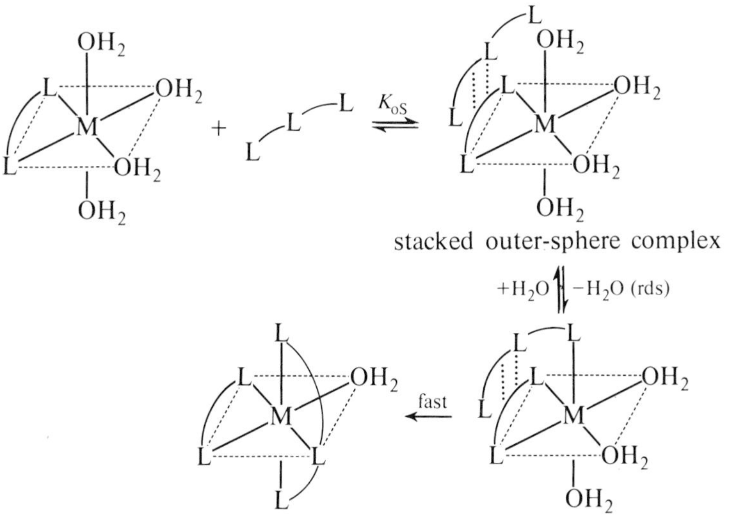

Journal of the Chemical Society, Chemical Communications

Figure 1-18 Proposed mechanism for ternary complex formation with stacking interaction

D. Conclusions

For metal ion electronic configurations with no crystal field effects the donor properties of the various coordinating ligands seem to have little effect on the substitution rate in ternary complex formation. For a series of complexes of magnesium(II)[285,287,290] it appears that the bound ligand simply reduces the number of positions at which substitution can take place. Similar behavior has been observed for Mn(II)[286,287] and Zn(II) even though the different zinc complexes do not have the same coordination number.[452,646]

On the other hand, metal ions where the d^n electronic configuration leads to crystal-field stabilization energies show much larger sensitivity to the donor properties of the coordinated ligands. By far the largest amount of data are available for complexes of nickel(II), and here the coordinated ligands can greatly affect the value of $k^{\text{Ni-H}_2\text{O}}$. Large effects are also found for Co(II) complexes. There are a number of qualitative observations that can be made concerning the effect of bound ligands on the value of $k^{\text{Ni-H}_2\text{O}}$. This rate constant increases with the σ electron-donating ability of the coordinated ligand while ligands which are able to accept electron density back from the metal ion, such as bipyridyl, terpyridyl, and phenanthroline, tend to have relatively small values of $k^{\text{Ni-H}_2\text{O}}$ with respect to their position in the spectrochemical series.

Attempts to make more general correlations of the effects of the coordinated ligand on the value of $k^{\text{M-H}_2\text{O}}$ have had only limited success. Funahashi and Tanaka[249,667] have attempted to correlate the changes in $k^{\text{Ni-H}_2\text{O}}$ with a ligand basicity scale (the E_A value) developed by Edwards.[668] They found for a series of ligands that a plot of log$^{NiL\text{-}H_2O}$/Ni$^{Ni\text{-}H_2O}$ vs. the E_A value of the ligand varies linearly for a number of ligands as shown in Figure 1-19. This relationship holds quite well when only one or two donor atoms, apart from water, are bound to the metal ion. It fails badly with aminocarboxylate and heterocyclic-type ligands. Hoffmann and Yeager[251] have attempted to correlate the lability of the malonate ligand in a series of nickel complexes with the observed spectral shift of the d–d transitions in the ternary complexes (Figure 1-20). The ligands can be divided into three classes: (a) ligands that labilize the malonate chelate in which the logarithm of the rate constant increases as a linear function of the spectral shift; (b) ligands which have little influence on the lability of the malonate chelate although producing a substantial spectral shift; and (c) ligands which labilize the malonate chelate without producing a spectral shift. Hoffmann and Yeager's attempt to correlate these observations with crystal field theory predicted that the dissociation rates should decrease rather than increase with a spectral shift to shorter wavelengths as experimentally observed. This is not too surprising since the crystal field arguments they use rely on equivalent donors, which is not the case for these ligands, and because the crystal field energies are small compared with the strength of the metal–ligand bonds.

For other metal ions, particularly Cu(II), the geometry of the complex is much more important than the nature of the donor groups. This is particularly evident

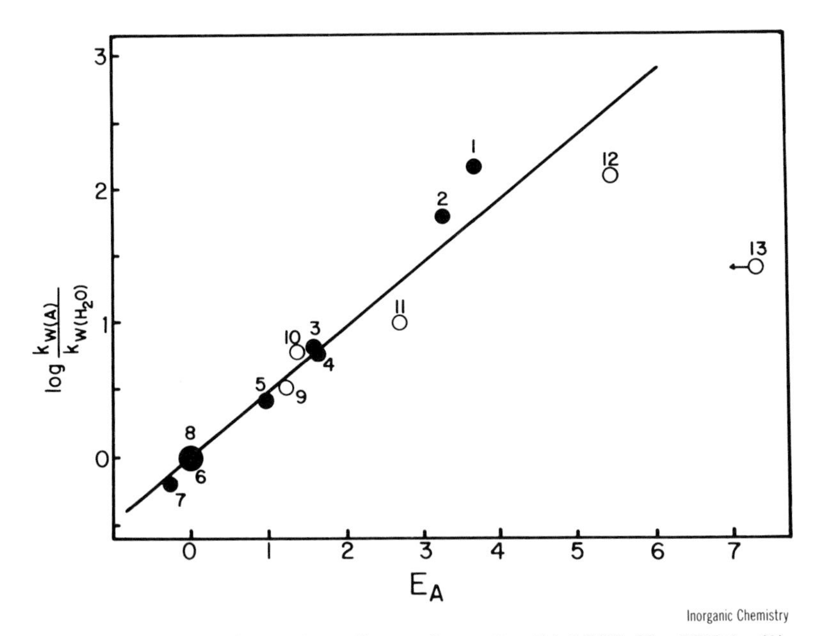

Figure 1-19 Plot of log ($k_{w(A)}/k_{w(H_2O)}$) vs. E_A: (1) $Ni(H_2O)_3$–EDDA; (2) $Ni(H_2O)_3NTA^-$; (3) $Ni(H_2O)_5OH^+$; (4) $Ni(H_2O)_5N_3^+$; (5) $Ni(H_2O)_5CH_3CO_2^+$; (6) $Ni(H_2O)_6^{2+}$; (7) $Ni(H_2O)_5F^+$; (8) $Ni(H_2O)_6^{2+}$; (9) $Ni(H_2O)_5Cl^+$; (10) $Ni(H_2O)_5NH_3^{2+}$; (11) $Ni(H_2O)_4en^{2+}$; (12) $Ni(H_2O)_2(en)_2^{2+}$; and (13) Ni-$(H_2O)_2(NCS)_4^{2-}$ [667]

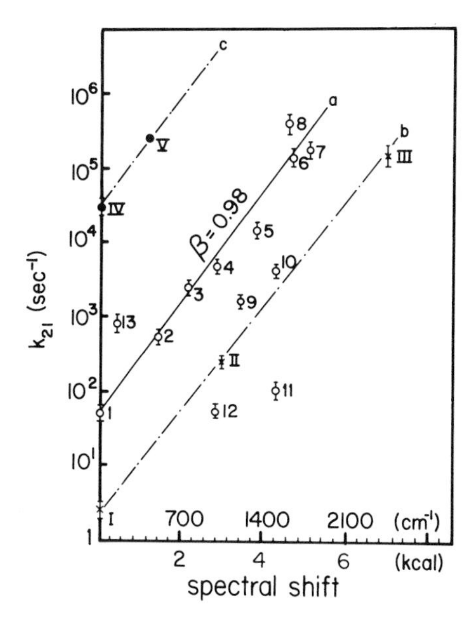

Berichte der Bunsengesellschaft fuer Physikalische Chemie

Figure 1-20 Rate constants vs. spectral shift of the first absorption peak of the nickel ion for the process

$$[NiaqXY] + aq \xrightarrow{k_{2,1}} [NiaqX] + Y \text{ at } 25°C$$

X =	Y =
1. aq	a) malonate
2. 2(pyridine)	b) NH_3
3. 2,2-aminoethylpyridine	c) H_2O
4. pn	
5. dien	
6. 3,3′-iminobispropylamine	
7. $(en)_2$	
8. $(pn)_2$	
9. (en)	
10. 2-(aminomethylpyridine)	
11. α,α'-bipyridine	
12. pyridinealdoxime	
13. acac	

The following points correspond to water or NH_3 exchange rates in the complexes: I., V.$[Ni(NH_3)(H_2O)_5]$; II. $[Ni(pn)(NH_3)(H_2O)_3]$; III. $[Ni(NH_3)_6]$; IV. $[Ni(H_2O)_6]$

for the Cutren$(H_2O)^{2+}$ complex where the rate of water exchange is $^1/_{10}{}^4$ that of the aquo metal ion. The geometry of Co(II) complexes is also thought to have considerable influence on the value of $k^{\text{Co-H}_2\text{O}}$, and it may be a reason why cobalt(II) can replace zinc(II) in many metalloenzymes to give only a slightly reduced enzyme activity while replacement by other metal ions gives unreactive enzymes.[669]

The study of ternary complexes in solution should be extremely useful in understanding the mechanisms of enzyme action, but it must be remembered that the large bulk of the protein in enzymes could have a significant effect on the water structure close to the active site, which could greatly affect the properties of the metal ion.[430]

VI. MONODENTATE LIGAND EXCHANGE REACTIONS

This section concerns the kinetics and mechanisms of complexes in which monodentate ligands exchange with monodentate and multidentate ligands or catalyze such exchange reactions. We have already considered reactions in which water is the monodentate ligand as well as the solvent, and we now examine the much more limited examples involving other monodentate ligands. With labile metal ions there are far more examples of multidentate ligand exchange kinetics, and these are discussed in section VII.

A. Monodentate Ligands Replacing Monodentate Ligands

One convenient method for determining monodentate ligand exchange rates is by nuclear magnetic resonance line-broadening techniques. Some of the systems that have been investigated by this method by Hunt and co-workers include Ni(II)–NH_3, Co(II)–NH_3, Ni(II)–NCS–H_2O, and Ni(II)–Cl–H_2O. The general reaction involves the replacement of a monodentate ligand in the inner-coordination sphere of the transition metal by one of the same type present in the solvent media. Table 1-29 lists the exchange rate constants for some Ni(II) and Co(II) systems; the water exchange values are included for comparison. The hexamine complexes of Ni(II) and Co(II) exchange ammonia in aqueous solution two to 30 times faster than the hexaaquo complexes exchange water.[671,673] As previously noted, however, the exchange of water in Ni$(NH_3)_5(H_2O)^{2+}$ is more than 100 times faster than in $Ni(H_2O)_6^{2+}$.

There are two basic mechanisms for monodentate ligand exchange in aqueous solution. Considering the nickel–hexaammine system, the first mechanism would be a dissociative type.

$$Ni(NH_3)_6^{2+} \xrightarrow{\text{rds}} [Ni(NH_3)_5^{2+}] + NH_3 \tag{53}$$

$$[Ni(NH_3)_5^{2+}] + NH_3 \rightarrow Ni(NH_3)_6^{2+} \tag{54}$$

TABLE 1-29
MONODENTATE LIGAND SELF-EXCHANGE RATE CONSTANTS

System	k (25°C) (sec^{-1})	$\Delta H^‡$ (kcal/mole)	$\Delta S^‡$ (cal/deg·mole)	Ref.
$Ni(NH_3)_6^{2+}$–NH_3 (liq. NH_3)	$1.0 \pm 0.1 \times 10^5$	11 ± 1	2 ± 3	670
$Ni(NH_3)_6^{2+}$–NH_3 (aq. NH_3)	$5.6 \pm 0.5 \times 10^4$	9.5 ± 1.1	-5 ± 4	671
$Ni(H_2O)_6^{2+}$–$H_2{}^{17}O$	2.7×10^4	11.6	0.6	33
$Ni(NCS)_4(H_2O)_2^{2-}$–NCS^-	5.3×10^5	9.2	−1	552
$Ni(NCS)_4(H_2O)_2^{2-}$–H_2O	1.1×10^6	6.0	−11	552
$NiCl(H_2O)_5^+$–Cl^- [a]	1.8×10^5 [b]	12 ± 1		551
$NiCl(H_2O)_5^+$–H_2O	6.9×10^5 [b]	8.0 ± 1		551
$Co(NH_3)_6^{2+}$–NH_3 (liq. NH_3)	$7.2 \pm 1.4 \times 10^6$	11.2 ± 0.4	10.2 ± 2	672
$Co(NH_3)_6^{2+}$–NH_3 (aq. NH_3)	$3.0 \pm 0.3 \times 10^7$	13 ± 1		673
$Co(H_2O)_6^{2+}$-$H_2{}^{17}O$	1.1×10^6	8.0	−4.1	33
$Co(NCS)(H_2O)_5^+$–NCS^- [c]	1.8×10^5	13.4		520
$Co(NCS)_2(H_2O)_2$–NCS^- [c]	$3.0 \pm 0.2 \times 10^6$ [b]	12.0 ± 0.6		520
$Co(NCS)_3(H_2O)^-$–NCS^- [c]	$4.7 \pm 0.1 \times 10^6$ [b]	9.6 ± 1.7		520
$Co(NCS)_4^{2-}$-NCS^- [c]	$> 1.0 \times 10^7$			520
$CoCl(H_2O)_5^+$–Cl^-	6.8×10^6			521
$CoCl_2(H_2O)_2$ [d]–Cl^-	$>5 \times 10^7$			521

[a] The actual nickel complex was not defined. Reactions were in 6.6–7.5M LiCl or HCl, and $NiCl_2$ and $NiCl^+$ are suggested as likely species.
[b] Units of M^{-1} sec^{-1}
[c] 27°C.
[d] Tetragonal complex.

The second would be a bimolecular mechanism involving water.

$$Ni(NH_3)_6^{2+} + H_2O \xrightarrow{\text{rds}} Ni(NH_3)_5H_2O^{2+} + NH_3 \qquad (55)$$

$$Ni(NH_3)_5H_2O^{2+} + NH_3 \rightarrow Ni(NH_3)_6^{2+} + H_2O \qquad (56)$$

It is very difficult to distinguish between such mechanisms in aqueous solution. However, the lack of NH_3 attack on $Ni(NH_3)_6^{2+}$ and the fact that a change of 30% in H_2O concentration produced no observable effect support a dissociative mechanism.[671]

The monodentate ligand substitution reactions of Co(III) complexes have been extensively investigated.[674] In general, the direct replacement of one anionic ligand by another does not occur, but the reactions proceed by an aquation path.

$$Co(NH_3)_5X^{2+} + H_2O \rightarrow Co(NH_3)_5OH_2^{3+} + X^- \qquad (57)$$

$$Co(NH_3)_5OH_2^{3+} + L^- \rightarrow Co(NH_3)_5L^{2+} + H_2O \qquad (58)$$

The log of the rate constants for the aquation reactions vs. pK_{eq} show a linear relationship of slope equaling 1.0, particularly for monodentate anions, indicating that the nature of the X group in the transition state is the same as that in the product.

For the aquation reaction of $Ru(NH_3)_5X^+$ where X is Br^- or Cl^- it has been found that the aquation rate does not depend on the leaving group.[675] This is in contrast to the other metal–pentaammine–halide complexes,[676] where in most cases the bromide complexes react faster than the chloride complexes because of the larger stability of the chloride complex. In the case of Ru(II) the stability of the chloride complex seems to be significantly larger than the bromide complex while the hydrolysis rates are the same. It is tempting to suggest that in these reactions bond making is more important.

Koren and Perlmutter-Hayman[677] have investigated the environmental effects on the aquation reaction of iron(III) chloride. For the reaction scheme shown below the rate constant corresponding to the *1* → *3* → *4* path depends on the nature of the added electrolyte, the difference becoming more pronounced the higher the concentration of added electrolyte. Existing theories would predict only small ionic strength effects on the rate of hydrolysis, k_{34}, while the effect of ionic strength on the value of K_4 would be expected to parallel the value of K_{OH}. Experimental evidence indicates that the effect on k_{34} outweighs that on K_4. For the acid-independent pathway only a weak ionic effect would be expected; however, a pronounced specific influence is observed. With the weakly basic chloride ligand it is unlikely that the pathway via $Fe(H_2O)_4OH^{2+}{\cdot}HX$ will be important, but the rate of hydrolysis may be increased by the ability of a neighboring water molecule to accept a proton. This would labilize the ligand and facilitate its exchange for a water molecule. Thus, two water molecules could play the role of the amine in the ICB effect.

$$
\begin{array}{ccc}
\underset{1}{Fe(H_2O)_5Cl^{2+}} & \xrightleftharpoons{H_2O} & \underset{2}{Fe(H_2O)_6^{3+}Cl^-} \\
\updownarrow {\scriptstyle -H^+} & & {\scriptstyle H^+} \updownarrow \\
\underset{3}{Fe(H_2O)_4ClOH^+} & \xrightleftharpoons{H_2O} & \underset{4}{Fe(H_2O)_5OH^{2+} + Cl^-}
\end{array}
\qquad (59)
$$

Koren and Perlmutter-Hayman[677] suggest that the effect of the added electrolytes can be explained if the hydrogen-bonded form of water accepts a proton from $FeCl(H_2O)_5^{2+}$ more rapidly and to a larger extent than the non-hydrogen-bonded form. Thus, the structure breaking potassium nitrate causes a reduction in the aquation rate compared with lithium nitrate.

The kinetics of substitution reactions of $Fe(CN)_5L^{3-}$ have recently been studied where L was an aromatic nitrogen heterocycle and the substituting ligand was *N*-methylpyrazinium.[678] The rate of substitution varied with the nature of L, and saturation kinetics typical of the rate-determining loss of L from the complex followed by rapid addition of the incoming ligand were observed. This is one of the few studies in which a limiting S_N1 ligand replacement mechanism (or D mechanism) has been observed and the first for monodentate ligand substitution reactions of simple low-spin Fe(II) complexes in aqueous solution. A similar study[679] of the reaction of $Fe(CN)_5SO_3^{5-}$ with cyanide ion has also been reported. The observed behavior is similar to that of the $Fe(CN)_5H_2O^{3-}$ complex mentioned earlier.

Complexes with lower coordination numbers frequently undergo monodentate-ligand–monodentate-ligand exchange. These reactions are most often found in the substitution reactions of square-planar complexes. There are extensive reviews of this area already in the literature[1] for Pt(II) complexes.

One example of monodentate–monodentate exchange with a more labile square-planar complex is the reaction of NH_3 with $Ni(CN)_4^{2-}$ where the released CN^- ion is scavenged by I_2 to drive the reaction.[680] In this case the reaction with one ammonia labilizes the complex, and all four cyanide ions are released to form ICN. The rate expression at 25°C is typical of square-planar complexes.

$$\frac{-d[Ni(CN)_4^{2-}]}{dt} = (4.8 \times 10^{-4} + 120[NH_3])[Ni(CN)_4^{2-}]$$

Additional reactions involving $Ni(CN)_4^{2-}$ are considered in subsequent sections. Associative mechanisms are found as is the case with Pt(II) complexes.

B. Monodentate Ligands Replacing or Catalyzing the Replacement of Multidentate Ligands

The thermodynamic feasibility of monodentate ligands' displacing multidentate ligands from metal ion complexes has limited the number of examples of direct-displacement reactions. However, cyanide ion forms very stable complexes with transition metal ions and displaces aminocarboxylate and polyamine complexes from Ni(II) and Co(II). In other reactions hydroxide ion and halide ions are able to initiate and to catalyze some multidentate ligand displacement reactions in which other species are present to force the release of the multidentate ligand.

1. Cyanide Ion Displacement. The general reaction in Eq. (60) has been studied for 10 different aminocarboxylate (L^{n-}) complexes of nickel. The rate of formation of $Ni(CN)_4^{2-}$ is first order in the nickel–aminocarboxylate concentration and varies from first to third order in cyanide ion concentration, the variation depending on the nature of the aminocarboxylate ion and the cyanide ion concentration.[502,681,682,683] Figure 1-21[684] shows the variable order of the observed first-order rate constant.

$$NiL^{2-n} + 4CN^- \rightarrow Ni(CN)_4^{2-} + L^{n-} \tag{60}$$

Although the observed rate constants differ by more than eight orders of magnitude, the same mechanism appears to hold for these complexes. In each

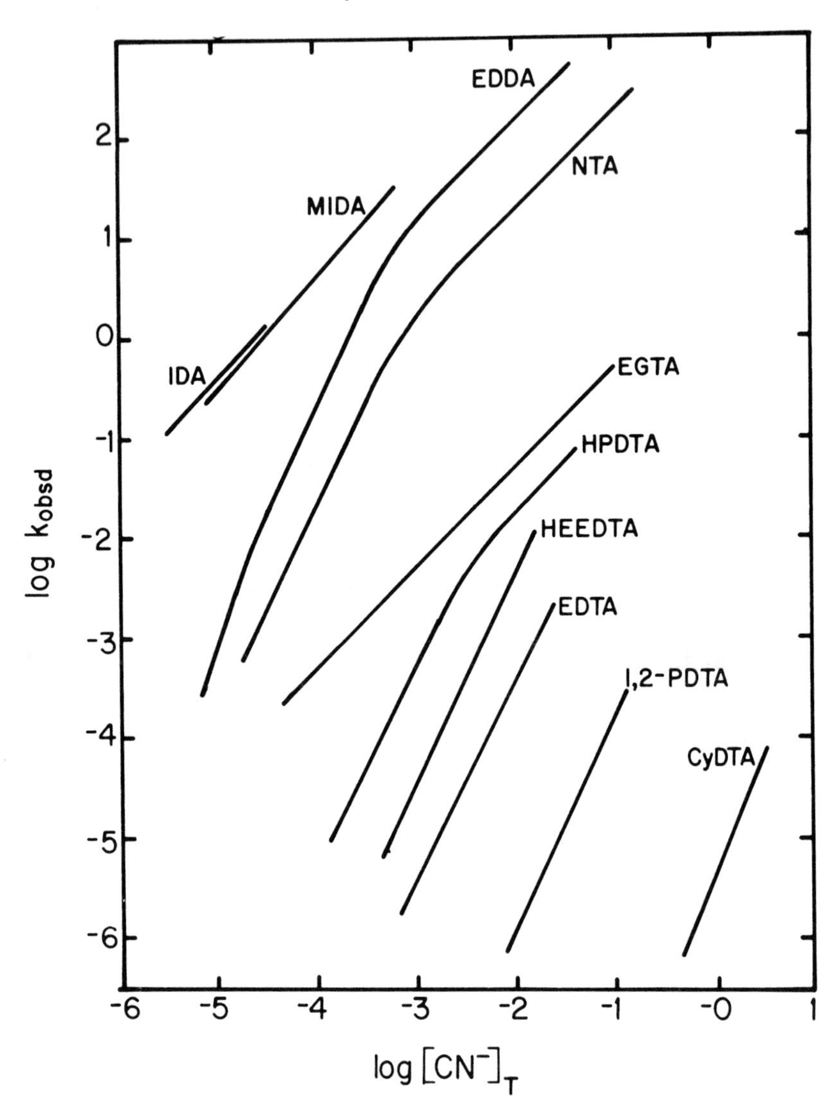

Analytical Chemistry

Figure 1-21 Dependence of the observed first-order rate constant (sec^{-1}) on the CN^- concentration (*M*) for the reaction of nickel–aminopolycarboxylate complexes with excess CN^- [684]

case, the transition state in the reaction mechanism has three cyanide ions, but the order dependence in cyanide ion depends upon the extent of formation of mixed cyanide complexes prior to the rate-determining step (rds). The mechanism is given in Eqs. (61)–(64) where $NiL(CN)^{1-n}$ and $NiL(CN)_2^{n-}$ may

$$NiL^{2-n} + CN^- \underset{}{\overset{K_1}{\rightleftarrows}} NiL(CN)^{1-n} \text{ (rapid)} \tag{61}$$

$$NiL(CN)^{1-n} + CN^- \overset{K_2}{\rightleftarrows} NiL(CN)_2^{n-} \text{ (rapid)} \tag{62}$$

$$NiL(CN)_2^{n-} + CN^- \xrightarrow{k_3} Ni(CN)_3L^{(n+1)-} \text{ (rds)} \tag{63}$$

$$Ni(CN)_3L^{(n+1)-} + CN^- \xrightarrow{k_4} Ni(CN)_4^{2-} + L^{n-} \text{ (rapid)} \tag{64}$$

be present in appreciable concentrations depending upon the cyanide ion concentration. Cyanide ion is present in excess so that the observed reaction rate is first order in the nickel–aminopolycarboxylate–cyanide complex [Eq. (65)],

$$\frac{d[Ni(CN)_4^{2-}]}{dt} = k_{obsd}[NiL(CN)_x{}^{2-n-x}] \tag{65}$$

which in general, forms rapidly and has a low absorbance compared with the final reaction product. The value of the first-order rate constant, k_{obsd}, depends on the cyanide ion concentration as expressed in Eq. (66) and shown in Figure 1-21.

$$k_{obsd} = \frac{K_1K_2k_3[CN^-]^3}{1 + K_1[CN^-] + K_1K_2[CN^-]^2} \tag{66}$$

The slope in Figure 1-21 can vary from unity at high cyanide ion concentrations (where $k_{obsd} = k_3[CN^-]$) to three at low cyanide ion concentrations (where $k_{obsd} = K_1K_2k_3[CN^-]^3$). Table 1-30 summarizes the values for the $K_1K_2k_3$ product, which change by nearly twenty orders of magnitude from the NiIDA complex to the $NiCyDTA^{2-}$ complex. For many of the more flexible aminocarboxylate ligands the $K_1K_2k_3$ term is proportional to the reciprocal of the stability constant of NiL^{2-n} after corrections are made for electrostatic repulsions in the transition state. This is consistent with the proposed structure for the intermediate species which has three CN^- coordinated as shown in Figure 1-22, where only one nitrogen remains coordinated, and where R represents the remainder of the aminocarboxylate molecule. The very small rate constants for the CyDTA complex can be attributed to its large stability constant ($\log K = 21.4$)[278] and to the fact that the cyclohexane ring does not allow the ligand to rotate away from the metal while a glycinate segment is still coordinated. Hence, a large negative charge and sheer bulk build up around the nickel ion in the transition state with three CN^- ions present. The EDDA complex may have two nitrogen atoms coordinated in the rate-determining step. The EDDS complex which contains two six-membered chelate rings has a larger rate constant.

In general, the aminocarboxylate complexes with nickel(II) that are high spin react with cyanide ion through a stepwise unwrapping mechanism where the rate-determining step occurs when three cyanide ions surround the nickel. Furthermore, $Co(EDTA)^{2-}$ and $Co(CyDTA)^{2-}$ have a similar rate dependence but react much faster as seen in the rate constants of Table 1-30.

TABLE 1-30
RATE CONSTANTS FOR THE FORMATION OF $Ni(CN)_4^{2-}$ AND STABILITY CONSTANTS FOR THE MIXED CYANIDE–AMINOCARBOXYLATE COMPLEXES OF Ni(II)

NiL^{2-n}	$K_1K_2k_3$ (M^{-3} sec^{-1})	Ref.
Ni(IDA)	8.1×10^{15}	502
Ni(MIDA)	1.1×10^{15}	502
$Ni(NTA)^{1-}$	1.0×10^{11}	682
Ni(EDDA)	2.4×10^{12}	682
$Ni(EGTA)^{2-}$	5.8×10^{9}	682
$Ni(EDDS)^{2-}$	2.2×10^{7}	683
$Ni(HPDTA)^{2-}$	2.6×10^{6}	682
$Ni(HEEDTA)^{1-}$	2.4×10^{6}	682
$Ni(EDTA)^{2-}$	2.3×10^{4}	681
$Ni(CyDTA)^{2-}$	$\sim 10^{-4}$	685
$Co(EDTA)^{2-}$	2.2×10^{6}	686
$Co(CyDTA)^{2-}$	3.7×10^{-1}	604

Recently the reaction of $MnCyDTA(OH)^-$ with cyanide has been investigated.[687] A mechanism to fit the observed data is shown below.

$$MnCyDTA(OH)^{2-} + CN^- \underset{k_{-1}}{\overset{k_1}{\rightleftharpoons}} MnCyDTA(CN)^{2-} + OH^- \quad (67)$$

$$MnCyDTA(CN)^{2-} + CN^- \xrightarrow{k_2} MnCyDTA(CN)_2^{3-} \quad (68)$$

$$MnCyDTA(CN)_2^{3-} + 4CN^- \xrightarrow{\text{fast}} Mn(CN)_6^{3-} + CyDTA^{4-} \quad (69)$$

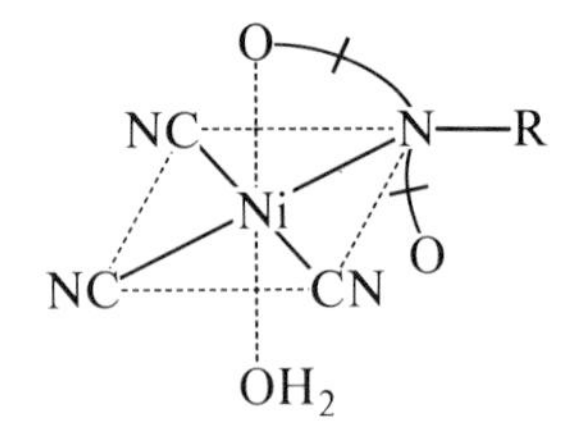

Inorganic Chemistry

Figure 1-22 Proposed structure of the $Ni(CN)_3L^{(n+1)-}$ intermediate formed in reaction of excess cyanide ion with nickel–aminopolycarboxylate complexes[682]

The values of k_1 and k_2 were $0.015M^{-1}$ sec^{-1} and $0.036M^{-1}$ sec^{-1}, respectively. In contrast to the analogous cobalt(II) and nickel(II) reactions only two cyanides are involved in the rate-determining step.

The presence of three cyanides does not appear to be a requirement, however, when the nickel(II) is already in the spin-paired state. In the reaction of cyanide[688,689] with deprotonated nickel(II)-triglycine ($NiH_{-2}GGG^{1-}$) and nickel(II)-diglycineamide ($NiH_{-2}GGa$) a two-path mechanism is observed. One path requires two cyanides, and the other requires three. The difference is attributed to the manner in which the ligands unwrap from the metal. These two examples in no way invalidate the previous discussion concerning reactions of high-spin complexes and in fact only dramatize the types of influence that a change in spin state can have in ligand exchange reactions.

The reaction of cyanide ion with (triethylenetetramine)nickel(II) differs from the aminocarboxylate reactions in that four cyanides are required in the rate-determining step.[476] The presence of four cyanides also is required to displace triethylenetetramminehexaacetate from Ni(II).[690] One CN^- adds rapidly to form $NiL(CN)^{5-}$, and there is a $[CN]_T{}^3$ dependence as well. In the reaction of cyanide with $Nitrien^{2+}$, HCN is an important reactant, and the rate depends on $[Nitrien^{2+}][CN^-]^2[HCN]^2$. At pH 5–8 $Nitrien^{2+}$ reacts faster with cyanide than does aquonickel, which is interesting because the dissociation of trien in this pH range is very slow in the absence of cyanide.[239] As with several of the nickel–aminocarboxylate complexes $Nitrien^{2+}$ forms an adduct with cyanide which is undoubtedly the first step in the reaction path leading to $Ni(CN)_4{}^{2-}$. The formation of $Nitrien(CN)^+$ reduces the free cyanide concentration and causes a novel behavior in which the rate of formation of $Ni(CN)_4{}^{2-}$ decreases as the amount of $Nitrien^{2+}$ in excess over cyanide increases.

For some reactions it is possible to control the experimental conditions so that the monodentate ligands only partially replace a multidentate ligand to give a mixed complex. This type of reaction occurs in the reaction of cyanide with $NiEDTA^{2-}$, $NiHEEDTA^-$, $NiCyDTA^{2-}$, and $CoCyDTA^{2-}$. The cobalt(II) reaction[604] which was observed at high pH where the HCN concentration is negligible has been discussed earlier. For the reactions of the nickel species under conditions where only the monocyanide mixed complex $[Ni(EDTA)CN]^{3-}$ is formed, a complex pH and buffer dependence was observed as shown in Figure 1-23.[319] Although $[Ni(EDTA)HCN]^{2-}$ is not a reaction product, it is obvious that HCN reacts faster with $NiEDTA^{2-}$ than CN^-, and the following mechanism has been proposed to explain the observed kinetics.

$$NiEDTA^{2-} + CN^- \rightleftarrows [Ni(EDTA)CN]^{3-} \tag{70}$$

$$NiEDTA^{2-} + HCN \rightleftarrows [Ni(EDTA)HCN]^{2-} \tag{71}$$

$$[Ni(EDTA)HCN]^{2-} + OH^- \rightleftarrows [Ni(EDTA)CN]^{3-} + H_2O \tag{72}$$

$$[Ni(EDTA)HCN]^{2-} + B \rightleftarrows [Ni(EDTA)CN]^{3-} + HB \tag{73}$$

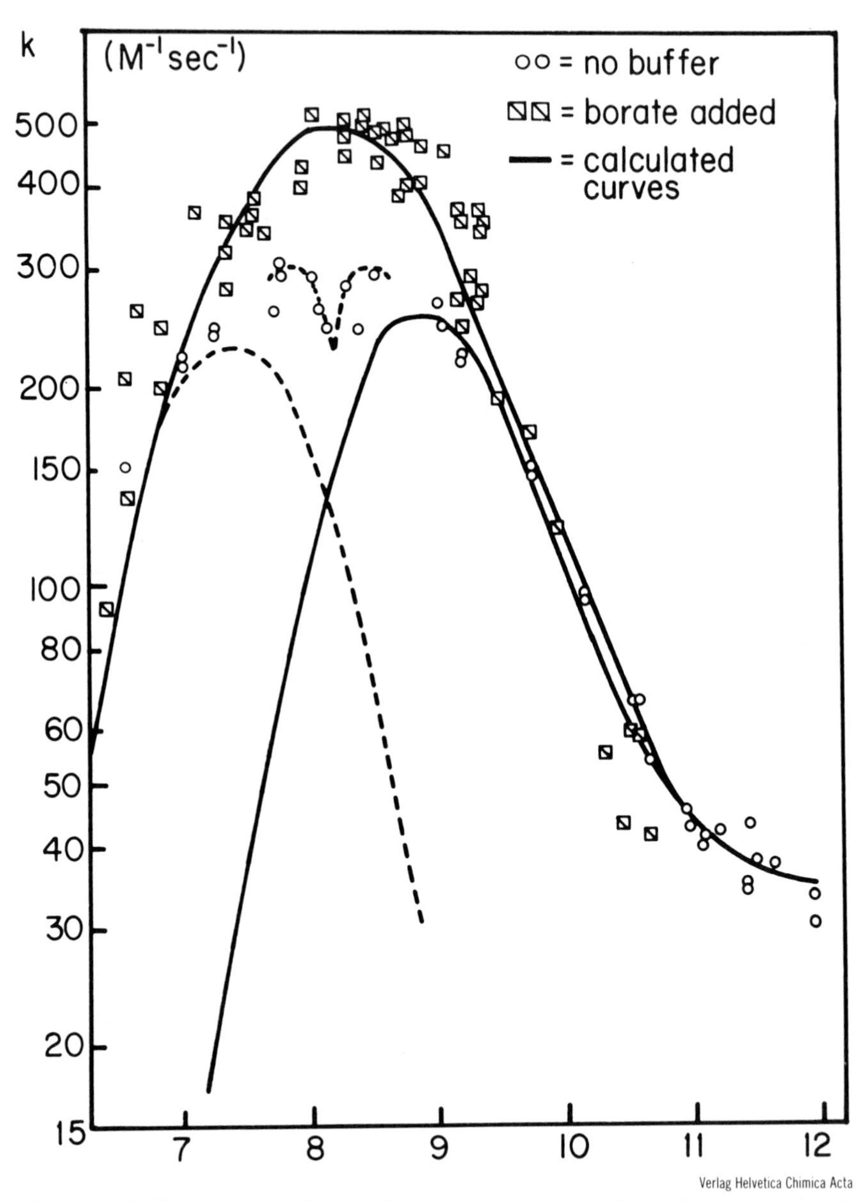

Verlag Helvetica Chimica Acta

Figure 1-23 Dependence of the second-order rate constant on pH for the reaction of total cyanide (CN^- + HCN) with $NiEDTA^{2-}$ to form $[Ni(EDTA)CN]^{3-}$ [319]

The observed data can be interpreted in terms of a mechanism where HCN reacts with $NiEDTA^{2-}$, but the intermediate must lose a proton to give products. Above pH 9 there is sufficient hydroxide to accept the proton from the intermediate, but below this pH the proton transfer step becomes rate limiting. From pH 8–10 added borate buffer catalyzes the reaction because it speeds the proton-transfer step as shown in Eq. (73). The reaction of cyanide with the square planar $Ni(daco)_2^{2+}$ proceeds via a stable intermediate $Ni(daco)(CN)_2$ before going to $Ni(CN)_4^{2-}$.[697]

2. Monodentate Ligand Assistance of Multidentate Ligand Displacement. Hydroxide ion can accelerate the displacement of multidentate ligands by coordinating to the metal ion and partially displacing the ligand. It also can accelerate ligand substitution reactions without entering the first coordination sphere. There are well-known examples of this behavior in the conjugate base (S_NICB) mechanism of Co(III) complexes[1] where the coordinated ligands must have acidic hydrogens—e.g., weakly acidic $Co^{III}—NH_3$ giving $Co^{III}—NH_2^-$—to experience the hydroxide ion catalysis. However, there are some metal complexes where the ligands have no acidic hydrogens and where hydroxide catalysis is still observed. This is particularly true of low-spin Fe(II) complexes with 1,10-phenanthroline, TPTZ, and bipyridyl derivatives. Thus, the dissociation rate of $Fe(phen)_3^{2+}$ is greatly accelerated by OH^- in the first step of a reaction sequence in which EDTA and O_2 later react to give Fe^{III}-$EDTA^-$.[691] The $Fe(phen)_3^{2+}$ complex also undergoes direct nucleophilic attack by CN^- and N_3^- as shown in Figure 1-24.[504,692] The electronic structure of the metal ion is important in these reactions because the $Fe(phen)_3^{2+}$ reactions

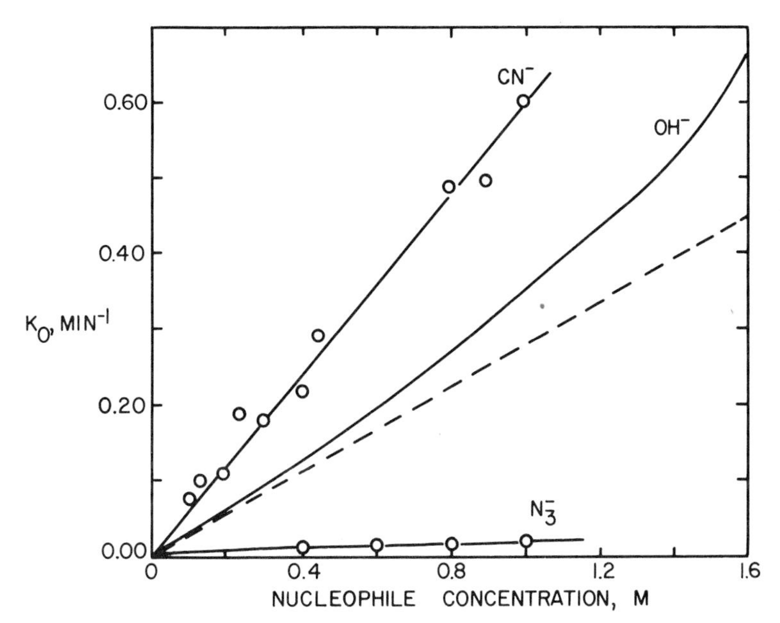

Advances in the Chemistry of Coordination Complexes

Figure 1-24 Observed first-order rate constant for the dissociation of $Fe(phen)_3^{2+}$ in the presence of excess nucleophiles (CN^-, OH^-, or N_3^-). The OH^- dependence is greater than first order.[692]

involve the formation of ion pairs between the incoming nucleophile and the metal complex, followed by an interaction between the nucleophile and the antibonding d-orbitals of the iron on the octahedral face. This weakens the other metal–ligand bonds and increases dissociation rates.

These reactions are not limited to $Fe(phen)_3^{2+}$ complexes since similar results have been obtained for the reaction of hydroxide, cyanide, and $S_2O_8^{2-}$ with tris[α-(2-pyridyl)benzylidene-4″-chloroaniline]iron(II)[693,694] and with some tris(bipyridyl) complexes[695] and for cyanide reactions with tris(bipyridyl) complexes[696] and bis(terpyridyl) complexes.[222] For the latter reactions the hydroxide ion rate is reduced with the 4,4′-dimethyl- and 5,5′-dimethylbipyridyl derivatives compared with the unsubstituted ligand. This is consistent with electron donation from the methyl groups to the nitrogens causing increased electron density in the vicinity of the metal, thus discouraging nucleophilic attack. Burgess and co-workers support this mechanism with the reactions of CN^- and OH^- with several Fe(II) complexes of substituted phenanthrolines, bipyridyls, and Schiff bases in aqueous alcohol mixtures.[699,700,701,702]

The dissociation of TPTZ from $Fe(TPTZ)_2^{2+}$ [297] is more complicated than the bipyridyl and terpyridyl systems. (Figure 1-25.) Between pH 5.4 and 7.5 no spectral or titrimetric evidence exists for the formation of $Fe(TPTZ)_2OH^+$; in the absence of EDTA and O_2 the complex does not dissociate in this pH range. The plateau at pH 7–8 must therefore be caused by a limiting dissociation step

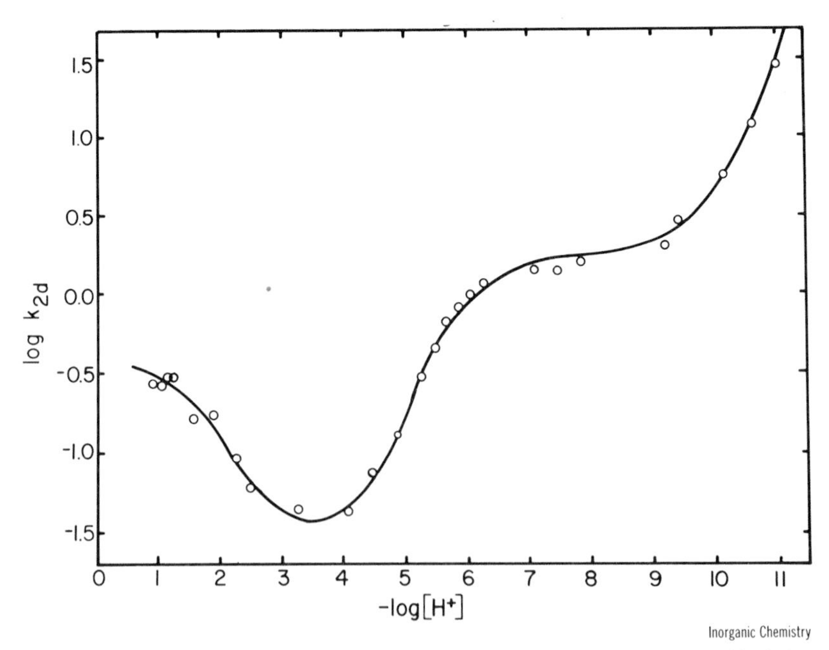

Inorganic Chemistry

Figure 1-25 Effect of acidity on the first-order dissociation rate constant (k_{2d}) for the loss of TPTZ from $Fe(TPTZ)_2^{2+}$ [297]

TABLE 1-31
FIRST-ORDER RATE CONSTANTS FOR THE DECOMPOSITION OF $[Fe(phen)_3]^{3+}$ IN CONCENTRATED SULFURIC ACID AT 100.7°C[a]

$[H_2SO_4]$ (wt %)	Activity of Water[b] (25°C)	$10^3 k_1$ (min^{-1})
98.1	3.1×10^{-6}	0.10
95.2	2.5×10^{-5}	0.26
91.0	2.1×10^{-4}	0.71
89.7	5.0×10^{-4}	0.92
86.9	8.9×10^{-4}	2.5
83.6	2.5×10^{-3}	4.9
82.9	3.0×10^{-3}	19.7
2.0*M*		50,000[c]

[a] Data from Ref. 703.
[b] Values interpolated from data in Ref. 704.
[c] Extrapolated from value measured at 25°C using experimentally determined activation energy (32.1 kcal/mol); data from Ref. 705.

prior to reaction with hydroxide ion, and since under other conditions the dissociation of TPTZ is slower, this must be only a partial dissociation of TPTZ. An intermediate such as (E) in Figure 1-26 is proposed between pH 7–8 while above pH 9 the reappearance of a hydroxide dependence suggests the direct reaction of hydroxide ion with $Fe(TPTZ)_2^{2+}$ as with $Fe(phen)_3^{2+}$.

The decomposition of $Fe(phen)_3^{3+}$ in concentrated sulfuric acid[703] is remarkable in that the dissociation rate decreases as the acid concentration increases (Table 1-31). The decrease in rate is too large to be explained by ion association or general medium effects. The authors have rationalized the results in terms of a bimolecular reaction in which the partial binding of water to the metal ion in the transition is essential to the progress of the reaction and in which the increased acid concentration decreases the concentration of unprotonated water. This reaction completely contrasts that for $Ni(phen)_3^{2+}$, which immediately dissociates in 98% sulfuric acid even though the nickel complex is 10 times more stable than the iron complex in 0.01*M* acid.

The exchange of CyDTA between mercury and copper depends on the anions present in solution.[706] The rate of exchange is independent of the copper concentration, and the same rates are observed if the scavenger is zinc or lead. This is typical of the exchange reaction of CyDTA in which the rate-determining step is usually the dissociation of the metal–CyDTA complex. For the exchange reactions of $HgCyDTA^{2-}$ halide and thiocyanate ions increase the rate of acid dissociation of $HgCyDTA^{2-}$ (Figure 1-27). These ions form mixed complexes with $HgCyDTA^{2-}$, and a change in reaction order from first to zero with bromide ion corresponds to the complete formation of the mixed, $Hg(CyDTA)Br^{3-}$,

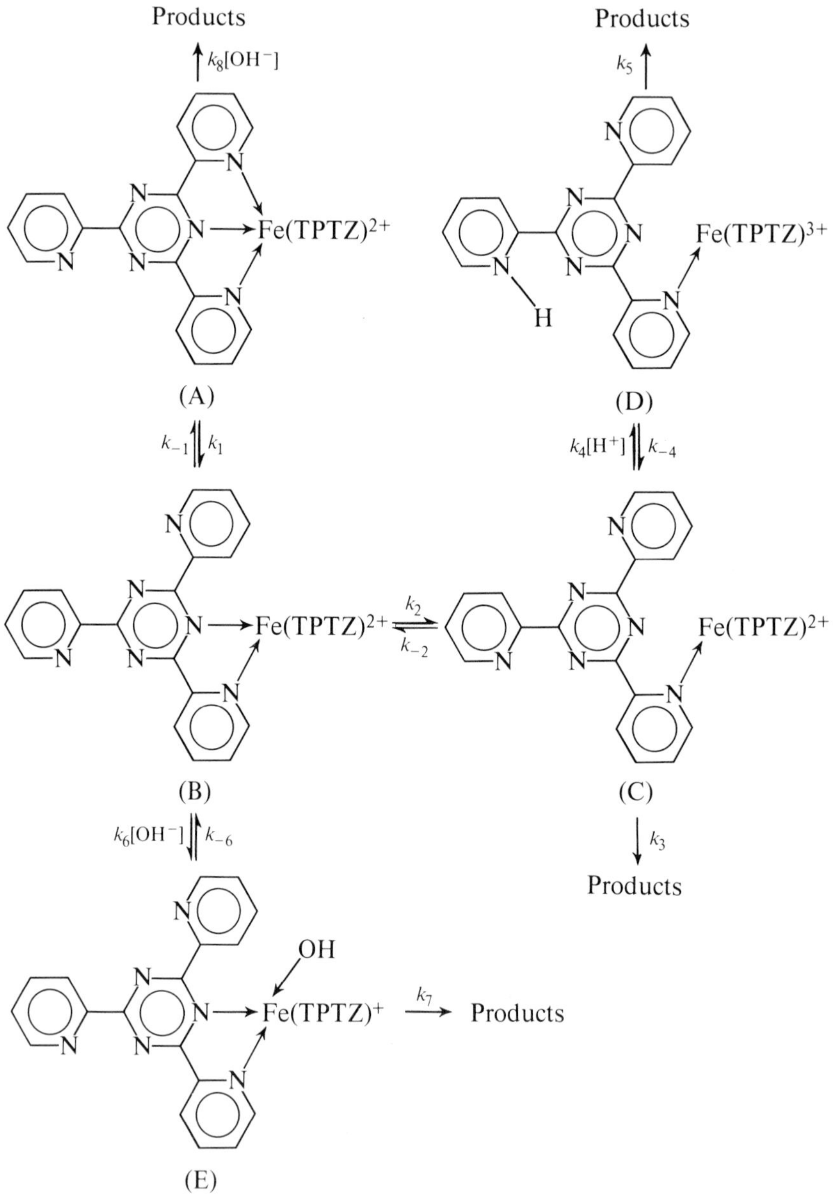

Products

$k_8[OH^-]$

N N N N N $\rightarrow Fe(TPTZ)^{2+}$

(A)

k_{-1} k_1

N N N N N $\rightarrow Fe(TPTZ)^{2+}$ $\underset{k_{-2}}{\overset{k_2}{\rightleftharpoons}}$ N N N N N $Fe(TPTZ)^{2+}$

(B)

$k_6[OH^-]$ k_{-6}

N N N N N OH $\rightarrow Fe(TPTZ)^{+}$ $\xrightarrow{k_7}$ Products

(E)

Products

k_5

N N N N N H $Fe(TPTZ)^{3+}$

(D)

$k_4[H^+]$ k_{-4}

(C)

k_3

Products

Inorganic Chemistry

Figure 1-26 Proposed mechanism for the dissociation of TPTZ from $Fe(TPTZ)_2^{2+}$ showing the direct OH^- attack (k_8) and the stepwise unwrapping of TPTZ(k_1, k_2, and k_3) which is assisted by OH^- (k_6, k_7) or by H^+ (k_4, k_5)[297]

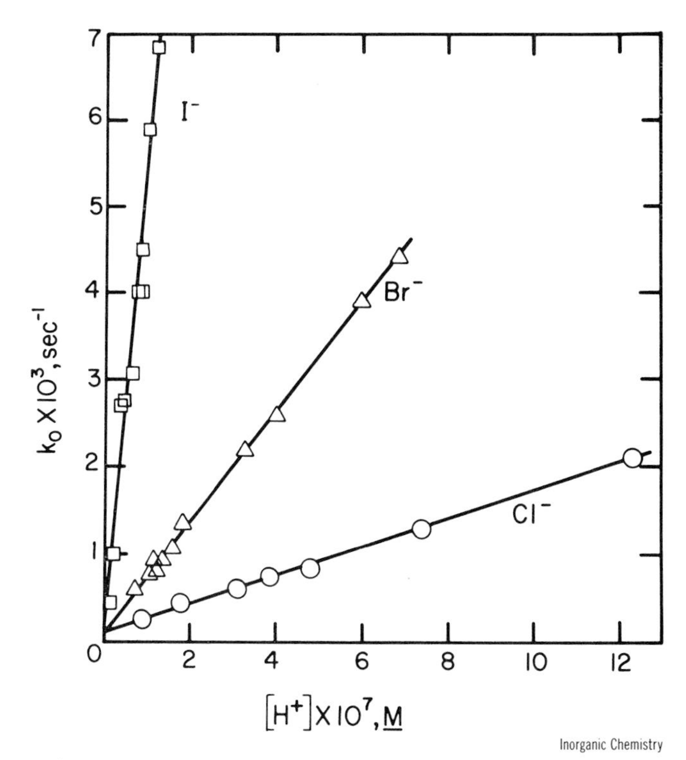

Inorganic Chemistry

Figure 1-27 Effect of hydrogen ion concentration on the first-order rate constant for $HgCy^{2-}$ dissociation in the presence of halide ions. The slopes are

$$k_H^{HgCyX}\left(\frac{K_{HgCyX}[X^-]}{1 + K_{HgCyX}[X^-]}\right).^{706}$$

TABLE 1-32
RATE CONSTANTS FOR THE MERCURY–CyDTA COMPLEXES AT 25.0°C, 0.10M $NaClO_4$

Complexing Anion	k_H^{HgCyX} $(M^{-1}\ sec^{-1})$[a]
None	3.1
Cl^-	2.6×10^3
Br^-	$(1.3 \pm 0.04) \times 10^4$[b]
I^-	5.9×10^4
SCN^-	2.3×10^3
(OH^-)	$<5 \times 10^6$

[a] Determined graphically except bromide ion.
[b] From least-squares analysis and with limits using 95% confidence level.

complex. The dissociation of $HgCyDTA^{2-}$ can proceed by several paths, the reaction in Eq. (74) making a major contribution.

$$H^+ + Hg(CyDTA)X^{3-} \xrightarrow{k_H^{HgCyX}} \text{products} \quad (74)$$

The rate constants given in Table 1-32 parallel the stability of the mixed $Hg(CyDTA)X^{3-}$ complex and the polarizability of the anion.

C. Multidentate Ligands Replacing Monodentate Ligands

Reactions of this type have been discussed extensively for the replacement of coordinated water concerning the formation reactions of multidentate ligand complexes in section III. There are relatively few studies of the replacement of other mondentate ligands despite the common occurrence of analytical procedures involving EDTA and other multidentate ligand displacements of NH_3, OH^-, or halide ions from metal complexes. Recent results of the reactions of $Cu(OH)_4^{2-}$ and $Cu(OH)_3^-$ with polyamines and of $Ni(CN)_4^{2-}$ with aminopolycarboxylate and polyamine ligands are summarized below. Another type of reaction meriting review is the ring-closure step of a partially bonded multidentate ligand in which a monodentate ligand is displaced. Examples are found for the reactions of Pt(II), Co(III), and Cr(III) and are considered here (despite our self-restricted limitation in regard to not reviewing the reactions of these sluggish metal ions) because even for these metal ions the reactions can be quite rapid.

1. Ring-Closure Reactions with Pt(II). Anchimeric assistance was reported[707] in the reactions of $[Pt(EDDA)Cl_2]^{2-}$ where the reactant has $EDDA^{2-}$ as a bidentate ligand with two of the carboxylate groups free and where the reaction sequence in Eq. (75) leads to the quadridentate EDDA complex.

$$[Pt(EDDA)Cl_2]^{2-} \xrightarrow{-Cl^-} [Pt(EDDA)Cl]^- \xrightarrow{-Cl^-} [Pt(EDDA)] \quad (75)$$

A much larger effect was found by Carter and Beattie[708] in the ring-closure reactions of *trans*-dichlorobis(ethylenediamine monohydrochloride)platinum(II) when the complex is mixed with base.

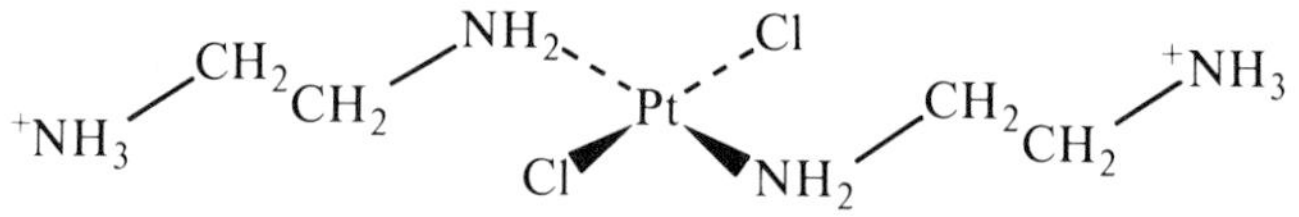

$$^+NH_3\!-\!CH_2\!-\!CH_2\!-\!NH_2\!-\!Pt(Cl)_2\!-\!NH_2\!-\!CH_2\!-\!CH_2\!-\!NH_3^+$$

The direct measurement of the closure of both ethylenediamine rings was possible using stopped-flow methods, and the following reaction scheme is given.

$$Pt(enH)_2Cl_2{}^{2+} \overset{K_{a_1}}{\rightleftharpoons} Pt(enH)(en\text{—})Cl_2{}^{+} + H^{+}$$

$$\downarrow k_1$$

$$Pt(en\text{—})(en)Cl^{+} + H^{+} \overset{K_{a_2}}{\rightleftharpoons} Pt(enH)(en)Cl^{2+} + Cl^{-} \qquad (76)$$

$$\downarrow k_2$$

$$Pt(en)_2{}^{2+} + Cl^{-}$$

For the second ring-closure step an acid-independent path attributed to a solvent-assisted path also exists.

$$Pt(en)(enH)Cl^{2+} + H_2O \rightarrow Pt(en)(enH)(H_2O)^{3+} + Cl^{-} \xrightarrow{\text{fast}} Pt(en)_2{}^{2+} + H_3O^{+} \qquad (77)$$

The rate constant for the solvent path in Eq. (77), $1.4 \times 10^{-5}\ sec^{-1}$, is similar to that for the reaction in Eq. (78)—i.e., $2.6 \times 10^{-5}\ sec^{-1}$, giving added support for the mechanism.

$$[Pt(NH_3)_3Cl]^{+} + H_2O \rightarrow [Pt(NH_3)_3H_2O]^{2+} + Cl^{-} \qquad (78)$$

The rate constant (k_1) for the first ring-closure step is 1800 times greater than it is for the reaction of $1M$ ammonia with *trans*-$[Pt(NH_3)_2Cl_2]$ while for the second reaction it is 1400 times greater than for the reaction of ammonia with $[Pt(NH_3)_3Cl]^{+}$. These large increases in ring-closure rates can be partly attributed to the increased nucleophilicity of ethylenediamine compared with ammonia, but this can only account for a maximum factor of five. Taking into account the statistical factor of two for the monodentate ligands, the chelate ring-closing reactions are still about 800 times faster than a normal bimolecular substitution in which the concentration of the entering ligand is $1M$.

Carter and Beattie used the statistical analysis of Cotton and Harris[273] for the distribution of the free end of a chelate to estimate its effective concentration to be about $30M$. Thus, the proximity effect accounts for a significant portion of the rate enhancement. However, this falls short of the factor of 800 needed. An explanation has not been given, but it is possible that during chelate ring closure the geometry of the complex or its secondary solvation is distorted in a manner favorable for Cl^{-} displacement.

2. Ring-Closure Reactions with Cr(III). Thorneley, Sykes, and Gans[709] have measured the ring-closure reactions of quadridentate $[Cr^{III}(EDTA)(H_2O)_2]^{-}$ to give the quinquidentate $[Cr^{III}(EDTA)(H_2O)]^{-}$ complex. This acetate ring-closure reaction has a rate constant of $330\ sec^{-1}$ at 25°C. Rate constants also are large for the protonated derivatives. The magnitude of the rate constants for substitution at a Cr(III) center is quite remarkable, and a high degree of S_N2 character or an anchimeric effect with neighboring-group participation is suggested. Another example of this type is the reaction $Cr(NH_3)_5OH_2^{3+}$ + oxal^{2-} → $[Cr(NH_3)_4(oxal)]^+$ in which ring closure and the displacement of NH_3 is rapid.[710]

3. Ring-Closure Reactions with Co(III). Reactions of the general type in Eq. (79) have been examined where L is $EDTA^{4-}$, $CyDTA^{4-}$, or 1,2-$PDTA^{4-}$ and

$$[Co^{III}(L)B] \xrightarrow{k_{rc}} [Co^{III}(L)] + B \tag{79}$$

where the ligand undergoes a change from quinquidentate to sexidentate coordination; B is H_2O, OH^-, Cl^-, or Br^-. The reactive species may be generated from redox reaction such as the reaction of $[Co^{II}EDTA]^{2-}$ with Br_2 or with OBr^-.[711] Table 1-33 summarizes some of the values obtained for k_{rc} by Higginson and co-workers[712,713] by Grossman,[714] by Morris and co-workers,[715] and by Busch and co-workers.[716] Higginson and co-workers[717,718] have shown very interesting specific cation catalysis of the ring-closure reactions. In Table 1-33 note the very large increase, by factors of 10^4, in ring-closure rate constants for the loss of Cl^- and of OH^- in going from EDTA to PDTA to CyDTA. The behavior of H_2O and the unusual variations of the $\Delta H^‡$ and $\Delta S^‡$ values (not given) merit additional study.

TABLE 1-33
RING-CLOSURE RATE CONSTANTS (25°C) FOR QUINQUIDENTATE-TO-SEXIDENTATE COORDINATION OF Co(III) AMINOPOLYCARBOXYLATE COMPLEXES

Complex	k_{rc} (sec^{-1})	Ref.
$[Co(EDTA)H_2O]^-$	1.78×10^{-3}	712
$[Co(EDTA)OH]^{2-}$	4.3×10^{-5}	712
$[Co(EDTA)Cl]^{2-}$	2.5×10^{-8}	713
trans-N-$[Co(EDTA)Br]^{2-}$	2.0×10^{-5}	714
trans-O-$[Co(EDTA)Br]^{2-}$	3.0×10^{-6}	714
$[Co(CyDTA)H_2O]^-$	7.16×10^{-4}	714
$[Co(CyDTA)OH]^{2-}$	1.5×10^{-1}	711
$[Co(CyDTA)Cl]^{2-}$	1.43×10^{-4}	714
$[Co(PDTA)H_2O]^-$	4.67×10^{-4}	716
$[Co(PDTA)OH]^{2-}$	1.3×10^{-4}	716
$[Co(PDTA)Cl]^{2-}$	2.24×10^{-6}	716
$[Co(PDTA)Br]^{2-}$	1.16×10^{-5}	716
$[Co(HEDTA)Br]^-$	3.09×10^{-6}, 5.83×10^{-5} (2 isomers present)	715
$[Co(HEDTA)Cl]^-$	3.3×10^{-6}	715

Some general points are that steric interactions appear to be quite important in ring-closure rates, that the proximity effect will cause relatively large rate constants, and that many of these reactions show strong signs that the entering dentate group participates.

4. $Ni(CN)_4^{2-}$ Reactions with Multidentate Ligands. With aminopolycarboxylate ligands the reaction with $Ni(CN)_4^{2-}$ is the reverse process of that given in Eqs. (61)–(64). A mixed $Ni(CN)_3L^{(n+1)-}$ complex forms before the rate-determining step in which an additional CN^- is lost. The behavior with trien is different, however, because the rate-determining step is the reaction of the ligand directly with $Ni(CN)_4^{2-}$ by an associative mechanism.[476] The initial reaction is followed by a series of rapid rearrangements where the polyamine fully coordinates and where all the cyanide ions are released. Ethylenediamine (in large excess or with I_2 present as a scavenger) also reacts directly with $Ni(CN)_4^{2-}$. We have already discussed the reaction of NH_3. The rate constants (M^{-1} sec^{-1} at 25°C) are 310, 425, and 120 for trien, en, and NH_3, respectively.[680] When EDTA rather than I_2 is used as a scavenger to force the dissociation of $Ni(CN)_4^{2-}$, the polyamines are much better catalysts than ammonia. This led to the earlier postulate[476] that chelation of the polyamine ligands contributed in their initial reaction with $Ni(CN)_4^{2-}$. However, the new results with I_2 as a scavenger for released CN^- indicate that this is not the case and that a five-coordinate transition state is sufficient to explain the kinetic results for the reaction of NH_3, en, and trien with this square-planar complex. Nevertheless, in the absence of I_2 ammonia reacts slowly because of the tendency of the released CN^- to reverse the displacement reaction.

5. $Cu(OH)_3^-$ and $Cu(OH)_4^{2-}$ Reactions with Polyamines. The reactions of open-chain and macrocyclic polyamines with Cu(II) in strongly basic solutions have been studied[295,298] to examine the kinetic behavior of the unprotonated ligands and to try to determine the relative number of hydroxide ions which must be displaced from the copper to reach the rate-determining step with various ligands. The main copper species are $Cu(OH)_3^-$ and $Cu(OH)_4^{2-}$.

The results in Table 1-34 show that secondary and tertiary amines are much less reactive than primary amines. The macrocyclic ligands and Et_4dien tend to displace at least one more hydroxide ion from Cu(II) than does 2,3,2-tet. This indicates that the rate-determining step takes place after the first nitrogen is coordinated. A shift to second-nitrogen bond formation seems most likely.

VII. MULTIDENTATE LIGAND EXCHANGE REACTIONS

Nearly all of the factors which have been reviewed in the previous sections come into play in considering the kinetics and mechanisms of multidentate ligand exchange reactions. Characteristic water exchange rate constants, outer-sphere association effects, steric effects, protonated ligands as reactants, metal ion stability constants, coordinated ligands' effects on additional exchange reactions, the partial dissociation of coordinated ligands, the proximity effect, and nucleophilicity all affect the speed and the location of the rate-determining step

TABLE 1-34
RATE CONSTANTS FOR THE REACTIONS OF POLYAMINES WITH $Cu(II)$ IN BASIC SOLUTION: $Cu(OH)_3^-$ (k_3) AND $Cu(OH)_4^{2-}$ (k_4) AT 25°C[a]

Ligand	Rate Constant (M^{-1} sec^{-1})		k_3/k_4
	k_3 ($Cu(OH)_3^-$)	k_4 ($Cu(OH)_4^{2-}$)	
2,3,2-tet	3.4×10^7	4.6×10^6	7.4
Et_2-2,3,2-tet	$>3 \times 10^6$	—	—
Et_4dien	3.8×10^5	$<10^3$	$>3.8 \times 10^2$
Cyclam	2.5×10^6	$\sim 2 \times 10^4$	$\sim 10^2$
Me_2cyclam	5×10^5	1.5×10^4	33
Tet *a*	$\geqslant 4 \times 10^3$	$<10^2$	>40

[a] Refs. 295 and 298.

in multidentate ligand kinetics. We have seen that the rate of removal of some multidentate ligands from metal ions can be exceedingly slow—e.g., some macrocyclic and CyDTA complexes—and yet other complexes of similar stability can react rapidly to transfer completely the metal ion to a new coordination environment. Furthermore, the resulting kinetics can be annoyingly complicated. As one of us wrote in an earlier publication:[22]

> The reaction between copper and nickel-EDTA appeared at first to be a kinetic maze—three apparent changes of reaction order with copper concentration and a pH dependence sometimes present and sometimes not. However, the very complexity of the kinetics imposes restrictions upon the possible explanations of the behavior at each condition. This, in turn, means increased confidence in any mechanism that can fit all the data. All the complexities can be unravelled both qualitatively and quantitatively by the consideration of a simple step-by-step mechanism of unwrapping EDTA from nickel and wrapping it around copper.

This system had simple kinetics compared with some others—e.g., the coordination chain reactions[23]—but the generalization holds. It should now be possible to predict both the rates and mechanisms of several types of multidentate ligand exchange reactions by careful consideration of stepwise reaction paths and the factors previously mentioned. The rate expressions need not be complicated. In fact, very often the reaction conditions and the nature of the exchanging ligands result in simple kinetic behavior even for ligands with many dentate groups. There are enough variables to make such predictions a challenging endeavor, and remember that the reaction path may shift with the concentration level of the reactants. We attempt to indicate in general terms what mechanistic predictions can be made to encourage the investigation of new systems and of specific interactions which are not well understood. Naturally, the better the predictions the less need there is to measure the reaction rates and kinetics of the almost infinite number of possible multidentate ligand exchange reactions. There are a few types of systems where additional rate measurements represent "turning the crank" as opposed to the elucidation of the mechanism or the achievement of a better understanding of the basic chemistry. On the other hand, there are many important aspects of this area of study which need additional study.

A. Exchanging the Ligand between Two Metal Ions

1. EDTA. About a decade ago the relative rate constants for the reaction of several metal ions with $NiEDTA^{2-}$ (and the reverse reaction) led to a proposed general mechanism for metal–EDTA exchange reactions.[22,494,495,498] Dinuclear reaction intermediates in which each metal is coordinated to an iminodiacetate segment of EDTA were proposed. The experimental exchange rate constants correlated very well with the stabilities of metal–iminodiacetate complexes.

The metal–metal exchange rate constants in Table 1-35 permit additional comparisons and serve as a basis for checking some predicted values using the proposed mechanism. Consider the relative rate constants for two sets of reactions:

$$\text{A.}\quad NiEDTA^{2-} + Cu^{2+} \xrightarrow{0.016} CuEDTA^{2-} + Ni^{2+} \tag{80}$$

$$CdEDTA^{2-} + Cu^{2+} \xrightarrow{2.2} CuEDTA^{2-} + Cd^{2+} \tag{81}$$

$$PbEDTA^{2-} + Cu^{2+} \xrightarrow{5.1} CuEDTA^{2-} + Pb^{2+} \tag{82}$$

$$CoEDTA^{2-} + Cu^{2+} \xrightarrow{15.3} CuEDTA^{2-} + Co^{2+} \tag{83}$$

$$ZnEDTA^{2-} + Cu^{2+} \xrightarrow{19} CuEDTA^{2-} + Zn^{2+} \tag{84}$$

$$\text{B.}\quad CoEDTA^{2-} + Zn^{2+} \xrightarrow{0.23} ZnEDTA^{2-} + Co^{2+} \tag{85}$$

$$CoEDTA^{2-} + Pb^{2+} \xrightarrow{0.1} PbEDTA^{2-} + Co^{2+} \tag{86}$$

$$CoEDTA^{2-} + Cu^{2+} \xrightarrow{15.3} CuEDTA^{2-} + Co^{2+} \tag{87}$$

Neither set A nor set B has a direct correlation for all of the rate constants with metal–IDA stabilities. This is expected because the rate-determining step shifts within each set. In Eq. (88) the general mechanism is written with several steps combined to emphasize that the reaction passes through an intermediate (or transition state) of dinuclear IDA segments. It can be seen that the rate-deter-

TABLE 1-35
METAL–METAL EXCHANGE REACTIONS

Complex	Metal	k (M^{-1} sec^{-1})	Ref.
$Mn(EDTA)^{2-}$	Co^{2+}	0.3	403
$Ni(NTA)^{-}$ (0°C)		8×10^{-4}	427
$Co(NTA)^{-}$ (0°C)	Ni^{2+}	1×10^{-3}	427
$Cu(NTA)^{-}$ (0°C)		2×10^{-4}	427
$Zn(EDTA)^{2-}$		6.7×10^{-4}	495
$Pb(CyDTA)^{2-}$		5×10^{-3}	719
$Pb(DTPA)^{2-}$		22	719
$Pb(EDTA)^{2-}$		2.7×10^{-2}	719
$Pb(EDTA{\cdot}OH)^{3-}$		6.7×10^{-2}	719
$Co(EDTA)^{2-}$	Cu^{2+}	15.3	449
$Co(EDTA)H^{-}$		1.5×10^{2}	449
$Ni(EDDA)^{2-}$		7.5×10^{-2}	720
$Ni(DPA)^{-}$		3.0×10^{-3}	721
$Ni(EDTA)^{2-}$		1.6×10^{-2}	494
$Ni(HEEDTA)^{-}$		1.5×10^{-2}	496
$Ni(IDA)$		2.0×10^{-2}	331
$Ni(NTA)^{-}$		1.39×10^{-3}	283
$Ni(BPEDA)$		4.9×10^{-4}	738
$Zn(EDTA)^{2-}$		19	448
$Zn(EDTA)H^{-}$		60	448
$Zn(NTA)^{-}$		5.4	446
$Cd(EDTA)^{2-}$		2.2	722
$Cd(EDTA)H^{-}$		45	722
$Pb(EDTA)^{2-}$		5.1	723
$Pb(PDTA)^{2-}$		8.9×10^{-1}	724
$Ce(EDTA)^{-}$		3.7×10^{-1}	725
$Sm(EDTA)^{-}$		1.2×10^{-1}	725
$Tb(EDTA)^{-}$		1.3×10^{-2}	725
$Er(EDTA)^{-}$		4×10^{-3}	725
Co(*meso*-BDTA)$^{2-}$	Zn^{2+}	$< 7 \times 10^{-4}$	726
$Co(EDTA)^{2-}$		2.3×10^{-1}	727
$Co(EDTA)H^{-}$		5.0×10^{-1}	727
$Co(PDTA)^{2-}$		2×10^{-3}	726
$Co(PDTA)H^{-}$		2×10^{-2}	726
$Co(TRDTA)^{2-}$		2.5	727
$Co(TRDTA)H^{-}$		1×10^{2}	727
$Ni(EDTA)^{2-}$		2.5×10^{-6}	495
$Co(EDTA)^{2-}$	Pb^{2+}	1×10^{-1}	728
$Co(EDTA)H^{-}$		15	728
$Zn(EDTA)^{2-}$		8.9	471

TABLE 1-35 (CONTINUED)
METAL–METAL EXCHANGE REACTIONS

Complex	Metal	k (M^{-1} sec^{-1})	Ref.
$Zn(EDTA)H^-$	Pb^{2+}	1×10^3	471
$Zn(PDTA)^{2-}$		6.6×10^{-1}	724
$Ca(EDTA)^{2-}$	Cd^{2+}	2.3×10^2	729
$Pb(EDTA)^{2-}$	Ga^{3+}	7.5	730
$Pb(EDTA \cdot OH)^{3-}$		1.4	730
$Pb(EDTA)^{2-}$	In^{3+}	50	731
$Pb(EDTA \cdot OH)^{3-}$		3.0×10^2	731
$Y(CyDTA)^-$	Y^{3+}	1.27×10^{-4}	732
$Y(DTPA)^{2-}$		4.0×10^{-3}	732
$Ce(CyDTA)^-$ (20°C)	Ce^{3+}	9.3×10^{-2}	733
$Ce(EDTA)^-$ (5°C)		2.0×10^{-2}	734
$Ce(EEDTA)^-$ (5°C)		5.3×10^{-2}	734
$Ce(EGTA)^-$		3.8	735
$Ce(EGTA)^-$ (5°C)		3.0	734
$Ce(PDTA)^-$ (5°C)		2.7×10^{-2}	734
$Nd(DPTA)^{2-}$	Nd^{3+}	5.8×10^{-3}	736
$Lu(EDTA)^-$	Lu^{3+}	3.7×10^{-4}	737

mining step in the forward direction will depend in part on the ratio $k_2k_3/(k_{-2} + k_3)$.

$$M' + M(EDTA)^{2-} \underset{k_{-1}}{\overset{k_1}{\rightleftharpoons}} M{<}^{O}_{O}{>}N\text{—}N{<}^{O}_{O} + M' \underset{k_{-2}}{\overset{k_2}{\rightleftharpoons}} M{<}^{O}_{O}{>}N\text{—}N{<}^{O}_{O}{>}M'$$

$$\Big\downarrow\!\uparrow \; k_3,\ k_{-3} \qquad (88)$$

$$M + M'(EDTA)^{2-} \underset{k_{-4}}{\overset{k_4}{\rightleftharpoons}} M + {}^{O}_{O}{>}N\text{—}N{<}^{O}_{O}{>}M'$$

Relative M(IDA) dissociation rates:

$$\log\left(\frac{k^{M-H_2O}}{K_{M(IDA)}}\right): \begin{array}{ccccccccccc} Ni & < & Cu & < & Co & < & Zn & < & Pb & < & Cd \\ (-3.7) & & (-1.1) & & (-0.7) & & (+0.2) & & (+2.0) & & (+3.0) \end{array}$$

A simple way to obtain a rough estimate of k_{-2}/k_3 is to compare the ratio of $k^{M'-H_2O}/K_{M'(IDA)}$ with $k^{M-H_2O}/K_{M(IDA)}$. These ratios show that the slowest segment to dissociate is NiIDA, and the fastest is CdIDA. Therefore, the rate constants for the reactions in set A will be equal to k_1k_2/k_{-1} for all the reactions except that of nickel, which will depend on $k_1k_2k_3/k_{-1}k_{-2}$. With this exception the relative rate constants for $k_{Cu}{}^{M(EDTA)}$ should be proportional to $k_1/k_{-1} \simeq K_{M(IDA)}/K_{M(EDTA)}$. The predicted ratios using Anderegg's stability constants[739] are in the correct sequence with rate constants for Cd < Pb < Co < Zn in the relative ratios of 1 < 1.4 < 24 < 55. This compares with the experimental ratio of rate constants of 1 < 2.3 < 7.0 < 8.7. Clearly the model might be refined, but the predictions are not too far astray as long as nickel–EDTA is excluded, which is proper because it has a different rate-determining step. Similarly, for the reactions in set B the rds for Eqs. (85) and (86) is the Co(II) bond dissociation, while for the reaction of $CoEDTA^{2-}$ with Cu^{2+} it is the copper bond formation. Hence, the rate constants for $CoEDTA^{2-}$ with Pb^{2+} and with Zn^{2+} should equal $k_1k_2k_3/k_{-1}k_{-2}$, and the relative rate constants are proportional to $k_2/k_{-2} \simeq K_{M'(IDA)}$. This predicts that Zn^{2+} and Pb^{2+} will react at nearly the same rate with $CoEDTA^{2-}$, which is correct, but this rough prediction gives the edge to the Pb^{2+} reaction by a factor of 1.5 while the experimental ratio actually favors Zn^{2+}. As expected the rate constant with Cu^{2+} does not fit the direct proportionality to its IDA stability constant because it has a different rds.

2. $NiEDTA^{2-}$ and Cu^{2+}. The apparent vagaries of the kinetics of the reaction between nickel–EDTA and copper ion have been mentioned. Depending on the pH and the copper ion concentration, it is possible for this reaction to exhibit three different first-order rate constants (independent of the $[Cu^{2+}]$) and two different second-order rate constants (dependent on $[Cu^{2+}]$). The proposed mechanism[22] given in Figure 1-28 is an abbreviated version of the stepwise process, but it readily explains the observations which qualitatively are as follows:

(*1*) 1st order; low $[Cu^{2+}]$, high $[H^+]$; $k_{2,3}$ rds.
(*2*) 2nd order; low $[Cu^{2+}]$, low $[H^+]$; $k_{5,6}$ rds.
(*3*) 1st order; moderate $[Cu^{2+}]$, low $[H^+]$; $k_{1,2}$ rds.
(*4*) 2nd order; high $[Cu^{2+}]$, low $[H^+]$; $k_{4,5}$ rds.
(*5*) 1st order; very high $[Cu^{2+}]$, low $[H^+]$; $k_{4,5}$ rds (species *4* in Figure 1-28 is now the reactant).

The actual reaction orders overlap a little, but it is clear that there are two distinctly different dinuclear reaction intermediates (species *4* and species *5* in Figure 1-28). An important conclusion is that changes in reaction order should be expected for multidentate ligand exchange reactions, and unless the complete stepwise mechanism is considered, care must be exercised in extrapolating rate data from one concentration level to another.

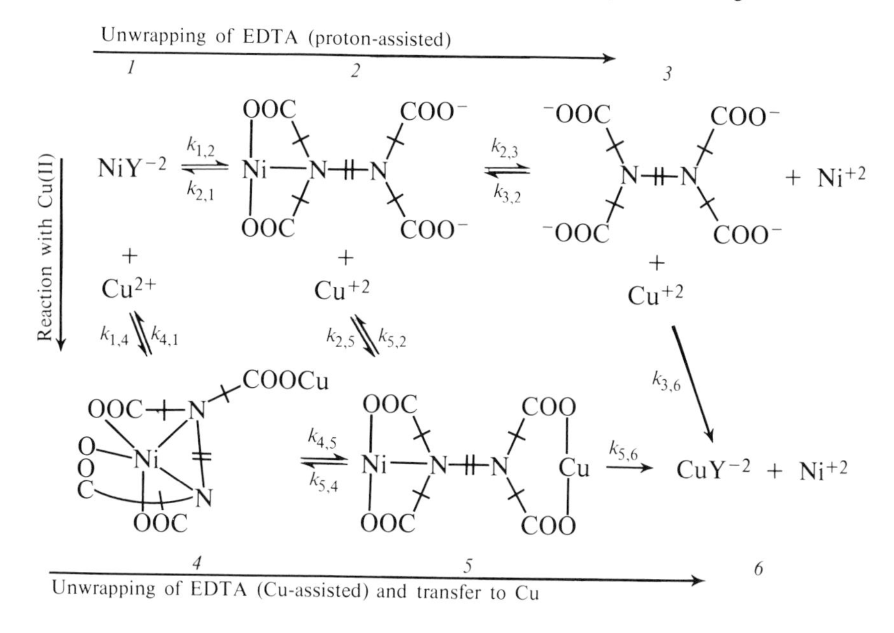

Journal of the American Chemical Society

Figure 1-28 Schematic of the transfer of EDTA(Y^{4-}) between nickel and copper. Species *4* is observed as a reactant at high Cu^{2+} concentrations while species *5* is a reaction intermediate only. H^+ assists the $k_{1,2}$ and $k_{2,3}$ steps.[22]

3. NiEDDA + Cu^{2+}. A recent study by Steinhaus[720] of the copper ion and hydrogen ion dependence in the reaction with NiEDDA shows that it undergoes the same type of reaction order changes as $NiEDTA^{2-}$ except for the last change. Even in 0.2*M* Cu^{2+} the reaction has a second-order dependence (first-order in [NiEDDA] and first-order in [Cu^{2+}]) because the dinuclear Ni(EDDA)Cu^{2+} complex is too weak to form in appreciable concentrations. Once again there are two distinctly different second-order rate constants with values (M^{-1} sec^{-1} at μ = 1.25*M*, 25°C) of 7.5×10^{-2} at low [Cu^{2+}] and 2.3×10^{-3} at higher (above 0.02*M*) [Cu^{2+}] concentrations. The ratio of second-order rate constants for NiEDDA compared with $NiEDTA^{2-}$ is 4.7 at low [Cu^{2+}] and is 0.16 at high [Cu^{2+}]. The rate-constant ratios fit the same type of mechanism proposed for EDTA. Relative stabilities of the bonded segments in the various rate-determining steps were used to give these ratios. Although the $NiEDTA^{2-}$ complex is 10^5 times more stable than the NiEDDA complex, the rate of transfer of the two ligands to Cu^{2+} is of the same order of magnitude as for EDTA and under some conditions can actually be faster.

4. Other Nickel–Aminocarboxylates with Cu^{2+}. Studies by Bydalek and coworkers[283,331] of the kinetics of NiIDA, $NiNTA^-$, and $NiHEEDTA^-$ as well as of the above reactions of $NiEDTA^{2-}$ and NiEDDA show that in every case Cu^{2+} can attack the complex directly and that the rate-determining step is the cleavage of the Ni—N bond of the last aminocarboxylate chelate bonded to nickel. When electrostatic factors and the stabilities of bonded segments relative to the reactants are taken into account, the calculated relative rate constants agree very well with the experimental ones. In contrast Cu^{2+} does not assist the transfer of $CyDTA^{4-}$ from Ni^{2+} and as discussed in section IV.C, it can inhibit the main reaction path which is via H^+ attack.

5. Other EDTA and Aminocarboxylate Exchange Reactions. The kinetics of exchange of EDTA between various metal ions have been a popular topic because of the importance of this ligand in many chemical applications. Some metal ion exchange rate constants are given in Table 1-35, but in many cases additional pathways involving hydrogen ion are known, and the original literature should be checked for these constants. Table 1-35 is not comprehensive for either EDTA or other aminocarboxylates. Additional isotopic and other exchange rate data are available, including studies by Choppin[740,741,742] Betts,[743] Fomin,[744] Kopanica,[745,746,747] and Ryhl and co-workers.[748,749] We selected a few of the resolved rate constants to show the relative behavior of different metal ions. The very recent work of Brücher and Boros[725] resolves the rate constants for a number of lanthanide–EDTA complexes with Cu^{2+}. In accord with our previous discussion the rate constants will be proportional to $K_{M(IDA)}/K_{M(EDTA)}$ if the reactions proceed via the IDA mechanism. Using IDA[750] and EDTA[751] stability constants, the predicted relative rate constants for Ce(III), Sm(III), Tb(III), and Er(III) are 1, 0.20, 0.045, and 0.011, respectively, while the experimental relative rate constants are 1, 0.32, 0.035, and 0.011, respectively. The agreement is almost good enough to discourage additional measurements. These authors[725] also show from proton magnetic resonance studies that the IDA group of $LaEDTA^-$ is labile.

In view of the previous discussion of the CyDTA reactions it may seem surprising to see direct metal ion exchange rate constants for isotopic exchange with cerium.[733] The contribution from this term is small compared with the acid dissociation rate constants, and it is observed at low acidity. Nyssen and Margerum[233] proposed a mechanism to account for this behavior and for the fact that there is an acid-independent (and copper-independent) rate constant, k_d^{MCy}, for the dissociation of the rare earth–CyDTA complexes which is much too large to represent the complete dissociation of the MCy^- complexes. The mechanism is given in Figure 1-29 where the k_d^{MCy} rate step involves a distortion of the initial complex with the metal moving away from the nitrogens. The resulting intermediate can react rapidly with Cu^{2+}, or it can react with another rare earth ion so that both metal ions are outside the coordination cage (bottom structure of Figure 1-29). In the subsequent study of the yttrium kinetics Glentworth and Newton[732] conclude that "the presence of a path involving direct attack of Y(III) on $YCyDTA^-$ does provide some support for the existence of a distorted intermediate which would allow two Y(III) ions to be coordinated to the same ligand." The diminished affinity of these ions for nitrogen donors helps to make such a pathway feasible.

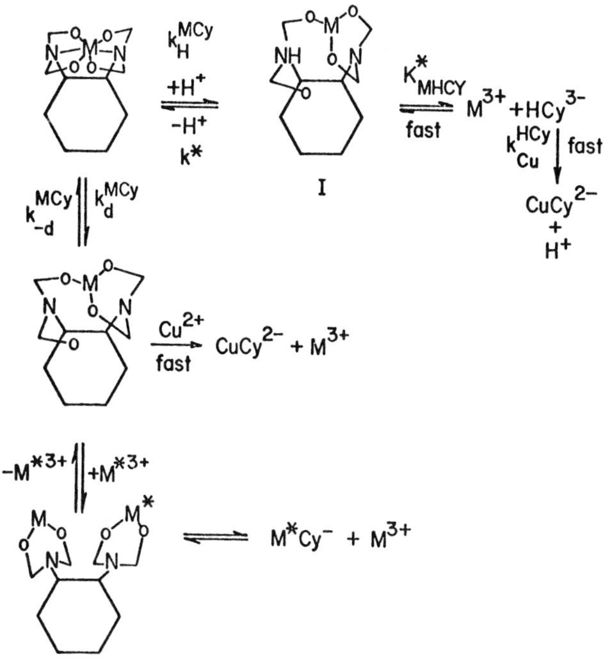

Inorganic Chemistry

Figure 1-29 Proposed mechanism for the formation and dissociation of CyDTA complexes of lanthanide ions. The acid path at the top of the figure predominates for most experimental conditions.[233]

Anions influence the rates of metal exchange reactions with both acceleration and retardation being encountered. Acetate slows the reaction of Cu^{2+} with $NiEDTA^{2-}$,[752] $ZnEDTA^{2-}$,[497,450] and $PbEDTA^{2-}$ [753] and also the reaction of Pb^{2+} with $ZnEDTA^{2-}$.[754] However, faster rates are observed for Co^{2+} and Ni^{2+} with metal EDTA and metal NTA complexes in the presence of acetate.[403,755,427] $CuOH^+$ reacts much more rapidly with $ZnEDTA^{2-}$ and $NiEDTA^{2-}$ than the aquo ion; however, CuN_3^+ reacts at about the same rate as $CuOAc^+$.[497] Hydroxide ion appears to have a special effect. Perhaps it is more effective in reducing the electrostatic repulsion between the two metal ions in the dinuclear intermediate.[497]

The isotopic exchange reactions of In^{3+} and $InEDTA^-$,[756] Ho^{3+} and $HoEDTA^-$,[757] the reaction of Zn^{2+} and Co(*meso*-BDTA)$^{2-}$,[726] the Fe^{3+}, Ni^{2+}, and Co^{2+} exchange with $AlEDTA^-$,[758] and all the metal exchange reactions of CyDTA complexes[281,278,233,499] proceed via the dissociation pathway. Most of these reactions were studied in the presence of high hydrogen ion concentration which favors the dissociative process; however, steric interactions control the reactions with Co(*meso*-BDTA)$^{2-}$ and all the CyDTA reactions. The $AlEDTA^-$

complex is believed to be a dimer,[758,759] which may account for the kinetic behavior. In strong base the In^{3+}–$InEDTA^-$ exchange reaction exhibits a third-order hydroxide dependence which cannot be explained by this mechanism.[757]

For the isotopic exchange reactions of the lanthanide–EDTA complexes, the low acidities at which the work has been performed have usually prevented the observation of a free metal-dependent pathway. However, for the isotopic exchange of Lu^{3+} with $LuEDTA^-$ a metal ion dependence has been observed in the pH range of 4–6[737] while other workers found no evidence for such a pathway at pH 2.5.[760] At first it might appear that Cu^{2+} is much more effective than Lu^{3+} in exchanging with $LnEDTA^-$, but Brücher[760] has shown that it is not easy to detect direct Cu^{2+} attack below pH 4. Since this is the upper pH range for which most of the $LnEDTA^-$ isotope exchange reactions have been studied, it might be expected that above pH 4 the other lanthanide complexes will exhibit direct lanthanide attack.[734,698]

6. Nickel–Polyamines and Copper. Table 1-36 gives the rate constants for the reaction of Cu^{2+} with a series of nickel–polyamines.[475] Whereas conventional wisdom might indicate that the larger the dentate number of the polyamine the slower the reaction, we find just the opposite. In this instance the greater the degree of polyamine coordination to nickel ion the faster the polyamine transfers to copper. Why? The suggested reaction intermediate is shown in Figure 1-30 for dien, and it is similar for the trien or tetren. When one nitrogen of the polyamine complex bonds to copper, the cleavage of the adjacent nickel–nitrogen bond is the rate-determining step. This is because the bonding to copper is rapid and is much stronger so that subsequent reaction intermediates are energetically more favorable. Hence, the largest energy barrier occurs early in the dissociation reaction of the nickel–polyamine. $Nitrien^{2+}$ reacts faster than $Nidien^{2+}$ because

TABLE 1-36
EXCHANGE RATE CONSTANTS FOR THE REACTION OF Cu(II) WITH NICKEL–POLYAMINE COMPLEXES AT 25°C, $\mu = 0.1$[a]

Complex	Log K_1 (Ref.)	k_H[b] (M^{-1} sec^{-1}) (Complete Dissoc.)	k_d^{NiHP}[c] (sec^{-1}) (2nd-Bond Dissoc.)	k_{Cu} (M^{-1} sec^{-1})
$Nien^{2+}$	7.8 (832)	6.8×10^4	rapid	no direct attack observed
$Nidien^{2+}$	10.7 (833)	3×10^2	2.8	1.8
$Nitrien^{2+}$	13.9 (834)	9.1×10^{-1}	4.0	2.7
$Nitetren^{2+}$	17.4 (835)	2×10^{-4}	—	8.0

[a] Ref. 475.
[b] Data from k_{Ni}^{Hen}, Ref. 384.
[c] Data from Table 1–20.

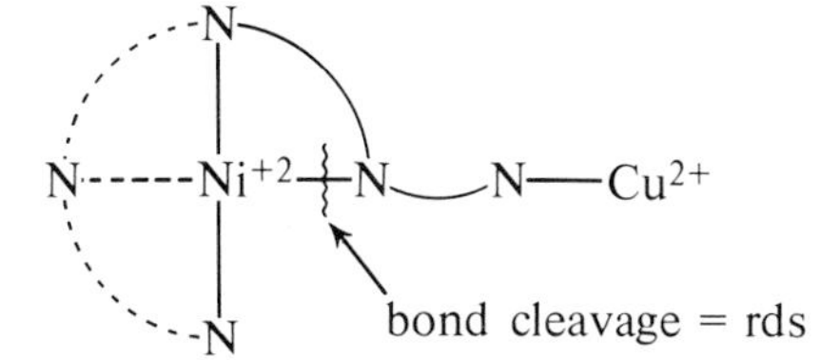

Figure 1-30 Proposed reaction intermediate in the reaction between Cu^{2+} and $Nidien^{2+}$. The corresponding intermediates for $Nitrien^{2+}$ and $Nitetren^{2+}$ are shown in italics.

its second Ni—N bond dissociation rate is faster, and the same should be true for $Nitetren^{2+}$ (*see* Table 1-20). Obviously the predicted rate constants for the transfer of the polyamines from nickel to zinc would be very much different, and dien would be much faster than trien, etc.

7. Metal Exchange Reactions of Porphyrins and Macrocycles. The stepwise transfer of macrocyclic or porphyrin ligands between metal ions would be expected to be much more difficult. Kinetic studies of these reactions indicate that they are slow.[761] The reactions of Zn(II) with Cd(II), Pb(II), and Hg(II) complexes of tetraphenylporphyrin (CdTPP, PbTPP, and HgTPP) have been studied in pyridine solution.[762] The CdTPP and HgTPP reactions are first order in [MTPP] and [Zn(II)]; however, the Zn(II) reaction with PbTPP is first order in [PbTPP] and second order in [Zn]. The following mechanism has been proposed to account for the kinetic data.[762]

$$M_1 + M'P \underset{k_{2,1}}{\overset{k_{1,2}}{\rightleftarrows}} M_1—P—M'$$

$$M_1—P—M' \underset{k_{3,2}}{\overset{k_{2,3}}{\rightleftarrows}} M_1P^* + M'$$

$$M_1P^* \underset{k_{4,3}}{\overset{k_{3,4}}{\rightleftarrows}} M_1P^{**}$$

$$M_1P^{**} + M_2 \underset{k_{5,4}}{\overset{k_{4,5}}{\rightleftarrows}} M_2P + M_1 \qquad (89)$$

A rate-limiting reaction in the third step for the CdTTP and HgTTP complexes and in the fourth step for PbTPP would account for the different Zn(II) dependence. The nature of the two complexes M_1P^* and M_2P^{**} is not clear, and the effect of pyridine is likely to prove important.[762,763]

In aqueous solution the metal exchange reactions of Cu(II) for Zn(II) in the Zn(II) complexes of TMTPyP [tetrakis(*N*-methyltetrapyridyl)porphine][764] and TENP (a tetraethylenediamine-substituted protoporphyrin IX)[451,452,765] have been studied. The reaction of Cu^{2+} with Zn(TMTPyP) is first order with respect to both Cu^{2+} and Zn(TMTPyP), pH independent from 3.0 to 4.4, and shows no suppression by added Zn^{2+}. A dinuclear [Zn(TMTPyP)Cu] reaction intermediate (activated complex) has been proposed to account for the observed kinetics.[764] The rate of the Cu^{2+} reaction with Zn(TENP) on the other hand decreases with increasing pH (from 4.6 to 5.1), tends toward a limiting value at high Cu^{2+} concentrations, and shows a slight suppression by added Zn^{2+}. The mechanism suggested for this reaction involves the dissociation of the Zn(TENP) complex, followed by Cu^{2+} incorporation into the porphyrin.[451] The above two systems display some similarities to the metal–metal aminocarboxylate exchange reactions (section VII.A and IV.C).

Exchange of the Metal Ion between Two Ligands

The multidentate ligand displacement of another multidentate ligand from a metal ion can, under favorable circumstances, be a very rapid process (Table 1-37). It is fascinating to find that under conditions where the solvent dissociation rate of a multidentate ligand complex is extraordinarily slow, another multidentate ligand can completely displace the initial ligand, forcing a whole series of bond cleavages in rapid sequence. The first example of such a system to be studied in detail was the exchange of Ni(II) between trien and EDTA.[282,326,498,802] The attack of $EDTA^{4-}$ on $Nitrien^{2+}$ is so rapid by comparison with the solvent dissociation that $k_{EDTA}{}^{Nitrien}/k_{d}{}^{Nitrien}$ approaches 10^{13}[trien]. The ligand attack is three orders of magnitude faster than the attack of Cu^{2+} on $Nitrien^{2+}$. Similarly, the attack of tetren on $CuEDTA^{2-}$ [25] is seven orders of magnitude faster than the attack of Ni^{2+} on $CuEDTA^{2-}$. It is one of the characteristics of multidentate ligand exchange reactions that ligand–ligand exchange rates are often much faster than metal–metal exchange rates.

1. Requirements and Possible Predictions. What are the requirements for rapid ligand–ligand exchange reactions, and to what extent can these rate processes be predicted? First, the attacking ligand needs an initial coordination site on the complex. If it has to displace a chelated dentate group before it can react, then its rate will be reduced by the degree to which that chelation stabilized the initial complex or less commonly, the rate will be limited by the speed of the chelate ring-opening step. On the other hand, if the attacking ligand gains a foothold by solvent displacement, there is an accompanying gain in accord with the stability of the mixed complex which offsets subsequent loss of stability resulting from the necessary ring-opening processes. Once the attacking ligand has a foothold, the proximity effect permits it to be chelated as rapidly as the initial ligand breaks dentate groups unless steric hindrance prevents this. Steric factors are very important in preventing rapid exchange reactions, but they can also assist some dissociation processes in the mixed complexes. These effects are not yet fully characterized, and this introduces some uncertainty into predicting reaction rates. The attacking ligand must become chelated to reduce the rate of its own dissociation from the mixed complex. In typical reactions the rate-determining step can occur while chelating the attacking ligand or any time between this step and breaking of the last chelate ring of the leaving ligand. This may still leave a choice of several steps to consider in a series of structures for the mixed complex. Choosing the appropriate step as the rds depends on the product of the relative stability of the mixed complex compared with the reactants and on the estimated rate constant for the next chelate ring closure or chelate ring opening. If these products are close together for a number of sequential steps, then the choice is not critical in predicting the rate constant. Conversely, if a *necessary* step gives a weak mixed complex and a small rate constant for the subsequent reaction, then the rds can be specifically assigned. (Naturally, if such a step is not necessary, this merely represents a reaction path which will not be taken. In cases such as this, steric and electrostatic factors are sometimes difficult to judge.)

The overall mechanism is one of nucleophilic attack by the multidentate ligand on the complex. However, this does not specify the intimate mechanism of the individual steps which, as implied in the preceding discussion, can occur by a series of dissociative or associative steps. In the metal–peptide reactions there is good evidence for associative reactions in multidentate ligand attack.[343]

The rate constants for ligand–ligand exchange will frequently be pH dependent, and a rate constant–pH profile is often needed to understand the mechanism or make use of the kinetics. It is helpful to resolve these profiles into a sum of rate constants for the various protonated reactants—i.e., $EDTA^{4-}$, $HEDTA^{3-}$, and H_2EDTA^{2-} reacting with $Nitrien^{2+}$—but this is only a convenient way of keeping track of the degree of protonation of the mixed complex and does not necessarily specify the location of protons in the transition state. Proton transfer is more rapid than most of the coordination rate steps, and a proton initially on H_2EDTA^{2-} can be transferred to the trien molecule before the rate-determining step. The resolved rate constant, $k_{H_2EDTA}{}^{Nitrien}$, refers to a transition state consisting of $[NiH_2(trien)(EDTA)]^0$, and in this particular case it probably is $[Ni(Htrien)(HEDTA)]^0$.[282] It also is possible for a proton to be shifted to a less basic but kinetically more favorable location in the EDTA ligand.

Another variable besides steric effects which hinders quantitative predictions of ligand–ligand exchange reactions is our incomplete knowledge of the effects of coordinated ligands on subsequent bond formation and dissociation rates. The data available in section V are helpful, but the amount of information is limited for cases where there is a high degree of ligand coordination, which is the case for these reactions.

Flexible tetradentate and pentadentate ligands with sterically open coordinating groups—i.e., primary amines and carboxylate groups—are very effective for rapid multidentate ligand exchange. So is EDTA because the sixth group is not strongly bound in many transition metal complexes. Reactions with favorable ΔG^0 changes tend to be faster because the rds is closer to the reactants. Favorable electrostatic interaction can be quite important too. At pH 5–7 $CuNTA^-$ actually reacts faster with tetren than does $Cu_{aquo}{}^{2+}$ despite the need to replace the strongly coordinated NTA group during the reaction. In this pH range the main species present for the polyamine is $H_3tetren^{3+}$, and the most kinetically reactive species is $H_2tetren^{2+}$.[325]

TABLE 1-37
MULTIDENTATE LIGAND EXCHANGE REACTIONS AT 25°C

Complex	Ligand	k (M^{-1} sec^{-1})	Ref.
$Ca(CyDTA)^{2-}$	$CyDTA^{4-}$	$<3 \times 10^{-5}$	334
$Ca(EDTA)^{2-}$	$EDTA^{4-}$ (30°C)	6.0×10^{1}	364
	$EDTA^{4-}$	7.4×10^{1}	334
		1.2×10^{2}	769
$Ca(EDTA)H^{-}$	$EDTA^{4-}$	2.5×10^{6}	364
	$H(EDTA)^{3-}$	3.0×10^{5}	364
	$H_2(EDTA)^{2-}$	$<3.0 \times 10^{5}$	364
$Ca(PDTA)^{2-}$	$PDTA^{4-}$	4.1×10^{1}	334
$Mn(CyDTA)H_2O^{-}$	$H(CyDTA)^{3-}$ (10.2°C)	22.8	766
	$H_2(CyDTA)^{2-}$ (10.2°C)	~0	766
$Mn(CyDTA)OH^{2-}$	$H(CyDTA)^{3-}$ (10.2°C)	0.63	766
	$H_2(CyDTA)^{2-}$ (10.2°C)	649	766
$Mn(EDTA)H_2O^{-}$	$H(EDTA)^{3-}$	65.2	766
	$H_2(EDTA)^{2-}$	0.564	766
$Mn(EDTA)OH^{2-}$	$H(EDTA)^{3-}$	2.63	766
	$H_2(EDTA)^{2-}$	9.03	766
$Co(CyDTA)^{2-}$	BT^{3-}	0.926	767
	$H(BT)^{2-}$	1.7×10^{-3}	767
$Co(dien)^{2+}$	$CyDTA^{4-}$ (0°C)	4.3×10^{4}	768
	$H(CyDTA)^{3-}$ (0°C)	4.53×10^{3}	768
$Co(DGITA)^{2-}$	BT^{3-}	2.28×10^{4}	769
	$H(BT)^{2-}$	7.37×10^{1}	769
	$DTPA^{5-}$	1.0×10^{4}	770
	$H(DPTA)^{4-}$	2.0×10^{3}	770
$Co(DTPA)^{3-}$	BT^{3-}	5.25×10^{-1}	767
	$H(BT)^{2-}$	1.1×10^{-2}	767
$Co(EDTA)^{2-}$	BT^{3-}	0.23	771
	$H(BT)^{2-}$	6.76×10^{2}	771
	CAL^{3-}	6.32×10^{2}	772
	$H(CAL)^{2-}$	7.01	772
$Co(HEEDTA)^{1-}$	$DTPA^{5-}$	21	770
	$H(DTPA)^{4-}$	3.8	770
	BT^{3-}	3.24×10^{4}	769
	$H(BT)^{2-}$	2.2×10^{1}	769
	CAL^{3-}	1.79×10^{4}	772
	$H(CAL)^{2-}$	9.65×10^{1}	772

TABLE 1-37 (CONTINUED)
MULTIDENTATE LIGAND EXCHANGE REACTIONS AT 25°C

Complex	Ligand	k (M^{-1} sec^{-1})	Ref.
$Co(EGTA)^{2-}$	H(PAR)	3.2	773
$Co(EGTA)H^{-}$	H(PAR)	2.1×10^4	773.
Co(GEDTA)	$DTPA^{4-}$	1.0×10^4	770
	$H(DTPA)^{3-}$	2.0×10^3	770
$Co(HIDA)^{+}$	$CyDTA^{4-}$	8.2×10^4	768
	$H(CyDTA)^{3-}$	2.3×10^2	768
Co(IDA)	$H(CyDTA)^{3-}$	3.17×10^3	768
$Co(N\text{-}Eten)^{2+}$	$CyDTA^{4-}$ (0°C)	8.7×10^5	768
	$H(CyDTA)^{3-}$ (0°C)	4.49×10^4	768
$Co(NTA)^{-}$	CAL^{3-}	6.5×10^5	774
	$H(CAL)^{2-}$	2.6×10^5	774
	$CyDTA^{4-}$	6.5×10^1	775
	$CyDTA^{4-}$	8.5×10^1	768
	$H(CyDTA)^{3-}$	0.34	775
	$H(CyDTA)^{3-}$	0.58	768
$Co(TDTA)^{2-}$	$H_2(PDTA)^{2-}$	0.89	799
Ni(Asp)	CAL^{3-}	3.5×10^6	774
	$H(CAL)^{2-}$	4.05×10^5	774
	$H(CyDTA)^{3-}$	2.5×10^3	776
	$DTPA^{5-}$	3.0×10^4	776
	$H(DTPA)^{4-}$	6.8×10^3	776
$Ni(CyDTA)^{2-}$	BT^{3-}	8.9×10^{-2}	777
	$H(BT)^{2-}$	4.25×10^{-3}	777
$Ni(DGITA)^{2-}$	BT^{3-}	3.30	769
	$H(BT)^{2-}$	0.16	769
	$DTPA^{5-}$	0.7	778
	$H(DTPA)^{4-}$	0.566	778
$Ni(dien)^{2+}$	CAL^{3-}	9.8×10^7	774
	$H(CAL)^{2-}$	2.5×10^5	774
	$CyDTA^{4-}$	3.8×10^4	779
	$H(CyDTA)^{3-}$	2.97×10^1	779
	$DTPA^{5-}$	7.2×10^5	779
	$H(DTPA)^{4-}$	9.80×10^4	779
	$H(EDTA)^{3-}$	6.3×10^4	296
$Ni(DTPA)^{3-}$	BT^{3-}	1.48×10^{-2}	767
	$H(BT)^{2-}$	2.9×10^{-4}	767
Ni(EDDA)	$EDTA^{4-}$	7.0	780
	$H(EDTA)^{3-}$	3.72	780

TABLE 1-37 (CONTINUED)
MULTIDENTATE LIGAND EXCHANGE REACTIONS AT 25°C

Complex	Ligand	k (M^{-1} sec^{-1})	Ref.
Ni(EDDA)[a]	H(PAR)	1×10^3	667
$Ni(EDMA)^+$	$DTPA^{5-}$	1.36×10^3	779
	$H(DTPA)^{4-}$	2.58×10^2	779
	$EDTA^{4-}$	1.5×10^3	781
	$H(EDTA)^{3-}$	3.9×10^2	781
	$HEEDTA^{3-}$	2.8×10^2	781
	$H(HEEDTA)^{2-}$	—	781
$Ni(EDTA)^{2-}$	BT^{3-}	1.25	782
	$H(BT)^{2-}$	—	782
	CAL^{3-}	1.82	772
	$H(CAL)^{2-}$	4.2×10^{-2}	772
	$EDTA^{4-}$	1.95×10^{-3}	783
$Ni(HEEDTA)^{1-}$	$DTPA^{5-}$	2.3×10^{-3}	770
	$H(DTPA)^{4-}$	3.76×10^{-3}	770
$Ni(EDTP)^{2-}$	$CyDTA^{4-}$	7.2×10^2	784
	$H(CyDTA)^{3-}$	9.45×10^{-1}	784
$Ni(en)^{2+}$	$CyDTA^{4-}$	7.2×10^4	779
	$H(CyDTA)^{3-}$	1.23×10^4	779
Ni(glut)	CAL^{3-}	4.9×10^7	774
	$H(CAL)^{2-}$	1.58×10^6	774
$Ni(HEEDTA)^-$	BT^{3-}	6.93	769
	$H(BT)^{2-}$	0.08	769
	CAL^{3-}	18.4	772
	$H(CAL)^{2-}$	0.43	772
Ni(HEIDA)	$CyDTA^{4-}$	3.7×10^3	785
	$H(CyDTA)^{3-}$	0.94	785
	$EDTA^{4-}$	1.7×10^4	785
	$H(EDTA)^{3-}$	7.35×10^3	785
Ni(HIDA)	$DTPA^{5-}$	1.1×10^5	770
	$H(DTPA)^{4-}$	1.5×10^4	770
Ni(IDA)	CAL^{3-}	4.8×10^5	774
	$H(CAL)^{2-}$	1.52×10^4	774
	$CyDTA^{4-}$	1.7×10^4	776
	$H(CyDTA)^{3-}$	3.3×10^2	776
	$DTPA^{5-}$	3.8×10^4	776
	$H(DTPA)^{4-}$	6.13×10^3	776
	$EDTA^{4-}$	2.1×10^4	775
	$H(EDTA)^{3-}$	7.33×10^3	775
$Ni(NDAIP)^-$	$H(EDTA)^{3-}$	1.73	786
	$H_2(EDTA)^{2-}$	0.284	786

TABLE 1-37 (CONTINUED)
MULTIDENTATE LIGAND EXCHANGE REACTIONS AT 25°C

Complex	Ligand	k (M^{-1} sec^{-1})	Ref.
$Ni(NDAP)^-$	$H(EDTA)^{3-}$	3.27	786
	$H_2(EDTA)^{2-}$	0.389	786
$Ni(NTA)^-$	BT^{3-}	3.27×10^4	782
	$H(BT)^{2-}$	5.8×10^1	782
	CAL^{3-}	2.5×10^5	774
	$H(CAL)^{2-}$	1.67×10^3	774
	$DTPA^{5-}$	9.1×10^2	787
	$H(DTPA)^{4-}$	3.06×10^2	787
	$H_2(DTPA)^{3-}$	1.78	787
	$EDTA^{4-}$	1.5×10^3	770
	$H(EDTA)^{3-}$	4.14	777
	$H(EDTA)^{3-}$	2.33×10^2	770
	$H_2(EDTA)^{2-}$	0.443	777
	$H(HEEDTA)^{2-}$	1.96	777
	$H_2(HEEDTA)^-$	0.427	777
	NTA^{3-}	3×10^{-5}	776
	H(PAR)[a]	4.2×10^2	667
$Ni(penten)^{2+}$	$TTHA^{6-}$	1.6×10^3	836
$Ni(tetren)^{2+}$	$EDTA^{4-}$	1.5	282
	$H(EDTA)^{3-}$	1.1	282
$Ni(tetren)H^{3+}$	$H(EDTA)^{3-}$	5.7×10^2	282
$Ni(tetren)H_2^{4+}$	$H(EDTA)^{3-}$	1.7×10^5	282
$Ni(trien)^{2+}$	$EDTA^{4-}$	$\sim 10^4$	282
	$H(EDTA)^{3-}$	5.0×10^2	282
$Ni(trien)H^{3+}$	$H(EDTA)^{3-}$	1.6×10^5	282
$Ni(TTHA)^+$	Dien	1.5×10^6	831
	H(dien)	5.46×10^5	831
	$EDMA^-$	4.3×10^3	831
	H(EDMA)	2.86×10^3	831
	$H_2(EDMA)^+$	4.36	831
	NTA^{3-}	2.8×10^3	831
	$H(NTA)^{2-}$	9.65×10^2	831
	$H_2(NTA)^{1-}$	1.06	831
$Cu(CyDTA)^{2-}$	En	0.12	788
	Dien	15	788
	$H(dien)^+$	3.4	788
	Tetren	5.0	788
	$H(tetren)^+$	4.7	788
	$H_2(tetren)^{2+}$	0.2	788
	$H_3(tetren)^{3+}$	0.007	788
	Trien	5.4	788

TABLE 1-37 (CONTINUED)
MULTIDENTATE LIGAND EXCHANGE REACTIONS AT 25°C

Complex	Ligand	k (M^{-1} sec^{-1})	Ref.
$Cu(CyDTA)^{2-}$	$H(trien)^{+}$	2.5	788
	$H_2(trien)^{2+}$	0.04	788
$Cu(dien)^{2+}$	Tetren	1.02×10^{8}	801
$Cu(EDTA)^{2-}$	$EDTA^{4-}$	0.174	789
	$H(EDTA)^{3-}$	1.5×10^{-2}	789
	MNT^{2-}	7.6×10^{3}	790
	Penten	1.7×10^{4}	788
	$H(penten)^{+}$	1.6×10^{4}	788
	$H_2(penten)^{2+}$	3.0×10^{3}	788
	$H_3(penten)^{3+}$	1.5×10^{3}	788
	Tetren	2.2×10^{5}	788
	$H(tetren)^{+}$	3.7×10^{5}	788
	$H_2(tetren)^{2+}$	6.7×10^{3}	788
	$H_3(tetren)^{3+}$	3.4×10^{1}	788
	Trien	4.3×10^{5}	788
		9.1×10^{5}	791
	$H(trien)^{+}$	3.5×10^{4}	788
		1.8×10^{4}	791
	$H_2(trien)^{2+}$	2.1×10^{4}	788
		3.6×10^{2}	791
	$H_3(trien)^{3+}$	44	791
Cu(EDTA)H	$EDTA^{4-}$	3.9×10^{2}	789
$Cu(EGTA)^{2-}$	PAR^{-}	3×10^{2}	792
	H(PAR)	1.8	792
$Cu(NTA)^{-}$	MNT^{2-}	1.6×10^{3}	790
	$H_2(tetren)^{2+}$	2.1×10^{8}	435
	$H_3(tetren)^{3+}$	6.1×10^{4}	435
$Cu(PAR)^{+}$	$EGTA^{4-}$	8.8×10^{2}	793
$Cu(tren)^{2+}$	Dien	1.06×10^{5}	794
	Tetren	5.41×10^{4}	794
	Trien	8.52×10^{4}	794
Cu(trien)	Tetren	7.47×10^{4}	801
$Cu(2M8Q)_2$	$H(CyDTA)^{3-}$	0.13	800
	$H_2(CyDTA)^{2-}$	0.26	800
Zn(EDDA)	$CyDTA^{4-}$	1.9×10^{1}	780
	$H(CyDTA)^{3-}$	0.60	780
$Zn(EDMA)^{+}$	$CyDTA^{4-}$	5.6×10^{2}	780
	$H(CyDTA)^{3-}$	3.2×10^{1}	780

TABLE 1-37 (CONTINUED)
MULTIDENTATE LIGAND EXCHANGE REACTIONS AT 25°C

Complex	Ligand	k (M^{-1} sec^{-1})	Ref.
$Zn(HIDA)^{+}$	$CyDTA^{4-}$	2.7×10^{4}	780
	$CyDTA^{4-}$ (0°C)	2.65×10^{3}	768
	$H(CyDTA)^{3-}$	2.6×10^{2}	780
	$H(CyDTA)^{3-}$ (0°C)	2.08×10^{1}	768
$Zn(IDA)$	$CyDTA^{4-}$ (0°C)	2.7×10^{4}	768
	$H(CyDTA)^{3-}$ (0°C)	5.90×10^{2}	768
$Zn(NTA)^{-}$	$CyDTA^{4-}$ (0°C)	7.5	768
	$H(CyDTA)^{3-}$ (0°C)	5.4×10^{-2}	768
	NTA^{3-}	2.0×10^{6}	337
	$H(NTA)^{2-}$	5.3×10^{2}	337
$Zn(PAR)^{+}$	$EGTA^{4-}$	2.5×10^{4}	795
$Zn(1,3\text{-}PDTA)^{2-}$	$H(1,3\text{-}PDTA)^{3-}$	6.5	459
	$H_2(1,3\text{-}PDTA)^{3-}$	25	459
$Sr(EDTA)^{2-}$	$EDTA^{4-}$	1.1×10^{3}	359
$Cd(EDDDA)^{2-}$	$EDDDA^{4-}$	1.5×10^{3}	459
$Cd(EDTA)^{2-}$	$EDTA^{4-}$ (28°C)	1.30×10^{2}	796
$Cd(NTA)^{-}$	NTA^{3-}	1.8×10^{7}	337
	$H(NTA)^{2-}$	2.9×10^{2}	337
$Cd(1,3\text{-}PDTA)^{2-}$	$H(1,3\text{-}PDTA)^{3-}$	3.5×10^{2}	459
		2.5×10^{2}	
	$H_2(1,3\text{-}PDTA)^{2-}$	26	459
		32	459
$Pb(l\text{-}CyDTA)^{2-}$	$d\text{-}CyDTA^{4-}$	3.77×10^{-2}	797
$Pb(EDDDA)^{2-}$	$EDDDA^{4-}$	2.0×10^{4}	459
$Pb(EDTA)^{2-}$	$d\text{-}(-)\text{-}CyDTA^{4-}$	0.408	798
	$H(d\text{-}(-)\text{-}CyDTA)^{3-}$	5.0×10^{-3}	798
$Pb(NTA)^{-}$	NTA^{3-}	6.6×10^{7}	339
	$H(NTA)^{2-}$	3.1×10^{3}	339
$Pb(l\text{-}PDTA)^{2-}$ $Pb(d\text{-}PDTA)^{2}$	$d\text{-}CyDTA^{4-}$	1.38	797
$Pb(l\text{-}PDTA)^{2-}$ $Pb(d\text{-}PDTA)^{2}$	$l\text{-}CyDTA^{4-}$	2.86	797
$Pb(l\text{-}PDTA)^{2-}$	$d\text{-}PDTA^{4-}$	1.58	797
$Pb(l,3\text{-}PDTA)^{2-}$	$1,3\text{-}PDTA^{4-}$	2.0×10^{4}	459
	$H(1,3\text{-}PDTA)^{3-}$	1.5×10^{3}	459
	$H_2(1,3\text{-}PDTA)^{2-}$	1.7×10^{2}	459

[a] Forms mixed complex.

2. Nickel–Polyamines and EDTA. The log k–pH profile for the reactions of EDTA with $Nitrien^{2+}$ and $Nitetren^{2+}$ is given in Figure 1-31.[282] The structure of the reaction intermediate preceding the rate-determining step can be deduced in the following manner. The exchange reactions are reversible with the forward and reverse reactions involving the same type of bond formation and breakage.

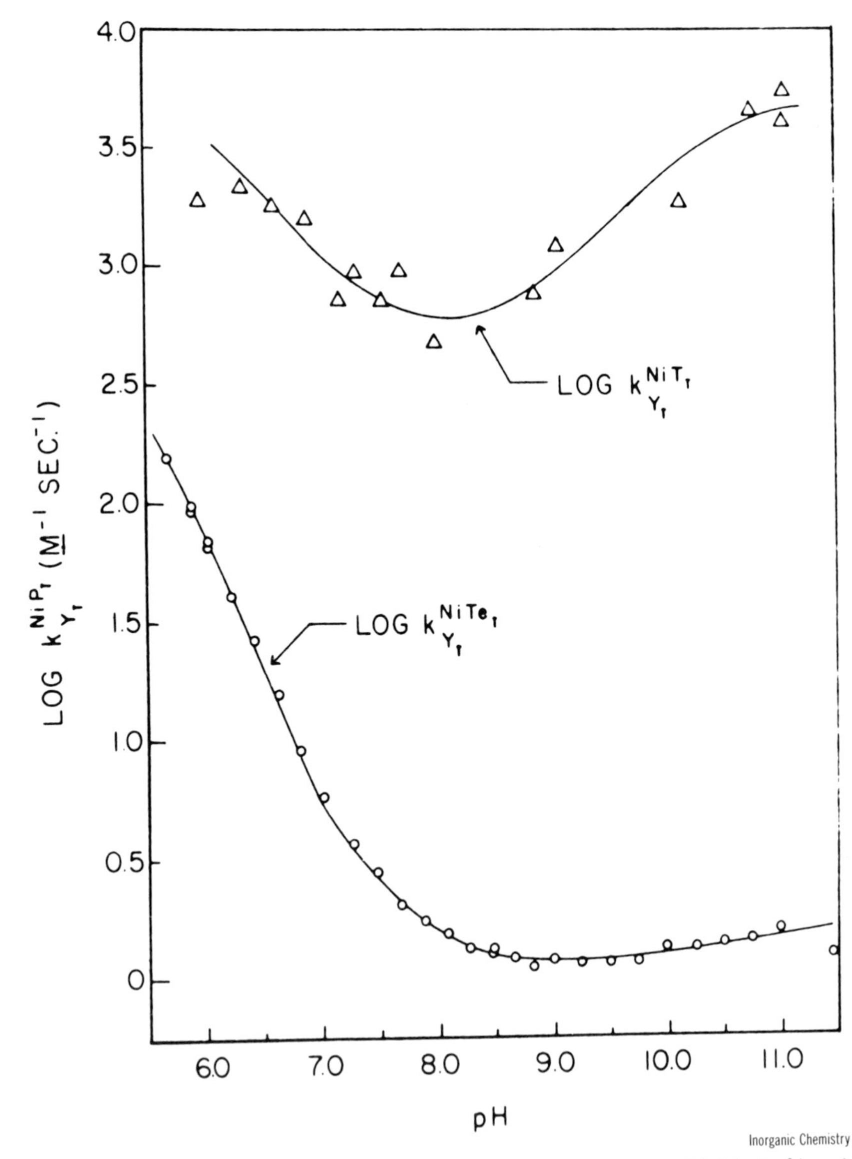

Figure 1-31 Rate constant–pH profiles for the reactions of EDTA with $Nitrien^{2+}$ and $Nitetren^{2+}$ [282]

The concentration of these intermediates prior to the rate-determining step in most cases is small and in equilibrium with the reactants. For these conditions the experimental rate constant can be equated to the relative stability of the intermediate preceding the rate-determining step:

$$k_{\text{expt}} = \frac{K_{(n\text{th intermediate})}}{K_{\text{reactants}}} \times k_n \tag{90}$$

where k_n is the rate constant for nickel–nitrogen or nickel–solvent bond rupture corrected for the coordinated groups and where

$$K_{(n\text{th intermediate})} = \frac{K_{(\text{polyamine segment})} \cdot K_{(\text{EDTA segment})}}{K_{(\text{electrostatic})}} \tag{91}$$

The ratio $K_{(n\text{th intermediate})}/K_{(\text{reactants})}$ represents the relative concentration of the nth intermediate compared with the concentration of reactants. The value of $K_{(\text{electrostatic})}$ takes into account the possibility of electrostatic attraction favoring one structure over another. Structural models show that it is difficult to have more than four of the six nitrogens in the Ni(trien)(EDTA)$^{2-}$ mixed complex coordinated at one time. The preceding treatment supports this so that a structural flow diagram of the type given in Figure 1-32 is suggested where the k_3 step is the most likely to determine the reaction rate. The Nitetren^{2+} re-

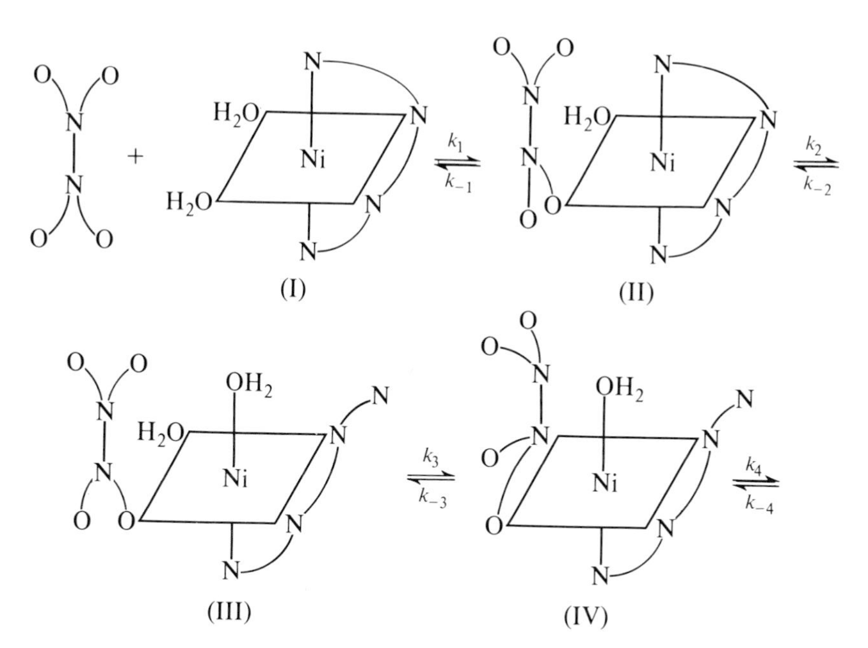

Figure 1-32 Proposed mechanism for EDTA exchange with Nitrien^{2+} where k_3 is the most probable rate-determining step[282]

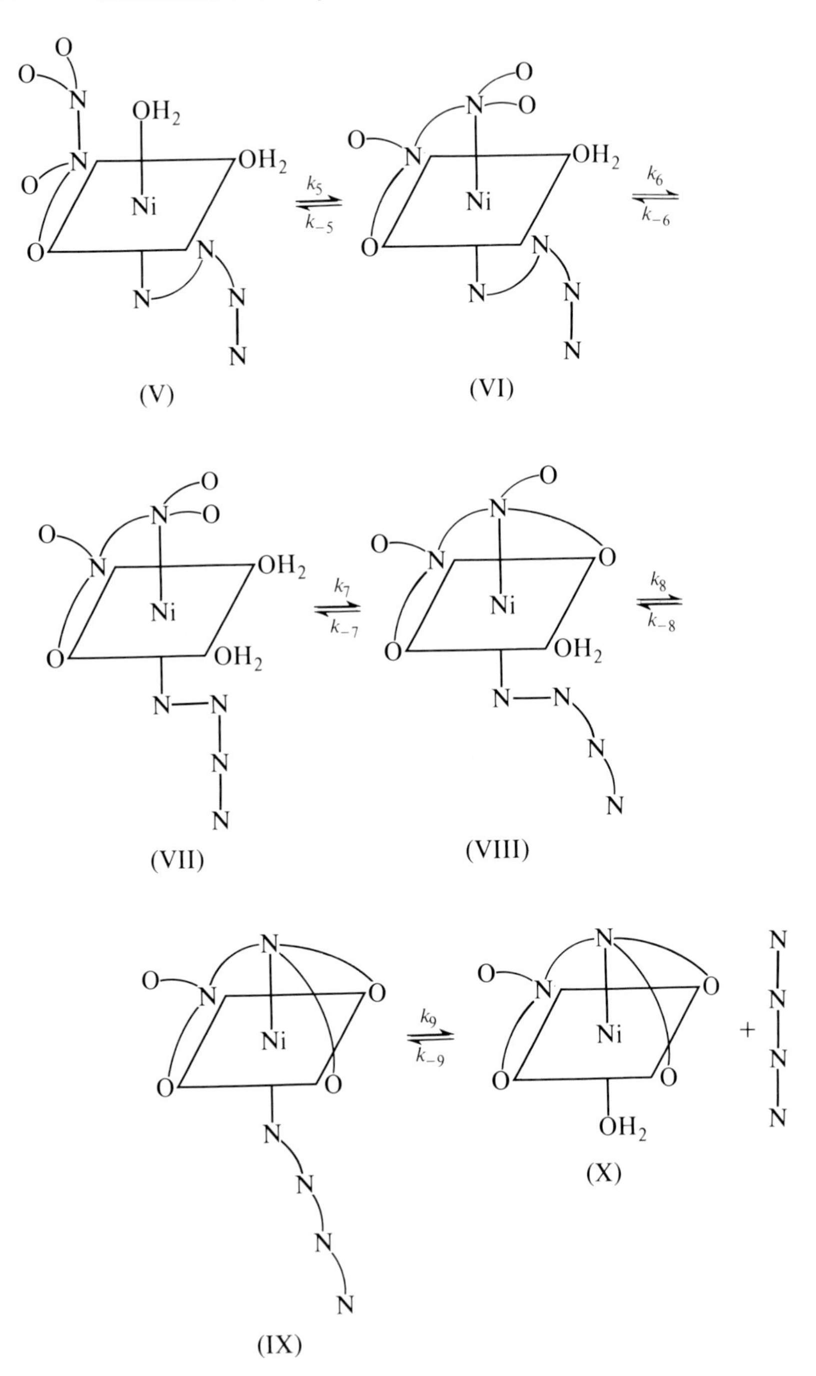

Figure 1-32 Continued

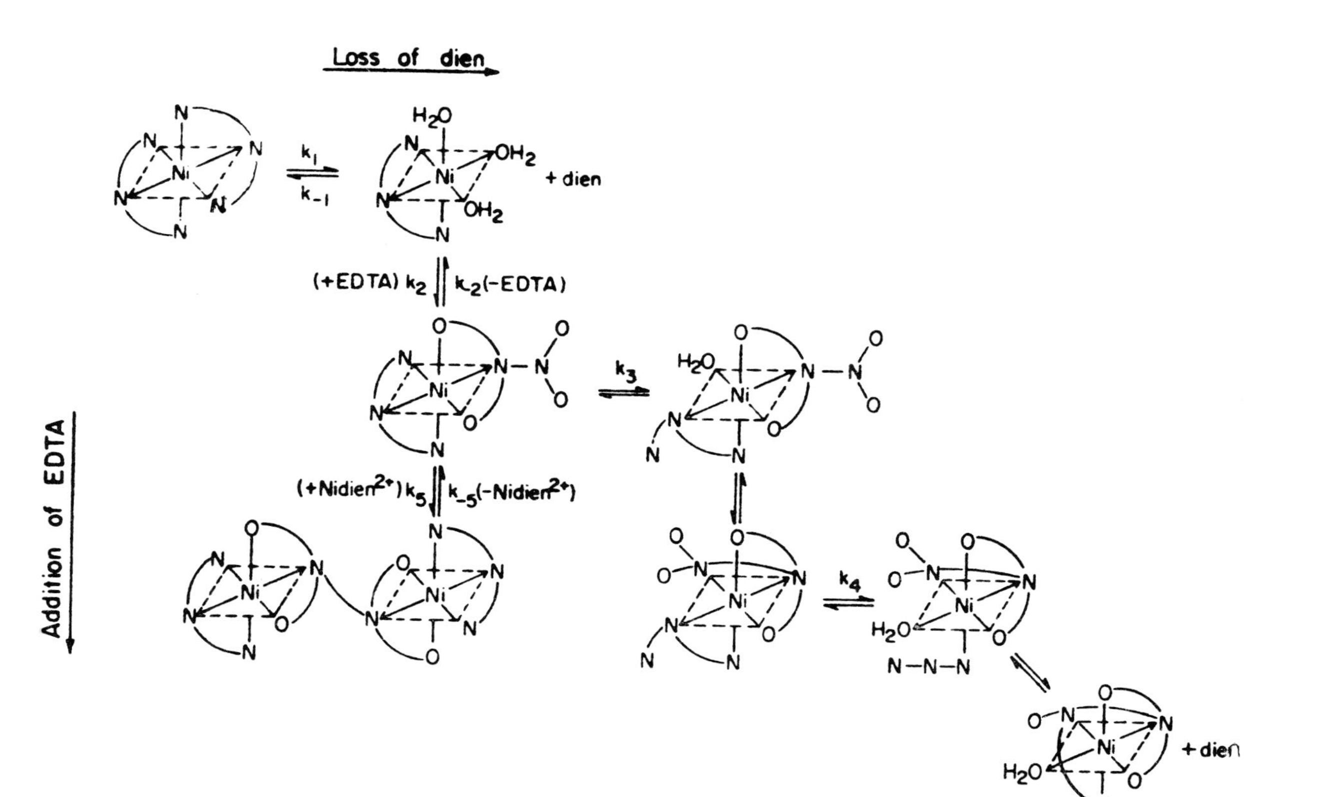

Figure 1-33 Stepwise unwrapping of dien with the addition of EDTA showing the steps identified in the kinetic study (k_1, k_2, k_3, k_4, and k_{-5})[296]

action appears to have a parallel path in which two of its amine groups are displaced before the rds, and hence it is slower to react.

Subsequent studies of the reactions of nickel–dien with EDTA supported these general conclusions.[296] The $Ni(dien)_2^{2+}$ complex does not react directly with EDTA but most completely lose one dien first. This agrees with the general behavior of the other polyamines in having only three nitrogens coordinated to nickel in the rds. As a result the reaction is relatively slow because the value of k_1 (see Figure 1-33) at pH 8 is only 0.012 sec^{-1}. However, the reaction of EDTA with the mono-dien complex is very rapid, and the mixed $Ni(dien)(EDTA)^{2-}$ complex is more stable than $Nidien^{2+}$ so it can be observed along with the two successive first-order steps (k_3 and k_4) which eventually lead to the complete release of dien from nickel. The values of $k_3 = 26\ sec^{-1}$ and $k_4 = 2.6\ sec^{-1}$ are strikingly similar to parallel chelate ring-opening steps of $Nidien^{2+}$ in strong acid as observed by Melson and Wilkins.[474] The entire reaction scheme is given in Figure 1-33 including the side product, $Ni(dien)_2(EDTA)$, which is formed when the $Nidien^{2+}$ concentration exceeds the EDTA concentration.

3. Copper–Aminocarboxylates and Polyamines. Figures 1-34 and 1-35 give the rate constant–pH profiles for the reactions of polyamines with $CuEDTA^{2-}$ [788] and of tetren with various copper–aminocarboxylate complexes.[435] Analysis of these kinetic data together with the comparative behavior of dien, en, and NH_3 reactions with $CuEDTA^{2-}$ indicate that three nitrogen donors of the polyamines are needed in the rate-determining step to have a rapid ligand–ligand displacement. The reactivities of the aminocarboxylate complexes vary inversely with their copper stability constants because of the significant degree of unwrapping of the aminocarboxylate which is necessary to accommodate three polyamine nitrogens around copper. The rate is altered by the number of basic sites available for protons transferred from the polyamine to the leaving ligand, and in some cases $Htetren^{+}$ is more reactive than tetren, but $H_2tetren^{2+}$ and $H_3tetren^{3+}$ are less reactive.

The rate of addition of EDTA to $Cudien^{2+}$ to form the mixed complex is too fast to measure using flow techniques and therefore must be greater than $1 \times 10^7 M^{-1}\ sec^{-1}$. An interesting example of steric hindrance exerted by the coordinated ligand is illustrated by the reaction of EDTA with $CuEt_4dien^{2+}$. The ligand Et_4dien has two ethyl groups on each of the terminal nitrogen atoms. These ethyl groups block coordination to the axial sites which reduces the reaction rate[803] as compared with the similar reaction with $Cudien^{2+}$.

$$k_{EDTA}^{CuEt_4dien} = 3.65 \times 10^6 M^{-1}\ sec^{-1}$$

$$k_{H(EDTA)}^{CuEt_4dien} = 2.62 \times 10^6 M^{-1}\ sec^{-1}$$

Square-planar complexes MEt_4dienX (M = Pd(II), Pt(II), Au(III)) undergo monodentate ligand substitution reactions; however, the rate is independent of the incoming ligand concentration.[804–808] In these reactions the four ethyl groups block the usual associative type mechanism of square-planar complexes. Roulet

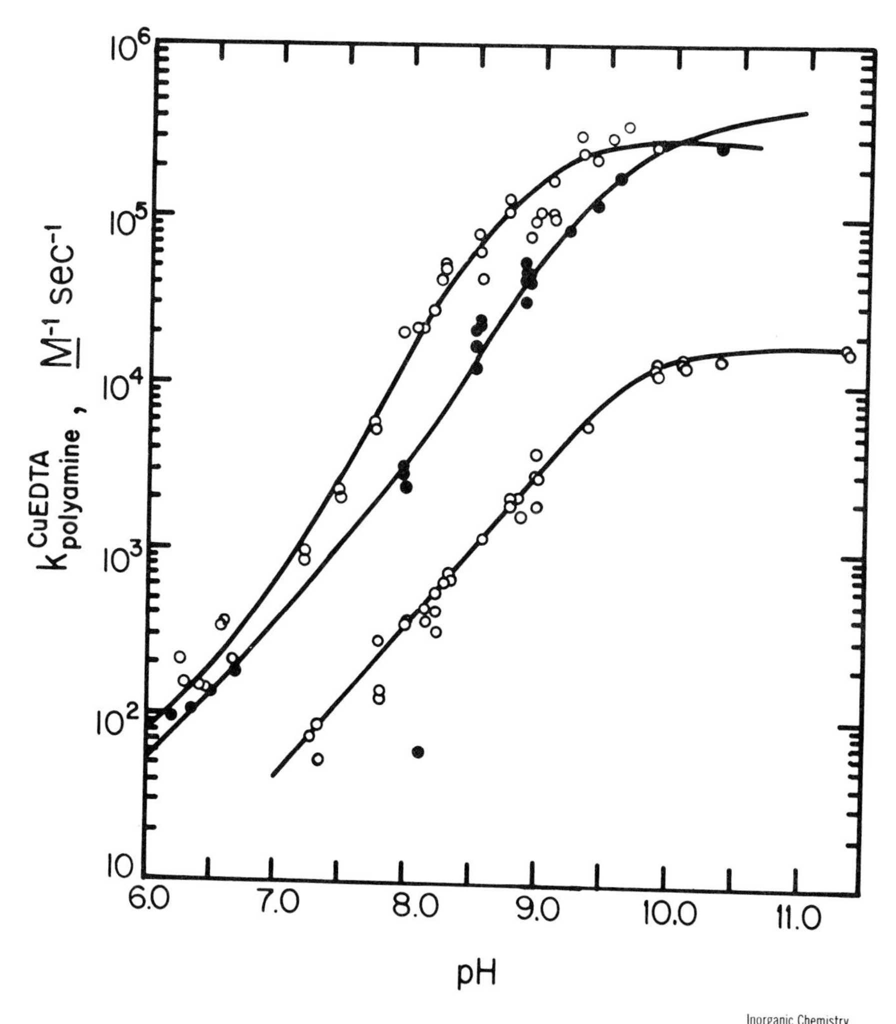

Figure 1-34 Rate constant–pH profile for the reaction of penten(Pe), tetren(Te) and trien(T) with copper EDTA (CuY)[788]

and Gray[809] have produced data which suggest an associative mechanism for the solvent pathway for attack on $PtEt_4dienX$.

4. Copper–Tren and Polyamines. Carr and Vasiliades[794] measured the rates of ligand exchange reactions of dien, trien, and tetren with $Cutren^{2+}$. The rate constants for the unprotonated ligands (M^{-1} sec^{-1} at 25°C, $\mu = 0.1$) are 1.06×10^5, 8.52×10^4, and 5.41×10^4 for dien, trien, and tetren, respectively. The pH dependence of the reaction rate is analyzed, and it is concluded that two unprotonated nitrogen donors on the incoming nitrogen are sufficient to reach the rds. Other rate constants are given in Table 1-37.

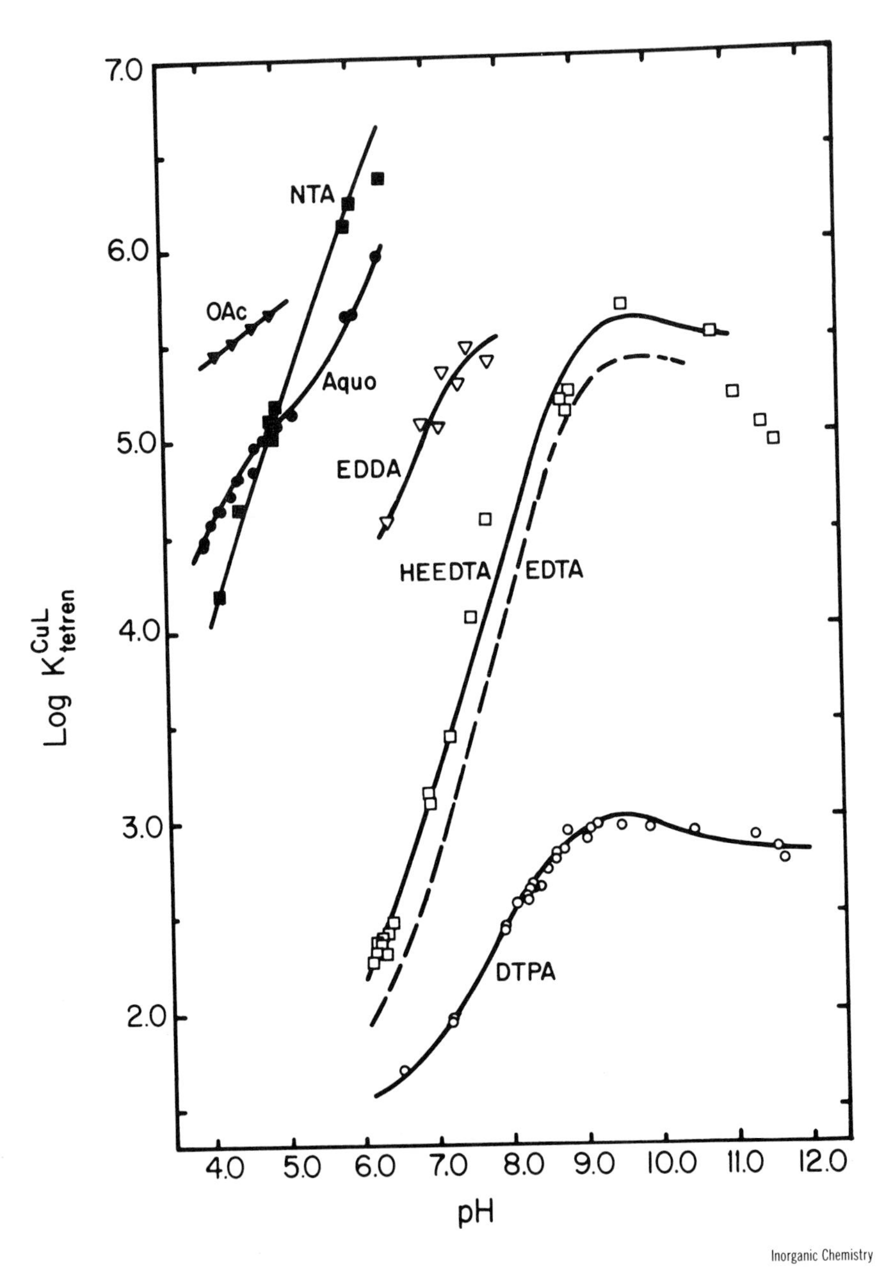

Figure 1-35 Rate constant–pH profile for the reaction of tetren with aquo copper(II) and copper complexes of DTPA, EDTA, HEEDTA, EDDA, NTA, and acetate[297]

5. Metal–Aminocarboxylate Exchange with Aminocarboxylates. Table 1-37 summarizes rate constants for the exchange of EDTA, NTA, and other aminocarboxylate complexes for a variety of systems. The exchange reactions have, in general, been studied using NMR techniques. The results obtained for Ca,[364,765] Sr,[359] Cd,[796] Cu,[783] Ni,[783] and Pb[783] aminocarboxylates are consistent with the stepwise unwrapping of the aminocarboxylate with simultaneous coordination of the exchanging ligand. The exchange rates tend to be much slower than those involving polyamines, partly because of the unfavorable buildup of the electrostatic charge and partly because the possible mixed intermediate complexes have fewer nitrogen donor groups and are less stable.

Some interesting comparisons can be made from the data in Table 1-37. The reactions of $NiEDTA^{2-}$ are much slower than the reactions of $NiNTA^{1-}$, which in turn are slower than NiIDA. CyDTA can act as a nucleophilic reagent with NiIDA but not with $NiNTA^{1-}$ or $CuNTA^{1-}$. These reactions proceed via the dissociation of the metal complex.[776,810] The hydroxyl group in HEEDTA has only a small effect on the reaction with $NiNTA^{1-}$ as compared with EDTA. A change of the acetate groups in NTA for propionate (NDAP) and 2-propionate (NDIAP) also has a small effect.

6. Metal-Oligopeptides Reacting with Multidentate Ligands. Metal-*N*-(peptide) complexes behave somewhat differently from other aminocarboxylate complexes because of the deprotonated-*N*-(peptide) bonding which occurs with Cu(II), Ni(II), Co(II), and Pt(II).[343] Proton-transfer limiting rates in the partial dissociation of the Ni(II) and Cu(II) complexes were discussed in section IV.D. These metal-*N*-(peptide) complexes tend to have square-planar coordination and tend to undergo nucleophilic substitution reactions as well as the protonation reactions. In fact, the $M(H_{-2}triglycine)^{-}$ (where M = Cu or Ni) reactions with polyamines are rapid,[811,812,813] and a ligand such as trien has second-order rate constants which are comparable with the proton-transfer rate constants. The nucleophile must be able to obtain an equatorial position in the metal–peptide complex. When this is is not possible, protonation or solvent dissociation steps may limit the rate. Thus, tetrapeptide complexes which have four nitrogen donors in planar positions are less subject to direct nucleophilic attack. For example, in the reaction of trien with $Cu(H_{-3}tetraglycine)^{2-}$ or with $Ni(H_{-3}tetraglycine)^{2-}$ the pathway in which cleavage of the terminal peptide bond to the metal is rate determining is more important than with the corresponding tripeptide complexes.[814] Steric factors have a large influence on the reaction rate and have been examined for tripeptides in regard to both peptide steric effects[506] and nucleophile steric effects.[812,813] The rate of the trien reaction with $Cu(H_{-2}tripeptide)^{-}$ is reduced by two orders of magnitude when L-leucyl residues replace glycyl residues in the middle position or in the carboxylate terminal group (positions 2 and 3 of structure **23**). Substitution of L-leucine for glycine at the amine terminal (position 1, structure **23**) is relatively ineffective in reducing the rate of the trien reaction. This behavior supports a mechanism in which the initial attack of trien is at the carboxylate end with the rate step being the cleavage of the Cu–*N*-(peptide) bond adjacent to the carboxylate terminal.[506]

Steric hindrance is very important in determining the relative reactivity of ligands with $Cu(H_{-2}triglycine)^-$ [812] and with $Ni(H_{-2}triglycine)^-$.[813] Polyamines and aminocarboxylates in which a nitrogen donor can coordinate to the metal ion in the plane of the metal–*N*-(peptide) bonds are much more reactive nucleophiles than those which are sterically hindered from doing this. Thus, trien, en, and $EDDA^{2-}$ are effective nucleophiles while $EDTA^{4-}$, NTA^{3-}, and Me_4en are not. The general order of reactivity is: polyamines $\gg$ aminocarboxylates $\gg NH_3 \gg CH_3COO^-$. The data support the mechanism discussed above. A recent review[343] examines in detail the kinetics and mechanisms of metal ion and proton-transfer reactions of oligopeptide complexes.

Three main reaction pathways have been found for the displacement of copper(II) from peptide complexes: (a) proton transfer to the peptide group, (b) nucleophilic attack on copper, and (c) a combination of proton transfer and nucleophilic attack.[508,509] In each case there is a concurrent or subsequent metal–*N*-(peptide) bond cleavage. The reaction rates for the three mechanisms differ in their pH dependence, their general acid dependence, and their nucleophilic dependence.

C. Double Exchange of Multidentate Ligand Complexes: Coordination Chain Reactions

An interesting situation exists when two multidentate complexes are mixed and the thermodynamics dictate a double exchange.[498] As we have seen, strong complexes in neutral or basic solution can dissociate exceedingly slowly, and therefore the question arises as to what mechanisms and what rates can be expected for a double exchange process [Eq. (92)].

$$ML + M'L' \rightarrow ML' + M'L \tag{92}$$

A number of pathways could be suggested.

(*1*) Dissociation of ML and of M′L′ followed by recombination of M + L′ and M′ + L.

(*2*) Direct collision of ML and M′L′ to form a mixed bridged MLM′L′ intermediate in which the ligands replace one another.

(*3*) Partial dissociation of one complex to give small amounts of free ligand which catalyzes the exchange.

(*4*) Partial dissociation of one complex to give small amounts of free metal ion which catalyzes the exchange.

The first pathway is most unfavorable and can be eliminated except in cases where steric hindrance prevents any metal ion or ligand attack on one of the complexes. The sluggishness of this pathway discourages its investigation, but macrocyclic ligand complexes are possible candidates for this behavior. The second pathway, in which one multidentate ligand slithers past another, seems a fascinating possibility for mutual assistance involving a combination of ligand–ligand and metal–metal exchange processes. However, there is no evidence yet for this pathway in double exchange. The third pathway is favored because of the rapidity of multidentate ligand–ligand exchange reactions and leads to the *coordination chain reactions* discussed below. The fourth pathway is, in general, not so favored as the third pathway, but examples of metal-catalyzed coordination chain reactions exist.

1. Coordination Chain Reactions: $Nitrien^{2+}$ and $CuEDTA^{2-}$. The exchange reaction in Eq. (93) is thermodynamically favorable, going essentially to com-

$$\mathrm{Nitrien^{2+} + CuEDTA^{2-} \rightarrow NiEDTA^{2-} + Cutrien^{2+}} \qquad (93)$$

pletion. In neutral solution the solvent dissociation processes for the complexes have half-lives between years and decades, but the exchange can take place in a few minutes. The reaction is very sensitive to trace catalysis and trace inhibition. Olson and Margerum[23] showed that this reaction proceeds by a chain mechanism [Eqs. (94)–(97)] in which the free ligands (EDTA and trien) are

$$\mathrm{Nitrien^{2+}} \underset{k_{-1}}{\overset{k_1}{\rightleftarrows}} \mathrm{Ni^{2+} + (trien)_T} \qquad (94)$$

$$\mathrm{(trien)_T + CuEDTA^{2-}} \underset{k_{-2}}{\overset{k_2}{\rightleftarrows}} \mathrm{Cutrien^{2+} + (EDTA)_T} \qquad (95)$$

$$\mathrm{(EDTA)_T + Nitrien^{2+}} \xrightarrow{k_3} \mathrm{NiEDTA^{2-} + (trien)_T} \qquad (96)$$

$$\mathrm{(EDTA)_T + Ni^{2+}} \xrightarrow{k_4} \mathrm{NiEDTA^{2-}} \qquad (97)$$

chain centers. Eq. (94), is the chain initiation step, which provides small concentrations of trien to the system (the subscript T for trien and EDTA in Eqs. (94)–(97) refers to the total concentration of the free ligands—i.e., the summation of the protonated and unprotonated species. Eqs. (95) and (96) are chain propagating reactions which maintain the chain centers. Eq. (97) and the reverse reaction of Eq. (94) are chain terminating, although Eq. (97) gives the reaction product. The reversibility of Eq. (95) also can affect the observed kinetics. In this system the parallel initiation reaction caused by the dissociation of $CuEDTA^{2-}$ and the possible metal–metal exchange reactions are too slow to contribute except under special conditions. There is no evidence of direct exchange between $Nitrien^{2+}$ and $CuEDTA^{2-}$. The coordination chain reaction system has many similarities to the classical chain reaction between H_2 and Br_2[815] except that the chain centers are not free radicals but are chemically stable free ligands. As a result the rate of the coordination chain reaction can be easily controlled by the addition of traces of free ligands which greatly accelerate the exchange. Under these circumstances the initiation and termination reactions are no longer important, and the rate is determined only by the two chain-propagating steps [Eqs. (95) and (96)] which maintain $(\text{trien})_T + (\text{EDTA})_T$ at a constant level, resulting in Eqs. (98). Any ligand which releases trien or

$$\text{rate} = k_2(\text{trien})_T[\text{CuEDTA}^{2-}] = k_3(\text{EDTA})_T[\text{Nitrien}^{2+}] \qquad (98)$$

EDTA will catalyze the reaction, and conversely any metal ion which complexes EDTA or trien will inhibit the reaction. These properties have led to the development of the coordination chain reaction system as a sensitive analytical method for trace concentrations (down to $10^{-8}M$) of metal ions or ligands.[24,816,817] Bidentate ligands such as oxalate, glycinate, and pyrophosphate act as blocking agents for the coordination chain reaction by forming mixed complexes with Nitrien^{2+}.[818] Thus, $[\text{Ni(trien)}(C_2O_4)]$ does not have a foothold for EDTA attack and does not participate in the chain-propagation reaction.

The rates of propagation of the chain reaction [Eqs. (95) and (96)] are increased by a change of solvent from water to methanol–water mixtures. The same change to methanol-rich solvent tends to decrease the rate of initiation. The sensitivity of the system to trace concentrations of the chain centers depends on the ratio of the rate of propagation to the rate of initiation. This ratio is termed the "chain length." The chain length is increased by factors of 10–100 in 90% $CH_3OH—H_2O$, and the kinetics indicate the formation of mixed-ligand complexes of [Ni(trien)EDTA] and of [Cu(EDTA)trien].[473] There is limited evidence for the formation of mixed-ligand dinuclear complexes of the type [(trien)Cu(EDTA)Ni(trien)] in methanol-rich solvents. The formation of such a reaction intermediate approaches pathway (2) but still falls under pathway (3) because the reaction is catalyzed by added free ligand. Finally, the exchange reaction in 90% $CH_3OH—H_2O$ will take place by a metal-catalyzed chain reaction [Eq. (97)] if traces of Ni^{2+} are added to the system.[819] The chain-propagating steps are those in Eqs. (99) and (100). Because the metal exchange reactions are slower than the ligand exchange reactions, this mechanism is observed only when the free metal ion concentrations are much greater than the free ligand concentrations and when the rate of the overall reaction is much slower than for the ligand-catalyzed system.

$$Ni^{2+} + \text{CuEDTA}^{2-} \rightarrow \text{NiEDTA}^{2-} + Cu^{2+} \qquad (99)$$

$$Cu^{2+} + \text{Nitrien}^{2+} \rightarrow \text{Cutrien}^{2+} + Ni^{2+} \qquad (100)$$

2. Other Coordination Chain Reaction Systems. The general requirements for the chain mechanism are that the catalytic pathways resulting from ligand–ligand or metal–metal exchange must be faster than the rate at which the solvent (or its conjugate acid or base) displaces the ligands from complexes capable of double exchange. The larger the ratio of these rates the longer the chain length and the more readily observable the coordination chain mechanism. The exchange reactions of the following reactant pairs have been shown to proceed through a chain mechanism.

$Nitetren^{2+} + CuEDTA^{2-}$ (Ref. 25)

$Znpenten^{2+} + CuEDTA^{2-}$ (Ref. 820)

$PbPDTA^{2-} + CuEDTA^{2-}$ (Ref. 789)

$Zntrien^{2+} + Cutren^{2+}$ (Ref. 794)

Reversibility of the chain-propagating steps and shifts of the identity of the rate-limiting step as a function of pH have major effects on the kinetic behavior which are considered in detail.[25,820] The all-polyamine coordination chain should have a built-in selectivity for transition metals in the presence of alkaline earths.[794]

D. Catalysis

The double exchange reactions described in the previous section are good examples of multidentate ligand catalysis of the substitution reactions of multidentate ligands. Many ligand–ligand exchange reactions are subject to catalysis when the main reactants have difficulty in undergoing direct reaction because of steric, electrostatic, or other effects. The metal–tripeptide reactions with EDTA provide several examples of this behavior [Eq. (101)] where the catalyst

$$\mathrm{Cu(H_{-2}GGG)^- + (EDTA)_T \xrightarrow{\text{catalyst}} CuEDTA^{2-} + GGG^-} \qquad (101)$$

can be trien,[811] buffer acids,[505] amino acids,[821] or GGG^- itself.[822] Catalysis by amino acids can be expected in the transfer of copper when it is complexed to proteins and may be important in the biological transport of copper.

Metal–metal exchange reactions can be catalyzed by other metal ions. The reaction of Zn^{2+} with $NiEDTA^{2-}$ is catalyzed by small amounts of Cu^{2+} because the Cu^{2+} rate constant is 6500 times larger than the Zn^{2+} rate constant with $NiEDTA^{2-}$.[495]

The reactions of EDTA with the square-planar nickel(II) complexes of bidentate α-amineoximes are catalyzed by bidentate ligands.[823] With a quadridentate ligand, $[Ni(enAO){-}H]^+$, no catalytic effects were observed. The bidentate ligands react with the square-planar complexes displacing an α-amineoxime ligand to form an octahedral nickel complex which reacts rapidly with EDTA. With a diamine which forms a square-planar nickel complex (tetraMeen) a large catalytic effect is not observed; a square-planar Ni(AO)(tetraMeen) complex is thought to be formed, which reacts slowly with EDTA.[823]

The exchange reaction between $Fe^{III}NTA$ and transferrin has been studied by stopped-flow techniques, and it has been shown that the breakup of the intermediate, which is rapidly formed, is catalyzed by the presence of bicarbonate[824] (the intermediate proceeds to products in about 2 min in the absence of bicarbonate but takes only 10 sec when bicarbonate is added). So far, the mechanism of this reaction has not been completely unravelled, and the binding site of the bicarbonate has not been determined. Several other exchange reactions between iron(III) chelates and transferrin are catalyzed by bicarbonate.[825,826]

Hydroxide ion catalyzes a number of multidentate ligand exchange reactions by virtue of its coordinating ability—e.g., the EDTA reaction with $Fe(TPTZ)_2^{2+}$.[297] It also abstracts protons from nitrogen donor atoms to cause conjugate base effects and to assist the racemization of asymmetric nitrogen centers. Examples of the latter include hydroxide ion catalysis of the racemization of $[Co^{III}(NH_3)_4Sar]^{2+}$,[827] $[Co(NH_3)_4(N\text{-Meen})]^{3+}$,[828] $[Pt^{II}(N\text{-Meen})(phen)]^{2+}$, and $[Pt^{II}(NH_3)_2(N\text{-Meen})]^{2+}$.[829] Racemization kinetics of macrocyclic tetramine complexes of Ni(II) and Cu(II) are also hydroxide-ion catalyzed.[830] Another type of hydroxide ion catalysis is observed in the isomerization reactions of macrocyclic complexes of Cu(II) where the blue complex of $Cu(tet\ b)^{2+}$ is converted to the red complex of $Cu(tet\ b)^{2+}$.[830]

$$Cu(tet\ b)^{2+}{}_{blue} \xrightarrow{OH^-} Cu(tet\ b)^{2+}{}_{red} \tag{102}$$

In this reaction the configurational changes leading to the blue-to-red conversion are caused by a coordinated hydroxide ion which accepts a proton from an adjacent nitrogen atom to permit its inversion in the complex. The blue complex can exist as a five-coordinate complex,[301] and $[Cu(tet\ b)X_{blue}]^+$ species inhibit the hydroxide ion catalysis. The unusual feature of this reaction is that coordinated hydroxide ion is a much more effective catalyst than free hydroxide ion in solution. Catalysis tends to be the rule rather than the exception in many reactions of multidentate ligand complexes, and the possibility of catalytic pathways should be kept in mind.

APPENDIX

abu	aminobutyric acid
ACA°	Apocarbonic Anhydrase
acac	acetylacetone
ADP	adenosine-5′-diphosphate
AEP	2-(2-aminoethyl)pyridine
Ala	alanine
ama	aminomalonic acid
AMP	2-aminomethylpyridine
AN	anthranilate
AO	α-amineoxime
Arg	arginine
Asp	aspartate
ATP	adenosine-5′-triphosphate

BCA	Bovine Carbonic Anhydrase
meso-BDTA	*meso*-butylenediaminetetraacetic acid
bipy	2,2′-bipyridine
bn	2,3-diaminobutane
BPEDA	*N,N′*-bis(2-picolyl)ethylenediamine
BT	Eriochrome Black T.
CAL	1-(hydroxy-4-methyl-2-phenylazo)-2-naphthol-4-sulfonate
CDP	cytosine diphosphate
Chel H^{2-}	4-hydroxypyridine-2,6-dicarboxylate
CH_2PEP	α-(dihydroxyphosphinylmethyl)-acrylate
chx	1,2-diaminocyclohexane
CR	2,12-dimethyl-3,7,11,17-tetraazabicyclo-[11.3.1]heptadeca-1(17),2,11,13,15-pentaene
$CRCH_3$	2,7,12-trimethyl-3,7,11,17-tetraazabicyclo-[11.3.1]heptadeca-1(17), 2,11,13,15-pentaene
CTP	cytosine triphosphate
cyclam	1,4,8,11-tetraazacyclotetradecane
CyDTA	cyclohexylenediaminetetraacetic acid
Cys	cystein
dapa	3,3′-diaminodipropylamine
αβDap	α,β-diaminopropionic acid
deuteroporphyrin	deuteroporphyrin-2,4-disulfonic acid dimethylester
1,3-ddmp	1,3-diamino-2,2-dimethylpropane
DGITA	2,2′-ethylenedioxybis[ethyliminodi(acetic acid)]
dien	diethylenetriamine
trans-[14]-diene	5,7,7,12,14,14-hexamethyl-1,4,8,11-tetraazacyclotetradeca-4,11-diene
N,N-diEten	*N,N*-diethylethylenediamine
N,N′-diEten	*N,N′*-diethylethylenediamine
N,N-diMeen	*N,N*-dimethylethylenediamine
*N,N′*diMeen	*N,N′*-dimethylethylenediamine
N,N′-di-*n*-pren	*N,N′*-di-*n*-propylethylenediamine
dopa	dihydroxyphenylalanine
α,β-DPA	α,β-diaminopropionate
DPA	2,6-dicarboxypiperidineacetic acid
DPEDA	*N,N′*-dibenzylethylenediamine
DTO	dithioxalate
DTPA	diethylenetriaminepentaacetic acid
DTZ	diphenylthiocarbazone
EDDA	*N,N′*-ethylenediaminediacetic acid
EDDDA	ethylenediamine-*N,N′*-diacetic-*N,N′*-dipropionic acid

EDMA	ethylenediaminemonoacetic acid
EDTA	ethylenediaminetetraacetic acid
EGTA	ethyleneglycolbis(2-aminoethyl ether)-tetraacetic acid
en	ethylenediamine
ENAO	2,2′-ethylenediaminobis(2-methyl-3-butanone) dioxime
enolase	phosphoenolpyruvatehydratase
N-Eten	*N*-ethylethylenediamine
$EtNH_2$	ethylamine
Et_2-2,3,2-tet	1,11-diethyl-1,4,8,11-tetraazaundecane
Et_4dien	1,1,7,7-tetraethyldiethylenetriamine
Ga	glycinamide
Glu	glutamate
Gly (G)	glycine
GlyGly, digly (GG)	diglycine
GlyLeu	glycylleucine
Glysa	glycylsarcosine
HCA	Human Carbonic Anhydrase
HEEDTA	hydroxyethylethylenediaminetetraacetic acid
HEIDA	*N*-hydroxyethyliminodiacetic acid
hematoporphyrin	hematoporphyrin IX
hm	histamine
His	histidine
HPDTA	1,3-diamino-2-hydroxypropanetetraacetic acid
IDA	iminodiacetic acid
IDP	iminodipropionic acid
imac	4(5)-imidazoleacetate
imid	imidazole
LeuGly	leucylglycine
mal	malonate
Me_2cyclam	5,12-dimethyl-1,4,8,11-tetraazacyclotetradecane
N-Meen	*N*-methylethylenediamine
MeHis	L-histidine methyl ester
$MeNH_2$	methylamine
Me_2NH	dimethylamine
$MePO_4^{2-}$	methylphosphate
metalphth	metalphthalein
MIDA	*N*-methyliminodiacetic acid
MNT	maleonitriledithiolate

2M8SQ	2-methyl-8-mercaptoquinoline
murexide	ammonium purpurate
NDAP	nitrilo(diacetic)propionate
NDAIP	nitrilo(diacetic)isopropionate
NSA	5-nitrosalicylic acid
NTA	nitrilotriacetic acid
OAc	acetate
OPr	propionate
4,7-$(OH)_2$phen	4,7-dihydroxy-1,10-phenanthroline
OX, oxine	8-hydroxyquinoline
oxal	oxalate
P	tetra(*N*-methyltetrapyridyl)porphine
OXS	8-hydroxyquinoline-5-sulfonic acid
pald	pyridine-2-aldoxime
pad	2-picolinamide
PADA	pyridine-2-azo-*p*-dimethylaniline
PAR	4-(2-pyridylazo)resorcinol
PC	pyruvate carboxylase
PDA	dicarboxypiperidineacetate
PDTA	propylenediaminetetraacetic acid
pen	D-L-$HSC(CH_3)_2CH(NH_2)COOH$
Pen	an ethylenediamine-substituted protoporphyrin IX (Ref. 403a)
penten	*N,N,N′,N′*-tetra(2-aminoethyl)ethylenediamine
penH_2	penicillamine
phen	1,10-phenanthroline
Phenala	phenylalanine
phencarb	phenanthroline-2-carboxylate
pidic	piperidine-2,6-dicarboxylate
pn	1,2-diaminopropane
N-pren	*N*-propylethylenediamine
ptn	1,2,3-triaminopropane
py	pyridine
pya	pyridine-2-acetate
pya_2tn	(*N,N′*-bis(2-pyridylmethylene)-1,3-diaminopropane
pyc	pyridine-2-carboxylate
pydic	pyridine-2,6-dicarboxylate
sal	salicylic acid
sali	salicylaldehyde
Sar	sarcosine
Ser	serine
8-SQ	8-mercaptoquinoline
SSA	5-sulfo-salicylate
SXO	semi-xylenol orange

cis,cis-tach	*cis,cis*-1,3,5-triaminocyclohexane
tart	tartrate
te	*C*-tetramethylethylenediamine
TEA	triethanolamine
te-Eteen	*N,N,N′,N′*-tetraethylethylenediamine
te-Meen	*N,N,N′,N′*-tetramethylethylenediamine
terpy	2,2′,2″-terpyridine
2,3,2-tet	*N,N′*-bis(2-aminoethyl)-1,3-propanediamine
tet *a*	5,7,7,12,14,14-hexamethyl-1,4,8,11-tetraazacyclotetradecane(*a* isomer)
tet *b*	5,7,7,12,14,14-hexamethyl-1,4,8,11-tetraazacyclotetradecane(*b* isomer)
tetraGly (GGGG)	tetraglycine
C-tetraMeen	2,3-diamino-2,3-dimethylbutane
tetren	tetraethylenepentamine
THPED	*N,N,N′,N′*-tetrakis(2-hydroxypropyl)ethylenediamine
tiron (ti)	4,5-dihydroxybenzene-1,3-disulfonic acid
TKED	*N,N,N′,N′*-tetrakis(2-hydroxyethyl)ethylenediamine
tn	1,3-diaminopropane(trimethylenediamine)
TP	tripolyphosphate
TPP	tetraphenylporphyrin
TPP^{3-}	thiamine pyrophosphate
TPTZ	2,4,6-tripyridyl-*s*-triazine
tren	2,2′,2″triaminotriethylamine
TRDTA	trimethylenediaminetetraacetic acid
trien	triethylenetetramine
tri-Eten	*N,N,N′*-triethylenediamine
triGly (GGG)	triglycine
N,N,N′-trimethylen	*N,N,N′*-trimethylethylenediamine
TTF	thenoyltrifluoroacetone
TTHA	triethylenetetraminehexaacetic acid
UDA	uramil-*N,N*-diacetate
XO	xylenol orange

Acknowledgment

This review was prepared with the assistance of many members of the Margerum research group whose assistance is greatly acknowledged. Special thanks are due to A. Graham Lappin and to Patty L. Shoemaker for the final reorganization and updating of the manuscript. We would like to thank the National Science Foundation Chemical Dynamics Program for their support of major portions of our research work which are presented here.

REFERENCES

1. Basolo, F., and R. G. Pearson. *Mechanisms of Inorganic Reactions*. 2nd Ed., John Wiley and Sons, Inc., New York, 1967.
2. Langford, C. H., and H. B. Gray. *Ligand Substitution Processes*. W. A. Benjamin, Inc., New York, 1965.
3. *Inorganic Reaction Mechanisms*. Vol. 1 (1971); Vol. 2 (1972); Vol. 3 (1974), J. Burgess, ed. Specialist Periodical Reports, Chemical Society, London.
4. Edwards, J. O., F. Monacelli and G. Ortaggi. *Inorg. Chim. Acta*. **11,** 47 (1974).
5. Wilkins, R. G. *Acc. Chem. Res.* **3,** 408 (1970).
6. Hague D. N. *Inorganic Reaction Mechanisms*. Specialist Periodical Reports, Vols. 1–4, Chemical Society, London (*see Ref.* 3).
7. Eigen, M. *Pure Appl. Chem.* **6,** 97 (1963).
8. McAuley, A., and J. Hill. *Q. Rev.* **23,** 18 (1969).
9. Hewkin, D. J., and R. H. Prince. *Coord. Chem. Rev.* **5,** 45 (1970).
10. Kustin, K., and J. Swinehart. *Prog. Inorg. Chem.* **13** (1970).
11. Diebler, H., M. Eigen, G. Ilgenfritz, G. Maass and R. Winkler. *Pure Appl. Chem.* **20,** 93 (1969).
12. Wilkins, R. G., and M. Eigen. "Mechanisms of Inorganic Reactions." *Adv. Chem. Ser.* **49,** 55 (1965).
13. Winkler, R. *Struct. Bonding*. **10,** 1 (1972).
14. Eigen, M., and L. C. de Maeyer. *Technique of Organic Chemistry*. Vol. VIII, Part II. S. L. Friess, E. S. Lewis, and A. Weissberger, eds. Interscience, New York, 1963, p. 895.
15. Caldin, E. F. *Fast Reactions in Solution*. John Wiley and Sons, Inc., New York 1964.
16. *Methods in Enzymology*. Vol. XVI. K. Kustin, ed. Academic Press, New York 1969.
17. Hague, D. N. *Fast Reactions*. Wiley-Interscience, London, 1971.
18. Strehlow, H. *Adv. Mol. Relaxation Processes*. **2,** 235 (1972).
19. Davies, M. H., J. R. Keefe and B. H. Robinson. *J. Chem. Soc. Ann. Rep., A*. **70,** 123 (1973).
20. Hammes, G. G., ed. *Techniques of Chemistry*. Vol. VI, Part II. Wiley-Interscience, New York, 1974.
21. Cayley, G. R., and D. W. Margerum. *J. Chem. Soc., Chem. Comm.* 1002, (1974).
22. Margerum, D. W., D. L. Janes and H. M. Rosen. *J. Am. Chem. Soc.* **87,** 4463 (1965).
23. Olson, D. C., and D. W. Margerum. *J. Am. Chem. Soc.* **85,** 297 (1963).
24. Margerum, D. W., and R. K. Steinhaus. *Anal. Chem.* **37,** 222 (1965).
25. Margerum, D. W., and J. D. Carr. *J. Am. Chem. Soc.* **88,** 1639 (1966).
26. Taube, H. *Electron Transfer Reactions of Complex Ion in Solution*. Academic Press, New York, 1970.
27. Silver, B. L., and Z. Luz. *Q. Rev. Chem. Soc.* **21,** 458 (1967).
28. Stengle, T. R., and C. H. Langford. *Coord. Chem. Rev.* **2,** 349 (1967).
29. Hunt, J. P. *Coord. Chem. Rev.* **7,** 1 (1971).
30. Lincoln, S. F. *Coord. Chem. Rev.* **6,** 309 (1971).
31. Hinton, J. F., and E. S. Amis. *Chem. Rev.* **71,** 627 (1971).

32. Fratiello, A. *Prog. Inorg. Chem.* **17(2)**, 57 (1972).
33. Swift, T. S., and R. E. Connick. *J. Chem. Phys.* **37**, 307 (1962).
34. Merideth, C. W. *U. S. At. Energy Comm. Rep. UCRL-11704, 1965.*
35. Poupko, R., and Z. Luz. *J. Chem. Phys.* **57**, 3311 (1972).
36. Connick, R. E., and D. N. Fiat. *J. Chem. Phys.* **39**, 1349 (1963).
37. Neely, J. W., Ph.D. Thesis, University of California, Berkeley, UCRL-20580 (1971).
38. Neely, J. W., and R. Connick. *J. Am. Chem. Soc.* **92**, 3476 (1970).
39. Olson, M. V., Y. Kanazawa and H. Taube. *J. Chem. Phys.* **51**, 289 (1969).
40. Zetter, M. S., M. W. Grant, E. J. Wood, H. W. Dodgen and J. P. Hunt. *Inorg. Chem.* **11**, 2701 (1972).
41. Chmelnick, A. M., and D. Fiat. *J. Chem. Phys.* **47**, 3986 (1967).
42. Hoggard, P. E., H. W. Dodgen and J. P. Hunt. *Inorg. Chem.* **10**, 959 (1971).
43. Connick, R. E., and D. Fiat. *J. Chem. Phys.* **44**, 4103 (1966).
44. Desai, A. G., H. W. Dodgen and J. P. Hunt. *J. Am. Chem. Soc.* **92**, 798 (1970).
45. Connick, R. E. *Symposium on Relaxation Techniques.* Buffalo, New York, June, 1965.
46. Lewis, W. B., M. Alei, Jr. and L. O. Morgan. *J. Chem. Phys.* **44**, 2409 (1966).
47. Fiat, D., and R. E. Connick. *J. Am. Chem. Soc.* **90**, 608 (1968).
48. Glass, G. E., W. B. Schwabacher and R. S. Tobias. *Inorg. Chem.* **7**, 2471 (1968).
49. Chmelnick, A. M., and D. Fiat. *J. Chem. Phys.* **51**, 4238 (1969).
50. Donham, D., and H. Taube, quoted in Ref. *10.*
51. Chmelnick A. M., and D. Fiat. *J. Magn. Reson.* **8**, 325 (1972).
52. Stranks, D. R., and T. W. Swaddle. *J. Am. Chem. Soc.* **93**, 2783 (1971).
53. Judkins, M. R. *U. S. At. Energy Comm. Rep. UCRL-17561 1967; Chem. Abstr.* **69**, 23496e (1968).
54. Judkins, M. R., unpublished data quoted in Ref. *37.*
55. Plumb, W., and G. M. Harris. *Inorg. Chem.* **3**, 542 (1964).
56. Reuben, J., and D. Fiat, unpublished data quoted in Ref. *45.*
57. Marianelli, R., *Lawrence Livermore Lab. Rep. UCRL-17069, 1966.*
58. Reuben, J., and D. Fiat. *J. Chem. Phys.* **51**, 4909 (1969).
59. Reuben, J., and D. Fiat. *Chem. Commun.* **1967**, 729.
60. Wüthrich, K., and R. E. Connick. *Inorg. Chem.* **7**, 1377 (1968).
61. Zeltman, A. H., and L. O. Morgan. *Inorg. Chem.* **10**, 2739 (1971).
62. Reuben, J., and D. Fiat. *Inorg. Chem.* **6**, 579 (1967).
63. Wüthrich, K., and R. E. Connick. *Inorg. Chem.* **6**, 583 (1967).
64. Miceli, J., and J. Stuehr. *J. Am. Chem. Soc.* **90**, 6967 (1968).
65. Crea J., and S. F. Lincoln, *J. Chem. Soc., Dalton Trans.,* **1973** 2075.
66. Frankel, L. *Inorg. Chem.* **10**, 813, 2360 (1971).
67. Swaddle, T. W. *Coord. Chem. Rev.* **14**, 217 (1974).
68. Roughton, F. J. W., and B. Chance. *Technique of Organic Chemistry.* Vol. VIII, Part II. S. L. Friess, E. S. Lewis, and A. Weissburger, eds., Interscience, New York, 1963, pp. 704–792.
69. Eigen, M., and K. Tamm. *Z. Elektrochem.* **66**, 93, 107 (1962).
70. Purdie, N., and M. M. Farrow. *Coord. Chem. Rev.* **11**, 189 (1973).
71. Jackopin, L. G., and E. Yeager. *J. Phys. Chem.* **74**, 3766 (1970).

72. Bechtler, A., K. G. Breitschwerdt and K. Tamm. *J. Chem. Phys.* **52,** 2975 (1970).
73. Jackopin, L. G., and E. Yeager. *J. Phys. Chem.* **70,** 313 (1966).
74. Atkinson, G., and S. K. Kor. *J. Phys. Chem.* **71,** 673 (1967).
75. Atkinson, G., and S. K. Kor. *J. Phys. Chem.* **70,** 314 (1966).
76. Atkinson, G., M. M. Emara and R. Fernandez-Prini, *J. Phys. Chem.* **78,** 1913 (1974).
77. Kor, S. K., Rai Gulshan and S. S. Bhatti, *Z. Phys. Chem.* (*Liepzig*). **238,** 388 (1968).
78. Atkinson, G., and S. K. Kor. *J. Phys. Chem.* **69,** 128 (1965).
79. Maass, G. *Z. Phys. Chem.* (*Frankfurt am Main*). **60,** 138 (1968).
80. Fittipaldi, F., and S. Petrucci. *J. Phys. Chem.* **71,** 3414 (1967).
81. Yasunaga, T., and S. Harada. *Bull. Chem. Soc. Jpn.* **44,** 848 (1971).
82. Reidler, J., and H. B. Silber. *J. Phys. Chem.* **77,** 1275 (1973).
83. Fay, D. P., D. Litchinsky and N. Purdie. *J. Phys. Chem.* **73,** 544 (1969).
84. Garza, V. L., and N. Purdie. *J. Phys. Chem.* **74,** 275 (1970).
85. Fay, D. P., and N. Purdie. *J. Phys. Chem.* **74,** 1160 (1970).
86. Farrow, M. M., and N. Purdie. *J. Solution Chem.* **2,** 513 (1973).
87. Silber, H. B. *J. Chem. Soc., Chem. Commun.* **1971,** 731.
88. Garnsey, R., and D. W. Ebdon. *J. Am. Chem. Soc.* **91,** 50 (1969).
89. Silber, H. B., N. Scheinin, G. Atkinson and J. J. Grecsek. *J. Chem. Soc., Faraday Trans. I.* **68,** 1200 (1972).
90. Farrow, M. M., and N. Purdie. *Inorg. Chem.* **13,** 2111 (1974).
91. Wang, H., and P. Hemmes. *J. Phys. Chem.* **78,** 261 (1974).
92. Periera, M. S., and J. M. Malin. *Inorg. Chem.* **13,** 386 (1974).
93. Geier, G. *Helv. Chim. Acta.* **51,** 94 (1968).
94. Geier, G. *Ber. Bunsenges. Phys. Chem.* **69,** 617 (1965).
95. Graffeo, A. J., and J. L. Bear. *J. Inorg. Nucl. Chem.* **30,** 1577 (1968).
96. Silber, H. B., J. H. Swinehart. *J. Phys. Chem.,* **71,** 4344 (1967).
97. Silber, H. B., R. D. Farina and J. H. Swinehart. *Inorg. Chem.* **8,** 819 (1969).
98. Yatsimirskii, K. B., and L. I. Budarin. *Dokl. Akad. Nauk SSSR.* **180,** 1383 (1968).
99. Beck, M. T. *Coord. Chem. Rev.* **3,** 91 (1968).
100. Fuoss, R. M. *J. Am. Chem. Soc.* **80,** 5059 (1958).
101. Eigen, M. *Z. Phys. Chem.* (*Frankfurt am Main*) NF **1,** 176 (1954).
102. Rorabacher, D. B., T. S. Turan, J. A. Defever and W. G. Nickels. *Inorg. Chem.* **8,** 1498 (1969).
103. Lin, C., and D. B. Rorabacher. *Inorg. Chem.* **12,** 2402 (1973).
104. Kodama, M., K. Namekawa and T. Horiuchi. *Bull. Chem. Soc. Jpn.* **47,** 2011 (1974).
105. Eisenstadt, M. *J. Chem. Phys.* **51,** 4421 (1969).
106. Rorabacher, D. B., and C. A. Melendez-Cepeda. *J. Am. Chem. Soc.* **93,** 6071 (1971).
107. Baldwin, W. G., and D. R. Stranks. *Aust. J. Chem.* **21,** 2161 (1968).
108. Patel, R. C., and R. S. Taylor. *J. Phys. Chem.* **77,** 2318 (1973).
109. Malin, J. M., and J. H. Swinehart. *Inorg. Chem.* **7,** 250 (1968).
110. Kruse, W., and D. Thusius. *Inorg. Chem.* **7,** 464 (1968).

111. Baker, B. R., M. Orhanovic and N. Sutin. *J. Am. Chem. Soc.* **89,** 722 (1967).
112. Hayes, R. G., and R. J. Myers. *J. Chem. Phys.* **40,** 877 (1964).
113. Kustin, K., I. A. Taub and E. Weinstock. *Inorg. Chem.* **5,** 1079 (1966).
114. Rorabacher, D. B. *Inorg. Chem.* **5,** 1891 (1966).
115. Hammes, G. G., and J. I. Steinfeld, *J. Am. Chem. Soc.* **84,** 4639 (1962).
116. Letter, J. E., and R. B. Jordan. *Inorg. Chem.* **10,** 2692 (1971).
117. Cassatt, J. C., W. A. Johnson, L. M. Smith and R. G. Wilkins. *J. Am. Chem. Soc.* **94,** 8399 (1972).
118. Letter, J. E., and R. B. Jordan. *Inorg. Chem.* **13,** 1152 (1974).
119. Melson, G. A., and R. G. Wilkins. *J. Chem. Soc.* **1962,** 4208.
120. Bulmer, R. S., E. F. Caldin and A. W. Walton. *Trans. Faraday Soc.* **67,** 3343 (1971).
121. Brumfitt, G., E. F. Caldin and R. H. Holyer, unpublished data quoted in Ref. *120.*
122. Kustin, K., and M. A. Wolff. *J. Chem. Soc., Dalton Trans.* **1973,** 103.
123. Davies, A. G., and W. MacF. Smith. *Proc. Chem. Soc., London.* **1961,** 380.
124. Dickert, F., H. Hoffman and T. Janjic. *Ber. Bunsenges. Phys. Chem.* **78,** 712 (1974).
125. Brintzinger, H., and G. G. Hammes. *Inorg. Chem.* **5,** 1286 (1966).
126. Hoffmann, H. *Ber. Bunsenges. Phys. Chem.* **73,** 432 (1969).
127. Diebler, H., and Ph. Rosen. *Ber. Bunsenges. Phys. Chem.* **76,** 1031 (1972).
128. Eisenstadt, M., and H. L. Friedman. *J. Chem. Phys.* **48,** 4445 (1968).
129. Harada, S., Y. Tsuji and T. Yasunaga. *Bull. Chem. Soc. Jpn.* **45,** 1930 (1972).
130. Hertz, H. G. *Z. Elektrochem.* **65,** 36 (1961).
131. Kallen, T. W., and J. E. Earley. *J. Chem. Soc. D.* **1970,** 851.
132. Postmus, C., and E. L. King. *J. Phys. Chem.* **59,** 1208, 1216 (1955).
133. Below, J. F., R. E. Connick and C. P. Coppel. *J. Am. Chem. Soc.* **80,** 2961 (1958).
134. Posey, F. A., and H. Taube. *J. Am. Chem. Soc.* **78,** 15 (1956).
135. Johansson, L. *Acta Chem. Scand.* **27,** 1637 (1973).
136. Johansson, L. *Acta Chem. Scand.* **27,** 2335 (1973).
137. Davies, G. *Coord. Chem. Rev.* **4,** 199 (1969).
138. Pouli, D., and W. MacF. Smith. *Can. J. Chem.* **38,** 567 (1960).
139. Seewald, D., and N. Sutin. *Inorg. Chem.* **2,** 643 (1963).
140. Cavasino, F. P., and E. Di Dio. *J. Chem. Soc. A.* **1970,** 1151.
141. Tamura, K. *Bull. Chem. Soc. Jpn.* **46,** 1581 (1973).
142. Nakamura, K., T. Tsuchida, A. Yamagishi and M. Fujimoto. *Bull. Chem. Soc. Jpn.* **46,** 456 (1973).
143. Gouger, S., and J. Stuehr. *Inorg. Chem.* **13,** 379 (1974).
144. Kalidas, C., W. Knoche and D. Papadopoulos. *Ber. Bunsenges. Phys. Chem.* **75,** 106 (1971).
145. Diebler, H., *Proc. XVI ICCC.* Dublin (1974).
146. Diebler, H. *Z. Phys. Chem. (Frankfurt am Main).* **68,** 64 (1969).
147. Petek, M., T. E. Neal, R. L. McNeely and R. W. Murray. *Anal. Chem.* **45,** 32 (1973).
148. Baker, B. R., N. Sutin and T. J. Welch. *Inorg. Chem.* **6,** 1948 (1967).
149. Patel, R. C., and H. Diebler. *Ber. Bunsenges. Phys. Chem.* **76,** 1035 (1972).
150. Espenson, J. H., and J. R. Pladziewicz. *Inorg. Chem.* **9,** 1380 (1970).

151. Tonat, R., and A. Rigo. *J. Inorg. Nucl. Chem.* **36,** 611 (1974).
152. Hale, C. F., and E. L. King. *J. Phys. Chem.* **71,** 1779 (1967).
153. Espenson, J. H., and E. L. King. *J. Phys. Chem.* **64,** 380 (1960).
154. Swaddle, T. W., and G. Guastalla. *Inorg. Chem.* **7,** 1915 (1968).
155. Swaddle, T. W., and E. L. King. *Inorg. Chem.* **4,** 532 (1965).
156. Sasaki, Y., and A. G. Sykes. *J. Chem. Soc., Dalton Trans.* **1975,** 1048.
157. Wells, C. F., and L. V. Kuritsyn. *J. Chem. Soc. A.* **1970,** 676.
158. Davies, G., L. J. Kirschenbaum and K. Kustin. *Inorg. Chem.* **7,** 146 (1968).
159. Davies, G., and K. Kustin. *Inorg. Chem.* **8,** 484 (1969).
160. Davies, G., and K. Kustin. *Trans. Faraday Soc.* **65,** 1630 (1969).
161. Conocchioli, T. J., G. H. Nancollas and N. Sutin. *Inorg. Chem.* **5,** 1 (1966).
162. Davies, G., and K. O. Watkins. *J. Phys. Chem.* **74,** 3388 (1970).
163. Thompson, R. C. *J. Phys. Chem.* **72,** 2642 (1968).
164. Murmann, R. K., J. C. Sullivan and R. C. Thompson. *Inorg. Chem.* **7,** 1876 (1968).
165. Connick, R. E., and C. P. Coppel. *J. Am. Chem. Soc.* **81,** 6389 (1959).
166. Yasunaga, T., and S. Harada. *Bull. Chem. Soc. Jpn.* **42,** 2165 (1969).
167. Rowley, J. K., and N. Sutin. *J. Phys. Chem.* **74,** 2043 (1970).
168. Davis, G. G., and W. MacF. Smith. *Can. J. Chem.* **40,** 1836 (1962).
169. Wendt, H., and H. Strehlow. *Z. Elektrochem.* **66,** 228 (1962).
170. Goodall, D. M., P. W. Harrison, M. J. Hardy and C. J. Kirk. *J. Chem. Educ.* **49,** 675 (1972).
171. Cavasino, F. P., and M. Eigen. *Ric. Sci. Parte 2: Sez. A.* **4,** 509 (1964).
172. Funahashi, S., S. Adachi and M. Tanaka. *Bull. Chem. Soc. Jpn.* **46,** 479 (1973).
173. Cavasino, F. P. *J. Phys. Chem.* **72,** 1378 (1968).
174. Baker, F. L., and W. MacF. Smith. *Can. J. Chem.* **48,** 3100 (1970).
175. Jayson, G. G., B. J. Parsons and A. J. Swallow. *J. Chem. Soc., Faraday I.* **69,** 1079 (1973).
176. Accascina, F., F. P. Cavasino and E. Di Dio. *Trans. Faraday Soc.* **65,** 489 (1969).
177. Pandey, R. N., and W. MacF. Smith. *Can. J. Chem.* **50,** 194 (1972).
178. Accascina, F., F. P. Cavasino and S. D'Alessandro. *J. Phys. Chem.* **71,** 2474 (1967).
179. Carlyle, D. W., and J. H. Espenson. *Inorg. Chem.* **6,** 1370 (1967).
180. Bauer, R. F., and W. MacF. Smith. *Can. J. Chem.* **43,** 2755 (1965).
181. Moorhead, E. G., and N. Sutin. *Inorg. Chem.* **5,** 1866 (1966).
182. Singleton, D. L., and J. H. Swinehart. *Inorg. Chem.* **6,** 1536 (1967).
183. Walker, R. G., and K. O. Watkins. *Inorg. Chem.* **7,** 885 (1968).
184. Espenson, J. H., and D. F. Dustin. *Inorg. Chem.* **8,** 1760 (1969).
185. Kuehn C., and W. Knoche. *Trans. Faraday Soc.* **67,** 2101 (1971).
186. Strelow, H., and H. Wendt. *Inorg. Chem.* **2,** 6 (1963).
187. Schlund, A., and H. Wendt. *Ber. Bunsenges. Phys. Chem.* **72,** 652 (1968).
188. Hurwitz, P., and K. Kustin, *J. Phys. Chem.* **71,** 324 (1967).
189. Ahrens, M. L., H. Diebler and H. Wendt, unpublished data quoted in Ref. *179.*
190. Sasaki, Y., R. S. Taylor and A. G. Sykes. *J. Chem. Soc., Dalton, Trans.* **1975,** 396.
191. Frei, V., and H. Wendt. *Ber. Bunsenges. Phys. Chem.* **74,** 593 (1970).

192. Eigen, M. *Advances in the Chemistry of Coordination Compounds.* S. Kirschner, ed., Macmillan, New York, 1961, p. 371.
193. Eigen, M., and J. S. Johnson. *Annu. Rev. Phys. Chem.* **11,** 307 (1960).
194. Stranks, D. R. *Pure Appl. Chem.* **38,** 303 (1974).
195. Caldin, E. F., M. W. Grant and B. H. Hasinoff. *Chem. Commun.* **1971,** 1351.
196. Caldin, E. F., M. W. Grant and B. H. Hasinoff. *J. Chem. Soc., Faraday Trans. I.* **68,** 2247 (1972).
197. Yu, A. D., M. D. Waissbluth and R. R. Grieger. *Rev. Sci. Instrum.* **44,** 1390 (1973).
198. Hunt, H. R., and H. Taube. *J. Am. Chem. Soc.* **80,** 2642 (1958).
199. Stanks, D. R., and N. Vanderhoek, *Inorg. Chem.,* **15,** 2639 (1976).
200. Swaddle, T. W., and D. R. Stranks. *J. Am. Chem. Soc.* **94,** 8357 (1972).
201. Tong, S. B., and T. W. Swaddle. *Inorg. Chem.* **13,** 1538 (1974).
202. Jones, W. E., and T. W. Swaddle. *Chem. Commun.* **1969,** 998.
203. Jones, W. E., L. R. Carey and T. W. Swaddle. *Can. J. Chem.* **50,** 2739 (1972).
204. Bott, H. L., A. J. Poë and K. Shaw. *Chem. Commun.* **1968,** 793.
205. Borghi, E., and F. Monacelli. *Inorg. Chim. Acta.* **5,** 211 (1968).
206. Monacelli, F. *Inorg. Chim. Acta.* **2,** 263 (1968).
207. Spiro, T. G., A. Revesz and J. Lee. *J. Am. Chem. Soc.* **90,** 4000 (1968).
208. Langford, C. H., and T. R. Stengle. *Annu. Rev. Phys. Chem.* **19,** 193 (1968).
209. Dickert, F., H. Hoffmann and W. Jaenicke. *Ber. Bunsenges. Phys. Chem.* **74,** 500 (1970).
210. Fischer, P., H. Hoffmann and G. Platz. *Ber. Bunsenges. Phys. Chem.* **76,** 1060 (1972).
211. Pearson, R. G., and P. Ellgen. *Inorg. Chem.* **6,** 1379 (1967).
212. Williams, J., S. Petrucci, B. Sesta and M. Battistini. *Inorg. Chem.* **13,** 1968 (1974).
213. Langford, C. H., and H. G. Tsiang. *Inorg. Chem.* **9,** 2346 (1970).
214. Frankel, L. S. *J. Chem. Soc. D.* **1969,** 1254.
215. Dickert, F., and H. Hoffmann. *Ber. Bunsenges. Phys. Chem.* **75,** 1320 (1971).
216. Chattopadhyay, P. K., and J. F. Coetzee. *Inorg. Chem.* **12,** 113 (1973).
217. Bennetto, H. P., and E. F. Caldin. *J. Chem. Soc. A.* **1971** 2191.
218. Bennetto, H. P., and E. F. Caldin. *J. Chem. Soc. A.* **1971,** 2211.
219. Bennetto, H. P., and E. F. Caldin. *J. Chem. Soc. A.* **1971,** 2207.
220. Bennetto, H. P., and E. F. Caldin. *J. Chem. Soc A.* **1971,** 2198.
221. Caldin, E. F., and H. P. Bennetto. *J. Solution Chem.* **2,** 217 (1973).
222. Burgess, J., and M. V. Twigg. *J. Chem. Soc., Dalton Trans.* **1974.** 2032.
223. Moore, P., and D. M. W. Buck. *J. Chem. Soc., Dalton Trans.* **1973,** 1602.
224. Benton, D. J., and P. Moore. *J. Chem. Soc., Dalton Trans.* **1973,** 399.
225. Buck, D. M. W., and P. Moore. *J. Chem. Soc., Chem. Commun.* **1974,** 60.
226. Chattopadhyay, P. K., and J. F. Coetzee. *Anal. Chem.* **46,** 2014 (1974).
227. Buck, D. M. W., and P. Moore. *J. Chem. Soc., Dalton Trans.* **1975,** 409.
228. Langford, C. H., J. P. K. Tong and A. Merbach. *Can. J. Chem.* **53,** 702, (1975).
229. Bradić, Z., and M. Pribanić and S. Asperger. *J. Chem. Soc., Dalton Trans.* **1975,** 353.
230. Caldin, E. F., and M. W. Grant. *J. Chem. Soc., Faraday Trans. I,* **1973,** 1648.

231. MacKellar, W. J., and D. B. Rorabacher. *J. Am. Chem. Soc.* **93,** 4379 (1971).
232. Shu, F. R., and D. B. Rorabacher. *Inorg. Chem.* **11,** 1496 (1972).
233. Nyssen, G. A., and D. W. Margerum. *Inorg. Chem.* **9,** 1814 (1970).
234. Geier, G. *Chimia.* **23,** 148 (1969).
235. Lin, C-T., and J. L. Bear. *J. Phys. Chem.* **75,** 3705 (1971).
236. Hamid, S. A., I. P. Alimarin and I. V. Puzdrenkova. *Vestn. Mosk. Univ., Khim.* **20,** 71 (1965).
237. Cavasino, F. P. *J. Phys. Chem.* **69,** 4380 (1965).
238. Hoffmann, H., and J. Stuehr. *J. Phys. Chem.* **70,** 955 (1966).
239. Margerum, D. W., D. B. Rorabacher and J. F. G. Clarke, Jr. *Inorg. Chem.* **2,** 667 (1963).
240. Taylor, R. W., H. K. Stepien and D. B. Rorabacher. *Inorg. Chem.* **13,** 1282 (1974).
241. Nancollas, G. H., and N. Sutin. *Inorg. Chem.* **3,** 360 (1964).
242. Cassatt, J. C., and R. G. Wilkins. *J. Am. Chem. Soc.* **90,** 6045 (1968).
243. Steinhaus, R. K., and D. W. Margerum. *J. Am. Chem. Soc.* **88,** 441 (1966).
244. Hammes, G. G., and M. L. Morell. *J. Am. Chem. Soc.* **86,** 1497 (1964).
245. Moss, D. B., C-T. Lin and D. B. Rorabacher. *J. Am. Chem. Soc.* **95,** 5179 (1973).
246. Kirschenbaum, L. J., and K. Kustin, *J. Chem. Soc., A.* **1970,** 684.
247. Rorabacher, D. B., and D. B. Moss. *Inorg. Chem.* **9,** 1314 (1970).
248. Turan, T. S. *Inorg. Chem.* **13,** 1584 (1974).
249. Funahashi, S., and M. Tanaka. *Inorg. Chem.* **8,** 2159 (1969).
250. Jones, J. P., and D. W. Margerum. *J. Am. Chem. Soc.* **92,** 470 (1970).
251. Hoffmann, H., and E. Yeager. *Ber. Bunsenges. Phys. Chem.* **74,** 641 (1970).
252. Platz, G., and H. Hoffmann. *Ber. Bunsenges. Phys. Chem.* **76,** 491 (1972).
253. Harada, S., H. Tanabe and T. Yasunaga. *Bull. Chem. Soc. Jpn.* **46,** 3125 (1973).
254. Harada, S., Y. Okuue, H. Kan and T. Yasunaga. *Bull. Chem. Soc. Jpn.* **47,** 769 (1974).
255. Bear, J. L., and C-T. Lin. *J. Phys. Chem.* **72,** 2026 (1968).
256. Calvaruso, G., F. P. Cavasino, and E. Di Dio. *J. Chem. Soc., Dalton Trans.* **1972,** 2632.
257. Calvaruso, G., F. P. Cavasino and E. Di Dio. *J. Inorg. Nucl. Chem.* **36,** 2061 (1974).
258. Harada, S., K. Amidaiji and T. Yasunaga. *Bull. Chem. Soc. Jpn.* **45,** 1752 (1972).
259. Hoffmann, H., and U. Nickel. *Ber. Bunsenges. Phys. Chem.* **72,** 1096 (1968).
260. Harada, S., and T. Yasunaga. *Bull. Chem. Soc. Jpn.* **46,** 502 (1973).
261. Harada, S., H. Tanabe and T. Yasunaga. *Bull. Chem. Soc. Jpn.* **46,** 2450 (1973).
262. Cavasino, F. P., E. DiDio and G. Locanto. *J. Chem. Soc., Dalton Trans.* **1973,** 2419.
263. Kustin, K., R. F. Pasternack and E. M. Weinstock. *J. Am. Chem. Soc.* **88,** 4610 (1966).
264. Kowalak, A., K. Kustin, R. F. Pasternack and S. Petrucci. *J. Am. Chem. Soc.* **89,** 3126 (1967).

265. Pasternack, R. F., and K. Kustin. *J. Am. Chem. Soc.* **90,** 2295 (1968).
266. Kustin, K., and R. F. Pasternack. *J. Am. Chem. Soc.* **90,** 2805 (1968).
267. Pasternack, R. F., E. Gibbs and J. C. Cassett. *J. Phys. Chem.* **73,** 3814 (1969).
268. Pasternack, R. F., M. Angwin, L. Gipp and R. Reingold, *J. Inorg. Nucl. Chem.* **34,** 2329 (1972).
269. Makinen, W. B., A. F. Pearlmutter and J. Stuehr. *J. Am. Chem. Soc.* **91,** 4083 (1969).
270. Schwarzenbach, G. *Helv. Chim. Acta.* **35,** 2344 (1952).
271. Ahmed, A. K. S., and R. G. Wilkins. *J. Chem. Soc.* **1960,** 2895.
272. Bertsch, C., W. C. Fernelius and P. B. Block. *J. Phys. Chem.* **62,** 444 (1958).
273. Cotton, F. A., and F. E. Harris. *J. Phys. Chem.* **60,** 1451 (1956).
274. Ahmed, A. K. S., and R. G. Wilkins. *J. Chem. Soc.* **1960,** 2901.
275. Davies, G., K. Kustin and R. F. Pasternack. *Inorg. Chem.* **8,** 1535 (1969).
276. Pearlmutter, A. F., and J. Stuehr. *J. Am. Chem. Soc.* **90,** 858 (1968).
277. Letter, J. E., and R. B. Jordan. *J. Am. Chem. Soc.* **97,** 2381 (1975).
278. Margerum, D. W., P. J. Menardi and D. L. Janes. *Inorg. Chem.* **6,** 283 (1967).
279. Hamm, R. E., *J. Am. Chem. Soc.* **75,** 5670 (1953).
280. Sykes, A. G., and R. N. F. Thornley. *J. Chem. Soc. A.* **1969,** 655.
281. Margerum, D. W., and T. J. Bydalek. *Inorg. Chem.* **2,** 683 (1963).
282. Rorabacher, D. B., and D. W. Margerum. *Inorg. Chem.* **3,** 382 (1964).
283. Bydalek, T. J., and M. L. Blomster. *Inorg. Chem.* **3,** 667 (1964).
284. Diebler, H. *Ber. Bunsenges. Phys. Chem.* **74,** 268 (1970).
285. Hague, D. N., and M. Eigen. *Trans. Faraday Soc.* **62,** 1236 (1966).
286. Hague, D. N., and M. S. Zetter. *Trans. Faraday Soc.* **66,** 1176 (1970).
287. Hague, D. N., S. R. Martin and M. S. Zetter. *J. Chem. Soc., Faraday Trans. I.* **68,** 37 (1972).
288. Johnson, W. A., and R. G. Wilkins. *Inorg. Chem.* **9,** 1917 (1970).
289. Holyer, R. H., C. D. Hubbard, S. F. A. Kettle and R. G. Wilkins. *Inorg. Chem.* **4,** 929 (1965); **5,** 622 (1966).
290. Cayley, G. R., and D. N. Hague. *J. Chem. Soc., Faraday Trans. I.* **68,** 2259 (1972).
291. Hubbard, C. D., and W. Palaitis. *Inorg. Chem.* **12,** 480 (1973).
292. Holmes, F., and F. Jones. *J. Chem. Soc.* **1960,** 2398.
293. Turan, T. S., and D. B. Rorabacher. *Inorg. Chem.* **11,** 288 (1972).
294. Rorabacher, D. B., and R. B. Cruz, to be published.
295. Cabbiness, D. K., and D. W. Margerum. *J. Am. Chem. Soc.* **92,** 2151 (1970).
296. Margerum, D. W., and H. M. Rosen. *Inorg. Chem.* **7,** 299 (1968).
297. Pagenkopf, G. K., and D. W. Margerum. *Inorg. Chem.* **7,** 2514 (1968).
298. Lin, C.-T., D. B. Rorabacher, G. R. Cayley and D. W. Margerum, *Inorg. Chem.,* **14,** 919 (1975).
299. Margerum, D. W., and R. A. Bauer, unpublished data.
300. Weaver, J., and P. Hambright. *Inorg. Chem.* **8,** 167 (1969).
301. Bauer, R. A., W. R. Robinson and D. W. Margerum. *Chem. Commun.* **1973,** 289.
302. Cabbiness, D. K., Ph.D. Thesis, Purdue University (1970).
303. Kaden, T. *Chimia.* **23,** 193 (1969).
304. Kaden, T., private communication.
305. Hertli, L., and T. A. Kaden. *Helv. Chim. Acta.* **57,** 1328 (1974).

306. Stanmann, W., and T. A. Kaden. *Helv. Chim. Acta.* **58,** 1358 (1975).
307. Buxtorf, R., and T. A. Kaden. *Helv. Chim. Acta.* **57,** 1035 (1974).
308. Smith, G. F., and D. W. Margerum, unpublished data.
309. Hambright, P. *Ann. N. Y. Acad. Sci.* **206,** 443 (1973).
310. Longo, F., E. M. Brown, D. J. Quinby, A. D. Adler and M. Meot-Ner. *Ann. N.Y. Acad. Sci.* **206,** 420 (1973).
311. Baker, H., P. Hambright and L. Wagner. *J. Am. Chem. Soc.* **95,** 5942 (1973).
312. Fleischer, E. B., and M. Krishnamurthy. *Ann. N. Y. Acad. Sci.* **206,** 32 (1973).
313. Hambright, P., and P. B. Chock. *J. Am. Chem. Soc.* **96,** 3123 (1974).
314. Stone, A., and E. B. Fleischer. *J. Am. Chem. Soc.* **90,** 2735 (1968).
315. Dearden, G. R., and A. H. Jackson. *J. Chem. Soc. D.* **1970,** 205.
316. Broadhurst, M. J., R. Grigg and G. Shelton. *J. Chem. Soc. D.* **1970,** 231.
317. Grigg, R., A. Sweeney, G. R. Dearden, A. H. Jackson and A. W. Johnson. *J. Chem. Soc. D.* **1970,** 1273.
318. Shears, B., B. Shah and P. Hambright. *J. Am. Chem. Soc.* **93,** 776 (1971).
319. Margerum, D. W., and L. I. Simándi. *Proc. Int. Conf. Coord. Chem., 9th.* W. Schneider, ed., *Verlag Helv. Chim. Acta,* Basel, Switzerland, 1966, p. 371.
320. Kolski, G. B., and D. W. Margerum. *Inorg. Chem.* **7,** 2239 (1968).
321. Eigen, M., and G. G. Hammes. *J. Am. Chem. Soc.* **82,** 5951 (1960).
322. Eigen, M., and G. G. Hammes. *J. Am. Chem. Soc.* **83,** 2786 (1961).
323. Roche, T. S., and R. G. Wilkins, unpublished data quoted in Ref. *117.*
324. Roche, T. S., and R. G. Wilkins. *J. Am. Chem. Soc.* **96,** 5082 (1974).
325. Shepherd, R. E., G. M. Hodgson and D. W. Margerum. *Inorg. Chem.* **10,** 989 (1971).
326. Margerum, D. W., and B. A. Zabin. *J. Phys. Chem.* **66,** 2214 (1962).
327. Ackerman, H., and G. Schwarzenbach. *Helv. Chim. Acta.* **35,** 485 (1952).
328. Sharma, V. S., and D. L. Leussing. *Inorg. Chem.* **11,** 138 (1972).
329. Sharma, V. S., and D. L. Leussing. *Inorg. Chem.* **11,** 1955 (1972).
330. Kaden, T. *Helv. Chim. Acta.* **53,** 617 (1970).
331. Bydalek, T. J., and A. H. Constant. *Inorg. Chem.* **4,** 833 (1965).
332. Litvak, H., and D. W. Margerum, unpublished data.
333. Pausch, J. B., and D. W. Margerum. *Anal. Chem.* **41,** 226 (1969).
334. Carr, J. D., and D. G. Swartzfager. *J. Am. Chem. Soc.* **97,** 315 (1975).
335. Eigen, M., W. Kruse, G. Maass and L. De Maeyer. *Prog. React. Kinet.* **2,** 287 (1964).
336. Eigen, M. *Angew. Chem., Int. Ed. Engl.* **3,** 1 (1964).
337. Rabenstein, D. L., and R. J. Kula. *J. Am. Chem. Soc.* **91,** 2492 (1969).
338. Maguire, J. *Can. J. Chem.* **52,** 4106 (1974).
339. Rabenstein, D. L. *J. Am. Chem. Soc.* **93,** 2869 (1971).
340. Sheinblatt, M., and H. S. Gutowsky. *J. Am. Chem. Soc.* **86,** 4814 (1964).
341. Karpel, R. L., K. Kustin and R. F. Pasternack. *Biochim. Biophys. Acta.* **177,** 434 (1969).
342. Cayley, G. R., and D. W. Margerum, unpublished data.
343. Margerum, D. W., and G. R. Dukes. *Metal Ions in Biological Systems.* Vol. 1, Chap. V. H. Sigel, ed., Marcel Dekker, New York, 1973.
344. Jaffe, M. R., D. P. Fay, M. Cefola and N. Sutin. *J. Am. Chem. Soc.* **93,** 2878 (1971).
345. Spedding J. H., et al. *J. Phys. Chem.* **70,** 2423, 2430, 2440, 2450 (1966).

346. Moeller, T., D. F. Martin, L. C. Thompson, R. Ferrús, G. R. Feistel and W. J. Randall. *Chem. Rev.* **65,** 1 (1965).
347. Geier, G., and U. Karlen. *Helv. Chim. Acta.* **54,** 135 (1971).
348. Geier, G., and C. K. Jørgensen. *Chem. Phys. Lett.* **9,** 263 (1971).
349. Geier, G., and U. Karlen. *Proc. XIII ICCC.* Cracowe, Poland, p. 45, 1970.
350. Cayley, G. R., Ph.D. Thesis, University of Kent, Canterbury (1972).
351. Lin, C-T., and J. L. Bear. *J. Inorg. Nucl. Chem.* **31,** 263 (1969).
352. Banyasz, J. L., and J. E. Stuehr. *J. Am. Chem. Soc.* **95,** 7226 (1973).
353. Frey, C. M., J. L. Banyasz and J. E. Stuehr. *J. Am. Chem. Soc.* **94,** 9198 (1972).
354. Kustin, K., and K. O. Watkins. *Inorg. Chem.* **3,** 1706 (1964).
355. Diebler, H., M. Eigen and G. G. Hammes. *Z. Naturforsch.* **15b,** 554 (1960).
356. Czerlinski, G., H. Diebler and M. Eigen. *Z. Phys. Chem.* (*Frankfurt am Main*). **19,** 246 (1959).
357. Eigen, M., and G. Maass. *Z. Phys. Chem. N.F.* **49,** 163 (1966).
358. Larsen, N. R., and A. Jensen. *Acta Chem. Scand.* **28A,** 638 (1974).
359. Kula, R. J., and D. L. Rabenstein. *J. Am. Chem. Soc.* **89,** 552 (1967).
360. Onodera, T., and M. Fuimoto. *Bull. Chem. Soc. Jpn.* **44,** 2003 (1971).
361. Boivin, G., and M. Zador, *Can. J. Chem.* **48,** 3053 (1970).
362. Farrow, M. M., N. Purdie and E. M. Eyring. *Inorg. Chem.* **13,** 2024 (1974).
363. Hammes, G. G., and S. A. Levison. *Biochemistry.* **3,** 1504 (1964).
364. Bryson, A., and I. S. Fletcher. *Aust. J. Chem.* **23,** 1095 (1970).
365. Pearson, R. G., and O. A. Gansow. *Inorg. Chem.* **7,** 1373 (1968).
366. Kruse, W., unpublished data, quoted in Ref. *12.*
367. Bennett, L. E., and H. Taube. *Inorg. Chem.* **7,** 254 (1968).
368. Ng, F. T. T., R. Pizer and R. G. Wilkins, quoted in Ref. *618.*
369. Wilkins, R. G., and K. P. Williams. *J. Am. Chem. Soc.* **96,** 2241 (1974).
370. Hague, D. N., and S. R. Martin. *J. Chem. Soc., Dalton Trans.* **1974,** 254.
371. Koryta, J. *Z. Elektrochem.* **64,** 196 (1960).
372. Kuntz, G. P. P., Y. F. Lam and G. Kotowycz. *Can. J. Chem.* **53,** 926 (1975).
373. Kimura, M. *Nippon Kagaku Zasshi.* **90,** 1246 (1969).
374. Ellis, K. J., and A. McAuley. *J. Chem. Soc., Dalton Trans.* **1973,** 533.
375. Fogg, P. G. T., and R. J. Hall, *J. Chem. Soc. A.* **1971,** 1365.
376. Saini, G., and E. Mentasti. *Inorg. Chim. Acta.* **4,** 210 (1970).
377. Saini, G., and E. Mentasti. *Inorg. Chim. Acta.* **4,** 585 (1970).
378. Bennetto, H. P., and Z. S. Imani. *J. Chem. Soc., Faraday Trans. I.* **71,** 1143 (1975).
379. Fay, D. P., A. R. Nichols, Jr. and N. Sutin. *Inorg. Chem.* **10,** 2096 (1971).
380. Hubbard, C. D. *Inorg. Chem.* **10,** 2340 (1971).
381. Wilkins, R. G. *Inorg. Chem.* **3,** 520 (1964).
382. Cobb, M. A., and D. N. Hague. *Trans. Faraday Soc.* **67,** 3069 (1971).
383. Cobb, M. A., and D. N. Hague. *Chem. Commun.* **1971,** 20.
384. Taylor, R. W., Ph.D. Thesis, Wayne State University (1973).
385. Hubbard, C. D. *Inorg. Nucl. Chem. Lett.* **7,** 139 (1971).
386. Hubbard, C. D. *J. Inorg. Nucl. Chem.* **36,** 1178 (1974).
387. Rudolf, S. A., and J. M. Sturtevant. *Biophys. Soc. Abstr.,* 189a (Feb. 1970); Rudolf, S. A., Ph.D. Thesis, Yale University (1971).
388. Davies, G., K. Kustin and R. F. Pasternack. *Int. J. Chem. Kinet.* **1,** 43 (1969).

389. Williams, J. C., and S. Petrucci. *J. Am. Chem. Soc.* **95,** 7619 (1973).
390. Pasternack, R. F., K. Kustin, L. A. Hughes and E. Gibbs *J. Am. Chem. Soc.* **91,** 4401 (1969).
391. Barr, M. L., K. Kustin and S-T. Liu. *Inorg. Chem.* **1,** 1486 (1973).
392. Barr, M. L., E. Baumgartner and K. Kustin. *J. Coord. Chem.* **2,** 263 (1973).
393. Kustin, K., and S-T. Liu. *J. Chem. Soc., Dalton Trans.* **1973** 278.
394. Grant, M. W. *J. Chem. Soc., Faraday Trans. I.* **69,** 560 (1973).
395. Kowalak, A., K. Kustin and R. F. Pasternack. *J. Phys. Chem.* **73,** 281 (1969).
396. Davies, G., K. Kustin and R. F. Pasternack. *Trans. Faraday Soc.* **64,** 1006 (1968).
397. Pasternack, R. F., L. Gipp and H. Sigel. *J. Am. Chem. Soc.* **94,** 8031 (1972).
398. Kustin, K., and R. F. Pasternack. *J. Phys. Chem.* **73,** 1 (1969).
399. Cavasino, F. P. *Ric. Sci.* **8,** 1120 (1965) (II-A).
400. Wendt, H. *Ber. Bunsenges. Phys. Chem.* **70,** 556 (1966).
401. Long, F. A., private communication quoted in Ref. *12.*
402. Tanaka, N. *Bull. Chem. Soc. Jpn.* **36,** 73 (1963).
403. Kimura, M. *Bull. Chem. Soc. Jpn.* **42,** 404 (1969).
404. Kimura, M. *Nippon Kaguku Zasshi.* **90,** 1246 (1969).
405. Steinhaus, R. K., and Z. Amjad. *Inorg. Chem.* **12,** 151 (1973).
406. Murmann, R. K. *J. Am. Chem. Soc.* **84,** 1349 (1962).
407. Wawro, R. G., and R. G. Wilkins, quoted in Ref. *5.*
408. Melvin, W. S., D. P. Rablen and G. Gordon. *Inorg. Chem.* **11,** 488 (1972).
409. Ellis, P., R. Hogg and R. G. Wilkins. *J. Chem. Soc.* **1959,** 3308.
410. Irving, H., and D. Mellor. *J. Chem. Soc.* **1962,** 5222.
411. Barrett, P. F., and W. MacF. Smith. *Can. J. Chem.* **42,** 934 (1964).
412. Margerum, D. W., R. I. Bystroff and C. V. Banks, *J. Am. Chem. Soc.* **78,** 4211 (1956).
413. Irving, H., and D. H. Mellor. *J. Chem. Soc.* **1962,** 5237.
414. Cobb, M. A., and D. N. Hague. *J. Chem. Soc., Faraday Trans. I.* **68,** 932 (1972).
415. Karpel, R. L., K. Kustin and M. A. Wolff. *J. Phys. Chem.* **75,** 799 (1971).
416. Boivin, G., and M. Zador. *Bull. Soc. Chim. Fr.* **1971,** 4279.
417. Pearson, R. G., and O. P. Anderson. *Inorg. Chem.* **9,** 39 (1970).
418. Hague, D. N., and K. Kinley, *J. Chem. Soc., Dalton Trans.* **1974,** 249.
419. Steinfeld, J. I., and G. G. Hammes. *J. Phys. Chem.* **67,** 528 (1963).
420. McClellan, B. E., and H. Frieser. *Anal. Chem.* **36,** 2262 (1964).
421. Frieser, H. *Solvent Extraction Chemistry, Proc. Int. Conf., Gottenburg, Sweden, 1966,* North Holland, Amsterdam (1967) p. 85, D. Dryssen, J. O. Liljenzin, J. Rydberg, eds.
422. Oh, J. S., and H. Frieser. *Anal. Chem.* **39,** 295 (1967).
423. Nakatini, H., and J. Osuji. *Nippon Kagaku Kaishi.* **1972,** 1816.
424. Kodama, M. *Bull. Chem. Soc. Jpn.* **46,** 3422 (1973).
425. Hoffmann, H., and W. Ulbricht. *Z. Naturforsch.* **25B** 1327 (1970).
426. Hoffmann, H., and U. Nickel. *Z. Naturforsch.* **26B,** 299 (1971).
427. Tanaka, N., and M. Kimura. *Bull. Chem. Soc. Jpn.* **40,** 2100 (1967).
428. Cook Jr., C. M., and F. A. Long. *J. Am. Chem. Soc.* **80,** 33 (1958); Tanaka, N., and Y. Sakuma, *Bull. Chem. Soc. Jpn.* **32,** 578 (1959).
429. Pearlmutter-Hayman, B., and F. Secco. *Isr. J. Chem.* **11,** 623 (1973).

430. Katz, H. B., and K. Kustin. *Biochim. Biophys. Acta.* **313,** 235 (1973).
431. Rechnitz, G. A., and Z. Lin. *Anal. Chem.* **39,** 1406 (1967).
432. Bhat, T. R., D. Raahama and J. Sankar. *Inorg. Chem.* **5,** 1132 (1966).
433. Kodama, M., and N. Oyama. *Bull. Chem. Soc. Jpn.* **45,** 2169 (1972).
434. Meyerstein, D. *Inorg. Chem.* **14,** 1716 (1975).
435. Shepherd, R. E., G. M. Hodgson and D. W. Margerum. *Inorg. Chem.* **10,** 989 (1971).
436. Roche, T. S., and R. G. Wilkins. *Chem. Commun.* **1970,** 1681.
437. Sharma, V. S., and D. L. Leussing. *Chem. Commun.* **1970,** 1278.
438. Boivin, G., and M. Zador, *Can. J. Chem.* **50,** 3117 (1972).
439. Ketesz, D., G. Rotilo, M. Brunori, R. Zito and E. Antonini. *Biochem. Biophys. Res. Commun.* **49,** 1208 (1972).
440. Feltch, S. M., J. E. Stuehr and G. W. Tin. *Inorg. Chem.* **14,** 2175 (1975).
441. Brubaker, J. W., A. F. Pearlmutter, J. E. Stuehr and T. V. Vu. *Inorg. Chem.* **13,** 559 (1974).
442. Karpel, R. L., K. Kustin, A. Kowalak and R. F. Pasternack. *J. Am. Chem. Soc.* **93,** 1085 (1971).
443. Pasternack, R. F., M. Angwin and E. Gibbs. *J. Am. Chem. Soc.* **92,** 5878 (1970).
444. Bewick, A., and P. M. Robertson. *Trans. Faraday Soc.* **63,** 678 (1967).
445. Osuji, J., and H. Nakatani. *Nippon Kagaku Kaishi* **1972,** 465.
446. Tanaka, N., and M. Kimura. *Bull. Chem. Soc. Jpn.* **41,** 2375 (1968).
447. Brundage, R. S., R. L. Karpel, K. Kustin and J. Weisel. *Biochem. Biophys. Acta.* **267,** 258 (1972).
448. Kato, K. *Bull. Chem. Soc. Jpn.* **33,** 600 (1960).
449. Tanaka, N., H. Osawa and M. Kamada. *Bull. Chem. Soc. Jpn.* **36,** 530 (1963) and references therein.
450. Kimura, M. *Bull. Chem. Soc. Jpn.* **43,** 1594 (1970).
451. Das, R. R. *J. Inorg. Nucl. Chem.* **34,** 1263 (1972).
452. Cayley, G. R., and D. N. Hague. *Trans. Faraday Soc.* **67,** 786 (1971).
453. Henkens, R. K., and J. M. Sturtevant. *J. Am. Chem. Soc.* **90,** 2669 (1968).
454. Cheung, S. K., F. L. Dixon, E. B. Fleischer, D. Y. Jetle and M. Krishnamurthy. *Bioinorg. Chem.* **2,** 281 (1973).
455. Miceli, J. A., and J. E. Stuehr. *Inorg. Chem.* **11,** 2763 (1972).
456. Reed, G. H., and R. J. Kula. *Inorg. Chem.* **10,** 2050 (1971).
457. Pearson, R., and D. G. DeWit. *J. Coord. Chem.* **2,** 175 (1973).
458. Bril, K., S. Bril and P. Krumholtz. *J. Phys. Chem.* **59,** 596 (1955); **60,** 251 (1956).
459. Fuhr, B. J., and D. L. Rabenstein. *Inorg. Chem.* **12,** 1868 (1973).
460. Tanaka, N., R. Tamamushi and M. Kodama, *Z. Phys. Chem.* (*Frankfurt am Main*) **14,** 141 (1958).
461. Kuempel, J. R., and W. B. Schaap. *Inorg. Chem.* **7,** 2435 (1968).
462. Koryta, J., and Z. Zabransky. *Collect. Czech. Chem. Commun.* **25,** 3153 (1960).
463. Aylward, G. H., and J. W. Hayes. *Anal. Chem.* **37,** 195 (1965).
464. Smith, G. F., and D. W. Margerum. *Inorg. Chem.* **8,** 135 (1969).
465. Taylor, R. S. *Talanta.* **21,** 1210 (1974).

466. Lukomskaya, N. D., T. V. Mal'kova and K. B. Yatsimirskii. *Zh. Neorg. Khim.* **12,** 2462 (1967).
467. Mal'kova, T. V., and V. D. Ovchinnikova. *Russ. J. Inorg. Chem.* **17,** 813 (1972).
468. Secco, F., and M. Venturini. *Inorg. Chem.* **14,** 1978 (1975).
469. Matušek, M., and H. Strehlow. *Ber. Bunsenges. Phys. Chem.* **73,** 982 (1969).
470. Kawai, Y., T. Takahashi, K. Hayashi, T. Imamura, H. Nakayama and M. Fujimoto. *Bull. Chem. Soc. Jpn.* **45,** 1417 (1972).
471. Tanaka, N., and K. Kato. *Bull. Chem. Soc. Jpn.* **33,** 1236 (1960).
472. Takahashi, T., T. Koiso and N. Tanaka. *Nippon Kagaku Zasshi.* **91,** 236 (1970).
473. James, M. R., D. W. Margerum, G. M. Hodgson and R. E. Shepherd, unpublished data.
474. Melson, G. A., and R. G. Wilkins. *J. Chem. Soc.* **1962,** 2662.
475. Weatherburn, D. C., E. J. Billo, J. P. Jones and D. W. Margerum. *Inorg. Chem.* **9,** 1557 (1970).
476. Kolski, G. B., and D. W. Margerum. *Inorg. Chem.* **8,** 1129 (1969).
477. Weatherburn, D. C., and D. W. Margerum, unpublished data.
478. Sigel, H., and D. B. McCormick. *Acc. Chem. Res.* **3,** 201 (1970).
479. Moore, P., Ph.D. Thesis, University of Sheffield (1964).
480. Ahmed, A. K. S., and R. G. Wilkins. *Proc. Chem. Soc.* **1959,** 399.
481. *Stability Constants of Metal Ion Complexes.* Supplement No. 1, The Chemical Society, 1970.
482. Wilkins, R. G. *J. Chem. Soc.* **1962,** 4475.
483. Childers, R. F., and R. A. D. Wentworth. *Inorg. Chem.* **8,** 2218 (1969).
484. Ahmed, A. K. S., and R. G. Wilkins. *J. Chem. Soc.* **1959,** 3700.
485. Lincoln S. F., and C. D. Hubbard. *J. Chem. Soc., Dalton Trans.* **1974,** 2513.
486. Margerum, D. W. "Mechanisms of Inorganic Reactions," *Adv. Chem. Ser.* **49,** 77 (1965).
487. Bosnich, B., M. L. Tobe and G. A. Webb. *Inorg. Chem.* **4,** 1109 (1965).
488. Busch, D. H., K. Farmery, V. Goedken, V. Katovic, A. C. Melnyk, C. R. Sperati and N. Tokel. "Bioinorganic Chemistry." *Adv. Chem. Ser.* **100,** 54 (1971).
489. Kodama, M., and E. Kimura. *J. Chem. Soc., Chem. Commun.* **1975,** 326.
490. Kodama, M., and E. Kimura, *J. Chem. Soc., Chem. Commun.,* **1975,** 891.
491. Jones, T. E., L. L. Zimmer, L. L. Diaddario and D. B. Rorabacher, *J. Am. Chem. Soc.* **97,** 7163 (1975).
492. Loyola, L. V., R. G. Wilkins and R. Pizer, *J. Am. Chem. Soc.,* **97,** 7382 (1975).
493. Rabenstein, D. L., G. Blakney and B. J. Fuhr. *Can. J. Chem.* **53,** 787 (1975).
494. Bydalek, T. J., and D. W. Margerum. *J. Am. Chem. Soc.* **83,** 4326 (1961).
495. Bydalek, T. J., and D. W. Margerum. *Inorg. Chem.* **1,** 852 (1962).
496. Margerum, D. W., and T. J. Bydalek. *Inorg. Chem.* **2,** 678 (1963).
497. Margerum, D. W., B. A. Zabin and D. L. Janes. *Inorg. Chem.* **5,** 250 (1966).
498. Margerum, D. W. *Rec. Chem. Prog.* **24,** 237 (1963).
499. Margerum, D. W., J. B. Pausch, G. A. Nyssen and G. F. Smith. *Anal. Chem.* **41,** 233 (1969).
500. Glentworth, P., and C. L. Wright. *J. Inorg. Nucl. Chem.* **31,** 1263 (1969).

501. Glentworth, P., and B. Wiseall. *Chemical Effects of Nuclear Transformations.* Vol. II. Int. At. Energy Agency, Vienna **1965,** 483. (Proceedings of the international conference held in Vienna, Dec. 1964).
502. Coombs, L. C., and D. W. Margerum. *Inorg. Chem.* **9,** 1711 (1970).
503. Morgenthaler, L. P., and D. W. Margerum. *J. Am. Chem. Soc.* **84,** 710 (1962).
504. Margerum, D. W., and L. P. Morgenthaler. *J. Am. Chem. Soc.* **84,** 706 (1962).
505. Pagenkopf, G. K., and D. W. Margerum. *J. Am. Chem. Soc.* **90,** 6963 (1968).
506. Hauer, H., G. R. Dukes and D. W. Margerum. *J. Am. Chem. Soc.* **95,** 3515 (1973).
507. Paniago, E. B., and D. W. Margerum. *J. Am. Chem. Soc.* **94,** 6704 (1972).
508. Cooper, J. C., L. F. Wong, D. L. Venezky and D. W. Margerum. *J. Am. Chem. Soc.* **96,** 7560 (1974).
509. Cooper, J. C., L. F. Wong and D. W. Margerum, unpublished data.
510. Brice, V. T., and G. K. Pagenkopf. *J. Chem. Soc., Chem. Commun.* **1974,** 75.
511. Pagenkopf, G. K., and V. T. Brice. *Inorg. Chem.* **14,** 3118 (1975).
512. Wong, L. F., and D. W. Margerum, unpublished data.
513. Libby, R. A., and D. W. Margerum. *Biochemistry.* **4,** 619 (1965).
514. Riepe, M. E., and J. H. Wang. *J. Biol. Chem.* **243,** 2779 (1968).
515. Vallee, B. L., and R. J. P. Williams. *Chem. Br.* **4,** 397 (1968).
516. Malmström, B. G., and A. Rosenberg. *Adv. Enzymol.* **21,** 131 (1959).
517. Margerum, D. W., and H. M. Rosen. *J. Am. Chem. Soc.* **89,** 1088 (1967).
518. Jones, J. P., E. J. Billo and D. W. Margerum. *J. Am. Chem. Soc.* **92,** 1875 (1970).
519. Pavelich, M. J. *Inorg. Chem.* **14,** 982 (1975).
520. Zeltmann, A. H., and L. O. Morgan. *Inorg. Chem.* **9,** 2522 (1970).
521. Zeltmann, A. H., N. A. Matwiyoff and L. O. Morgan. *J. Phys. Chem.* **73,** 2689 (1969).
522. Glass, G. E., and R. S. Tobias. *J. Am. Chem. Soc.* **89,** 6371 (1967).
523. Eigen, M., G. Geier and W. Kruse. *Essays in Coordination Chemistry.* Birkhauser Verlag, Basel, Switzerland, p. 164, 1964.
524. Doyle, M., and H. B. Silber. *J. Chem. Soc., Chem. Commun.* **1972,** 1067.
525. Tomiyasu, H., K. Dreyer and G. Gordon. *Inorg. Chem.* **11,** 2409 (1972).
526. Hoffmann, H., and W. Ulbricht. *Ber. Bunsenges. Phys. Chem.* **76,** 1052 (1972).
527. Kustin, K., and R. Pizer. *Inorg. Chem.* **9,** 1536 (1970).
528. Fitzgerald, J. J., and N. D. Chasteen. *Biochemistry.* **13,** 4338 (1974).
529. Whittaker, M. D., J. Asay and E. M. Eyring. *J. Phys. Chem.* **70,** 1005 (1966).
530. Kustin, K., and D. L. Toppen. *J. Am. Chem. Soc.* **95,** 3564 (1973).
531. Knowles, P. F., and H. Diebler. *Trans. Faraday Soc.* **64,** 977 (1968).
532. Diebler, H., and R. E. Timms. *J. Chem. Soc. A.* **1971,** 273.
533. Kustin, K., S. T. Liu, C. Nicolini and D. L. Toppen. *J. Am. Chem. Soc.* **96,** 7410 (1974).
534. Kustin, K., and S-T Liu, *J. Am. Chem. Soc.* **95,** 2487 (1973).
535. Honig, D. S., and K. Kustin, *J. Am. Chem. Soc.* **95,** 5525 (1973).
536. Kustin, K., and S. T. Liu. *Inorg. Chem.* **12,** 2362 (1973).
537. Mentasti, E., E. Pelizzetti and G. Saini. *J. Chem. Soc., Dalton Trans.* **1973,** 2605.

538. Gilmour, A. D., and A. McAuley. *J. Chem. Soc. A.* **1969,** 2345.
539. Cavasino, F. P., and E. Di Dio. *J. Chem. Soc. A.* **1971,** 3176.
540. Mentasti, E., E. Pelizzetti and G. Saini. *Gazzetta.* **104,** 201 (1974).
541. Mentasti, E., E. Pelizzetti and G. Saini. *J. Chem. Soc., Dalton Trans.* **1974,** 1944.
542. Ong, W. K., and R. H. Prince. *J. Chem. Soc. A.* **1966,** 458.
543. Matthies, P., and H. Wendt. *Z. Phys. Chem. (Frankfurt am Main).* **30,** 137 (1961).
544. Espenson, J. H., and S. R. Helzer. *Inorg. Chem.* **8,** 1051 (1969).
545. Carlyle, D. W. *Inorg. Chem.* **10,** 761 (1971).
546. Toma, H. E., J. M. Malin and E. Giesbrecht. *Inorg. Chem.* **12,** 2080 (1973).
547. Espenson, J. H., and S. G. Woleneuk. *Inorg. Chem.* **11,** 2034 (1972).
548. Hoggard, P. E., H. W. Dodgen and J. P. Hunt. *Inorg. Chem.* **10,** 959 (1971).
549. Griesser, R., and H. Sigel. *Inorg. Chem.* **10,** 2229 (1971).
550. Pearson, R. G., and D. A. Sweigart. *Inorg. Chem.* **9,** 1167 (1970).
551. Lincoln, S. F., F. Aprile, H. W. Dodgen and J. P. Hunt. *Inorg. Chem.* **7,** 929 (1968).
552. Jordan, R. B., H. W. Dodgen and J. P. Hunt. *Inorg. Chem.* **5,** 1906 (1966).
553. Farrow, M. M., N. Purdie and E. M. Eyring. *Inorg. Chem.* **14,** 1584 (1975).
554. Kirschenbaum, L. J., J. H. Ambros and G. Atkinson. *Inorg. Chem.* **12,** 2832 (1973).
555. Bekker, P. van Z., W. J. Louw and W. Robb. *Inorg. Chim. Acta.* **6,** 564 (1972).
556. Louw, W. J., and W. Robb. *Inorg. Chim. Acta.* **3,** 303 (1969).
557. Valleau, J. P., and S. J. Turner. *Can. J. Chem.* **42,** 1186 (1964).
558. Koryta, J. *Z. Elektrochem.* **61,** 423 (1957).
559. Gerischer, H. *Z. Phys. Chem. (Frankfurt am Main).* **2,** 79 (1954).
560. Gerischer, H. *Z. Elektrochem.* **64,** 29 (1960).
561. Geier, G. *Chimia.* **27,** 636 (1974).
562. Simpson, R. B. *J. Chem. Phys.* **46,** 4775 (1967).
563. Eigen, M., and E. M. Eyring. *Inorg. Chem.* **2,** 636 (1963).
564. Williams, M. N., and D. M. Crothors. *Biochemistry.* **14,** 1944 (1975).
565. O'Rielly, D. E., G. E. Schacher and K. Schug. *J. Chem. Phys.* **39,** 1756 (1963).
566. Schuster, R. E., and A. Fratiello. *J. Chem. Phys.* **47,** 1554 (1967).
567. Fratiello, A., R. E. Lee, V. M. Nishida and R. E. Schuster. *J. Chem. Phys.* **47,** 4951 (1967).
568. Lincoln, S. F., A. Sandercock and D. R. Stranks. *J. Chem. Soc., Chem. Commun.* **1972,** 1069.
569. Friedman, H. L., J. P. Hunt, R. A. Plane and H. Taube, *J. Am. Chem. Soc.* **73,** 4028 (1951).
570. Armor, J. N., and H. Taube. *J. Am. Chem. Soc.* **92,** 6170 (1970).
571. Itzkovitch, I. J., and J. A. Page. *Can. J. Chem.* **46,** 2743 (1968).
572. Duffy, N. V., and J. E. Earley. *J. Am. Chem. Soc.* **89,** 272 (1967).
573. Rutenberg, A. C., and H. Taube. *J. Chem. Phys.* **20,** 285 (1952).
574. Monacelli, F., and E. Viel. *Inorg. Chim. Acta.* **1,** 467 (1967).
575. Armor, J. N., and H. Taube. *J. Am. Chem. Soc.* **91,** 6874 (1969).
576. Davies, G., and B. Warnqvist. *Coord. Chem. Rev.* **5,** 349 (1970).
577. Johnson, D. A., and A. G. Sharpe. *J. Chem. Soc. A.* **1966,** 798.

578. Haim, A., and W. K. Wilmarth. *Inorg. Chem.* **1,** 573 (1962).
579. Haim, A., R. G. Grassie and W. K. Wilmarth. "Mechanisms of Inorganic Reactions," *Adv. Chem. Ser.* **49,** 31 (1965).
580. Basolo and Pearson. *Mechanisms of Inorganic Reactions,* p. 207 (Ref. 1).
581. Rablen, D. P., H. W. Dodgen and J. P. Hunt, *J. Am. Chem. Soc.* **94,** 1771 (1972).
582. Fenton, D. E. *J. Chem. Soc. A.* **1971,** 3481.
583. Geier, G., M. Furrer and R. Gehrig, *Proc. Int. Conf. Coord. Chem., 16th,* **1974,** 3, 28.
584. Funahashi, S., Y. Ito and M. Tanaka. *J. Coord. Chem.* **3,** 125 (1973).
585. Grant, M., H. W. Dodgen and J. P. Hunt. *Inorg. Chem.* **10,** 71 (1971).
586. Lanir, A., and G. Navon. *Biochemistry.* **11,** 3536 (1972).
587. Tomiyasu, H., S. Ito and S. Tagami. *Bull. Chem. Soc. Jpn.* **47,** 2843 (1974).
588. Scrutton, M. C., and A. S. Mildvan. *Biochemistry.* **7,** 1490 (1968).
589. Scrutton, M. C., and A. S. Mildvan. *Arch. Biochem. Biophys.* **140,** 131 (1970).
590. Nowak, T., A. S. Mildvan and G. L. Kenyon. *Biochemistry.* **12,** 1690 (1973).
591. Zetter, M. S., H. W. Dodgen and J. P. Hunt. *Biochemistry.* **12,** 778 (1973).
592. Ward, R. L., and M. D. Cull. *Arch. Biochem. Biophys.* **150,** 436 (1972).
593. Miziorko, H., and A. S. Mildvan. *J. Biol. Chem.* **249,** 2743 (1974).
594. Villafranca, J. J., and F. C. Wedler. *Biochemistry.* **13,** 3286 (1974).
595. Kiriyak, L. G., and E. G. Chikryzova. *Izv. Akad. Nauk Mold. SSR, Ser. Biol. Khim. Nauk.* **1973,** 59.
596. Kolski, G. B., and R. A. Plane. *J. Am. Chem. Soc.* **94,** 3740 (1972).
597. Gupta, R. K., and A. G. Redfield. *Biochem. Biophys. Res. Commun.* **41,** 273 (1970).
598. Erman, J. E. *Biochemistry.* **13,** 34 (1974).
599. Huchital, D. H., and A. E. Martell. *Inorg. Chem.* **13,** 2966 (1974).
600. Sigel, H., P. R. Huber and R. F. Pasternack. *Inorg. Chem.* **10,** 2226 (1971).
601. Pasternack, R. F., P. R. Huber, U. M. Huber and H. Sigel. *Inorg. Chem.* **11,** 420 (1972).
602. Nakatani, H., and J. Osuji. *Nippon Kagaku Kaishi.* **1972,** 1809.
603. Rosen, H. M., Ph.D. Thesis, Purdue University (1967).
604. Jones, J. P., and D. W. Margerum. *Inorg. Chem.* **8,** 1486 (1969).
605. Taylor, P. W., J. Feeney and A. S. V. Burgen. *Biochemistry.* **10,** 3866 (1971).
606. Pasternack, R. F., and M. A. Cobb. *J. Inorg. Nucl. Chem.* **35,** 4327 (1973).
607. Gerber, K., F. T. T. Ng and R. G. Wilkins. *Bioinorg. Chem.* **4,** 153 (1975).
608. Gerber, K., F. T. T. Ng, R. Pizer and R. G. Wilkins. *Biochemistry.* **13,** 2663 (1974).
609. Taylor, P. W., R. W. King and A. S. V. Burgen. *Biochemistry.* **9,** 3894 (1970).
610. Thusius, D. *J. Am. Chem. Soc.* **93,** 2629 (1971).
611. Hasinoff, B. B. *Can. J. Chem.* **52,** 910 (1974).
612. Pasternack, R. F., and M. A. Cobb. *Biochem. Biophys. Res. Commun.* **51,** 507 (1973).
613. Waldmeier, P., and H. Sigel. *Inorg. Chem.* **11,** 2174 (1972).
614. Fleischer, E. B., S. Jacobs and L. Mestichelli. *J. Am. Chem. Soc.* **90,** 2527 (1968).
615. Ivin, K. J., R. Jamison and J. J. McGarvey. *J. Am. Chem. Soc.* **94,** 1763 (1972).

616. Margerum, D. W., D. C. Weatherburn and G. W. Hoffmann, unpublished data.
617. Hoffmann, G. W. *Rev. Sci. Inst.* **42,** 1643 (1971).
618. Wilkins, R. G. *Pure Appl. Chem.* **33,** 583 (1973).
619. Apel, M., and H. Hoffmann. *Proc. Symp. Coord. Chem., 3rd.* **1,** 302 (1970).
620. Grant, M., H. W. Dodgen and J. P. Hunt. *J. Am. Chem. Soc.* **92,** 2321 (1970).
621. Desai, A. G., H. W. Dodgen and J. P. Hunt. *J. Am. Chem. Soc.* **91,** 5001 (1969).
622. Rablen, D., quoted in Ref. *29.*
623. Smith, L. M., Ph.D. Thesis, State University of New York, Buffalo (1971).
624. Rusnak, L. L., and R. B. Jordan. *Inorg. Chem.* **10,** 2686 (1971).
625. Farrar, D. T., J. Stuehr, A. Moradi-Araghi, F. L. Urbach and T. G. Campbell. *Inorg. Chem.* **12,** 1847 (1973).
626. Rablen, D., and G. Gordon. *Inorg. Chem.* **8,** 395 (1969).
627. Seyse, R. J., Ph.D. Thesis, University of Iowa (1971).
628. Steinhaus, R. K., and T. A. Boersma. *Inorg. Chem.* **11,** 1505 (1972).
629. Margerum, D. W., and R. K. Steinhaus. *J. Am. Chem. Soc.* **87,** 4643 (1965).
630. Grant, M., H. W. Dodgen and J. P. Hunt. *Chem. Commun.* **1970,** 1446.
631. Belluco, U., L. Cattalini, F. Basolo, R. G. Pearson and A. Turco. *J. Am. Chem. Soc.* **87,** 241 (1965).
632. Pasternack, R. F., P. R. Huber, U. M. Huber and H. Sigel. *Inorg. Chem.* **11,** 276 (1972).
633. Cobb, M. A., and D. N. Hague. *Chem. Commun.* **1971,** 193.
634. Wilkins, R. G. *J. Chem. Soc.* **1957,** 4520.
635. Pearson, R. G., and R. D. Lanier. *J. Am. Chem. Soc.* **86,** 765 (1964).
636. Boivin, G., and M. Zador, *Can. J. Chem.* **51,** 3322 (1973).
637. Taylor, P. W., R. W. King and A. S. V. Burgen. *Biochemistry.* **9,** 2638 (1970).
638. Taylor, P. W., and A. S. V. Burgen. *Biochemistry.* **10,** 3859 (1971).
639. Chen, R. F., A. N. Schechter and R. L. Berger. *Anal. Biochem.* **29,** 68 (1969).
640. Olander, J., and E. T. Kaiser. *Biochem. Biophys. Res. Commun.* **45,** 1083 (1971).
641. Rasmussen, S. E. *Acta Chem. Scand.* **13,** 2009 (1959).
642. Jørgenson, C. J. *Acta Chem. Scand.* **10,** 887 (1956).
643. Anbar, M., Z. B. Alfasse and H. Bregman-Reisler. *J. Am. Chem. Soc.* **89,** 1263 (1967).
644. Jain, Prem. C., and E. C. Lingafelter. *J. Am. Chem. Soc.* **89,** 724 (1967); **89,** 6131 (1967).
645. Marongiu, G., E. C. Lingafelter and P. Paoletti. *Inorg. Chem.* **8,** 2763 (1969).
646. Cayley, G. R., and D. N. Hague. *Trans. Faraday Soc.* **67,** 2896 (1971).
647. Phillips, C. S. G., and R. J. P. Willians. *Inorganic Chemistry.* Vol. II. Oxford University Press, 1966, p. 487.
648. Marongiu, G., M. Cannas and G. Carta. *J. Coord. Chem.* **2,** 167 (1973).
649. Levanon, H., and Z. Luz. *J. Chem. Phys.* **49,** 2031 (1968).
650. Fiat, D., Z. Luz and B. L. Silver. *J. Chem. Phys.* **49,** 1376 (1968).
651. Luz, Z., and S. Meiboom. *J. Chem. Phys.* **40,** 1066 (1964).
652. Vriesenga, J. R., and R. Gronner. *Inorg. Chem.* **12,** 1112 (1973).
653. Luz, Z. *J. Chem. Phys.* **51,** 1206 (1969).
654. Plotkin, K., J. Copes and J. R. Vriesenga. *Inorg. Chem.* **13,** 1494 (1974).

655. Vriesenga, J. R. *Inorg. Chem.* **11,** 2724 (1972).
656. Williams, J., and S. Petrucci. *J. Phys. Chem.* **77,** 130 (1973).
657. Hodgkinson, J., and R. B. Jordan. *J. Am. Chem. Soc.* **95,** 763 (1973).
658. West, R. J., and S. F. Lincoln. *Inorg. Chem.* **12,** 494 (1973).
659. Lincoln, S. F. *J. Chem. Soc., Dalton Trans.* **1973,** 1896.
660. Lincoln, S. F., and R. J. West. *Aust. J. Chem.* **27,** 97 (1974).
661. West, R. J., and S. F. Lincoln. *J. Chem. Soc., Dalton Trans.* **1974,** 281.
662. Lincoln, S. F., and R. J. West. *J. Am. Chem. Soc.* **94,** 400 (1974).
663. Sanduja, M. L., and W. MacF. Smith. *Can. J. Chem.* **47,** 3773 (1969).
664. Caldin, E. F., and P. Godfrey. *J. Chem. Soc., Faraday Trans. I.* **1974,** 2260.
665. Seyse, R. J. *Diss. Abstr.* **3225868** (1971); Ph.D. Thesis, University of Iowa (1971).
666. Williams, M. J. G., and R. G. Wilkins. *J. Chem. Soc.* **1957,** 4514.
667. Funahashi, S., and M. Tanaka. *Inorg. Chem.* **9,** 2092 (1970).
668. Edwards, J. O. *J. Am. Chem. Soc.* **76,** 1540 (1954); **78,** 1819 (1956).
669. Vallee, B. L., and R. J. P. Williams. *Proc. Nat. Acad. Sci.* **59,** 458 (1968).
670. Glaeser, H. H., G. A. Lo, H. W. Dodgen and J. P. Hunt. *Inorg. Chem.* **4,** 206 (1965).
671. Hunt, J. P., H. W. Dodgen and F. Klanberg. *Inorg. Chem.* **2,** 478 (1963).
672. Glaeser, H. H., H. W. Dodgen and J. P. Hunt. *Inorg. Chem.* **5,** 1061 (1965).
673. Murray, R., S. F. Lincoln, H. H. Glaeser, H. W. Dodgen and J. P. Hunt. *Inorg. Chem.* **8,** 554 (1969).
674. Basolo and Pearson. *Mechanisms of Inorganic Reactions,* chap. 3 (Ref. *1*).
675. Coleman, G. N., J. W. Gesler, F. A. Shirley and J. R. Kuempel. *Inorg. Chem.* **12,** 1036 (1973).
676. Basolo and Pearson. *Mechanisms of Inorganic Reactions,* p. 164 (Ref. *1*).
677. Koren, R., and B. Perlmutter-Hayman. *Inorg. Chem.* **11,** 3055 (1972).
678. Toma, H. E., and J. M. Malin. *Inorg. Chem.* **12,** 1039 (1973).
679. Bradić, Z., D. Pavlović, I. Murati and S. Asperger. *J. Chem. Soc., Dalton Trans.* **1974,** 344.
680. Crouse, W. C., and D. W. Margerum, *Inorg. Chem.* **13,** 1437 (1974).
681. Margerum, D. W., T. J. Bydalek and J. J. Bishop. *J. Am. Chem. Soc.* **83,** 1791 (1961).
682. Coombs, L. C., D. W. Margerum and P. C. Nigam. *Inorg. Chem.* **9,** 2081 (1970).
683. Pagenkopf, G. K. *J. Coord. Chem.* **2,** 129 (1972).
684. Coombs, L. C., J. Vasiliades and D. W. Margerum. *Anal. Chem.* **44,** 2325 (1972).
685. Young, D. C., J. Vasiliades and D. W. Margerum, unpublished data.
686. Nakamura, S., Ph.D. Thesis, University of Chicago (1964).
687. Hamm, R. E., and J. C. Templeton. *Inorg. Chem.* **12,** 755 (1973).
688. Pagenkopf, G. K. *J. Am. Chem. Soc.* **94,** 4359 (1972).
689. Pagenkopf, G. K., *Inorg. Chem.* **13,** 1591 (1974).
690. Stará, V., and M. Kopanica. *Collect. Czech. Chem. Commun.* **37,** 2882 (1972).
691. Margerum, D. W. *J. Am. Chem. Soc.* **79,** 2728 (1957).
692. Margerum, D. W., and L. P. Morgenthaler. *Advances in the Chemistry of Coordination Compounds.* S. Kirschner, ed. Macmillan, New York, p. 481, 1961.
693. Burgess, J. *J. Chem. Soc. A.* **1968,** 497.

694. Gardener, E. R., F. M. Mekhail and J. Burgess. *Int. J. Chem. Kinet.* **6,** 133 (1974).
695. Burgess, J., and R. H. Prince. *J. Chem. Soc.* **1965,** 6061.
696. Nord, G. *Acta Chem. Scand.* **27,** 743 (1973).
697. Billo, E. J. *Inorg. Chem.* **12,** 2783 (1973).
698. Balcombe, C. I., and B. Wiseall. *J. Inorg. Nucl. Chem.* **36,** 881 (1974).
699. Burgess, J. *J. Chem. Soc. Dalton Trans.* **1972,** 1061.
700. Burgess, J. *J. Chem. Soc. A.* **1967,** 955.
701. Burgess, J. *Inorg. Chim. Acta.* **5,** 133 (1971).
702. Burgess, J., G. E. Ellis, D. J. Evans, A. Porter, R. Wane and R. D. Wyvill. *J. Chem. Soc. A.* 44 (1971).
703. Richards, A. F., J. H. Ridd and M. L. Tobe. *Chem. Ind.* (*London*). **1963,** 1726.
704. Deno, N. C., and R. W. Taft, Jr. *J. Am. Chem. Soc.* **76,** 244 (1954).
705. Dickens, J. E., F. Basolo and H. M. Neumann. *J. Am. Chem. Soc.* **79,** 1286 (1957).
706. Janes, D. L., and D. W. Margerum. *Inorg. Chem.* **5,** 1135 (1966).
707. Tanner, S. P., F. Basolo and R. G. Pearson. *Inorg. Chem.* **6,** 1089 (1967).
708. Carter, M. J., and J. K. Beattie. *Inorg. Chem.* **9,** 1233 (1970).
709. Thorneley, R. N. F., A. G. Sykes and P. Gans. *J. Chem. Soc. A.* **1971,** 1494.
710. Nor, D., and A. G. Sykes. *J. Chem. Soc., Dalton Trans.* **1973,** 1232.
711. Woodruff, W. H., D. W. Margerum, M. J. Milano, H. L. Pardue and R. E. Santini. *Inorg. Chem.* **12,** 1490 (1973).
712. Shimi, I. A. W., and W. C. E. Higginson. *J. Chem. Soc.* **1958,** 260.
713. Dyke, R., and W. C. E. Higginson. *J. Chem. Soc.* **1960,** 1998.
714. Grossman, B., Ph.D. Thesis, State University of New York at Buffalo (1969).
715. Morris, M. L., M. E. Banasik and J. Knoeck. *Inorg. Nucl. Chem. Lett.* **10,** 331 (1974).
716. Swaminathan, K., and D. H. Busch. *Inorg. Chem.* **1,** 256 (1962).
717. Higginson, W. C. E., and M. P. Hill. *J. Chem. Soc.* **1959,** 1620.
718. Dyke, R., and W. C. E. Higginson. *J. Chem. Soc.* **1963,** 2788.
719. Nozaki, T., K. Kasuga and N. Kagaua. *Nippon Kagaku Kaishi.* **1973,** 718.
720. Steinhaus, R. K., and R. L. Swann. *Inorg. Chem.* **12,** 1855 (1973).
721. Bydalek, T. J., T. M. Stohich and D. M. Coleman. *Inorg. Chem.* **9,** 29 (1970).
722. Tanaka, N., and M. Kamada. *Bull. Chem. Soc. Jpn.* **35,** 1596 (1962).
723. Tanaka, N., K. Kato and R. Tamamushi. *Bull. Chem. Soc. Jpn.* **31,** 283 (1958).
724. Ogino, H., and N. Tanaka. *Bull. Chem. Soc. Jpn.* **40,** 852 (1967).
725. Brücher, E., and L. Boros. *Proc. XV ICCC.* Moscow, U.S.S.R. (1973).
726. Ogino, H., T. Baba and N. Tanaka. *Bull. Chem. Soc. Jpn.* **42,** 1578 (1969).
727. Ogino, H., and N. Tanaka. *Bull. Chem. Soc. Jpn.* **40,** 857 (1967).
728. Tanaka, N., H. Osawa and M. Kamada. *Bull. Chem. Soc. Jpn.* **36,** 67 (1963).
729. Kuempel, J. R., and W. B. Schaap. *Inorg. Chem.* **7,** 2435 (1968).
730. Nozaki, T., and K. Kasuga. *Nippon Kagaku Zasshi.* **1973,** 2117.
731. Nozaki, T., and N. Suemitsu. *Bull. Chem. Soc. Jpn.* **49,** 445 (1976).
732. Glentworth, P., and D. A. Newton. *J. Inorg. Nucl. Chem.* **33,** 1701 (1971).
733. Glentworth, P., B. Wiseall, C. L. Wright and A. J. Mahmood. *J. Inorg. Nucl. Chem.* **30,** 967 (1968).
734. Balcome, C. I., and B. Wiseall. *J. Inorg. Nucl. Chem.* **35,** 2859 (1973).

735. Wiseall, B., and C. Balcombe. *J. Inorg. Nucl. Chem.* **32,** 1751 (1970).
736. Asano, T., S. Okada and S. Taniguchi. *J. Inorg. Nucl. Chem.* **32,** 1287 (1970).
737. Asano, T., S. Okada, K. Sakamoto, S. Taniguchi and Y. Kobayashi. *J. Inorg. Nucl. Chem.* **31,** 2127 (1969).
738. Steinhaus, R. K., and C. L. Barsuhn. *Inorg. Chem.* **13,** 2922 (1974).
739. Anderegg, G. *Helv. Chim. Acta.* **47,** 1801 (1964).
740. D'Olieslager, W., and G. R. Choppin. *J. Inorg. Nucl. Chem.* **33,** 127 (1971).
741. Choppin, G. R., and K. R. Williams. *J. Inorg. Nucl. Chem.* **35,** 4255 (1973).
742. Williams, K. R., and G. R. Choppin. *J. Inorg. Nucl. Chem.* **36,** 1849 (1974).
743. Betts, R. H., O. F. Dahlinger and D. M. Munro. *Radioisotopes in Scientific Research.* Vol. 2. Pergamon Press, Oxford, p. 236.
744. Fomin, V. V., E. P. Maiorova and G. A. Leman. *Russ. J. Inorg. Chem.* **14,** 1000 (1969).
745. Neubauer, L., and M. Kopanica. *Collect. Czech. Chem. Commun.* **36,** 1121 (1971).
746. Kopanica, M., and V. Stará, *Collect. Czech. Chem. Commun.* **37,** 80 (1972).
747. Kopanica, M. *Talanta.* **15,** 1457 (1968).
748. Rhyl, T. *Acta Chem. Scand.* **26,** 3955, 4001 (1972).
749. Rhyl, T. *Acta Chem. Scand.* **27,** 303 (1973).
750. Thompson, L. C. *Inorg. Chem.* **1,** 490 (1962).
751. Schwarzenbach, G., R. Gut and G. Anderegg. *Helv. Chim. Acta.* **37,** 937 (1954).
752. Bydalek, T. J. *Inorg. Chem.* **4,** 232 (1965).
753. Tanaka, N., and K. Kato. *Bull. Chem. Soc. Jpn.* **32,** 1376 (1959).
754. Tanaka, N., and H. Ogino. *Bull. Chem. Soc. Jpn.* **36,** 175 (1963).
755. Kimura, M. *Bull. Chem. Soc. Jpn.* **42,** 407 (1969).
756. Saito, K., and M. Tsuchimoto. *J. Inorg. Nucl. Chem.* **25,** 1245 (1963).
757. Makashev, Yu. A., I. E. Makasheva and V. A. Mel'nikov. *Radiokhimiya.* **8,** 371 (1966).
758. Das, R. R., T. R. Bhat and J. Shankar. *J. Inorg. Nucl. Chem.* **30,** 1691 (1968).
759. Sawyer, D. T., and J. E. Tackett. *J. Am. Chem. Soc.* **85,** 2390 (1963).
760. Brücher, E., and P. Szarvas. *Inorg. Chim. Acta.* **4,** 632 (1970).
761. Barnes, J. W., and G. D. Dorough. *J. Am. Chem. Soc.* **72,** 4045 (1950).
762. Grant, C., and P. Hambright. *J. Am. Chem. Soc.* **91,** 4195 (1969).
763. Baum, S. L., and R. A. Plane. *J. Am. Chem. Soc.* **88,** 910 (1966).
764. Baker, H., P. Hambright, L. Wagner and L. Ross. *Inorg. Chem.* **12,** 2200 (1973).
765. Kula, R. J., and G. H. Reed. *Anal. Chem.* **38,** 697 (1966).
766. Shirakashi, T., and N. Tanaka. *Nippon Kagaku Zasshi.* **91,** 142 (1970).
767. Kodama, M., C. Sasaki and K. Miyamoto. *Bull. Chem. Soc. Jpn.* **42,** 163 (1969).
768. Kodama, M., and K. Hagiya. *Bull. Chem. Soc. Jpn.* **46,** 3151 (1973).
769. Kodama, M., C. Sasaki and T. Noda, *Bull. Chem. Soc. Jpn.* **41,** 2033 (1968).
770. Kodama, M. *Bull. Chem. Soc. Jpn.* **47,** 2200 (1974).
771. Kodama, M. *Bull. Chem. Soc. Jpn.* **40,** 2575 (1967).
772. Kodama, M. *Nippon Kagaku Zasshi.* **91,** 134 (1970).
773. Funahashi, S., and M. Tanaka. *Bull. Chem. Soc. Jpn.* **43,** 763 (1970).
774. Kodama, M., T. Sato and S. Karasawa. *Bull. Chem. Soc. Jpn.* **45,** 2757 (1972).

775. Kodama, M. *Bull. Chem. Soc. Jpn.* **42,** 3330 (1969).
776. Kodama, M., and T. Ueda. *Bull. Chem. Soc. Jpn.* **43,** 419 (1970).
777. Kimura, M. *Nippon Kagaku Zasshi.* **89,** 1209 (1968).
778. Kodama, M., and T. Ueda. *Nippon Kagaku Zasshi.* **91,** 138 (1970).
779. Kodama, M., Y. Fujii and T. Ueda. *Bull. Chem. Soc. Jpn.* **43,** 2085 (1970).
780. Kodama, M. *Bull. Chem. Soc. Jpn.* **47,** 1430 (1974).
781. Kodama, M., M. Hashimoto and T. Watanabe. *Bull. Chem. Soc. Jpn.* **45,** 2761 (1972).
782. Kodama, M., C. Sasaki and M. Murata. *Bull. Chem. Soc. Jpn.* **41,** 1333 (1968).
783. Carr, J. D., and C. N. Reilley. *Anal. Chem.* **42,** 51 (1970).
784. Kodama, M. *Bull. Chem. Soc. Jpn.* **43,** 2259 (1970).
785. Kodama, M., S. Karasawa and T. Watanabe. *Bull. Chem. Soc. Jpn.* **44,** 1815 (1971).
786. Kimura, M. *Bull. Chem. Soc. Jpn.* **42,** 2844 (1969).
787. Kodama, M. *Bull. Chem. Soc. Jpn.* **42,** 2532 (1969).
788. Carr, J. D., R. A. Libby and D. W. Margerum. *Inorg. Chem.* **6,** 1083 (1967).
789. Carr, J. D., K. Torrance, C. J. Cruz and C. N. Reilley. *Anal. Chem.* **39,** 1358 (1967).
790. Sweigart, D. A., and D. G. DeWit. *Inorg. Chem.* **9,** 1580 (1970).
791. Katsuyama, T., and T. Kumai. *Bull. Chem. Soc. Jpn.* **48,** 3581 (1975).
792. Funahashi, S., S. Yamada and M. Tanaka. *Inorg. Chem.* **10,** 257 (1971).
793. Funahashi, S., S. Yamada and M. Tanaka. *Bull. Chem. Soc. Jpn.* **43,** 769 (1970).
794. Carr, J. D., and J. Vasiliades. *Inorg. Chem.* **11,** 2104 (1972).
795. Tanaka, M., S. Funahashi and K. Shirai. *Inorg. Chem.* **7,** 573 (1968).
796. Sudmeier, J. L., and C. N. Reilley. *Inorg. Chem.* **5,** 1047 (1966).
797. Carr, J. D., and D. R. Baker. *Inorg. Chem.* **10,** 2249 (1971).
798. Reinbold, P. E., and K. H. Pearson. *Inorg. Chem.* **9,** 2325 (1970).
799. Ogino, H., T. Watanabe, J. J. Chung and N. Tanaka. *Bull. Chem. Soc. Jpn.* **46,** 2460 (1973).
800. Haraguchi, K., K. Yamada and S. Ito. *J. Inorg. Nucl. Chem.* **36,** 1611 (1974).
801. Carr, J. D., and V. K. Olsen. *Inorg. Chem.* **14,** 2168 (1975).
802. Olson, D. C., and D. W. Margerum. *J. Am. Chem. Soc.* **84,** 680 (1962).
803. Luthy, J., and D. W. Margerum, unpublished observations.
804. Goddard, J. B., and F. Basolo. *Inorg. Chem.* **7,** 936 (1968).
805. Basolo, F., and W. H. Baddley. *J. Am. Chem. Soc.* **86,** 2075 (1966).
806. Baddley, W. H., and F. Basolo. *J. Am. Chem. Soc.* **88,** 2944 (1968).
807. Hewkin and Prince. *Coord. Chem. Rev.,* p. 81 (Ref. 9).
808. Weick, C. F., and F. Basolo. *Inorg. Chem.* **5,** 576 (1966).
809. Roulet, R., and H. B. Gray. *Inorg. Chem.* **11,** 2101 (1972).
810. Kimura, M. *Bull. Chem. Soc. Jpn.* **42,** 2841 (1969).
811. Pagenkopf, G. K., and D. W. Margerum. *J. Am. Chem. Soc.* **90,** 502 (1968).
812. Pagenkopf, G. K., and D. W. Margerum. *J. Am. Chem. Soc.* **92,** 2683 (1970).
813. Billo, E. J., G. F. Smith and D. W. Margerum. *J. Am. Chem. Soc.* **93,** 2635 (1971).
814. Youngblood, M. P., and D. W. Margerum, unpublished results.
815. Laidler, K. J. *Chemical Kinetics.* McGraw Hill, New York, 1965, p. 356.

816. Stehl, R. H., D. W. Margerum and J. J. Latterell. *Anal. Chem.* **39,** 1346 (1967).
817. Margerum, D. W., and R. H. Stehl. *Anal. Chem.* **39,** 1351 (1967).
818. Crouse, W. C., and D. W. Margerum, unpublished data.
819. Blancher, J. A., D. W. Margerum and M. R. James, unpublished data.
820. Carr, J. D., and D. W. Margerum. *J. Am. Chem. Soc.* **88,** 1645 (1966).
821. Dukes, G. R., and D. W. Margerum. *Inorg. Chem.* **11,** 2952 (1972).
822. Dukes, G. R., G. K. Pagenkopf and D. W. Margerum. *Inorg. Chem.* **10,** 2419 (1971).
823. Murmann, R. K. *Inorg. Chem.* **2,** 116 (1962).
824. Bates, G. W., and J. Wernicke. *J. Biol. Chem.* **246,** 3679 (1971).
825. Bates, G. W., C. Billups and P. Saltman. *J. Biol. Chem.* **242,** 2810 (1967).
826. Bates, G. W., C. Billups and P. Saltman. *J. Biol. Chem.* **242,** 2816 (1967).
827. Halpern, B., A. M. Sargeson and K. R. Turnbull. *J. Am. Chem. Soc.* **88,** 4630 (1966).
828. Buckingham, D. A., L. G. Marzilli and A. M. Sargeson. *J. Am. Chem. Soc.* **89,** 825 (1967).
829. Sledziewska, E., L. Plachta, D. Vonderschmitt and K. Bernauer. *Chimia.* **25,** 330 (1971).
830. Chung, C., and D. W. Margerum, unpublished data.
831. Kodama, M. *Bull. Chem. Soc. Jpn.* **48,** 2813 (1975).
832. Ciampolini, M., P. Paoletti and L. Sacconi. *J. Chem. Soc.* **1960,** 4553.
833. Ciampolini, M., P. Paoletti and L. Sacconi. *J. Chem. Soc.* **1961,** 2994.
834. Sacconi, L., P. Paoletti and M. Ciampolini. *J. Chem. Soc.* **1961,** 5115.
835. Paoletti, P., and A. Vacca. *J. Chem. Soc.* **1964,** 5051.
836. Stará, V., and M. Kopanica. *Collect. Czech. Chem. Commun.* **37,** 3545 (1972).

2

Ligand Reactions—the Effect of Metal Complexes on Chemical Processes

Christian J. Hipp
E. I. du Pont de Nemours & Co.
Brevard, N.C. 28712

Daryle H. Busch
Department of Chemistry
Ohio State University
140 W. 18th Ave.
Columbus, Ohio 43210

I. DEFINITION OF TERMS

A ligand reaction is a chemical process wherein the central metal atom in a complex plays a determining role in effecting the chemical reaction of a ligand. The chemical efficacy of metal ions as they occur in coordination and organometallic compounds is widespread, ranging from basic inorganic and organic chemical reactions to a number of neighboring areas of science. It is our purpose to focus attention on the purely chemical aspects of the subject. An attempt has been made to organize this vast area with a structural and mechanistic outline, proceeding by way of basic classes of interactions to certain broad areas of ligand reactions. Parts of the field that have previously been reviewed in detail are included at a depth and with a thoroughness consistent with the intent to present an overview of the whole field. More detailed treatments are accorded to subjects that have not been summarized earlier. Thus, we consider general elemental chemical processes and the relationships that exist between structure and function rather than review, in a complete way, any part of this vast subject.

0-8412-0292-3/78/36-174-221$30.00/1

A single question underlies the following discussions: how can the metal atom effect this process? In particular, how can $HCo(CO)_4$ put three small molecules (C_2H_4, CO, and H_2) together to form a large molecule,

$$CH_3CH_2C(=O)H \quad ?$$

How do low valent Ti or V compounds produce high molecular weight, stereo-regular polyolefins at low temperature and pressure? How do metal activated peptidases, lyases, and esterases use metal ions as they hydrolyze carboxyl derivatives? How do molybdenum catalysts convert propylene into mixtures containing butylene and ethylene? Why are some reagents specific in cleaving *cis*-glycols? How do Mo- and Fe-activated enzymes fix nitrogen? How does the cobalt in vitamin B_{12} participate in the stereospecific rearrangement of C—H and C—C bonds? How do transition metal catalysts run a single olefin linkage in a linear hydrocarbon moiety up or down the C—C chain like the slide of a zipper? These and many more questions come from areas of application, and millions of words have been written about them. Some of them have been more or less answered. From the perspective of ligand reactions, these questions imply a great number of chemical processes and specific metal atom–substrate relationships that must be first identified, then explored, and eventually understood. Reviews have been written on the general subject of ligand reactions (Refs. 1–10), and numerous ones have been written on specific aspects of the field. Many of these will be cited in the discussion to follow.

The questions raised by the needs of applied areas by no means limit the range of ligand reactions that have come under study in recent years. In any specific instance of a ligand reaction, we are really inquiring about a chemical property or set of chemical properties that relate to the metal atom in the reacting compounds.

In the course of a chemical reaction between two (or more) species and involving metal complexing as an integral part, three stages of the process are often recognizable. These are (a) the reactant stage or intermediate complex-forming stage; (b) the intermediate complex stage; and (c) the product-forming stage. In (a), the metal complex serves as a reagent toward a separate entity, and the latter may, or may not, become a ligand during this process. As we study this phase of the reaction between the metal complex and substrate, we think in terms of a nucleophilic, electrophilic, or radical attack of the substrate by the metal atom. Very often, bond forming between the metal atom and an atom(s) of the ligand will be critically involved. In (b), the metal atom is bound to the substrate, and it exerts its effect on the reactions of that substrate by serving as a substituent. The ligand, for example, may be rendered more strongly electrophilic as a result of the action of the metal complex substituent. Attention is often directed toward the effect of the metal atom on the electron density of the substrate. The intermediate complex may, or may not, interact with external reagents. In (c), the altered substrate is liberated by the metal complex. In some cases the substrate may not be altered until this separation occurs. Generally, this aspect of the process involves bond breakage, and we must think in terms such as heterolytic cleavage, homolytic cleavage, and reductive elimination. Often the metal-containing fragment should be considered to be a leaving group. In any given ligand reaction, the relative importance of these three stages must be realized before progress can be made in understanding the process. In some cases all three stages will be important; in many cases, two stages will dominate, while in others only one stage will be of consequence. The term ligand reactions may be taken to involve all these instances, although it was defined[2] for the purpose of focusing attention on the advantages of studying processes wherein the binding between substrate and metal atom persists throughout the course of the reaction under investigation.

Whether or not a process is catalytic has contrasting pay-offs depending on the need that is to be served. Obviously, catalytic processes are desirable for the large scale operation of industrial syntheses; however, noncatalytic systems more often reveal the nature of the intermediate, thereby illuminating those elemental steps that govern the ligand reactions involved. For example, there is no question that the existence of easily studied alkyl and acyl derivatives of manganese,

$$CH_3Mn(CO)_5 \text{ and } CH_3\overset{\overset{\displaystyle O}{\|}}{C}Mn(CO)_5,$$

facilitated the development of an understanding of the role of the catalyst, $HCo(CO)_4$, in the oxo process.[7,11,12,13]

It is particularly helpful to elaborate on the reagent, substituent, leaving group scheme outlined above. Each provides useful insight into the various roles a metal atom may play in ligand reactions, and this review begins by using this as a part of its organization. Many ligand reactions function because of the intrinsic polyfunctional character of metal atoms. Since this polyfunctionality derives from the fact that metal atoms and ions have several sites for coordination, we classify the attendant ligand reaction phenomena as multisite effects. This set of relationships appears next in our organization, and summaries of large but specific areas of ligand reactions follow. These include systems involving unsaturated and strained organic molecules (hydrogenation, olefin isomerization, oligomerization, polymerization, disproportionation, and oxo-reppe chemistry); systems in which nucleophilic substitution is the central issue (amino acid derivatives, phosphate derivatives, and Schiff bases); systems involving electrophilic reactions (β-diketones and their analogs, cyclic tetrapyrroles and other macrocyclic ligands, and reactions of coordinated nucleophiles); and substrate oxidations (promoted autoxidation, promoted peroxide oxidation, and oxidation by metal reagent).

II. METAL COMPLEXES AS REAGENTS IN LIGAND REACTIONS

A. Outer-Sphere Electron Transfer

Perhaps the least complicated involvement of a metal complex in ligand reactions depends on outer-sphere electron transfer, as exemplified by the oxidation of a mercaptan by the ferricyanide ion.[14]

$$2Fe^{III}(CN)_6^{3-} + 2RSH \rightarrow 2Fe^{II}(CN)_6^{4-} + RSSR + 2H^+ \qquad (1)$$

The conclusion that these processes do not involve bond formation with any part of the complex during intermediate stages of the reaction rests, in part, on the knowledge of the types of reactions into which the reagent enters. Ferricyanide generally undergoes reduction via an outer-sphere mechanism. Tracer experiments ruled out all possibility of substitution at Fe^{3+} during the reaction described by Eq. (1). Perhaps the role of the iron center in the heme protein cytochrome c is equally simple,[15] involving outer-sphere electron transfer.

B. Lewis Acid Effects

In contrast to the class of reagent just illustrated, the central metal atom of a complex usually becomes involved as the site of the action when the complex serves as a reagent. In the most obvious mode of interaction, the positive metal ion acts as an electron pair acceptor, a Lewis acid; consequently, much attention is devoted to this function of the metal complex. There are a variety of notable examples of some generality.

Various Friedel-Crafts reactions[16] involve activation as a consequence of the Lewis acidity of the catalyst. Recent crystallographic results[17] have proven that the active adducts in Friedel-Crafts acylations may have either structure **1** or structure **2** depending on the polarity of the solvent in which they are generated (MX_n is the Lewis acid).

$$R{-}C{\equiv}O + MX^-_{n+1}$$

1

$$R{-}C(X){=}O{-}MX_n$$

2

The Lewis acid properties of metal ions are instrumental in many processes involving nucleophilic substitution reactions of various carboxylate derivatives. The early observation by Kroll[18] that the hydrolysis of α-amino acid esters is accelerated by solutions of divalent metal ions may serve as an illustration [Eq. (2)].

$$NH_2CH_2C(=O)OR + H_2O \xrightarrow{M^{2+}} NH_2CH_2C(=O)O^{\ominus} + ROH + H^+ \quad (2)$$

The reagent function of the metal ion is exercised in the act of complex formation in processes of this kind. During the course of the hydrolysis, the metal ion and substrate are bound together so that part of the process belongs in section III on metal ions as substituents. This chemistry underlies the function of several classes of enzymes.

Similarly, the hydrolysis of alkyl halides is greatly accelerated by those metal ions that we now think of as labile, b-type species—e.g., Ag^+ and Hg^{2+}.

$$CH_3I + H_2O + Ag^+ \rightarrow CH_3OH + AgI + H^+ \quad (3)$$

This example emphasizes the selectivity toward substrates (bases) of metal ions (Lewis acids). Again, the effects of Lewis acids on the reactivities of the bases with which they combine are more properly subjects for other sections of this chapter.

C. Oxidation-Reduction Reactions Involving Group Transfer

Often systems that involve initial Lewis acid-base interactions are dominated by other more subtle and interesting interrelationships. The abstraction of a bromine atom from benzyl bromide by cobalt in $Co(CN)_5^{3-}$ [Eq. (4)] surely involves at least incipient complexing between the cobalt atom and the benzyl bromide molecule. However, the important processes are the radical abstraction reaction [Eq. (4)] and the following radical combination reaction [Eq. (5)].[19,20]

$$Co(CN)_5^{3-} + C_6H_5CH_2Br \rightarrow BrCo(CN)_5^{3-} + C_6H_5CH_2\cdot \quad (4)$$

$$C_6H_5CH_2\cdot + Co(CN)_5^{3-} \rightarrow C_6H_5CH_2Co(CN)_5^{3-} \quad (5)$$

Similarly, although some weak complexing of the H_2 molecule by $Ir(CO)[P(C_6H_5)_3]_2Cl$ must precede the oxidation-reduction process that is called oxidative addition, it is the latter [Eq. (7)] that must interest us as we try to understand catalytic hydrogenation.[20,21,22]

$$Ir(CO)[P(C_6H_5)_3]_2Cl + H_2 \rightleftarrows Ir(CO)[P(C_6H_5)_3]_2Cl(H)_2 \quad (6)$$

These two examples act to return our discussion to the broad area of substrate oxidation-reductions by metal complexes. In beginning these discussions of metal complexes as reagents in ligand reactions, we first considered redox processes in which the metal complex merely acted as a source, or receiver, of electrons. The examples just cited emphasize that there is a much larger field of substrate oxidation or reduction by metal complexes in which relatively profound interactions occur between the metal ion and the substrate. In the first example, the cobalt(II) atom may be thought to have reduced the bromine radical, or the benzyl radical, thereby being oxidized to cobalt(III). In the second example, oxidation of the iridium(I) to iridium(III) is accompanied by reduction of two hydrogen atoms to hydride ligands.

D. Analogy to Carbon Chemistry

Halpern[20,23] has provided a correlation between the electronic structure of the central metal atom of the reagent complex and the kinds of processes into which the metal complex enters. The classification scheme focuses on low-spin transition metal complexes and provides an insight into behavior by drawing analogies between these species and familar reactive carbon species (Table 2-1). Whereas the stable reference structure in the case of carbon includes such saturated tetrahedral molecules as methane or CCl_4, for the complexes, it is a low-spin d^6 octahedral ion or molecule such as $Co(NH_3)_6^{3+}$ or $Cr(CO)_6$. In

TABLE 2-1
ANALOGIES BETWEEN REACTIVE SPECIES IN CARBON CHEMISTRY

Carbon Species			
Species	Coordination No.	Characteristic Structural Feature	Reactions
Saturated molecule R_3CX	4	all orbitals and electron pairs used	substitution
Free Radical $R_3C\cdot$	3	one unpaired electron	dimerization, abstraction, addition
Carbanion R_3C^-	3	an unshared pair of electrons, minus charge	addition of electrophile
Carbonium ion R_3C^+	3	a vacant low energy orbital	addition of nucleophile
Carbene R_2C	2	(singlet)· vacant orbital and lone pair of electrons	addition, insertion

carbon chemistry, the homolytic breaking of a bond produces a carbon radical:

$$CCl_4 \rightarrow \cdot CCl_3 + Cl\cdot \qquad (7)$$

Such a process for a related transition metal complex produces a derived radical—a low-spin d^7 complex having a coordination number of 5 [Eq. (8)].

$$Co(CN)_5Br^{3-} \rightleftharpoons \cdot Co(CN)_5^{3-} + Br\cdot \qquad (8)$$

With many soft and strong ligands, d^7 metal ions are often low-spin and five-coordinate. Their oxidation reduction reactions may be understood in general terms by following up on this analogy. Eq. (4) describes a radical abstraction reaction.

AND TRANSITION METAL CHEMISTRY

		Transition Metal Counterpart		
Coordination No.	No. of d electrons	Characteristic Structural Feature	Examples	Reactions
6	6	all orbitals filled	$Co(III)(NH_3)_5Cl^{2+}$ $Cr^0(CO)_6$	substitution
5	7	one unpaired electron	$Co(II)(CN)_5^{3-}$	$2Co(CN)_5^{3-} \rightleftharpoons (CN)_5Co{-}Co(CN)_5^{6-}$ $Co(CN)_5^{3-} + CH_3I \rightarrow Co(CN)_5I^{3-} + CH_3\cdot$ $2Co(CN)_5^{3-} + HC{\equiv}CH \rightarrow (NC)_5Co{-}CH{=}CHCo(CN)_5^{6-}$
5	8	a high energy pair of electrons	$Mn(CO)_5^-$	$Mn(CO)_5^- + H^+ \rightleftharpoons HMn(CO)_5$
5	6	a vacant low energy orbital	$Cl(III)(CN)_5^{2-}$	$Co(CN)_5^{2-} + H_2O \rightarrow Co(CN)_5H_2O^{2-}$
4	8	vacant orbital and high energy lone pair	$Ir(I)(CO)[P(C_6H_5)_3]_2Cl$	$Ir(CO)[P(C_6H_5)_3]_2Cl + CO \rightleftharpoons Ir(CO)_2[P(C_6H_5)_3]_2Cl$ $Ir(CO)[P(C_6H_5)_3]_2Cl + H_2 \rightleftharpoons Ir(CO)[P(C_6H_5)_3]_2Cl(H)_2$

If the d^7 complex is reduced, the process is analogous to reduction of a carbon radical to produce a carbon anion.

$$\cdot Co(CN)_5^{3-} + e^- \rightarrow :Co(CN)_5^{4-} \qquad (9)$$

$:Co(CN)_5^{4-}$ is indeed a carbon anion analog. It is a superb nucleophile[24] [Eq. (10)] and

$$CH_3I + :Co(CN)_5^{4-} \rightarrow CH_3Co(CN)_5^{3-} + I^- \qquad (10)$$

acts as a Brønsted base [Eq. (11)].

$$H_3O^+ + :Co(CN)_5^{4-} \rightarrow HCo(CN)_5^{3-} + H_2O \qquad (11)$$

One electron oxidation of our carbon radical analog (a d^7 complex) produces a carbonium ion analog—a five-coordinate complex having a low-spin d^6 electronic configuration [Eq. (12)].

$$\cdot Co(CN)_5^{3-} \rightarrow Co(CN)_5^{2-} + e^- \qquad (12)$$

Such species usually act as highly reactive Lewis acids, a behavior to be expected in view of their coordinately unsaturated structures.

Central metal ions having the low-spin d^8 electronic configuration commonly exist in the square planar, four-coordinate structure. Compounds having this structure can be moderately stable or they may be highly reactive. Besides substitution, their reactions[25-29] include ligand addition, more or less of the usual Lewis acid-base kind [Eq. (13)], and oxidative addition, a process wherein both the oxidation number and the coordination number increase by two units [Eq. (14)].

$$Cl{-}Ir(P(C_6H_5)_3)_2{-}CO + C_2H_4 \longrightarrow (CH_2{=}CH_2)Ir(Cl)(CO)(P(C_6H_5)_3)_2 \qquad (13)$$

$$[PtCl_4]^{2-} + Cl_2 \longrightarrow [PtCl_6]^{2-} \qquad (14)$$

This behavior has been compared to that of carbenes (Table 2-1) and either reaction (Lewis addition or oxidative addition) leads to a relatively stable closed-shell configuration. The five-coordinate product may still undergo an oxidative addition reaction coupled to substitution [Eq. (15)].[28]

$$Ru^{0}(CO)_3[P(C_6H_5)_3]_2 + HCl \rightarrow Ru^{II}(H)(Cl)(CO)_2[P(C_6H_5)_3]_2 + CO \quad (15)$$

These materials are intrinsically ambiphilic. Simplistically, however, the mode of reaction we identify as Lewis addition for these compounds involves their combination with nucleophiles—i.e., the d^8 complexes are acting as electrophiles. In a similar context, oxidative addition involves a reaction with an electrophile, so one might expect the complex to participate in such reactions as a nucleophile. Despite these appealing relationships, considerable ambiguity exists in just this area, for a number of the molecules that commonly react by "Lewis addition" with low-spin, four-coordinate d^8 complexes are easily recognized as electrophiles—e.g., O_2 and SO_2. For the purpose of electron counting, it is probably still best to presume that all such species are electron pair donors since this leads to the sums associated with stable configurations. If one does not adhere to some such convention, it is difficult to rationalize the coordination numbers in some cases.[26,28] The tendency of d^8 ions to add a fifth ligand and become coordinately saturated decreases upon descending the periodic table ($Fe^0 > Ru^0 > Os^0$) and from left to right [$Fe^0 > Co(I) > Ni(II)$].[27,28] The mechanisms assigned to some oxidative addition reactions are discussed in more detail later.

In a very real sense, the ultimate oxidative addition reactions are those into H—H, C—C, and C—H bonds, for these processes are instrumental in the catalysis of many organic reactions. H_2 activation is a common phenomenon but the insertion of metal atoms into C—C and C—H bonds remains a relatively rare phenomenon. The oxidative cleavages of C—H bonds by metal atoms, as first reported by Chatt and Davidson[30] and by Cope and Siekman,[31] are summarized in Eqs. (16) and (17), respectively.

$$Ru[(CH_3)_2P\frown P(CH_3)_2]_2 \rightleftharpoons Ru(H)[(CH_3)_2P\frown P(CH_3)_2][(CH_3)_2P\frown P(CH_3)CH_2] \quad (16)$$

$$C_6H_5—N{=}N—C_6H_5 + PdCl_2 \rightarrow [Pd(C_6H_4N{=}NC_6H_5)(\mu\text{-}Cl)]_2 + 2HCl \quad (17)$$

Note that the second example [Eq. (17)] actually includes an oxidative addition into H—C followed by elimination of HX. The subject has been reviewed by Parshall.[32]

Compounds having metal ions with the d^{10} electronic configuration may also undergo oxidative addition with accompanying changes in coordination number from two to four [Eq. (18)].[29,33]

$$(C_6H_5)_3P{-}Pt^0{-}P(C_6H_5)_3 + CH_3I \longrightarrow (C_6H_5)_3P{-}\underset{\displaystyle P(C_6H_5)_3}{\overset{\displaystyle CH_3}{Pt}}{-}I \qquad (18)$$

Since the product of oxidative addition given in Eq. (18) is of the same electronic and geometric structure as the reactant in the oxidative addition reaction described in Eq. (6), one must conclude that the relative orbital energies of the two reactants and of the product are crucial in such processes. One can, in fact, offer a series of related oxidative addition processes starting with an elemental $d^{10}s^2$ atom of coordination number zero and going step by step to the six-coordinate d^6 structure.

Configuration	$d^{10}s^2 \rightarrow$	$d^{10} \rightarrow$	$d^8 \rightarrow$	d^6
Coordination No.	0	2	4	6

When one considers the σ-bonding to ligands, all these species have filled d-levels, and each has what is probably the most common coordination number for its particular electronic configuration. An example of the first class of process ($d^{10}s^2$) is the oxidation of gaseous Hg atoms by Cl_2 to produce molecular $HgCl_2$. Perhaps the unusual reaction products (unsolvated Grignards, RMgX) observed to form upon warm-up of a frozen matrix of alkyl halide molecules and magnesium atoms represent nontransition metal analogs.[34] For d^{10} atoms, Ni^0, Pd^0, and Pt^0 are relatively active while Zn^{2+} is not. Au^+, of course, is more reactive than Ag^+. For d^8 atoms, the sequence should be:

$$Os^0 > Ru^0 > Fe^0 > Ir(I) > Rh(I) > Co(I) > Pt(II) > Pd(II) \gg Ni(II), Au(III).$$

Ugo[33] emphasizes the fact that the zero valent complexes of Ni, Pd, and Pt provide the closest relatives to the free metals, and that they should, therefore, provide the best models for those long-used catalysts. Paradoxically, these zero valent complexes are generally unreactive toward molecular hydrogen.

Note that reductive elimination is the reverse of oxidative addition and, of course, results in a decrease in both the coordination number and the oxidation state by two units. Both forward and reverse reactions figure heavily in many catalytic processes. These will be encountered later. Also, Halpern and others have commonly discussed the radical abstraction and radical combination reactions of low-spin d^7 (and related) ions [Eqs. (4) and (5)] as oxidative additions. We choose not to do so here but refer to such processes by the alternate labels just stated.

E. Mechanisms of Oxidative Addition Reactions

Mechanistic studies of the oxidative addition reaction have centered on d^8 ions with Ir(I) and Rh(I) receiving much early attention. Vaska's compounds $IrCO(PR_3)_2X$, where X is Cl, Br, I, NCS, undergo a variety of oxidative addition reactions with such substrates as molecular hydrogen, alkyl halides, and hydrogen halides.[25,35,36] In addition, certain so-called Lewis acid-base reactions of these molecules have been compared with the more obvious oxidative addition since these may also be considered to be oxidative addition[26,28] processes. We concentrate on those cases where the linkage is cleaved between two atoms in the reacting molecule. A number of mechanisms have been proposed for these reactions. Chock and Halpern[37] proposed that

$$Ir(CO)(PR_3)_2X + AB \rightarrow Ir(CO)(PR_3)_2(A)(B)X \tag{19}$$

the oxidative addition of H_2 (and the kinetically similar addition of O_2) may proceed via a concerted process involving a relatively nonpolar transition state.

The O—O bond distances in the oxygen adducts vary from 1.30 Å for the X = Cl complex to 1.51 Å for the X = I complex. In a related structure, the O—O distance is 1.65 Å.[38,39,40] The shorter linkage may be compared with that in free O_2 (1.21) or in superoxide (1.28), while the longer distances are similar to, or exceed, that of the single O—O bond in peroxides (~1.5). The apparent variation in the O—O bond order requires a variation in the degree and possibly the nature of the interaction between the O_2 grouping and the iridium atom. An especially simple view considers the adducts of short O—O distance to involve a metal-ligand linkage strictly analogous to that with ethylene but involving a singlet form of oxygen (Figure 2-1a). This concept can be described in terms of a model involving the mixing of ground and excited states of O_2 upon complexing as developed by Mason.[41] Again, most simply, the species containing the longer O—O distances may be considered to be chelates of the O_2^{2-} ion (Figure 2-1b).

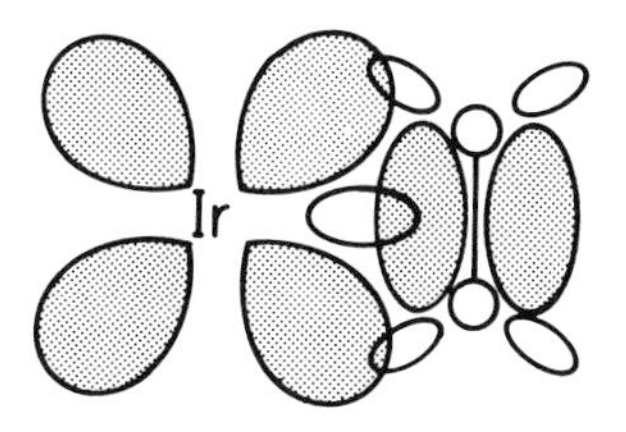

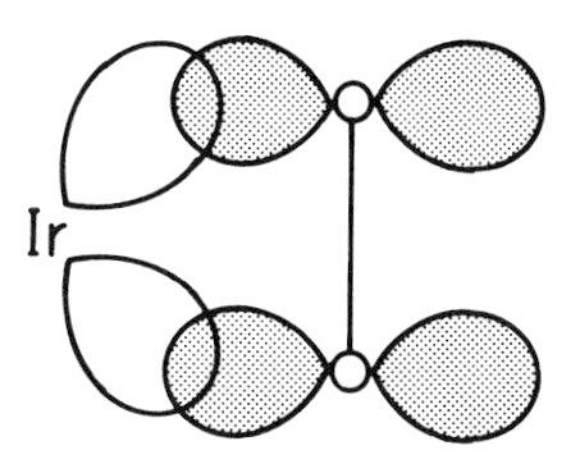

Figure 2-1 Bonding in O_2 adducts of Ir(I) complexes

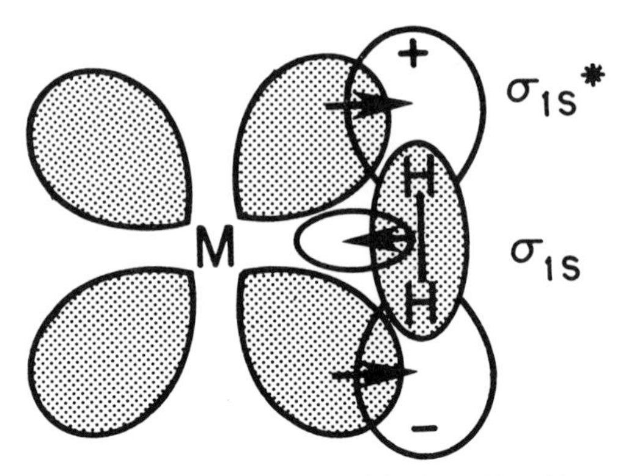

Figure 2-2 The cleavage of the H—H bond by a metal atom

In any case it seems reasonable to conclude that the bond order of the O_2 unit varies depending on the nature of its interaction with the metal atom.

The bonding relationships in the Chatt-Dewar model for ethylene complexes are particularly helpful in rationalizing the cleavage of the H—H bond in a three-centered or concerted transition state.[44,45] As Figure 2-2 shows, the overlap of the filled bonding molecular orbital of H_2 with a vacant metal orbital removes electron density from the H—H bond. At the same time, overlap of the vacant $\sigma_{1s}*$ antibonding orbital of H_2 with a filled d-π orbital of the metal puts electron density into an antibonding orbital. Clearly, no H—H bond exists when the antibonding electron density exceeds the bonding electron density. As pointed out by Nyholm, alternate concerted processes can be conceived; however, the most obvious would involve a highly polar transition state (simultaneous development of a hydride and a proton).[4]

In contrast, the oxidative addition of alkyl halides has been ascribed to various mechanisms, including S_N2 in which the metal acts as a nucleophile,[37] concerted addition,[46] and free radical combination, sometimes requiring initiation.[47] Kinetic activation parameters, solvent effects, and substituent effects for substituted benzyl halides clearly parallel those of known S_N2 displacements.[29,37,48,49] Such studies have generally been restricted to CH_3I and benzyl halides.

The use of optical isomers provides definitive evidence to test alternate mechanisms when such a diversity of possibilities is on hand. Investigations of oxidative addition reactions have been reported. The earlier studies are generally nondefinitive for one reason or another; however, Stille and associates[50] provide unequivocal results. These investigations established that inversion of carbon atom configurations occurs in certain oxidative addition reactions such as Eq. (20).

$$S(+)\text{-}C_6H_4\text{-}CHDCl + Pd[P(C_6H_5)_3]_4 \longrightarrow C_6H_5CHDPd[P(C_6H_5)_3]_2Cl \quad (20)$$

The conclusion is that there exists an S_N2 pathway for the oxidative addition of some alkyl halides.

The reactions of $Pt(PEt_3)_3$ with isopropyl iodide and benzyl bromide demonstrate the existence of radical pathways for oxidative addition reactions.[51] The behavior of these and related systems has led to a general formulation of multiple mechanisms for oxidative addition as indicated in Scheme 2-1.

Scheme 2-1

$$Pt^0 + RX \xrightarrow{S_N2} R—Pt^{II}—X$$

$$k_{-1} \nwarrow\searrow k_1 \qquad \nearrow k_c$$

$$\overline{Pt^I—XR\cdot}$$

$$\downarrow k_d$$

$$Pt^I—X + R\cdot$$

$$RX \swarrow k_a \qquad k_b \searrow Pt^0, RX$$

$$X—Pt^{II}—X + R\cdot \qquad \text{chain process} \quad R—Pt^{II}—X,\ H—Pt^{II}—X$$

This scheme brings together the results from many studies. The S_N2 process is incorporated in agreement with the conclusions stated above. The second path is visualized as involving the abstraction of halogen, possibly preceded by electron transfer to form a radical pair ($Pt(I) - X + \dot{R}$). This pair can either collapse to form the expected adduct $R - Pt(II) - X$ or separate to form $Pt(I) - X$ and $\cdot R$. The subsequent steps then vary depending on the reactivity of the alkyl halide. $Pt(I) - X$ can rapidly abstract halogen from very reactive RX molecules producing $X - Pt(II) - X$ and $\cdot R$. For less reactive alkyl halides, the organic radical can initiate a chain process to produce the usual adduct $R - Pt(II) - X$. The dominant path, or paths, in any given case depends on many factors, including the nature of the alkyl halide, the nucleophilicity of the metal complex, the ability of the metal complex to undergo one-electron processes, steric effects, and ligand exchange effects.

Parshall has offered suggestions on the possible mechanism of oxidative addition of C—H (Figure 2-3).[32] Concerning himself exclusively with the reaction at the ortho position in aromatic substrates, he predicts (a) arene complex formation, (b) conversion to a σ-complex not unlike those ascribed to some Friedel-Crafts systems, and (c) proton transfer. A preliminary reaction may involve dissociation of a ligand to make a coordination site available on the complex. Although arene complexing may occur, one might question whether this has pertinence to the oxidative addition reaction since what is needed is electrophilic attack at the ortho-carbon or a biphilic attack on the C—H bond (like on H_2

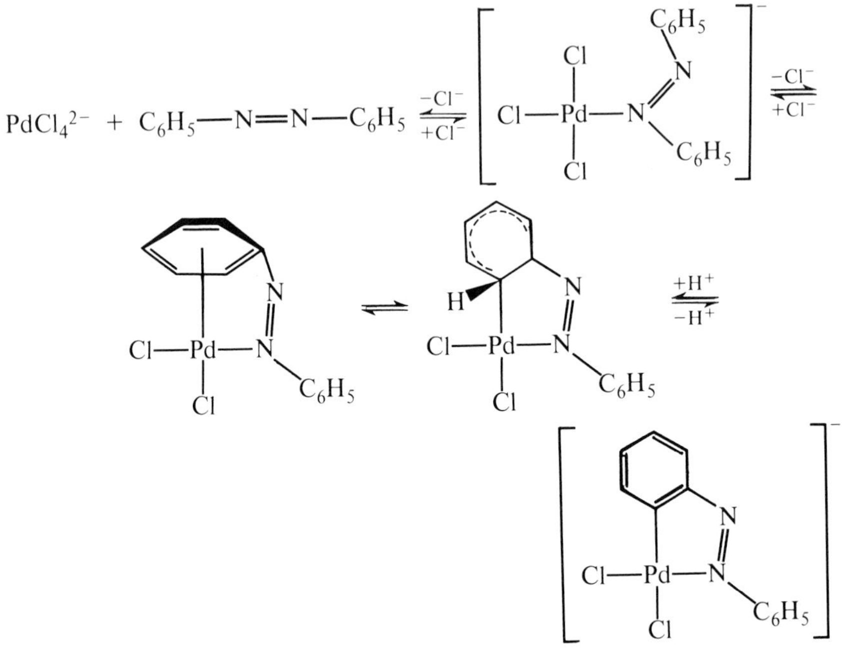

Figure 2-3 Proposed mechanism of oxidative addition of a C—H bond

in Figure 2-2). The stereochemistries for the latter two interactions would be very different from that of the arene complex.

Since four-coordinate d^8 species are simultaneously either nucleophiles or electrophiles, and potentially diradicals, they should be expected to operate on substrates via multiple mechanisms. These processes are important elemental steps in a wide range of ligand reactions, many of which involve catalytic systems, and they arise repeatedly in sections to follow. Insertions into C—C and C—H bonds are perhaps the most interesting examples and certainly among those requiring the strongest action by the metal atom.

III. CENTRAL METAL ATOMS OF COMPLEXES AS SUBSTITUENTS ON THE SUBSTRATE IN LIGAND REACTIONS

The various modes of bonding that a metal ion (atom) can employ in its complexes may serve to delineate the ways in which a metal ion can alter the electron distribution in a ligating substrate.[9] Depending on the orbitals and electrons available in a given complex, the metal–ligand linkage may be formed by either accepting or donating electrons from σ- and π-electron systems. Of course, the simplest mode of interaction is σ-electron pair donation from the ligand to a positively charged metal ion—i.e., the classic Lewis acid-base adduct formation.

$$M^{n+}L_n + :B \rightarrow L_nM{:}B^{n+} \quad (21)$$

However, even in this case, the electron-pair donor may use delocalized π electrons (ethylene, etc). In general, the metal ion, when held to the ligand by such a mode of bonding, will withdraw electron density from the ligand. The metal atom in its complex may also act as the electron pair donor in some cases—e.g., when a σ-bond is formed as a consequence of the nucleophilic attack of the metal complex on a substrate.

$$L_nM: + A \rightarrow L_nM{:}A \qquad (22)$$

In the product, one may choose to view the metal atom as donating electron density to the ligand through its σ-bond. More generally, it is found that metal atoms donate electron density to ligands by means of π-back bonding using their filled d-orbitals. This interaction is most important among low oxidation state derivatives of transition elements. In the case of very high oxidation state derivatives, the central metal ion may accept electron density from the ligand via the π-system as well as the σ-system (oxyanions such as MnO_4^-, etc.). The interaction of the π-orbitals of the metal atom with the π-electron systems of ligands leads to conjugative effects as well, and these are also of importance in ligand reactions.

A. Decreased Substrate Electron Density—Masking of Nucleophiles

When a nonbonded pair of electrons on a free ligand (NH_3) becomes a σ-bonding pair of electrons in a metal complex [$Ni(NH_3)_6^{2+}$, Eq. (23)], a profound change in the reactivity of the ligand occurs. Classically, "a valence is saturated," and a site of nucleophilic reactivity on the ligand has been used up.

$$Ni^{2+} + 6{:}NH_3 \rightarrow Ni({:}NH_3)_6^{2+} \qquad (23)$$

The conversion of the electron pair from nonbonding to σ-bonding constitutes both a lowering in energy of the electrons and a site alteration for their electron distribution from a region of relative void to a closely confined region between two atoms. Such a change literally covers up the electron pair; any alternate electrophile would have to compete directly with the metal ion for the electron pair in a displacement reaction, whereas the original process was a relatively simple addition reaction. Interactions of this kind often lead to extreme changes in chemical reactivity. For example, coordinated NH_3 is completely nonbasic. In fact, it may become acidic. Such extreme changes in reactivity are called masking, and although these effects have been reviewed earlier,[2,9,10,52] a few examples are presented here.

The selective formation of isonitriles when cyanide complexes act as nucleophiles toward alkyl halides and related electrophilic alkylating agents [Eq. (24)] is a direct consequence of the selective prior coordination of the more strongly nucleophilic carbon atom to the metal atom.[2,53]

$$M:C{\equiv}N:^{\ominus} + RX \rightarrow M:C{\equiv}N{-}R + X^{\ominus} \quad (24)$$

The corresponding reactions of ionic cyanide produce nitriles, R—C≡N.

The oxidation potentials of metal atom couples are greatly altered by changing the ligands to which they are bound. For example, the oxidation potential for the Co^{2+}–Co^{3+} couple changes from −1.82 *v* to +0.83 *v* when CN^- is substituted for the water ligands.[54] In this transformation, the Co^{3+} species ceases to be a powerful oxidizing agent. In a very real sense, CN^- masks the reactivity of Co^{3+}. Indeed, in the presence of oxygen donors, Co^{3+} is a very useful oxidant[55] and catalyst for air oxidations.[56] This activity vanishes when the cobalt is bound to strong donors.

Returning to the masking of nucleophiles, the effect may be considerably smaller when the metal ion interacts with a different electron pair from the site that may serve as a nucleophile toward other substrates. This is most simply seen in the usual effect of coordination on the basicity of a remote site. Free

N⬡NH$^+$

is a moderately strong acid having a pK of 0.6. However, coordination to Ru(III) in

$(NH_3)_5$RuN⬡NH^{4+}

increases its acidity by two orders of magnitude, the pK becoming −0.8. This is considered to be a result of the normal inductive effect of the metal cation,[57] and it is presumed to operate through the σ-electron system of the ligand.

The withdrawal of electron density from the π-electron system of the substrate also leads to partial masking of basicity. This is clear from the data of Table 2-2 which shows that the formation of a π-complex by a $Cr(CO)_3$ group reduces

TABLE 2-2
ACIDITIES OF SUBSTITUTED BENZOIC ACIDS

Compound	pK
$C_6H_5CO_2H$	5.68
$(CO)_3Cr(C_6H_5CO_2H)$	4.77
$O_2NC_6H_4CO_2H$	4.48

the basicity of the benzoate anion almost as much as attaching a nitro group in the para position.[58]

B. Decreased Substrate Electron Density—Promotion of the Reactivity of Electrophiles

When a metal cation binds to a substrate by using a substrate electron pair, it reduces the electron density on the substrate. If the latter contains an electrophilic reactive site, the reactivity of that site should be enhanced. On this there seems to be universal agreement, and, in fact, there is sometimes a tendency to view this as the sole route whereby a metal atom can promote reactivity in a substrate. While so extreme a view must be considered myopic, this category of ligand activation remains most obvious and among the more common. Schiff base linkages that are not parts of chelate rings serve to illustrate the phenomenon [Eq. (25)].[59,60,61]

$$M^{n+}(N{=}C(R)R')_2 + 2H_2O \longrightarrow M^{n+}(NH_2)_2 + 2RC(=O)R' \qquad (25)$$

Their rates of hydrolysis are substantially increased by coordination. Similarly, the action of halophiles to promote the solvolysis of C—X bonds is a direct result of the promotion of electrophile reactivity [Eqs. (26) and (27)].

$$RX + H_2O + Hg^{2+} \longrightarrow ROH + HgX^+ + H^+ \qquad (26)$$

$$(X = Cl,\ Br,\ I)$$

$$RO\text{---}P(O)(O)\text{---}F + H_2O \xrightarrow{Cu^{2+}} RO\text{---}P(O)(O)\text{---}OH + HF \qquad (27)$$

In the first example, polarization of the azomethine function by the cation enhances the electrophilic character of the carbon atom, thereby facilitating attack by water[59,60] (structure **3**).

$$M^{n+}—\overset{\delta^-}{N}=C^{\delta+} \leftarrow OH_2$$

3

In the remaining two examples, coordination of the metal to the halogen in the reactant molecule presumably plays the same role.[62,63,64] Thioesters[65,66,67,68] and even thioethers[69,70,71] suffer metal ion-promoted electrophilic cleavage of a C—S bond, at least in certain cases.

C. Decreased Substrate Electron Density—Formation of New Nucleophiles

A variety of ligands containing ionizable hydrogen become stronger Brønsted acids upon coordination to a metal ion. Consequently, their conjugate bases exist in complexes in aqueous media at relatively low pH values. The formation of such species [Eqs. (28) and (29)] constitutes an additional example of the promotion of the reactivity of an electrophile through coordination, for ionization of QH_n in a polar medium occurs by the nucleophilic attack of a solvent molecule on a proton.

$$M^{m+} + :QH_n \rightleftarrows M:QH_n{}^{m+} \tag{28}$$

$$M:QH_n{}^{m+} \rightleftarrows M:\ddot{Q}H_{n-1}{}^{(m-1)+} + H^+ \tag{29}$$

Coordination of $:QH_n$ to a metal is an example of masking of a nucleophile. In contrast, the subsequent ionization of coordinated $:QH_n$ produces a new nucleophile. The loss of a proton makes available an electron pair on the ligand. Examples of such processes are given in Eqs. (30)–(33) below.

$$M(H_2O)_6{}^{n+} + H_2O \rightarrow M(H_2O)_5(OH)^{(n-1)+} + H_3O^+ \tag{30}$$

$$Pt(NH_3)_6{}^{4+} + H_2O \rightarrow Pt(NH_3)_5(NH_2)^{3+} + H_3O^+ \tag{31}$$

$$2H_2O + Ni^{2+} + 2NH_2CH_2CH_2SH \longrightarrow Ni\left(S-CH_2-CH_2-NH_2\right)_2 + 2H_3O^+ \tag{32}$$

$$3H_2O + Fe^{3+} + CH_3C(=O)CH_2C(=O)CH_3 \longrightarrow Fe\left(O-C(CH_3)-CH-C(CH_3)-O\right)_3 + 3H_3O^+ \tag{33}$$

The reactions of Eq. (30) merely represent the Brønsted acid ionization of hydrated metal ions. Some common hydrated-ion pK_a values[72] are: Mg^{2+}, 11.4; Cu^{2+}, 6.8; Hg^{2+}, 3.7; Cr^{3+}, 3.8; Fe^{3+}, 2.2; Co^{3+}, 0.7. Working with a hydroxy derivative of the glycyl-glycine complex of Cu(II), $(Cu(GG)OH^-)$, Koltun, Fried, and Gurd[73] were able to compare the nucleophilic strength of the coordinated hydroxide to those of water and free hydroxide. The sequence is $OH^- > Cu^{2+}—OH^- > H—OH$ with the coordinated OH^- only some 60 times stronger as a nucleophile than H_2O. In the reaction studied [hydrolysis of *p*-nitrophenyl acetate, Eq. (34)] $Cu^{2+}—OH^-$ is quite effective because it is present in much greater concentrations than OH^- at the neutral pH of water.

$$CH_3C(=O)O-C_6H_4-NO_2 + H_2O \xrightarrow{Cu(GG)OH^-} HO_2CCH_3 + HO-C_6H_4-NO_2 \quad (34)$$

Eq. (31) illustrates that coordinated amines undergo Brønsted acid ionization only when the metal ion to which they are coordinated has a sufficiently great polarizing power. The fact that the resulting amide complexes can act as nucleophiles has been amply demonstrated[74] [Eq. (35)].

$$[(H_2N\frown NH_2)Pd^{2+}(\ddot{N}H^{\ominus}\frown NH_2)] + CH_3I \longrightarrow [(H_2N\frown NH_2)Pd^{2+}(NH(CH_3)\frown NH_2)] + I^- \quad (35)$$

As shown[75,76] in Eq. (32), it is most common that mercaptan groups ionize upon coordination to metal ions. Such coordinated mercaptans have long been known to act as nucleophiles (see section VIII). Finally, Eq. (33) demonstrates promoted ionization of a proton from a carbon atom, and again, the coordinated conjugate base is subject to a range of electrophilic reactions.[5,6,77] The actions of these various functions as nucleophiles are discussed at length later in this chapter.

D. Increased Substrate Electron Density—Masking of Electrophiles

Metal atoms are often presumed to enter into π-bonding with appropriate ligands by donating shares in their d-π electron pairs. This should increase the electron density on the ligand. In turn, if the ligand were normally subject to nucleophilic attack, this susceptibility should be decreased. The reported sensitivity of the trivalent halides of phosphorus to hydrolysis when coordinated to various metal ions appears to support this point of view. Whereas $PtCl_2(PCl_3)_2$[78,79] and $AuCl(PCl_3)$[80] hydrolyze rapidly upon exposure to moisture, $IrCl_3(PCl_3)_3$[81] does not react with a variety of nucleophiles. The fact that $Ni(PF_3)_4$ can be steam distilled without hydrolysis reflects the same phenomenon.[82] A quantitative study of the effects of coordinating Ru(II) and Ru(III) to the pyridine nitrogen on the hydrate formation of 4-formyl pyridine[83] has shown a similar effect [Eq. (36)].

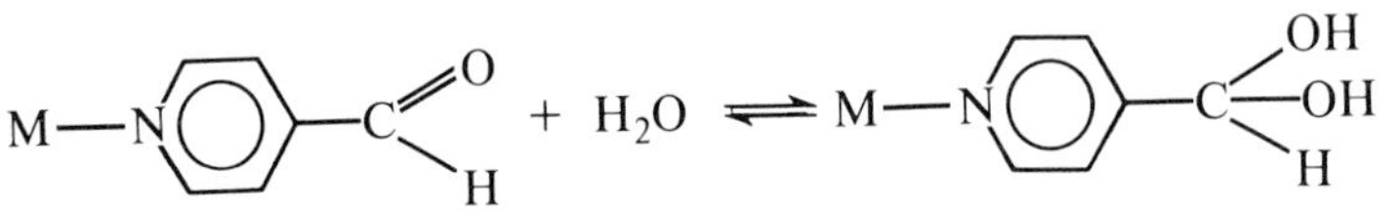

$$\text{M—N}\langle C_5H_4\rangle\text{—C(=O)H} + H_2O \rightleftharpoons \text{M—N}\langle C_5H_4\rangle\text{—C(OH)(OH)H} \quad (36)$$

The electrophilic character of the carbonyl function is diminished at equilibrium by coordination to the π-electron donating Ru(II) atom.

E. Increased Substrate Electron Density—Promotion of Nucleophiles

Metal atoms that form strong π-bonds by back-donation should enhance the electron densities on coordinated substrates that still can act as nucleophiles. Ru^{2+} is a strongly back-bonding ion, and when it binds to

$$\text{N}\langle C_4H_4\rangle\text{NH}^+$$

to form the complex

$$(NH_3)_5Ru\text{—N}\langle C_4H_4\rangle\text{NH}^{3+},$$

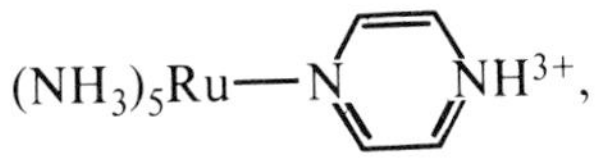

it decreases the acidity of the ligand by almost two orders of magnitude (pKa = 2.5 compared with 0.6 for the free ligand).[57] This can only be ascribed to the increased basicity of the protonated nitrogen atom when the other nitrogen is bound to Ru(II).[57]

In keeping with the rationalization given above for the interaction of low-valent metal ions with O_2 and H_2 molecules, the activation of some small (diatomic) molecules upon coordination is the result of weakening the bond uniting the small molecule by donation of d-π electron density. Thus, the coordinated O_2 in $Ir(CO)(PR_3)_2X(O_2)$ becomes more reactive as the O—O distance increases and O—O bond order decreases. The metal ion substituent lowers the bond order from two to about one, or even less.

F. Increased Electron Density—Formation of New Electrophiles

Pursuing analogies and contrasts to the extreme provides a maximum range of types. This section and the preceding section are direct counterparts to categories relating to electron withdrawal.

Electron Withdrawal	Electron Donation
(*1*) Masking of nucleophiles	(*1*) Masking of electrophiles
(*2*) Promotion of electrophiles	(*2*) Promotion of nucleophiles
(*3*) Formation of new nucleophiles by loss of a cation (H^+)	(*3*) Formation of new electrophiles by loss of an anion (X^-)

The third category under electron donation deserves rationalization and, if possible, illustration. Category *3* under electron withdrawal occurs because the polarization of the ligand makes it more positive, and this can be relieved by loss of a positive ion. Conversely, electron donation to a ligand should make it more negative, and dissociating an anion could counter this. Sulfite is a good ligand, but so is sulfur dioxide. In fact, the SO_3^{2-} ion in $Ru(NH_3)_5SO_3$ is supposed to be transformed to $Ru(NH_3)_5SO_2^{2+}$ under acidic conditions[84]—i.e., it loses an anion (O^{2-} or OH^-).

An alternative mode of reaction is probably more common under those circumstances in which electron donation by the metal atom is considerable. This is the simple electrophilic addition implied by the enhanced nucleophilicity of the substrate. For example, the activated O_2 molecule will add SO_2 to form SO_4^{2-} within the complex [Eq. (37)].[85]

$$Ir(CO)(PR_3)_2X(O_2) + SO_2 \rightarrow Ir^{III}(CO)(PR_3)_2X(SO_4) \qquad (37)$$

Also, the formation of alkyl-cobalt(III) derivatives from olefins and Co(I) complexes in protic media may be viewed in these terms. The reaction proceeds thru an olefin complex, and electron donation by the cobalt(I) presumedly enhances the nucleophilicity of the olefin, which then abstracts a proton from an available donor, solvent, or other acid [Eq. (38)].[86,87,88,89]

$$\left[L_5Co^{I}\!-\!\overset{\textstyle CRR'}{\underset{\textstyle CH_2}{\|}} \right]^{n+} + H^+ \rightleftharpoons [L_5Co^{III}CH_2\!-\!CRR'H]^{(n+1)+} \qquad (38)$$

Such processes occur with vitamin B_{12}[86,87] and other cobalt complexes containing only one available reactive site at the cobalt atom.[88,90] It is unlikely that the conversion of olefin complexes to alkyl derivatives in these systems occurs by the common class of insertion reaction in which both the hydrogen and olefin are first bound to the metal atom.

G. Intramolecular Oxidation-Reduction Reactions

The preceding sections have dealt with the polarizing effects of metal ions by showing how the metal ion substituent alters the reactivity of the ligating substrate. In every case, reference is made to the influence of this partial displacement of electron density on reactivity toward some third external reagent, most often an electrophile or a nucleophile. The charge displacement in certain cases is countered by a detectable slow intramolecular process that appears to involve oxidation-reduction, with the metal ion and its ligand acting as the reactants. Such processes may be triggered by the action of an external reagent, a change in medium, or if a structural rearrangement is involved, they may be unimolecular processes.

Complexes of highly unsaturated macrocyclic ligands provide likely examples of unimolecular redox processes in which electrons appear to be transferred from molecular orbitals confined predominantly on the metal ion to molecular orbitals delocalized over some substantial part of the ligand. Wolberg and Manassen[91] found that the product of the first one-electron step in electrochemical oxidation of nickel(II) tetraphenylporphyrin [$Ni(II)(TPP^{2-})$] contains nickel(III) [$Ni(III)(TPP^{2-})^+$]. This substance is not stable in the medium in which it is produced (Ph-CN), and it is transformed back to the starting material, Ni(II)-(TPP^{2-}). However, evidence has been presented for the initial rearrangement of $Ni(III)(TPP^{2-})^+$, by an internal redox reaction, to a nickel(II)–anion radical complex $Ni(II)(TTP^{-}\cdot)^+$. The system's high reactivity made detailed study difficult so that only weak esr signals for $Ni(II)(TTP^{-}\cdot)^+$ were obtained. However, this example raises the very real question "why should an internal redox reaction be slow?" Since intramolecular electron transfer will not be rate dermining, some skeletal rearrangement must occur.

A related process involving the cobalt complexes of 1,19-diethoxycarbonyl-tetradehydrocorrin (R_2Cn) has been reported[92] (structure **4**).

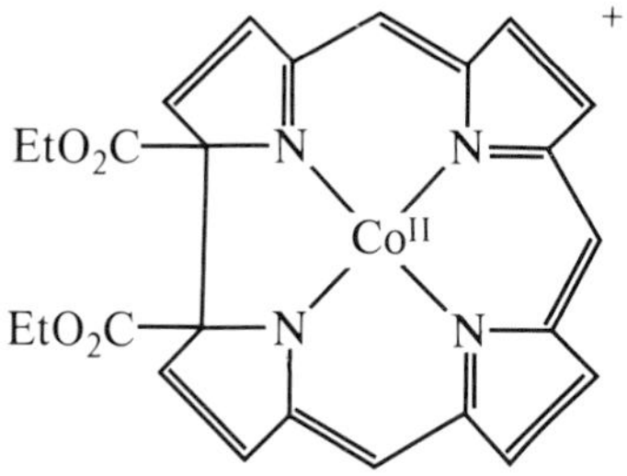

4

The one-electron reduction of the cobalt(II) complex, $Co(II)(R_2Cn^-)^+$, produces a stable cobalt(I) derivative, $Co(I)(R_2Cn^-)$. Two-electron reduction of the starting material, however, produces two-electron reduction of the ligand, leaving the metal ion in the divalent state [Eq. (39)].

$$Co^{II}(R_2Cn^-)^+ + 2e^- \rightarrow Co^{II}(R_2Cn^{3-})^- \qquad (39)$$

The two-electron reduction process carried out in steps is:

$$Co^{II}(R_2Cn^-)^+ + e^- \rightarrow (CO^{I}(R_2Cn^-) \qquad (40)$$

$$Co^{I}(R_2Cn^-) + e^- \rightarrow Co^{II}(R_2Cn^{3-})^- \qquad (41)$$

As Eqs. (40) and (41) show, the addition of one electron to the cobalt(I) complex causes oxidation, not reduction, of the cobalt. Here there is no evidence for a slow rearrangement, but the experiments reported were not designed to detect such a process.

One of the most compelling examples presently available to illustrate internal oxidation-reduction between a metal ion and its own ligand involves the transformation between the annulene-like ligand known as TAAB and its two-electron reduction product. The latter is a porphyrin analog.[93,94,95]

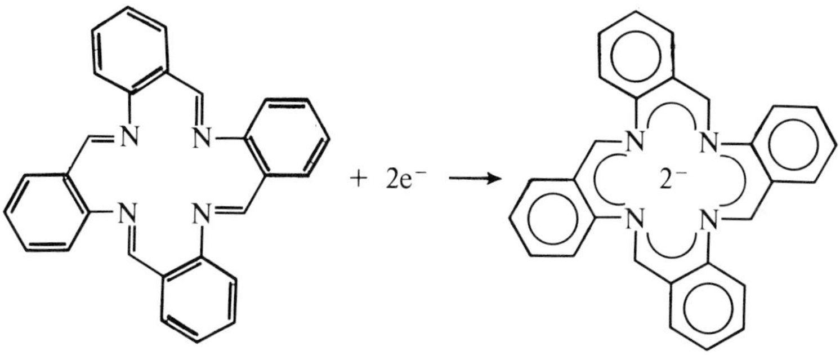

$$\text{TAAB} + 2e^- \longrightarrow \text{TAAB}^{2-} \qquad (42)$$

TAAB TAAB^{2-}

For Cu(II)(TAAB)$^{2+}$, Ni(II)(TAAB)$^{2+}$, and Co(II)(TAAB)$^{2+}$, the number of electrons that can be added to the complex at moderate potentials varies with the electronic configuration. This supports the view that a stable electronic structure occurs for the general unit M(TAAB)$^{n+}$, when M is a metal ion having 8 d-electrons, and the ligand has acquired two more electrons and become the aromatic anion TAAB^{2-}.[95] Eq. (42) could well be slow since the annulene-like ligand TAAB exists,[96] in so-called square-planar complexes, in a rather strongly ruffled-saddle conformation (Figure 2-4), while the aromatic dianion TAAB^{2-} should be more nearly planar. It is particularly interesting that the one-electron reduction product of Co(II)TAAB^{2+} undergoes a slow rearrangement forming a complex that has been isolated and characterized as a derivative of Co(III)-(TAAB^{2-})$^+$. The sequence of events, as outlined in Eqs. (43) and (44), is supported by detailed electrochemical studies.[95]

$$\text{Co}^{\text{II}}(\text{TAAB})^{2+} + e^- \rightleftharpoons \text{Co}^{\text{I}}(\text{TAAB})^+ \qquad (43)$$

$$\text{Co}^{\text{I}}(\text{TAAB})^+ \xrightarrow{\text{slow}} \text{Co}^{\text{III}}(\text{TAAB}^{2-})^+ \qquad (44)$$

It is concluded that the cobalt(I) formed in the electrode reaction performs a two-electron reduction on its own ligand so that the net effect of adding one electron to the cobalt complex is to induce oxidation of the central metal ion.

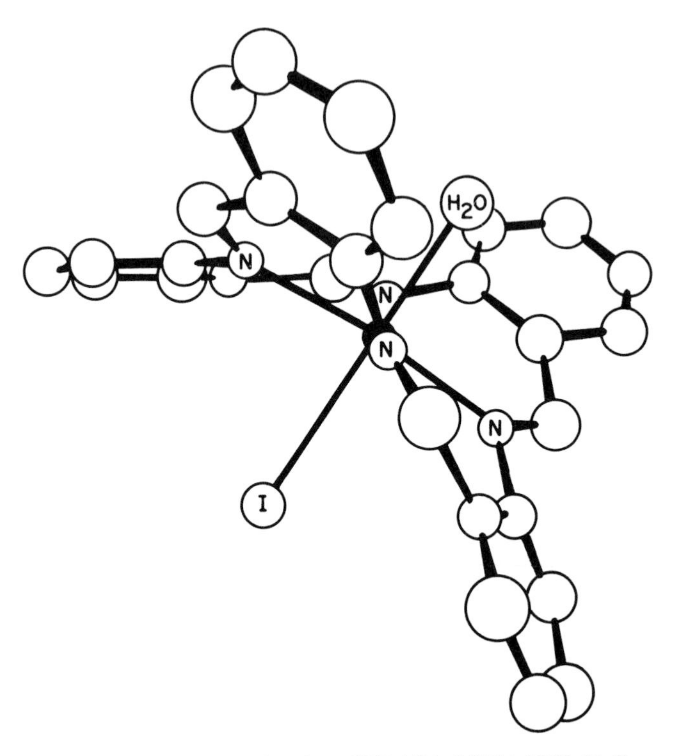

Figure 2-4 Perspective view of TAAB in $Ni(TAAB)I_2 \cdot H_2O$

Taube has identified and studied induced electron-transfer reactions involving the reduction of cobalt(III) to cobalt(II) [Eq. (45)].[97,98]

$$(NH_3)_5Co^{III}N\langle\bigcirc\rangle\text{—}CH_2OH^{3+} + [O^{I}] \xrightarrow{H^+}$$

$$Co^{2+} + 5NH_4^{+} + HN\langle\bigcirc\rangle\text{—}C\begin{matrix}H \\ \\ O\end{matrix} + [R^{I}] \quad (45)$$

[O(I)] is a one-electron oxidizing agent such as Ce(IV), Mn(III), or $Co(H_2O)_6^{3+}$. It is presumed that the one-electron oxidant produces a Co(III) complex of a highly reactive radical,

$$[(NH_3)_5CoN\langle\bigcirc\rangle\text{—}\dot{C}(H)OH]^{3+}$$

which then undergoes internal redox, leading to the products of Eq. (45). The central metal ion does not participate in the ligand oxidation when two-electron oxidants [O(II)] are used [Eq. (46)].

$$(NH_3)_5Co^{III}N\bigcirc\!\!-CH_2OH + [O^{II}] \xrightarrow{H^+} \left[(NH_3)_5Co^{III}N\bigcirc\!\!-C\begin{matrix}\nearrow O\\ \searrow H\end{matrix}\right]^{3+} + H^+ + [R^{II}] \quad (46)$$

Since Co(II) is very labile, the novel induced-oxidation is easily identified by the products formed. The ratio of Co(III) to Co(II) in the products, and other results, indicate that the reaction follows two pathways. Two different radicals may be formed in this process, and they would differ in their rates of internal electron transfer compared with their rates of formation under the action of the external oxidant. The second intermediate may be generated by electron abstraction from the oxidizable ligand [Eq. (47)].

$$-N\bigcirc\!\!-CH_2OH \longrightarrow -N\bigcirc\!\!-CH_2OH^+ + e^- \quad (47)$$

This second intermediate might then react relatively rapidly with an external oxidant by hydrogen atom transfer.

IV. METAL COMPLEXES AS LEAVING GROUPS IN LIGAND REACTIONS

A. Heterolytic Ligand Dissociation

This phrase has a precise meaning in coordination chemistry. Ligand dissociation involves the heterolytic cleavage of the metal-donor bond so that the σ-bonding pair of electrons goes with the departing ligand atom. The metal atom does not change its oxidation state in such a process. Indeed, this kind of process is extremely common for those metal complexes that undergo ligand substitution by S_N1 mechanisms [Eq. (48)].

$$(NH_3)_5CoCl^{2+} \xrightarrow{\text{slow}} (NH_3)_5Co^{3+} + Cl^- \quad (48)$$

The metal-containing product is a highly electrophilic carbonium ion analog in terms of the scheme described above (Table 2-1). Although oxidation states are defined so that this kind of process does not lead to an alteration in this parameter for either the ligand or the metal ion, the overall process of which it is part may leave either or both species very much changed. For example, cobalt(I) might serve as a nucleophile with CH_3I to form a Co-CH_3 grouping [Eq. (49)].

$$\ast Co^I\!: + CH_3I \longrightarrow \ast Co^{III}-CH_3 + I^- \quad (49)$$

If an electrophilic reagent is subsequently used to remove CH_3 as $:CH_3^-$ [Eq. (50)], the cobalt has clearly been oxidized.

$$H_2O + -Co^{III}-CH_3 + Hg^{2+} \longrightarrow -Co^{III}-OH_2 + CH_3Hg^+ \quad (50)$$

B. Nucleophilic Ligand Abstraction

This reaction class involves the removal of a ligating atom from the coordination sphere of a metal ion leaving the σ-bonding electron pair from the metal-donor bond with the metal [Eq. (51)].

$$HCo^{III}(CN)_5^{3-} + OH^- \rightarrow :Co^{I}(CN)_5^{4-} + H_2O \quad (51)$$

This leads, by definition, to two-electron reduction of the metal atom. For example, cobalt in $HCo(CN)_5^{3-}$ is assigned an oxidation state of positive three while the product obtained upon loss of H^+, $Co(CN)_5^{4-}$ contains cobalt(I). The rules of oxidation state assignment are formal and arbitrary, but the electron bookkeeping associated with such changes is very real.

The transfer of Cl^+ to a substrate by the iron-heme enzyme chloroperoxidase has been presumed to operate by such a mechanism[99] [Eq. (52)].

$$Cl^{\ominus}\text{-}Fe^{IV}(N_4) + SH \longrightarrow Fe^{III}(N_4) + SCl + H^+ \quad (52)$$

(SH = substrate)

The apparent imbalance of Eq. (52) comes from a further complication of the model. It is assumed that the iron-containing reagent exists as a radical anion complex of Fe(IV). In the course of transferring Cl^+, both the porphyrin ligand and the iron are assumed to be reduced. This metal-ligand collaboration (actually induced-reaction) would facilitate the involvement of the metal ion in two-electron processes.

The Wacker process seems to involve this class of bond breaking in the critical oxidation step. In this process, olefins are oxidized to aldehydes by a $PdCl_2$ catalyst in the presence of copper salts and oxygen [Eq. (53)].

$$C_2H_4 + \frac{1}{2}O_2 \xrightarrow[CuCl_2]{PdCl_2} CH_3CHO \quad (53)$$

A carbon-palladium σ-bonded species appears to be a critical intermediate, and Eq. (54) shows a simplified version of the step of immediate interest. This system is discussed in more detail later.

$$HOCH_2CH_2^{\ominus}\text{—}Pd^{II}\text{—} \longrightarrow HOCH_2CH_2^{+} + Pd^0 \longrightarrow H^+ + O{=}CHCH_3 \quad (54)$$

C. Homolytic Ligand Dissociation

Homolytic cleavage of the metal-ligand bond constitutes a one-electron redox process; the ligand is oxidized and the metal ion reduced. Such processes are not uncommon, one of the most celebrated being the cleavage of the Co—C bond in vitamin B_{12} co-enzyme when that species is combined with both enzyme and substrate[100,101] [Eq. (55)].

$$RCH_2\text{—}Co^{III}(N_4)(N) + E + S \longrightarrow RCH_2 \cdot \; Co^{II}(N_4)(N) \cdot S \cdot E \quad (55)$$

(E = enzyme, S = substrate)

A classical reaction for the coupling of alkyl radicals utilizes the addition of such transition metal salts as $CoCl_2$ or $CuCl_2$ to Grignard reagents. These processes probably involve the formation of M—C bonds which rapidly undergo homolytic cleavage. Such a homolytic cleavage also provides an obvious possibility for the mechanism of reduction of metal ions by hydrogen peroxide [Eq. (56)].[102]

$$M^{n+} + H_2O_2 \rightleftharpoons [M\text{—}O(H)\text{—}O\text{—}H]^{n+} \longrightarrow M^{(n-1)+} + HO_2\cdot + H^+ \quad (56)$$

Perhaps the most general reaction of this class is the oxidation of alcohols to carbonyl compounds by metal ions that act as one-electron oxidants [Eq. (57)].

$$\mathrm{RCH_2OH \xrightarrow{M^{n+}} RCHO} \tag{57}$$

It is well established that alcohol complexes form as intermediates. A critical step further on in the process[103] involves homolytic cleavage to generate a radical [Eqs. (58) and (59)].

$$\mathrm{R_2CHOH + VO_2^+ + H_3O^+ \rightleftharpoons R_2CH\underset{\underset{H}{|}}{O}{-}V(OH)_3^{2+}} \tag{58}$$

$$\mathrm{R_2CH\underset{\underset{H}{|}}{O}{-}V(OH)_3^{2+} \xrightarrow{slow} R_2\dot{C}OH + V^{IV}} \tag{59}$$

D. Reductive Elimination

The principle of microscopic reversibility assures us that this class of reaction will occur. In fact, the reverse of our first example of oxidative addition [Eq. (14)] illustrates the process of reductive elimination. The relationship is generalized in Eq. (60) where emphasis is placed on the simultaneous change by two units in oxidation number and coordination number.

$$\mathrm{M^{(N)}X_n + X_2 \underset{\text{reductive elimination}}{\overset{\text{oxidative addition}}{\rightleftharpoons}} M^{(N+II)}X_{(n+2)}} \tag{60}$$

In any given system, the value of the equilibrium constant for the total system will reflect the relative facility of the two opposing rate processes. In the case of $PtCl_4^{2-} + Cl_2$, the equilibrium obviously lies far to the right while the similar equilibrium for the $PdCl_2 + Cl_2$ or $PCl_3 + Cl_2$ system is not so extreme.

Reductive elimination processes are critical in a number of important catalytic processes, some of which are considered in detail in sections to follow. For example, a metal ion containing both coordinated hydrogen atoms and alkyl groups may produce alkanes by reductive elimination [Eq. (61)]. This may be the final step in some catalytic hydrogenations.

$$\mathrm{{>}M{<}^{H}_{R} \longrightarrow {>}M + HR} \tag{61}$$

V. SPECIAL EFFECTS DEPENDENT ON THE POLYFUNCTIONALITY OF METAL IONS

Many of the elemental reactions that are involved in ligand reactions require simultaneous use of two or more coordination sites on a single metal ion. Three categories of effects of this kind are: (a) Intramolecular combination-fragmentation processes, wherein monodentate groups react with each other while bound to a single metal ion; (b) Ligand reactions requiring chelate ring formation; and (c) Ligand reactions controlled by geometric relationships within the coordination sphere, including stereospecific processes of both the asymmetric and template types. The geometric constraints of (a) and (b) require that the sites binding the reactive ligands to the given metal ion need only be in suitable proximity. More stringent geometric requirements on both the ligating reagents and the activating metal ion accompany (c).

A. Intramolecular Combination-Fragmentation Processes

In this widely ranging class of metal-ion effects reactive groups are brought together and united within the coordination sphere of a metal ion. Further, this class of processes must include the microscopic reverse, the splitting of single ligands into two or more bound fragments. For a system involving two starting ligands, the process is envisioned most simply in Eqs. (62a) and (62b).

$$\text{M}\langle^{\text{X}}_{\text{Y}} \;\underset{\text{fragmentation}}{\overset{\text{(a) combination}}{\rightleftharpoons}}\; \text{M}\langle^{\text{X}}_{\text{Y}}| \qquad \text{M}\langle^{\text{X}}_{\text{Y}} \;\underset{\text{fragmentation}}{\overset{\text{(b) combination}}{\rightleftharpoons}}\; \text{M—X—Y} \tag{62}$$

A new chelate ring is formed in Eq. (62a) with no change in coordination number at the central metal ion. A chelate is cleaved in the reverse process. The combination process in Eq. (62b) is called an insertion reaction, and the two originally monodentate ligands are converted into a single monodentate ligand; the coordination number is lowered by one. The corresponding fragmentation process produces two ligands where only one existed; the coordination number is increased.

The most thoroughly documented examples of intramolecular combination processes leading to the in-situ production of new chelate ligands are those involving the reactions of α-amino acid esters.[104,105,106] In basic media, the reactions shown in Eqs. (63)–(65) occur. The reaction depends on the acid ionization of a coordinated NH_3 group in order to provide a nucleophile adjacent to the bound amino acid ester [Eq. (63)]. The coordinated NH_2^- group then displaces the alkoxy group from the ester to produce a chelated amide in Eq. (64). At the pH of this reaction the product amide is sufficiently acidic to ionize a proton [Eq. (65)].

$$[(H_3N)_5Co^{3+}-NH_2-CH_2C(=O)OR] + OH^- \overset{\text{fast}}{\rightleftharpoons} [(H_3N)_4(\overset{\ominus}{N}H_2)Co^{3+}-NH_2-CH_2C(=O)OR] \quad (63)$$

$$[(H_3N)_4(\overset{\ominus}{N}H_2)Co^{3+}-NH_2-CH_2-C(=O)OR] \overset{\text{slow}}{\rightleftharpoons} [(H_3N)_4Co^{3+}(NH_2-CH_2-C(=O)-NH_2)] + OR^- \quad (64)$$

$$[(H_3N)_4Co^{3+}(NH_2-CH_2-C(=O)-NH_2)] + OH^- \overset{\text{fast}}{\rightleftharpoons} [(H_3N)_4Co^{3+}(NH_2-CH_2-C(=O)-\overset{\ominus}{N}H)] + H_2O \quad (65)$$

The corresponding ring opening or fragmentation process would be the reverse of Eq. (64). A chelate ring is cleaved by nucleophilic displacement of the coordinated donor atom from a carboxylate function. Such processes are illustrated by the ring opening reaction of chelated oxalate in basic media. Using ^{18}O-

labeled water, Andrade and Taube showed that the C—O, and not the Co—O bond, is cleaved in this process [Eq. (66)].[107]

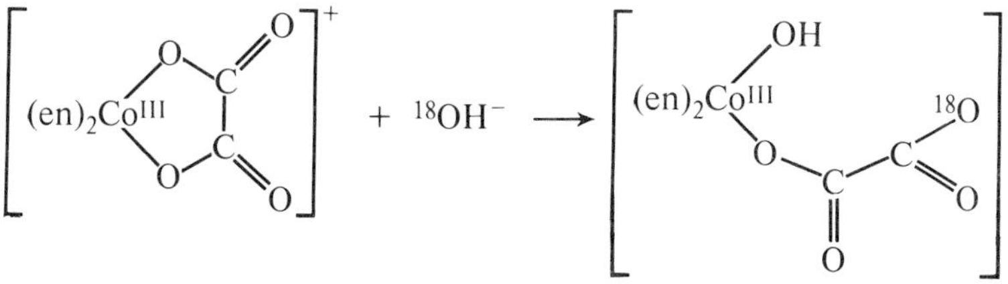

$$\left[(en)_2Co^{III}\text{(oxalato)}\right]^+ + {}^{18}OH^- \longrightarrow \left[(en)_2Co^{III}(OH)(O-C(=O)-C(=O)-{}^{18}O)\right] \qquad (66)$$

The mode of cleavage has been studied for a variety of systems.[107,108,109] As is true of many ligand reactions, this set of reversibly related processes has not been fully exploited. For example, it appears that monodentate ester complexes have *not* been prepared by the reaction described in Eq. (67).

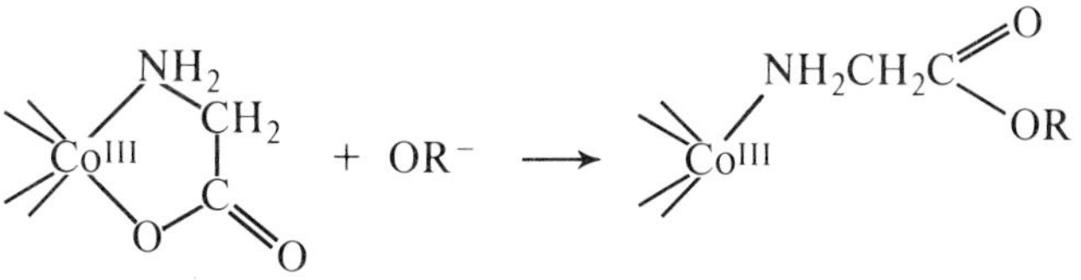

$$Co^{III}(NH_2CH_2C(=O)O) + OR^- \longrightarrow Co^{III}-NH_2CH_2C(=O)OR \qquad (67)$$

The discovery by Leussing and his associates[110,111,112] that only certain metal ions (Zn^{2+}, Mg^{2+}, Pb^{2+}, Cd^{2+}, Mn^{2+}), and not others (Co^{2+}, Cu^{2+}, Ni^{2+}), are active in promoting the formation of Schiff bases has been interpreted in terms of a process that is readily placed in the present category. Metal-ion promotion is presumed to occur along a path [Eqs. (68) and (69)] that serves to convert the carbonyl-amine condensation process from a second order to a first order process.

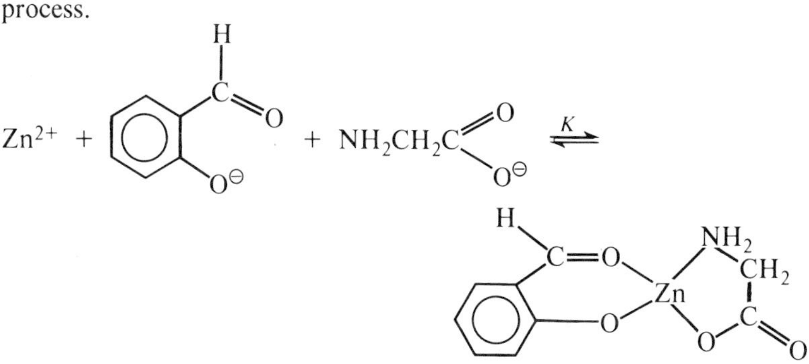

$$Zn^{2+} + C_6H_4(CHO)(O^\ominus) + NH_2CH_2C(=O)O^\ominus \overset{K}{\rightleftharpoons} \text{(salicylaldehyde)Zn(glycinate)} \qquad (68)$$

$$\text{(salicylaldehyde)Zn}(NH_2CH_2C(=O)O) \xrightarrow{k_1} \text{Zn(}HC{=}N-CH_2-C(=O)O\text{)} + H_2O \qquad (69)$$

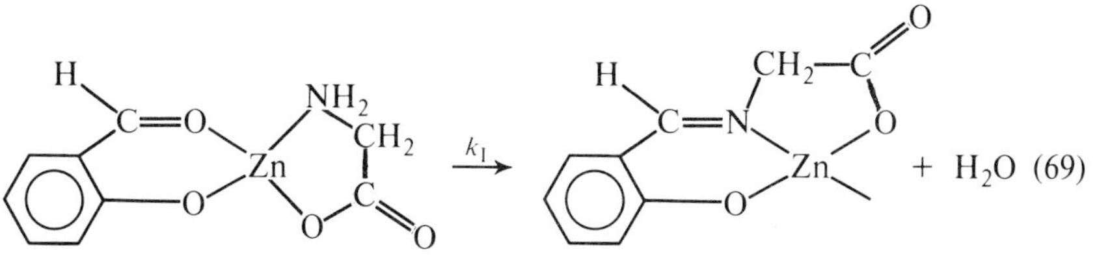

The commonly accepted dynamic behavior of metal chelates, in which metal-donor bonds of chelate rings break and reform with great frequency, is assumed to provide a sufficiently large concentration of uncoordinated amine groups to allow the addition of the amine to the carbonyl group while the latter remains bound. The efficacy of the active metal ions increases as the bond to the amino group weakens. These investigators have also pointed out the likelihood that the inactivity of the strongly complexing metal ions in this process may be related to the stringent geometric constraints deriving from crystal field effects. It is an impressive fact that all the metal ions showing the effect are devoid of the influence of crystal field stabilization energy. It has been pointed out that ligands constrained strictly to certain geometric sites may remain separated and be prevented from reacting unless additional energy is expended to allow for less stable configurations. The more effective ions are free from such influences, and the stereochemistry can adjust relatively readily to an arrangement that is suitable for the intramolecular process. This propensity for propituous paresis has been dubbed the *promnastic* effect—from root meaning "match maker".[111]

Insertion reactions are of extreme importance in the catalysis or promotion of reactions of unsaturated organic molecules by metal ions.[11,12,13,113,114] As used here, the name implies processes whose general stoichiometric properties conform to Eq. (62b). The classic example is carbonylation, a reaction that is an essential part of the oxo process [Eq. (70)].

$$CH_2{=}CH_2 + H_2 + CO \xrightarrow{HCo(CO)_4} CH_3CH_2C(=O)H \tag{70}$$

The catalyst is $HCo(CO)_4$ but the classic example of insertion involves an intermediate, an alkyl compound, $RCo(CO)_4$. During the insertion reaction, one mole of CO is absorbed, and a single CO molecule inserts between the σ-bonded cobalt and carbon atoms [Eq. (71)] converting the alkyl–cobalt moiety into an acyl–cobalt function.[11]

$$RCo(CO)_4 + CO \longrightarrow RC(=O)Co(CO)_4 \tag{71}$$

A less important analog of the cobalt–carbonyl system, that based on manganese carbonyl, has proven useful in providing information concerning the mechanism of the insertion reaction.[12,113,114]

Coffield, Kozikowski, and Closson[115,116] showed that alkyl manganese pentacarbonyls absorb CO to form acyl manganese pentacarbonyls. They used ^{14}C-labeled CO to show that the CO used to form the acyl group is not the external CO molecule but one already coordinated to the metal atom [Eq. (72)].

$$*CO + CH_3Mn(CO)_5 \longrightarrow CH_3C(=O)\text{—}Mn(CO)_4(*CO) \quad (72)$$

Questions concerning the mechanism of this reaction have been asked. Is the insertion process concerted with the entry of the additional mole of CO as might be suspected if the stoichiometric equation were given a literal mechanistic interpretation? Alternatively, is insertion a separate process preceding ligand addition [Eq. (73)]?

$$RMn(CO)_5 \rightleftharpoons RC(=O)\text{—}Mn(CO)_4 \underset{-CO}{\overset{+CO}{\rightleftharpoons}} RC(=O)\text{—}Mn(CO)_5 \quad (73)$$

It is significant that a variety of other ligands besides CO will cause the insertion reaction to occur. These include phosphines and amines. An early kinetic study suggested a simple second-order mechanism involving concerted carbonyl addition and insertion.[117] However, more extensive studies[118,119,120] strongly support the model given in Eq. (73). The general mechanism proposed to describe the insertion reactions of CO[113] involves two paths. The first path involves a first-order process [Eqs. (74) and (75)] while the second involves a second-order process [Eq. (76)].

$$M(CO)R \underset{k_{-1}}{\overset{k_1}{\rightleftharpoons}} \text{intermediate} \quad (74)$$

$$\text{Intermediate} + L \xrightarrow{k_2} M(\overset{O}{\overset{\|}{C}}R)L \quad (75)$$

$$M(CO)R + L \xrightarrow{k_3} M(\overset{O}{\overset{\|}{C}}R)L \quad (76)$$

L is the incoming ligand that returns the coordination number to its original value. The general rate law is expressed in Eq. (77). This expression shows how the observed orders of such processes can vary depending on the relative mag-

$$\text{rate} = \frac{k_1 k_2[M(CO)R][L]}{k_{-1} + k_2[L]} + k_3[M(CO)R][L] \quad (77)$$

nitudes of the various terms in the rate law—e.g., if $k_2[L] \gg k_{-1}$ and the first term of Eq. (77) dominates, the rate is simply first-order. The intermediate is most simply viewed as a coordinately unsaturated acyl derivative, such as

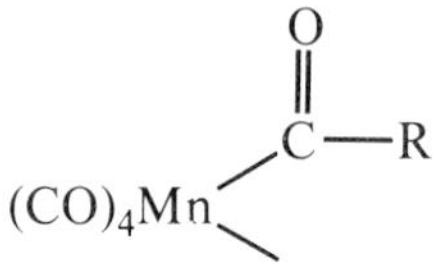

$$(CO)_4Mn\text{—}C(=O)\text{—}R$$

However, it might involve solvent coordination,

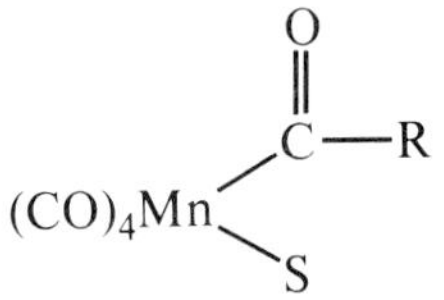

$$(CO)_4Mn(S)\text{—}C(=O)\text{—}R$$

or a different binding mode of the acyl group,

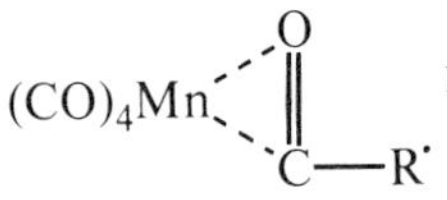

$$(CO)_4Mn(\eta^2\text{-}O{=}C\text{—}R') \qquad 121$$

The effects of the solvent, the varying nature of the incoming ligand L, and variation in the alkyl group have been studied for a number of metal centers.

Studies on the stereochemistry of the reverse reaction, decarbonylation, using

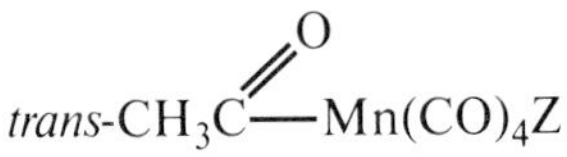

$$trans\text{-}CH_3C(=O)\text{—}Mn(CO)_4Z$$

have been revealing.[119,120,123] The question under examination is "which group moves, the CH_3 or the CO?"

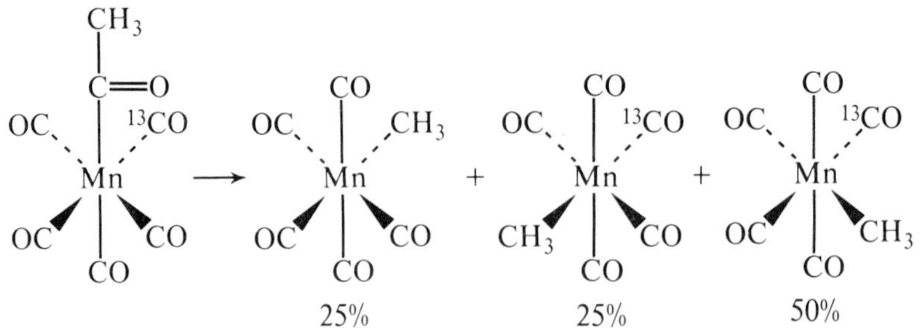

$$cis\text{-}(CH_3CO)Mn(CO)_4(^{13}CO) \longrightarrow \underset{25\%}{cis\text{-}CH_3Mn(CO)_5} + \underset{25\%}{CH_3Mn(CO)_4(^{13}CO)\ (trans\text{ to }CH_3)} + \underset{50\%}{CH_3Mn(CO)_4(^{13}CO)\ (cis\text{ to }CH_3)} \qquad (78)$$

$$(CO)_4(^{13}CO)Mn{-}C(=O)CH_3 \longrightarrow CH_3Mn(CO)_5\ (25\%) + CH_3Mn(CO)_4(^{13}CO)\ (75\%) \qquad (79)$$

Eq. (78) shows the distribution of isomers obtained upon decarboxylation of stereospecifically ^{13}C-labelled

$$(CO)_5Mn\overset{O}{\overset{\|}{C}}CH_3$$

if we assume that the CH_3 group migrates. We assume that the starting complex first loses one of the four CO molecules that are cis to the acyl group. This forms a tetragonal pyramidal intermediate, and the migration of CH_3 would produce three isomers (based on ^{13}CO) in a ratio of 1:1:2. If the CO of the acyl group migrates [Eq. (79)], two isomers will be formed in a ratio of 1:3. Experiment shows that three isomers are formed, so decarboxylation involves methyl migration.[119,120] Invoking the principle of microscopic reversibility, one concludes that the process we traditionally call insertion is actually better described as an alkyl migration. The radicals that form the most stable carbanions migrate most slowly. This may imply stronger metal–carbon bonding in these cases or that the migration process is favored by alkyl (aryl) groups having electrophilic character.

Whitesides and Boschetto[124] have shown that the insertion reaction proceeds with complete (>95%) retention of configuration when the alkyl group is optically active. Their technique was based on the predicted magnitudes of coupling constants for the *threo* and *erythro* isomers of the organic group in the example shown in Eq. (80).

$$C(CH_3)_3CHDCHD{-}Fe(CO)_2C_5H_5 \xrightarrow[THF,\ 65^\circ]{PPh_3} C(CH_3)_3CHDCHD{-}C(=O){-}Fe(CO)(PPh_3)C_5H_5 + C(CH_3)_3CHDCHD{-}C(=O){-}Fe(PPh_3)(CO)C_5H_5 \qquad (80)$$

The complete retention of configuration requires that the alkyl group essentially remain bound during the migration process. Additional stereochemical studies have been reported.[113] These relationships are consistent with three-center mechanisms of the type given in Eq. (81).

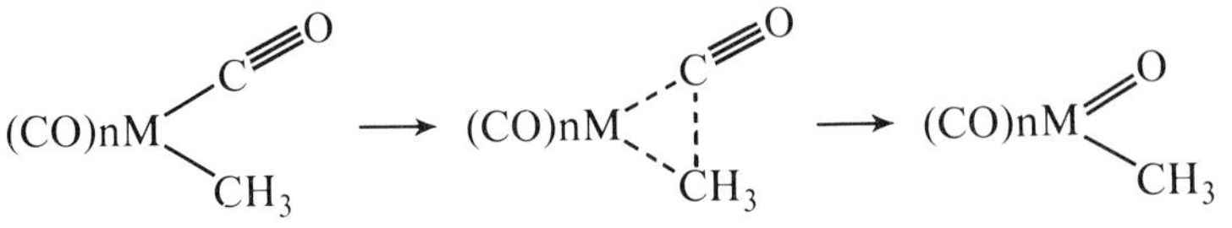

$$(CO)_nM(C{\equiv}O)(CH_3) \longrightarrow (CO)_nM\cdots(C{\equiv}O)\cdots CH_3 \longrightarrow (CO)_nM{-}C({=}O)CH_3 \qquad (81)$$

Orchin and Rupilius[11] have suggested that the alkyl migration of the alkyl metal carbonyls is analogous to the 1,2 migration within carbonium ions that is generalized as the Wagner-Meerwein rearrangement (Figure 2-5a). In an organometallic system, the empty orbital in the alkyl compound is the low-lying antibonding orbital of a coordinated CO molecule. This orbital is largely carbon in character. Also, it is usually presumed to be involved in back-bonding with the filled metal d-orbitals; however, such back-bonding to any single CO must be small in tetra- and pentacarbonyls. The arrangement is shown in Figure 2-5b.

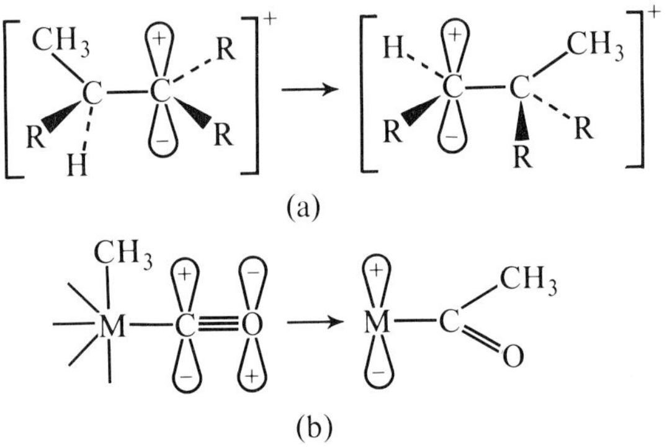

(a)

(b)

Figure 2-5 Analogy between the Wagner-Meerwein rearrangement and the carbonyl insertion reaction

Insertion reactions cover a broad range, and a number of examples are discussed in sections that follow. Reflecting on our earlier definition, such processes merely require that some unsaturated species X insert between M and Y in the unit M—Y. M—Y linkages that have been found to participate in such reactions include, among others, metal–alkyl, metal–hydrogen, metal–halogen, metal–metal, and metal–oxygen. Some of the more common inserting molecules are CO, SO_2, olefins, and acetylenes. Recent studies have also involved organic isocyanides,[125] germanium(II) chloride,[126] tetracyanoethylene,[127] oxygen,[128] S_4,[129] and CS_2.[130]

Brief consideration of a few SO_2-insertion processes provides a useful comparison to the behavior observed in CO-insertion systems.[113,131,132] The first such reaction involved dissolving and keeping π-$C_5H_5Fe(CO)_2R$ in liquid SO_2 for several hours at ~$-40°C$[133] [Eq. (82)].

$$\pi\text{-}C_5H_5Fe\begin{matrix}\diagup CO\\ -R\\ \diagdown CO\end{matrix} + SO_2 \longrightarrow \pi\text{-}C_5H_5Fe\begin{matrix}\diagup CO\\ -SO_2R\\ \diagdown CO\end{matrix} \quad (82)$$

The products contain S-bonded sulfinates

$$(Fe—\overset{\overset{\displaystyle O}{\|}}{\underset{\underset{\displaystyle O}{\|}}{S}}—R)$$

when the metallic element is iron, manganese, and, presumably, other transition elements.[132,134,135] However, nontransition elements such as Zn and Hg[137] show a marked tendency to generate O-bonded sulfinates.[138] The possibility that SO_2 might emulate CO in undergoing insertion reactions was first suspected[132] because SO_2 was shown to act as a good π-back-bonding ligand toward transition metals.[139] Thus, the coordinated SO_2 has a vacant low-energy orbital which could be used in the alkyl migration process (Figure 2-5). Rate studies show[140] that more strongly electron-releasing R groups react more rapidly with SO_2. This also agrees with the model. The molecules π-$C_5H_5Fe(CO)(L)CH_3$ react with SO_2 in chloroform with rates increasing in the order $CO \ll PPh_3 < P(n\text{-}Bu)_3$.[141] Thus, the more basic the ligand L, the faster the reaction of an adjacent Fe—C bond with SO_2. The results of recent studies on the reverse process

$$(M—SO_2R \longrightarrow M\begin{matrix}\diagup SO_2\\ \\ \diagdown R\end{matrix})$$

are consistent with these observations.[142]

A number of observations contrast the SO_2 insertion reaction with that of CO. The thermal parameters associated with the rate processes provide the first strong contrast. The values of $\Delta S^{\ddagger}$ for the SO_2 insertion are remarkably large negative numbers.[132] (-43 with -62 eu as compared with -20 to -30 eu for CO) Alexander and Wojcicki first showed that SO_2 insertion is a stereospecific process,[143] however, it was left for Whitesides and Boschetto to demonstrate that the insertion reaction [Eq. (83)] proceeds with an inversion of configuration.[150]

$$\textit{threo-}[(CH_3)_3C(CHD)_2Fe(CO)_2(\pi\text{-}C_5H_5) + SO_2 \xrightarrow{\sim 80\%}$$

$$\textit{erytho-}(CH_3)_3C(CHD)_2SO_2Fe(CO)(\pi\text{-}C_5H_5) \quad (83)$$

A most significant conclusion from many studies on these systems is that the inserting SO_2 molecule is not bound prior to the insertion process in certain cases. From this we conclude that insertion mechanisms need not all be the same.[113,143]

B. Ligand Reactions Requiring Chelate Ring Formation

The chelate effect[144] relates to the special stability and kinetic effects that result when a ligand forms chelate rings of favorable sizes. Several kinds of ligand reactions are strongly affected by the chelate effect. The most obvious cases are those in which the formation of chelate rings of a favored geometry play a determining role. We consider four such categories: (1) masking of specific groups because of chelate ring formation, (2) the activation of specific groups because of chelate ring formation, (3) chelate stabilization of the transition state for a reaction, and (4) chelate stabilization of the product structure. The fifth category of chelate ring effect on ligand reactions involves the reactions of aromatic chelate rings.

1. Chelate Effect and Specific Group Masking. Kurtz[145,146,147] provided the classic examples of specific masking of the reactions of certain functional groups by means of the chelate effect, and these were confirmed and extended.[148,149,150] The studies provided a technique for site-specific reaction of one of the amino groups in such α,ω-diamino carboxylic acids as lysine and ornithine. The α-amino group is ideally located for chelation when the carboxylate ion is bound to a metal ion [Eq. (84)].

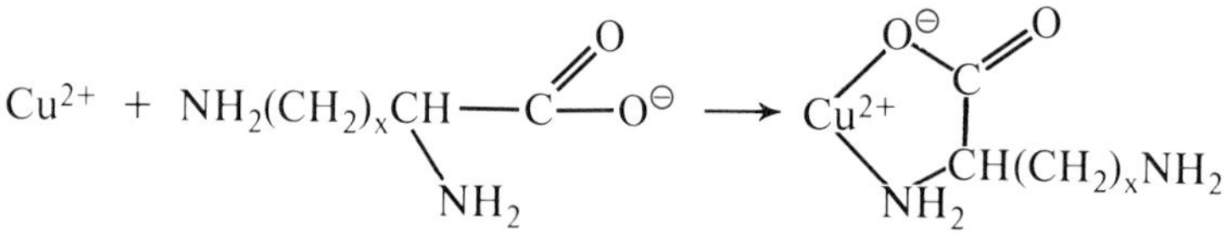

$$Cu^{2+} + NH_2(CH_2)_xCH(NH_2)-C(=O)-O^{\ominus} \longrightarrow Cu^{2+}[O^{\ominus}-C(=O)-CH(CH_2)_xNH_2-NH_2] \quad (84)$$

The nucleophilicity of the α-amino group is destroyed, as discussed above. The reactivity of the ω-amino group depends on the number of carbon atoms separating it from the α-amino group and the nature of the available metal ion. The copper(II) complex is readily converted to the related arginine derivative, Eq. (85).

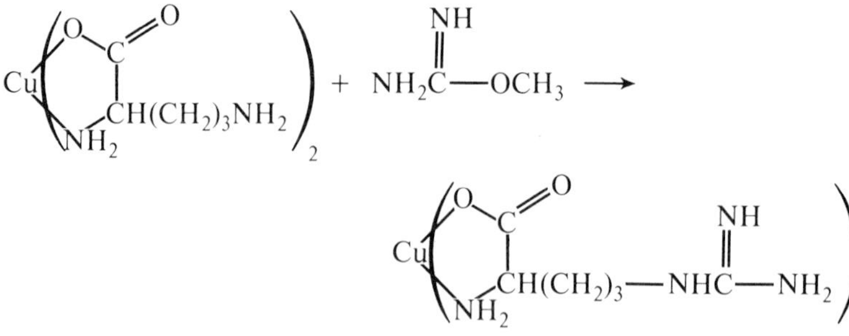

$$Cu[O-C(=O)-CH(CH_2)_3NH_2-NH_2]_2 + NH_2C(=NH)-OCH_3 \longrightarrow Cu[O-C(=O)-CH(CH_2)_3-NHC(=NH)-NH_2-NH_2]_2 \quad (85)$$

A similar chelate selectivity can be found in the masking of the near carboxylate group in the reactions of aminodicarboxylates. This is illustrated for the case of glutamic acid in Eq. (86).[152]

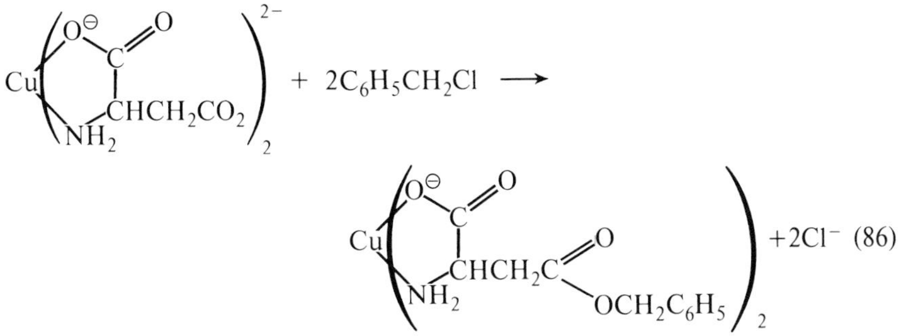

$$\left(\mathrm{Cu}\begin{matrix}\mathrm{O}^{\ominus}\text{—}\mathrm{C}(=\mathrm{O})\\ |\\ \mathrm{NH_2}\text{—}\mathrm{CHCH_2CO_2}\end{matrix}\right)_2^{2-} + 2\mathrm{C_6H_5CH_2Cl} \longrightarrow \left(\mathrm{Cu}\begin{matrix}\mathrm{O}^{\ominus}\text{—}\mathrm{C}(=\mathrm{O})\\ |\\ \mathrm{NH_2}\text{—}\mathrm{CHCH_2C}(=\mathrm{O})\mathrm{OCH_2C_6H_5}\end{matrix}\right)_2 + 2\mathrm{Cl}^- \quad (86)$$

Considerable difficulty has been encountered in attempts to utilize the nucleophilic powers of uncoordinated alcohol functions in structures where other functions were masked. The complex of structure **5** failed to undergo acetylation under a variety of stringent conditions.[153]

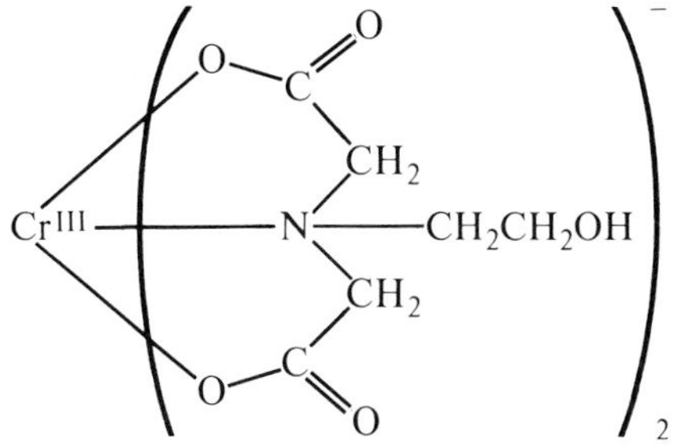

$$\left(\mathrm{Cr^{III}}\begin{matrix}\mathrm{O}\text{—}\mathrm{C}(=\mathrm{O})\text{—}\mathrm{CH_2}\\ \mathrm{N}\text{—}\mathrm{CH_2CH_2OH}\\ \mathrm{O}\text{—}\mathrm{C}(=\mathrm{O})\text{—}\mathrm{CH_2}\end{matrix}\right)_2^{-}$$

5

Substantial success was achieved only by reaction of ketene with the compound in acetonitrile solution. Very low yields of the ester were obtained from refluxing glacial acetic acid–acetic anhydride mixed solvent. Attempts by others to run reactions of free hydroxyethyl functions in cationic complexes were unsuccessful.[154]

An interesting case of masking by electronic means occurs among complexes of acyl- or formyl-substituted β-diketones.[155] The carbonyl group in these substances (structure **6**) fails to form essentially all of the electrophilically generated adducts of normal ketones and aldehydes—e.g., oximes, hydrazones, and imines.

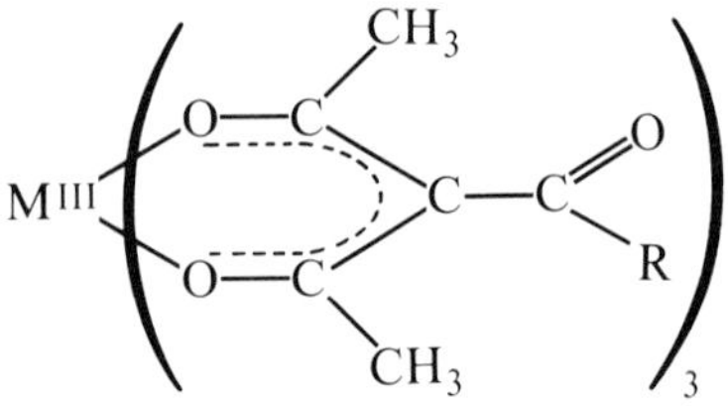

6

The complexes even fail to react with Grignard reagents. The fact that such functions are unusually electron rich has recently been demonstrated by the study of nucleophilic reactions of such a function bound to a macrocyclic ligand[156] [Eq. (87)].

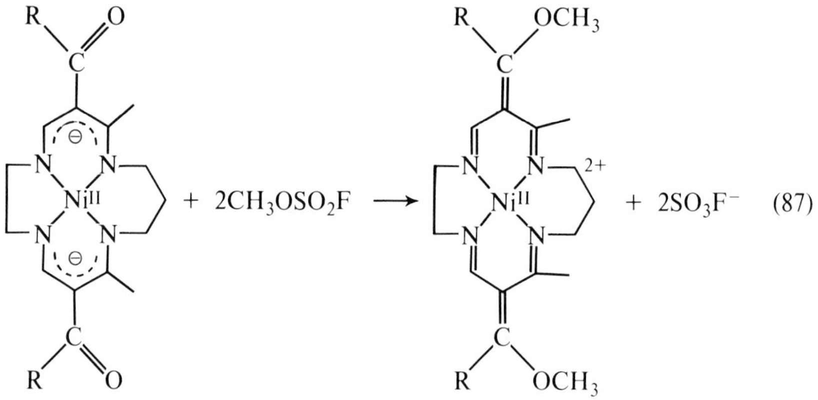

$$+ 2CH_3OSO_2F \longrightarrow \quad ^{2+} + 2SO_3F^- \qquad (87)$$

2. Chelate Effect and Specific Group Activation. Just as favorable chelate ring size may cause coordination and concommitant masking of a functional group, so may the same cause affect the coordination of groups that may be activated in the process. The modes of activation may be any of those discussed earlier. The novel point here is that chelation facilitates coordination of the active group thereby bringing it under the influence of the metal ion. The best example of this probably involves promoting the reactivity of carboxylate functions by coordination of the carbonyl oxygen atom. The occurrence of such activated intermediates was established by Alexander and Busch[157] [Eq. (88a)] using inert cobalt (III) derivatives.

$$\left[(en)_2Co^{III}(Cl)(NH_2CH_2C(=O)OR)\right]^{2+} + Hg^{2+} \longrightarrow \left[(en)_2Co^{III}(O{=}C(OR)CH_2NH_2)\right]^{3+} + HgCl^{+} \quad (88a)$$

$$(en)_2Co^{III}(O{=}C(OR)CH_2NH_2) + H_2O \longrightarrow (en)_2Co^{III}(OC(=O)CH_2NH_2) + H^{+} + ROH \quad (88b)$$

The necessity for chelation in such activation steps is apparent since the hydrolysis of such unsubstituted esters as ethyl acetate is unaffected by metal ions in aqueous solutions.[167] A variety of other functional groups also lead to metal-ion promotion of these nucleophilic processes, an anionic carboxylate group being most common.[158,159] The intermediate of Eq. (88a) was first identified unequivocally by spectroscopic means. The most obvious alternative structure would have been the aquo complex (or hydroxo complex depending on pH) (structure **7**), and the electronic spectrum of this species could be predicted very accurately. The C=O stretching frequency of the ester grouping occurred in the starting material at 1740 cm^{-1}. In the intermediate, it occurred at 1610 cm^{-1} and at 1640 cm^{-1} in the solvolyzed product [structure **8,** Eq. (88b)].

$$\left[(en)_2Co(OH_2)(NH_2CH_2C(=O)OR)\right]^{3+} \qquad \left[(en)_2Co(OC(=O)CH_2NH_2)\right]^{2+}$$

7 **8**

No drastic change in this vibrational mode is expected for structure **7,** while it is expected to occur at reduced frequency if the carbonyl oxygen is bound to a metal ion.[157] The intermediate was subsequently isolated and thoroughly characterized by Buckingham and his associates.[104,160] The ligand reactions of amino acids and their derivatives are considered in detail in section VIII.

The relatively rapid cleavage of *cis*-glycols by a number of metal containing reagents provides a second example in which chelation promotes a chemical reaction. The mechanism for such a cleavage by lead tetraacetate is outlined in Eq. (89).[161,162]

$$R_2C\text{—}OH\;|\;R_2C\text{—}OH + Pb(OAc)_4 \longrightarrow R_2C(O)\text{—}C(O)R_2\,Pb(OAc)_2 \longrightarrow 2R_2C{=}O + Pb(OAc)_2 \quad (89)$$

Two electron oxidizing agents are most often implicated in such processes, and the evidence for the cyclic intermediate is most extensive for oxidation of glycols by periodate.[161] Chromate also seems to participate in this mode of reaction with highly substituted glycols—e.g., *cis*-1,2-dimethyl-1,2-cyclopentanediol is oxidized to 2,6-heptanedione 17,000 times more rapidly than the trans isomer [Eq. (90)].[163] This is not, however, the usual manner of oxidation of alcohols by chromate (see section IX).

$$\text{diol} + Cr^{VI}O_4^{2-} \longrightarrow \text{cyclic ester } Cr^{VI}O_2 \longrightarrow \text{2,6-heptanedione} + Cr(IV) \quad (90)$$

3. Stabilization of a Transition State through Chelation. It is sometimes clear that the structure of the substrate is better suited for chelation in the transition state of its reaction than in its usual or initial state. In these instances, chelation should greatly accelerate the substrate reaction. In the most general sense, many of the examples that are readily recognized as belonging to the immediately preceding category also involve stabilization of the transition state. The decarboxylation of oxaloacetate exemplifies these relationships [Eqs. (91) and (92)].

$$M^{n+}\text{(oxaloacetate chelate)} \longrightarrow M^{n+}\text{(transition state)} \longrightarrow M\text{(enolate chelate, } CH_2) + CO_2 \quad (91)$$

$$M^{n+}\text{(enolate chelate, } CH_2) + H^+ \longrightarrow M^{n+}\text{(pyruvate chelate, } CH_3) \longrightarrow M^{n+} + CH_3\overset{O}{\overset{\|}{C}}CO_2^{\ominus} \quad (92)$$

Presumably, the transition state closely resembles the intermediate

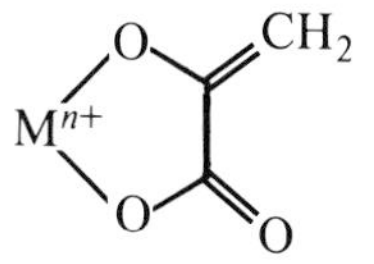

which is a close structural analog of an oxalate complex. In accordance with this conclusion, the rate constants for metal-ion catalysis of the reaction bear a linear relationship to the stability constants of the corresponding oxalate complexes.[164,165,166] Ionization of a proton from the ligand prevents decarboxylation [Eq. (93)].

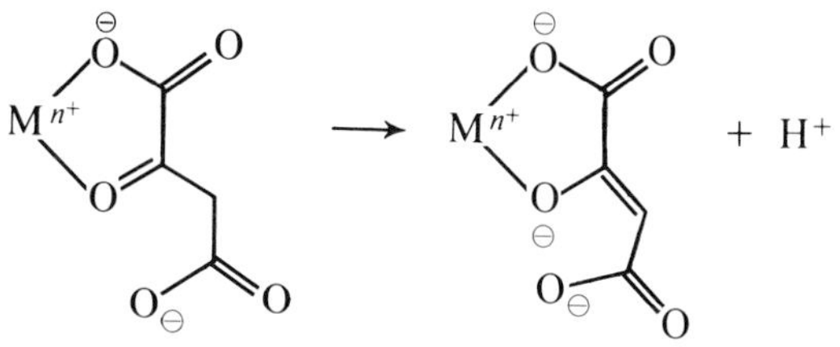

$$\longrightarrow \quad + H^+ \qquad (93)$$

α,α-Dimethyloxaloacetate cannot undergo the inactivation reaction given in Eq. (93). Consequently, the study of its decarboxylation reaction has been highly informative.[167-170] These and some related reactions have been summarized.[171]

It is reasonable to assume that the elimination of the phosphate from a serine phosphate ester[172] proceeds through an intermediate closely related to those just discussed [Eq. (94)].

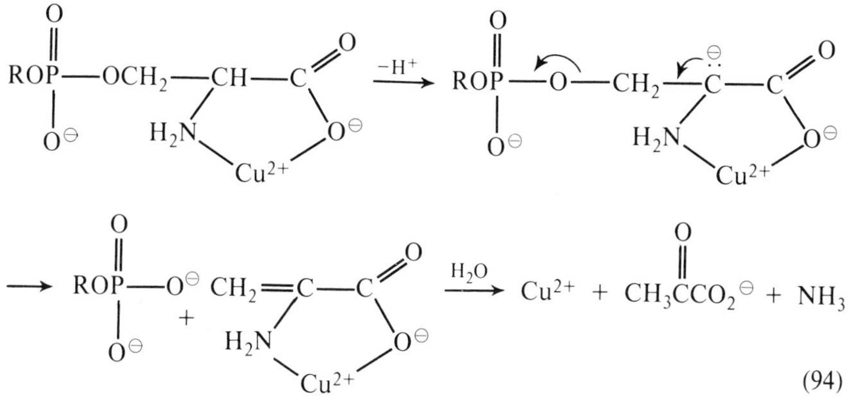

$$\xrightarrow{H_2O} Cu^{2+} + CH_3CCO_2^{\ominus} + NH_3 \qquad (94)$$

A second particularly useful illustration of the stabilization of the transition state because of favorable chelation involves the hydrolysis of a nitrile group substituted in the 2-position on 1,10-phenanthroline.[173] Substitution of bulky groups in the 2, or 2 and 9 positions of 1,10-phenanthroline prevents the formation of tris(substituted phenanthroline) complexes of the usual high stability and substitution inertness. The presence of the CN group in the 2-position may prevent coordination of one water molecule or, through repulsion, may generate a very long metal–oxygen bond. It is appealing to assume that a water molecule has been displaced, and that the CN group, though not ideally oriented for coordinating, interacts strongly with the metal ion. The strain inherent in this structure is largely removed upon the addition of OH^- to the nitrile carbon atom to produce a trigonal amide carbon (tautomer) well suited to coordinate its nitrogen atom to the metal ion [Eq. (95)].

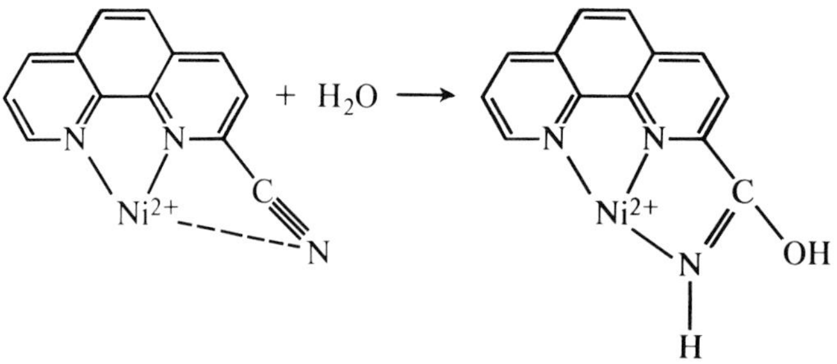

$$+ H_2O \longrightarrow \qquad (95)$$

This example has been related to the so-called "rack-mechanism" for some enzymatic processes in which the binding of the substrate to the enzyme is presumed to induce a distortion toward the geometry of the transition state.

4. Stabilization of the Product through Chelation. Under many different conditions, the formation of chelate rings enhances the stability of a structure. This often acts as a driving force in ligand reactions, and, not infrequently, ligands that are relatively unstable in the free state are readily formed in their complexes. An often quoted example is that of cyclobutadiene. The free ligand is extremely reactive and not well characterized, while it is quite stable in its complexes with low valent metal atoms (structure **9**). Theoretical prediction[174] preceded synthesis in this case.[175,176]

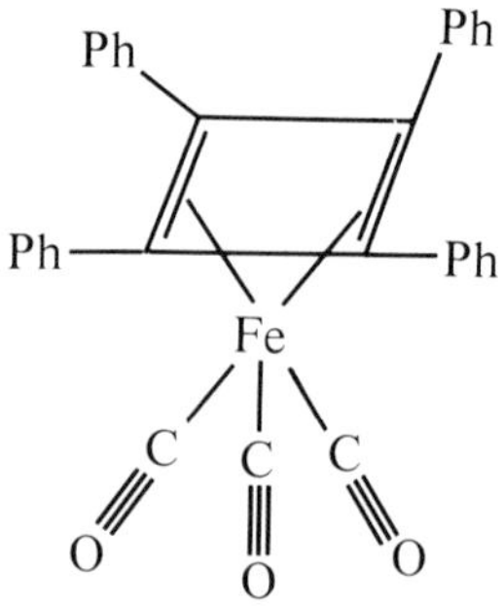

9

Eichhorn and co-workers used Schiff base derivatives to show that chelation may stabilize or activate the same linkage depending on the specific structural relationships involved. In the molecule bis(thiophenal)-ethylenediimine, only the nitrogen atoms coordinate to the Cu^{2+} ion, and the metal ion greatly accelerates the hydrolysis of the C=N group [Eq. (96)].[159,160]

$$(C_4H_3S)CH{=}N{-}CH_2CH_2{-}N{=}CH(C_4H_3S)\cdot Cu^{2+} + 2H_2O \longrightarrow H_2N{-}CH_2CH_2{-}NH_2\cdot Cu^{2+} + 2\,(C_4H_3S)CHO \qquad (96)$$

In contrast, the C=N group in the Schiff base formed between salicylaldehyde and glycine is stabilized by chelate formation at pH values where the free ligand is readily hydrolyzed[161,177] [Eq. (97)].

$$C_6H_4(CHO)OH + NH_2CH_2CO_2H + Cu^{2+} \xrightarrow{pH^3} C_6H_4(O^-)CH{=}N{-}CH_2{-}C({=}O)O^-\cdot Cu \qquad (97)$$

The coordination of a metal ion to the Schiff base linkage is expected to enhance the electrophilicity of the imine carbon atom. The isolation of a number of α-hydroxyamines and related addition products has confirmed this expectation (structures **10** and **11**).[178,179,180]

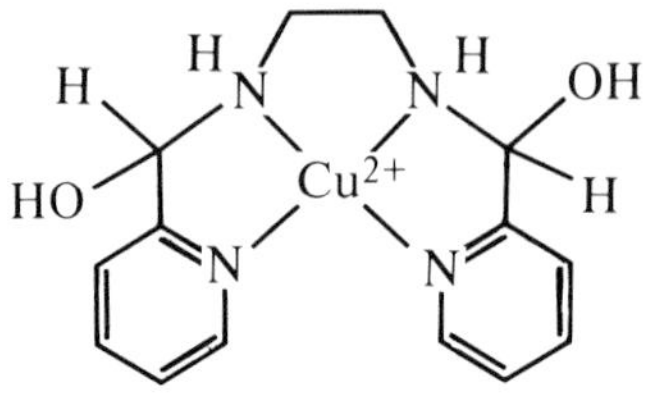

10

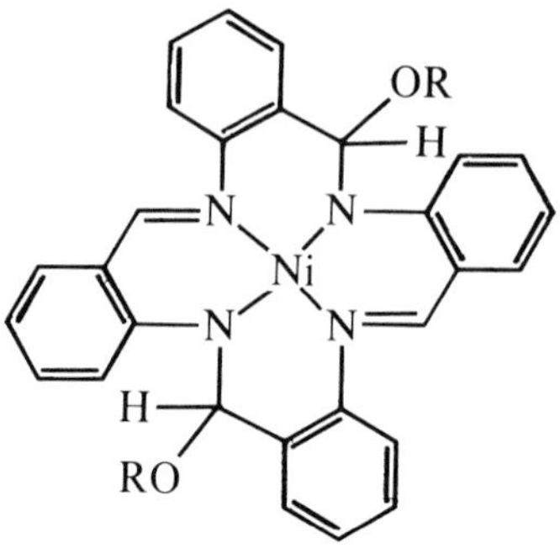

11

Such structures for uncoordinated Schiff bases are usually not very stable and readily revert either to the hydrolysis fragments or the unsaturated Schiff base.

2,2′-Pyridil (structure **12**) rearranges under the influence of Co(II) or Ni(II) to form complexes of 2,2′-pyridilic acid (structure **13**) coordinated in a tridentate fashion (structure **14**).[181,182,183]

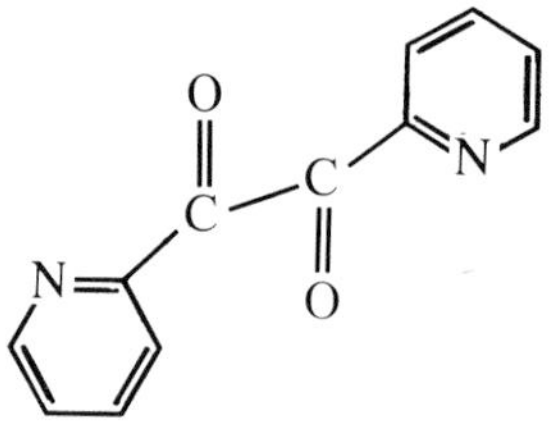

12

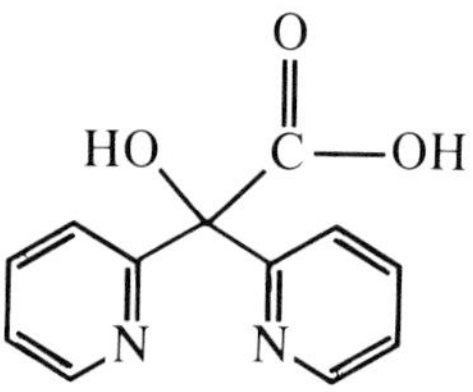

13

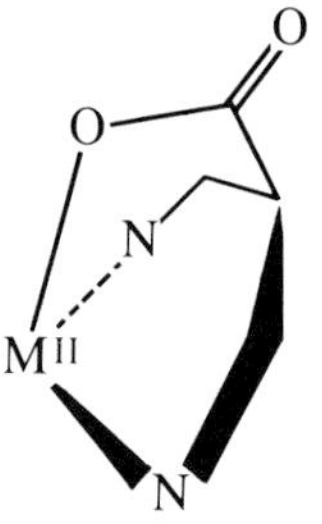

14

Although the metal ion almost certainly alters the mechanism of this reaction, it is equally likely that the rearrangement is promoted by the stability of the product. In the absence of the metal ion, strong alkali and heat are required to drive the process while it proceeds under mild conditions in the presence of Co(II) or Ni(II). When Cu(II) is the catalyst, the product reacts further, undergoing decarboxylation and yielding the copper complex of 2,2′-dipyridyl-methanol (structure **15**).

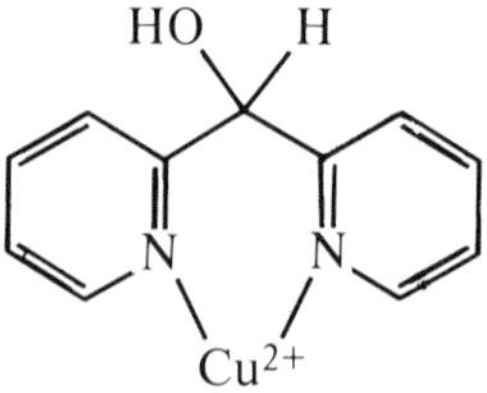

15

The α-diimine-iron(II) chelate ring

provides an additional example of the dramatic effect of product stability on ligand reactions. As was pointed out long ago,[178,184] this particular ring constitutes an aromatic system of exceptional stability. This grouping of atoms has now been produced by at least three independent ligand reaction pathways, all of which reflect the influence of product stability. First [Eq. (98)], the condensation of α-diketones with aliphatic primary amines leads to the formation of the tris(α-diimine) chelates, although the free ligands are unknown.[178,184]

$$Fe^{2+} + 3\,(CH_3)C(=O)\text{–}C(=O)(CH_3) + 6NH_2CH_3 \longrightarrow [Fe(CH_3N{=}C(CH_3)\text{–}C(CH_3){=}NCH_3)_3]^{2+} + 6H_2O \quad (98)$$

Recently, coordinated 1,2-diamines have been converted to α-diimines by oxidative dehydrogenation reactions of the coordinated ligands.[185,186] [Eq. (99)].

$$Fe^{II}(RNH\text{–}CH_2\text{–}CH_2\text{–}NHR) \xrightarrow{[O]} Fe^{II}(RN{=}CH\text{–}CH{=}NR) \quad (99)$$

This reaction has been applied both to bidentate ethylenediamine derivatives and to tetradentate macrocyclic complexes. Still more remarkably, it has been shown that tautomerization[186] of complexes containing iron(II) and unconjugated imines will produce α-diimine rings [Eq. (100)].

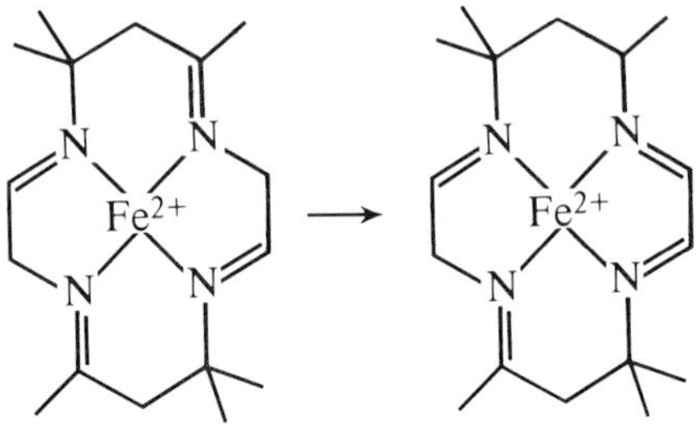

Fe^{2+} → Fe^{2+} (100)

5. Electrophilic Reactions of "Aromatic" Chelate Rings. The complexes of β-diketonate anions provide the best known examples of this class of reaction and the subject ligand reactions have been reviewed extensively.[5,6,187] The centermost carbon is the point of reactivity [Eq. (101)], and the observed reactions include halogenation, nitration, thiocyanation, acylation, formylation, chloromethylation, and aminomethylation.

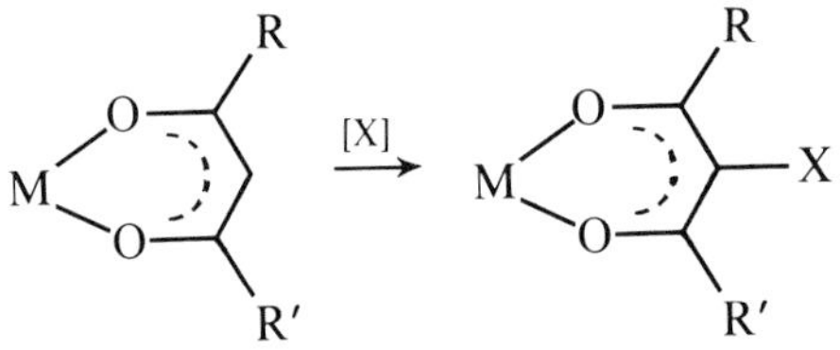

$$\text{M(O,O-chelate; R, R')} \xrightarrow{[X]} \text{M(O,O-chelate; R, X, R')} \tag{101}$$

The usual reactants were the tris(acetylacetonates) of such trivalent ions as Cr(III), Co(III), and Rh(III).

The α-diimine chelate ring with iron(II) is less thoroughly studied, but it has provided a striking example of electrophilic substitution[188,189] [Eq. (102)].

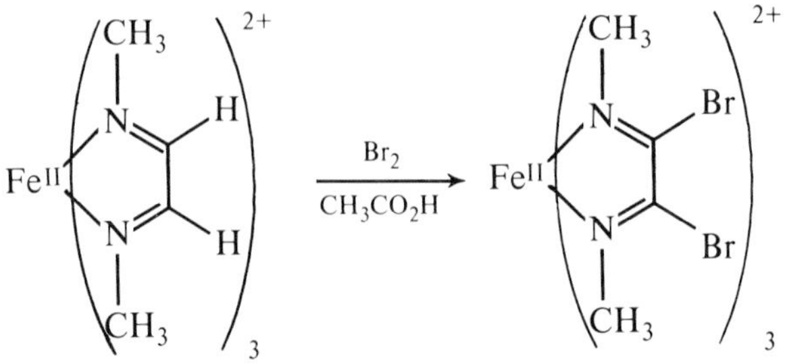

$$\left[Fe^{II}(CH_3N{=}CH{-}CH{=}NCH_3)_3 \right]^{2+} \xrightarrow[CH_3CO_2H]{Br_2} \left[Fe^{II}(CH_3N{=}CBr{-}CBr{=}NCH_3)_3 \right]^{2+} \tag{102}$$

It is impressive that the bromine serves to brominate the ring rather than oxidize the more often vulnerable iron(II) atom. Whereas it was originally believed that the resistance to oxidation of iron(II) was associated with octahedral structures having three α-diimine ligands, it has recently been shown that even one such ring has a profound effect on the electrode potential of the iron(II)–iron(III) couple.[186]

Of course, coordination to a metal ion is most often expected to decrease the electron density on a ligand so that electrophilic processes are often not enhanced by coordination. We have already discussed two ways in which coordination effects may contribute electron density to the ligand. These are the promotion of a ligand's acidity—e.g., acetylacetone loses a proton and coordinates as an anion, and also metal to ligand π back-bonding. The latter effect is presumed to operate in the case of Fe^{2+}-α-diimine chelates with the result that coordination enhances the electron density on the ring.

Additional evidence for the quasi-aromatic character of metal-α-diimine chelate rings comes from recent electrochemical and esr studies.[190] Characteristically, the reduction of nickel(II) complexes containing α-diimine linkages produces a metal-ion-stabilized ligand radical anion [Eq. (103)].

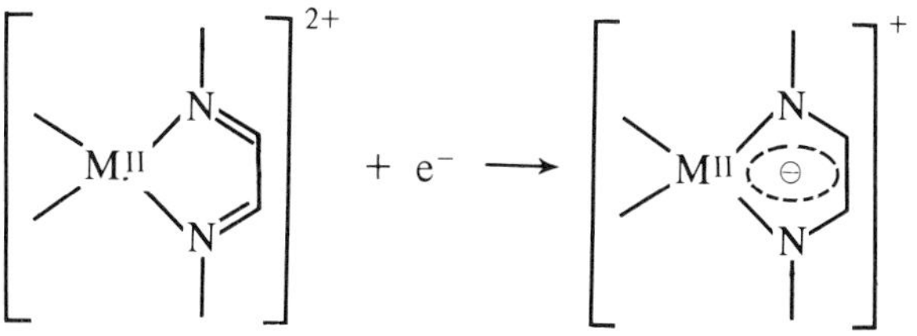

$$[\mathrm{M^{II}}(\alpha\text{-diimine})]^{2+} + e^- \longrightarrow [\mathrm{M^{II}}(\alpha\text{-diimine}^{\ominus})]^{+} \quad (103)$$

It is suggested that the added electron goes into a low lying π antibonding orbital that is localized largely on the ligand. In contrast, complexes containing unconjugated azomethine linkages in their chelate rings undergo reduction at the metal ion [Eq. (104)].

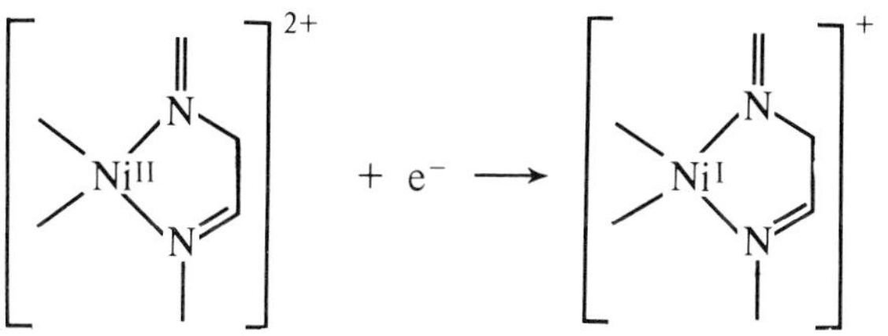

$$[\mathrm{Ni^{II}}]^{2+} + e^- \longrightarrow [\mathrm{Ni^{I}}]^{+} \quad (104)$$

It is an interesting fact that the rates of nitration of the iron(II) and cobalt(III) complexes of 1,10-phenanthroline are more rapid than the rate of the protonated 1,10-phenanthroline.[191] The rate difference may reflect a difference in the abilities of the three cations to polarize the substrate molecule via the σ-electron system, or alternatively, it might derive from some π-electron back donation.

C. Stereospecificity

Optically active complexes are selective toward complexation with the enantiomers of optically active ligands. This selectivity also comes into play when a ligand reaction in an enantiomeric complex involves either an optically active ligand as a substrate or produces an antipodal ligand. An example of the phenomenon is found in the formation of α-alanine within an optically active cobalt(III) complex by decarboxylation of the symmetrical ligand 2-methyl-2 amino-malonate [Eq. (105)].[192,193] The alanine is produced preferentially in one antipodal form.

$$\xrightarrow{-CO_2} \qquad (105)$$

The first example of this phenomenon involves the distinctly novel process represented by performing the Knoevenagel reaction on an optically active metal complex containing chelated amino acid anions. Murakami and Takahashi[194] first showed that such reactions produce optically active amino acids when the substrate is chelated glycine [Eq. (106)].

$$\xrightarrow{RCHO} \qquad (106)$$

The metal ion exerts several modes of influence in this process. Coordination of the glycinate-NH_2 group masks it, thereby preventing competition by a Schiff base reaction. Coordination of the carboxyl function provides the polarization of the entire glycine moiety that is necessary in order to enhance the acidity of its methylene hydrogens.[195,196] Finally, the nonbonded repulsions in the complex favor formation of a specific enantiomer of the serine or threonine that is produced by the ligand reaction. It is not clear whether the selectivity is exercised in the transition state or is strictly a thermodynamic phenomenon.

1. Coordination Template Effects. When the geometric relationships between the polyhedral arrangement of the coordination sphere of a metal ion and the orientation of donor atoms in a substrate lead to the formation of geometrically specific ligand reaction products, the phenomenon is called a coordination template effect.[197-200] Stated more simply, the metal ion serves as a template to organize the course of complex multistep reactions.

Two general categories of such reactions are recognized. In one, the metal ion influences the position of chemical equilibrium by sequestering a species that is especially suited to chelating strongly; such examples are said to involve thermodynamic coordination template effects. In the other, the metal ion controls the geometric course of a multistep process; these cases are described as involving kinetic coordination template effects. These kinetic effects enjoy the twin advantages of having figured most heavily in the early history of the area and in containing examples in which the role of the metal ion is relatively well-established. For these reasons, attention will be given first to examples of the kinetic coordination template effect. Many synthetic applications of these principles have been made, but only limited effort has been directed toward their mechanistic study. It is worth indicating early in the discussion, however, that these reactions have produced many new cyclic ligands, important routes to natural products, and exotic new kinds of structures.

2. Kinetic Coordination Template Effect. The central idea of the kinetic coordination template effect is shown in the theoretical examples of Eqs. (107), (108), (109), and (110). Eqs. (107) and (108) show how polymerization might

$$nA \rightarrow \sim A \sim A \sim A \sim A \sim A \sim A \sim A \tag{107}$$

$$A_4M^{n+} \longrightarrow \text{(cyclic A–A linked around } M^{n+}) \tag{108}$$

result in ring formation when such a reaction occurs within a coordination sphere. Since the propagation process in many condensation polymerization reactions can also serve as the chain-termination step, the possibilities are real. The second class of process, as outlined in Eqs. (109) and (110) has been adequately demonstrated. The general example given below shows clearly that the essential feature of a process of this type is really new chelate ring formation via a ligand reaction. In order for such a process to occur, both geometric and chemical constraints must be met. Chemically, the proposed site of attachment on the bound ligand (the donor atom in the example) must be reactive. Geometrically, the reagent X—X must be the right size and shape to span the reactive sites.

$$n(\text{B}\sim\text{B}) + n(\text{X}\sim\text{X}) \rightarrow \sim\text{B—B}\sim\text{B—B}\sim\text{B}\ldots + 2n\chi \qquad (109)$$

$$\text{B}_2\text{M}^{n+}\text{B}_2 \text{ (chelate)} + 2\text{XX} \longrightarrow \text{(macrocyclic B}_4\text{M}^{n+}\text{)}\,\text{X}_4 \qquad (110)$$

Although chelate ring formation in these cases involves the making of new intraligand bonds rather than metal–donor bonds, the geometric arguments applied to the chelate effect are germane.[144]

Earlier it was deemed most reasonable to attempt to establish the kinetic coordination template effect in a system that required formation of only one additional chelate ring in order to produce a new cyclic ligand. The venture also required identification and development of a suitable coordinated functional group to serve as the reactive site on the coordinated ligand. Because the geometry of the coordination sphere is most obviously felt by the donor atoms, first choice was given to these species. Ligand reactions were recognized[2] for coordinated amines, phosphorus halides, and mercaptans. Because coordination more often masks them, amines were not considered further. Mercaptans were chosen over phosphorus halides for two reasons. Early literature[69,201] showed the general ability of the coordinated mercaptide ion to function as a nucleophile. Further, the reagents necessary to produce nonlabile, new chelate rings are especially simple to obtain and manipulate—dihalohydrocarbons. In contrast, coordinated phosphorus halides would require nucleophiles as reagents; these either produce relatively labile linkages (phosphite esters) or are more difficult to handle (dilithio-hydrocarbons). In either case, it would have been necessary to produce new ligands and characterize new classes of complexes in order to test the template hypothesis. The model chosen for the test is given in Eq. (111) and is

$$\text{(A}_2\text{S}_2)\text{M}^{\text{II}} + \text{X}\sim\text{X} \longrightarrow \left[\text{(A}_2\text{S}_2)\text{M}^{\text{II}}\text{—S}\sim\text{X}\right]\text{X} \longrightarrow \text{(cyclic A}_2\text{S}_2)\text{M}^{\text{II}}\,\text{X}_2 \qquad (111)$$

based on a proposed planar complex of a tetradentate ligand containing *cisoid*-coordinated mercaptide functions. The principal stumbling block in obtaining suitable reactants was the fact that the mercaptide ion tends to form bridges between metal ions [e.g., Eq. (112)].

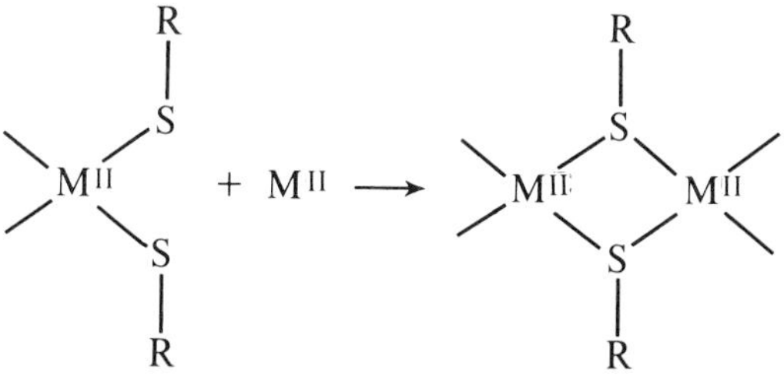

$$R_2S_2M^{II} + M^{II} \longrightarrow M^{II}(\mu\text{-SR})_2M^{II} \qquad (112)$$

The most obvious solution to this synthetic problem failed in some cases because the insolubilities of some bridged products seemed to drive the coordination reactions in the undesirable direction. The obvious starting materials for the template reaction, derivatives of the ligand given in structure **16,** showed a tendency to form trinuclear adducts[202], analogous to those formed by β-mercaptoethylamine, $NH_2CH_2CH_2SH_2$ (structure **17**). Conveniently, the Schiff-

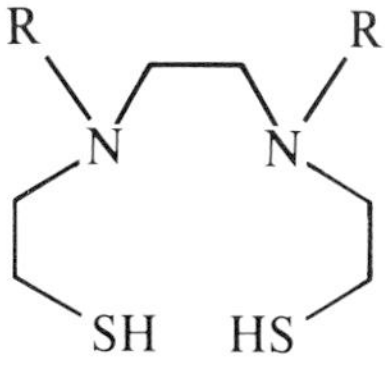

16

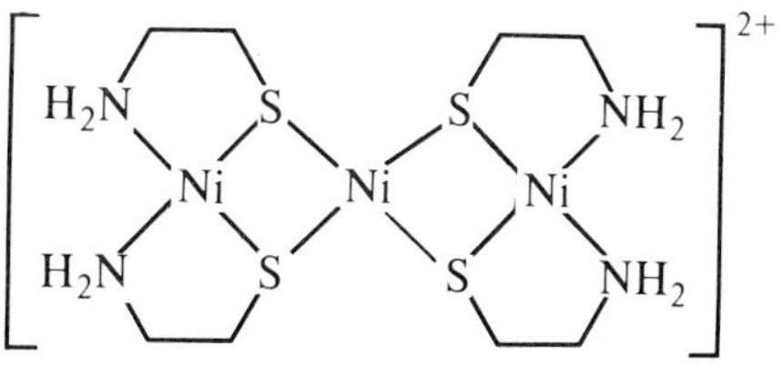

17

base derivatives of α-diketones and β-mercaptoethylamine[197,198,205] formed monomeric square planar nickel(II) complexes with the required pair of cis mercaptide functions (structure **18**).[206] The results of an X-ray study confirm this structure and show that the S—Ni—S angle exceeds substantially the ideal 90° as a result of the compression of the remaining donor–metal–donor angles. The coordinated mercaptides in this structure show a reduced tendency to chelate to other metal ions.

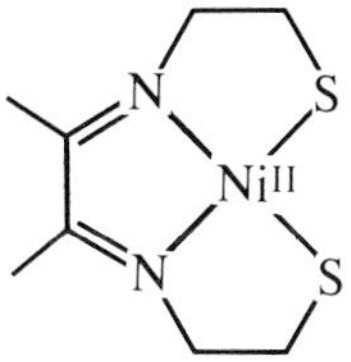

18

Extensive studies on the alkylation of the coordinated S atoms in the metal chelates of mercaptoamines and related ligands have shown that most unbridged coordinated mercaptides can act as nucleophiles. These investigations have involved both synthetic and kinetic studies, and they are discussed in some detail in section VIII of this review. Three conclusions from those studies are helpful at this point. Alkylation converts a very soft, strong mercaptide donor into a relatively weak thio-ether ligand. Secondly, in the course of reactions between alkyl halides and planar nickel complexes containing mercapto functions, the alkyl halide coordinates to the metal ion at its vacant site prior to the reaction between the sulfur nucleophile and the electrophilic carbon center [Eq. (113)].

$$\text{Ni(S,N)} + RCH_2X \longrightarrow \text{Ni(X–CH}_2\text{R)(S,N)} \longrightarrow \text{XNi(S(CH}_2\text{R),N)} \qquad (113)$$

Finally, and most important, the mercapto function remains bound to the metal ion during the course of its reaction with electrophiles. Clearly this is a minimum requirement for a kinetic coordination template effect.

Although the reaction was studied[197,199] prior to the reported crystal structure[206] of the compound indicated in structure **18,** it was expected that the distance between the cis sulfur atoms would be great. Consequently, α,α'-dibromo-*o*-xylene was chosen as the reagent for ring closure. It not only forms a large chelate ring (seven members), but also its reactive sites are directed in space. Consequently, it was to be expected that reaction would lead cleanly to a single product of a precisely predicted geometry [Eq. (114)].[197,199]

$$\text{Ni(N}_2\text{S}_2) + \text{Br–CH}_2\text{C}_6\text{H}_4\text{CH}_2\text{–Br} \longrightarrow \text{NiBr}_2(\text{N}_2\text{S}_2\text{-macrocycle}) \qquad (114)$$

Note that the ring closure of Eq. (114) was preceded by studies on the reaction of the same complexes (several closely related structures were used) with monofunctional alkyl halides[207] [Eq. (115)] and by chelate ring-forming reactions with α,α'-dibromo-*o*-xylene in which macrocycles were not formed[199] [Eq. (116)]. Thus, the coincidence of the physical properties of the product of Eq. (114) with those expected left little doubt that the desired process had taken place.

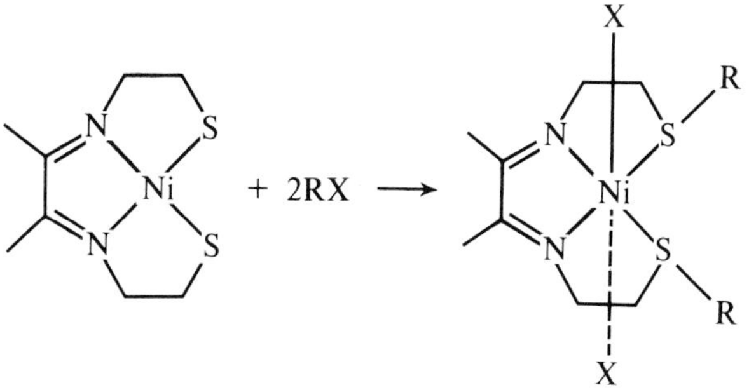

(115)

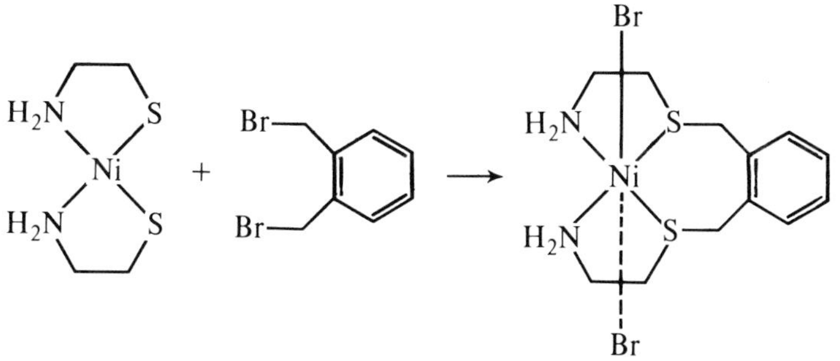

(116)

Despite the apparent success of reaction (114), the magnetic moment of the product was disconcerting. Samples handled somewhat casually appeared to contain low-spin nickel(II) while those carefully protected showed values corresponding to a fraction of an unpaired electron. It has since been shown that the room temperature moment of the carefully protected compound corresponds to 1.57β, and the susceptibility obeys the Curie-Weiss law with a small Weiss constant (1.8).[208] This implies nonequivalent complexes in the unit cell with either 3 of 4, or 4 of 5, of the units in the low-spin form and the remaining in high-spin form. On exposure to air, about 0.5 mol H_2O is absorbed, and the high-spin nickel(II) atoms become low-spin. This change is presumed to be caused by strong hydrogen bonding between the absorbed water and the previously coordinated bromide ions. This, in turn, would weaken the extraplanar ligand field, permitting the stronger in-plane field to force electron pairing on the nickel atom. The general problem of such magnetic anomalies as these has been reviewed[209], as has the subject of solvent and hydrogen bonding effects on spin state.[210] The slight distraction from this complicated physical property was completely obviated by metathesis of the bromide to produce a series of pseudotetragonal complexes of the formula Ni(macrocycle)X_2, where X included NCS^{-1}, Br^{-1}, Cl^{-1}, and N_3^- [208]. All of the new derivatives displayed physical properties in complete agreement with their predicted geometries.

Similar macrocycles can be derived from the same starting complex and 1,3-dibromopropane or 1,4-dibromobutane, but 1,2-dibromoethane is too short to span the intramolecular S—S distance.[199,211] An interesting extension of this work involves the reaction in chloroform of the complex of structure **18** with sulfur monochloride to produce a substance having a tetrasulfide chelate ring as part of a macrocycle (structure **19**).[212] An aromatic analog of structure **18** undergoes corresponding alkylation reactions in 1,2-dichloroethane solvent, producing a series of related octahedral macrocyclic complexes (structure **20**)[213].

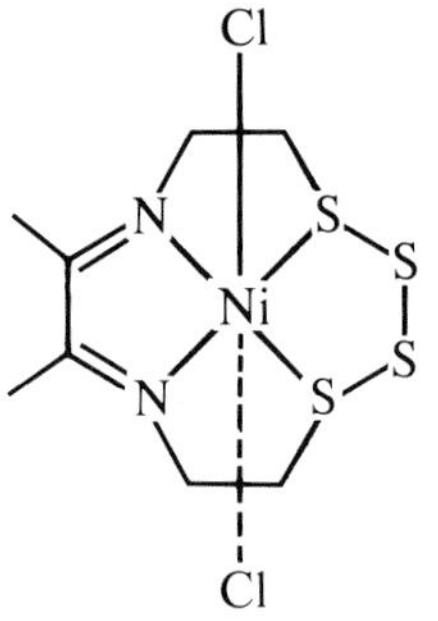

Cl N S S Ni S N S Cl

19

Br N S Ni N S Br

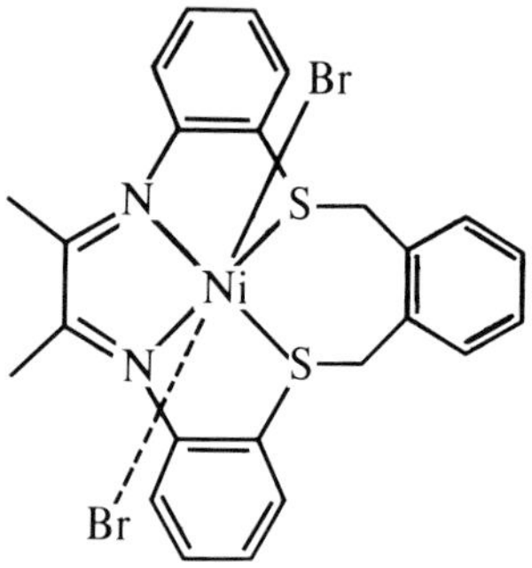

20

The reaction of α,α'-dibromo-*o*-xylene has also been used to produce an unsaturated macrocycle containing only sulfur donors [Eq. (117)][214].

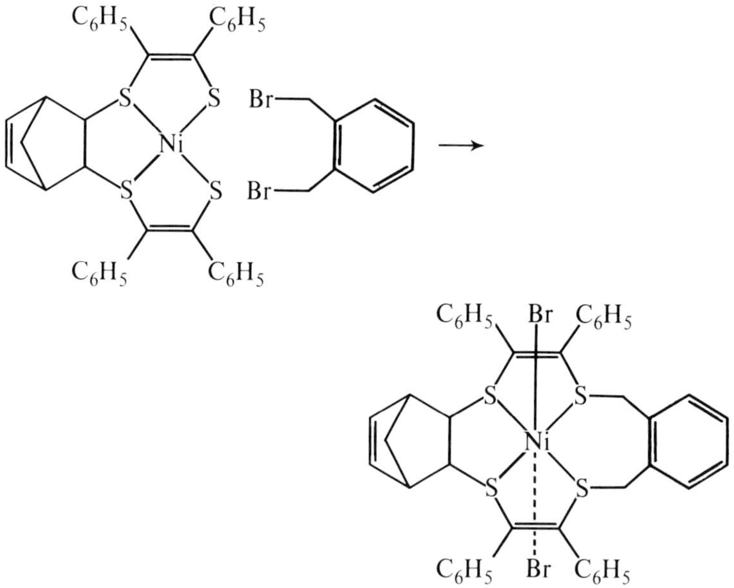

C_6H_5 C_6H_5 S S Br Ni S S Br C_6H_5 C_6H_5 →

C_6H_5 Br C_6H_5 S S Ni S S C_6H_5 Br C_6H_5 (117)

Kinetic studies[215] have confirmed the existence of the template effect in the case of the first system studied [Eq. (114)]. The interpretation of these studies was aided by broad-ranging investigations of the reaction mechanism of the coordinated mercaptide nucleophile (see section VIII).

A simple demonstration of the kinetic nature of the template effect is shown in Figures 2-6 and 2-7. These figures reproduce spectrophotometer scans taken at different times for the reaction in 1,2-dichloroethane of the complex having structure **18** with α,α'-dibromo-*o*-xylene [Eq. (114)] and with benzyl bromide [Eq. (115)], respectively. The most obvious distinction between the two figures lies in the simplicity of the scan sequence for the α,α'-dibromo-*o*-xylene reaction. The well-developed isosbestic point near 4000Å suggests that only two colored species exist at significant concentrations during the reaction. The spectra for the benzyl bromide reaction show an increase in optical density for the first 3.5 min within wave length ranges between 6000Å and 6400Å and between 3800Å and 4900Å. The optical density then decreases so that no persistent isobestic points occur. Obviously, an intermediate is formed in a sufficiently large concentration to dominate the spectrum. At relatively short reaction times, an isosbestic does occur between the intermediate and the starting material at 5900Å. Significantly, many other monofunctional alkylating agents also give spectrophotometer scans like benzyl bromide (Figure 2-7). All contrast with α,α'-dibromo-*o*-xylene in that only that reagent fails to give evidence for significant concentrations of intermediates. It was also shown that α,α'-dibromo-*p*-xylene behaves in essentially the same manner as benzyl bromide. This, of course, focuses attention on the necessity for the two reactive centers of the dialkylating agent to be capable of forming a new chelate ring.

Clearly benzyl bromide reacts in two steps, as shown in Eqs. (118) and (119), and the second-order rate constants (25°C, dichloroethane) for these steps are

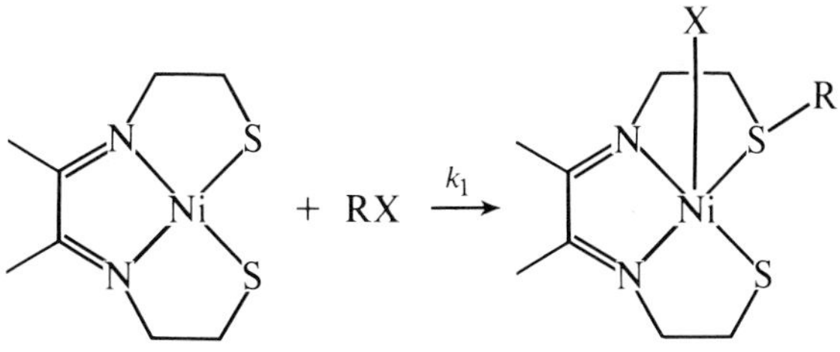

N N S S Ni + RX $\xrightarrow{k_1}$ X N N S S R Ni (118)

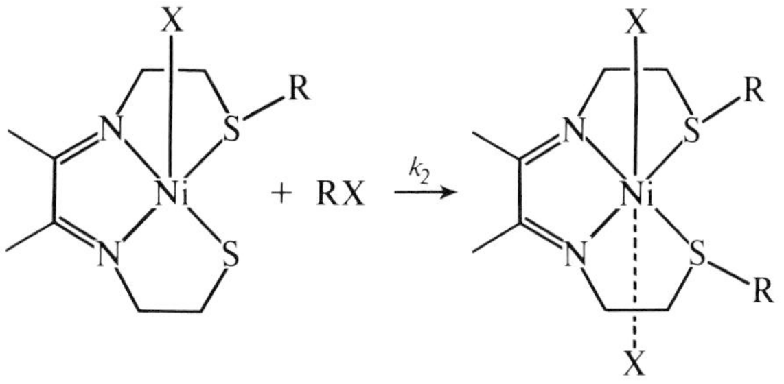

X N N S S R Ni + RX $\xrightarrow{k_2}$ X X N N S S R R Ni (119)

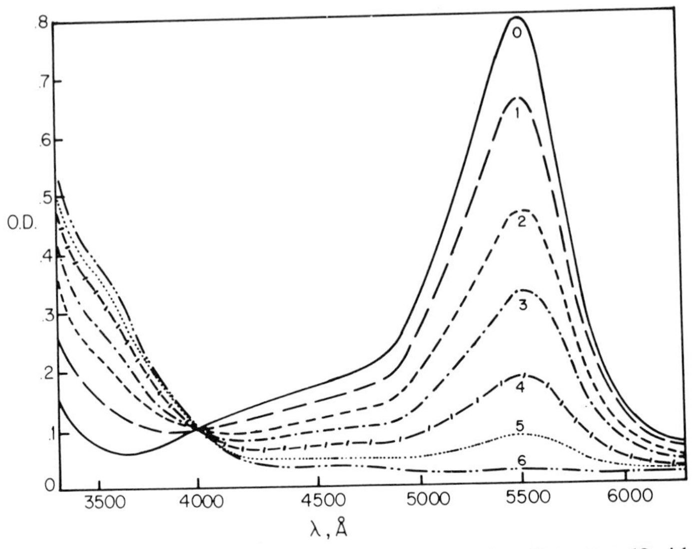

Figure 2-6 Visible spectra obtained during the reaction of structure **18** with α,α'-dibromo-*o*-xylene in 1,2-dichloroethane at 25°C. [**18**] = 2.12×10^{-4} *M*; [$C_8H_8Br_2$] = 3.61×10^{-3} *M*; chart speed = 25 Å/sec. Spectrum: 0, t = 0; 1, t = 1.5 min; 2, t = 5 min; 3, t = 9 min; 4, t = 17 min; 5, t = 28 min; 6, t = 50 min.

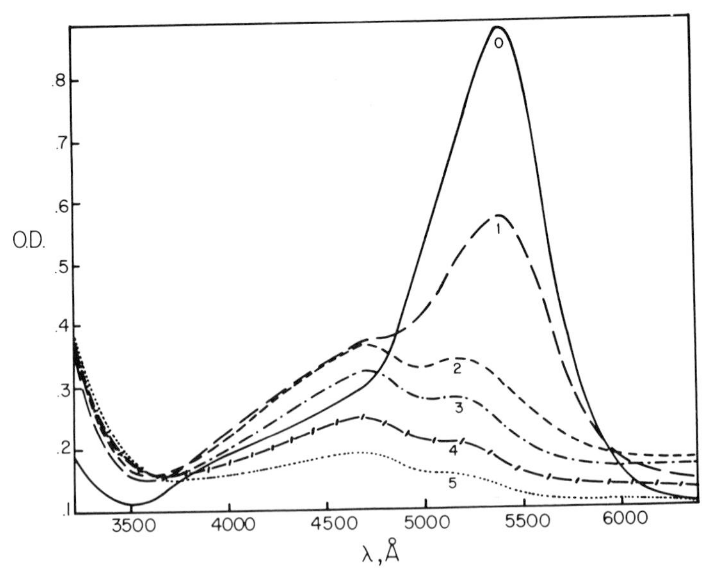

Figure 2-7 Visible spectra obtained during the reaction of structure **18** with benzyl bromide in 1,2-dichloroethane at 25°C. [**18**] = 2.2×10^{-1} *M*, [$C_6H_4CH_2Br$] = 4.48×10^{-2} *M*, chart speed = 25 Å/sec. Spectrum: 0, t = 0; 1, t = 30 sec; 2, t = 3.5 min; 3, t = 7 min; 4, t = 14 min; 5, t = 27 min.

$k_1 = 0.35_3(\pm 0.026)M^{-1}\text{sec}^{-1}$ and $k_2 = 9.8_5(\pm 0.15) \times 10^{-3}M^{-1}\text{sec}^{-1}$. Thus, the rate of the first step exceeds that of the second by a factor of approximately 36. Under corresponding conditions, the rate constant for the single measureable kinetic process between structure **18** and α,α'-dibromo-*o*-xylene is $0.57_8(\pm 0.050)M^{-1}$. Since this is about 1.6 times as fast as the first step for benzyl bromide, it must conform to the displacement of the first bromide from α,α'-dibromo-*o*-xylene. Clearly, the second step is too fast to observe in this case. It seemed plausible to assume that detectable amounts of an intermediate would build up if the second rate exceeded the first by less than a factor of about 10. This and the fact that the second step for benzyl bromide is only $1/36$ of the first rate suggests that the relative rates of the first and second consecutive steps are changed by a factor of at least 300–400 in going from the reaction with a monofunctional reagent to a directed bifunctional reagent.

In the reaction of structure **18** with α,α'-dibromo-*o*-xylene, a sterically directed mechanism is imposed by the cis configuration of the sulfurs and the ortho position of the α-bromomethylene groups. Apparently, the first step is slow, and the second step is fast. After the combining of the first mercapto group with the reagent, the remaining sulfur and the α-bromomethylene group are oriented in position for the cyclization reaction (structure **21**).

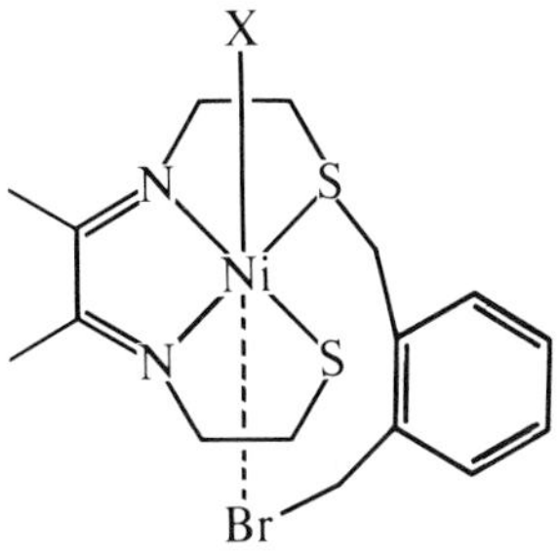

21

This relationship explains the conclusion that measurable concentrations of the intermediate are not formed and constitutes the essence of the kinetic coordination template effect.

The kinetic behavior of this template effect resembles the normal kinetic behavior for simple chelation of bidentate ligands[216] as, indeed, it should. The behavior of flexible α-ω-dibromo hydrocarbons toward structure **18** is also interesting. At a high concentration, 1,4-dibromobutane (100:1) gives a spectral scan vs. time profile resembling that of a monoalkylating agent while at lower concentrations it behaves like α,α'-dibromo-*o*-xylene. This is interpreted in terms of two competing second steps, one being ring closure and the other, reaction with a second mole of alkylating agent.

A number of distinctly different systems gave early evidence for the kinetic coordination template effect.[200] Two separate groups[217,218] discovered that a number of reagents, foremost among which is $BF_3 \cdot Et_2O$, will react with the bridged oxygen atoms of planar bis-α-dioxime complexes to produce new chelate rings (structure **22**). Because of the chemical nature of the starting complex, it seems almost certain that the reaction occurs without breaking any metal–nitrogen bonds. Incidently, this reaction was discovered in experiments designed to produce a metal–boron bond.[218]

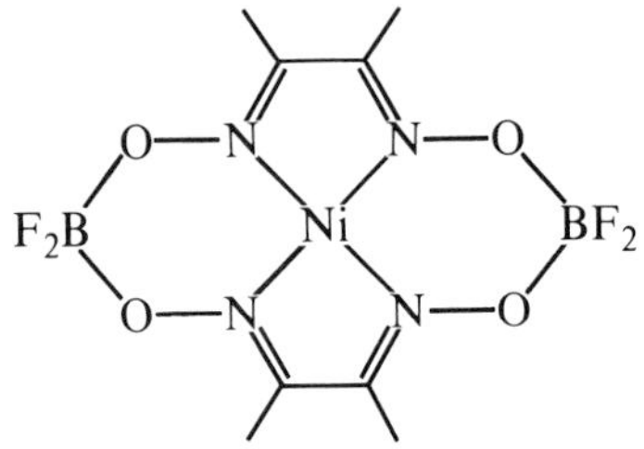

22

Rings closely related to natural products were closed in such reactions by investigators who made early applications of template effects. In the presence of coordinated palladium(II), the tetrapyrrole (structure **23**) was closed to form the palladium complexes of the derivatives (structure **24**) having X=O, NCH_3,

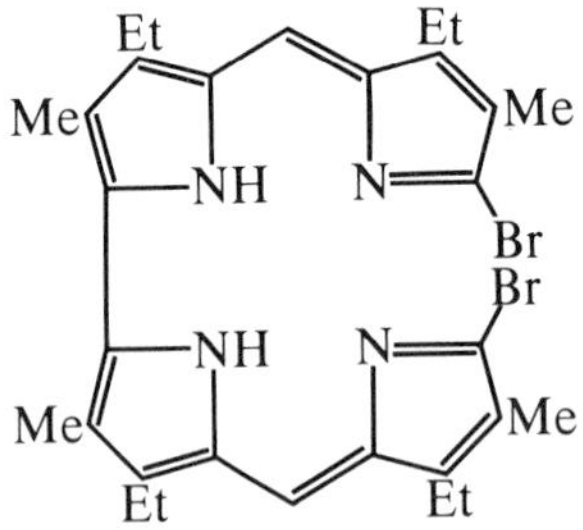

23 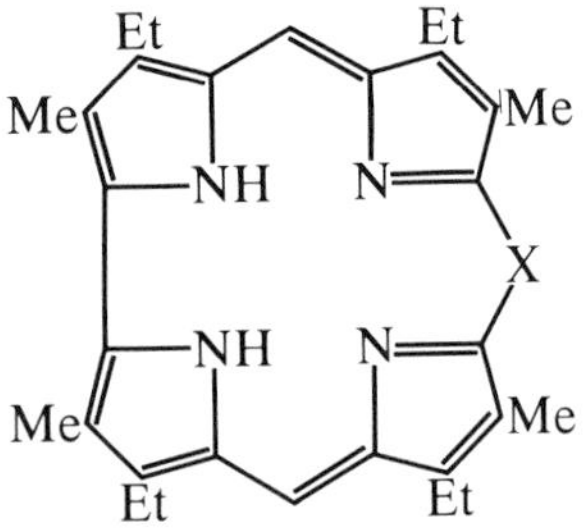 **24**

or S.[219,220] Eschenmoser et al.[221] used the templating action of Ni(II) in order to prepare the corrin ring of vitamin B_{12}.[221] Ring closure involved the treatment of the nickel perchlorate complex of structure **25** with *t*-butoxide in *t*-butyl alcohol to form structure **26,** where R = CN. Heating structure **26** in dilute hydrochloric acid at 22°C leads to acid hydrolysis of the cyano group along with subsequent decarboxylation, yielding nickel corrin (structure **26,** R = H). These studies have been extended.[222–225]

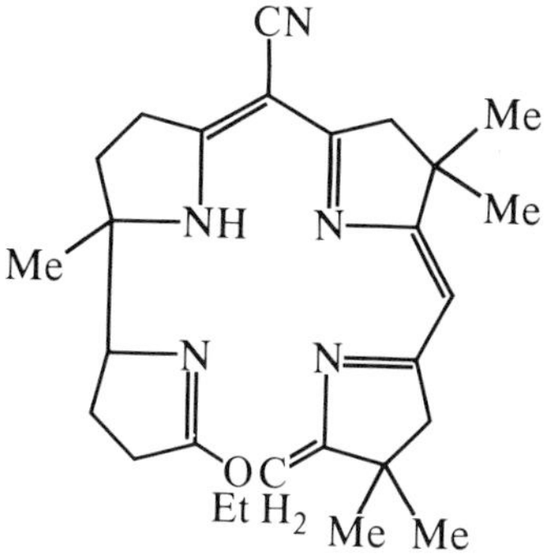

25

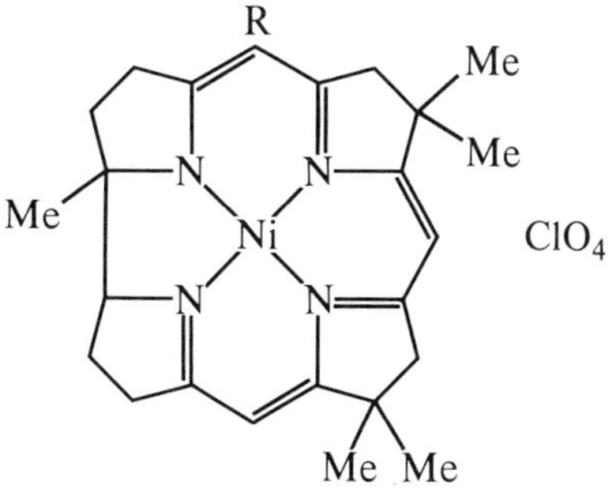

26

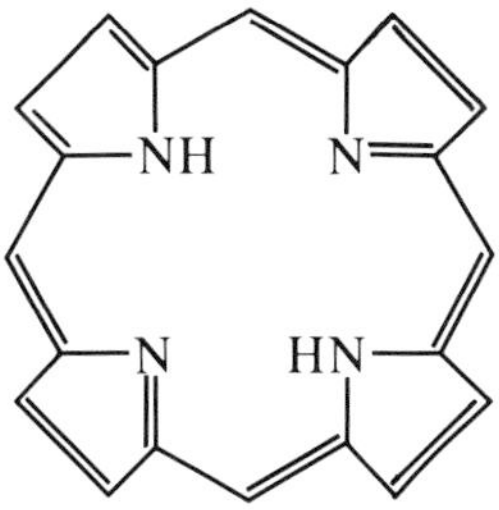

27

Though there are recurrent reports[226,227,228] of the influence of metal ions on reaction yields used to produce the porphin ring (structure **27**), convincing examples of the importance of coordination template effects are by no means abundant. Perhaps the most likely examples involve cyclization of tetrapyrrole derivatives,[226,229] reactions analogous to those just described for the synthesis of the corrin ring and its congenors. The related phthalocyanine (structure **28**) provides a more convincing example of a template synthesis.[230] In the stepwise

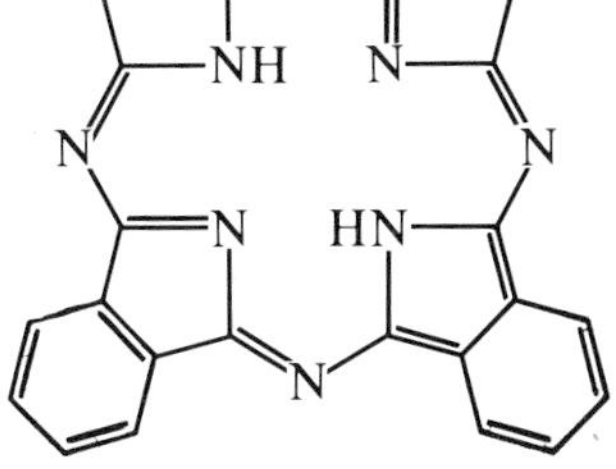

28

production of nickel phthalocyanine from nickel(II) chloride and 1,3-diiminoisoindoline, an intermediate, bisalkoxyiminoisoindolinenine (structure **29**), is isolated. This intermediate converts readily into the nickel phthalocyanine upon heating.

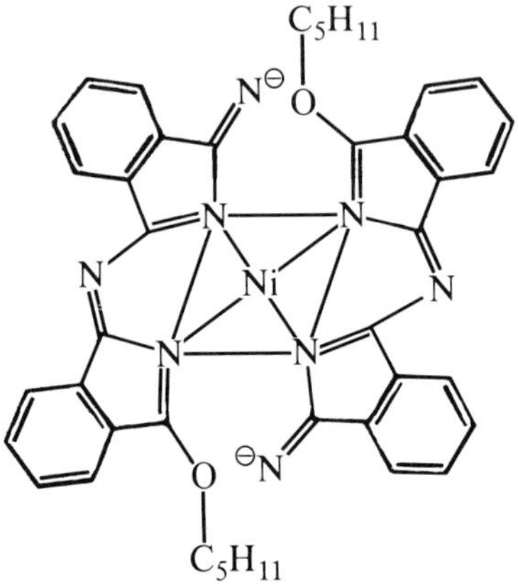

29

3. Equilibrium Template Effects. The basic concept of an equilibrium or thermodynamic template effect is illustrated in the metal-promoted rearrangement of oxazolines and thiazolines to produce Schiff-base complexes.[212,231,232,233] Equilibria[231,234,235] exist between the heterocyclic and Schiff-base structures in metal-free systems, and these favor the cyclic forms. Consequently, the metal ion causes shifts in these equilibria favoring the Schiff-base forms[231,198,200] [Eq. (120)]. Kinetic studies[233] show that the rates of certain of these reactions are dependent on the concentration of metal ion. This poses the real possibility that a metal ion can serve in both the kinetic and thermodynamic roles defined here for the purpose of distinguishing limiting classes of metal ion effects.

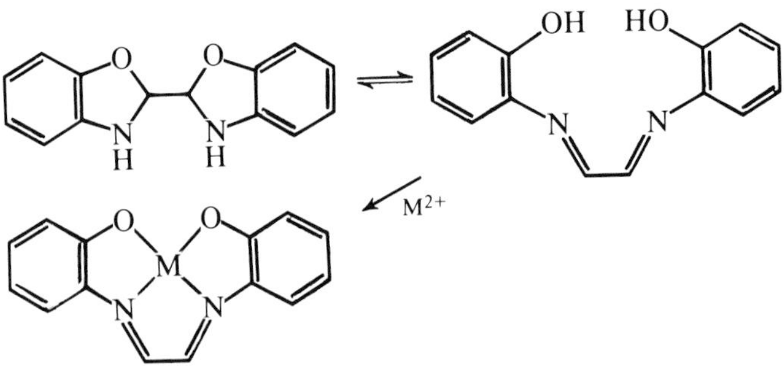

(120)

The ligand TRI (structure **30**) is known only in its metal complexes. It is produced[236] by the rearrangement of a very different molecule (structure **31**).[237]

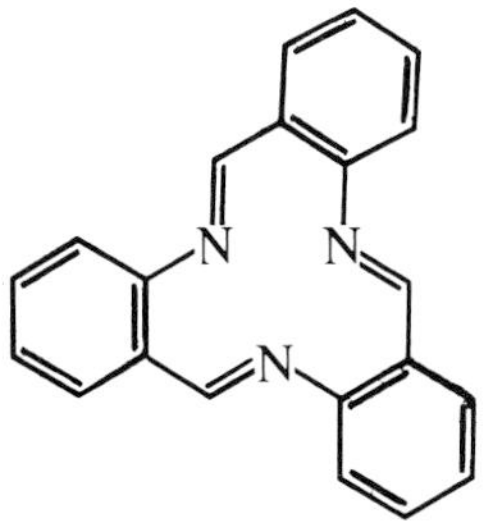

30 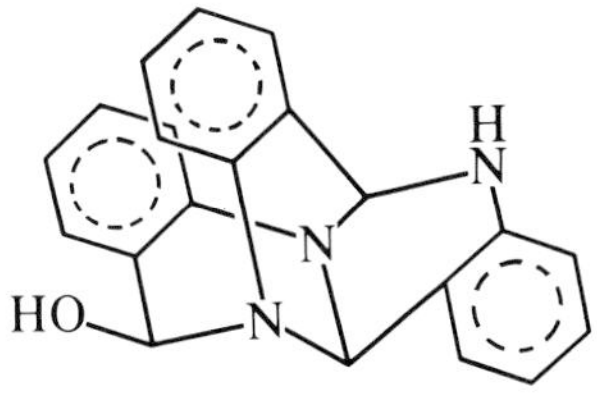 **31**

The stability of the complex is responsible for the existence of the ligand structure, but there is no information available on the kinetics of the rearrangement. TRI (structure **30**) is a trimeric Schiff base of *o*-aminobenzaldehyde. The monomer polymerizes in the presence of many metal ions producing TRI or a tetrameric condensate called TAAB (structure **32**).[238–242] These reactions provide

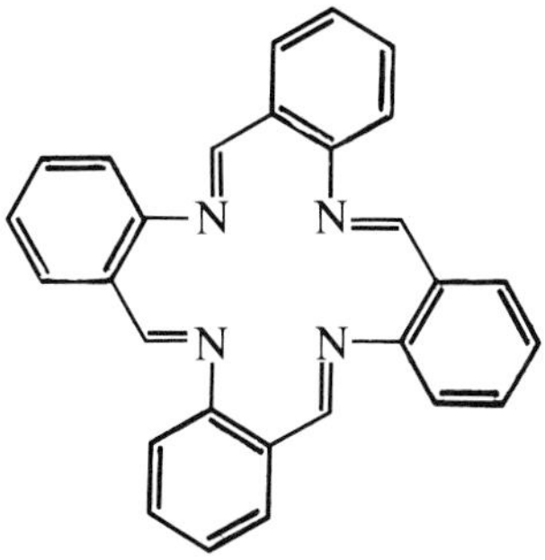

32

close approximation to a cyclic polymerization reaction as idealized in Eqs. (107) and (108). It is probably more nearly accurate to assume, however, that the principal roles of the metal ion are (a) to serve as a thermodynamic template, stabilizing the cyclic Schiff-base polymers and possibly (b) as a kinetic template in causing rearrangement of precondensed aminobenzaldehyde rather than promotion of the initial polymerization process. These relationships probably best explain the influence of metal ions on many cyclic ligands that are produced in their presence by Schiff-base reactions and some other condensation processes.[200,211,228] For example, although the ring of structure **33** was first discovered and has often been synthesized by reactions carried out in the presence of metal ions,[228,243] it can be prepared quite easily from the same starting materials (acetone and ethylenediamine) in the absence of metal ions.[244] Under severe conditions, such systems produce dark brown tars so that structure **33** may be considered to be a product of modest stability that forms quickly but whose presence is not favored by the system at its long-time equilibrium. Clearly, a metal ion can sequester the ligand as it forms and protect it from further reaction. The success of the metal-free synthesis depends in a similar way on

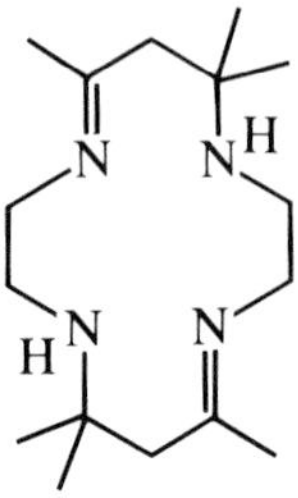

33

separation of this somewhat fragile intermediate product. The latter separation occurs by crystallization of various onium salts of the ligand.

4. Template Synthesis. It is especially true in the case of tetradentate macrocycles that the synthetic applications of the template effects and other techniques, including additional ligand reactions, are rapidly reaching a stage of maturity such that one can set out to synthesize almost any previously unknown ligand of the class with a reasonable expectation of success. The reactions, in the presence of metal ions, of mono- and dicarbonyl compounds with pairs of adjacent amine groups account for much of this generality.[94] The reaction of tris(ethylenediamine) nickel(II) salts with acetone produces the nickel(II) complex of structure **33** and its position isomer having the C = N groups mutually cis. The product was first reported to exist as the tetrakis(isopropylidene) derivative[245] (structure **34**), but shortly thereafter, the correct structure (**33**)

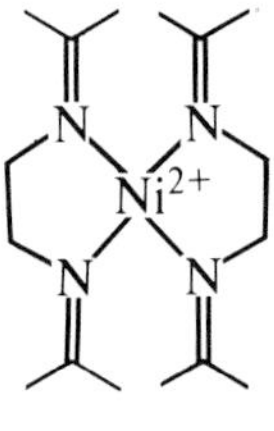

34

was deduced, largely on the basis of its infrared spectrum (an N—H stretching mode is evident), its remarkable inertness toward acid (the complex persists under conditions where any ordinary amine would be stripped from the metal), and the fact that when decomposition is accomplished, mesityl oxide $(CH_3)_2$-$C{=}CHCOCH_3$ is a major product.[243]

There are many examples of the condensation reactions of α-diketones with pairs of amines to produce new chelate rings containing the α-diimine group.[184,212,231,232,233,197,198,246] However, the first example of a ring closure utilizing such a reagent was reported by Baldwin and Rose[247] [Eq. (121)].

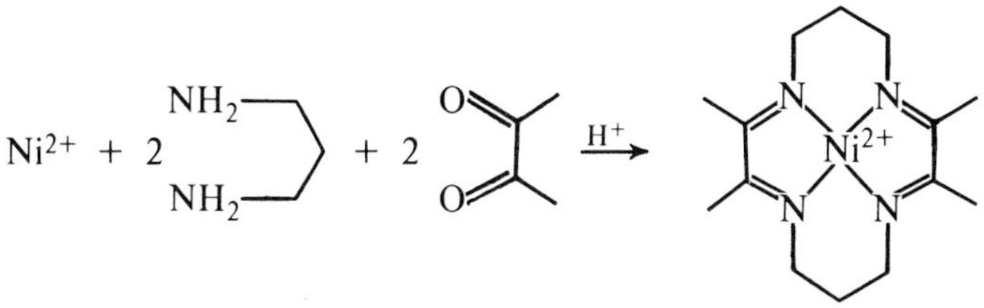

$$Ni^{2+} + 2\ H_2N(CH_2)_3NH_2 + 2\ CH_3COCOCH_3 \xrightarrow{H^+} [Ni(macrocycle)]^{2+} \quad (121)$$

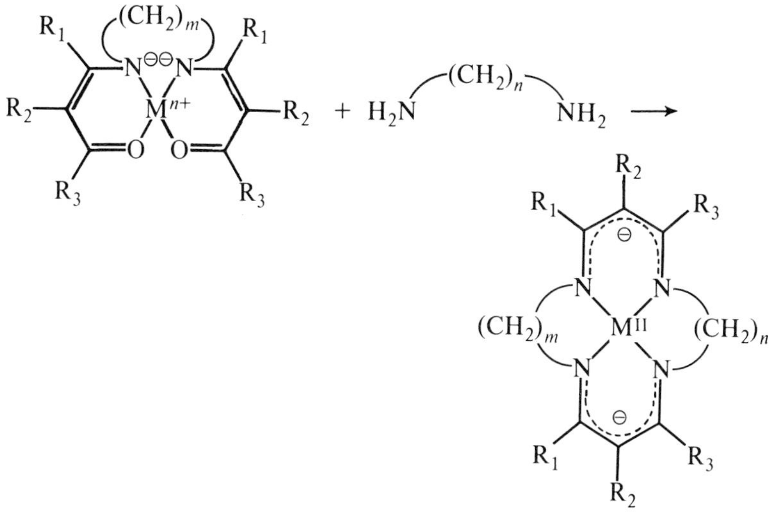

$$\text{M}^{n+}\text{ complex} + H_2N(CH_2)_nNH_2 \longrightarrow \text{M}^{II}\text{ macrocycle} \quad (122)$$

R1, R2, R3, $(CH_2)_m$, $N^{\ominus\ominus}N$, M^{n+}, O, O; $(CH_2)_n$, NH_2; M^{II}, N, N, N, N

$m,n = 2,3$

Jaeger first incorporated the β-diketone grouping into a macrocyclization reaction [Eq. (122)].[248,249,250] The formation of the new chelate rings is accompanied by ionization so that the unit assumes a negative charge. The products are given in structure **35** where M is Ni or Cu, and both X_1 and X_2 can be $(CH_2)_2$, $(CH_2)_3$, or *o*-phenylene, or $X_1 = (CH_2)_2$ and $X_2 = (CH_2)_3$.

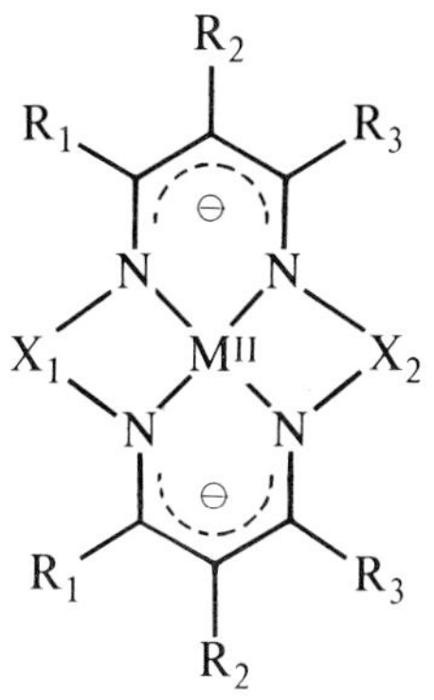

R2, R1, R3, N, N, X_1, M^{II}, X_2, N, N, R1, R3, R2

35

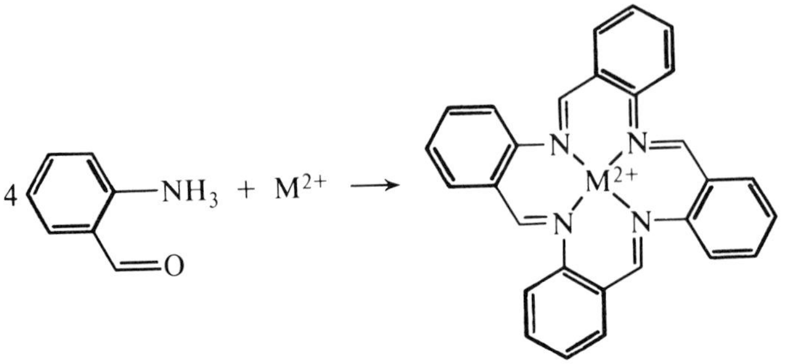

(123)

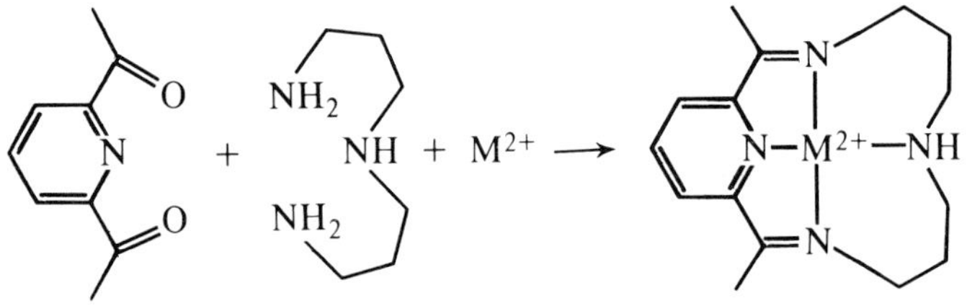

(124)

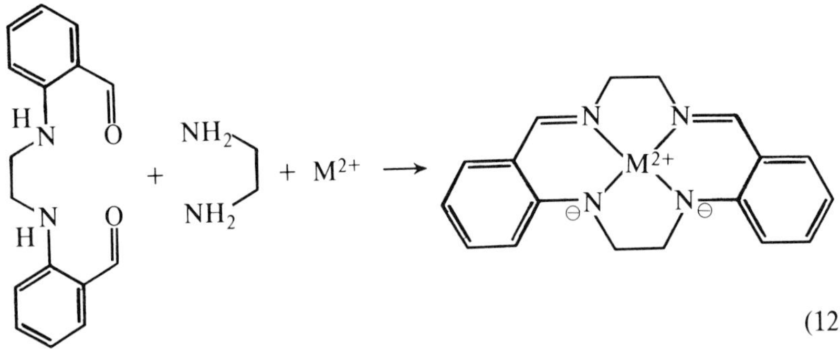

(125)

Reagents that contain carbonyl groups as well as amine donors have been particularly useful in the synthesis of certain cyclic structures via template processes. Three prominent examples are given in Eqs. (123), (124), and (125). Eq. (123) involves the cyclic polymerization of *o*-aminobenzaldehyde and was discussed above. The reaction of Eq. (124) utilizes a dicarbonyl reagent having a pyridine nitrogen strategically placed to facilitate coordination of its nitrogen and both carbonyl oxygens.[251–255] From earlier discussions it should be obvious that coordination of the carbonyl oxygen atoms should promote the Schiff-base reaction (structure **36**).

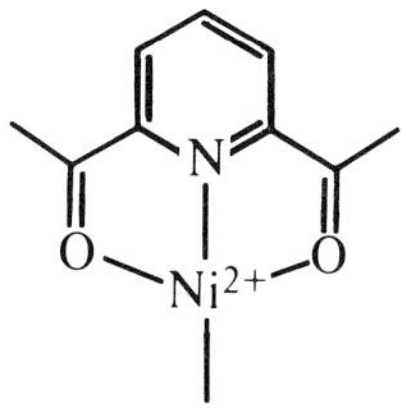

36

This principle of placing additional coordinating groups in the structure of the carbonyl-containing reactant has been used in other metal-ion, template-controlled[78] Schiff-base condensations. Eq. (125) provides a second example of this kind.[256] Variations of this structure include replacing the amines in the structure with thioether groups,[257,258,259] and the synthesis of a complex of a corrin analog (structure **37**).[260] Also, the product of the reaction described in

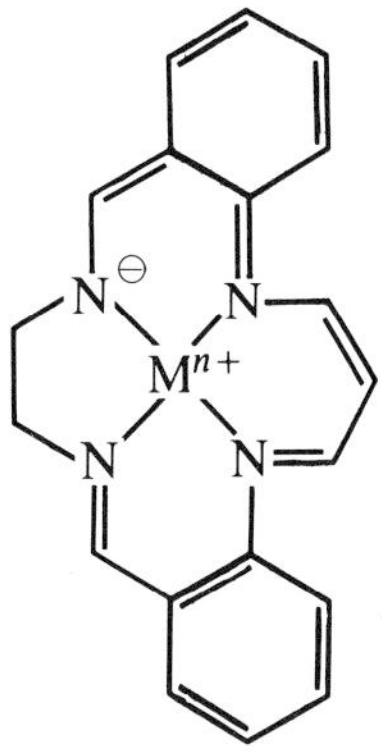

37

Eq. (125) can be prepared by the reaction of ethylene dibromide with the complex formed by the Schiff base of ethylenediamine and aminobenzaldehyde (structure **38**).[261]

Larger rings have also been produced by various template reactions. Pentadentate macrocycles include structures **39**[251,262,263,264,265] and **40**.[266] Sexadentate structures are equally abundant and include examples having structures **41**[251] and **42**[257,258] Permutations of the donor atoms in structure **42**[267] have greatly expanded that series of derivatives. The sexadentate rings of structure **42** serve as prime examples of how the coordination template effects may serve chemical synthesis. Attempts to prepare these large, polyfunctional cyclic molecules in the absence of metal ions have failed. Further, some of the rings are very easily removed from the metal ions.

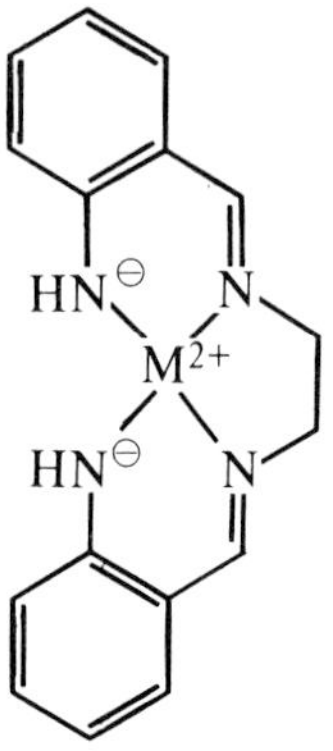

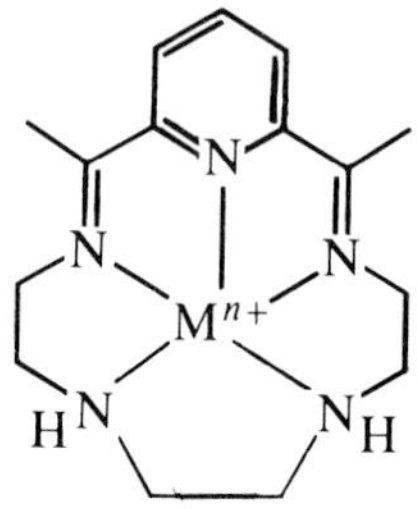

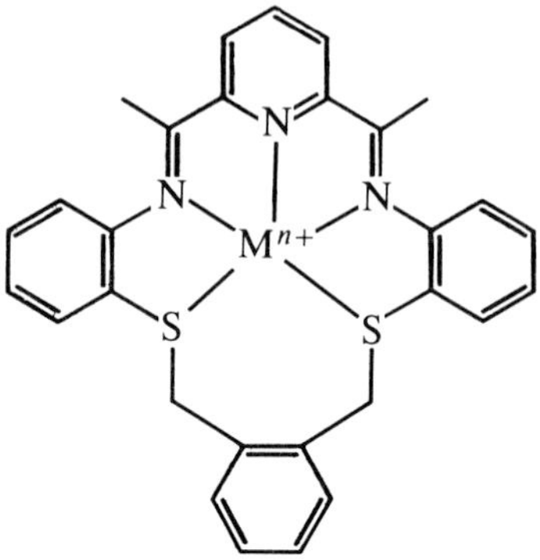

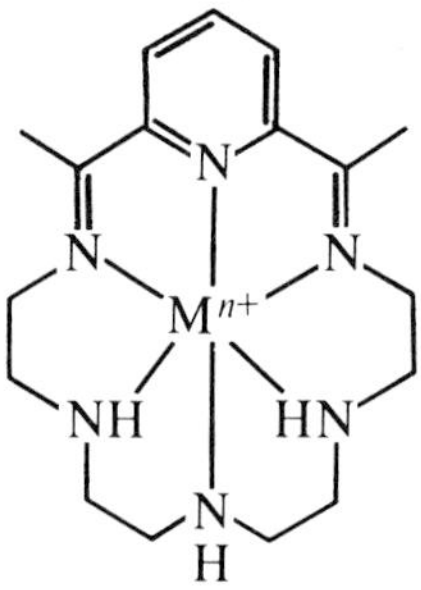

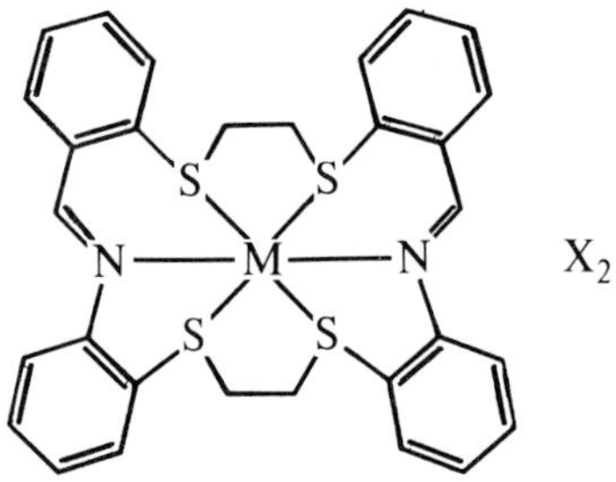

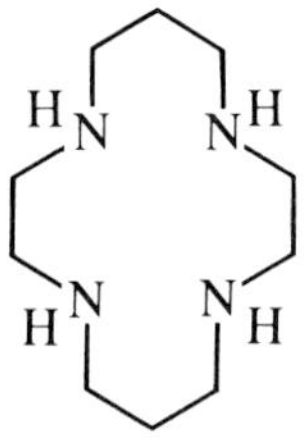

A second example shows in modest detail how template syntheses can provide extreme improvements over standard syntheses of certain substances. The 14-membered tetradentate ligand called cyclam (structure **43**) plays a central role in the chemistry of the synthetic macrocyclic ligands.[94,200,211] Two synthetic routes have generally been used to prepare the free ligand. The first is very simple but makes sacrificial use of most of the starting materials.[268,269] This involves reflux of 1,3-bis-(2′-aminoethylamino) propane with 1,3-dibromopropane followed by co-distillation from the reaction mass with additional unreacted 1,3-bis(2′-aminoethylamino)propane. It is fortuitous that the cyclam is only very slightly soluble in the entraining linear tetraamine and separates from the distillate as a white crystalline solid. The yield varies from 0–3%. The alternate route to cyclam provides a more deliberate but far more tedious synthesis with little or no improvement in yield.[270] Barefield[271] has recently reported a template synthesis in which the yield is as high as 30%. This is outlined in Eq. (126).

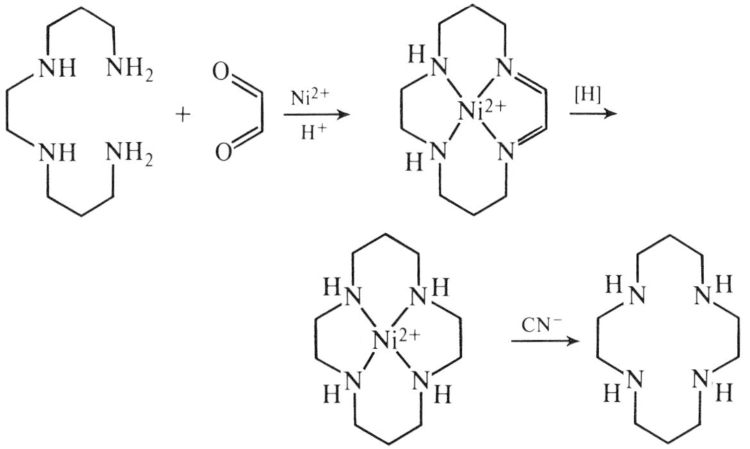

NH NH_2 NH NH_2 + O O $\xrightarrow[H^+]{Ni^{2+}}$ Ni^{2+} complex $\xrightarrow{[H]}$ Ni^{2+} complex $\xrightarrow{CN^-}$ cyclam (126)

Still more striking examples of template syntheses involve the formation of large-ring bicyclic structures. The simplest are those where a basket-like structure is produced by formation of new chelate rings (structure **44**).[186,272]

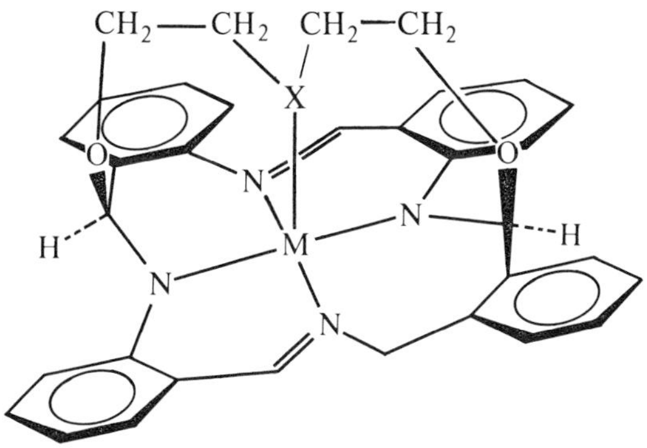

M = Ni and Cu
X = S and NCH_3

44

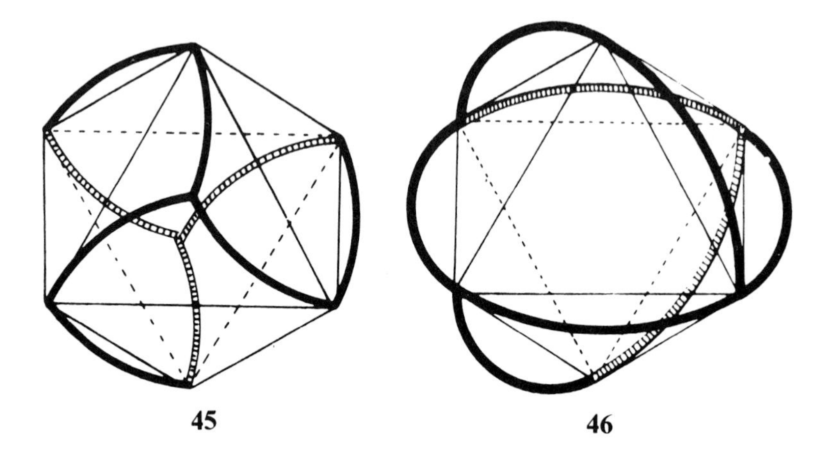

45 46

Some time ago, two stereochemical arrangements (structures **45** and **46**) were suggested to be appropriate for complete encapsulation of a metal ion via template reactions.[200] The first example of such an encapsulation of a metal ion by a template reaction was reported by Boston and Rose.[273] Their system utilized the reaction of $BF_3 \cdot Et_2O$ with potassium tris(dimethylglyoximato)cobaltate(III). The structure of the products conforms to the general structure, **45**, and is represented schematically in structure **47**. More recently, Parks and Holm have used the same ligand reaction to prepare new clathro chelates whose ligand geometry forces trigonal prismatic coordination on the enclosed metal ion[274] (structure **48**).

47

48

VI. Homogeneous Transition Metal-Catalyzed Olefin Reactions

Many homogeneous, transition metal-catalyzed olefin reactions are known. Here we examine the better known catalyst systems in terms of their mechanistic details and relate them to the recognized chemical features of the transition metal atom in the catalytic species. This provides a framework upon which to accumulate the bewildering variety of seemingly disparate reactions confronted in the ever-expanding area of transition metal catalysis.

Our approach progresses from simple to complex. Thus, a relatively uncomplicated intramolecular atom-transfer reaction, the reversible insertion into a metal–hydride bond leading to olefin isomerization, is considered first. Following this, we examine catalytic hydrogenation of olefins, in which hydrogen activation followed by olefin insertion into the M—H bond and subsequent hydrogenolysis of the metal—alkyl linkage is a dominant mechanism. Competitive insertion of carbon monoxide into and hydrogenolysis of metal–alkyl bonds (hydroformylation of olefins) is the next topic. Our interest is in the mechanisms of acyl formation and hydrogenolysis.

Finally, we examine a number of strained hydrocarbon rearrangements catalyzed by transition metal-containing species. These reactions, which all appear to be initiated by the Lewis acid action of the catalyst, display a range of products depending on the nature of the catalytic species. Our intention is to correlate the observed products with behavior patterns known for the catalytic transition metal atom and its complexes.

A. Olefin Isomerizations by Homogeneous Transition Metal Catalysis

Among the simplest reactions undergone by ligands coordinated to transition metal ions is that of hydrogen addition to or abstraction from coordinated olefins. In the former case, the resultant ligand can be an alkyl group [Eq. (127)] or a π-allyl ligand [Eq. (128)], depending on the nature of the original olefinic ligand.

$$RCH_2CH{=}CH_2 + ML_nH \longrightarrow L_nM{-}CH_2CH_2CH_2R + L_nM{-}\underset{\displaystyle CH_3}{\underset{|}{\overset{\displaystyle CH_2R}{\overset{|}{C}H}}} \qquad (127)$$

$$RCH{=}CH{-}CH{=}CHR' + ML_nH \longrightarrow RCH_2{-}\overbrace{CH \cdots CH \cdots CH}^{\pi\text{-allyl}}{-}R' \;(\text{with } ML_n \text{ bonded to the allyl}) \qquad (128)$$

Proton abstraction, however, leads exclusively to the formation of a π-allyl ligand [Eq. (129)].

$$RCH_2CH{=}CH_2 + ML_n \longrightarrow RCH\overset{CH}{\cdots}CH_2 \;(\text{with } HML_n \text{ bonded to the } \pi\text{-allyl}) \quad (129)$$

In the event that these reactions are reversible, a mechanism for olefin isomerization is possible [Eqs. (130–132)]. If the rearranged olefin is displaced from the coordination sphere of the catalyst by the original olefinic ligand, a homogeneous catalytic process has occurred.

$$L_nM{-}CH_2CH_2CH_2R + L_nM{-}CH(CH_3)(CH_2R) \rightleftharpoons CH_2{=}CHCH_2R + CH_3CH{=}CHR + ML_nH \quad (130)$$

$$RCH_2{-}CH\overset{CH}{\cdots}CH{-}CH_2R' \;(ML_n) \rightleftharpoons RCH_2CH{=}CH{-}CH{=}CHR' + RCH{=}CH{-}CH{=}CH{-}CH_2R' + ML_nH \quad (131)$$

$$R{-}CH\overset{CH}{\cdots}CH_2 \;(HML_n) \rightleftharpoons RCH_2CH{=}CH_2 + RCH{=}CHCH_3 + ML_n \quad (132)$$

We are primarily concerned with Eqs. (130) and (132) since the reaction of transition metal hydrides with conjugated olefins to form π-allyl species usually results in catalyst poisoning. This is further discussed in section B.

Deciding upon the mechanisms operative in such transition metal-catalyzed olefin isomerizations is a matter of determining if a preformed metal hydride species reacts with the olefin. In this case, overwhelming evidence points to a reaction via Eqs. (127) and (130), a 1,2-addition–elimination mechanism. In those cases where the existence of a metal hydride species is questionable, the π-allyl or proton abstraction mechanism may be involved [Eqs. (129) and (132)].[8,275,276]

As demonstrated by Cramer and Lindsey[277,278] [Eqs. (127) and (131)], 1,2-hydride addition or elimination is the most common mechanism observed for coordinated olefins. Either prior to, or subsequent to, hydride formation, an olefinic substrate is coordinated to a transition metal and followed by a reversible insertion of the olefin into the M—H bond [Eq. (133)], yielding a metal-alkyl

$$\rangle C{=}C\langle + L_mMH \rightleftarrows L_mM(H)(\eta^2\text{-}{>}C{=}C{<}) \rightleftarrows L_mM{-}\overset{|}{\underset{|}{C}}{-}\overset{|}{\underset{|}{C}}{-}H \qquad (133)$$

species. Evidence for this type of mechanism is obtained from deuterium scrambling in the isomerization of partially deuterated substrates, as well as the deuterium distribution found in use of M—D as the catalytic species and in the observation of metal-alkyl complex formation by transition metal hydride species. Thus, in the presence of a variety of transition metal hydrides, *trans*-ethylene-d_2 was found to undergo deuterium redistribution in benzene solution.[279] The π-allyl isomerization mechanism [Eqs. (129) and (132)] has been suggested to occur with specific catalyst systems and substrates and will be discussed in this context.

Of the transition metal carbonyls known to effect olefin isomerization, the carbonyls of iron and cobalt are of particular significance. Olefin isomerization is a side reaction in the hydroformylation of olefins in the "oxo" process[11] [Eq. (134)], particularly under conditions of low carbon monoxide pressure.[280] This

$$RCH_2CH{=}CH_2 + CO + H_2 \xrightarrow{Co(CO)_4H} RCH_2CH_2CH_2\overset{O}{\overset{\|}{C}}H + RCH_2\underset{CH_3}{\underset{|}{C}H}{-}\overset{O}{\overset{\|}{C}}H \ldots \qquad (134)$$

is indicated by the similar aldehyde product distribution formed from 1-pentene and 2-pentene under similar conditions.[281] Several mechanisms for isomerization reactions have been advanced for this system,[11] but the data reported can be rationalized on the basis of an olefin-alkyl mechanism as originally suggested by Heck and Breslow.[282] Evidence in support of this is found in the reaction of

$Co(CO)_4H$ with propylene-d_6 [Eq. (135)] yielding $CD_3CH{=}CD_2$ as one of the products.[283] Also, the reaction of $Co(CO)_4D$ with dimethyl malate gave deuterodimethyl fumarate.[284]

$$CD_3CD{=}CD_2 \xrightarrow{Co(CO)_4H} CD_3CH{=}CD_2 + \ldots \qquad (135)$$

The observed inverse relationship between CO partial pressure and the extent of isomerization[282,285] is rationalized by the need for a CO dissociation step in the isomerization sequence [Eq. (136)].

$$\underset{\substack{|\\ H-Co(CO)_3}}{RCH_2CH{=}CH_2} \underset{-CO}{\overset{+CO}{\rightleftharpoons}} \underset{\substack{|\\ Co(CO)_4}}{(CH_3)(CH_2R)CH} \underset{CO}{\overset{-CO}{\rightleftharpoons}} \underset{\substack{|\\ Co(CO)_3\\ |\\ H}}{CH_3CH{=}CH_2} \qquad (136)$$

Markovnikov addition is reported for the reaction of $Co(CO)_4H$ with terminal olefins in which a strong electron-withdrawing group is attached to one of the olefinic carbons. This is shown by the formation of the branched acyl complex from methyl acrylate[286] [Eq. (137)], and the π-allyl species from butadiene[11] [Eq. (138)].

$$CH_2{=}CHC({=}O){-}OCH_3 + Co(CO)_4H + CO \longrightarrow CH_3CH(C({=}O){-}Co(CO)_4){-}C({=}O){-}OCH_3 \qquad (137)$$

$$CH_2{=}CH{-}CH{=}CH_2 + Co(CO)_4H \longrightarrow \underset{\substack{|\\ Co(CO)_3}}{CH_3CH{-}CH{=}CH_2} \longrightarrow$$

$$\pi\text{-}(CH_3CH{\cdots}CH{\cdots}CH_2)Co(CO)_3 + \pi\text{-}(CH_2{\cdots}CH{\cdots}CH(CH_3))Co(CO)_3 \qquad (138)$$

The consequences of this 1,2-addition–elimination mechanism is further examined in section C.

Catalytic and stoichiometric olefin isomerizations by the iron carbonyls $Fe_3(CO)_{12}$, $Fe_2(CO)_9$, and $Fe(CO)_5$ have been studied extensively.[287] Terminal and internal mono-olefins are catalytically isomerized, and nonconjugated dienes are converted to conjugated species. Stoichiometric isomerization of conjugated dienes also occurs.

Two reaction mechanisms for the isomerization of mono-olefins have been proposed.[288,289] Evidence is found to support both a 1,2-addition–elimination mechanism [Eqs. (127) and (130)] and a π-allyl mechanism [Eqs. (129) and (132)]. Thus, in the presence of iron carbonyl, *cis*-stilbene is converted to *trans*-stilbene,[288] while for several olefins, isomerization proceeds to form products that are considered to be beyond an intermediate species, which is shown to be inactive from the standpoint of the 1,2-addition–elimination model [an example is shown in Eq. (139)]. These data suggest that isomerization of

$$\underset{\mathbf{50}}{CH_3-\overset{\overset{CH_3}{|}}{C}=CHCH_2CH_3} \xrightarrow{Fe_3(CO)_{12}} \text{no change}$$

$$\underset{\mathbf{49}}{CH_3CH=CH-\overset{\overset{CH_3}{|}}{C}HCH_3} \xrightarrow{Fe_3(CO)_{12}} CH_3CH=CHCH(CH_3)_2 + \underset{\mathbf{51}}{CH_3CH_2CH_2-\overset{\overset{CH_3}{|}}{C}=CH_2} + \underset{\mathbf{50}}{CH_3CH_2CH=\overset{\overset{CH_3}{|}}{C}-CH_3} \quad (139)$$

structure **49** does not occur via a 1,2-addition–elimination mechanism since the reaction would then proceed to structure **50** and stop, thereby not accounting for the presence of structure **51.** With the π-allyl mechanism, however, the transfer of hydrogen need not generate an olefin. An inverse dependence upon CO pressure in the rate of isomerization was observed, suggesting displacement of CO by the olefinic ligand.

Pettit[290] suggested a π-allyl mechanism for the catalytic conversion of allyl alcohol to propionaldehyde [Eq. (140)]. This has been substantiated by studies on isomerization of $CH_2=CHCD_2OH$ to CH_2DCH_2CDO by this catalyst.[291] Further, the presence of deuterium on the aldehyde carbon shows that the isomerization step is essentially irreversible. Evidence for this mechanism is also found in the catalytic isomerization of other α,β-unsaturated olefins[291,292] [Eqs. (141) and (142)].

$$CH_2{=}CHCH_2OH + Fe(CO)_5 \longrightarrow \underset{\displaystyle Fe(CO)_3H}{CH_2 \overset{CH}{\cdots} CH{-}OH} \longrightarrow$$

$$\underset{\displaystyle Fe(CO)_3}{CH_3CH{=}CHOH} \longrightarrow CH_3CH{=}CHOH \longrightarrow CH_3CH_2\overset{O}{\overset{\|}{C}}H \quad (140)$$

$$\mathbf{52} \xrightarrow{Fe(CO)_5} \text{no reaction} \quad (141)$$

$$\mathbf{53} \xrightarrow{Fe(CO)_5} \mathbf{54} \longrightarrow \mathbf{55} \quad (142)$$

Structure **52**, in which the migrating hydrogen is not sterically available to the $Fe(CO)_5$ catalyst, is not changed even under severe conditions, while structure **53** is catalytically transformed to structure **55** presumably via structure **54**.[292] Other evidence to support a π-allyl mechanism is found in the facile formation of such species from allylic halides[293] [Eq. (143)].

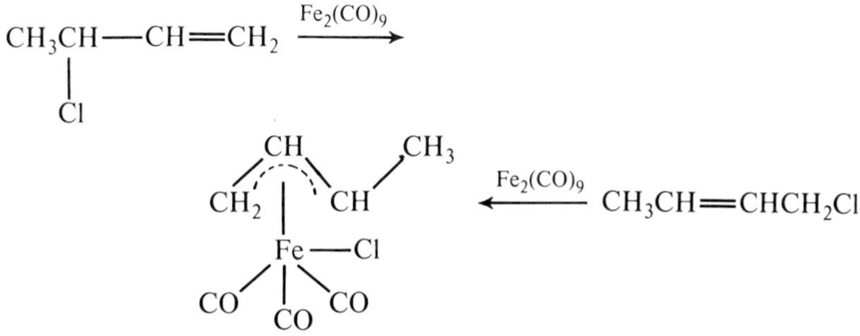

$$CH_3CH(Cl)—CH{=}CH_2 \xrightarrow{Fe_2(CO)_9} (\pi\text{-}CH_2CHCHCH_3)Fe(CO)_3Cl \xleftarrow{Fe_2(CO)_9} CH_3CH{=}CHCH_2Cl \quad (143)$$

A similar π-allyl mechanism is invoked to rationalize the catalytic isomerization of a nonconjugated diene to a conjugated species and the stoichiometric rearrangement of conjugated species. Pettit et al.[295] proposed a mechanism for the isomerization of a nonconjugated diene involving initial formation of one or two iron–olefin π bonds. The iron then deprotonates the diene at an allylic site to form an intermediate complex. Rearrangement gives the conjugated diene–iron carbonyl complex [Eq. (144)].

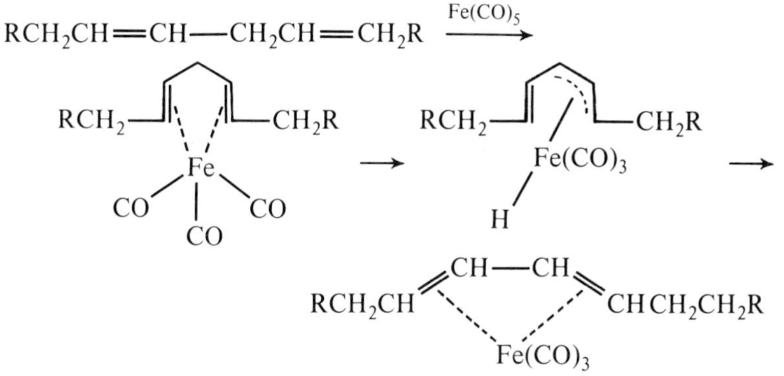

$$RCH_2CH{=}CH—CH_2CH{=}CH_2R \xrightarrow{Fe(CO)_5} (\text{diene})Fe(CO)_3 \rightarrow (\pi\text{-allyl})Fe(CO)_3H \rightarrow RCH_2CH{=}CH—CH{=}CHCH_2CH_2R\cdot Fe(CO)_3 \quad (144)$$

The original diolefin displaces the conjugated species and completes the catalytic cycle. This is observed in the catalytic conversion of 1,4- and 1,5-cyclooctadiene to 1,3-cyclooctadiene by $Fe(CO)_5$[295] and several diolefin substituted-$Fe(CO)_3$ species[294,296] [Eq. (145)].

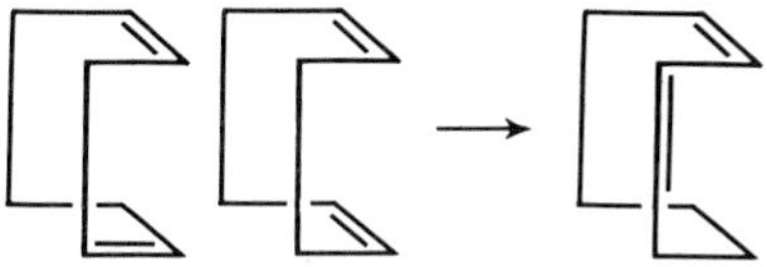

(145)

$$\text{C}_6\text{H}_8 \xrightarrow{Fe(CO)_5} \text{C}_6\text{H}_8 \cdot Fe(CO)_3 \tag{146}$$

A variety of nonconjugated diolefins which are isomerized to stable $Fe(CO)_3$ adducts,[294,297–300] however, [Eq. (146)] have been shown by deuterium labelling to occur by a 1,3-hydrogen shift.[301] These reactions indicate that the cissoid conjugated arrangement, of double bonds, which can more readily coordinate to the $Fe(CO)_3$ moiety, is achieved in the isomerization process. This is a general feature of diolefin rearrangement reactions of iron carbonyls. The same mechanism is also invoked to rationalize the isomerism of conjugated dienes to less sterically strained isomers.[294,297,298,299] Recently, however, it has been suggested that the isomerization of conjugated dienes by $Fe(CO)_5$ is controlled by kinetic rather than thermodynamic factors.[302]

The generally accepted mechanism for olefin isomerizations catalyzed by group VIII transition metal complexes involves reversible 1,2-hydrogen addition–elimination [Eqs. (127) and (131)] with the exception of palladium (II) catalysis, where there is evidence for a possible reversible π-allyl complex formation mechanism [Eqs. (129) and (132)].

Cramer and Lindsey[277] have suggested the mechanism shown in Scheme 2-2 to account for the products obtained in the isomerization of 1-butene by complexes of rhodium, palladium, platinum, nickel, and iron in CH_3OD (only those steps which result in the observed products are given).

The key features in this catalytic scheme necessary to account for the observed products (deuterated 1-butenes and 2-butenes) are:

(1) deuteration of a preformed M^{a+}–olefin complex;
(2) reversible insertion of the coordinated olefin into the M^{a+2}—D bond leading predominantly to a M^{a+2}–n-alkyl species;
(3) rapid olefin exchange with the deuterated product;
(4) occasional reversible formation of an isoalkyl complex;
(5) elimination of a proton from the M^{a+2} (rearranged olefin)–π-complex;
(6) subsequent olefin exchange.

The variation in rate for these individual steps was then used to rationalize the isomerization behavior of the catalytic species examined.

The distribution of deuterium in homogeneous $RhCl_3 \cdot 3H_2O$ and bis(benzonitrile) $PdCl_2$-catalyzed isomerization of vinyl-and allyl-deuterated 1-olefins supported reversible alkyl complex formation for Rh catalysis.[303] In the case of Pd(II), however, the low rate of deuterium exchange with other olefins and the deuterium distribution in the product, isomerized olefins, was interpreted as being in accord with a π-allyl mechanism. Catalytic hydrogen exchange with D_2, C_2D_4, and 1-butene by a variety of transition metal hydrides has been in-

Scheme 1-2

$$MI_n^{a+} = M^a$$

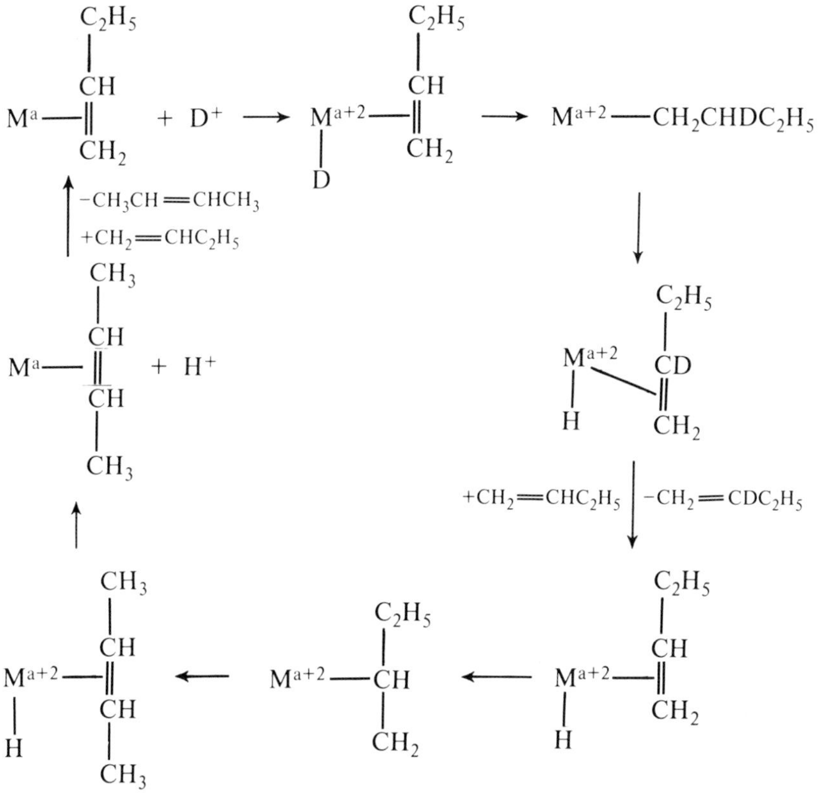

$$M^a\text{—}\overset{\displaystyle C_2H_5 \atop \displaystyle |\atop \displaystyle CH}{\|}\underset{\displaystyle CH_2}{} + D^+ \longrightarrow M^{a+2}(D)\text{—}(CH_2{=}CHC_2H_5) \longrightarrow M^{a+2}\text{—}CH_2CHDC_2H_5$$

$$\downarrow$$

$$M^{a+2}(H)\text{—}(CH_2{=}CDC_2H_5)$$

$$\downarrow\ {+CH_2{=}CHC_2H_5}\ \ {-CH_2{=}CDC_2H_5}$$

$$M^{a+2}(H)\text{—}(CH_2{=}CHC_2H_5) \longrightarrow M^{a+2}\text{—}CH(CH_2)C_2H_5 \longrightarrow M^{a+2}(H)\text{—}(CH_3CH{=}CHCH_3)$$

$$\uparrow$$

$$M^a\text{—}(CH_3CH{=}CHCH_3) + H^+$$

$$\uparrow\ {-CH_3CH{=}CHCH_3}\ \ {+CH_2{=}CHC_2H_5}$$

$$M^a\text{—}(CH_2{=}CHC_2H_5)$$

terpreted in terms of a reverse alkyl mechanism.[304] Also, catalytic deuterium redistribution in ethylene-d_2[279] mentioned earlier, supports this mechanism for several transition metal hydride complexes and $Ru[P(C_6H_5)_3]_3Cl_2$. However, $Pd(C_6H_5CN)_2Cl_2$ is found to catalyze cis-trans rearrangement but not intermolecular deuterium exchange. The mechanism of olefin isomerization by a $Pd(II)/SnCl_2/H_2$ catalytic system is also found consistent with a reversible insertion of olefin into a Pd—H bond in the isomerization of 1-butene.[305]

B. Catalytic Hydrogenation of Olefins

The homogeneous catalytic hydrogenation of olefinic substrates by transition metal ions and their complexes has been the subject of numerous investigations since the pioneering work of Calvin[306] dealing with Cu(II)-catalyzed reduction of benzoquinone to quinol [Eq. (147)]. General reviews of the literature[8,21,276,307,309] and more detailed topical reviews have recently appeared. We examine here a few well-known homogeneous catalytic systems and

$$\text{O=C}_6\text{H}_4\text{=O} + H_2 \xrightarrow{Cu^{II}} \text{HO-C}_6\text{H}_4\text{-OH} \quad (147)$$

consider the mechanisms of hydrogen and substrate activation, as well as the mechanism of the hydrogenation process itself. Attention is given to the consequences of these mechanistic details on the substrate and product selectivities of homogeneous hydrogenation catalyst.

The catalytic process can be considered to occur in a stepwise manner, involving activation of molecular hydrogen, activation of a reducible substrate, and finally, transfer of hydrogen with liberation of the reduced substrate and regeneration of the catalyst. Three different patterns can be envisioned for this process: hydrogen activation followed by hydrogenation of the substrate [Eq. (148)]; substrate activation followed by rapid hydrogen activation and subsequent hydrogenation [Eq. (149)]; and sequential activation of both hydrogen and substrate to form a discrete intermediate(s) followed by hydrogen transfer [Eq. (150)].

$$L_xM \xrightarrow{H_2} L_xMH_y \xrightarrow{S} L_xM + SH_2 \quad (148)$$

$$L_xM \xrightarrow{S} L_xMS \xrightarrow{H_2} L_xM + SH_2 \quad (149)$$

$$L_xM \begin{array}{l} \xrightarrow{H_2} L_xMH_y \xrightarrow{S} \\ \xrightarrow{S} L_xMS \xrightarrow{H_2} \end{array} L_xMH_yS \longrightarrow L_xM + SH_2 \quad (150)$$

Since each sequence outlined requires addition of at least one ligand during some step in the catalytic process, the catalyst must be either coordinately unsaturated[20,23,27,29,310] (or readily oxidizible to a species which is) or substitution labile.

With respect to the first step, hydrogen activation, three mechanisms have been recognized[310] and examples are given below:

$$Ru^{III}Cl_6^{3-} + H_2 \rightleftarrows Ru^{III}Cl_5H^{3-} + H^+ \quad (151)$$

$$2Co^{II}(CN)_5^{3-} + H_2 \rightleftarrows 2Co^{III}(CN)_5H^{3-} \quad (152)$$

$$Ir[P(C_6H_5)_3]_2(CO)Cl + H_2 \rightleftarrows Ir[P(C_6H_5)_3]_2(CO)Cl(H)_2 \quad (153)$$

The heterolytic cleavage of the H_2 molecule (or hydride abstraction) [Eq. (151)] is essentially a ligand substitution reaction in which the metal is not oxidized. It is dependent on the ease of ligand substitution, the stability of the resultant M—H bond, and the presence of a base in the system to retard the reverse reaction. The reaction described in Eq. (152) is an example of homolytic cleavage of the hydrogen molecule (or radical abstraction) by two catalyst molecules. During this process, the metal atom in the catalyst undergoes a one-electron oxidation. Eq. (153) is example of oxidative addition, in which the catalyst is simultaneously oxidized by two equivalents and gains two hydride ligands.

Activation of the substrate involves formation of a π-complex with the catalyst, and this may occur before or after reaction of the catalyst with hydrogen. This step, when observed, requires the presence of coordination unsaturation in the catalyst species and thus should be slowed down or completely inhibited by the presence of good coordinating ligands in the reaction solution. Absence of an inhibitory effect on the catalytic process in the presence of such ligands is suggestive of a reaction sequence in which the substrate is not activated prior to hydrogen transfer.

The transfer of two hydrogen atoms to an olefin molecule can be either stepwise of simultaneous. If stepwise, the addition of hydrogen to a coordinated olefin or addition of olefin to a metal hydride species generates an alkyl group bound to the catalyst. If this alkyl formation is reversible (*see* preceding section), then the olefinic substrate can be isomerized[275] [Eq. (154)]. Occurrence of

$$\text{M(olefin)} \underset{-\text{H}}{\overset{+\text{H}}{\rightleftharpoons}} \text{M-alkyl} \underset{+\text{H}}{\overset{-\text{H}}{\rightleftharpoons}} \text{M(olefin}') \qquad (154)$$

isomerization during hydrogenation is taken as evidence for a stepwise addition of hydrogen. In catalytic deuterogeneration, deuteration of residual olefin or extensive H-D scrambling in the product is explained by this mechanism. Other processes are known to allow deuterium scrambling, however, including oxidative addition of vinyl groups[311] and exchange of metal hydrides with solvent deuterium in deuterated hydroxylic solvents.[277,278,312]

1. Homolytic Cleavage in Hydrogen Activation. The classic example of homolytic cleavage of hydrogen by homogeneous transition metal catalysts is found in the chemistry of pentacyanocobaltate(II), a subject that has been reviewed extensively.[19,29,276,308,313,314] This system is selective for the hydrogenation of conjugated dienes to mono-olefins and the reduction of α,β-olefins conjugated to activating groups. Isolated olefins and acetylenes are not reduced.

When $HCo(CN)_5^{3-}$, the active catalytic species,[19,313,314] reacts with activated olefins ($CH_2{=}CRX$ and $XCR{=}CRX$, where R is H, an alkyl, or a substituted alkyl group, ànd X is allyl, acyl, CN, COR, COOR, or COO^-), a variety of products are formed, depending on R and X. The olefin may be hydrogenated

$$CH_2{=}CRX + Co(CN)_5H^{3-} \rightleftarrows CH_3CRX\text{-}Co(CN)_5{}^{3-} \qquad (155)$$

or isomerized, or a stable organocobalt(III) complex may be formed[315] [Eq. (155)]. It was also shown that formation of alkyl complexes occurs to give stereospecific cis addition but that the reduction and isomerization reactions are not stereospecific. For example, the deuterium content of isomerized acids and the saturated acid products indicate that the initial hydrogen transfer step is reversible.

$$-d[\text{olefin}]/dt = k[Co(CN)_5H^{3-}][\text{olefin}] \qquad (156)$$

The rate of the reaction given by Eq. (155) for a variety of olefins obeys the rate law [Eq. (156)]. The absence of an inverse term in the cyanide ion concentration suggests that this catalyst does not have an available coordination site to activate the olefin and hence that only hydrogen is activated.[315,316] The value of k decreases for the following olefins in this order.

butadiene > styrene > 2-vinylpyridine > acrylonitrile > acrylic acid

Substituting an alkyl group on the α-carbon for an α-hydrogen atom increases k, thus indicating an electrophilic attack on the olefin.[316]

A mechanism for the first step in hydrogenation by $Co(CN)_5H^{3-}$ [involving initial hydrogen transfer to the activated substrate, Eq. (157)] now seems generally accepted. The intermediate, structure **56,** can then either rapidly collapse to give a stereospecific cis addition of Co—H to the olefin, an organic radical plus $Co(CN)_5{}^{3-}$, or by dissociation, return to the original reactants.

Reduction involving an organic radical intermediate is reported for cinnamic acid,[19,313,314,317] *trans*-($C_6H_5CH{=}CHCOOH$), styrene,[318] and sorbic acid,[318,319] along with several other anions of unsaturated acids.[315] These reactions are thought to occur in accord with the pathways of Eq. (157), generating

$$Co(CN)_5H^{3-} + CH_2{=}CRX \rightleftharpoons \left[\begin{array}{l} CH_2{\cdots}\dot{C}RX \\ \vdots \\ H{\cdots}\dot{C}o \\ \quad (CN)_5 \end{array}\right]^{3-}$$

$$\underset{\mathbf{56}}{\left[\begin{array}{l} CH_2{\cdots}\dot{C}RX \\ \vdots \\ H{\cdots}\dot{C}o \\ \quad (CN)_5 \end{array}\right]^{3-}} \begin{array}{l} \rightleftharpoons \dot{C}o(CN)_5{}^{3-} + \underset{\mathbf{57}}{\overset{\beta}{C}H_3\overset{\alpha}{\dot{C}}RX} \\ \rightleftharpoons CH_3{-}CRX^{3-} + Co(CN)_5 \end{array} \qquad (157)$$

a stable intermediate radical species, structure **57**, by elimination of $Co(CN)_5^{3-}$. This radical intermediate can then abstract a hydrogen atom from a second $Co(CN)_5H^{3-}$ to produce a saturated product, or it can lose hydrogen via the reverse of [Eq. (157)].

A special case of this reaction sequence is seen in the hydrogenation of butadiene. Kwiatek[19,314,315] has shown that butadiene is converted to a mixture of butene isomers by way of the butenyl intermediate [Eqs. (158) and (159)].

$$Co(CN)_5H^{3-} + C_4H_8 \rightleftharpoons C_4H_7Co(CN)_5^{3-} \tag{158}$$

$$C_4H_7Co(CN)_5^{3-} + HCo(CN)_5^{3-} \rightleftharpoons C_4H_8 + C_4H_8 + 2Co(CN)_5^{3-} \tag{159}$$

A more detailed kinetic study has established this reaction sequence and suggests that two σ-butenyl complexes and a π-complex are in equilibrium[320] (Scheme 2-3).

Scheme 2-3

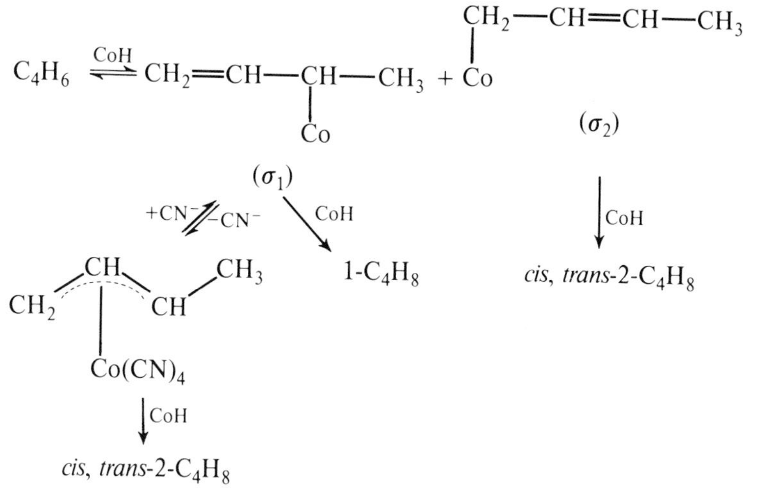

$$C_4H_6 \xrightleftharpoons{CoH} CH_2{=}CH{-}\underset{\displaystyle Co}{\underset{|}{CH}}{-}CH_3 \ (\sigma_1) + \underset{\displaystyle Co}{\underset{|}{CH_2}}{-}CH{=}CH{-}CH_3 \ (\sigma_2)$$

(σ_1) $\xrightarrow{CoH}$ $1\text{-}C_4H_8$

(σ_2) $\xrightarrow{CoH}$ *cis, trans*-2-C_4H_8

(σ_1) $\xrightleftharpoons[-CN^-]{+CN^-}$ π-allyl complex $CH_2\cdots CH\cdots CH{-}CH_3$ bound to $Co(CN)_4$ $\xrightarrow{CoH}$ *cis, trans*-2-C_4H_8

The initial butenyl complex formation is consistent with the rate law, given in Eq. (156), and is independent of cyanide ion concentration. This suggests initial formation of a σ-complex (probably the branched isomer σ_1). The composition of the resulting butenes has been explained on the basis of an equilibrium between σ- and π-allyl species at low 1cn0;1co] ratios and formation of the two σ species (σ_1, σ_2) at high [CN]:[Co] ratios.

For the addition of the second hydrogen atom, these authors suggest a transition state in which the incipient carbon–hydrogen bond bridges the two cobalt atoms. This yields the product shown in Scheme 2–2 directly. Kwiatek,[19,314,315] however, suggests a different intermediate in the second hydrogen addition step for butadiene (Scheme 2–4). This model assumes an attack on the allylic system

Scheme 2-4

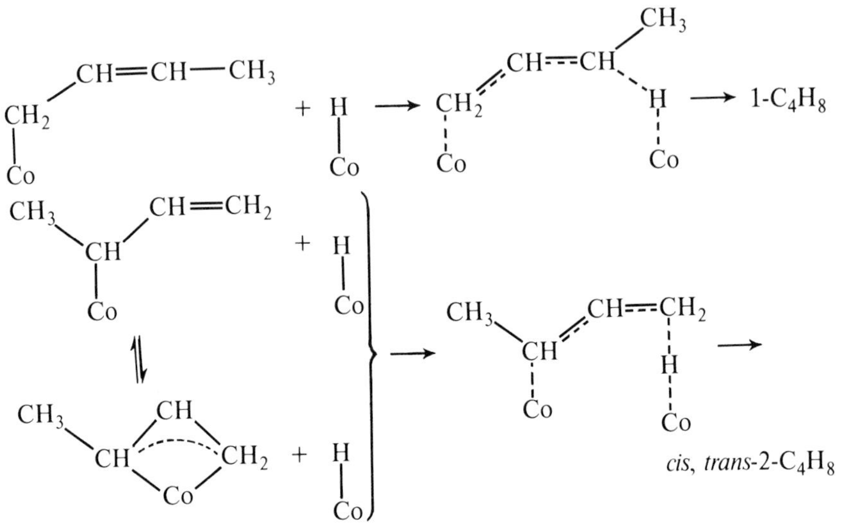

by a second $Co(CN)_5H^{3-}$ molecule. The mixture of butenes formed for several [CN]:[Co] ratios is given in Table 2-3.[307] This catalytic system can be modified by replacing some of the cyanide ligands with various chelating species. For example, dipyridyl, 1,10-phenanthroline, and amino acids have been used in the presence of cobaltous cyanide solutions (CN:Co < 5) to produce catalytically active hydride species.[321-324] It is found that the replacement of cyanide by amine or heterocyclic aromatic ring nitrogens weakens the Co—H bond, producing more reactive hydrogenation catalysts for activated olefins than $Co(CN)_5H^{3-}$.[321,322] The most active catalyst appears to be $Co(dipy)(CN)_3H^-$. The kinetics of hydrogenation and hydrogen exchange indicate that, for this system, these processes involve homolytic cleavage of molecular hydrogen and reversible

TABLE 2-3
DEPENDENCE OF PRODUCT DISTRIBUTION ON THE CN/Co RATIO IN THE HYDROGENATION OF BUTADIENE

CN/Co	%	%	%
4.5	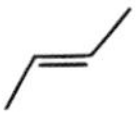 86	 1	 13
5.5	70	1	29
6.0	12	3	85
8.5	19	1	80

alkyl formation.[322] Hydrogenolysis then takes place by attack of a second molecule of the hydrido cobalt species. Further evidence for the presence of an intermediate Co–alkyl complex is seen in the hydrogenation of optically active α,β-unsaturated acids with chiral-ligand-substituted cobaltous cyanide systems to produce optically active products.[323,324]

Homolytic cleavage of hydrogen is also exhibited by dicobalt octacarbonyl [Eq. (160)].[325] The reactions of this catalytically active hydride species are of particular interest in the hydroformylation (oxo) reaction [Eq. (161)].[11,326]

$$Co_2(CO)_8 + H_2 \rightleftharpoons 2Co(CO)_4H \qquad (160)$$

$$RCH{=}CH_2 \xrightarrow{H_2,\ CO} RCH_2CH_2\overset{\overset{\displaystyle O}{\|}}{C}H \qquad (161)$$

Olefin hydrogenation is an observed side reaction which becomes more important in the presence of olefins which are branched, conjugated, or carry electronegative substituents. Very recently,[327] it has been shown that the polycyclic aromatic hydrocarbons (PAH) almost certainly are hydrogenated by a free radical mechanism [Eqs. (160), (162), (163), and (164)]. This mechanism explains observations that are anomalous when the mechanism is presumed to involve Co—C bond formation.

$$\mathbf{58} + HCo(CO)_4 \underset{k_{-2}}{\overset{k_2}{\rightleftharpoons}} \mathbf{59} + {\cdot}Co(CO)_4 \qquad (162)$$

(58: 9-R,10-R′-anthracene; 59: 9-H,9-R-9,10-dihydro-10-anthryl radical bearing R′)

$$\mathbf{59} + HCo(CO)_4 \xrightarrow{fast} \text{cis-9,10-dihydro (H, R)} + \text{trans-9,10-dihydro (H, R)} + {\cdot}Co(CO)_4 \qquad (163)$$

$$2{\cdot}Co(CO)_4 \xrightarrow{fast} Co_2(CO)_8 \qquad (164)$$

2. Hydrogen Activation by Oxidative Addition. Perhaps the most interesting and important class of homogeneous hydrogenation catalysts are Osborn and Wilkinson's phosphine-stabilized rhodium and ruthenium complexes. In addition to being some of the most active catalytic species, they also display remarkable selectivity in their hydrogenation reactions. Chlorotris(triphenylphosphine)-rhodium(I), $Rh(I)[P(C_6H_5)_3]_3Cl$, whose structure has been established by X-ray crystallography to be distorted from a planar array of ligands toward a tetrahedral arrangement,[342] was observed to be an excellent catalyst for the hydrogenation of olefins at ambient temperature and pressure.[328,329,330]

Osborn et al.[330,331,332] have proposed the mechanism for hydrogenation shown in Scheme 2-5 (P is phosphine). Note that this catalyst is thus suggested to activate both molecular hydrogen and the olefinic substrate. The details of this activation are in accord with the known reactivity patterns of rhodium, which involve the oxidative addition of molecular hydrogen to Rh(I) and produce either a Rh(III) olefin–dihydride complex directly, or a reactive Rh(III) dihydride which then coordinates an olefin. The validity of this mechanism is established by the numerous reports dealing with catalysis by this system.

Scheme 2-5

$$\begin{array}{ccccc} RhP_3Cl & \rightleftharpoons & RhP_2Cl & \overset{H_2}{\rightleftharpoons} & RhP_2ClH_2 \\ & {\scriptstyle -olefin} \swarrow\!\!\nearrow {\scriptstyle olefin} & & & \downarrow {\scriptstyle k'\ olefin\ (fast)} \\ & RhP_2Cl(olefin) & & \xrightarrow{H_2} & RhP_2Cl + paraffin \end{array}$$

These systems have created much interest and the resultant literature has revealed a number of complexities associated with it. James[309] has reviewed the literature through 1972, and the complexities are demonstrated and clarified substantially in recent work.[333] The complex $RhCl(PR_3)_3$ remains largely intact[334,335,336] in relatively concentrated solutions, and the equilibrium constant for the equilibrium of Eq. (165) is small (1.4×10^{-4} *M* for R = C_6H_5).[337]

$$RhP_3Cl \rightleftharpoons RhP_2Cl + P \qquad (165)$$

A dimerization equilibrium has been identified and studied.[330,333] The dimer $Rh_2P_4Cl_2$ (as well as the monomer) reacts with H_2 and acts as a hydrogenation catalyst.[333]

Evidence for the existence of reversible dihydride–rhodium complex formation under the reaction conditions encountered in the kinetic studies comes from proton magnetic resonance studies of $Rh[P(C_6H_5)_3]_3Cl$ solutions in the presence of molecular hydrogen.[330,333,338] Thus, on exposure to hydrogen, high field resonances corresponding to two hydride ligands are observed in a variety of solvents. Further, on flushing these solutions with nitrogen gas, the original spectrum is obtained. The structure of this dihydride species in dichloromethane has been deduced from its phosphorus and proton magnetic resonance spectra to be **60**.[333] The nuclear magnetic resonance spectrum for this species further demonstrated that at room temperature the only ligand exchange occurring is that of the phosphorus trans to the hydride; at lower temperatures no exchange was observed. Complex **60**[21,338] and a number of similar Rh(III) hydride derivatives have been isolated.[21,330,332]

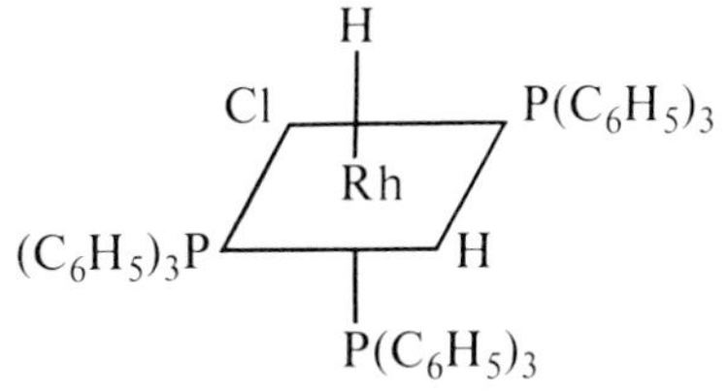

H
Cl — Rh — $P(C_6H_5)_3$
$(C_6H_5)_3P$ — H
$P(C_6H_5)_3$

60

Evidence has been reported on the reaction of ethylene and other simple mono-olefins with $Rh[P(C_6H_5)_3]_3Cl$ under the reaction conditions used in hydrogenation which leads to the equilibrium [Eq. (166)]. For mono-olefins larger than ethylene, however, the equilibrium constant is small, and no spectral properties corresponding to olefin complex formation have been reported.[330,333] Kinetic studies have been reported on these systems.[333,339,340]

$$\text{olefin} + Rh(PR_3)_3Cl \rightleftarrows Rh(PR_3)_2(\text{olefin})Cl + PR_3 \qquad (166)$$

The hydrogen transfer step in catalytic olefin hydrogenations by these catalysts was initially thought to be a simultaneous cis addition of both hydrogens[330] on the basis of the stereochemistry found in the reaction products. Thus, internal alkynes gave predominately *cis*-olefins, and deuterium addition to maleic acid and fumaric acids gave *meso*-1,2-dideuterosuccinic and DL-*syn*-1,2-dideuterosuccinic acids, respectively.[330]

Evidence now indicates the presence of a reversible rhodium–alkyl bond formation in the reaction of $Rh[P(C_6H_5)_3]_3ClH_2$ with olefins (Figures 2–8). Hex-1-ene gave $C_6H_{13}D$ (27%) upon hydrogenation with mixed hydrogen–deuterium[330] and partial reduction of cyclohexene with hydrogen. Tritium gave tritiated cyclohexene as well as cyclohexane.[341] Study of the deuteriogenation of substituted cyclohexenes also gave as products cyclohexanes-d_0-d_4 and some cyclohexene-d_1.[342] Isomerization of substrates during hydrogenation also offers evidence for reverse alkyl formation.[343,344] Such a process presumably would involve an intramolecular insertion.

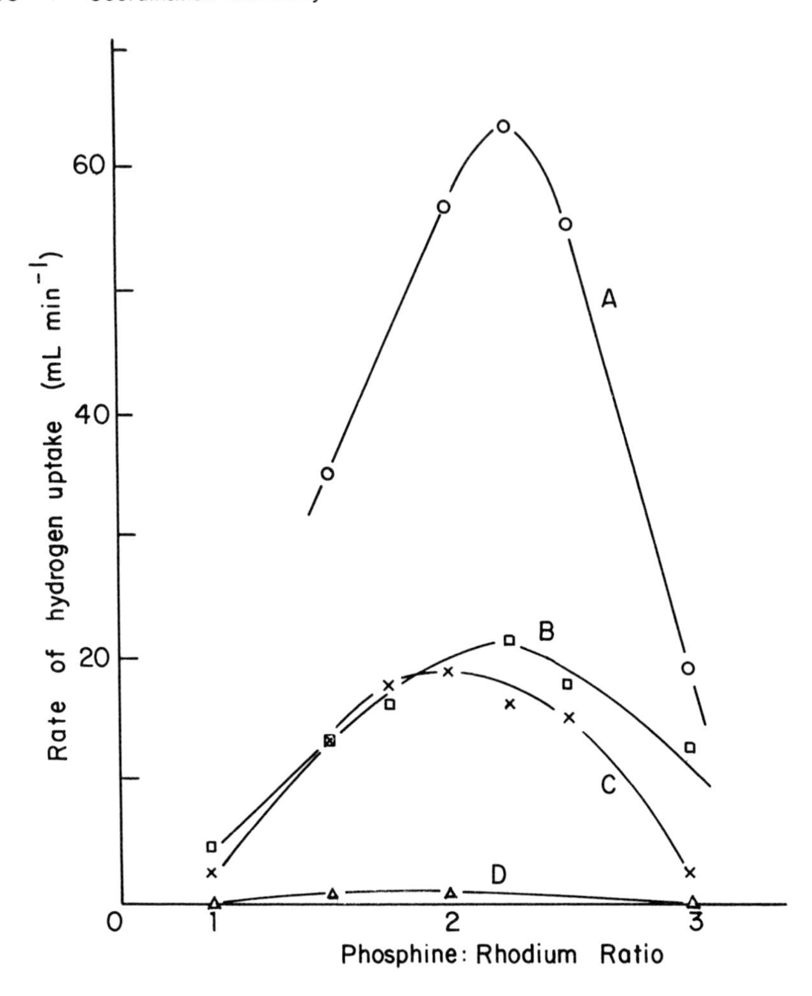

Figure 2–8. Hydrogenation of cyclohexene (0.6*M* benzene) at 25°C at catalyst concentration 1.25m*M* with various phosphine:rhodium ratios; A, $(MeOC_6H_4)_3P$; B, $(C_6H_5)_3P$; C, $(C_6H_5)_2EtP$; D, PEt_3

Some substances are known to poison or inhibit this catalyst system. Oxidation of the catalyst to Rh(III) in a form which is not readily reduced accounts for the effect of chlorinated hydrocarbons[21,345] and acyl halides.[346] Oxygen or peroxides are accelerators at low concentrations but inhibit the reaction at higher concentrations.[330] It has been suggested that at low oxygen concentrations, phosphine is converted to phosphine oxide, thus facilitating the dissociation of the RhL_3Cl complexes, but other studies indicate that this acceleration in catalytic rate may be a result of formation of a more active species.[347] Evidence for this is found in the catalytic reduction of benzene to cyclohexane and the isomerization of substrates observed for this system after exposure to oxygen[348] in contrast to the behavior of the oxygen free system. Competition with substrate for a site on the metal ion is observed. Pyridine and tetrafluoroethylene form $Rh[P(C_6H_5)_3]_2LCl$ complexes,[330] while chelating olefins form $RhP(C_6H_5)_3$-(diene)Cl, and stable hydrides are formed in good donor solvents. Carbon monoxide and organic compounds from which it can be abstracted are also strong poisons since the complex $Rh[P(C_6H_5)_3]_2(CO)Cl$ is formed: it is not catalytically active under the reaction conditions.[330]

Catalysis by $Rh[P(C_6H_5)_3]_3Cl$ only hydrogenates olefinic and acetylenic groups. The order of activity[21] is:

terminal olefin $\gg$ disubstituted olefin $>$ trisubstituted olefin
$\gg$ tetrasubstituted olefin

Ketones and aldehydes are not hydrogenated. The carbon–carbon double bond of α,β-unsaturated acids, esters, aldehydes, ketones, nitriles, and nitro compounds can also be hydrogenated with $Rh[P(C_6H_5)_3]_3Cl$.[350]

$Rh[P(C_6H_5)_3]_3Cl$ is also a good catalyst for the synthesis of stereospecifically labelled deutero compounds. The addition of two hydrogen (deuterium) atoms occurs quickly enough to minimize isomerization.[342,351,352] On the other hand, olefins which can be converted to tetra-substituted olefins by one double bond shift do not give high yields of dideuterated compounds, and it appears that the stereospecificity of dideuteriogenation decreases with increasing substitution on the double bond.[21]

Use of optically active phosphine ligands in RhL_3Cl catalytic hydrogenations has been reported to lead to asymmetric reductions with small optical yield.[353,354] Recently, however, use of chiral phosphine with the optical activity on the carbon in such systems gave asymmetric reduction products including a 61% optical yield in the reduction of β-methylcinnamic acid to 3-phenylbutanoic acid.[355] RhL_3Cl catalysts have been incorporated into polymeric matrices, thus producing catalytically active materials with the properties of both homogeneous and heterogeneous catalysts [Eq. (167)].[356]

P ≡ polymeric matrix

$$\text{P}\!-\!C_6H_4\!-\!CH_2P(C_6H_5)_2 \xrightarrow{RhL_nCl} \text{P}\!-\!C_6H_4\!-\!CH_2\!-\!P(C_6H_5)_2(\text{RhL}_n\text{Cl}) \quad (167)$$

Another well-characterized rhodium–phosphine hydrogenation catalyst with properties similar to $Rh[P(C_6H_5)_3]_3Cl$ is $Rh[P(C_6H_5)_3]_3(CO)H$.[357,358,359] The mechanism suggested for this catalyst is similar to the olefin activation–hydride formation pathway of Scheme 2-5 involving ligand dissociation, sequential activation of both substrate and hydrogen (by oxidative addition), and hydrogen transfer (Scheme 2-6).[360] The rate law for this mechanism is then given by Eq.

Scheme 2-6

$P \equiv P(C_6H_5)_3$

$$RhP_3(CO)H \underset{+P}{\overset{-P}{\rightleftharpoons}} RhP_2(CO)H \underset{-\text{olefin}}{\overset{K_1 \atop \text{olefin}}{\rightleftharpoons}} RhP_2(\text{olefin})(CO)H$$

$$-\text{paraffin} \uparrow \qquad\qquad \updownarrow$$

$$RhP_2(CO)H_2(\text{alkyl}) \underset{-H_2}{\overset{k_1 \atop H_2}{\rightleftharpoons}} RhP_2(CO)(\text{alkyl})$$

(168). Assuming that the other reactions are fast compared with the rate-determining activation of hydrogen, then Eq. (168) rearranges to Eq. (169). In support of this kinetic analysis, plots of $1/R$ against $1/[H_2]$ and $1/R$ against $1/[\text{olefin}]$ are straight, with the former passing through the origin. Also, a deuterium isotope effect ($k'_H/k'_D = 1.47$) was observed, which is of the order expected for the oxidative addition of hydrogen as the rate determining step.

$$R = \frac{-d[\text{olefin}]}{dt} = \frac{k'K_1[H_2][\text{olefin}][Rh]}{1 + K_1[\text{olefin}]} \tag{168}$$

$$\frac{1}{R} = \frac{1}{[\text{olefin}]} \cdot \frac{1}{k'K_1[H_2][Rh]} + \frac{1}{k'[H_2][Rh]} \tag{169}$$

Again, for $Rh[P(C_6H_5)_3]_3(CO)H$, the activity of the catalyst increased while the selectivity decreased as the catalyst concentration decreased, indicating further ligand dissociation from the catalytic species.

$Rh[P(C_6H_5)_3]_3(CO)H$ is as active as $Rh[P(C_6H_5)_3]_3Cl$ for the hydrogenation of terminal olefins but is much less active for internal olefins, conjugated dienes, and acetylenes. This, combined with evidence for the reversible formation of a rhodium–alkyl species from the catalysis of hydrogen–deuterium exchange in olefins,[361] indicates a strong preference for the formation of Rh-(*n*-alkyl) bonds. The half-life of hydrogen–deuterium exchange, measured by the formation of $Rh[P(C_6H_5)_3]_3(CO)H$ form $Rh[P(C_6H_5)_3]_3(CO)D$, is 20 sec for pent-1-ene and 1 hr for pent-2-ene; the half-life of isomerization is greater than 1 hr. *n*-Alkyl rhodium species are thus formed much more readily than secondary alkyl ligands.

An explanation for this has been advanced which suggests that the coordinatively unsaturated $Rh[P(C_6H_5)_3]_2(CO)H$ is planar with *trans*-phosphine groups, thus maximizing the steric requirements for coordination of an ole-

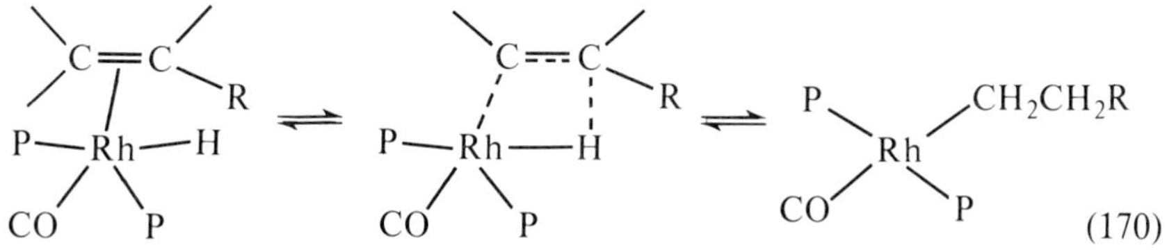

(170)

fin.[360,361] This, in turn, tends to produce a preponderance of *n*-alkyl rhodium complexes [Eq. (170)]. Recent studies concerning the isomerization of pentenes and hexenes support this suggestion and indicate that exchange reactions are much faster than hydrogenation for 1-pentene but occur at comparable rates for *cis*-2-pentene (*trans*-2-pentene is inert).[362] The nuclear magnetic resonance spectrum of this catalyst shows that it is not highly dissociated, although rapid ligand exchange occurs at room temperature.[358,363]

The compounds Rh(diene)L_n^+, where diene is nonbornadiene (NBD) or 1,5-cyclooctadiene (COD), L is a phosphine or arsine, and $n = 2, 3$, react with hydrogen in appropriate solvents (acetone, 2-methoxyethanol, THF) to form cationic catalytic species[364,365] that are characterized as $RhH_2L_2S_2^+$ and $RhHL_xS_y$ (S is solvent).[366] Both species are catalytically active, but the former causes hydrogenation with a minimum of olefin isomerization while that competing reaction is highly competitive in the case of the monohydride. The overall scheme for the process is similar to Schemes 2-5 and 2-6, involving both olefin coordination and hydrogen activation via oxidative addition. The limitation on olefin isomerization in the case of $RhH_2L_2S_2^+$ presumably is associated with very rapid reductive elimination of paraffin from the intermediate R-$RhHL_2S_2^+$. The alkyl intermediate derived from $RhHL_xS_y$ is expected to be longer-lived

Scheme 2-7

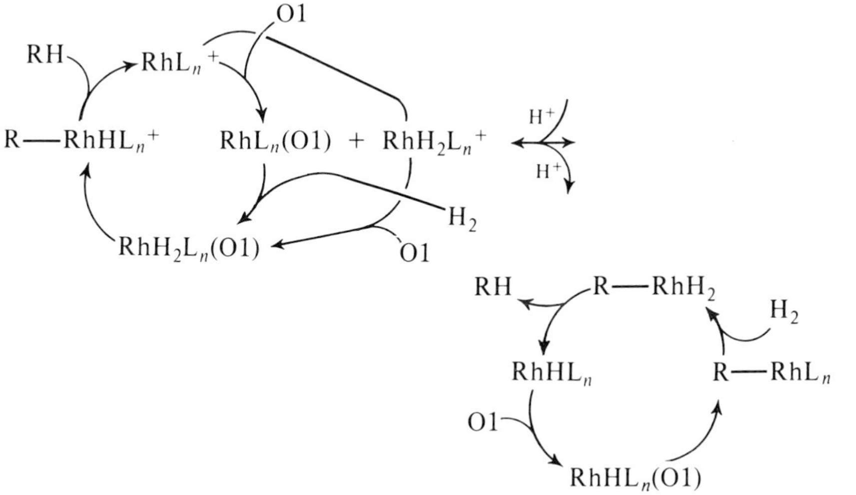

since it cannot eliminate paraffin prior to further oxidative addition of H_2. These relationships are shown in Scheme 2-7.[366] The relative abundances of the two catalytic species can be controlled by varying the H^+ activity—i.e., in acid solution, hydrogenation can be carried out without olefin isomerization. Some of these catalyst systems will reduce alkynes specifically to *cis* olefins,[367,368] chelating dienes specifically to monoenes[368] and ketones to alcohols.[368] A dramatic example of asymmetric olefin hydrogenation has been reported which involves the synthesis of optically active amino acids with up to 95% enantiomeric excess on a commercial scale.[369]

Iridium–phosphine complexes are of particular interest in that, although they are weak hydrogenation catalysts, their chemical reactivity patterns have provided much of the information known concerning the mechanisms of hydrogenation of group VIII phosphine complexes.

$IrClCO[P(C_6H_5)_3]_2$,[26] and its analogs catalyze the isomerization and hydrogenation of but-1-ene in isomeric butenes and butane and the deuterogenation of ethylene leading to deuterated ethylene and ethane. The results are consistent with a reversible alkyl formation in the catalytic process.[370] Hydrogenation of maleic and fumaric acids in dimethylacetamide has been reported.[371] This catalytic hydrogenation is best regarded as occurring by the mechanism in Scheme 2-8. Spectroscopic evidence indicates that the equilibria

Scheme 2-8

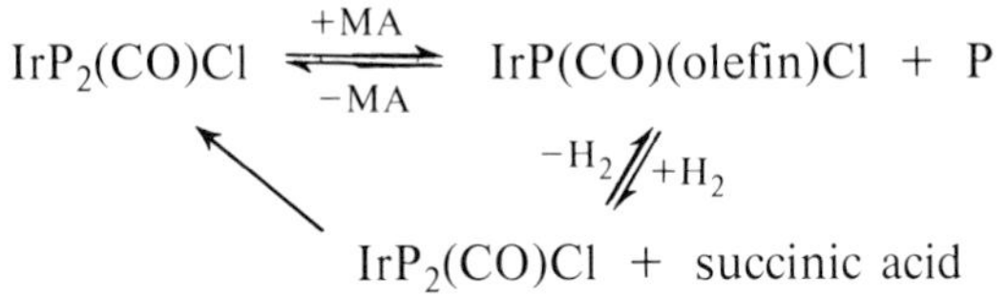

$$IrP_2(CO)Cl \underset{-MA}{\overset{+MA}{\rightleftharpoons}} IrP(CO)(olefin)Cl + P$$

$$\xrightleftharpoons[-H_2]{+H_2}$$

$$IrP_2(CO)Cl + \text{succinic acid}$$

$$P = P(C_6H_5)_3$$

$$MA = \text{maleic acid}$$

between $Ir[P(C_6H_5)_3]_2(CO)Cl$ and the hydride $Ir[P(C_6H_5)_3]_2(CO)Cl(H)_2$ occurs quickly, followed by slow formation of a maleic anhydride complex. Also, the rate was accelerated by two orders of magnitude in the presence of traces of oxygen, again suggesting that an active catalytic species is formed by a dissociation step. Since there is no evidence for the formation of $IrP(C_6H_5)_3$-$(CO)Cl(H)_2$, and since the existence of an $IrP(C_6H_5)_3(CO)$(maleic anhydride)Cl complex was indicated, the hydrogenation is presumed to occur predominately via addition of hydrogen to this latter species.

3. Activation of Hydrogen by Heterolytic Cleavage. A variety of noble metal salts catalyze the hydrogenation of olefinic substrates in polar solvents, using heterolytic cleavage of molecular hydrogen for hydrogen activation. Solutions of noble metal chlorides in aqueous HCl[20,23,310] or amide solutions[372] react with hydrogen to give hydrides thru hydride abstraction [Eq. (171)]. In the absence of stabilizing ligands, however, the hydride decomposes, eventually resulting in precipitation of the metal.

$$M^{X+} + H_2 \rightleftarrows MH^{(X-1)+} + H^+ \qquad (171)$$

The activity toward hydrogenation of aqueous solutions of Ru(II) in hydrochloric acid (which demonstrates the features common to noble metal salt catalysis) has been investigated by Halpern et al.[20,23,310,373] The mechanism proposed is given in Scheme 2-9. The rapidly formed, stabilized ruthenium ion

Scheme 2-9

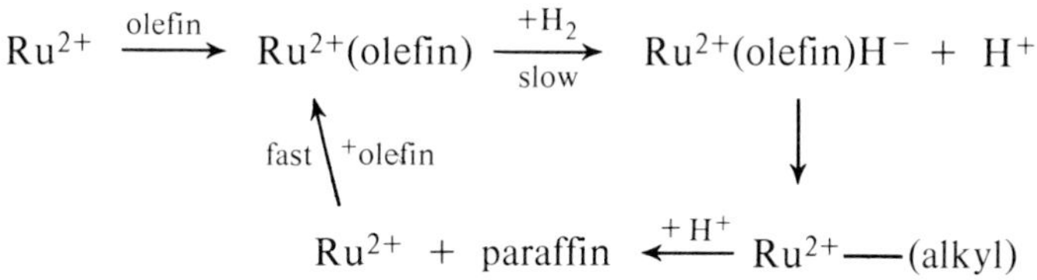

$$Ru^{2+} \xrightarrow{\text{olefin}} Ru^{2+}(\text{olefin}) \xrightarrow[\text{slow}]{+H_2} Ru^{2+}(\text{olefin})H^- + H^+$$

$$Ru^{2+}(\text{olefin})H^- \longrightarrow Ru^{2+}\text{—(alkyl)} \xrightarrow{+H^+} Ru^{2+} + \text{paraffin} \xrightarrow[\text{fast}]{+\text{olefin}} Ru^{2+}(\text{olefin})$$

(Ru^{2+}[olefin]) heterolytically cleaves molecular hydrogen to give a hydride species which then forms a ruthenium–alkyl complex, followed by protonation of the alkyl group. Evidence for this mechanism is found in the observed dependence of the reaction rate on olefin concentration, hydrogen–deuterium exchange by the hydride complex, and the stereospecificity of the products obtained. Spectroscopic data indicate that ethylene and α,β-unsaturated acids form complexes with Ru(II), but only the activated olefins are hydrogenated. Hydrogen–deuterium exchange is found for the ethylene complex, however. The rate of hydrogenation is given by Eq. (172) and is independent of olefin except at low concentrations. The use of deuterium and deuterium oxide shows that the addition of hydrogen to maleic acid is cis and that the hydrogen is derived from the solvent. These observations indicate that the exchange [Eq. (173)] is fast compared with alkyl formation and that the protonation of the alkyl group is also fast.

$$\frac{-d[H_2]}{dt} = k[H_2][\text{Ru(II)olefin}] \qquad (172)$$

$$\text{Ru(II)(olefin)H} + D^+ \rightleftarrows \text{Ru(II)(olefin)D} + H^+ \qquad (173)$$

A particularly exciting hydrogenation reaction involving the catalyst η^3-C_3H_5Co-$[P(OCH_3)_3]_3$ has recently been reported.[374,375] This catalyst causes the specific hydrogenation of benzene and many of its derivatives to cyclohexane at 25°C and < 1 atm H_2. Remarkably, benzene is hydrogenated more rapidly than is cyclohexene, and the rate for 1,3-hexadiene is very similar to that for benzene. The reaction is stereospecific; for example, *m*-xylene is reduced to

Scheme 2-10

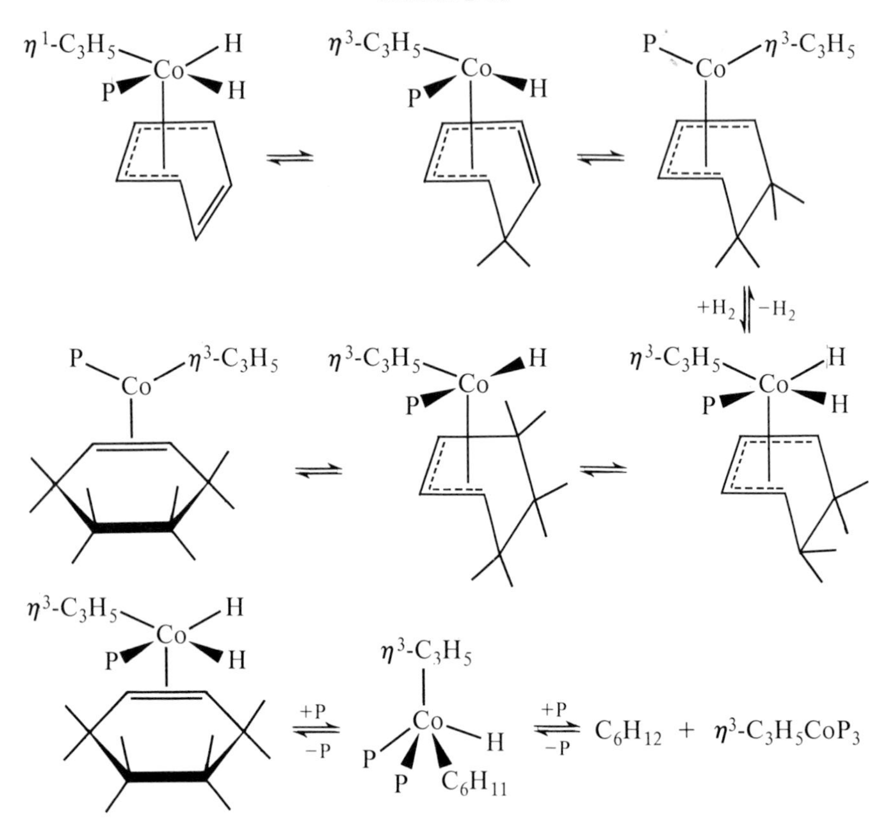

cis-1,3-dimethylcyclohexane and mesitylene to *cis,cis*-1,3,5-trimethylcyclohexane. These and other observations lead to the mechanism suggested in Scheme 2-10. The polyhapto binding of benzene and the derivatives derived from it are an essential feature of the process.

The systems discussed above serve to illustrate and highlight the subject of homogeneous catalytic hydrogenation by transition metal compounds. However, the range of catalytic systems is far broader than indicated here, and reference should be made to the reviews cited in order to appreciate the magnitude of field.

2. Catalytic Hydroformylation of Olefins by Cobalt Carbonyls. The commercially important hydroformylation reaction ("oxo" process)[11,326,376] [Eq. (174)] is catalytic at 150–300°C and at 100–300 atm of synthesis gas (hydrogen–carbon monoxide) in the presence of cobalt salts. Cobalt(II) is reduced to form $Co_2(CO)_8$, which then reacts with hydrogen to form the catalytic species [Eq. (175)]. General agreement is found concerning the steps in the reaction sequence [Eqs. (176–181)] leading to the observed products.

$$RCH{=}CH_2 + CO + H_2 \xrightarrow{Co(CO)_4H} RCH_2CH_2\overset{O}{\overset{\|}{C}}H \qquad (174)$$

$$Co_2(CO)_8 + H_2 \rightleftarrows 2Co(CO)_4H \qquad (175)$$

$$Co(CO)_4H \rightleftarrows Co(CO)_3H + CO \qquad (176)$$

$$RCH{=}CH_2 + Co(CO)_3H \rightleftarrows Co(CO)_3(\text{olefin})H \qquad (177)$$

$$Co(CO)_3(\text{olefin})H \rightleftarrows Co(CO)_3CH_2CH_2R + Co(CO)_3CH\begin{matrix}CH_3\\R\end{matrix} \qquad (178)$$

$$Co(CO)_3CH_2CH_2R + Co(CO)_3{-}CH\begin{matrix}CH_3\\R\end{matrix} \overset{CO}{\rightleftarrows} Co(CO)_4CH_2CH_2R + Co(CO)_4CH\begin{matrix}CH_3\\R\end{matrix} \qquad (179)$$

$$Co(CO)_4CH_2CH_2R + Co(CO)_4CH\begin{matrix}CH_3\\R\end{matrix} \rightleftarrows Co(CO)_3\overset{O}{\overset{\|}{C}}CH_2CH_2R + Co(CO)_3\overset{O}{\overset{\|}{C}}{-}CH\begin{matrix}CH_3\\R\end{matrix} \qquad (180)$$

$$Co(CO)_3\overset{O}{\overset{\|}{C}}{-}CH_2CH_2R + Co(CO)_3\overset{O}{\overset{\|}{C}}{-}CH\begin{matrix}CH_3\\R\end{matrix} \xrightarrow{H_2} Co(CO)_3H + RCH_2CH\overset{O}{\overset{\|}{C}}H + R\underset{CH_3}{\underset{|}{C}}H{-}\overset{O}{\overset{\|}{C}}H \qquad (181)$$

Much of the knowledge available concerning this reaction has been gathered from study of stoichiometric olefin hydroformylation by preformed $Co(CO)_4H$ in the presence of varying partial pressures of carbon monoxide where the hydrogen source for formation of product aldehydes from cobalt acyl complexes is cobalt hydrocarbonyl [Eq. (182)].

$$Co(CO)_m{-}\overset{\overset{O}{\|}}{C}R + Co(CO)_nH \longrightarrow Co_2(CO)_{m+n} + R\overset{\overset{O}{\|}}{C}H \qquad (182)$$

The inverse dependence on carbon monoxide observed in reaction rate for the stoichiometric hydroformylation reaction[282,285,286] supports the reversible dissociation of carbon monoxide from the hydrocarbonyl [Eq. (176)] as does kinetic data on the decomposition of $Co(CO)_4H$ in nonpolar solvents.[377] Olefin coordination prior to insertion into the Co—H bond [Eq. (177)] is assumed on the basis of the inhibition effects of carbon monoxide and since this is consistent with previously established alkyl formation reactions. The rapid isomerization of olefins by this catalyst under conditions where olefin dissociation does not occur also supports this intermediate.[11] Powerful evidence for the rapid isomerization of olefins by the previously established 1,2 addition–elimination mechanism is seen in the catalytic hydroformylation of optically active 3-methyl-1-hexene to yield structures **61, 62**, and **63** [Eq. (183)].[378] The formation of structure **63** in small yield with 70% retention of configuration was thought to occur by direct insertion of carbon monoxide into a C—H bond of the methyl group. It was subsequently demonstrated by the equivalent reaction of structure **64** (racemic), [Eq. (184)] that this reaction must involve a series of 1,2-hydride shifts.[379] In order to account for the observed retention of optical activity during

$$CH_2{=}CH{-}\underset{*}{\overset{\overset{CH_3}{|}}{C}H}{-}C_3H_7 \longrightarrow \underset{\mathbf{61}}{\underset{93\%}{CH_3CH_2CH_2{-}\underset{\underset{H^*}{|}}{\overset{\overset{CH_3}{|}}{C}}{-}CH_2CH_2\overset{\overset{O}{\|}}{C}H}} +$$

$$\underset{\mathbf{62}}{\underset{4\%}{CH_3CH_2CH_2\underset{\underset{H}{|}}{\overset{\overset{CH_3}{\backslash}}{C}}{-}\underset{\underset{H}{|}}{\overset{\overset{CH_3}{|}}{C}}{-}\overset{\overset{O}{/\!/}}{C}{-}H}} + \underset{\mathbf{63}}{\underset{3\%}{CH_3CH_2CH_2\underset{*}{\overset{\overset{C_2H_5}{|}}{C}H}{-}CH_2\overset{\overset{O}{\|}}{C}{-}H}} \qquad (183)$$

$$\mathrm{CH_2{=}CH{-}\overset{\displaystyle CH_3}{\underset{\displaystyle D}{C}}{-}C_3H_7} \longrightarrow \mathrm{CH_3CH_2CH_2{-}\overset{\displaystyle H}{\underset{\displaystyle CHD(CH_3)}{C}}{-}CH_2{-}\overset{\displaystyle O}{\overset{\|}{C}}H} +$$

64

$$\mathrm{CH_3CH_2CH_2{-}\overset{\displaystyle D}{\underset{\displaystyle CH_3}{C}}{-}CH_2CH_2\overset{\displaystyle O}{\overset{\|}{C}}H} + \mathrm{CH_3CH_2CH_2\overset{\displaystyle D}{\underset{\displaystyle CH_3}{C}}{-}\overset{\displaystyle H}{\underset{\displaystyle CH_3}{C}}{-}\overset{\displaystyle O}{\overset{\|}{C}}{-}H} \quad (184)$$

this isomerization process, the catalytic hydroformylation reaction must involve σ to π conversions in which the π-bonded intermediate is tightly bound and has a very short lifetime, and such that the addition and elimination from the olefinic linkage is largely stereospecific.

The addition of a molecule of CO to the $Co(CO)_3R$ species is indicated by the previously mentioned inhibition of isomerization with an increasing partial pressure of CO [Eq. (179)]. The carbonyl insertion reaction is assumed to proceed by an alkyl migration to a carbonyl ligand cis to the migrating alkyl group as is observed for $Mn(CO)_5CH_3$ [Eq. (185)].[120] This type of reaction has been reported for $Co(CO)_4CH_2C_6H_5$, where again the inserted carbonyl is bound to the cobalt atom prior to the reaction.[380]

$$\mathrm{CO} + \mathrm{Mn(CO)_5CH_3} \rightleftarrows \mathrm{Mn(CO)_5{-}\overset{\displaystyle O}{\overset{\|}{C}}{-}CH_3} \qquad (185)$$

The formation of the product aldehyde is then suggested to occur by hydrogenolysis in the catalytic reaction [Eq. (181)]. This is indicated by the reduction of $Co(CO)_4CO\text{-}CH_3$ with hydrogen under conditions where any $Co(CO)_4H$ would be destroyed.[381]

The factors governing the product distribution in the hydroformylation reaction include:

(1) the initial direction of addition of the hydrocarbonyl to the olefin [Eq. (178)];
(2) subsequent isomerization of the alkyl groups so formed [reverse of Eq. (178)];
(3) the rate of insertion of carbon monoxide into the various Co—R groups present in the reaction mixture [Eq. (180)];

(4) isomerization of the Co–acyl groups prior to their hydrogenolysis [reverse of Eq. (180)];
(5) the rate of hydrogenolysis of the various Co–acyl complexes present in the reaction mixture.

It was previously noted that the addition of the hydrocarbonyl to activated olefins is predominately Markovnikov. This is also found for isolated olefins; thus, the initial stoichiometric gas phase addition of $Co(CO)_4H$ to propylene-d_6 was shown to be 70% Markovnikov in the absence of carbon monoxide in a nitrogen atmosphere.[283] Under these conditions, the major reaction was isomerization, but the small yield of aldehyde observed also corresponded to the initial direction of addition to the olefin bond (70% $(CH_3)_2CHCHO$, 30% $CH_3CH_2CH_2CHO$). This indicated that in the absence of CO the initial direction of addition to the olefinic substrate determined the aldehyde distribution. Under a carbon monoxide atmosphere (which retards the exchange reaction with the olefin, preventing a measure of the initial direction of addition of $Co(CO)_4H$ to the olefin), the ratio of straight chain to branched aldehyde was much higher. This is a general phenomenon; increasing the carbon monoxide pressure in either the stoichiometric or catalytic hydroformylation reaction leads to greater yields of straight-chain aldehydes at the expense of branched products.[11] An explanation has been advanced[382] based on the equilibrium formation of $Co(CO)_4R$ with increasing carbon monoxide partial pressure, resulting in more stringent steric requirements on the coordinated alkyl groups, and thus leading to the formation of a more linear aldehyde. Orchin and Rupilius, however, on the basis of work with the stoichiometric hydroformylation reaction at varying pressures of CO, suggest that the actual catalytic species is different, being $Co(CO)_3H$ at low carbon monoxide pressure and $Co(CO)_4H$ at high pressure.[11] Either explanation is in accord with the observed effect of phosphine addition to the reaction mixture which leads to increased yields of linear aldehydes.[383] The catalytic species $Co(CO)_3$-$P(n\text{-}C_4H_9)_3H$ is more sterically demanding and promotes the formation of very high yields of *n*-aldehydes, although at a slower rate.

Isomerization of Co–acyl groups in $Co(CO)_4$-COR complexes, along with disproportionation to yield equimolar amounts of aldehyde and olefin under an inert atmosphere, has been reported[384] (Scheme 2-11). Further evidence[11] for this Co–acyl complex isomerization is seen in Eq. (186). The mixture of butenes obtained from equilibration of the isomeric pentanoyl cobalt carbonyls was determined to be the same. This equilibrium distribution of isomeric butenes was determined to be formed prior to olefin dissociation from the Co-$(CO)_3$(olefin)H species. Although in the catalytic reaction under normal conditions acyl complex formation is irreversible, these results do indicate the complexity of this system.

Other substrates which undergo hydroformylation catalyzed by this system include epoxides [Eq. (187)] and orthoformic esters [Eq. (188)].[376,326]

Scheme 2-11

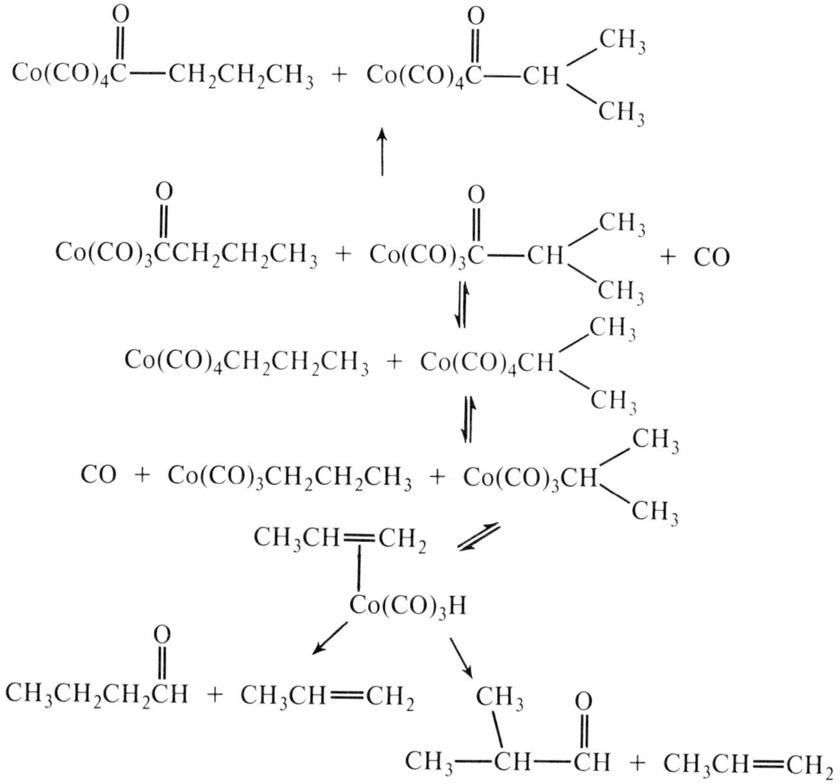

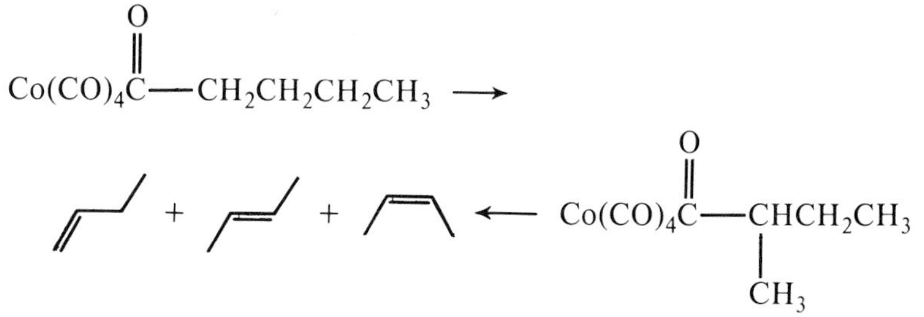

$$Co(CO)_4C(=O)CH_2CH_2CH_3 + Co(CO)_4C(=O)CH(CH_3)_2$$

$$\uparrow$$

$$Co(CO)_3C(=O)CH_2CH_2CH_3 + Co(CO)_3C(=O)CH(CH_3)_2 + CO$$

$$\rightleftharpoons$$

$$Co(CO)_4CH_2CH_2CH_3 + Co(CO)_4CH(CH_3)_2$$

$$\rightleftharpoons$$

$$CO + Co(CO)_3CH_2CH_2CH_3 + Co(CO)_3CH(CH_3)_2$$

$$\rightleftharpoons$$

$$CH_3CH{=}CH_2 \cdot Co(CO)_3H$$

$$\swarrow \qquad \searrow$$

$$CH_3CH_2CH_2CH{=}O + CH_3CH{=}CH_2 \qquad CH_3{-}CH(CH_3){-}CH{=}O + CH_3CH{=}CH_2$$

$$Co(CO)_4C(=O){-}CH_2CH_2CH_2CH_3 \longrightarrow$$

$$1\text{-butene} + \text{trans-2-butene} + \text{cis-2-butene} \longleftarrow Co(CO)_4C(=O){-}CH(CH_3)CH_2CH_3 \quad (186)$$

$$\text{epoxide } >C{-}C< \text{ (bridged by O)} + CO + H_2 \xrightarrow{Co_2(CO)_8} {-}C(OH){-}C{-}CH{=}O \quad (187)$$

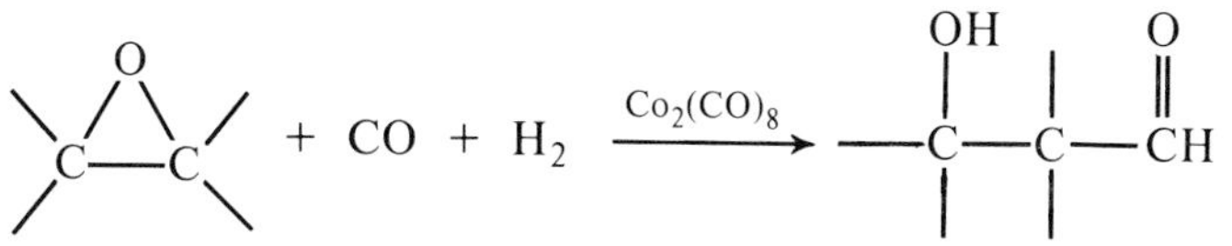

$$2HC(OR)_3 + CO + H_2 \xrightarrow{Co_2(CO)_8} RCH(OR)_2 + ROH + 2HCOOR \quad (188)$$

Other products are observed in the presence of active hydrogen sources [Eq. (189)]. The reported catalytic synthesis of *N*-acylamino acids from olefins or aldehydes under "oxo" conditions represents a reaction of this type[385] [Eq. (190)].

$$\mathrm{Co(CO)_4\overset{\overset{\displaystyle O}{\|}}{C}R + R'H \longrightarrow R\overset{\overset{\displaystyle O}{\|}}{C}R' + Co(CO)_4H} \qquad (189)$$

$$\mathrm{(R'H = H_2O,\ RSH,\ NHR_2,\ R\overset{\overset{\displaystyle O}{\|}}{C}OH)}$$

$$\mathrm{R^1CH{=}CH_2 + R^2\overset{\overset{\displaystyle O}{\|}}{C}NH_2 + 2CO + H_2 \xrightarrow{Co_2(CO)_8} R^1CH_2CH_2\underset{}{\overset{\overset{\displaystyle NH\overset{\overset{\displaystyle O}{\|}}{C}R^2}{|}}{C}H}{-}CO_2H} \qquad (190)$$

VII. PROMOTED REARRANGEMENT OF STRAINED HYDROCARBONS

The existence of stable polycyclic hydrocarbons possessing high ring-strain energies is rationalized by the orbital symmetry conservation concepts of Woodward and Hoffman.[386] In the presence of a catalyst which allows a low-energy pathway for isomerization, however, these high energy species readily rearrange.

The mechanisms proposed to account for the remarkable ability of certain transition metals (notably Ag,[29,387] Rh,[29] and Pd[29]) to homogeneously catalyze σ-bond rearrangements in such strained organic molecules are of two types: concerted, in which the transition metal serves to relax the constraints of orbital symmetry such that a pathway is open to concerted rearrangements; and stepwise reactions in which the rearrangement is postulated to go through one or more intermediates or transition states involving C—C-bond scission by the transition metal ion. The latter mechanism can involve radical formation, carbene–carbonium ion formation, or oxidative addition (insertion of the metal atom into the C—C bond). In the latter case, the transition metal serves by lowering an otherwise insurmountable energy barrier required for cleavage of a C—C σ bond.

Of particular interest are the skeletal rearrangements undergone by cubane derivatives, bicyclobutanes, and other organic molecules in which the product obtained varies with the choice of the transition metal catalyst.

A. Bicyclobutane Rearrangements

Tricyclo[4.1.0.0^{2,7}]heptane, structure **65,** and its methyl and dimethyl derivatives, **66** and **67**, are catalytically rearranged by a variety of transition metal catalysts.[388-398] Some products formed, for selected catalysts, are shown in Table 2-4. Other compounds which have been catalytically rearranged are a series ef methyl substituted bicyclobutanes and yield a variety of products (Table 2-5).

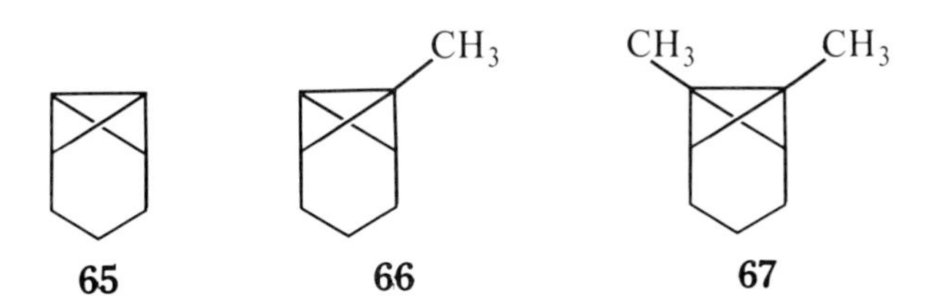

There is a striking variation in rearrangement product distribution with catalyst species which provides insight into the rearrangement mechanisms. Intervention of both carbene and carbonium ion transition metal intermediates with subsequent rearrangement yielding the observed products has been suggested for Rh(I),[388,389,394,400,401] Pd(II),[393,394] and Ag(I)[387,395] catalysis [Eq. (191).

M^{n+} → $\dot{M}^{n+}$ → M^{n-1} (⊕) (191)

Pd

68 **65** → **69** (CH_2) (192)

Also, structure **68**, an intermediate in the catalytic rearrangement of **65** to **69** [Eq. (192)] in deuterochloroform solution[440] by $PdCl_2(C_6H_5CN)_2$, has been detected. Other evidence indicating transition metal-containing intermediates is found in cyclopropylcarbonyl–mesylate elimination reactions[399] and in catalytic decomposition of diazoalkenes by these catalyst species.[392] Thus, the mesylate, structure **70**, in benzene solution undergoes an elimination reaction to yield 1,3-cycloheptadiene, the product of Ag^+-catalyzed rearrangement of structure **65**. This suggests an intermediate similar to structure **71**. Also, dia-

TABLE 2-4
TRANSITION METAL-CATALYZED REARRANGEMENTS OF TRI-

Catalyst	Conditions	Substrate
	(°C, hr, solvent)	
$AgBF_4$	~40°, < 1, $CDCl_3$[c,d]	
$(C_6H_5CN)_2PdCl_2$	25°, 20, CH_3CN[d]	
	25°, . . . $CHCl_3$[e]	
$[(\pi\text{-}C_3H_5)PdCl]_2$	25°. 0.5, $CHCl_3$[e]	
$[Rh(CO)_2Cl]_2$	25°, 0.25, CH_3CN[b]	
$[Ir(CO)_3Cl]_2$	25°, 14, $CHCl_3$[b]	
PtO_2	65°, 48, CH_3CN[b]	
		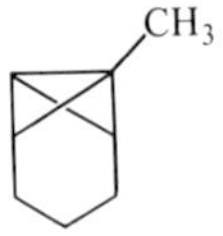 CH_3
$AgClO_4$	40°, . . . benzene[f]	
$[(\pi\text{-}C_3H_5)PdCl]_2$	25°, 0.5, $CHCl_3$[b]	
$[Rh(CO)_2Cl]_2$	25°, 0.25, $CHCl_3$[b]	
$[Ir(CO)_3Cl]_2$	25°, 14, $CHCl_3$[b]	
		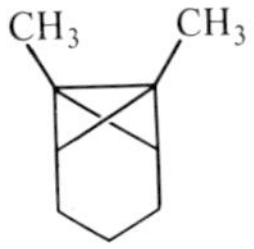 CH_3 CH_3
$AgClO_4$	. . . . . . benzene[f]	

[a] Refs. 3,7.
[b] Ref. 388.
[c] Ref. 390.

$$R\text{-}CHN_2 + M \rightarrow RCH{:}M + N_2 \qquad (193)$$

CYCLO[4.1.0.0.a] HEPTANE AND ITS DERIVATIVES

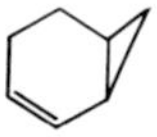 Products (%)	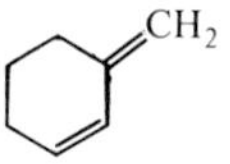	
...	...	100
...	69	...
...	90	...
...	94	...
...	98	...
...	91	...
24	62	...

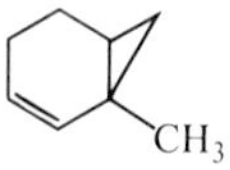	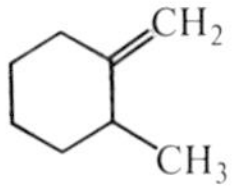	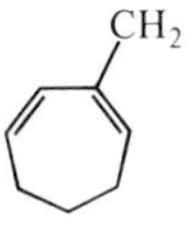	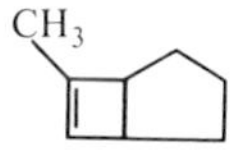	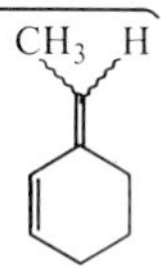
...	...	29	42	29
...	93	...	...	...
...	96	...	...	...
...	93	...	...	...

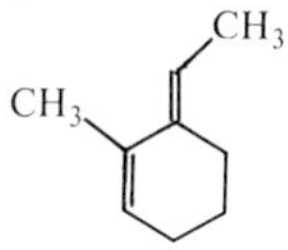	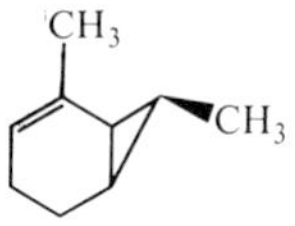
80	20

[d] Ref. 391.
[e] Ref. 392.
[f] Ref. 395.

zoalkanes have been observed to react with transition metal compounds leading to the formation of complexes consisting of the metal and the carbene species derived from the diazoalkane[404] [Eq. (193)]. Reactions of diazoalkenes with the various transition metal catalysts have thus been suggested to produce, initially, the carbene species. The subsequent fate of the diazoalkene substrate has been examined to determine the reaction sequence for a possible transition metal carbene intermediate in the rearrangements of the analogous bicyclobutanes.[392,393,394,402] Thus, Ag^+ acting on structure **72** yields **73** in contrast to its effect on structure **75**; Pd(II) produces structure **73** with both substrates.[399]

TABLE 2-5
TRANSITION METAL-CATALYZED REARRANGEMENT PRODUCTS

Substrate	Catalyst	Conditions (°C, solvent)
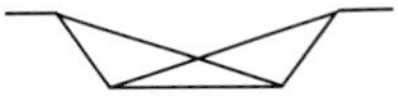	$AgClO_4$[a]	25°, C_6H_6
	$Pd(C_6H_5CN)_2Cl_2$[a]	25°, $CHCl_3$
	$[Rh(CO)_2Cl]_2$[b]	$CHCl_3$
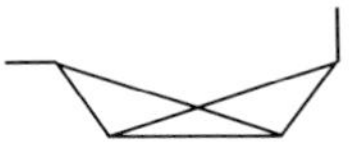	$AgClO_4$[a]	25°, C_6H_6
	$Pd(C_6H_5CN)_2Cl_2$[a]	25°, $CHCl_3$
	$[Rh(CO)_2Cl]_2$[b]	$CHCl_3$
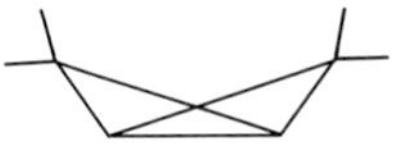	$AgClO_4$[a]	25°, C_6H_6
	$Pd(C_6H_5CN)_2Cl_2$[a]	25,° $CHCl_3$
	$[Rh(CO)_2Cl]_2$[c]	25°, $CHCl_3$ (5 min)
	$[Ir(CO)_3Cl]_2$[c]	25°, $CHCl_3$ (12 hr)
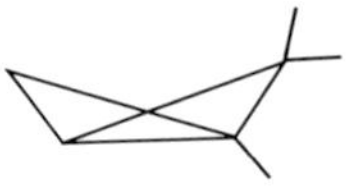	$AgBF_4$[d]	40,° $CDCl_3$
	$[C_6F_5Cu]_4$[g]	0,° $CHCl_3$
	$Pd(C_6H_5CN)_2Cl_2$[h]	25,° $CHCl_3$
	di-μ-chloro-bis-[(π-allyl)palladium (II)][h]	$CHCl_3$
	$[Rh(CO)_2Cl]_2$[f]	$CHCl_3$
	$[Ru(CO)_3Cl_2]_2$[f]	
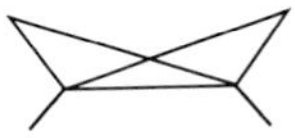	$AgClO_4$[e]	25,° C_6H_6
	$Pd(C_6H_5CN)_2Cl_2$[b]	$CHCl_3$
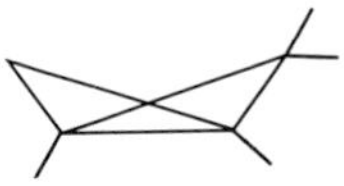	$[Rh(CO)_2Cl]_2$[c]	$CHCl_3$

[a] Ref. 392.
[b] Ref. 393.
[c] Ref. 394.
[d] Ref. 395.

FOR METHYL-SUBSTITUTED BICYCLOBUTANES

Distribution of Products (%)

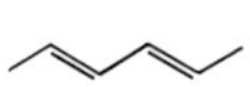	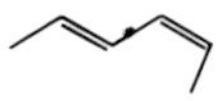	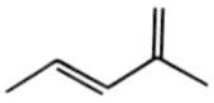
78	22	. . .
37	16	47
75	3	22
5	95	. . .
51	24	24
60	8	32

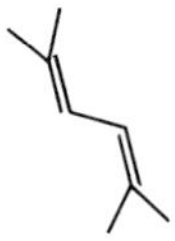

90
95
83
87

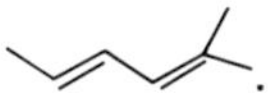	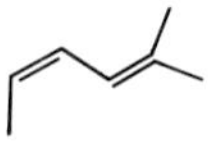	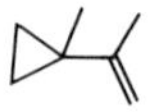	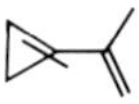	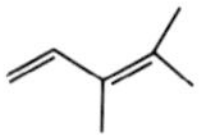
74	16	8	. . .	. . .
39	33	. . .	11	. . .
. . .	. . .	7	. . .	93
37	33	. . .	3	2
. . .	. . .	30	. . .	58
. . .	. . .	22	. . .	46

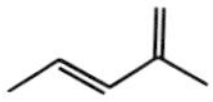

95
80

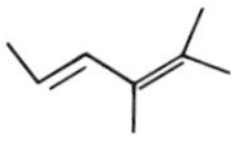	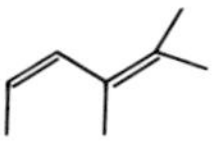
50	46

[e] Ref. 399.
[f] Ref. 400.
[g] Ref. 402.
[h] Ref. 403.

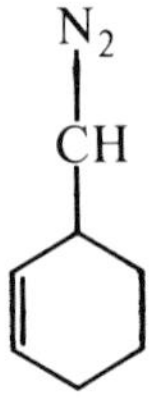

72

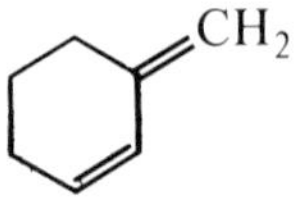

73

The decomposition products of cyclopropyl mesylates analogous to structure **70** are shown in Table 2-6 to indicate the products expected for carbonium ion-type intermediate rearrangements. Isomerization products for several diazoalkenes similar to structure **72** are also given to indicate the products expected from carbene–transition metal species (Table 2-7).

The similarities between the products of Ag^+ catalysis and those resulting from the decomposition of analogous mesylates, as well as the striking differences between products of Ag(I)-catalyzed bicyclobutane rearrangements and Ag(I)-catalyzed diazoalkene decomposition suggest an unstable polar intermediate. The specifics of the rearrangement are then dominated by formation of the most stable carbonium ion (Scheme 2-12).

Scheme 2-12

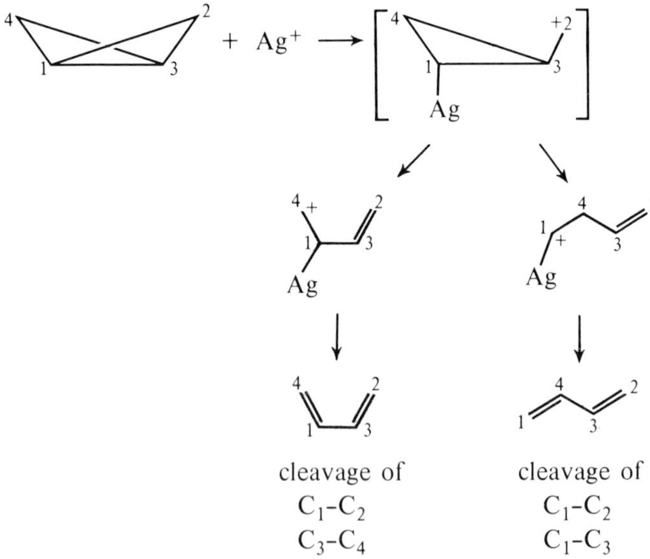

cleavage of C_1-C_2 C_3-C_4

cleavage of C_1-C_2 C_1-C_3

Thus, structure **74** is formed from **75** [Eq. (194)] but not from **76** and **77** (see also Table 2-5).

TABLE 2-6
PRODUCT DISTRIBUTION OF THE ELIMINATION REACTION OF MESYLATES[a]—CARBONIUM ION REARRANGEMENT PRODUCTS

Substrate	Products (%)		
OMs	95	CH_2 0	
MsO, H, CH_3	75	20	0
MsO, CH_3, H	8	86	0
CH_2OMs	. . .	. . .	95

[a] Ref. 399

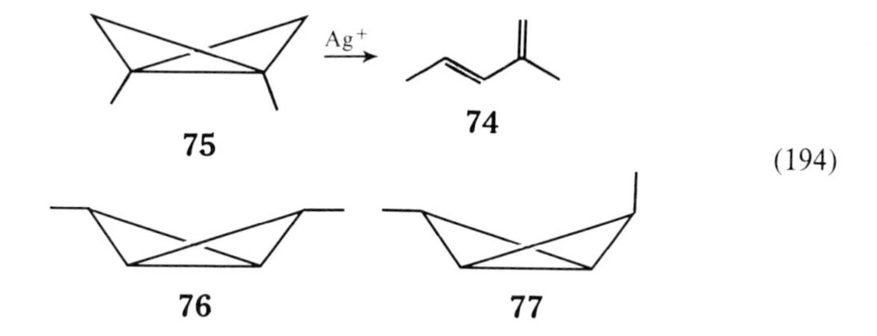

(194)

Recent work has shown that for Ag^+-catalyzed rearrangement of structure **75** and its derivative, the catalytic species approaches the substrate near the central carbon–carbon bond of the strained ring system. The product distribution is then influenced by both steric and electronic factors.[396,397,398]

For Rh(I) and Pd(II), carbene–carbonium intermediates are indicated by the similarities in results for catalyzed decomposition of bicyclobutanes and the analogous diazoalkenes. Product distributions observed in Tables 2-4 and 2-5 can be rationalized by competitive 1,2-hydrogen and vinyl migrations (Scheme 2-13). In the case where one of the central carbons is substituted and one of the terminal carbons doubly substituted, Ag(I) and Rh(I) appear to attack different carbon atoms (Scheme 2-14). Ag(I), on the basis of the product distribution,

TABLE 2-7
PRODUCT DISTRIBUTION FOR METAL-CATALYZED REACTIONS OF DIAZOALKENES[a]—DERIVATIVES OF TRANSITION METAL–CARBENE INTERMEDIATES

Substrate	Reagent	Products (%)		
(cyclohexenyl)CH_2N_2			(methylenecyclohexene)	
	$AgClO_4$		95	
	$Pd(C_6H_5CN)_2Cl_2$		95	
CHN_2 substrate		(diene)	(diene)	(diene)
	$AgClO_4$	17	38	45
	$Pd(C_6H_5CN)_2Cl_2$	26	14	60
CHN_2 substrate			(diene)	
	$AgClO_4$		90	
	$Pd(C_6H_5CN)_2Cl_2$		90	
	$[C_6F_5Cu]_4$[b]		"good yield"	
	$[Rh(CO)_2Cl]_2$		only volatile product	

[a] Ref. 392.
[b] Ref. 403.

Scheme 2-13

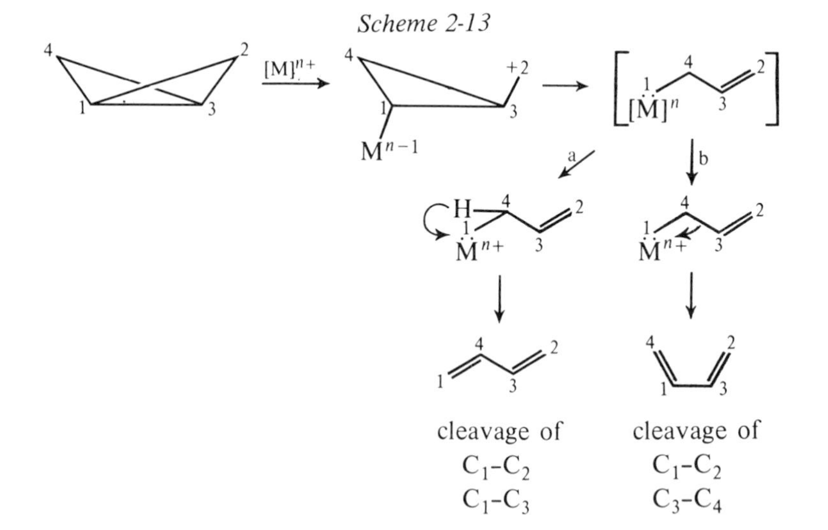

Scheme 2-14

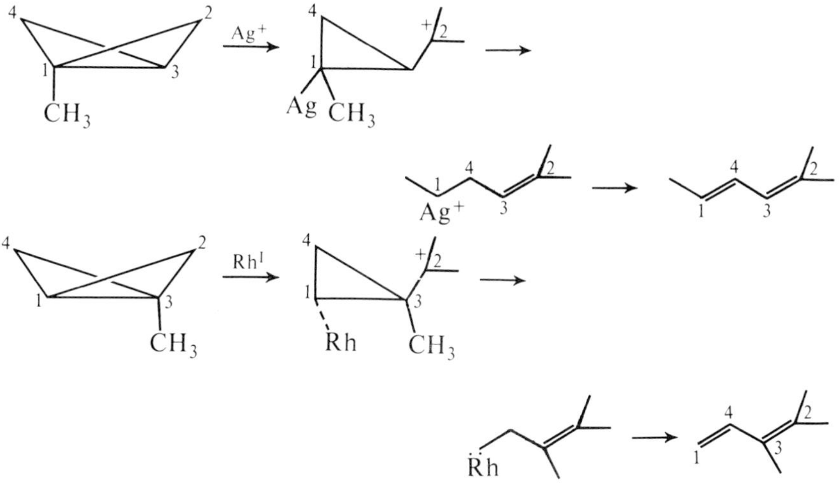

would seem to react at the substituted central carbon atom. Rh(I), however, appears to attack the unsubstituted central carbon atom.

The formation of a metal-bound carbene–carbonium ion as an intermediate in bicyclobutane rearrangement catalyzed by Rh(I) and Pd(II) in aprotic solvents is further supported by work with the phenyl substituted compounds **78** and **79.**

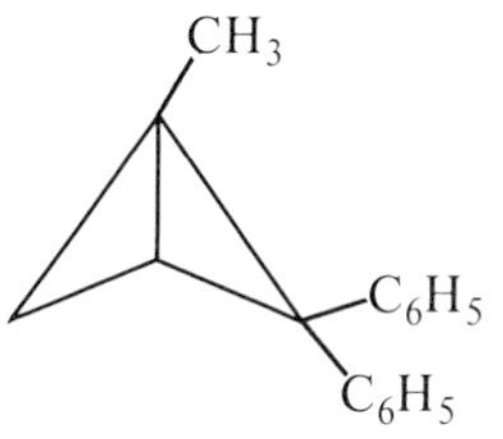

78 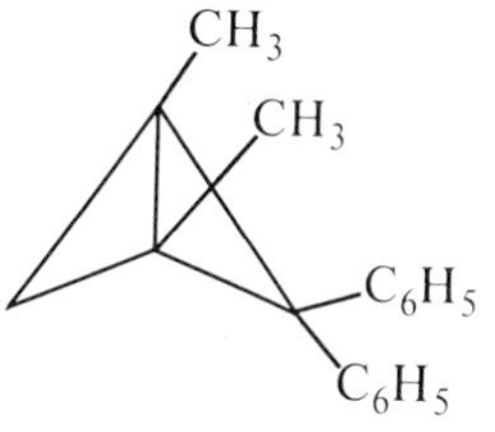 **79**

The mechanisms suggested for Rh(I) and Ag(I) catalysis readily account for the observed products as summarized in Eqs. (195) and (196). Initial cleavage of peripheral bicyclobutane C—C bond leads to the intermediate shown in Scheme 2-15. In the case of structure **78,** Rh(I) and Ag(I) again attack different central carbon atoms and give the products shown in Eq. (195). In Scheme 2-15, a hydrogen shift would produce structure **80,** and carbene addition to a benzene ring would give **82;** reaction at a benzene C—H bond would produce **81** [Eq. (195)]. With Cu^+ and Ag^+ catalysis (Scheme 2-15a), a hydrogen shift would

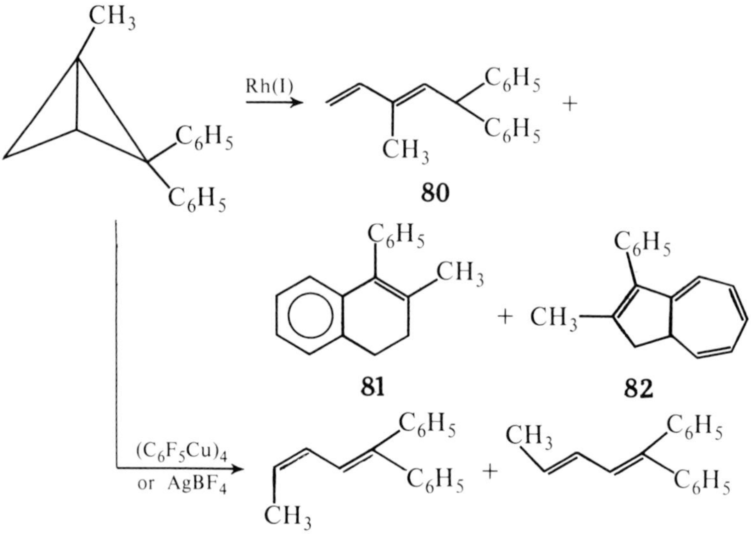

$$\text{(195)}$$

80

81 82

give the observed products in Eq. (195). Rearrangements of **89** are then assumed to involve the same type of intermediate. The addition of another methyl substituent should not alter the situation with respect to $[C_6F_5Cu]_4$ catalysis as is observed [Eq. (196)]. In $[Rh(CO)_2Cl]_2$ catalysis, however, the carbene–carbonium ion generated (Scheme 2-15b, $R_1 = R_2 = CH_3$) is now secondary; therefore, the reactivity of the intermediate and consequently the distribution of the products is expected to be altered [Eq. (196)]. The absence of a dehy-

Scheme 2-15

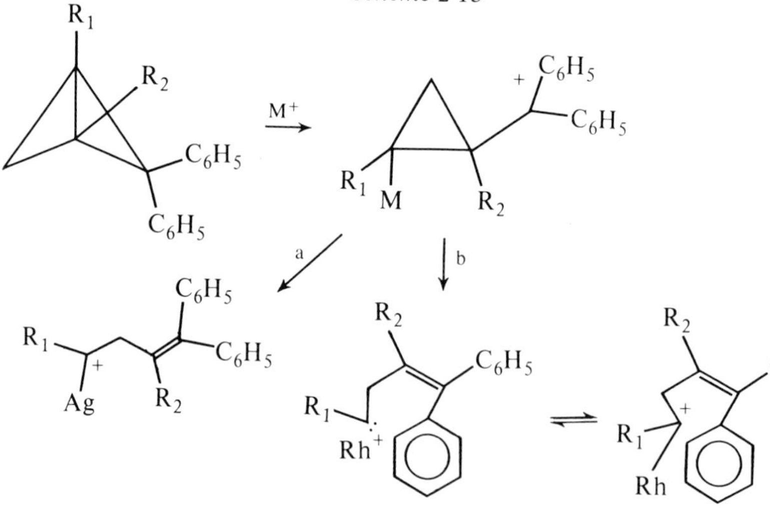

droazulene analogous to structure **81** in the reaction product mixture indicates more cationic character in the reactive intermediate.

The data presented with regard to catalytic rearrangement of bicyclobutanes by Rh(I), Pd(II), and Ag(I) indicate that the catalysts function initially as Lewis acids. The specifics of the reaction hereafter appear to be governed by the nature of the metal–carbon bond formed. Thus Rh(I) and to a lesser extent Pd(II), which are known to be ambiphilic in their reactions with electron deficient compounds, stabilize carbenoid reaction intermediates, while Ag^+, which is a better σ acceptor, forms more polar carbon bonds. The catalytic properties of Pd(II) in fact appear to be largely determined by the ligands present in the catalytically active species.[403] The modifying effects of ligands on these tendencies will become more apparent in the section following.

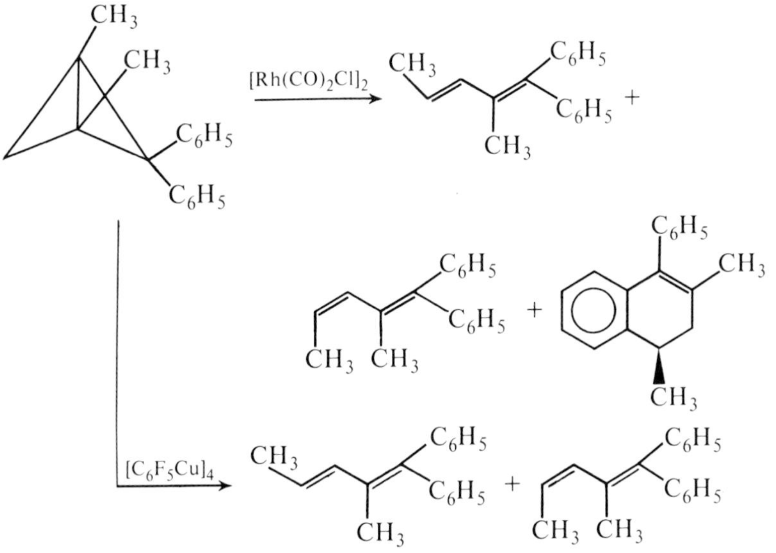

(196)

B. Cubane and Secocubane Rearrangements

Ag(I) catalyzes the rearrangement of a number of highly strained cage molecules under mild conditions to produce structures different from those generated by thermolysis. Reports of the synthesis of structure **83** resulted in a disagreement about the physical properties of the products.[407,408,409] Column chromatography of the intermediate **84** on silica gel–silver nitrate inadvertently resulted in its conversion to **85** [410–413] [Eq. (197)]. This rearrangement was also catalyzed by $AgBF_4$ in deuterochloroform solution.

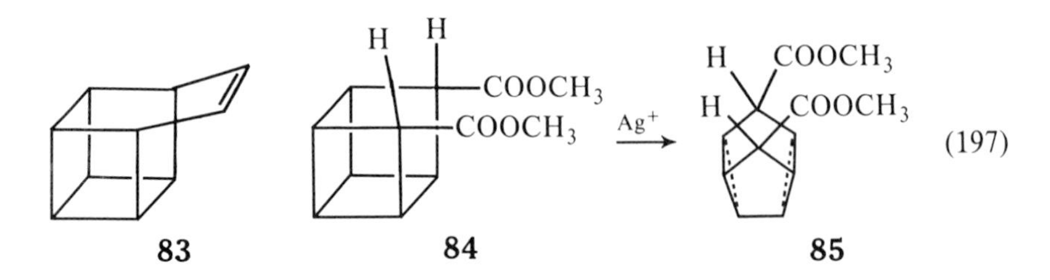

This isomerization is now known to exemplify a general type of reaction following a second-order rate law:[387]

$$\frac{-d[\text{cage cpd}]}{dt} = k_{\text{Ag}}[\text{cage cpd}][\text{Ag salt}]$$

The reaction of Eq. (198) is catalyzed by $AgBF_4$ in dilute $CDCl_3$ solution, and analogous structures give similar rearrangements.[411,412,413] Cubane **86** and substituted cubanes are transformed to **87** and its corresponding substitution products[414] [Eq. (199)]. The rearrangement of Eq. (198) is closely related.[411,412,413]

(198)

(199)

86 **87**

The presence of a complete cage structure appears unnecessary since a number of less strained derivatives are also similarly transformed in the presence of Ag(I), for example, Eq. (200).[388,415] There appears to be no correlation of the rate of Ag^+-catalyzed isomerization with steric strain in these cage compounds.[387]

R R R R

(200)

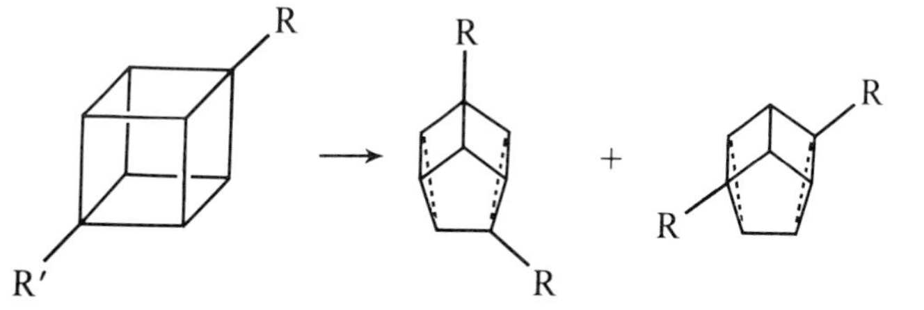

(201)

The product distribution for the Ag(I)-catalyzed rearrangement of mono- and disubstituted cubanes [Eq. (201)] are particularly interesting in that they indicate the effect of electron withdrawing groups on this type of reaction (Table 2-8).[414] The kinetic evidence, available suggests that the rate of Ag^+-catalyzed rearrangement of cage molecules depends primarily on the electron withdrawing ability of the substituents on the substrate. Increasing substitution does, however, lead to a reduction in rate of isomerization. Further, the ratios of products for the isomerization (Table 2-8) are in the order expected for the bond switching occurring at the termini remote from the corner(s) of the cube carrying the substituents (Scheme 2-16).

TABLE 2-8
PRODUCT DISTRIBUTION FOR AG(I)-CATALYZED REARRANGEMENTS OF MONO- AND DISUBSTITUTED CUBANES

Compound	Product Ratio		
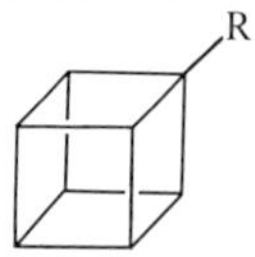 $R = CH_2OAc$	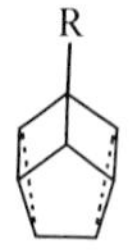 6	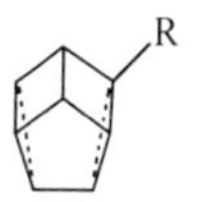 2	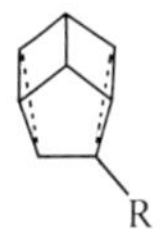 1
$R = COOCH_3$	2.5	1	0
$R = CH_2OH$	1	1	2
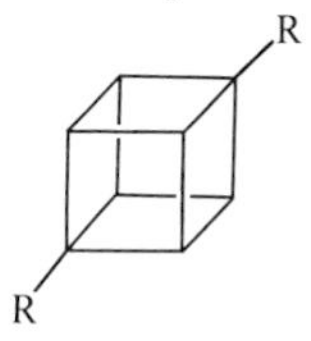 $R = CH_2OAc$	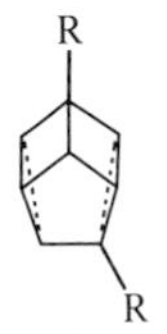 5	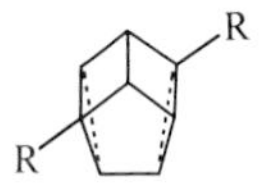 1	
$R = COOCH_3$	1	20	

Scheme 2-16

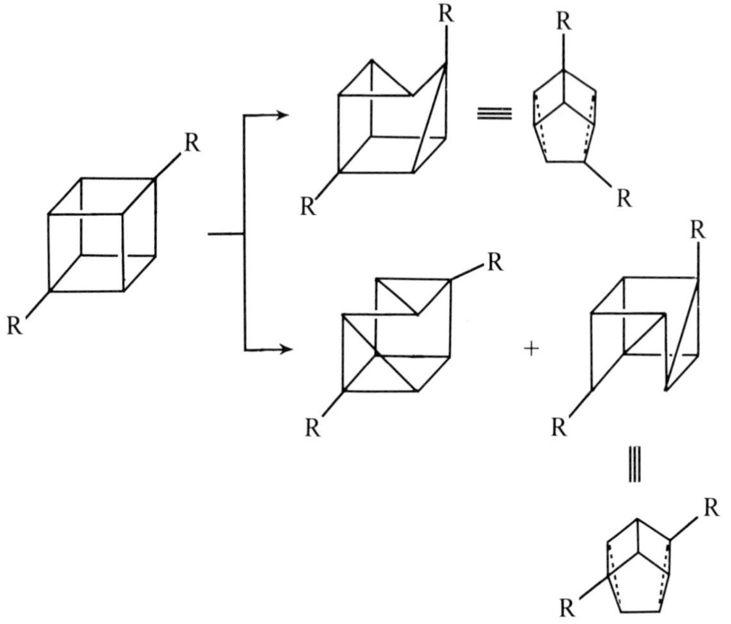

R R R R R R R R R R R R

A mechanism for these rearrangements has been suggested by analogy to the thallium(III), mercury(II), and palladium(II) oxidation of 1-methylcyclobutene. The product, cyclopropylmethyl ketone, is presumed to form thru a cyclobutyl carbonium ion to cyclopropyl carbinyl rearrangement (Scheme 2-17). So it has been suggested that the Ag^+ ion heterolytically cleaves a C—C bond, producing a carbonium ion which rearranges [Eq. (202)].[416]

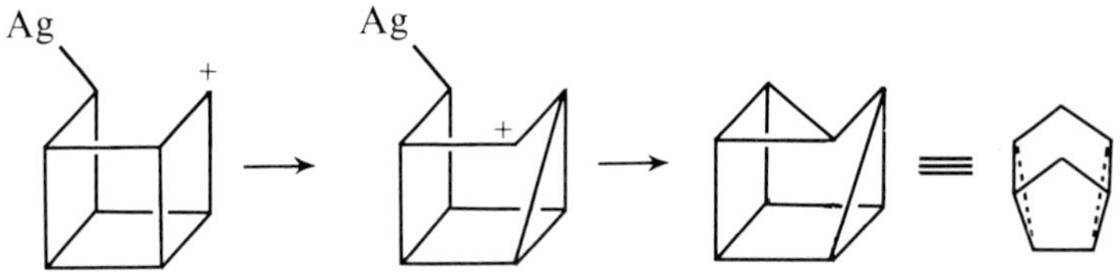

Ag + Ag + (202)

The data for hydroxymethylcubane (Table 2-8) suggest that the hydroxyl group coordinates to the Ag(I), permitting approach to an area otherwise sterically shielded and consequently affecting the product distribution. In support of this suggestion, the rate of isomerization for this compound is 13 times faster than that having R = CH_2OAc and approaches that of unsubstituted cubane.

Palladium(II)[414] and Rhodium(I)[417] have also been examined for their catalytic action upon cubane and its derivatives. Pd(II) was observed to catalytically

Scheme 2-17

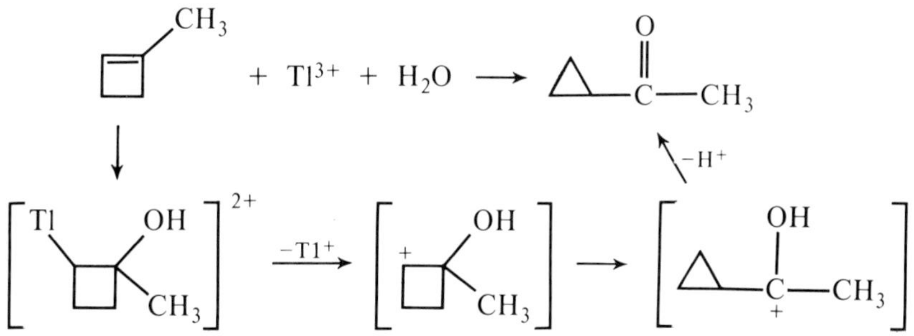

CH_3 + Tl^{3+} + H_2O ⟶ O, C—CH_3

$-H^+$

[Tl OH CH_3]$^{2+}$ $\xrightarrow{-Tl^+}$ [+ OH CH_3] ⟶ [OH C—CH_3 +]

rearrange cubane to give the same product as Ag^+ would give. Rh(I) complexes, however, have been reported to produce different compounds from cubane and substituted cubanes.[417] Some Rh(I) complexes in chloroform and carbon tetrachloride rapidly and quantitatively convert cubanes, as shown in Eqs. (203) and (204). Carbomethoxycubane is similarly transformed in Eq. (205).

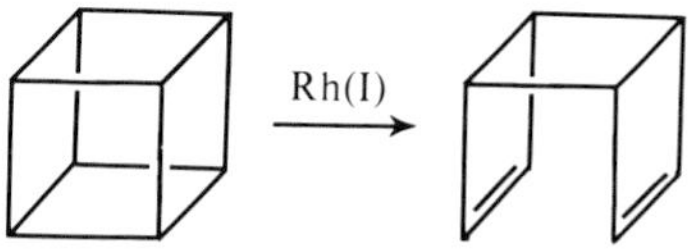

Rh(I) (203)

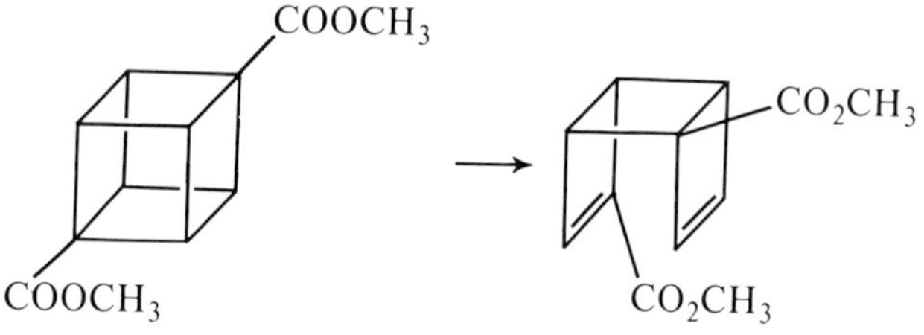

$COOCH_3$ $COOCH_3$ ⟶ CO_2CH_3 CO_2CH_3 (204)

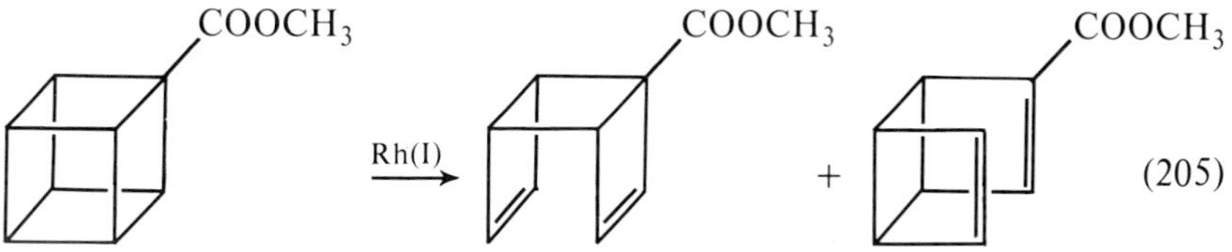

$COOCH_3$ Rh(I) $COOCH_3$ + $COOCH_3$ (205)

On the basis of the reaction intermediates isolated, the kinetic data, and the ratio of products obtained, the mechanism for these Rh(I) catalyses appears to differ from that for Ag(I) and Pd(II). Use of stoichiometric quantities of $[Rh(CO)_2Cl]_2$ with cubane and mono- and dicarbomethoxy substituted cubanes

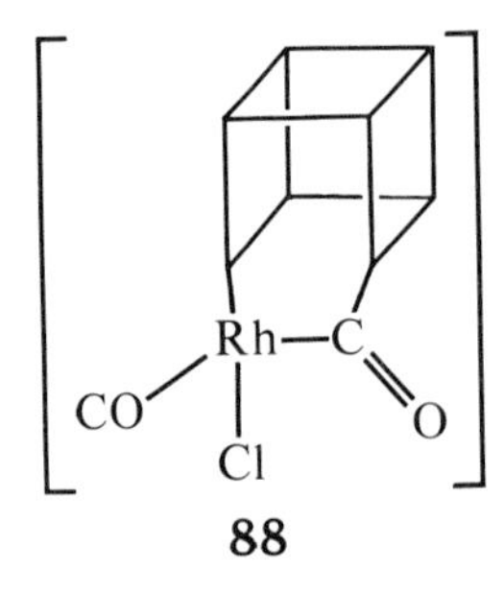

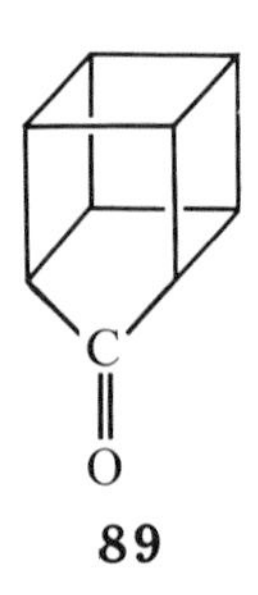

results in the isolation of intermediate complexes of structure **88**. Subsequent treatment of these intermediate species with triphenylphosphine yielded cyclic ketones (structure **89**).

The kinetic data reveal that in all cases examined the reactions follow the second-order rate law:

$$\frac{-d[\text{cubane}]}{dt} = k[\text{cubane}][\text{Rh(I) complex}]$$

The ratios of the rates observed for different catalysts with a series of substrates were found to be constant, although the rates varied by several orders of magnitude.

The product ratio for Eq. (205) for all Rh(I) catalysts examined, as well as the ratio of the products for all intermediates decomposed with triphenylphosphine (structures **90** and **91**) was approximately 1:2. This suggests that the initial

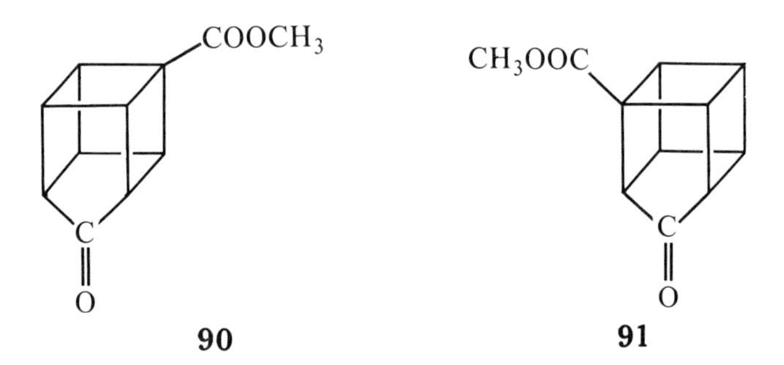

oxidative addition of the Rh(I) catalyst to carbomethoxycubane is statistical with respect to the remote carbon–carbon bonds sterically available for reaction. The oxidative addition is followed by carbonyl insertion into the rhodium–carbon bonds (in the case of the intermediates isolated) or catalytic generation of the products of Eq. (205), respectively (Scheme 2-18). The key feature of the Rh(I) cubane rearrangement reaction is thus suggested to involve insertion of the metal atom into a C—C bond.

Scheme 2-18

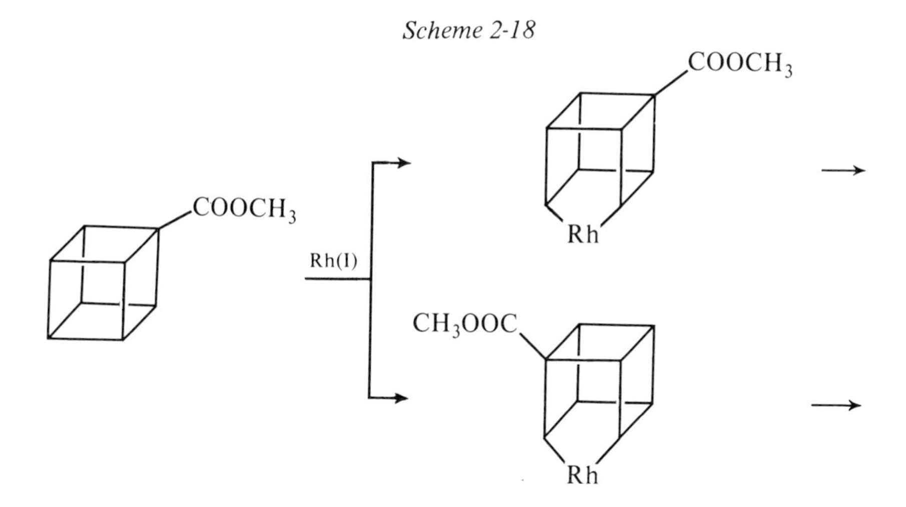

A slower catalytic process was observed to convert the product of cubane rearrangements to cyclooctatetraenes under the reaction conditions [Eq. (206)]. This reaction is also known for Ag(I),[418] and the rates for Rh(I) and Ag(I) are comparable.[417] Appropriately the rearrangement given in Eq. (207) produces semibullvalene.[414] More recent work shows that the product formed is a function of the ligand donor properties and the solvent employed [Eq. (208)].[419] Palladium(II) complexes give varying ratios of the three products. Ligands which are good σ donors and π acceptors enhance the ease of oxidative addition of Pd(II) and hence yield more of the dienes; complexes that have strong π-accepting ligands but are weak σ-donors yield less of the dienes. Increasing the polarizibility of the ligands also gives more of the dienes. For complexes having ligands which are weak σ donors and π acceptors and which are non-polarizible, the product distribution approaches that of the non-complexed metal ion (no diene). For the catalyst $PdCl_2[(C_6H_5)_3P]_2$, more polar solvents, which presumably stabilize carbonium ion intermediates, yield greater concentrations of the cage compound.

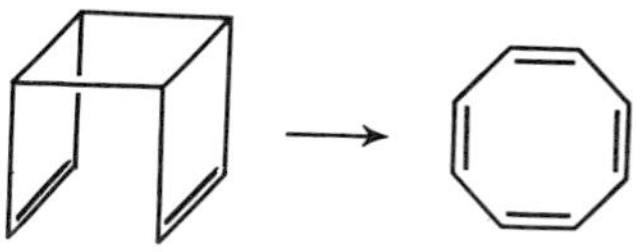

(206)

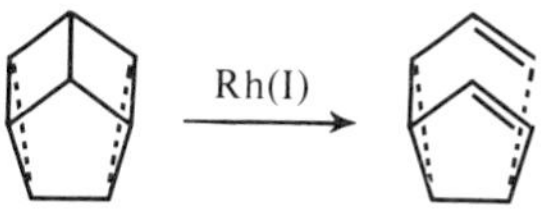

Rh(I)

(207)

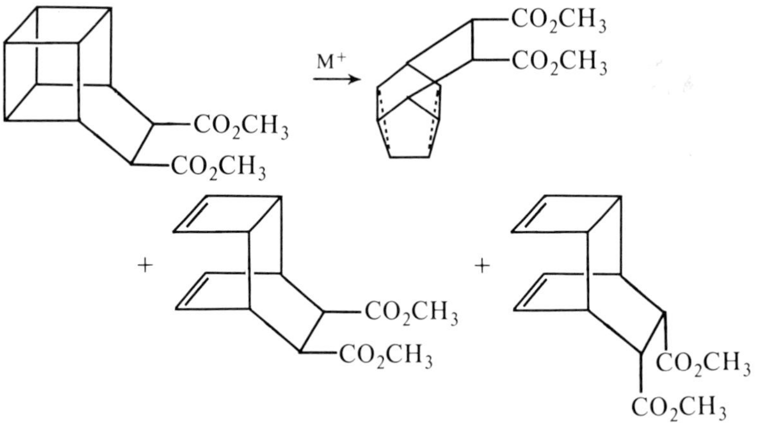
M+
CO2CH3
CO2CH3
CO2CH3
CO2CH3
+
CO2CH3
CO2CH3
+
CO2CH3
CO2CH3

(208)

C. Other Metal-Catalyzed Strained Hydrocarbon Rearrangements

A number of other transition metal-catalyzed strained hydrocarbon isomerizations are known, including the conversion of prismanes to Dewar benzenes,[420,421,422] [Eq. (209)] and quadricyclene to norbornadiene[423,424] [Eq. (210)]. Organo-transition metal complexes which shed light on the mechanism

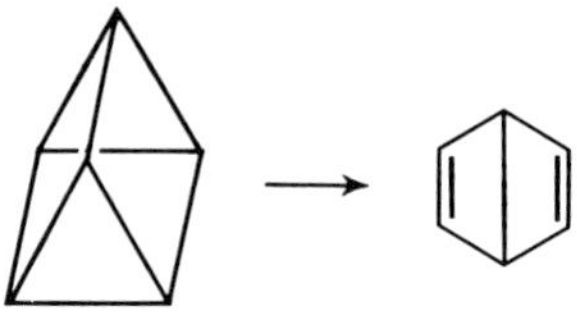

(209)

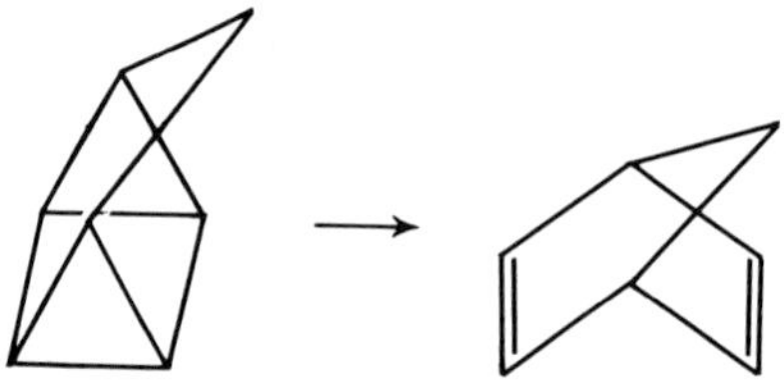

(210)

of these reaction have been reported. Platinum(II)[425] and rhodium(I)[426,427] have been observed to insert into cyclopropanes and substituted cyclopropanes yielding chelated σ-bonded organometallics; of course, these react by insertion of a carbonyl group into the rhodium–carbon bond. An analogous insertion product of rhodium(I) and quadricyclene has been reported (structure **92**).[424] Therefore, the isomerization of prismanes, quadricyclenes, and other structures by Pd(II), Pt(II), and Rh(I) catalysts is suggested to occur through formation of intermediates similar to those occurring in cubane rearrangements.

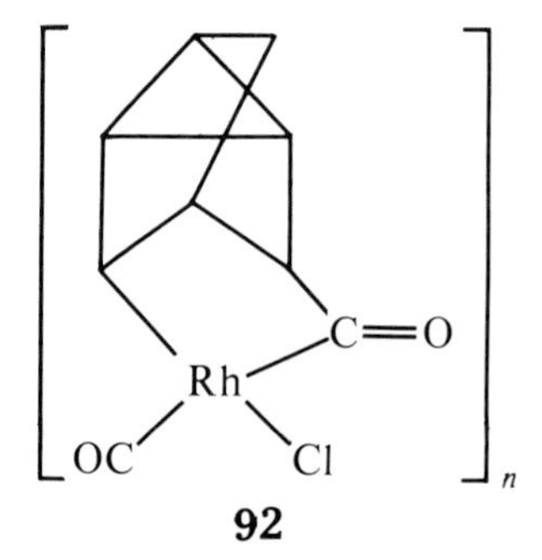

92

The isomerization of tri-*tert*-butyl-substituted prismane to the corresponding Dewar benzene[421] by a variety of catalysts, including those which are simply electron acceptors, seems reasonable by assuming removal of an electron from the substrate, thereby generating a radical cation which rearranges prior to reduction [Eq. (211)].

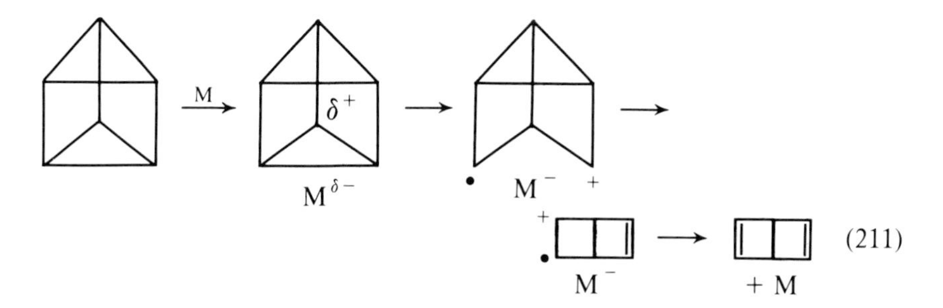

(211)

A similar redox process may be operative in the catalytic isomerization of quadricyclene to norbornadiene by various M(II) tetraphenylporphyrins.[428]

The Ag(I)- and Cu(I)-catalyzed isomerization of cyclobutene derivatives[418,429,430] provides yet another type of strained bond rearrangement. Catalytic amounts of Ag^+ and Cu^+ react as shown in Eq. (212) in acetone and methanol solutions to form cyclooctatetraene. An intermediate in this and similar reactions is indicated from the reaction of Eq. (213) in the presence and absence of maleic anhydride. The formation of the Diels-Alder adduct strongly implies the intermediate of structure **94,** which is consistent with a concerted disrotatory ring opening catalyzed by Ag^+. Benzotricyclooctadiene (structure **95**) undergoes an analogous $AgBF_4$-catalyzed reaction sequence, and the Diels-Alder adduct has been isolated. More recent studies of interest deal with the Ag(I) and Rh(I)

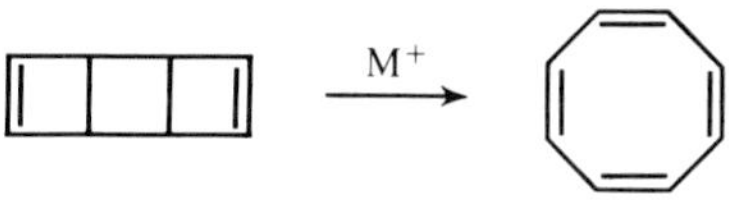

M^+ (212)

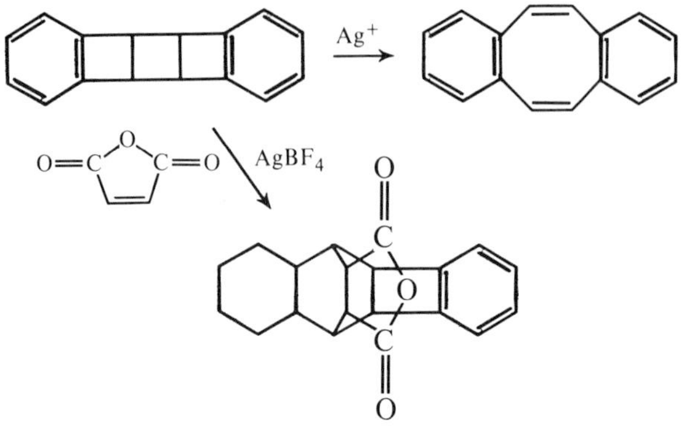

Ag$^+$

AgBF$_4$

(213)

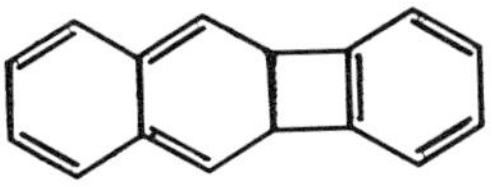

94 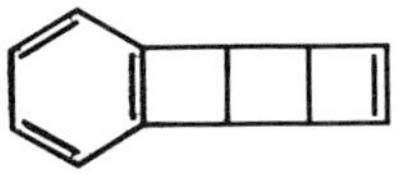 **95**

catalyzed isomerizations of *syn*-and *anti*-tricyclooctane.[431] The geometry of the tricyclooctane is of importance since both Ag(I) and Rh(I) catalyze the isomerization of the *syn* isomer [Eq. (214)] while the *anti* isomer is inert under the reaction conditions. In view of the similarities between this reaction and the cage compound isomerizations mentioned earlier, plausible mechanisms analogous to Eq. (202) (Ag(I) catalysis) or oxidative addition at the central cyclobutane ring (Rh(I) catalysis) could be suggested.

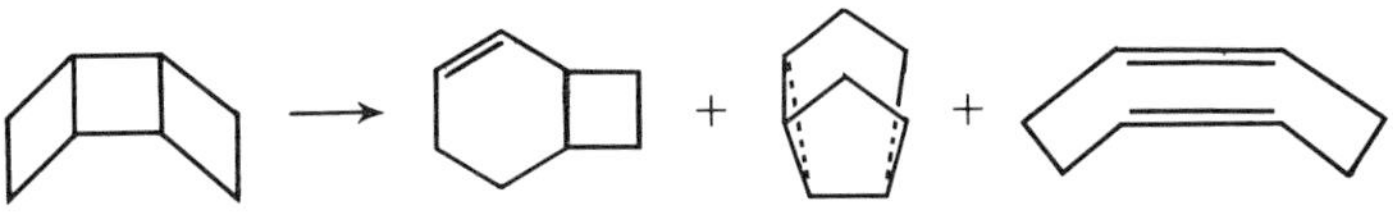

(214)

1. Olefin Metathesis. The olefin metathesis reaction[431] is a truly remarkable process which can be expressed in the general form given in Eq. (215). Such reactions were first discovered in systems using heterogeneous catalysts;[432,433]

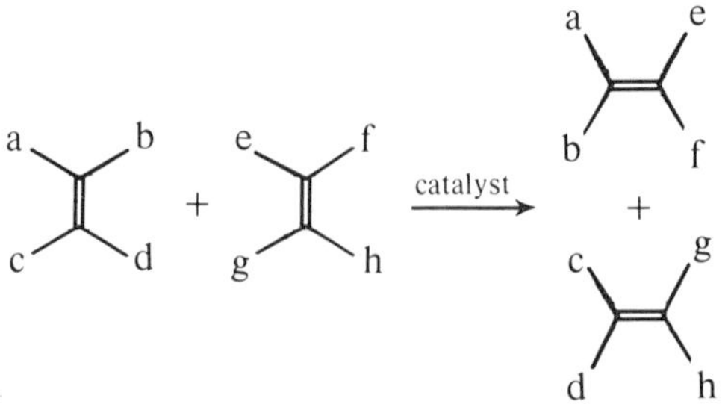

(215)

however, much activity has focused more recently on homogeneous catalysts based on tungsten, molybdenum, or rhenium.[434–442] The products isolated indicate cleavage of the substrate–olefin double bond and recombination of the fragments to produce different olefins. This cleavage has been confirmed by deuterium labeling in systems using tungsten catalysts,[438,439] as well as by ^{14}C labeling in propylene substrates with a variety of heterogeneous catalysts.[443,444] It is also characteristic of these reactions that cycloalkenes form macrocyclic polymers [Eq. (216)].[445] Since this process would be expected to involve no large

$$(CH_2)_n\left\langle\begin{matrix}CH\\ \Vert\\ CH\end{matrix}\right. + (CH_2)_m\left\langle\begin{matrix}CH\\ \Vert\\ CH\end{matrix}\right. \xrightleftharpoons{\text{catalyst}} (CH_2)_n\left\langle\begin{matrix}CH{=}CH\\ \\ CH{=}CH\end{matrix}\right\rangle(CH_2)_m \qquad (216)$$

enthalpy changes, the distribution of products should be determined by statistical and stereochemical relationships. This appears to be true. Metathesis experiments using varying proportions of 2-pentene and 6-dodecene and a tungsten catalyst gave results in excellent agreement with predictions based on random scrambling of constituents.[438] This equilibrium situation was also found for several molybdenum–nitrosyl catalysts.[442]

Kinetic data suggest that the rate-determining step in the reaction varies with the metal catalyst. For tungsten, accumulation of relatively high molecular weight macrocyclic polymers in the early stages of cyclooctene and cyclopentene polymerization suggests that the metathesis is faster than substrate exchange.[431] With molybdenum–nitrosyl catalysts and linear olefins, however, the kinetic data indicate metathesis as the rate-determining step.[442]

Intermediates proposed for reactions involving two simultaneously coordinated olefins include coordinated cyclobutane[446,447] [Eq. (217)] and four bound alkylidene groups [Eq. (218)].[448] These and related concerted mechanisms have fallen into disfavor for a number of reasons. The "cyclobutane" model would seem to require unusual theories for bonding[448,449,450] since cyclobutane itself has no accessible valencies for bonding to metal atoms. No cyclobutane has been found either to enter into the reaction or to evolve from it. Such mechanisms fail to explain known alkyne metathesis reactions[451] unless a presumably stable

$$\begin{matrix}a{-}C{=}C{-}b\\ +\\ c{-}C{=}C{-}d\end{matrix} \underset{}{\overset{M}{\rightleftharpoons}} \begin{matrix}a & & b\\ & C{-}{-}{-}C & \\ & \vdots \quad \vdots & \\ & C{-}{-}{-}C & \\ c & & d\end{matrix} \rightleftharpoons \begin{matrix}a{-}C{=}C{-}b\\ +\\ c{-}C{=}C{-}d\end{matrix} \text{ or } \begin{matrix}a{-}C{=}C{-}c\\ +\\ b{-}C{=}C{-}d\end{matrix} \qquad (217)$$

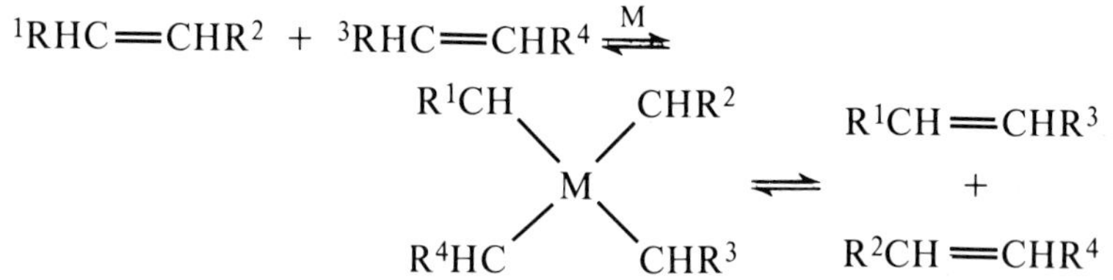

$$^{1}RHC{=}CHR^{2} + {}^{3}RHC{=}CHR^{4} \overset{M}{\rightleftharpoons} \begin{matrix} R^{1}CH & & CHR^{2} \\ & M & \\ R^{4}HC & & CHR^{3} \end{matrix} \rightleftharpoons \begin{matrix} R^{1}CH{=}CHR^{3} \\ + \\ R^{2}CH{=}CHR^{4} \end{matrix} \qquad (218)$$

cyclobutadiene intermediate occurs. Also, observations indicate that the process is not concerted but that it involves sequential steps.[451-456] The occurrence of the C_{14} hydrocarbon during the early stages of the reaction shown in Eq. (219) provides an example of this kind.[452,455]

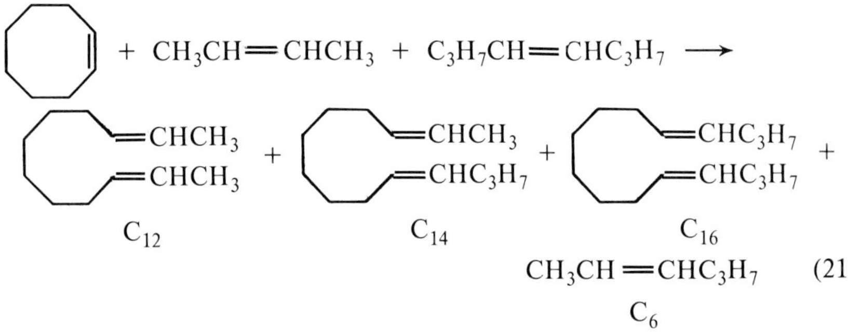

$$\text{cyclooctene} + CH_3CH{=}CHCH_3 + C_3H_7CH{=}CHC_3H_7 \longrightarrow$$

$$\underset{C_{12}}{(\text{ring}){=}CHCH_3,\ {=}CHCH_3} + \underset{C_{14}}{(\text{ring}){=}CHCH_3,\ {=}CHC_3H_7} + \underset{C_{16}}{(\text{ring}){=}CHC_3H_7,\ {=}CHC_3H_7} + \underset{C_6}{CH_3CH{=}CHC_3H_7} \qquad (219)$$

Opinion[453-456] favors a mechanism in which a single carbene exists in the active catalyst molecule (Scheme 2-19). Coordination of an olefin is presumed to be followed by an addition reaction between the carbene and the olefin to form a 4-membered metal-containing ring. When this ring opens in a way different from the reverse of its formation, olefin metathesis has occurred. Carbenes are well-established as ligands (see section VIII, D.), and recently, examples of metal

Scheme 2-19

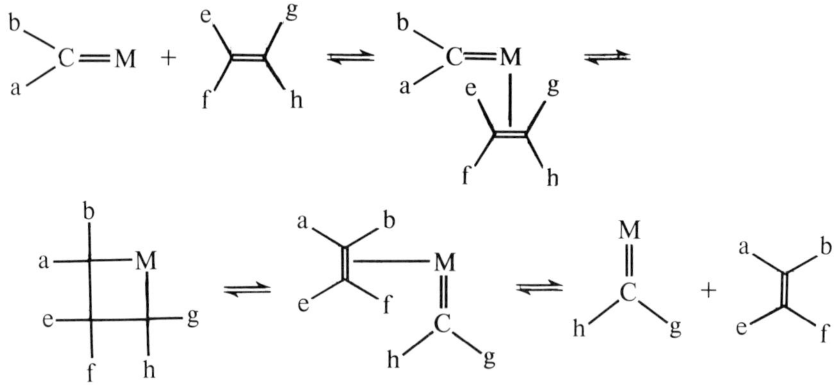

$$(a)(b)C{=}M + (e)(f)C{=}C(g)(h) \rightleftharpoons (a)(b)C{=}M \cdot [(e)(f)C{=}C(g)(h)] \rightleftharpoons$$

$$\begin{matrix} & b & \\ a- & C & -M \\ e- & C & -C-g \\ & f & h \end{matrix} \rightleftharpoons (a)(b)C{=}C(e)(f) \cdots M{=}C(h)(g) \rightleftharpoons M{=}C(h)(g) + (a)(b)C{=}C(e)(f)$$

complexes with carbenes analogous to those that would result from olefin cleavage have been reported.[457,458,459] A strong case can be made therefore for the carbene mechanism of Scheme 2-19, but as Katz and McGinnis[455] observed, "the major difficulty is finding a way to prove it."

A provocative step toward more firmly establishing the carbene mechanism has been taken through the work of Farona and Greenlee.[460] These authors have pursued the problem of identifying the carbene *first formed*, which by Scheme 2-19 is *not* a priori a fragment of the olefin or olefins undergoing metathesis. Studies were conducted with catalysts generated from rhenium carbonyl halides and aluminum alkyl halides. For the case where the halide is chlorine and the alkyl group is ethyl, the metathesis of 1-butene, the propagating carbenes are methylene and propylidene, and the products are ethylene and 3-hexene [Eq. (220)]. If the initiating carbene is formed from the alkyl group of C_2H_5-

$$2CH_2{=}CHCH_2CH_3 \xrightleftharpoons{\text{catalyst}} CH_2{=}CH_2 + CH_3CH_2CH{=}CHCH_2CH_3 \qquad (220)$$

$AlCl_2$, then it should be ethylidene, and the *first-formed* products should be propylene and/or 2-pentene (the amounts of these products formed depend on the amount of active catalyst present). In fact, the initial carbene is not ethylidene but propylidene. Its presence was explained as the result of insertion of carbon monoxide into a carbon–rhenium bond, followed successively by carbonyl alkylation and hydrogen–oxygen interchange.

VIII. ELECTROPHILIC AND NUCLEOPHILIC REACTIONS OF COORDINATED LIGANDS

Reactions of coordinated ligands which principally involve the modification of ligand electron density and distribution by the metal ion and the consequent enhancement of the susceptibility of this species to either electrophilic or nucleophilic attack are considered in this section. In view of the large amount of data available with these effects which precludes a systematic review, we include brief coverage of those areas which have been extensively reviewed, including illustrative examples to indicate dominant mechanisms and concepts. More comprehensive treatment is reserved for areas which are rapidly developing and have not been included in earlier reviews.

A. Electrophilic Addition at Coordinated Mercaptides

A number of donor atoms still possess unshared electron pairs when coordinated to metal ions. Many such species have the capability of entering into ligand reactions with electrophiles. Examples are given in Eqs. (221), (222),[461] and (223).[462]

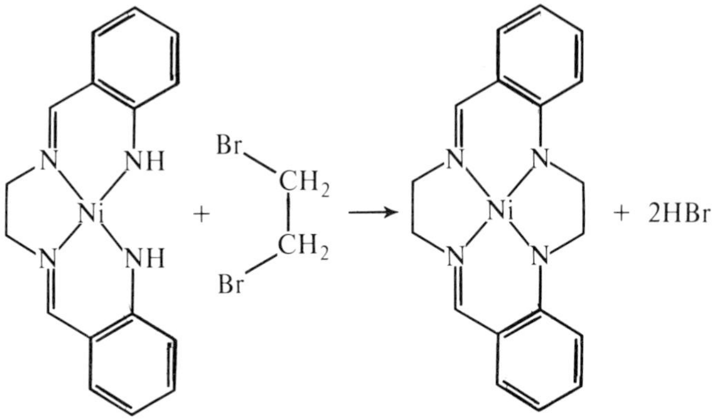

N NH Ni N NH + Br CH_2 CH_2 Br → N N Ni N N + 2HBr (221)

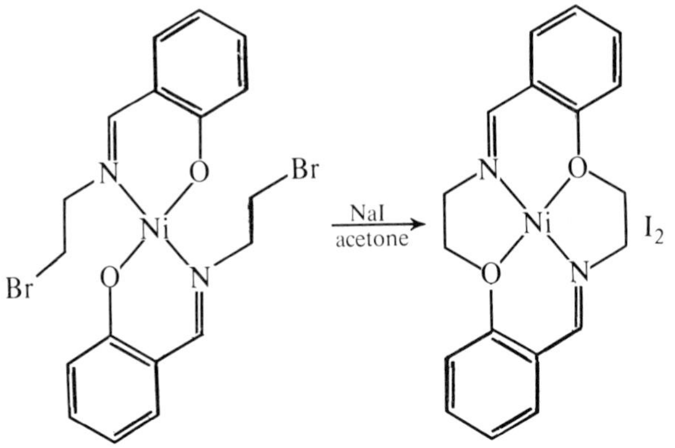

N O Ni O N Br Br $\xrightarrow[\text{acetone}]{\text{NaI}}$ N O Ni O N I_2 (222)

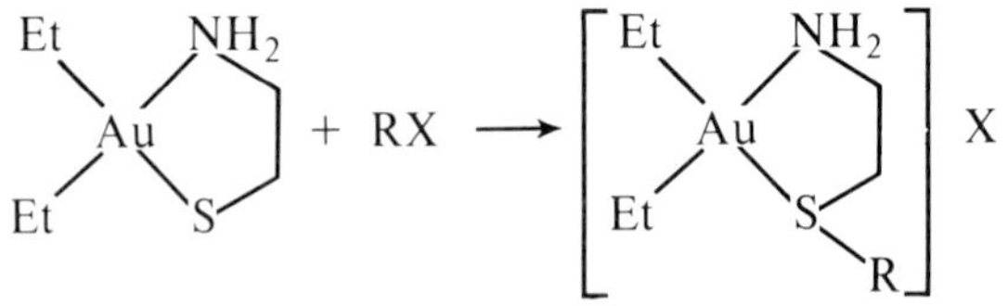

Et Et Au NH_2 S + RX → [Et Et Au NH_2 S R] X (223)

Early studies on the coordination template effect (*see* section V,C.) were based on considerations of the various reactive donor atoms.[199,463,464] Sulfur donors were chosen for a detailed investigation because history has shown many examples in which coordinated divalent sulfur acts as a nucleophile. Reports of the formation of coordinated thioethers by alkylation extend from the 1880s when Blomstrand reported his studies on mercaptides and sulfides of platinum.[465,466] He predicted the reaction of Eq. (224). In 1897 Hofman and Rabe[467] reported that $Pt(SEt)_2$ reacts with EtI and MeI in sealed tubes at 70°–80° to

$$Pt(SR)_2 + 2R'I \rightarrow Pt(SRR')_2I_2 \quad (224)$$

form the thioether complexes predicted by Blomstrand. Related reactions with HgS[468] and $Hg(SEt)_2$[467,469,470] were reported in the same era.

In the years 1916–19 Rây and co-workers published work on the alkylation of the mercaptides and thioethers of mercury with ethyl iodide and methyl iodide.[471,472,473] Much of the potential value of that work was lost because the reagents were not free of iodine, which resulted in competing oxidation reactions. Also, all reaction products were not characterized, and the concepts applied to the discussion of the product structures did not reflect an awareness of Werner's development of modern inorganic stereochemistry.

Modern studies on the subject began with the alkylation of the sulfur atom in diethyl-β-mercaptoethylaminegold(III) by Ewens and Gibson in 1949 [Eq. (223)].[462] Subsequently, copper complexes of alkyl and aryl mercaptans were used to prepare thioethers via alkylation reactions.[474,475] A number of electrophilic reactions of phenyl-substituted $C_6H_5HgSC_6H_5$ were reported by Spinelli and Salvemini.[476] It was suggested that the reactions with benzyl chlorides proceed via a four-center transition state [Eq. (225)].

$$XC_6H_4HgSC_6H_5 + C_6H_5CH_2Cl \downarrow XC_6H_4HgSC_6H_5 \cdots Cl\text{–}CH_2C_6H_5 \longrightarrow XC_6H_4HgCl + C_6H_5SCH_2C_6H_5 \quad (225)$$

By incorporating the mercaptide function into the structures of chelating ligands Busch and co-workers have been able to study the mechanism of the alkylation reaction in detail[215,464,477,478] and to exploit it in purposeful synthesis.[199,207,464,479]

The initial exploratory synthetic work quickly showed that the alkylation process may be accompanied by other changes in the coordination sphere of the metal ion. Whereas the mercaptoamine complexes of Ni(II) are low spin and square planar, the products of their reactions with alkyl halides are invariably high spin and six-coordinate [Eq. (226)].[207,464]

$$Ni(SCH_2CH_2NH_2)_2 + 2RX \longrightarrow NiX_2(RSCH_2CH_2NH_2)_2 \quad (226)$$

The reaction of the corresponding complex of Pd(II) with an alkyl halide is often accompanied by ligand displacement [Eq. (227)].

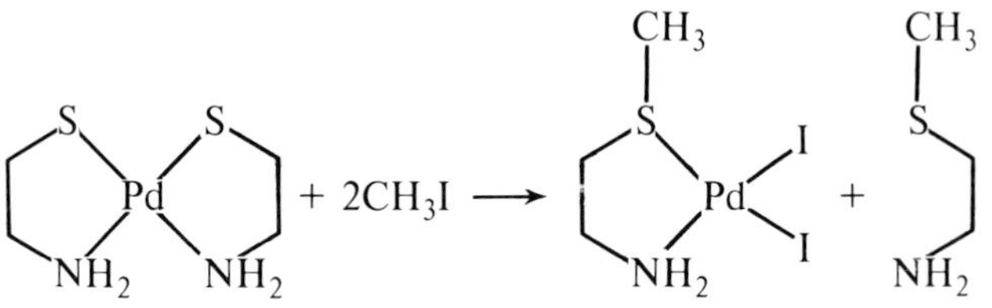

(227)

The dimeric complex of the ligand methyl-2,2′-dimercaptodiethylamine was among the earliest available complexes of a mercaptoamine having solubility characteristics suitable for detailed study.[464] The reactions of this substance clearly demonstrated the difference in electrophilic reactivity of terminal and bridged mercaptides. Only the former react under mild conditions [Eq. (228)].

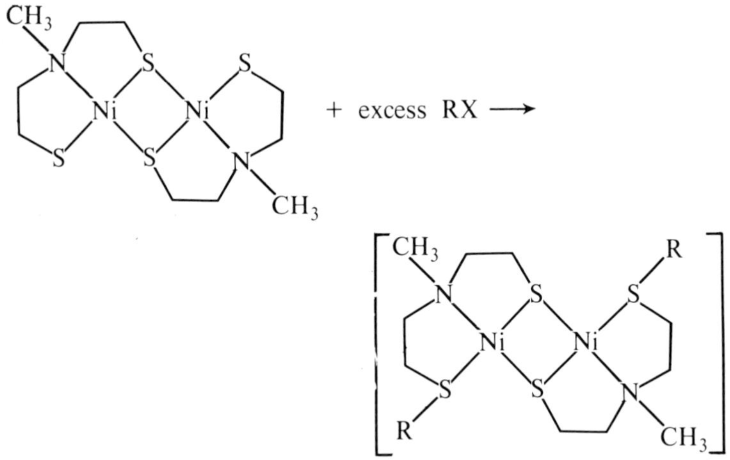

(228)

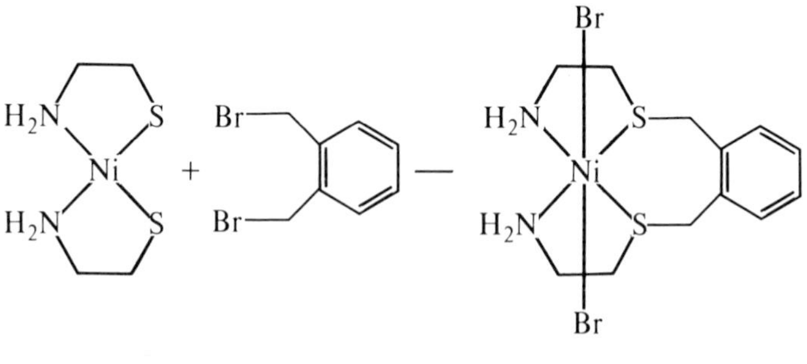

(229)

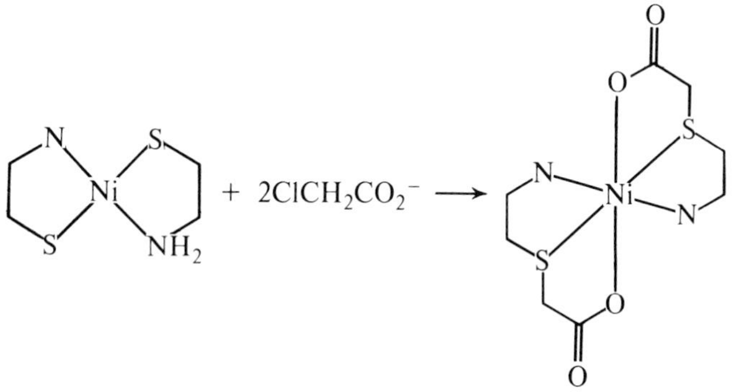

$$\text{Ni(NCH}_2\text{CH}_2\text{S)(SCH}_2\text{CH}_2\text{NH}_2\text{)} + 2ClCH_2CO_2^- \longrightarrow \text{Ni complex} \quad (230)$$

Synthetic application of these systems involved the addition of chelate rings [Eq. (229)],[199] the addition of functional groups that are capable of coordinating, [Eq. (230)],[479] the demonstration that certain structures do not contain bridged mercaptide groups,[480] and the synthesis of various macrocyclic complexes (*see* discussion of coordination template effect in section V,C).[199,212,463,481,482]

The well known dithiene or dithiolene complexes

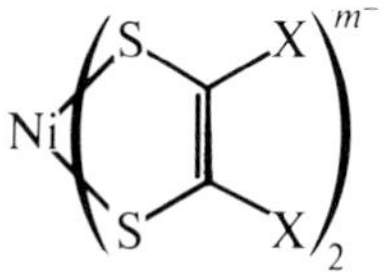

$$\mathrm{Ni}\left(\mathrm{S_2C_2X_2}\right)_2^{m-}$$

have also been subjected to alkylation studies.[478,483] Only the dianion ($m = 2$) reacts, apparently because the electron densities at the sulfur atoms in the less negative complexes are too small. Also, a significant selectivity is observed in the second alkylation step which occurs on the same bidentate ligand as the first step [Eq. (231)].

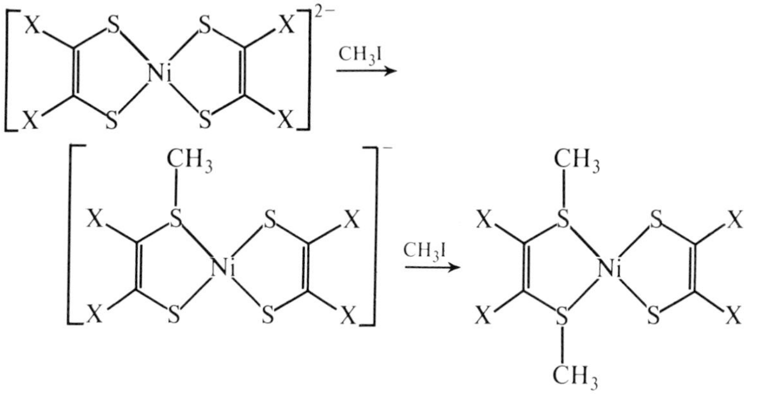

$$[\mathrm{Ni}(\mathrm{S_2C_2X_2})_2]^{2-} \xrightarrow{CH_3I} [\mathrm{Ni}(\mathrm{CH_3SSC_2X_2})(\mathrm{S_2C_2X_2})]^- \xrightarrow{CH_3I} \mathrm{Ni}((\mathrm{CH_3S})_2\mathrm{C_2X_2})(\mathrm{S_2C_2X_2}) \quad (231)$$

For the case where $X = C_6H_5$, the final product of Eq. (231) can be isolated, but the dimethylated ligand is too readily displaced from the coordination sphere for the case where X = CN.[483] Reaction with the difunctional alkylating agent α,α'-dibromo-*o*-xylene leads to cyclization but with a single ligand serving as reactant [Eq. (232)].

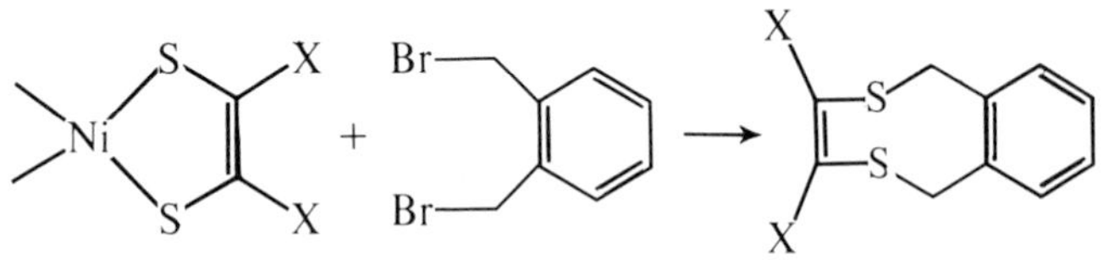

(232)

The study of the mechanisms of these reactions was initiated with the processes shown in Eqs. (226) and (228); however, in the former case, the reactant was not sufficiently soluble to permit its direct study.[464] The reactant bis-β-mercaptoethylamine nickel(II), $Ni(MEA)_2$, was solubilized by reaction with hydrated Ni^{2+} according to Eq. (233).[464] This necessitated the study of the reac-

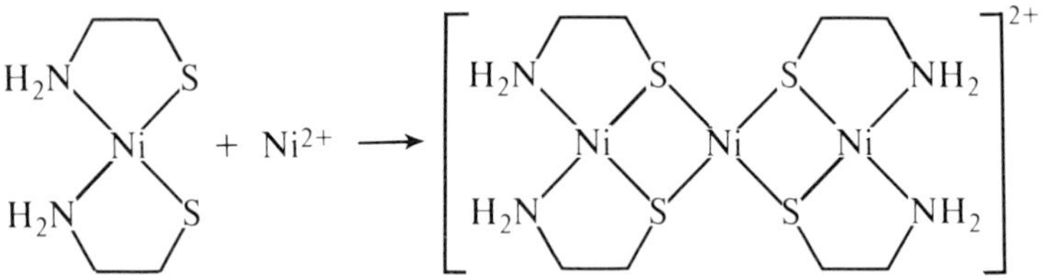

(233)

tions with alkyl halides in polar solvents; methanol was chosen. The alkylation reaction rate is pseudo first order in the presence of excess alkyl halide, but it depends parametrically on the first order of the alkyl halide concentration. The rate dependences on the nature of RX are not unusual—i.e., the rate decreases as the alkyl chain length increases and in the order X = (I > Br > Cl). Two observations support the view that the reactive complex is the monomer $Ni(NH_2CH_2CH_2S)_2$, which is produced by dissociation of $Ni_3(NH_2CH_2CH_2S)_4{}^{2+}$. The rate shows an inverse one-half-power dependence on the concentration of added Ni^{2+} ion; if Pd^{2+} is used in place of Ni^{2+} in the reaction of Eq. (233) (to form $Pd[Ni(NH_2CH_2CH_2S)_2]_2{}^{2+}$), the alkylation of the coordinated sulfur atoms is virtually stopped. The dimeric complex of structure **96** provided the first opportunity to investigate the mechanism of alkylation of a coordinated mercaptide ion under relatively uncomplicated conditions.[464] The reaction in question [Eq. (228)] was studied in chloroform solution. On the basis of the simple rate law, activation energies, and substituent effects on *p*-substituted benzyl bromides, the mechanism in Scheme 2-20 was suggested. All evidence supported bimolecular nucleophilic reaction between

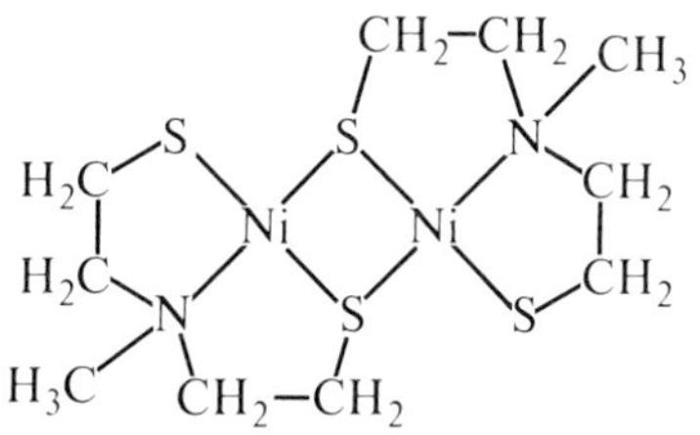

96

the coordinated sulfur and the carbon atom of the alkyl halide. At the same time, it strongly disagreed with models requiring dissociation of a coordinated RS^- group prior to reaction with the alkyl halide.

A family of complexes of the formula $Ni(X\text{-}MEA)_2$ (where X-MEA is a substituted mercaptoethylamine) was developed for the purpose of investigating the mechanism of reaction of coordinated mercaptides.[76,484] In order to establish definitely the coordination template effect, it was necessary to show that the sulfur nucleophile remains bound to a metal ion during addition of an electrophilic group. From among the several ligands studied, *N,N*-dimethyl-β-mercaptoethylamine (Me_2L), *n*-*N*-propyl-β-mercaptoethylamine (prL), and ethyl cystineate ($NH_2CH(COOEt)CH_2SH$, EtCy) illustrate the results.[477] For the reactions of $Ni(Me_2L)$ with various *p*-substituted benzyl bromides in 1,2-dichloroethane,[477] a simple second-order rate law is observed:

$$\text{rate} = k'[Ni(Me_2L)_2][RX]$$

However, the reactions of $Ni(PrL)_2$ with the same reagents do not exhibit so simple a behavior. Instead, the experimental rate law [Eq. (234)] corresponds to a pre-equilibrium complexation step [Eq. (235)] in which the alkyl halide bonds to the metal ion and is followed by an intramolecular rate-determining

Schemes 2-20

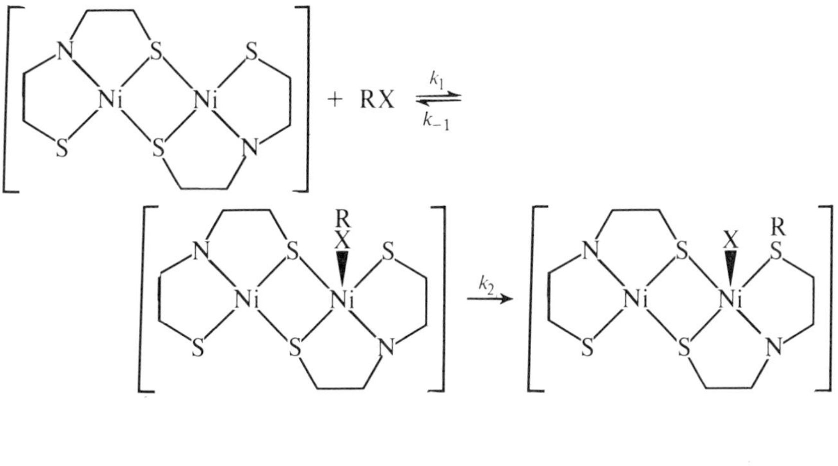

$$\text{rate} = \frac{kK[\mathrm{Ni(Me_2L)_2}][\mathrm{RX}]}{1 + K[\mathrm{RX}]} \tag{234}$$

$$\mathrm{Ni(PrL)_2 + RX \overset{K}{\rightleftharpoons} Ni(PrL)_2{\cdot}RX} \tag{235}$$

$$\mathrm{Ni(PrL)_2 \cdot RX \xrightarrow{k} Ni(PrL)(PrLR)X} \tag{236}$$

nucleophilic displacement step [Eq. (236)]. Electronic spectral data were also presented in support of the formation of the alkyl halide complex.[477]

The proposed mechanism is shown in Scheme 2-21. This model presumes that the complex probably exists in the trans configuration. An x-ray structure determination[485] has shown this to be true in the case of $Ni(Me_2L)_2$ in the solid state. The crystal structure also confirms the expectation that one of the pair of alkyl groups on the nitrogen atom of Me_2L is equatorially oriented and the other is axially oriented in the nickel complex. The axial alkyl group effectively blocks the region above the plane of coordination of the nickel(II) atom, thereby preventing çoordination of any fifth ligand in the case of $Ni(Me_2L)_2$. Thus, $Ni(Me_2L)_2$ reacts with alkyl halides by a straightforward S_N2 mechanism. In contrast, replacing an alkyl group by a hydrogen atom, as in $Ni(PrL)_2$, makes the metal atom vulnerable to attack by ligands from above and below the plane. This accounts for the difference in behavior between the secondary and the teritiary amine derivatives.

Schemes 2-21

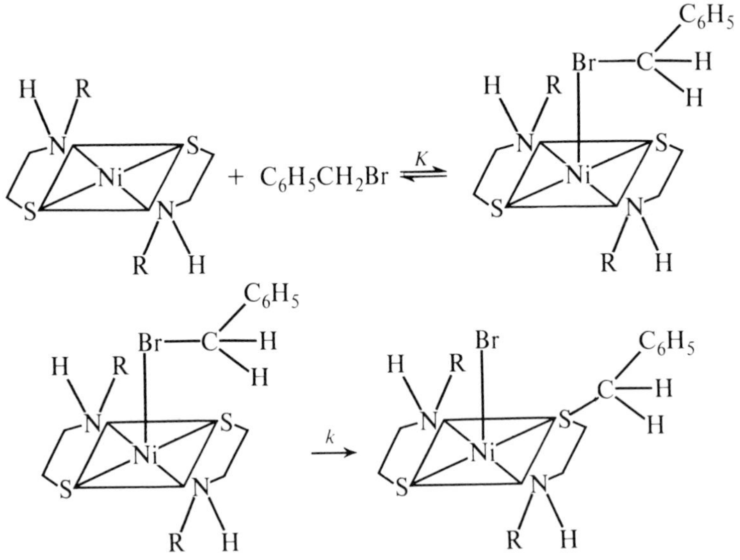

Further investigations also revealed that the coordination of the RX molecule accelerates the rate substantially. Most simply, the complex $Ni(RL)_2$ reacts with benzyl bromide three times as fast as does $Ni(Me_{2i}L)_2$, and the rate is independent of the length of the R group. The results of studies on substituent effects are more important in this regard.[478] Application of the Hammett equation showed significant variations in the extent of polarization of the carbon–halogen bond depending on X in $X\text{-}C_6H_4CH_2Br$. The rate data for the complex of ethyl cysteinate gave the same substituent effect sequence as simple sulfur-containing nucleophiles. This implies that the alkylation process proceeds via an S_N2 mechanism in this case. Since the carbethoxy groups of ethyl cysteinate are expected to hinder greatly the approach of any ligand above or below the plane of coordination, $Ni(EtCy)_2$ is a good basis for comparison. In contrast, the rates for $Ni(RL)_2$ follow an order that is typical of S_N1 reactions. It therefore seems clear that the substituted benzyl bromides are not only coordinated to Ni(II) via their bromide atoms prior to the reaction between the carbon and sulfur atoms, but this coordination substantially activates the carbon–bromine bond so that bond breaking plays a major role in the replacement process. It is also significant that the rate of reaction of $Ni(Me_2L)_2$ is relatively insensitive to the nature of the substituent in $p\text{-}X\text{-}C_6H_4CH_2Br$. It is concluded that although the steric requirements of Me_2L prevent the formation of detectable concentrations of the complex $Ni(Me_2L)_2 \cdot RX$, the alkyl halide still interacts weakly with the Ni(II) atom, through its halogen atom, in the transition state.

The results summarized here are particularly important from the standpoint of the use of the reactions of coordinated mercaptide functions in template processes. These detailed studies strongly indicate that the mercaptide ion remains bound to the metal ion throughout the course of its reaction with electrophiles.

B. Electrophilic Addition and Substitution on Charge-Delocalized Six-Membered Chelate Rings

Electron-rich ligands coordinated to metal ions are known to undergo electrophilic attack. Despite the large body of data dealing with π-bonded species (cyclopentadienyl,[486,487] cyclobutadiene,[488] etc.), we wish to restrict our attention to those systems involving σ-bonded chelate rings and to concentrate on those which yield reactive products.

Electrophilic attack on the central carbon in charge-delocalized six-membered chelate rings (*see* section V,B.) resembles the process of electrophilic substitution on aromatic rings and is thought to occur by an analogous mechanism. Thus, a cationic electrophile is suggested to add to the central carbon in the chelate ring, leading to the formation of a tetrahedral carbon atom [Eq. (237)]. This product may be stable, or by subsequent nucleophilically assisted elimination of the original substituent on the reactive carbon, electrophilic substitution may occur [Eq. (238)].

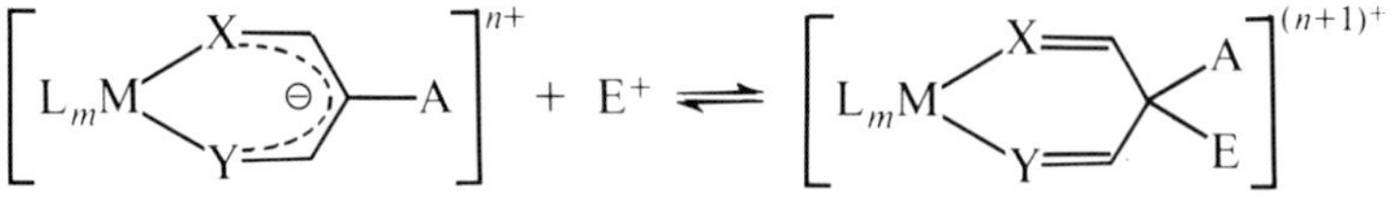

$$\left[L_mM \langle X, Y \rangle \ominus\text{—}A \right]^{n+} + E^+ \rightleftharpoons \left[L_mM \langle X{=}, Y{=} \rangle C(A)(E) \right]^{(n+1)+} \tag{237}$$

$$\left[L_mM \langle X{=}, Y{=} \rangle C(A)(E) \right]^{(n+1)+} + N^- \longrightarrow \left[L_mM \langle X, Y \rangle \ominus\text{—}E \right]^{n+} + NA \tag{238}$$

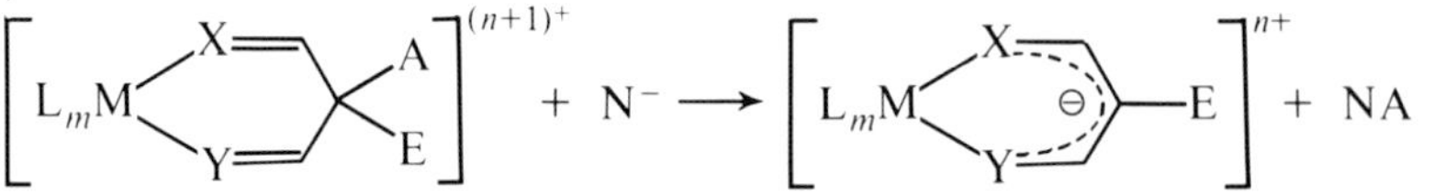

Much of the early work reported in this area has involved the reactions of enolate anions of 1,3-diones coordinated to nonlabile transition metal ions, investigated extensively by Collman.[187,489] The unsubstituted central carbon atom in a variety of such nonlabile acetylacetone complexes was found to undergo substitution with an enormous number of electrophiles [Eq. (239)].

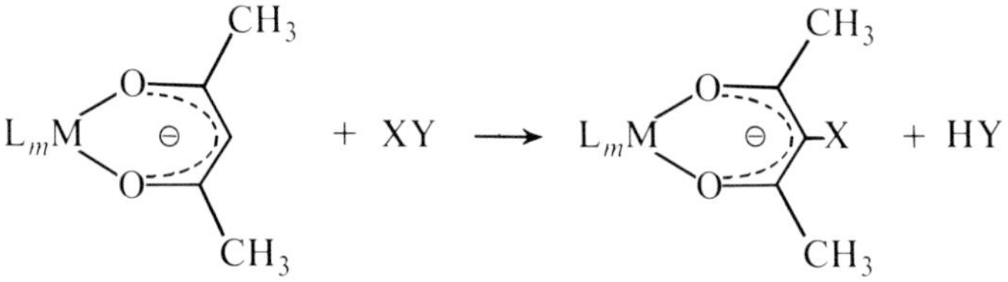

$$L_mM(O,O\text{-acac}, CH_3, CH_3) + XY \longrightarrow L_mM(O,O\text{-acac}, CH_3, CH_3)\text{—}X + HY \tag{239}$$

M = Cr, Co, Rh, Fe, Be, La, Al, Eu

X = Cl, Br, I, SCN, S-aryl, NO_2, $CH_2N(CH_3)_2$, $\overset{O}{\overset{\|}{C}}H$, $\overset{O}{\overset{\|}{C}}R$

Evidence in support of the ligand remaining coordinated during the reaction comes from exchange studies with free ^{14}C-labelled acetylacetone,[490] as well as from retention of optical activity by partially resolved trisacetylacetone chelates upon electrophilic substitution.[491] In addition to the relatively straightforward reactions with halogens and *N*-halosuccinimides leading to nucleophilically inert halo-substituted chelate rings,[489] several substituents have been introduced which offer reactive sites in the resultant complexes. Thus, nitration of $Cr^{III}(acac)_3$ with $Cu(NO_3)_2{\cdot}3H_2O$ in acetic anhydride leads to the mononitro species [Eq. (240)].

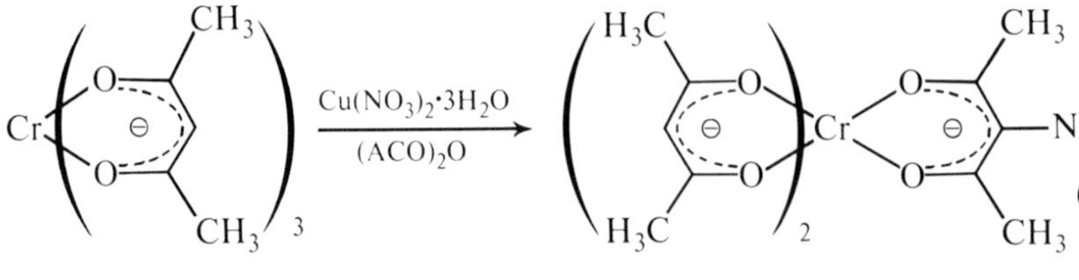

$$Cr(acac)_3 \xrightarrow[(ACO)_2O]{Cu(NO_3)_2\cdot 3H_2O} (acac)_2Cr(3\text{-}NO_2\text{-acac}) \tag{240}$$

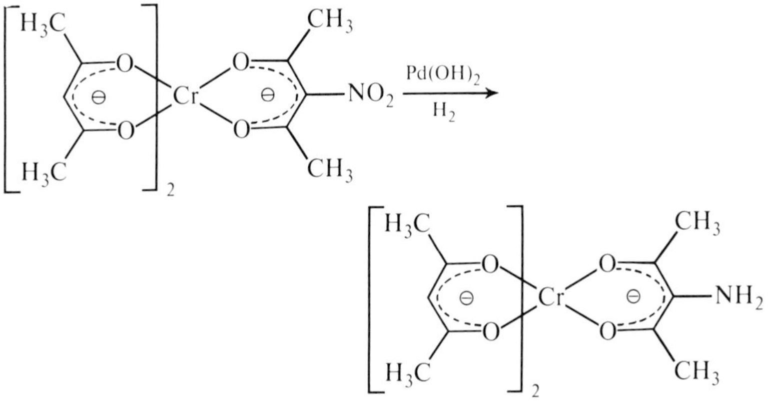

$$[\text{Cr(acac)}_2](\text{3-NO}_2\text{-acac}) \xrightarrow[H_2]{Pd(OH)_2} [\text{Cr(acac)}_2](\text{3-NH}_2\text{-acac}) \quad (241)$$

Catalytic hydrogenation of the product formed the amino derivative [Eq. (241)]. Reaction of the amino complex with acidic sodium nitrate then produced the diazonium salt, demonstrating the aromatic nature of acetylacetone chelate rings [Eq. (242)].

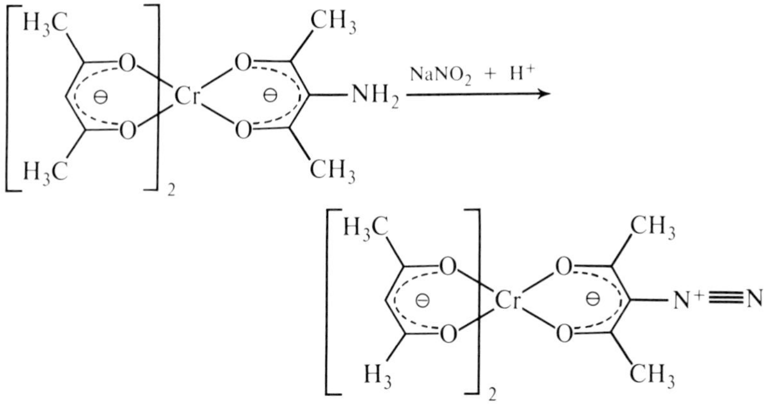

$$[\text{Cr(acac)}_2](\text{3-NH}_2\text{-acac}) \xrightarrow{NaNO_2\ +\ H^+} [\text{Cr(acac)}_2](\text{3-N}^+{\equiv}\text{N-acac}) \quad (242)$$

Other reactive intermediate species formed by electrophilic displacement reactions of M^{III} $(acac)_3$ complexes include acyl,[187,492] chlorosulfenyl,[493] dimethylaminomethyl,[494] and chloromethyl[494] derivatives. These complexes react further to yield a variety of products. In fact, the chloromethyl species proved to be too reactive to isolate and was characterized on the basis of its derivatives [Eq. (243)].

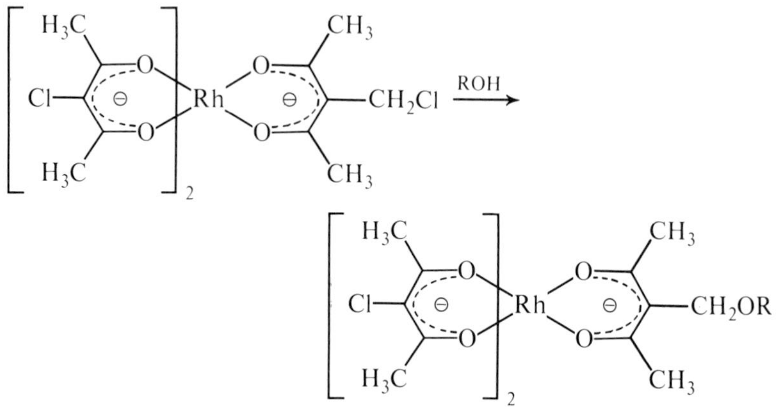

$$[(\text{3-Cl-acac})_2\text{Rh}(\text{3-CH}_2\text{Cl-acac})] \xrightarrow{\text{ROH}} [(\text{3-Cl-acac})_2\text{Rh}(\text{3-CH}_2\text{OR-acac})] \quad (243)$$

Formylation of these M^{III} $(acac)_3$ complexes [Eq. (244)] resulted in the monoformyl derivatives predominantly.[495] These materials have proved to be inert to both nucleophilic and electrophilic reagents. Thus, the unsubstituted rings could be nitrated and halogenated, the formyl group being unreactive with both oxidants and typical nucleophilic species known to attack aldehyde carbon atoms. Also, no reaction was observed in the presence of Grignard reagents.

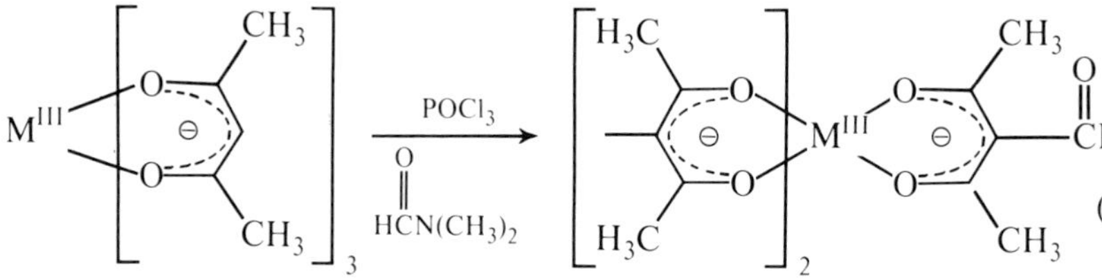

$$M^{III}(acac)_3 \xrightarrow[\text{HC(O)N(CH}_3)_2]{\text{POCl}_3} [(acac)_2 M^{III}(\text{3-CHO-acac})] \quad (244)$$

M = Cr, Co, Rh

Electrophilic attack on charge-delocalized six-membered chelate rings followed by chelate ring isomerization has also been reported for labile transition metal complexes. Thus, exposure of acetylacetoneimine complexes of nickel(II) and Cu(II) to either nitric oxide or acidic solutions of sodium nitrite leads to the formation of isonitroso species [Eq. (245)].[496]

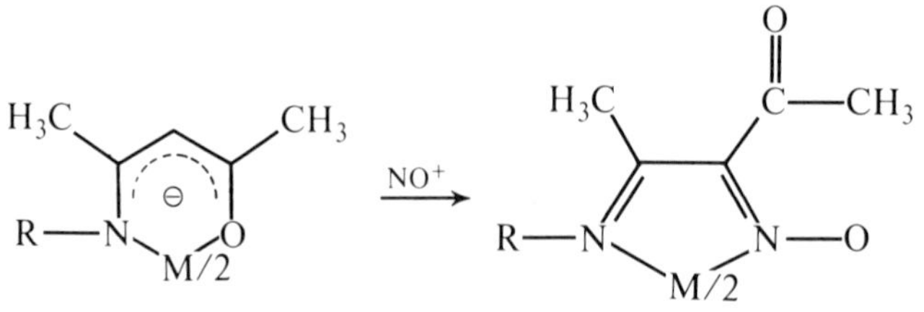

$$\text{R—N(M/2)O chelate (H}_3\text{C, CH}_3) \xrightarrow{\text{NO}^+} \text{R—N(M/2)N—O chelate (H}_3\text{C, C(O)—CH}_3) \quad (245)$$

Here, the reaction presumably involves electrophilic attack by NO^+ on the central carbon of the chelate ring, followed by rearrangement to form an oxime and subsequent coordination of the oxime nitrogen donor atom to the transition metal ion.

Incorporation of the charge-delocalized chelate ring into a macrocyclic ligand structure significantly alters the chemistry associated with such species by stabilization of otherwise unstable intermediates. This is clearly seen in the protonation of such charge-delocalized chelate rings leading to stable products under conditions where substitution-inert acetylacetone complexes are rapidly decomposed [Eq. (246)].[249,497,498]

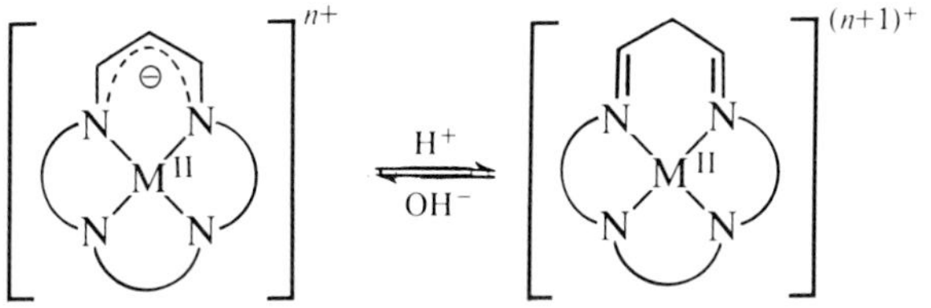

$$[M^{II}N_4]^{n+} \underset{OH^-}{\overset{H^+}{\rightleftharpoons}} [M^{II}N_4]^{(n+1)+} \quad (246)$$

M = Ni, Cu

A further difference in chemical behavior imposed by the constraints of a macrocyclic ligand system is seen in the reactions of complex **97** with sources of NO^+.[499] On exposure of **97** to conditions of oxidative nitrosation, the observed product is the dinitro species [Eq. (247)].

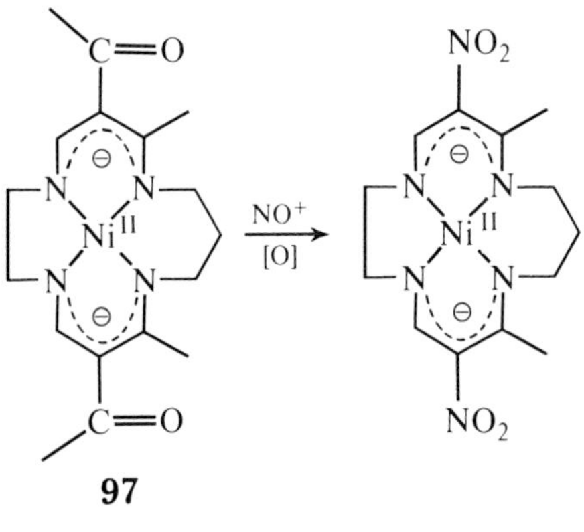

$$\mathbf{97} \xrightarrow[\text{[O]}]{NO^+} Ni^{II}(NO_2)_2 \text{ complex} \quad (247)$$

97

The charge-delocalized chelate rings cannot rearrange subsequent to attack by NO^+ and hence are readily oxidized to the observed nitro complexes.

Another type of intermediate isolated from the attack of electrophiles on charge-delocalized chelate rings in macrocyclic ligand systems is formed on treatment of complex **97** with bromine [Eq. (248)].[156]

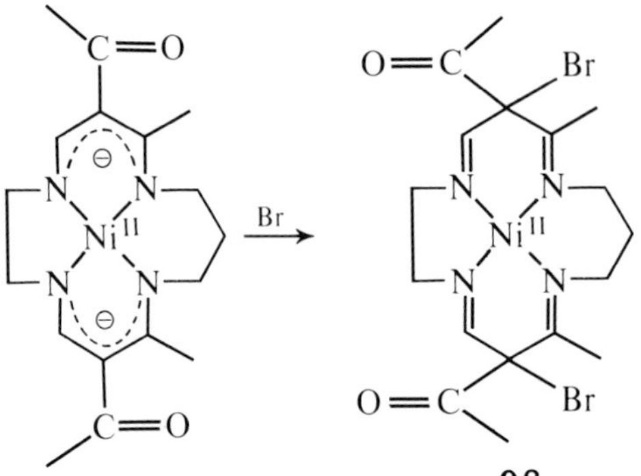

(248)

98

Complex **98** is readily convertible to **97** in the presence of suitable Br^+ acceptors, thus affording a source of Br^+. This reversible binding of electrophiles by such chelate rings in macrocyclic complexes may suggest a behavior associated with iron porphyrin-containing species. Dolphin et al.[500] have isolated several materials corresponding to the addition of cationic electrophiles to the framework of metal porphyrin species at the meso position [Eq. (249)].

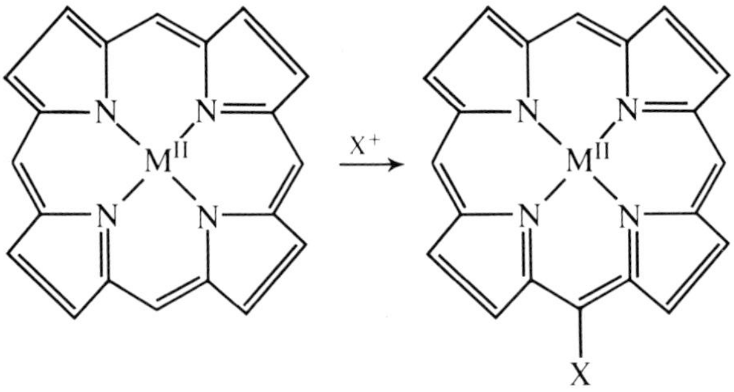

(249)

This, together with the reported behavior of the heme enzyme chloroperoxidase, which halogenates organic substrates in the presence of X^- and hydrogen peroxide [Eq. (250)] and which has been shown to incorporate into its active intermediate the constituents of OH^+,[99,501] may indicate the binding of electrophiles at the central carbon atom in charge delocalized six-membered chelate rings to be a significant role in the biochemistry of heme enzymes.

$$H_2O_2 + X^- + S \xrightarrow{CPO} S\text{-}X + H_2O + OH^- \qquad (250)$$

X = Cl, Br, I

Electrophilic attack on the central carbon atom of such charge-delocalized chelate rings is, however, influenced by substituents on this carbon atom. Thus, complex **97** reacts with active alkylating agents to form **99** in which the formerly inert acetyl substituents have been converted to reactive alkoxyethylidene groups. In contrast, the deacylated species **100** is readily alkylated to form, after deprotonation, the ring-substituted product [Eq. (251)].[156]

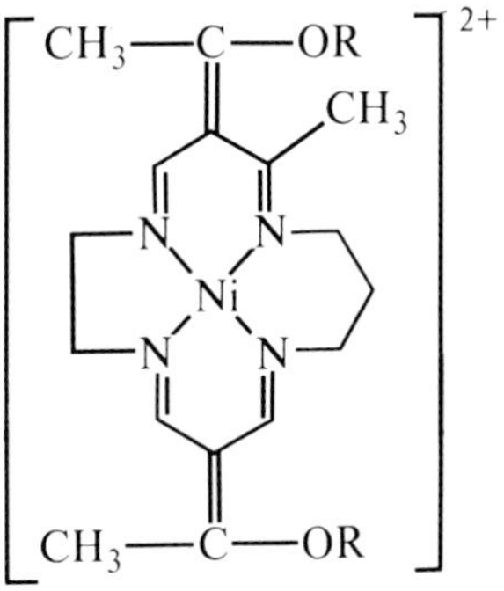

CH_3—C—OR
CH_3
N N
Ni
N N
CH_3—C—OR
$2+$

99

CH_3
N N
Ni
N N
CH_3

100

1. CH_3SO_3F
2. NaOEt

CH_3
CH_3
N N
Ni
N N
CH_3
CH_3

(251)

Macrocyclic complex **100** has been subjected to an extensive series of electrophilic substitution reactions[503] including: mono- and diacetylation with acetic anhydride; diacylation with *p*-substituted benzoyl chlorides; Michael addition of methyl vinyl ketone, ethyl acrylate, and 2-vinylpyridine; condensation with α-napthylisocyanate; and coupling with *p*-nitrobenzenediazonium hexafluorophosphate. Electrochemical and spectroscopic studies have shown that the presence of the various substituents at the 6- and 14-positions on the macrocyclic ligand have a profound effect on the ligand field experienced by the metal ion. For example, the potential for the Ni(II)/Ni(III) couple is caused to vary over a range of approximately 0.9 volt by changing the substituent.

C. Electrophilic Addition to σ-Alkyls, Alkenyls, and Alkynyls

A variety of organotransition metal σ-bonds undergo electrophilic attack, leading to insertion reactions. Neglecting the vast areas of unsaturated hydrocarbon and fluorocarbon electrophiles, we examine the chemistry of several electrophilic species which have been reported to insert into σ-metal–alkyl and –aryl bonds and which exhibit more complicated reactivity patterns with σ-bonded, unsaturated hydrocarbons.

Sulfur dioxide insertions[113,132,504] into transition metal–alkyl and –aryl bonds, most of which have been performed in liquid SO_2 on π-$C_5H_5Fe(CO)_2R$ complexes,[113] are of particular interest because of the relatively large amount of data available and because they strongly contrast with carbon monoxide insertion (alkyl migration) reactions. The most evident difference is that SO_2 undergoes facile insertion into transition metal–alkyl and –aryl bonds without prior or concommitant coordination to the transition metal center (*see* section VI,E.). SO_2 appears to attack electrophilically on the back side of the α-carbon (S_E2) (at least in the case of π-$C_5H_5Fe(CO)_2R$ in liquid SO_2) leading to a M-OS(O)R[505] species which then rearranges to the observed M—$S(O)_2R$ product. This reaction, which is generally irreversible, has been demonstrated for many low-valent L_mM—R organometallic species.

The mechanism proposed[113] to account for these observations (for coordinatively saturated systems) is shown in Eq. (252).

$$M{-}CR_3 + SO_2 \longrightarrow \left[\overset{\delta+}{M} \cdots CR_3 \cdots \overset{\delta-}{S}O_2 \right] \longrightarrow M^+O_2SCR_3^- \longrightarrow M{-}S(=O)_2{-}CR_3 \quad (252)$$

$$M^+O_2SCR_3^- \longrightarrow M{-}OS(=O)CR_3 \longrightarrow M{-}S(=O)_2{-}CR_3$$

In support of this are the abundant examples of M—R species which are reported to react without prior ligand dissociation,[113,132,504] as well as clear substituent and ligand effects which indicate initial electrophilic attack by the sulfur atom on the M—R bond. Thus, kinetic data for SO_2 insertion into π-$C_5H_5Fe(CO)_2R$ (R = alkyl, aryl) compounds in liquid SO_2, as measured by infrared spectroscopy, indicate that the rate decreases as electron-withdrawing groups are introduced on R.[140] This trend is also found for the reactions of π-$C_5H_5Mo(CO)_3R$

complexes.[506] Substitution of more strongly basic phosphine ligands for carbon monoxide, however, is observed to increase the rate of reaction in the π-$C_5H_5Fe(CO)_2R$[141] and π-$C_5H_5Mo(CO)_3R$[506] systems. Other information supporting this mechanism includes large negative entropies of activation in SO_2 solution which suggest a polar transition state, and second-order kinetics when the reactions are carried out in organic solvents. Also, SO_2 insertion was determined to be largely stereospecific in the case of optically active π-$C_5H_5Fe(CO)_2CH(CH_3)C_6H_5$[507] and to involve predominant inversion of configuration[508] for π-$C_5H_5Fe(CO)_2CHDCHDC(CH_3)_3$.

Introduction of carbon–carbon unsaturation at the 2-position of R groups of transition metal organometallic species produces more diverse reaction patterns. In the case of 2-alkenyls, SO_2 attack appears to occur on the remote olefinic carbon atom [Eq. (253)][113,509] in addition to the normal reaction sequence proceeding, in the former case, via a coordinated olefinic intermediate to the rearranged SO_2 insertion product [Eq. (254)]. In support of this, intermediates

$$M—CH_2—CH{=}CHR + SO_2 \longrightarrow M\cdots\left(\begin{matrix}CH_2\\ \| \\ CH\end{matrix}\right)—CHR—\overset{\overset{O}{\|}}{S}—O— \tag{253}$$

$$M\cdots\left(\begin{matrix}CH_2\\ \| \\ CH\end{matrix}\right)—CH_2—S(=O)—O— \longrightarrow M\overset{\overset{O}{\|}}{\underset{\underset{O}{\|}}{S}}CHRCH{=}CH_2 \tag{254}$$

corresponding to the product of Eq. (253) have been detected during SO_2 insertion into the M—R bonds of π-$C_5H_5Fe(CO_2)_2CH_2CH{=}CH_2$ and π-$C_5H_5Mo(CO)_3CH_2CH{=}CH_2$, and reactions with cationic electrophiles (H^+, $(C_6H_5)_3C^+$)[509] have been reported to lead to structure **101,** isolated as its BF_4^- salt.

$$\left[\pi\text{-}C_5H_5Fe(CO)_2\left(\begin{matrix}CH_2\\ \| \\ CH—CH_2R\end{matrix}\right)\right]^+$$

101

R = H, $(C_6H_5)_3C$

A different reaction sequence is reported for the interaction of SO_2 with π-$C_5H_5Fe(CO)_2CH\langle CH_2CH_2\rangle$ (cyclopropyl).[509] Here the initial electrophilic attack occurs at the C_1—C_2 bond leading to the carbenoid species of Eq. (255) as shown by the deuterium-labelled starting material. This intermediate can then form the cyclic product by nucleophilic attack on the α-carbon or, by hydrogen migration, lead to the olefin.

$$\pi\text{-}C_5H_5Fe(CO)_2\text{—}C_1\langle C_2C_3\rangle + SO_2 \longrightarrow \pi\text{-}C_5H_5Fe^{+}(CO)_2\text{—}CH\text{—}CH_2\text{—}CH_2SO_2^{-} \tag{255}$$

$$\nearrow\ \pi\text{-}C_5H_5Fe(CO)_2\text{—}C(\text{cyclic }CH_2CH_2S(=O)O)$$

$$\downarrow\ \pi\text{-}C_5H_5Fe^{+}(CO)_2\text{—}\|(CH_2{=}CH\text{—}CH_2SO_2^{-})$$

Tetracyanoethylene (TCNE), an electrophilic olefin, which has been reported to react with transition metal–alkyl and –aryl bonds leading to insertion products of Eq. (256),[127] also reacts with the Fe–σ-cyclopropyl complex to form structure **102,** presumably via an intermediate analogous to that of Eq. (255). The reaction of TCNE given in Eq. (257) can be rationalized in a similar way.[510] Also, both

$$\pi\text{-}C_5H_5Fe(CO)_2R + (NC)_2C{=}C(CN)_2 \longrightarrow$$

$$\pi\text{-}C_5H_5Fe(CO)_2C(CN)_2\text{—}C(CN)_2\text{—}R + C_5H_5Fe(CO)_2\text{—}N{=}C{=}C(CN)\text{—}C(R)(CN)_2 \tag{256}$$

CO
π-C_5H_5Fe
CN
CN
CO CN CN

102

CO
π-C_5H_5Fe—$CH_2C{=}CH_2$
CO R
TCNE →
CO
CH_2 CN CN
π-$C_5H_5Fe \cdots \|$
C—CH_2C—$O^{\ominus}$
CO
R CN CN
→

CO
CN
CN
π-C_5H_5Fe
CN
CN
CO R
(257)

CO
O
π-C_5H_5Fe
S—O
CO
↑

CO
π-C_5H_5Fe—CH_2—CH(CH_2)(CH_2)
CO
93
E →
CO
CH_2 E^-
π-$C_5H_5Fe^+ \cdots \|$
CH_2 CH_2
CO CH_2
(258)
↓

CO
CN
CN
CN
π-C_5H_5Fe
CN
CO

TCNE and SO_2 react with structure **93** to form analogous products, presumably through a similar intermediate [Eq. (258)]. 2-Alkynyl groups in these systems also react with SO_2 to form cyclic products. The mechanism proposed[113,511,512] is shown in Eq. (259) for π-$C_5H_5Fe(CO)_2(CH_2C{\equiv}CCH_3)$, where the structure

$$\pi\text{-}C_5H_5Fe(CO)_2\text{—}CH_2C\equiv CCH_3 + SO_2 \longrightarrow \pi\text{-}C_5H_5Fe^{+}(CO)_2\cdots\|CH_2{=}C{=}C(CH_3)\text{—}SO_2^{\ominus} \longrightarrow \pi\text{-}C_5H_5Fe(CO)_2\text{—}\overline{C{=}C\text{—}S({=}O)\text{—}O\text{—}CH_2} \quad (259)$$

of the SO_2 insertion product has been determined by x-ray crystallography.[513] Again, electrophilic attack on the remote unsaturated carbon atom is suggested to lead to the bound allene intermediate which then undergoes nucleophilic attack by SO_2^-. Evidence supporting this sequence is found in the protonation of various M-$CH_2C{\equiv}CR$ species generating allene[512] and the insertion of several carboxylic acids into these alkynyl metal bonds producing substituted vinyl complexes [Eq. (260)].

$$(CO)_4P(C_6H_5)_3MnCH_2C\equiv CH \xrightarrow{RCOOH} R\text{—}\overset{O}{\overset{\|}{C}}OCH_2\text{—}\underset{Mn(CO)_4P(C_6H_5)_3}{\underset{|}{C}}{=}CH_2 \quad (260)$$

SO_3, which has been reported to insert into transition metal–alkyl and –aryl bonds,[514] also reacts with several 2-alkynyl complexes [Eq. (261)].[515]

$$Mn(CO)_5(CH\equiv CCH_3) + SO_3 \longrightarrow (CO)_5Mn\text{—}\overline{C{=}C(CH_3)\text{—}S(O_2)\text{—}O\text{—}CH_2} \quad (261)$$

These products could also be synthesized by oxidizing the corresponding SO_2 insertion product [Eq. (262)].

$$(CO)_5Mn{-}\overset{\displaystyle H_2C{-}O}{C{=}C(C_6H_5){-}S{=}O} \xrightarrow[CH_3COOH]{KMnO_4} (CO)_5Mn{-}\overset{\displaystyle H_2C{-}O}{C{=}C(C_6H_5){-}S(=O)_2} \quad (262)$$

Significantly, SO_2 insertion into 3-alkynyl complexes leads to the formation of S-sulfinites [Eq. (263)][512] according to the proposed mechanism.

$$\pi\text{-}C_5H_5Fe(CO)_2{-}CH_2CH_2C{\equiv}CCH_3 + SO_2 \longrightarrow \pi\text{-}C_5H_5Fe(CO)_2{-}S(=O)_2{-}CH_2CH_2C{\equiv}CCH_3 \quad (263)$$

D. Nucleophilic Addition Reactions at Coordinated Carbon Donors

A number of unsaturated ligands are known to be activated toward attack by nucleophiles at a carbon atom that is bonded to the central metal in a complex [Eq. (264)].

$$L_mM{-}A{\equiv}B + X \rightleftharpoons L_mM{-}\overset{\displaystyle B}{\overset{\|}{A}}X \quad (264)$$

If the nucleophile is anionic or has an ionizable proton, the resultant ligand species is negatively charged [Eq. (265)].

$$L_mM^{n+}{-}A{\equiv}B + YH(\text{or } Y^-) \rightleftharpoons L_mM^{(n-1)+}{-}\underset{\displaystyle B}{\underset{\|}{A}}{-}Y + H^+ \quad (265)$$

Reactions of this type are found for carbon monoxide and such isoelectronic species as isonitriles and lead to carbamoyl and alkoxycarbonyl complex formation [Eqs. (266) and (267)], as well as to their analogs.[516]

$$L_mM^{n+}CO + NHR_2 \rightleftharpoons L_mM^{(n-1)+}C(=O)—NR_2 + H^+ \quad (266)$$

R = H, alkyl, aryl

$$L_mM^{n+}CO + ROH(\text{or } OR^-) \rightleftharpoons L_mM^{(n-1)+}—C(=O)—OR + H^+ \quad (267)$$

Subsequent reaction with an electrophile can then generate carbene ligand species [Eq. (268)].[516,517]

$$[L_mM^{(n-1)+}—A(=B)—Y] + Z^+ \rightleftharpoons L_mM^{n+}{:}A(BZ)(Y) \quad (268)$$

E. Formation of Carbamoyl and Alkoxycarbonyl Complexes

Many cationic transition metal-carbonyl complexes are known to undergo the reactions shown in Eqs. (266) and (267), producing stable products, and subsequent electrophilic attack yielding carbene ligands [Eq. (268)] is observed for formally neutral carbonyl complexes. Thus, monocationic complexes[516,518–526] $M^{II}(CO)_4(\pi\text{-}C_5H_5)^+$ and $M^{II}(CO)_3NH_3\text{-}(\pi\text{-}C_5H_5)^+$ (M = Mo, W), $M^I(CO)_5L^+$ and $M^I(CO)_4L_2^+$ (M = Mn and Re), $Fe^{II}(CO)_3(\pi\text{-}C_5H_5)^+$ and $Fe^{II}(CO)_2L(\pi\text{-}C_5H_5)^+$, $Co(CO)_4P(C_6H_5)_3^+$, *trans*-$Pt[P(C_6H_5)_3]_2(CO)Cl^+$, and neutral $Fe(CO)_5$ have been reported to react with alkylamines forming carbamoyl complexes. The electron density localized on the carbonyl ligands is important in determining the reactivity of such species toward amines.[522] In complexes having as ligands good σ donors or poor π acceptors, the formation of carbamoyl complexes is either readily reversible or not observed at all, even in alkylamine solvent.[516] This reactivity has been correlated with the force constants for the carbon–oxygen stretching vibrations in the ir spectra of these complexes[527] (carbonyl complexes with C—O stretching frequencies below ~2000 cm^{-1} do not yield carbamoyl complexes). In this regard, $Fe(CO)_5$ seems to be the only neutral carbonyl complex which reacts with amines to form a carbamoyl complex[522] [Eq. (269)].

$$Fe(CO)_5 + 2\,C_4H_8NH \rightleftharpoons Fe(CO)_4[C(=O)—NC_4H_8]^- + C_4H_8NH_2^{\oplus} \quad (269)$$

An analogous reaction with alcohol or water has been reported to produce alkoxy-carbonyl ligand complexes [Eq. (267)], carboxylic acid complexes, or hydride complexes [Eq. (270)].

$$L_mM^+\text{—}CO + H_2O \rightleftharpoons L_mM\text{—}\overset{\overset{O}{\|}}{C}\text{—}OH \longrightarrow L_mM\,H + CO_2 \quad (270)$$

Examples include the reaction of $Ir(CO)_2L_2I_2^+$ cations with alcohols leading to the formation of the corresponding alkoxycarbonyl complexes,[528] as well as the reactions $Pt(CO)L_2X^+$ species with water and alcohols producing either PtL_2XH or $PtL_2X(\text{-}COOR)$ depending on the nature of X and L.[529] A stable carboxylic acid ligand species has also been reported.[530] Reactions of carbon monoxide and alcohols in the presence of transition metal complexes have also been reported to yield alkoxycarbonyl ligands [Eqs. (271) and (272)].

$$Pt[(C_6H_5)_3]_2(NCO)X + CO + ROH \longrightarrow Pt[P(C_6H_5)_3]_2(NCO)(\overset{\overset{O}{\|}}{C}\text{—}OR) \quad (271)$$

$$Co^{III}(\text{corrinoid}) + ROH + CO \longrightarrow Co^{III}(\text{corrinoid})(\overset{\overset{O}{\|}}{C}\text{—}OR) \quad (272)$$

Several $Pt^{II}[P(C_6H_5)_3]_2XX'$ complexes in alcohol solution react with carbon monoxide as shown in Eq. (271).[531] A series of cobalt(III) correnoids was also observed to react with carbon monoxide in alcoholic solution to produce alkoxycarbonyl ligands [Eq. (272)] and in aqueous solution to be reduced to Co(II) in the presence of carbon monoxide.[532] The latter process presumably involves initial formation of a bound carboxylic acid intermediate, subsequent decomposition to a hydride, followed by homolytic cleavage of the Co—H bond to generate the Co(II) species. The most widely used synthesis for alkoxycarbonyl complexes, however, involves reaction of cationic carbonyl complexes with alkoxides [Eq. (267)].[516] Of particular note in this regard is the reaction of $Fe(CO)_2(CS)\,(\pi\text{-}C_5H_5)^+$ with methoxide to form both the thiocarboxylate and carboxylate complexes.[525] Methoxide reacts with $Ir(CO)_2[P(C_6H_5)_3]_2(CS)^+$ to form only the carboxylate, however.[533] Evidence for the absence of migration of alkoxy groups from one carbonyl ligand to another is found in the reaction of the optically active mentholate anion with $Fe(CO)_2[P(C_6H_5)_3](\pi\text{-}C_5H_5)^+$ to give a diastereoisomeric mixture of products. Separation of this mixture produced materials which are stable toward racemization.[534] The reaction of platinum(II)–isonitrile complexes with OH^- or SH^- leads to the formation of analogous products.[535]

The formal insertion of alkyl isocyanates into the M—H bond of $W(CO)_3(\pi\text{-}C_5H_5)H$ and $Fe(CO)_2(\pi\text{-}C_5H_5)H$ to form the corresponding carbamoyl complex has been reported [Eq. (273)].[536]

$$L_mM\text{—}H + RNCO \longrightarrow L_mM\text{—}\overset{\overset{\displaystyle O}{\|}}{C}NHR \qquad (273)$$

These reactions are suggested to occur by nucleophilic attack of the anion (formed by deprotonation of the hydride) on the alkyl isocyanate followed by protonation of the nitrogen [Eqs. (274) and (275)].

$$L_mM\text{—}H + B \rightleftharpoons L_mM^- + BH \qquad (274)$$

$$L_mM^- + RNCO \longrightarrow L_mM\text{—}\overset{\overset{\displaystyle O}{\|}}{C}\text{—}NR^- \xrightarrow{H^+} L_mM\text{—}\overset{\overset{\displaystyle O}{\|}}{C}NHR \qquad (275)$$

F. Reactions of Carbamoyl and Alkoxycarbonyl Complexes

The reactivity patterns of alkoxycarbonyl and carbamoyl complexes include sensitivity to both nucleophiles and electrophiles. In the presence of amines, alkoxycarbonyl complexes undergo a reaction analogous to amide formation. An example is seen in the reaction of $Re(CO)_5CO_2CH_3$ with methylamine in Eq. (276).[516]

$$Re(CO)_5(CO_2CH_3) + CH_3NH_2 \longrightarrow Re(CO)_5(\overset{\overset{\displaystyle O}{\|}}{C}\text{—}NHCH_3) + CH_3OH \qquad (276)$$

The reverse reaction[537] of Eq. (276), reactions analogous to ester exchange,[529] and hydride formation reactions (via intermediate carboxylic acid complexes)[538] have been reported.

Certain carbamoyl complexes also react with bases by deprotonation of the amide nitrogen leading to the formation of either isocyanates or, in the case of primary amines, substituted ureas[521] [Eqs. (277) and (278)].

$$W(CO)_3(\pi\text{-}C_5H_5)(\overset{\overset{\displaystyle O}{\|}}{C}NHCH_3) + (C_2H_5)_3N \rightleftharpoons$$
$$W(CO)_3(\pi\text{-}C_5H_5)^- + CH_3NCO + (C_2H_5)_3NH^+ \qquad (277)$$

$$W(CO)_3(\pi\text{-}C_5H_5)(\overset{O}{\overset{\|}{C}}NHCH_3) + CH_3NH_2 \longrightarrow W(CO)_3(\pi\text{-}C_5H_5)^- + CH_3NH_3^+ + (CH_3NH)_2\overset{O}{\overset{\|}{C}} \quad (278)$$

$$L_mM^+(CO) + N_2H_4 \longrightarrow [L_mM{-}\overset{O}{\overset{\|}{C}}NHNH_2] + H^+ \longrightarrow L_mMN{=}C{=}O + NH_3 \quad (279)$$

A closely related reaction is that of carbonyl complexes with hydrazine, producing isocyanato complexes [Eq. (279)]. This is also observed for several cationic species[525,539] including $Fe(CO)_3(\pi\text{-}C_5H_5)^+$ and $Fe(CO)_2(CS)(\pi\text{-}C_5H_5)^+$ [Eq. (280)].

$$Fe(CO)_2(CS)(\pi\text{-}C_5H_5)^+ + N_2H_4 \longrightarrow Fe(CO)_2(\pi\text{-}C_5H_5)(N{=}C{=}S) + NH_3 \quad (280)$$

Another reaction which is probably related is the dehydration of $Mn(CO)_{(5\text{-}n)}L_m(CONH_2)$ to give cyano species [Eq. (281)].[540]

$$Mn(CO)_4NH_3(\overset{O}{\overset{\|}{C}}NH_2) + NH_3 \longrightarrow Mn(CO)_3(NH_3)_2CN + H_2O + CO \quad (281)$$

Electrophiles also react with carbamoyl and alkoxycarbonyl complexes. Protonation of these complexes leads to the reverse of Eqs. (268) and (267), regenerating the carbonyl complex.[516] Alkylation of the carbamoyl and alkoxycarbonyl complexes, however, yields carbene complexes [Eq. (268)].

G. Formation of Carbene Complexes

The initial report of a transition metal–carbene complex dealt with the reaction of hexacarbonyltungsten with LiR and isolation of the product of Eq. (282).

$$W(CO)_6 + LiR \longrightarrow [W(CO)_5(\overset{O^-}{\overset{\|}{C}}{-}R)] \quad (282)$$

$$[W(CO)_5(\overset{O^-}{\overset{\|}{C}}—R)] \xrightarrow{H^+} [W(CO)_5(C(OH)—R)] \xrightarrow{CH_2N_2} [W(CO)_5(C(OCH_3)R)] \quad (283)$$

Subsequent protonation and reaction with diazomethane produced the carbene species[541] [Eq. (283)]. This reaction was extended to the analogous hexacarbonyls $M(CO)_6$ (M = Cr, Mo), mono-substituted species $M(CO)_5L$ (M = Cr, Mo, W), $M^I(CO)_3(\pi\text{-}C_5H_5)$ (M = Mn, Re), and to $Cr(CO)_3$(arene).[542] Other sources of R^+ which proved useful in place of $H^+ + CH_2N_2$ include oxonium salts,[543] $(CH_3)_3SiCl$,[544] and acyl halides.[545] These reactions are not limited to simple alkyllithium reagents and mononuclear carbonyl complexes, however, as is shown in Eqs. (284)[546] and (285).[547] Other nucleophiles which have been

$$(LiC_5H_4)Fe(C_5H_5) + Cr(CO)_6 \longrightarrow (CO)_5Cr—\overset{O^-}{\overset{\|}{C}}—C_5H_4FeC_5H_5 \xrightarrow{(C_2H_5)_3OBF_4} (CO)_5CrC(OC_2H_5)(C_5H_4FeC_5H_5) \quad (284)$$

$$Cr(CO)_6 + LiN(C_2H_5)_2 \longrightarrow [Cr(CO)_5(\overset{O^-}{\overset{\|}{C}}—N(C_2H_5)_2] \xrightarrow{(C_2H_5)_3OBF_4} (CO)_5Cr—\left[C\begin{matrix} OC_2H_5 \\ N(C_2H_5)_2 \end{matrix}\right] \quad (285)$$

used include amides[547] [Eq. (285)] and Grignard reagents.[548] Neutral acyl and carbamoyl complexes have also been alkylated, resulting in cationic carbene species.[549,550] A novel reaction type which probably occurs by this mechanism is shown in Eq. (286).[551] Insertion of carbon monoxide into the Mo–alkyl bond results in formation of an $Mo\text{-}C(O)(CH_2)_3Br$ species which rapidly rearranges to the observed product.

$$Mo(CO)_3(\pi\text{-}C_5H_5)(CH_2)_3Br + L \longrightarrow Mo(CO)_2L(\pi\text{-}C_5H_5[\overset{\displaystyle O}{\overset{\|}{C}}—(CH_2)_3Br]$$

$$\longrightarrow Mo(CO)_2L(\pi\text{-}C_5H_5)\overline{C{<}^{O—CH_2}_{CH_2—CH_2}{|}} + Br^- \quad (286)$$

Addition of an oxygen–hydrogen or nitrogen–hydrogen bond across the carbon–nitrogen bond of a coordinated isonitrile also forms amino carbene complexes [Eq. (287)].[517,518] This has been reported for a number of neutral Pt(II) and Pd(II) complexes, an example of which is shown in Eq. (288).[552] Similar reactions are known for cationic isonitrile complexes[553] as well as dicationic species.[554] Examples of these are given in Eqs. (289) and (290). This type of reaction has also been reported for Hg^{2+}.[550]

$$L_mMC{\equiv}NR + Q—H \longrightarrow L_mM—C{<}^{Q}_{NHR} \quad (287)$$

$$Q = {>}N— \text{ or } —O—$$

$$cis\text{-}Pt(CNPh)(P(C_2H_5)_3)Cl_2 + CH_3OH \longrightarrow cis\text{-}Pt(P(C_2H_5)_3)\left[C{<}^{OCH_3}_{NHPh}\right]Cl_2 \quad (288)$$

$$trans\text{-}[Pt(CNCH_3)(P(C_2H_5)_3)_2Cl]^+ + C_2H_5NH_2 \longrightarrow trans\text{-}Pt(P(C_2H_5)_3)_2\left[C{<}^{NHC_2H_5}_{NHCH_3}\right]Cl \quad (289)$$

$$[Pt(P(CH_3)_2Ph)_2(CNC_2H_5)_2]^{2+} + C_6H_5CH_2SH \longrightarrow trans\text{-}Pt(P(CH_3)_2C_6H_5)_2(CNC_2H_5)\left[C{<}^{NHC_2H_5}_{SCH_2C_6H_5}\right] \quad (290)$$

Transition metal isonitrile complexes form chelating carbene ligands under appropriate conditions. Thus, methylamine reacts with the octahedral isonitrile complex $Fe(CNCH_3)_6^{2+}$ to give the chelating carbene shown in Eq. (291).

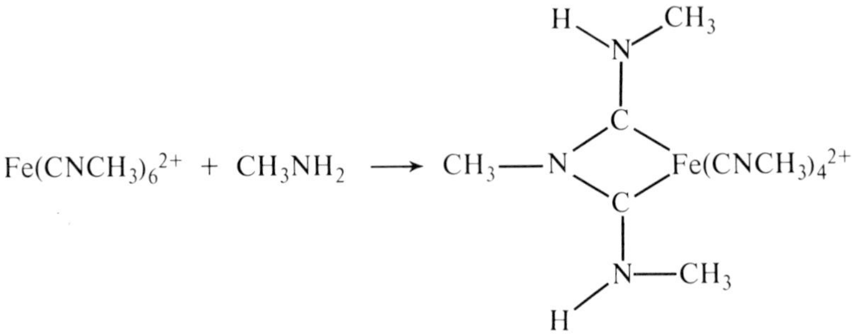

$$Fe(CNCH_3)_6^{2+} + CH_3NH_2 \longrightarrow CH_3{-}N\begin{matrix} C(=N(H)CH_3) \\ C(=N(H)CH_3) \end{matrix}Fe(CNCH_3)_4^{2+} \quad (291)$$

Methylamine reacts with tetrakis platinum and palladium isonitrile complexes giving monodentate carbenes as shown in Eq. (292).[555]

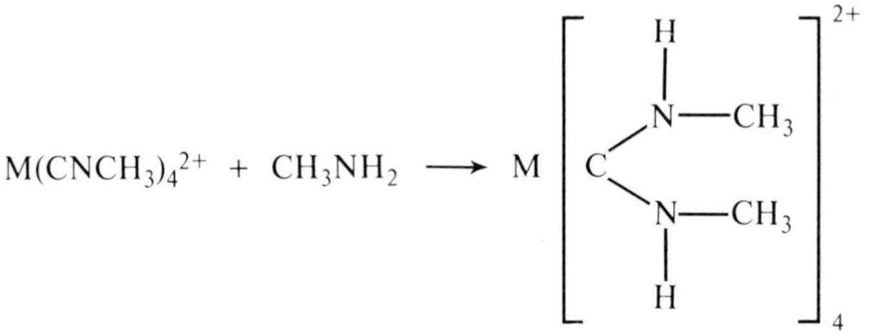

$$M(CNCH_3)_4^{2+} + CH_3NH_2 \longrightarrow M\left[C\begin{matrix} N(H){-}CH_3 \\ N(H){-}CH_3 \end{matrix}\right]_4^{2+} \quad (292)$$

$$M = Pd,\ Pt$$

Hydrazine or substituted hydrazines react with the Pd(II) or Pt(II) tetrakis isonitrile complexes to form chelating carbenes[556] [Eq. (293)].

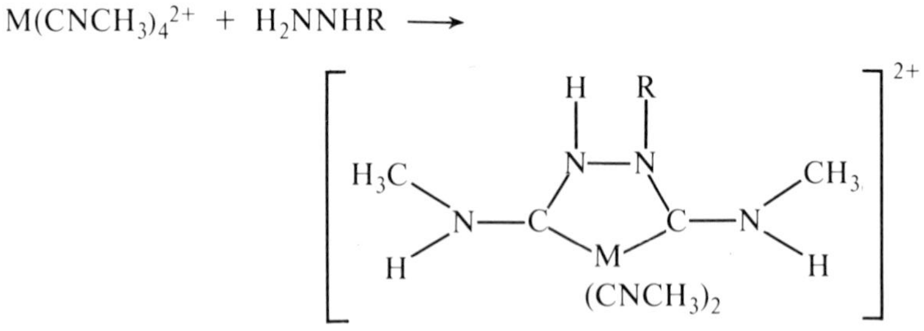

$$M(CNCH_3)_4^{2+} + H_2NNHR \longrightarrow$$

$$\left[(H_3C)(H)N{-}C\begin{matrix} N(H){-}N(R) \\ M(CNCH_3)_2 \end{matrix}C{-}N(CH_3)(H)\right]^{2+} \quad (293)$$

The latter reaction was also observed for hexakis isonitrile Fe(II) complexes.[557] A similar sequence is observed in the reaction of sodium borohydride with cationic isonitrile complexes [Eq. (294)] where the structure **103** was proposed for the bidentate complexes.[558]

$$ML_m(CNCH_3)_2^+ + BH_4^- \longrightarrow ML_m(CHNCH_3)_2BH_2 \qquad (294)$$

$$\begin{array}{ccc} H & & CH_3 \\ & \diagdown C{=}N \diagup & \\ M & & BH_2 \\ & \diagup C{=}N \diagdown & \\ H & & CH_3 \end{array}$$

103

Nucleophilic attack on acetylene coordinated to cationic Pt(II) complexes by solvent alcohol has also been reported[559,560] to lead to the formation of carbene species [Eq. (295)].

$$trans\text{-}[PtQ_2(HC{\equiv}CR)CH_3]^+ + CH_3OH \longrightarrow trans\text{-}\left[PtQ_2\left(C\begin{matrix} OCH_3 \\ CH_2 \end{matrix}\right)CH_3\right]^+ \qquad (295)$$

The mechanism for this reaction has been suggested[560] to involve nucleophilic attack on a coordinated carbonium ion formed by a hydride shift [Eq. (296)].

$$\left[L_mPt\cdots\begin{matrix} CR \\ ||| \\ CH \end{matrix}\right]^+ \longrightarrow \left[L_mPt{-}\overset{+}{C}{=}C\begin{matrix} R \\ H \end{matrix}\right]^+ \xrightarrow{CH_3OH} \left[L_mPtC\begin{matrix} CH_2R \\ OCH_3 \end{matrix}\right]^+ \qquad (296)$$

In fact, this reaction is but one of several possibilities arising from the attack of nucleophiles on acetylene coordinated to cationic Pt(II) complexes.[560]

Another area which involves formation of transition metal carbene complexes is the interaction of transition metals with diazo species. Evidence for the intervention of a Cu(II)-carbene species in the reaction of an optically active Cu(II) *N*-substituted bis-salicylideneimine complex is seen in the optically active product resulting from insertion of the carbene into styrene [Eq. (297)].[561]

$$N_2CHCOOCH_3 + CH_2CHC_6H_5 \xrightarrow{\text{Cu(II) bis-salicylideneimine (N-R*)}} \text{cyclopropane-}C^{*}H(COOCH_3)\text{, }C_6H_5 \quad (297)$$

Other variations which indicate the formation of reactive carbene intermediates are known. Formation of an intermediate carbene complex in the reaction of $Ir(CO)[P(C_6H_5)_3]_2Cl$ with diazomethane [Eq. (298)] is suggested on the basis of observed chloromethyl product and the activity of the complex in cationic decomposition of excess diazomethane.[562] Similar reactions have been reported for $Fe(CO)_2(\pi\text{-}C_5H_5)CH_2OCH_3$ as well as the corresponding $M^{II}(CO)_3$-$(\pi\text{-}C_5H_5)CH_2Y$ (M = Mo, W) complexes.[563,564] A possible anionic carbene intermediate is suggested in the reaction shown in Eq. (299). The methyl group found in the product has been determined by isotopic labelling to arise from the methylene group originally attached to the silicon.[565]

$$Ir(CO)[P(C_6H_5)_3]_2Cl + CH_2N_2 \rightarrow Ir(CO)[P(C_6H_5)_3]_2CH_2Cl \quad (298)$$

$$[Mo(CO)_3(\pi\text{-}C_5H_5)]^- + (CH_3)_3SiCH_2I \longrightarrow Mo(CO)_3(\pi\text{-}C_5H_5)CH_2Si(CH_3)_3 \xrightarrow{\text{THF}} Mo(CO)_3(\pi\text{-}C_5H_5)CH_3 \quad (299)$$

H. Reaction of Carbene Complexes

Among the reactions of carbene species with nucleophiles, the simplest involve deprotonation of the carbene moiety at either a carbon or a nitrogen atom[566] [Eqs. (300) and (301)].

$$Cr(CO)_5\left[C\begin{matrix}OR\\CH_2R'\end{matrix}\right] \xrightarrow[CH_3ONa]{CH_3OD} Cr(CO)_5\left[C\begin{matrix}OR\\CD_2R'\end{matrix}\right] \quad (300)$$

$$Pd[(P(C_6H_5)_3)]\left[C\begin{matrix}OCH_3\\NHC_6H_5\end{matrix}\right]Cl_2 \underset{H^+}{\overset{OH^-}{\rightleftharpoons}}$$

$$[(C_6H_5)_3P](H_5C_6{-}N{=}C(OCH_3))Pd(\mu\text{-}Cl)_2Pd[C(OCH_3){=}N{-}C_6H_5][P(C_6H_5)_3] \quad (301)$$

An extension of the former reaction leads to the formation of substituted alkyl groups[542] [Eq. (302)].

$$Cr(CO)_5\left[C\begin{matrix}OCH_3\\CH_3\end{matrix}\right] \xrightarrow[2.\ (CH_3)_3OBF_4]{1.\ NaOCH_3}$$

$$Cr(CO)_5\left[C\begin{matrix}OCH_3\\CH_2CH_3\end{matrix}\right] + Cr(CO)_5\left[C\begin{matrix}OCH_3\\CH(CH_3)_2\end{matrix}\right] \quad (302)$$

The anions resulting from deprotonation of tungstencarbonyl carbene complexes with alkyllithium reagents have reacted with electrophiles including benzaldehyde [Eq. (303)] and acetyl chloride [Eq. (304)] leading to novel structures.[567] As there exists a considerable partial π-bond between the carbene carbon and the electronegative substituents on this carbon, isomerization about this bond is possible. Base-catalyzed isomerization of carbenes formed from primary amines has been reported, the product ratio depending upon the base employed [Eq. (305)].[568]

$$W(CO)_5\left[C{<}^{OCH_3}_{CH_2^-}\right] + C_6H_5C(=O)H \longrightarrow W(CO)_5\left[C{<}^{OCH_3}_{CH{=}CHC_6H_5}\right] + OH^- \quad (303)$$

$$W(CO)_5\left[C{<}^{OCH_3}_{CH_2^-}\right] + 2CH_3\overset{O}{\overset{\|}{C}}Cl \longrightarrow W(CO)_5\left[C{<}^{CH{=}C{<}^{CH_3}_{O_2CCH_3}}_{OCH_3}\right] \quad (304)$$

$$cis\text{-}Cr(CO)_5\left[C{<}^{N(H){-}CH_3}_{CH_3}\right] \xrightarrow{\text{base}} cis\text{-}Cr(CO)_5\left[C{<}^{N(H){-}CH_3}_{CH_3}\right] + trans\text{-}Cr(CO)_5\left[C{<}^{N(CH_3){-}H}_{CH_3}\right] \quad (305)$$

$$Cr(CO)_5\left[C{<}^{OR}_{R}\right] + AH \rightleftharpoons Cr(CO)_5\left[C{<}^{R}_{A}\right] + ROH \quad (306)$$

$AH = NH_3,\ NH_2R,\ NHR_2,\ RO^-,\ RSH$

Nucleophilic attack on carbene ligands also leads to the formation of new carbene species. So reaction with primary and secondary amines,[569–572] alkoxides,[573] and thiols[542] yield as products carbene complexes [Eq. (306)]. This type of product has been reported in the reactions of acyloxy carbene complexes with nucleophiles.[545] A similar reaction with benzophenoneimine forms an imino carbene species.[574]

The mechanism of aminolysis of $Cr(CO)_5(C(OCH_3)(C_6H_5))$ has been studied, and the rate law observed for a variety of amines and solvents is given in Eq. (307).[575]

$$d\left[Cr(CO)_5\left(C{\begin{smallmatrix}NHR\\ Ph\end{smallmatrix}}\right)\right]/dt = k_n\left[Cr(CO)_5\left(C{\begin{smallmatrix}OCH_3\\ C_6H_5\end{smallmatrix}}\right)\right][RNH_2][HX][Y] \quad (307)$$

In the rate expression, HX is a proton donor, and Y is a proton acceptor. The overall mechanism is given in Scheme 2-22.

Scheme 2-22

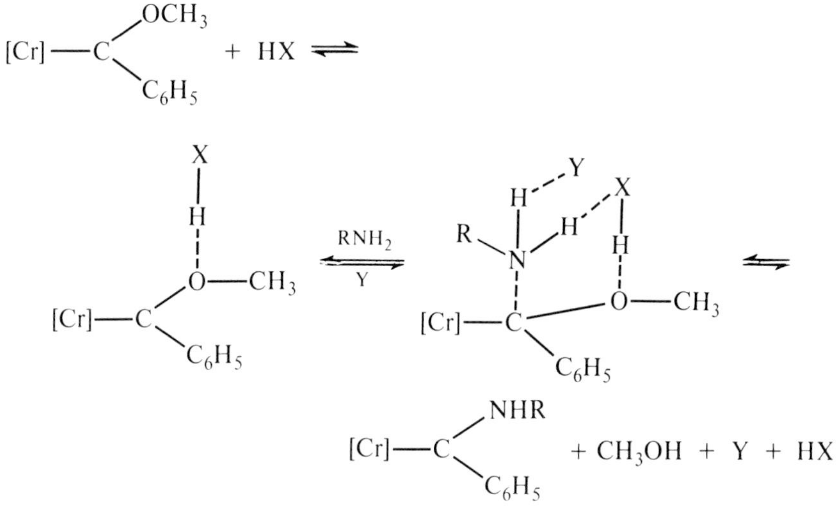

$$[Cr]{-}C(OCH_3)(C_6H_5) + HX \rightleftharpoons [Cr]{-}C(C_6H_5){=}O(CH_3){\cdots}H{-}X \underset{Y}{\overset{RNH_2}{\rightleftharpoons}} [Cr]{-}C(C_6H_5)(O(CH_3){\cdots}H{-}X)(N(R)(H{\cdots}Y)(H{\cdots}X)) \rightleftharpoons$$

$$[Cr]{-}C(NHR)(C_6H_5) + CH_3OH + Y + HX$$

$[Cr] = Cr(CO)_5$

In accord with this mechanism the entropy of activation for this reaction has a large negative value (−55–−82 esu).

Additional reactions have been observed. These include: reduction of coordinated carbene ligands to substituted alkyl compounds;[576] reaction with aldoximes and ketoximes;[574] reaction with hydroxylamine, substituted hydrazines, and other bifunctional nitrogen nucleophiles;[577] reaction with phosphorus ylides;[578] and reaction with olefins having electron-withdrawing groups as substituents on the olefinic carbon and which lead to displacement of the carbene ligand and stereospecific formation of cyclopropanes [Eq. (308)].[579]

$$Cr(CO)_5(C(OCH_3)(C_6H_5)) + (H)(H_3C)C{=}C(COOCH_3)(H) \xrightarrow{C_5H_5N}$$

$$\text{cyclopropane}[COOCH_3, H, CH_3;\ C_6H_5, OCH_3] + \text{cyclopropane}[COOCH_3, H, CH_3;\ OCH_3, C_6H_5Cr(CO)_3] \qquad (308)$$

I. Palladium-Catalyzed Olefin–Substituent Exchange

Following the initial report by Smidt[580] of palladium(II)-catalyzed vinylic olefin–substituent exchange, [Eq. (309), X, Y = carboxylates], a number of

$$CH_2{=}CHX + Y^- \xrightarrow{Pd^{II}} CH_2{=}CHY + X^- \qquad (309)$$

related reactions involving nucleophilic substituent exchange on coordinated olefins have appeared in the literature.[581,582,583] In view of the mechanistic similarities between these reactions and the oxidation of olefins catalyzed by Pd(II) in the presence of nucleophiles (*see* section X,C.), and their possible synthetic utility, the kinetics and stereochemical course of these exchange reactions have been investigated. It was determined that in the case of ester exchange, the actual process is trans-vinylation [Eq. (310)] as shown by isotopic labelling experiments.[584] In addition, use of 2,2-dideuterovinyl chloride [Eq. (311)] demonstrated that the carbons in the ethylene moiety retained their identity during the exchange process.[585] Further, ester exchange[584] and alkoxide exchange[586] reactions were shown to be accompanied by cis-trans isomerization.

$$CH_2{=}CHOC(=O)C_2H_5 + CH_3C(={}^{18}O){-}OH \longrightarrow CH_2{=}CH{-}OC(={}^{18}O)CH_3 + C_2H_5C(=O)OH \quad (310)$$

$$CD_2{=}CHCl + {}^{-}OC(=O)CH_3 \xrightarrow{Pd(II)} CD_2{=}CHOC(=O)CH_3 \quad (311)$$

A mechanism was proposed for these reactions which could account for the above observations: the nucleophile attaches on the coordinated olefin at the hetero-substituted carbon leading to σ-alkyl species [Eq. (312)], then, rearrangement occurs by rotation about the C—C bond; elimination of the original hetero substituent follows, generating the bound, substituent-exchanged olefin [Eq. (313)]. Two features of this generally accepted mechanism merit comment. First, rotation about the C—C bond in the σ-alkyl species is required to allow substituent exchange. Also, if the stereochemistry of Y^- addition (*cis* or *trans*-L_mPd^-, Y^- addition) is different from that of X^- elimination, then olefin configuration is retained. If both processes are of the same stereochemistry, olefin configuration inversion occurs as shown in Eq. (313).

$$L_mPd^{II} + (R)(H)C{=}C(X)(H) \rightleftharpoons L_mPd{\cdots}[(H)(R)C{=}C(H)(X)] + Y^- \rightleftharpoons L_mPd{-}C(R)(H){-}C(X)(Y)(H) \quad (312)$$

$$[\text{Newman projection: } L_mPd \text{ (front)}, H, X, R, H, Y] \rightleftharpoons [\text{Newman projection: } PdL_m, Y, H, R, H, X] \longrightarrow L_mPd{-}[(H)(R)C{=}C(Y)(H)] + X^- \quad (313)$$

A number of kinetic and stereochemical investigations into these processes by Henry et al.[581,583] have now been reported which support this mechanism and suggest different modes of addition for different nucleophiles. These studies, which were carried out in acetic acid with $Li_2Pd_2Cl_6$ as the catalytic species in the presence of varying amounts of LiCl and LiOAc, all display a reduction in rate as the olefinic carbons are progressively substituted. This is consistent with the addition of L_mPd and Y^- as a step in the catalytic sequence. Exchange of the substituent on vinyl hetero-substituted olefins has been examined in three systems which illustrate the major mechanistic considerations: vinyl ester exchange with acetate;[587,588] vinyl chloride exchange with acetate;[589] and vinyl chloride exchange with isotopically labelled chloride.[590] The stereochemistry of L_mPdCl and acetate addition (and by microscopic reversibility, elimination) has been inferred from the product distributions in the Pd-catalyzed oxidation of various deuterated cyclohexenes in acetic acid. The trans structures indicate attack by uncoordinated acetate from the solution on the bound olefin [Eq. (312)].[591]

Further support for this contention is seen by the lack of evidence for complexation of OAc^- by the $Li_2Pd_2Cl_6$ systems.[581] Addition of L_mPd and Cl^- is, however, suggested to involve prior coordination of the halide to Pd(II) and, hence, to be cis addition (and elimination) [Eq. (314)].

$$\text{>Pd(Cl)} \cdots (R_2)(R_1)C{=}C(X)(R_3) \rightleftharpoons \text{>Pd}{-}C(R_1)(R_2){-}C(R_3)(X){-}Cl \qquad (314)$$

This is supported by the rate expression for reactions involving Pd–Cl addition or elimination, by the observed stereochemistry for these reactions, and by analogy with L_mPd–phenyl addition, in which the phenyl group is generally accepted to be coordinated, and for which cis addition has been reported.[592]

Pd(II)-catalyzed exchange of $CH_2{=}CHOC(O)CD_3$ and $CH_3CH{=}CHOAc$,[586] as well as $CH_3CH{=}CHOC(O)C_2H_5$ with acetate,[588] follow the rate law [Eq. (315)]. $CH_3CH{=}CHOAc$ and $CH_3CH{=}CHOC(O)C_2H_5$ ex-

$$\frac{-d[\text{vinyl ester}]}{dt} = \left(\frac{[Li_2Pd_2Cl_6][\text{vinyl ester}]}{[LiCl]}\right)(k' + k''[LiOAc]) \qquad (315)$$

change only with isomerization. The observed occurrence of isomerization with accompanying exchange supports the mechanism given in Eqs. (312) and (313) since both addition and elimination are suspected to be trans; the rate expression

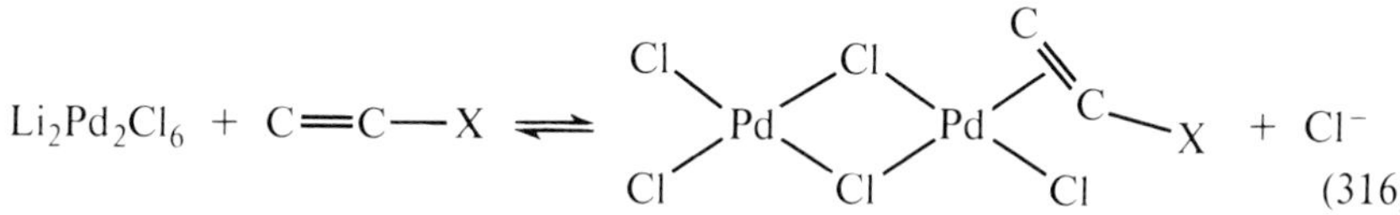

$$Li_2Pd_2Cl_6 + C{=}C{-}X \rightleftharpoons \text{Cl}_2\text{Pd}(\mu\text{-Cl})_2\text{Pd(Cl)}(\text{C}{=}\text{C}{-}\text{X}) + Cl^- \quad (316)$$

provides further evidence. Presuming that the first step in the reaction is olefin binding by $Li_2Pd_2Cl_6$ [Eq. (316)], then the inverse chloride dependence in the rate law is rationalized, and (excluding five-coordinate Pd(II)) the absence of a $[LiCl]^{-2}$ term argues against coordination of acetate to the catalyst prior to nucleophilic attack on the olefin. This point is reaffirmed by the observed failure of —$OC(O)CD_3$ to exchange with 1-acetoxy-1-cyclopentene under the reaction conditions. This is rationalized by the assumption of pure trans addition leading to structure **104** which cannot rearrange by rotation about the C—C σ-alkyl bond and hence reacts only by trans elimination to give exclusively the starting material.

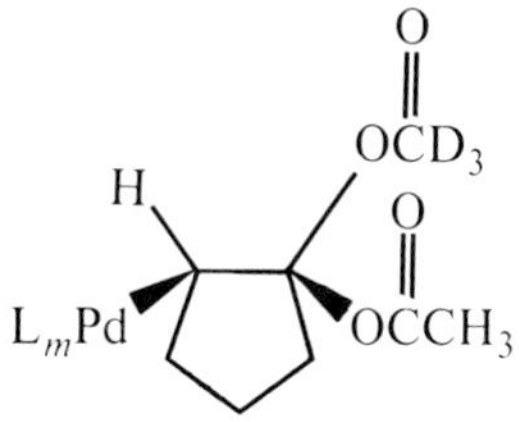

104

In the exchange of acetate for the chloride in vinyl chloride,[589] which has the rate expression of Eq. (317), and for exchange of acetate for chloride in *cis*- and *trans*-1-chloro-1-propene,

$$\frac{-d[\text{vinyl chloride}]}{dt} = \left(\frac{[Li_2Pd_2Cl_6][\text{olefin}]}{[LiCl]^2}\right)(k' + k''[LiOAc]) \quad (317)$$

predominant retention of configuration was found. These observations are compatable with initial olefin binding by the Pd(II) catalyst [Eq. (316)] followed by fast trans acetoxypallidation [Eq. (318)] and slow cis dechloropallidation involving the prior formation of an empty site on the catalyst [Eq. (319)]. Thus, the two chloride dissociation steps [Eqs. (316) and (319)] rationalize the $[LiCl]^{-2}$ term in the rate law [Eq. (317)]. The retention of configuration for propenyl chlorides supports a different stereochemistry for acetoxypallidation and dechloropallidation, the latter of which is suggested to require coordinative unsaturation in the catalytic intermediate.

Exchange of labelled chloride[590] with vinyl chloride presents a more complicated case since rapid olefin isomerization is observed in the absence of exchange. The available information, however, can be satisfactorily accommodated

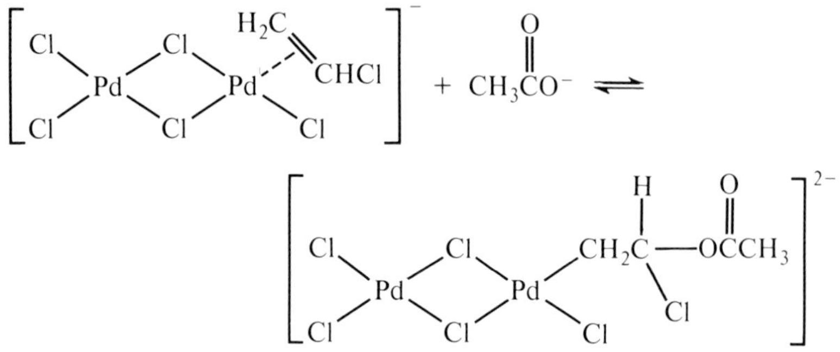

$$\left[\begin{matrix} Cl & & Cl & & H_2C \\ & Pd & & Pd & \Vert \; CHCl \\ Cl & & Cl & & Cl \end{matrix} \right]^- + CH_3\overset{O}{\overset{\Vert}{C}}O^- \rightleftharpoons$$

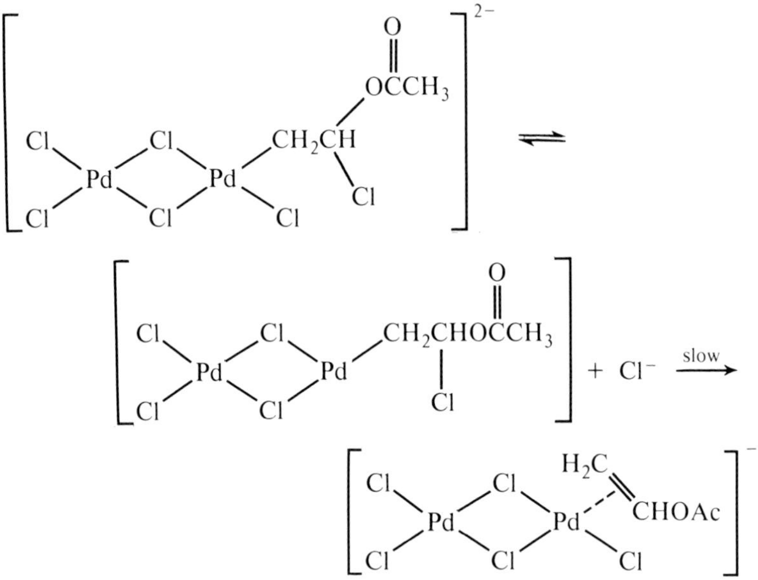

$$\left[\begin{matrix} Cl & & Cl & & CH_2\overset{H}{\overset{|}{C}}—O\overset{O}{\overset{\Vert}{C}}CH_3 \\ & Pd & & Pd & \quad \diagdown Cl \\ Cl & & Cl & & Cl \end{matrix} \right]^{2-} \qquad (318)$$

$$\left[\begin{matrix} & & & & O\overset{O}{\overset{\Vert}{C}}CH_3 \\ Cl & & Cl & & CH_2CH \\ & Pd & & Pd & \quad \diagdown Cl \\ Cl & & Cl & & Cl \end{matrix} \right]^{2-} \rightleftharpoons$$

$$\left[\begin{matrix} Cl & & Cl & & CH_2\underset{Cl}{\underset{|}{C}}HO\overset{O}{\overset{\Vert}{C}}CH_3 \\ & Pd & & Pd & \\ Cl & & Cl & & \end{matrix} \right] + Cl^- \xrightarrow{slow}$$

$$\left[\begin{matrix} Cl & & Cl & & H_2C \\ & Pd & & Pd & \Vert \; CHOAc \\ Cl & & Cl & & Cl \end{matrix} \right]^- \qquad (319)$$

by a dechloropalladation reaction mechanism involving an empty donor site on the catalyst.

Further evidence for an intermediate involving a σ-alkyl-Pd(II) species resulting from acetoxy-pallidation is obtained from Pd(II)-catalyzed allylic ester–acetate exchange.[593,594] The rate law observed is the same as found for vinyl ester–acetate exchange [Eq. (315)] again suggesting coordination of the allylic ester to the Pd(II) catalyst, addition of L_mPd and acetate to the double bond, and elimination of the original carboxylate group to complete the reaction. The key requirement of this mechanism is that unsymmetric allylic esters should exchange only with isomerization. This is the observed result as shown in Eqs. (320) and (321) for acetate exchange with crotyl propionate and 3-buten-2-yl

$$CH_3CH{=}CHCH_2OC(=O)C_2H_5 + L_mPd^+ + CH_3C(=O)O^- \rightleftharpoons CH_3CH(OC(=O)CH_3)CH(PdL_m)CH_2OC(=O)C_2H_5 \quad (320)$$

$$CH_3CH(OC(=O)C_2H_5)CH{=}CH_2 + L_mPd^+ + CH_3C(=O)O^- \rightleftharpoons CH_3CH(OC(=O)C_2H_5)CH(PdL_m)CH_2OC(=O)CH_3$$

$$\longrightarrow CH_3CH{=}CHCH_2OC(=O)CH_3 \quad (321)$$

propionate. This same type of Pd(II) isomerization with exchange was reported for the reaction of allylic trifluoroacetates with chloride.[595]

Vinyl ester exchange with acetate has also been studied in chloride-free media using vinyl propionate.[596] In the acetate medium, three Pd(II) complexes are present: an unreactive monomer and reactive dimers and trimers, the dimer being the most effective catalyst. The predominant rate term is given by Eq. (322).

$$\text{rate} = k[Na_2Pd_2(OAc)_6][C_2H_3OAc] \quad (322)$$

In contrast to the chloride systems, the reaction is not stereospecific in this medium. This has been attributed to the presence of acetate in the coordination sphere of the Pd(II).

The hydrolysis of vinyl acetate in wet acetic acid is of interest because it might involve steps analogous to those in the Wacker process (*see* section X,B.). The reaction is shown in Eq. (323).[597]

$$CH_2CHOAc + H_2O + Pd(II) \xrightarrow{HOAc} Pd(II)—CH_2CH\begin{matrix}OAc\\OH\end{matrix}$$

$$\downarrow \qquad (323)$$

$$Pd(II) + CH_3CHO \xleftarrow{H^+} Pd(II)—CH_2CHO + HOAc + H^+$$

A kinetic study is consistent with reaction between water and a dimeric complex containing the vinyl ester [Eq. (324)].

$$^{-}Cl_5Pd_2\text{----}\begin{matrix}CH_2\\ \| \\ CH—OAc\end{matrix} + H_2O \longrightarrow {}^{2-}Cl_5Pd_2—CH_2CH\begin{matrix}OH\\OAc\end{matrix} + H^+ \qquad (324)$$

These results indicate that the mechanism is different from that for ethylene in water.

Thus, a variety of Pd(II)-catalyzed, heteroatom-substituted vinylic and allylic olefins are observed to undergo Pd(II)-catalyzed nucleophilic substitution reactions with similar mechanisms. Initial coordination of the olefin is observed, followed by attack of the nucleophile, either trans (from solution) or cis, via coordination to the catalyst prior to attack. This leads to a Pd(II)-σ-alkyl species. Elimination of the original nucleophile then completes the reaction.

J. Nucleophilic Addition and Substitution on Coordinated Aromatic Ligands

Coordination of arenes to transition metals in positive oxidation states has long been observed to promote nucleophilic attack by hydride ions on the arene ring, exo- to the metal ion [Eq. (325)].[598-603]

X–C₆H₅(–ML_m^{n+}) + H⁻ ⇌ X–C₆H₆⁻(–$ML_m^{(n-1)+}$) (325)

The resulting ligand is a planar pentahapto-6π-electron moiety with nearly equivalent M—C bond distances and a methylene above this plane (for η^5-$C_6(CH_3)_6HRe(CO)_3$, the dihedral angle between the planar ring segment and the plane containing the methylene carbon is approximately 50°).[603] This reaction has been reported for mono- and bis-π-arene complexes,[604,605] yielding both mono- and bis-η^5-cyclohexadienyl species [Eqs. (325a), (326), and (327)].

$$\eta^6\text{-}C_6H_6Mn(CO)_3^+ + H^- \rightleftharpoons \quad (325a)$$

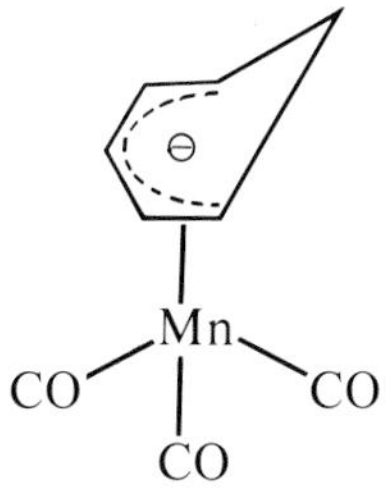

$$(\eta^6\text{-}C_6H_6)_2Re^+ + H^- \rightleftharpoons \quad (326)$$

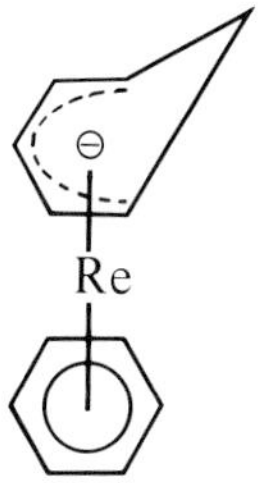

$$(\eta^6\text{-}C_6H_6)_2Ru^{2+} + 2H^- \longrightarrow \quad (327)$$

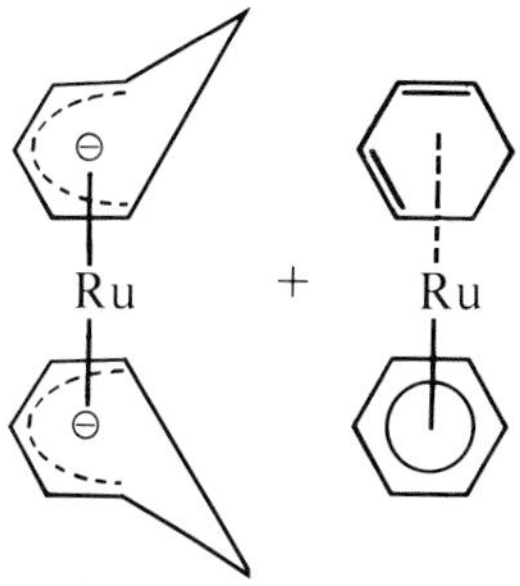

The π-acid properties of the ancillary ligands are found to affect the course of the hydride addition, as shown in Eqs. (325), (326), and (327).

Extension of this work to include reactions with alkyl- and aryllithium compounds gave analogous products. The cyclohexadienyl ligand could then be converted into a substituted arene by hydride extraction in favorable cases [Eq. (328)]. Recently, utilization of bis(π-arene)Fe(II) complexes in these reactions followed by Ce(IV) oxidation has been demonstrated to be a superior synthesis for certain substituted arene rings.[606]

$$\text{Fe}(\eta^5\text{-C}_5\text{H}_5)(\eta^5\text{-C}_6\text{H}_6\text{R}^-) + (C_6H_5)_3C^+ \longrightarrow \text{Fe}^+(\eta^5\text{-C}_5\text{H}_5)(\eta^6\text{-C}_6\text{H}_5\text{R}) + (C_6H_5)_3CH \qquad (328)$$

Other nucleophiles have also been reported to react with π-arene M^{n+} species leading to cyclohexadienyl products including $(\eta^6\text{-}C_6H_6)_2RuCl_2$ [Eq. (329)][607] and $\eta^6\text{-}C_6(CH_3)_nH_{(6-n)}]Mn(CO)_3{}^+$.[608] The hexadienyl complex produced by the reaction of Eq. (329) could be oxidized with Ce(IV) in sulfuric acid liberating pure $C_6(CH_3)_nH_{(5-n)}CN$.

$$(\eta^6\text{-}C_6H_6)_2RuCl_2 + CN^- \longrightarrow \text{Ru}(\eta^5\text{-C}_6\text{H}_6\text{CN}) \qquad (329)$$

K. Nucleophilic Substitution Reactions of Amino Acid Esters and Amides and Polypeptides

Early investigators called attention to the hydrolysis of amino acid esters and amides and peptides by aqueous metal derivatives.[609–612] For a summary *see* Ref. 67. Bender and Turnquest[612] studied the hydrolysis of glycine ethyl ester and DL-phenylalanine ethyl ester in aqueous solution at a pH of 7.3, using the amino acid as the buffer. Under these conditions the main complexes of Cu(II) are those with the glycinate anion, and a small extent of ester hydrolysis does not change the buffer concentration significantly. Consequently, the reaction follows the rate law

$$\frac{-d(\text{ester})}{dt} = k_{\text{obs}}(\text{ester})(\text{Cu(II)buffer}) \qquad (330)$$

The equations appropriate to this rate law are Eqs. (331) and (332).

$$\text{ester} + \text{Cu(II)buffer} \xrightarrow{K} \text{Cu(II)ester} + \text{buffer} \qquad (331)$$

$$\text{Cu(II)ester} \xrightarrow[H_2O]{k} \text{Cu(II)(buffer)}_2 + \text{ROH} \qquad (332)$$

The use of other buffer systems—e.g., tris(hydroxymethyl)-aminomethane—leads to more complex behavior. The acceleration of hydrolysis of phenylalanine ethyl ester by Cu(II) over the base hydrolysis was estimated to be much larger than that produced by a positive charge on the amino group.[613,614] This was taken as evidence for direct interaction of the ester group with the metal ion center. The role of coordination via the amino group was confirmed by complete absence of a promotional effect by Cu(II) in the case of ethyl acetate.[615] Oxygen tracer experiments (carbonyl-^{18}O) showed that the ratio of the hydrolysis rate to the ^{18}O exchange rate was 3.9 ± 0.4. These studies suggested the formation of a symmetrical intermediate, and the following reaction sequence was proposed [Eq. (333)].

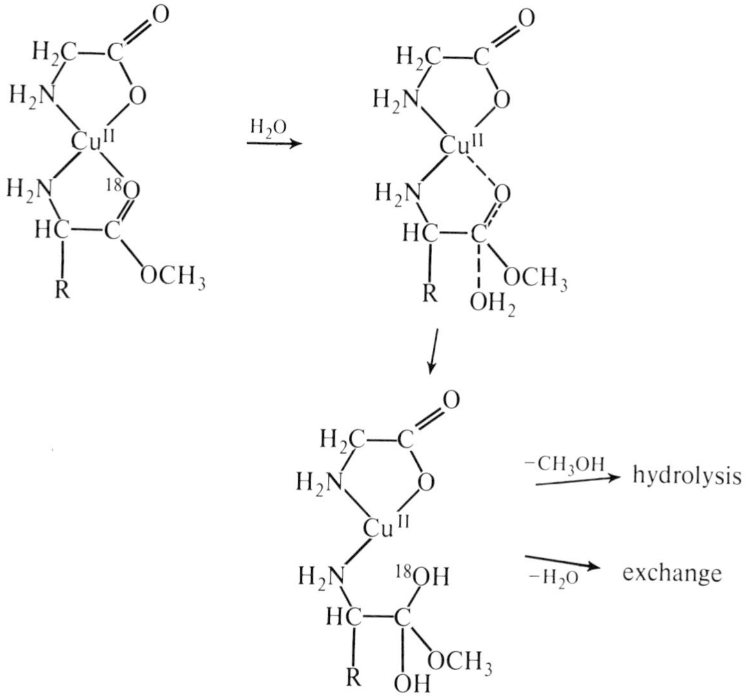

H_2O →

$-CH_3OH$ → hydrolysis

$-H_2O$ → exchange

(333)

Bender and Turnquest believed that it was reasonable to assume that the metal ion activates the ester by coordination to the carbonyl group in analogy to accepted points of view on the path of proton-promoted ester hydrolysis. They recognized, however, the possibility of an alternative mode of attack on the alkoxy oxygen atom. In fact, a third mode of activation could not be ruled out by early studies. This would involve a neighboring group effect due to a coordinated OH^- group (structure **105**).

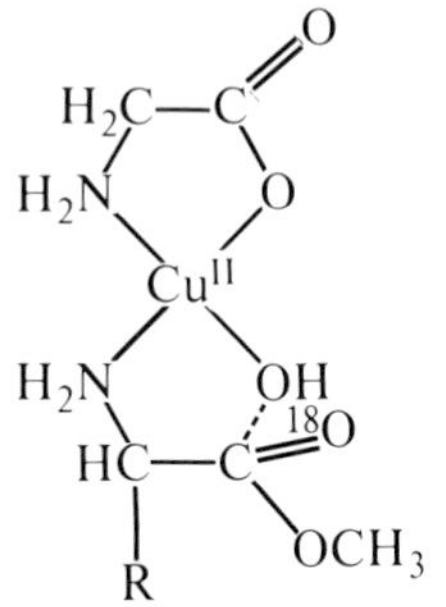

105

Conly and Martin[616] studied the acid-, base-, and metal ion-catalyzed hydrolysis of glycine ethyl ester by a pH-stat method to eliminate the effects of buffers. For the metal-free systems, four processes were considered [Eqs. (334)–(337)].

$$\text{ester H}^+ + H_2O \xrightarrow{k_w{}^+} \text{products} \tag{334}$$

$$\text{ester H}^+ + OH^- \xrightarrow{k^+} \text{products} \tag{335}$$

$$\text{ester} + H_2O \xrightarrow{k_w{}^0} \text{products} \tag{336}$$

$$\text{ester} + OH^- \xrightarrow{k^0} \text{products} \tag{337}$$

The overall rate was taken to be the sum of contributions from these four processes and, from the values deduced for the separate constants, the reactions with hydroxide dominate.

Consideration of the formation constants of the Cu(II)-glycine ester complexes led to the conclusion that over 96% of the complexed Cu(II) exists as the 1:1 complex under the conditions used by Conly and Martin in their rate measurements. Two reactions were considered [Eq. (338) and (339)].

$$\text{Cu(II)(ester)} + H_2O \xrightarrow{k_{1w}{}^{2+}} \text{products} \tag{338}$$

$$\text{Cu(II)(ester)} + OH^- \xrightarrow{k_1{}^{2+}} \text{products} \tag{339}$$

From the value of $k_{1w}{}^{2+}$ and $k_1{}^{2+}$, it was concluded that the reaction with OH^- predominates above pH 5. An interesting comparison between rates can be made. Whereas OH^- reacts with the protonated ester 41 times as fast as with the neutral ester, it reacts with the Cu(II) complex by a factor of 3200 times that involved in the protonated ester case. Thus, the overall effect of binding Cu(II) to the neutral ester is to accelerate its rate of base hydrolysis by about 10^5. This large augmentation in rate has been taken as an example of *super-acid catalysis*,[617] and, again, chelation to the carbonyl oxygen was assumed. A Zn(II) (mercaptoethylamine) complex was also found to give a super-acid effect, though much less than that produced by Cu(II). These authors[618] also report that amino acid esters that strongly bind metal ions through chelation at two sites, neither of which is the reactive site (histidine methyl ester,[619] cysteine methyl ester, and aspartic acid β-methyl ester), are much less susceptible to metal ion promotion of hydrolysis. For example, the second order rate constant for hydroxide attack on the 2:1 histidine methyl ester complex of Cu(II) is 265, but it is 101000 a^{-1} sec^{-1} in the case of Cu(II)(glycine ethyl) ester.

Jones and coworkers[620,621] have studied the hydrolysis of ethyl glycinate by the metal ions Co(II), Ni(II), Cu(II), and Zn(II) and essentially agree with the interpretation of Conly and Martin. The effectiveness of the metal ions in promoting OH^- hydrolysis of glycine ethyl ester is

$$Cu^{II} \gg Zn^{II} > Co^{II} > Ni^{II}$$

This is the same sequence as the stability constants of the 1:1 metal–hydroxide complexes. The rate constants for attack by water which are

$$Co^{II} > Zn^{II} > Ni^{II} \gg Cu^{II}$$

vary inversely with the stabilities of the 1:1 metal ester complexes.

In order to simplify the equilibrium relationship in systems of this class containing labile metal ions, Angelici and co-workers[622] studied the hydrolysis of the lone ester group in a series of *N,N*-diacetic acid derivatives of amino acids (structure **106**).

$$RO\overset{\overset{\displaystyle O}{\|}}{C}(CHR)_nN(CH_2CO_2H)_2$$

106

These ligands generally chelate in a tridentate fashion using the tertiary amine and carboxylate ions. The question of binding of the ester carbonyl has been investigated but generally with negative results. The simplest ligand of this class, ethylglycine-*N,N*-diacetic(EGDA) acid, gives the rate law[622] [Eq. (340)] in

$$\text{rate} = k[M(EGDA)][OH^-] \tag{340}$$

pH range of 5–7. From the ionic strength dependence of the rates with M = Cu(II) or Sm(III), it was believed that the mechanism involved coordination of the carbonyl oxygen of the ester group to the metal ion in the transition state.

The rate of ester hydrolysis increases in the sequences given below[623] for the divalent and trivalent metal ions studied:

$$Cd^{II} < Ni^{II} < Mn^{II} < Co^{II} < Fe^{II} < Zn^{II} < La^{III} < Cu^{II} < Pb^{II} < Nd^{III} < Gd^{III} < Sm^{III} < Dy^{III} < Er^{III} < Yb^{III} < Lu^{III}$$

The range of second order rate constants is from 2.1×10^2 to 3.5×10^5 M^{-1} sec^{-1}. For the lanthanides, the rates parallel the formation constants for the complexes with the related ligand iminodiacetic acid, generally increasing with increasing atomic number but with a slight decrease at gadolinium. The divalent ions do not show a related trend; instead, a weak correlation again exists between catalytic rate and the magnitude of the formation constant of the hydroxo complex. From limited thermal data it has been shown that the trend for the divalent ions is determined by changes in $\Delta H^{\ddagger}$, and the variations observed among the lanthanides are the result of changes in $\Delta S^{\ddagger}$.

Rate changes with ligand–substrate structure were also interesting. Replacing a hydrogen atom α to the ester group with a methyl group is without significant effect (ethyl glycinate-*N,N*-diacetic acid → ethyl-α-alanine-*N,N*-diacetic acid). However, the bulkier groups present in the same location in the leucine and valine derivatives had a great retarding effect on the catalysis.[623,624] Consequently, these derivatives were convenient for more detailed studies. Formation constants could be determined accurately, and it was possible to study the effects on their hydrolysis rates of a number of alternate nucleophilic catalysts. From the ratio of the first (and only) two formation constants, it was determined that the ester carbonyl group was not extensively coordinated. The dependence of the rate on a general set of nucleophiles gave strong support to the mechanism postulated earlier and discussed above. The effectiveness of nucleophiles varied in the sequence observed for their hydrolysis of uncoordinated esters and not with their basicities.

The variation of the length of the chain linking the ester group (*n* in structure **106**) to the tertiary nitrogen had a profound effect on rates. The effect was variable (depending on the metal ion) and small in going from one to two carbons, but the catalytic effect disappeared entirely for all metal ions reported upon going from two to three carbons. Thus, the seven-membered chelate ring shows so little tendency to form that no metal ion promotion of hydrolysis occurs.

By using nitrilotriacetic acid (H_3NTA) to block four coordination sites, Hopgood and Angelici[625,626] produced a range of catalysts with labile metal ions that have only two sites for interaction with a substrate. They then investigated the bonding of amino acids and their esters at these sites[625] and the catalysis of base hydrolysis of the esters for the Cu(II) case.[636] For the amino acid anions, the formation constants followed the Irving-Williams order,

$$Mn^{II} < Co^{II} < Ni^{II} < Cu^{II} > Zn^{II}$$

but the ethyl valinate ester gave a distinctly different sequence. The anomalously large constants for bonding to Mn(II), Pb(II), and Co(II) were attributed to some chelation through the ester carbonyl; the other metal ions were presumed to coordinate only to the amino group of the ester. In the case of Co(II), this suggestion was supported by a substantial Cotton effect. The rate of base hydrolysis of $H_2NCHRCO_2Et$ is mainly responsive to the steric effect of the group R both for the free ester and the catalytic process in which it is bound to $Cu(NTA)^-$. The ratio of catalyzed to uncatalyzed rates is approximately constant at 200 with the exceptions of ethyl-β-alaninate and methyl histidinate, which may not bind their ester carbonyls to the metal in the catalytic process.

Similar studies with substituted iminodiacetates (RIMDA) indicated that steric interaction occurs between the R group on RIMDA and large groups on the esters.[627] The rates for Cu(IMDA)-ester decrease in the order:

methyl glycinate > methyl-α-alaninate > ethyl sarcosinate
> ethyl glycinate > butyl glycinate > ethyl leucinate

Both these and the NTA systems show evidence for hydrolysis by both H_2O and OH^-, and the rate is insensitive to the nature of R in RIMDA unless R contains a coordinating group. In all cases, there is no evidence for the participation of a bound OH^- group in the rate process. Indeed, the concentrations of $Cu(R\text{-}IMDA)OH^-$ are vanishingly small under the conditions of the studies. Interestingly, a crystal structure[627] on $Cu[EtOC(O)CH(i\text{-}Pr)N(CH_2CO_2)_2]\cdot H_2O$ shows weak interaction between the metal ion and alkoxy oxygen of the ester group rather than the carbonyl oxygen. Of course, the solution structure need not correspond closely to the structure in the solid state.

The complex of Cu(II) with diethylenetriamine(dien) is a slightly poorer catalyst for the hydrolysis of amino acid esters than $Cu(NTA)^-$ despite its positive charge.[628] An interesting parallel between the formation constant for ester complexing and rate of catalysis has been drawn:

$Cu(EtGly)^{2+}$ > Cu(IMDA)(MeGly) > $Cu(NTA)(MeGly)^-$
> $Cu(dien)(MeGly)^{2+}$ >MeGly

It is reasonable to assume that the ability of the catalyst to polarize the C=O unit of the ester would parallel its ability to bind to the ester; however, as indicated above, this has not always been observed.

The foregoing discussion illustrates the need for a different class of information in order to obtain a more intimate understanding of the mechanism, or mechanisms, by which metal ions can activate amino acid esters and related functions. Perhaps the most compelling question that cannot be confronted adequately in the study of labile systems is the nature of the interaction between the metal ion and the ester function. The three possibilities that have attracted the attention of investigators are shown in structures **107, 108,** and **109.**

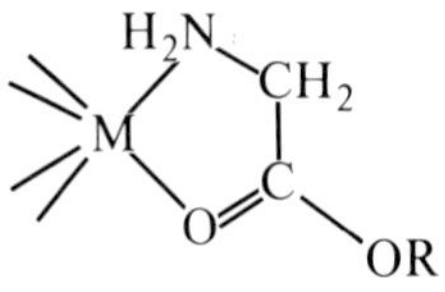

107 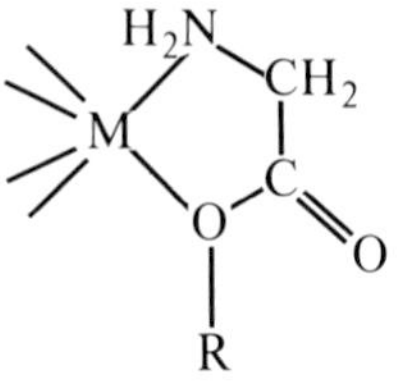 **108** 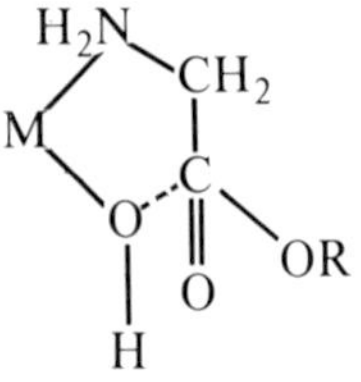 **109**

Alexander and Busch[157,629] initiated studies with systems well-adapted to the investigation of this question and related mechanistic considerations by synthesis of *N*-bonded amino acid ester complexes of the inert metal ion, cobalt(III). *trans*-Dichlorobis(ethylenediamine)-cobalt(III) reacts with glycine esters to form *cis*-$[Co(en)_2(NH_2CH_2CO_2R)X]^+$ (X = Cl, Br) which were isolated and characterized by spectral studies and resolution (R = Et, *i*-Pr) into optical isomers. The kinetics of the Hg^{2+}-catalyzed hydrolysis to $[Co(en)_2(NH_2CH_2CO_2)]^{2+}$ was studied in detail. Hg^{2+} electrophilically removes the halide ion [Eq. (341)], and the intermediate (IM) so generated hydrolyzes to the Co(III) complex of glycinate ion [Eq. (342)]. Of central importance is the

$$[Co(en)_2(RGly)X]^{2+} + Hg^{2+} \rightarrow [IM] + HgX^+ \qquad (341)$$

$$[IM] \xrightarrow{H_2O} [Co(en)_2(Gly)]^{2+} + ROH \qquad (342)$$

structure of the intermediate. Most of the studies were carried out in acidic media. The rate of reaction [Eq. (341)] is faster for X = Br than for Cl while that for the reaction in Eq. (342) decreases in the usual sequence with increasing size of R,

$$CH_3 > C_2H_5 > i\text{-}C_3H_7$$

By proper choice of the system, the critical intermediate could be generated in near stoichiometric quantities, and its structure was studied by spectroscopic methods. The simplest change might involve replacement of X by H_2O to yield $[Co(en)_2(H_2O)(RGly)]^{3+}$. Such a complex could lose a proton, forming a coordinated hydroxide ion, which could then act as an intermolecular nucleophile toward the ester group. The aquo species would have an electronic spectrum closely related to those of the corresponding amine complexes $[Co(en)_2(H_2O)(NH_2R)]^{3+}$; however, such was not the case. Also the aquo–ester complex should retain its ester function in an unperturbed condition so that a C=O stretching mode should be observed above 1700 cm^{-1} in the infrared spectrum of the complex. The most popular model for intermediates in ester hydrolysis has been structure **107**, involving coordination of the ester carbonyl oxygen. From

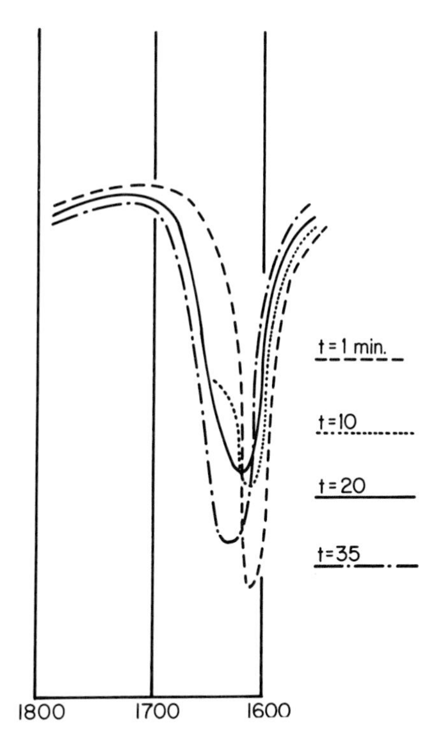

Figure 2-9 Infrared spectrum of intermediate in reaction of *cis*-$[Co(en)_2(i\text{-PrGly})Br]^{2+}$ with aqueous Hg^{2+}.

the work of Springer and Curran[630] such a structure is expected to show a C=O stretching mode in its ir spectrum that is much reduced in energy. Finally, structure **109** with a coordinated alkoxy oxygen should show an enhanced frequency in the C=O mode because of the electron-withdrawing effect of complexation. Figure 2-9 shows the infrared spectrum in the C=O region of a D_2O solution of *cis*-$[Co(en)_2(i\text{-PrGly})Br]^+$ after addition of Hg^{2+}. The rate of the reaction of Hg^{2+} with the coordinated Br^- is fast, so that the process is complete before the first spectrum is measured. The C=O of the intermediate [IM] absorbs at 1610 cm^{-1}, much lower than a free ester group. This requires the assignment of structure **107** to the active intermediate and confirms the expectation that metal ion activation of the hydrolysis of amino acid esters can proceed via carbonyl oxygen coordination.

The accelerating influence resulting from this mode of interaction with Co(III) is about 10^6 over protonated ethyl glycinate. The rate for the latter is reported[631] to be 6×10^{-9} sec^{-1}; the rate of reaction (342) for

$$\left[(en)_2Co\langle H_2N{-}CH_2{-}C(OC_2H_5){=}O\rangle\right]^{3+}$$

is 7.2×10^{-3} sec^{-1}.[157]

The hydrolysis of the intermediate identified in these studies was also shown to be subject to general nucleophilic catalysis. The rate law is therefore of the general form [Eq. (343)] where N_i are the nucleophiles available in a given reaction system. Under the conditions of these studies a hydroxide term is not involved in the rate law although evidence for such a term was obtained at higher pH.

$$\frac{-d[\mathrm{IM}]}{dt} = k_{\mathrm{H_2O}} + \sum_i k_{\mathrm{N}_i}[\mathrm{N}_i] \qquad (343)$$

Buckingham, Sargeson, and co-workers have clarified the subject of metal ion activation of amino acid esters and related functions in a series of highly detailed studies. Among their more striking results are the isolation and characterization of the intermediate containing the chelated ester

$$\left[(\mathrm{en})_2\mathrm{Co}\begin{matrix}\diagup \mathrm{H_2N}-\mathrm{CH_2} \\ \diagdown \mathrm{O}=\mathrm{C}-\mathrm{OR}\end{matrix}\right]^{3+}$$

($R = CH_3$, C_2H_5, *t*-Bu, benzil, and *n*-Bu).[160] The syntheses of these species are reported in a later paper.[632] The hydrolytic behavior of the intermediate has been examined and compared with the behavior of other reactions that generate the same intermediate [Eqs. (341), (344), and (345)]. The rates of formation

$$[\mathrm{Co(en)_2}(i\text{-PrGly})\mathrm{Br}]^{2+} + \mathrm{HOCL} \rightarrow \text{products} \qquad (344)$$

$$[\mathrm{Co(en)_2}(i\text{-PrGly})\mathrm{Br}]^{2+} + \mathrm{Cl_2} \rightarrow \text{products} \qquad (345)$$

of the chelated glycine product $[Co(en)_2(Gly)]^{2+}$ are the same in all cases. The identity of the intermediate generated in the studies of Alexander and Busch[629] with $[Co(en)_2(i\text{-}PrGly)]^{3+}$ has been confirmed by ir studies. Competition studies with such potential ligands as NO_3^-, Cl^-, Br^-, and HSO_4^- show that the intermediate, once formed, does not undergo ring opening reactions. Also, the Hg^{2+}-induced path [Eq. (446)] involves formation only of the chelated intermediate even in the presence of large excesses of competing anions. When HOCl is used to activate the starting material, sulfate competes successfully with the ester carbonyl for the vacant coordination site on the cobalt ion.

Oxygen tracer experiments identify the point of cleavage of the ester function as the C=O bond of the alkoxyl group. They also show that the carbonyl oxygen atom is completely retained in the product, thus confirming the sequence of events given in Eq. (346). The obvious alternative routes predict erroneous exchange patterns.

$$\left[(en)_2Co^{III}\begin{matrix} H_2N-CH_2 \\ O=C-OR \end{matrix}\right]^{3+} + OH_2 \longrightarrow \left[(en)_2Co^{III}\begin{matrix} H_2N-CH_2 \\ O-C=O \end{matrix}\right]^{2+} + ROH \qquad (346)$$

Studies on the roles of other nucleophiles have substantiated and expanded on the report that the hydrolysis of $[Co(en)_2(RGly)]^{3+}$ is subject to general nucleophilic catalysis.[632,633] In fact, nitrogen bases form chelated amides where the observed rate constant k_{obs} obeys Eq. (347).

$$k_{obs} = k_{N_1}[NHR_2] + k_{N_2}[NHR_2]^2 \qquad (347)$$

The first term represents the nucleophilic attack on the chelated ester carbon by the nitrogen nucleophile; the second is associated with general base catalysis of the nucleophilic process.

Extensive product analysis confirmed the presence of multiple pathways in aqueous solutions containing such bases. Thus, carrying out the solvolysis in the presence of glycine isopropyl ester yielded $[Co(en)_2(i\text{-PrGlyGly})]^{3+}$, and the similar reaction in the presence of dimethylamine yielded $[Co(en)_2\text{-}(NH_2CH_2CO{-}N(CH_3)_2)]^{3+}$. A linear correlation was found between pK_a and $\log k_{N_1}$ for the various nitrogen bases studied. Although the question of whether such oxygen bases as substituted acetates act as nucleophiles or as general bases presents a more difficult problem, it seems to have been resolved in the case of acetate by competition experiments.[633] In these experiments, the hydrolysis of $[Co(en)_2(i\text{-PrGly})]^{3+}$ was carried out in the presence of both acetate and aminoacetonitrile. The rate is given by Eq. (348)

$$k_{obs} = k_{H_2O}[H_2O] + k_{OH^-}[OH^-] + k_N[N] + K_{Ac}[Ac] \qquad (348)$$

where N refers to aminoacetonitrile and Ac refers to acetate. The dependences of the rate on acetate and aminoacetonitrile were mutually independent—i.e., there was no cross term $k[Ac][N]$ as might have been expected for general base catalysis. Of greater importance, the distribution of products depended on the concentration of aminoacetonitrile and *not* on the concentration of acetate. This cannot be accommodated by a general base model because as the concentration of acetate increased, the amount of

$$\left[(en)_2Co^{III}\begin{matrix} H_2N-CH_2 \\ O-C=O \end{matrix}\right]^{2+}$$

would then increase since the process [Eq. (349)] cannot incorporate the amine.

$$\left[(en)_2Co\langle H_2N{-}CH_2{-}C(OR)(\cdots OH{-}OCCH_3)=O\rangle\right] \rightarrow \left[(en)_2Co\langle H_2N{-}CH_2{-}C(=O){-}O\rangle\right]^{2+} + ROH + HC_2H_3O_2 \quad (349)$$

In contrast, if the acetate acts as a nucleophile, an anhydride is formed as an intermediate, and the amine, H_2O, and OH^- may compete for reaction with it with success comparable with that they enjoy in the case of ester itself [Eq. (350)].

$$\left[(en)_2Co\langle H_2N{-}CH_2{-}C(OR)=O\rangle\right]^{3+} + {}^-O{-}C(=O){-}CH_3 \rightarrow \left[(en)_2Co\langle H_2N{-}CH_2{-}C(=O){-}O{-}C(=O){-}CH_3\rangle\right]^{3+} + OR^-$$

$$\xrightarrow{NH_2R} \left[(en)_2Co\langle H_2N{-}CH_2{-}C(NHR)=O\rangle\right]^{3+} \qquad \xrightarrow{H_2O,\ OH^-} \left[Co\langle H_2N{-}CH_2{-}C(=O){-}O\rangle\right]^{2+} + CH_3COOH \quad (350)$$

This provides substantial support for the conclusion that the chelated ester is subject to general nucleophilic attack by a broad range of nucleophiles.

In the immediately preceding discussion, it has been shown that metal ion activation of amino acid esters in acidic media occurs by chelation of the carbonyl oxygen. This constitutes a well-established example of metal ion promotion of the reactivity of a coordinated electrophile by the displacement of electron density from the substrate toward the metal ion. Studies on base hydrolysis in systems of this class have also established an alternate process, the formation of a new nucleophile and its role as an intramolecular reactant. The ideal reactant, $[Co(NH_3)_5(EtGly)]^{3+}$, was designed and prepared by Buckingham, Foster, and Sargeson.[105] Since the Co(III) is attached only to inert amine nitrogen donors, no site for binding to the metal ion becomes available during the reaction. In acidic media, the free ester function hydrolyzes to form the carboxylic acid group, and the metal ion functions merely as a distant, charged substituent [Eq. (351)].

$$\left[(NH_3)_5Co^{III}NH_2CH_2C(=O)OC_2H_5\right]^{3+} + H_2O \longrightarrow (NH_3)_5Co^{III}NH_2CH_2C(=O)OH + C_2H_5OH \quad (351)$$

The hydrolysis of $[Co(NH_3)_5(EtGly)]^{3+}$ proceeds by two pathways in basic media, and the rate law follows Eq. (352).

$$\text{rate} = (k_1[OH^-] + k_2[OH^-]^2)[Co(NH_3)_5(EtGly)^{3+}] \quad (352)$$

The pathway characterized by the constant k_1 produces the monodentate glycinate ion [Eq. (353)], and that corresponding to k_2 leads to formation of the chelated imide (structure **110**).

$$\left[(NH_3)_5Co^{III}NH_2CH_2C(=O)OC_2H_5\right]^{3+} + OH^- \longrightarrow (NH_3)_5Co^{III}NH_2CH_2C(=O)O^- + ROH \quad (353)$$

$$\left[(H_3N)_4Co(NH_2CH_2C(=O)NH)\right]^{2+}$$

110

The reaction corresponding to k_1 is given in Eq. (353). The rate constant $k_1 = 50M^{-1}sec^{-1}$, $\mu = 0.1M$, compares to that for base hydrolysis of the protonated ester $NH_3^+—CH_2CO_2C_2H_5$, $24M^{-1}sec^{-1}$, $\mu = 0.16M$,[616] showing that the $[Co(NH_3)_5]^{3+}$ entity plays a role similar to that of the proton in these processes.

Scheme 2-23

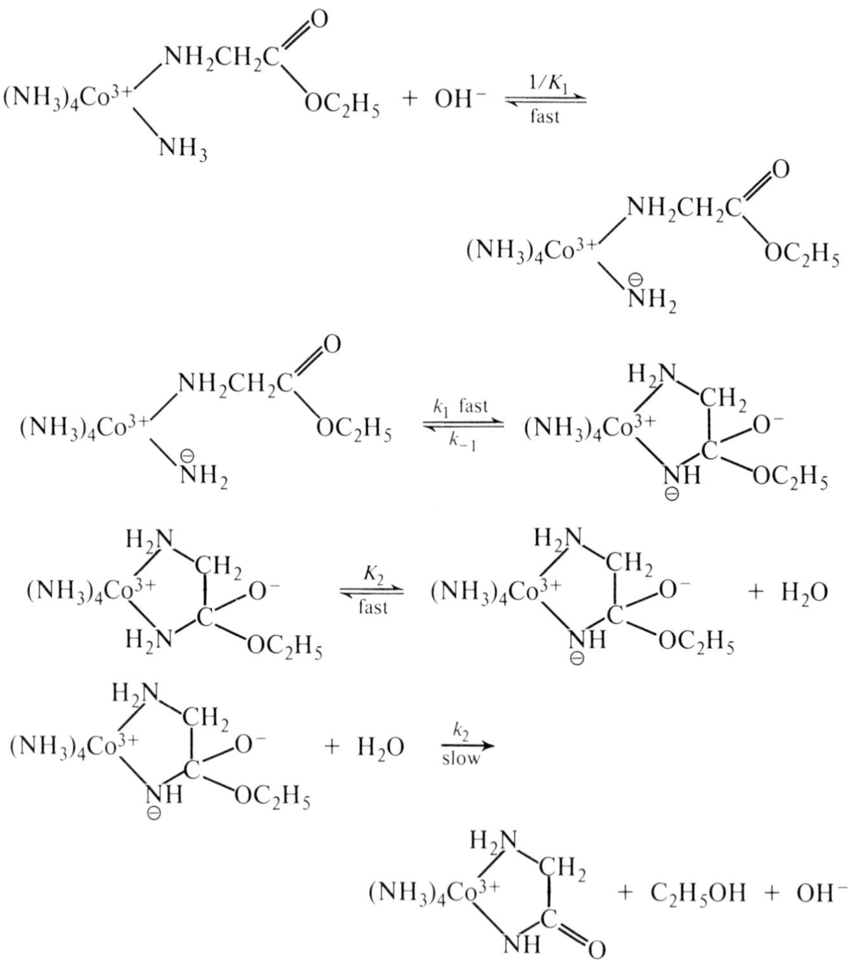

$$(NH_3)_4Co^{3+}(NH_3)(NH_2CH_2C(=O)OC_2H_5) + OH^- \underset{\text{fast}}{\overset{1/K_1}{\rightleftharpoons}} (NH_3)_4Co^{3+}(\overset{\ominus}{N}H_2)(NH_2CH_2C(=O)OC_2H_5)$$

$$(NH_3)_4Co^{3+}(\overset{\ominus}{N}H_2)(NH_2CH_2C(=O)OC_2H_5) \underset{k_{-1}}{\overset{k_1 \text{ fast}}{\rightleftharpoons}} (NH_3)_4Co^{3+}[H_2N\text{—}CH_2\text{—}C(O^-)(OC_2H_5)\text{—}\overset{\ominus}{N}H]$$

$$(NH_3)_4Co^{3+}[H_2N\text{—}CH_2\text{—}C(O^-)(OC_2H_5)\text{—}NH_2] \underset{\text{fast}}{\overset{K_2}{\rightleftharpoons}} (NH_3)_4Co^{3+}[H_2N\text{—}CH_2\text{—}C(O^-)(OC_2H_5)\text{—}\overset{\ominus}{N}H] + H_2O$$

$$(NH_3)_4Co^{3+}[H_2N\text{—}CH_2\text{—}C(O^-)(OC_2H_5)\text{—}\overset{\ominus}{N}H] + H_2O \xrightarrow[\text{slow}]{k_2} (NH_3)_4Co^{3+}[H_2N\text{—}CH_2\text{—}C(=O)\text{—}NH] + C_2H_5OH + OH^-$$

Scheme 2-23 was proposed[105] to reconcile the rate law with the formation of the chelated imide. Key features in this scheme are the internal nucleophilic attack of the coordinated NH_2 on the carbonyl carbon of the ester group and the formation of a tetrahedral carbon intermediate. It is the solvent-assisted dissociation of that intermediate that is rate determining. Assuming steady state conditions and rapid proton exchanges, the rate law given in Eq. (354) is derived.

$$\frac{-d[(Co(NH_3)_5(EtGly))^{3+}]}{dt} = \left(\frac{k_1k_2}{K_1K_2k_{-1}}\right) \times [(Co(NH_3)_5(EtGyl))^{3+}][OH^-]^2 \quad (354)$$

The base hydrolysis of $[Co(en)_2(RGly)X]^{2+}$, for X = Br or Cl and R = CH_3, C_2H_5, $CH(CH_3)_2$, $(CH_2)_3CH_3$, $C(CH_3)_3$, and $CH_2C_6H_5$, has been studied and interpreted in terms of a related series of reactions. The reactions are generally complicated and give numerous side products in addition to the expected $[Co(en)_2(Gly)]^{2+}$. Rate measurements, extensive product analyses, ^{18}O tracer studies, and racemization measurements support the interpretations. In general, the rate of the process leading to formation of $[Co(en)_2(Gly)]^{2+}$ is independent of R. The process proceeds with incorporation of one ^{18}O atom from solvent, and about one-half of the ^{18}O originally in the carbonyl oxygen position is retained. Optically active starting material $[(-)_{589}$-$Co(en)_2(MeGly)Br]^{2+}$ reacts to give 50% racemic $[Co(en)_2(Gly)]^{2+}$, the other 50% being formed with retention of configuration. These and other results can be rationalized by assuming that the hydrolysis process quickly follows halide loss but that the process proceeds by the two pathways (Scheme 2-24). The two pathways may reflect a competition between the ester carbonyl and solvent for the vacated position in a five-coordinate intermediate.

A related complex of triethylenetetraamine (trien), $[\beta_2$-Co(trien)-$Cl(EtGly)]^{2+}$, has provided a useful second system to confirm the conclusions derived from works with the $[Co(en)_2(RGly)X]^{2+}$ compounds.[634] The Hg^{2+}-catalyzed acid hydrolysis reaction of $[\beta_2$-Co(trien)(EtGly)Cl$]^{2+}$ resembles that for the ethylenediamine complex in several ways: (a) the same rate law and similar rates are observed; (b) the glycinato complex is the only product; (c) the optically active complex reacts with full retention of chirality; (d) added anions fail to compete successfully with the ester carbonyl for a position in the coordination sphere of the Co(III); and (e) tracer experiments show that the carboxylate oxygen bound to Co(III) in the product is the ester carbonyl oxygen of the reactant. This all favors the mechanism given for this metal-promoted reaction.

Unlike the Hg^{2+}-induced reaction, the base hydrolysis of $[\beta_2$-Co(trien)-(EtGly)Cl$]^{2+}$ differs in a number of ways from that of $[Co(en)_2(EtGly)X]^{2+}$: (a) The rate laws are the same, but the trien complex hydrolyzes more rapidly by a factor of about 3000; (b) in contrast to the en derivative, the trien complex yields a single product, $[\beta_2$-Co(trien)(Gly)$]^{2+}$, upon base hydrolysis; (c) $[\beta_2$-Co(trien)(Gly)$]^{2+}$ is formed from optically active material with complete retention of chirality. (d) N_3 competition is observed for both starting materials. (e) base hydrolysis of $[\beta_2$-Co(trien)(EtGly)Cl$]^{2+}$ results in retention of ap-

Scheme 2-24

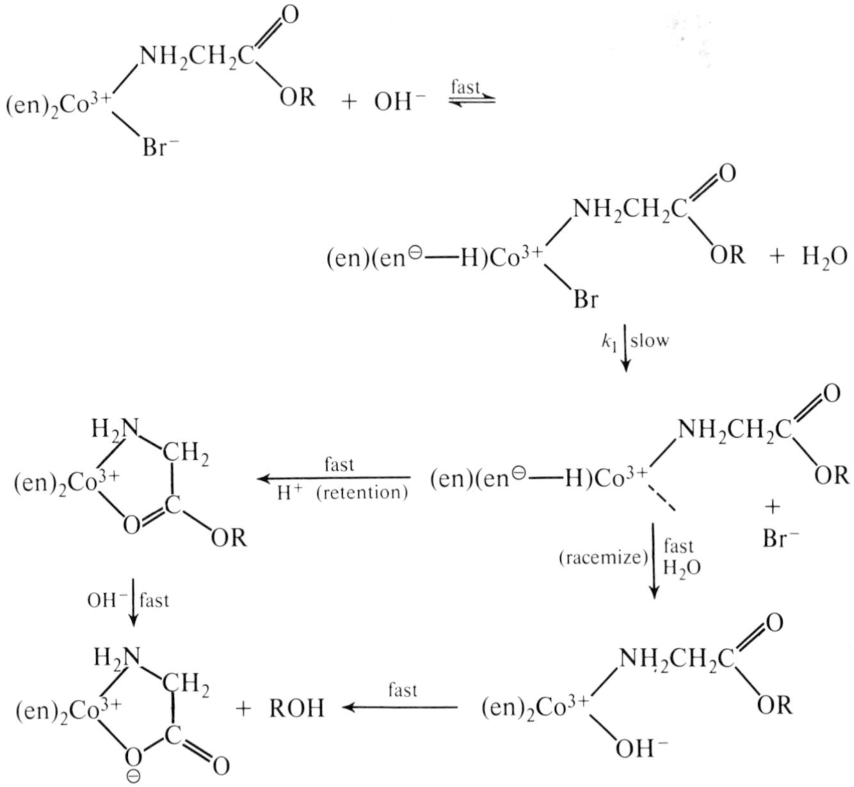

$(en)_2Co^{3+}$ ($NH_2CH_2C(=O)OR$)(Br^-) + OH^- $\xrightleftharpoons{\text{fast}}$ $(en)(en^{\ominus}{-}H)Co^{3+}$ ($NH_2CH_2C(=O)OR$)(Br) + H_2O

k_1 ↓ slow

$(en)(en^{\ominus}{-}H)Co^{3+}$ ($NH_2CH_2C(=O)OR$) + Br^-

fast, H^+ (retention) → $(en)_2Co^{3+}$ (H_2N–CH_2–$C(OR)$=O chelate)

fast, H_2O (racemize) ↓ $(en)_2Co^{3+}$ ($NH_2CH_2C(=O)OR$)(OH^-)

OH^- ↓ fast; fast →

$(en)_2Co^{3+}$ (H_2N–CH_2–$C(=O)$–$O^{\ominus}$ chelate) + ROH

proximately 84% of the ester carbonyl oxygen, and about 16% of the solvent oxygen appears in the Co—O position in the product. Scheme 2-24, worked out to explain the behavior of the $[Co(en)_2(RGly)X]^{2-}$ systems, applies even more easily to the trien complex. The principal differences are that the latter system is associated with fewer troublesome side reactions, and it displays a different fractionation along the two pathways of reaction of the five-coordinate intermediate produced by chloride expulsion.

Hay and Bennett have studied the contrasting case of base hydrolysis of a long-chain amino acid ester, methyl-6-aminohexanoate, in its Co(III) complex [*cis*-$Co(en)_2(NH_2(CH_2)_5COOCH_3)^{3+}$].[635] The ester hydrolyzes at a slow rate compared with the rate of chloride loss. The role of the metal ion is minor in this case because of the great distance between it and the ester function.

Since it has long been known that the hydrolysis of *tert*-butyl esters proceeds largely by alkyl–oxygen bond breakage, the metal ion-promoted hydrolyses of the *tert*-butyl esters of amino acids were expected to be exceptional; this has been found to be the case. The Hg^{2+}-promoted hydrolysis of $[Co(en)_2(t\text{-BuGly})Br]^{2+}$ yields $[Co(en)_2(Gly)]^{2+}$ with the acceleration of the hydrolysis rate due to

Co(III)'s being about 100-fold.[636] This is smaller by a factor of 10^4 than the effect on EtGly. Since ir spectral studies on reacting solutions of the *t*-BuGly complex confirm the formation of the chelated ester intermediate, this variation in the promotional effect suggests a substantial alteration in the hydrolysis mechanism. The rates also increase as the solvent polarity decreases for the *tert*-butyl complex, the converse behavior is observed for MeGly, EtGly, and *i*-PrGly complexes. Oxygen-18 tracer experiments clearly showed the point of cleavage to involve the alkyl–oxygen linkage.[636,637] The essential aspects of the process are shown in Eq. (355).

$$\left[(en)_2Co(Br)(NH_2CH_2C(=O)OR)\right]^{2+} \xrightarrow[H^+]{-Br^-} \left[(en)_2Co\langle O{=}C(O{-}R)CH_2NH_2\rangle\right]^{3+} \xrightarrow{H_2{}^{18}O} \left[(en)_2Co\langle OC(=O)CH_2NH_2\rangle\right]^{2+} + R^{18}OH \quad (355)$$

Stereoselectivity in the hydrolysis of optically active amino acid esters by optically active metal complexes was first reproted by Murakami et al.[638] They found that the hydrolysis of L-(−)-phenylalanine ethyl ester to form the chelated derivative is faster with *trans*-$[Co(D(-)\text{-}pn)_2Cl_2]^+$ than with *trans*-$[Co(L(+)\text{-}pn)_2Cl]^+$. Hix and Jones[639] studied the hydrolysis of (*R*)-(−)- and (*S*)-(+)-histidine methyl esters in the presence of the Ni(II) complexes of (*R*)-(−)-histidine and (*S*)-(+)-histidine. The rate of hydrolysis of the (*R*)-(−)-ester is greater in the presence of Ni[(*S*)-(+)-histidinate]$^+$ than in the presence of Ni([*R*]-(−)-histidinate)$^+$, the rate constants differing by about 40%. When the components (ester, amino acid anion, and metal ion) are present in equimolar amounts, the predominant species is Ni(ester)(amino acidate). Further, the formation constant for this species is the same regardless of which isomers of ester and amino acid anion are present. It follows that the selectivity occurs in the hydrolysis step not in the relative stabilities of the reacting complexes Ni(ester)(amino acidate). Note that in other systems the formation constants for the diastereoisomers[640,641,642] shown in Eqs. (356) and (357) are not always

$$M(\text{L-amino acidate})^{n+} + \text{L-(amino acidate)}^- \rightleftharpoons M(\text{L-amino acidate})_2{}^{(n-1)+} \quad (356)$$

$$M(\text{L-amino acidate})^{n+} + \text{D-(amino acidate)}^- \rightleftharpoons M(\text{L-amino acidate})(\text{D-amino acidate})^{(n-1)+} \quad (357)$$

the same. Leach and Angelici used the Cu(II) complex of L-valine-*N*-monoacetate as a catalyst for the hydrolysis of the antimers of MeLeu, MPhe, MeAla, and MeSer. A selectivity was found only with the first two, and its extent, and even its sign, appeared to depend on the conditions of the experiment.[641]

Considerable attention has been given to the metal ion-promoted hydrolysis of amide linkages and peptide bonds and to peptide bond formation in metal complexes. This is, in part, because of the possible significance of such studies to the understanding of the role of metal ions in such enzymes as carboxypeptidase A, leucine aminopeptidase, and serine proteinases. A limited treatment of the subject is offered here.

In early studies Meriwether and Westheimer[610] showed that Cu(II), Ni(II), and Co(II) promote the hydrolysis of glycine amide. They also found that Cu(II) promotes the cleavage of both the amide and peptide linkages in phenylalanylglycine amide over ring closure to 3-benzyl-2,5-diketopiperazine. Later studies have shown that the hydrolysis of glycine amide is promoted by Cu(II) over the pH range of 6 to 10.5 but that Cu(II) inhibits the hydrolysis at still higher pH values.[616] This behavior has been attributed to the known isomerization of the point of coordination of amide functions upon ionization [Eq. (358)].[643] The N-terminal-bonded species does not hydrolyze, and the protonated, C-terminal-bonded species is readily hydrolyzed. The fact that Cu(II) is less effective at promoting amide hydrolysis than ester hydrolysis is attributed to a larger k_2/k_3 in the former case [Eq. (359)].

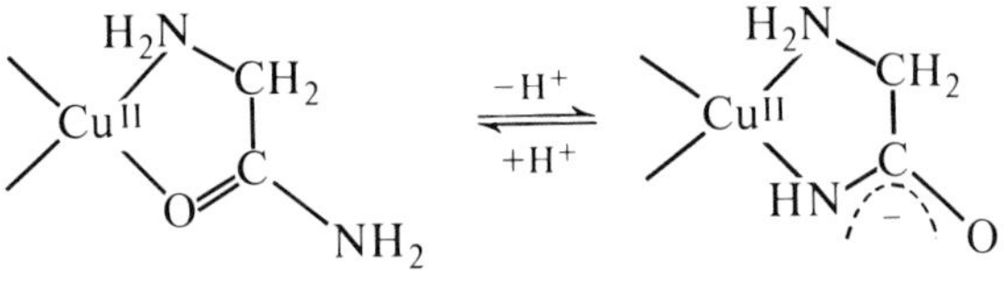

$$\mathrm{Cu^{II}(H_2NCH_2C(=O)NH_2)}\ \text{(N,O-chelate: } \mathrm{H_2N{-}CH_2{-}C(NH_2){=}O\rightarrow Cu}) \underset{+\mathrm{H^+}}{\overset{-\mathrm{H^+}}{\rightleftharpoons}} \mathrm{Cu^{II}(H_2NCH_2C(O)NH)}\ \text{(N,N-chelate: } \mathrm{H_2N{-}CH_2{-}C(=O){-}HN\rightarrow Cu}) \qquad (358)$$

$$\mathrm{NH_2CH_2C(=O)R} \underset{k_2}{\overset{k_1}{\rightleftharpoons}} \mathrm{NH_2CH_2C(OH)(O^-){-}R} \xrightarrow{k_3} \mathrm{NH_2CH_2C(=O)O^-} + \mathrm{RH} \qquad (359)$$

$$[\mathrm{Co(trien)(H_2O)OH}]^{2+} + \mathrm{H_2NCHR\overset{O}{\overset{\|}{C}}NHP} \longrightarrow \left[\mathrm{(trien)Co}\ \text{(chelate: } \mathrm{H_2N{-}CHR{-}C(=O){-}O\rightarrow Co})\right]^{2+} + \text{peptide residue} \qquad (360)$$

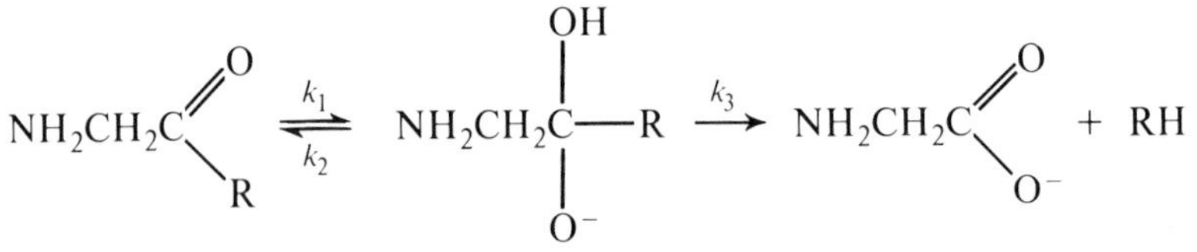

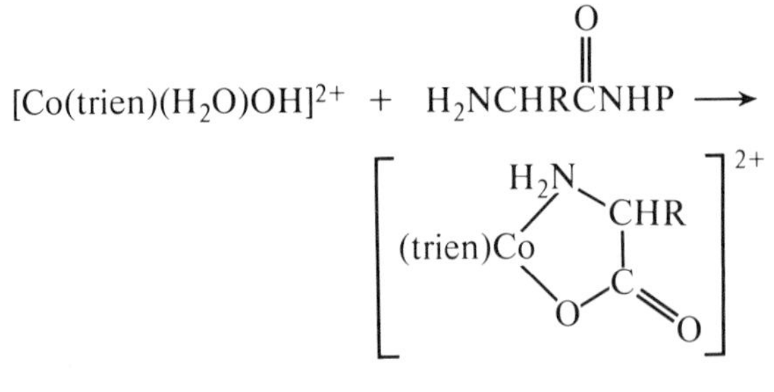

In 1963 Collman and Buckingham[644] reported the hydrolytic cleavage of N-terminal peptide bonds by [*cis*-Co(trien)(H_2O)OH]$^{2+}$ according to Eq. (360). These studies were expanded in a later publication.[645] Many [*cis*-β-Co(trien)-(amino acidate)]$^{2+}$ complexes were prepared as reference compounds, and the products of the reaction of [Co(trien)(H_2O)OH]$^{2+}$ with many amino acid amides and low molecular weight peptides were investigated. The rate of Eq. (361) was shown to be independent of the nature of the amide at 40°C and pH

$$[\beta\text{-Co(trien)(H}_2\text{O)OH}]^{2+} + \text{NHRGly} \rightarrow [\beta\text{-Co(trien)(Gly)}]^{2+} + \text{NH}_2\text{R} \quad (361)$$

7.35, indicating that coordination of the substrate, not its hydrolysis, was rate determining. The rate varies with pH in a way that indicates the most reactive cobalt(III) species is [Co(trien)(H_2O)OH]$^{2+}$ and not the diaquo or dihydroxo complex. The nature of the reaction products [Co(trien)(amino acidate)]$^{2+}$ from such related reactants as glycyl-*d,l*-phenylalanine and *d,l*-phenylalanylglycine firmly established the N-terminal specificity of the process. Less conclusive but similar studies were conducted with [Co(en)$_2$(H_2O)OH]$^{2+}$ as the reagent.[646] The fact that C-terminal binding to the cobalt is ineffective in peptidasis is also of much interest. The complex of structure **111** did not undergo peptide cleavage.[645]

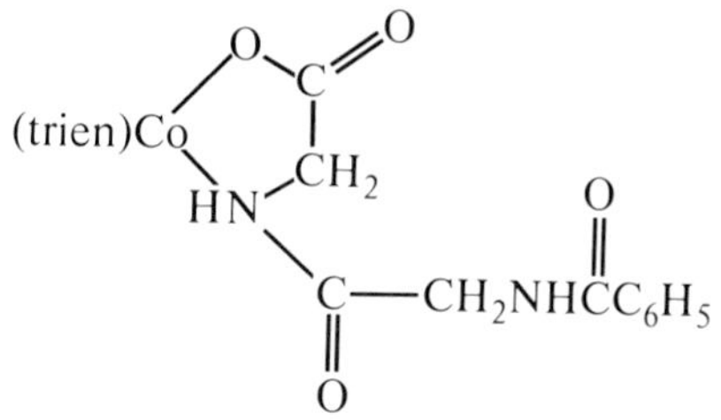

(trien)Co[–O–C(=O)–CH$_2$–NH–] chelate; HN–C(=O)–CH$_2$NHC(=O)C$_6$H$_5$

111

The reagent [Co(trien)(H_2O)OH]$^{2+}$ proved to be still more versatile when it was found that it promoted the condensation of amino acid esters to form dipeptides [Eq. (362)].[647,648] A crystal structure confirmed the nature of the product,[649] and the identical material could be made directly from the glycylglycine ester. It was also possible to prepare[648] the Gly-Gly complex from previously prepared ester complex [Eq. (363)]. It was proposed that the sequence of events involved loss of chloride, coordination of the carbonyl oxygen of the ester, and then nucleophilic attack on the carbonyl carbon by the nitrogen atom of the second mole of ester. The variation in mode of chelation of amides with ionization that relates to these processes has been considered by a number of investigators.[650,653–656]

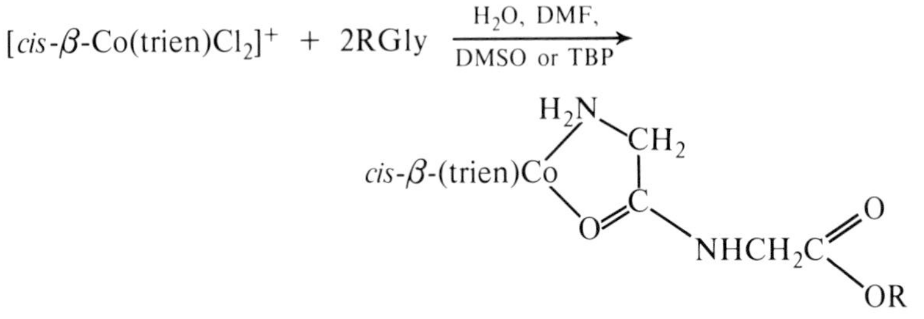

$$[cis\text{-}\beta\text{-Co(trien)Cl}_2]^+ + 2\text{RGly} \xrightarrow[\text{DMSO or TBP}]{\text{H}_2\text{O, DMF,}} cis\text{-}\beta\text{-(trien)Co}(\text{H}_2\text{NCH}_2\text{C(=O)O})\text{-C(NHCH}_2\text{C(=O)OR)} \qquad (362)$$

$R = CH_3,\ C_2H_5,\ i\text{-}C_3H_7$

$$[\beta_2\text{-Co(trien)(EtGly)Cl}]^{2+} + \text{EtGly} \rightarrow [\beta_2\text{-Co(trien)(EtGlyGly)}]^{3+} \qquad (363)$$

$$\text{rate} = \left(\frac{k_3 k_1}{k_2 + k_3}\right) \frac{K_b[\text{Co}][\text{OH}^-]}{K_b + [\text{OH}^-]} \qquad (364)$$

The investigation of the kinetics of base hydrolysis of the chelated peptide complex and related amide derivatives revealed a rate law [Eq. (364)] that suggested that ionization of the amide function inhibits the hydrolysis (Scheme 2-25).[651] This is consistent with the expectation that ionization of that group

Scheme 2-25

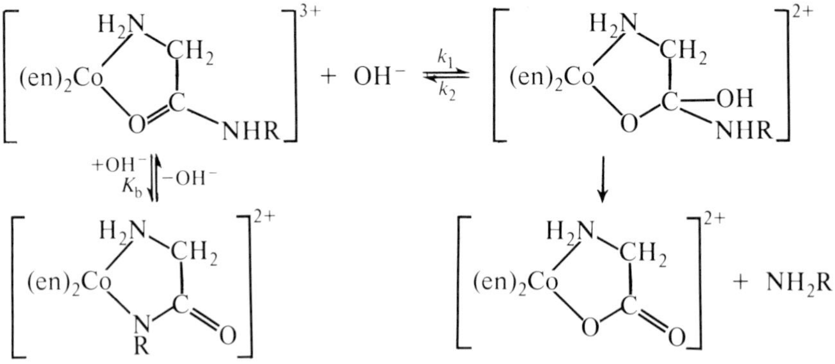

$$[(\text{en})_2\text{Co}(\text{H}_2\text{NCH}_2\text{C(=O)NHR})]^{3+} + \text{OH}^- \underset{k_2}{\overset{k_1}{\rightleftharpoons}} [(\text{en})_2\text{Co}(\text{H}_2\text{NCH}_2\text{C(OH)(NHR)O})]^{2+}$$

$$\downarrow \qquad\qquad [(\text{en})_2\text{Co}(\text{H}_2\text{NCH}_2\text{C(=O)O})]^{2+} + \text{NH}_2\text{R}$$

$$[(\text{en})_2\text{Co}(\text{H}_2\text{NCH}_2\text{C(=O)NHR})]^{3+} \underset{-\text{OH}^-}{\overset{+\text{OH}^-,\ K_b}{\rightleftharpoons}} [(\text{en})_2\text{Co}(\text{H}_2\text{NCH}_2\text{C(=O)NR})]^{2+}$$

would produce the hydrolytically stable N—N-chelated amide. Glycine dimethylamide did not give this behavior. Oxygen tracer experiments showed that the chelate ring remained intact during the hydrolytic process as required by the suggested mechanism. The rate of hydrolysis of the amide is 10^5–10^6 slower than that of the related esters, but the metal ion promotional effect is comparable for the two functional groups. Dangling amide groups show little metal ion effect on their rates of base hydrolysis (structure **112**).[652]

112

Surprisingly, the precursor to the chelated amide complex [*cis*-$Co(en)_2$-($NR_1R_2Gly)Br]^{2+}$ gave a relatively complicated base hydrolysis pattern with strong evidence for a different, very rapid hydrolysis pathway.[106] The rate of loss of Br^- from these complexes in basic media is much faster than the rate of hydrolysis of the chelated amide that is expected to be the product of that reaction [Eq. (365)].

$$[cis\text{-}Co(en)_2(NR_1R_2Gly)Br]^{2+} + OH^- \xrightarrow[\text{pH 9–13}]{} (en)_2Co(OC(NR_1R_2)CH_2NH_2) + Br^- \quad (365)$$

However, a substantial amount of $[Co(en)_2(Gly)]^{2+}$ is also produced in a reaction for which the loss of Br^- appears to be rate determining [Eq. (366)].

$$[cis\text{-}Co(en)_2(NR_1R_2Gly)Br]^{2+} + OH^- \xrightarrow[\text{pH 9–13}]{} [Co(en)_2(Gly)]^{2+} + NHR_1R_2 \quad (366)$$

The relative amounts of products resulting from competing reactions in Eqs. (365) and (366) depend on the nature of the amide; the percentage of chelated amide [Eq. (365)] is 46% for NH_2, 66% for —NHMe, and 82% for Me_2. Experiments with $[Co(en)_2(NR_1R_2Gly)]^{3+}$ show that all of its carbonyl oxygen is retained in its base hydrolysis product $Co(en)_2(Gly)$, and the corresponding experiment shows that the $[Co(en)_2(Gly)]^{2+}$ formed by the process of Eq. (366) contains solvent oxygen in its Co—O bond. These facts and others are consistent with Scheme 2-26. The loss of bromide is presumed to occur via the usual S_N1cb mechanism forming a five-coordinate intermediate. This intermediate then reacts via two pathways with the competing nucleophiles, solvent, and the amide oxygen. Path *A* then reflects the now familiar mechanism for Co(III)-promoted hydrolysis via carbonyl oxygen coordination. Path *B* is the less familiar but previously invoked route whereby a coordinated OH^- acts as an intramolecular

Scheme 2-26

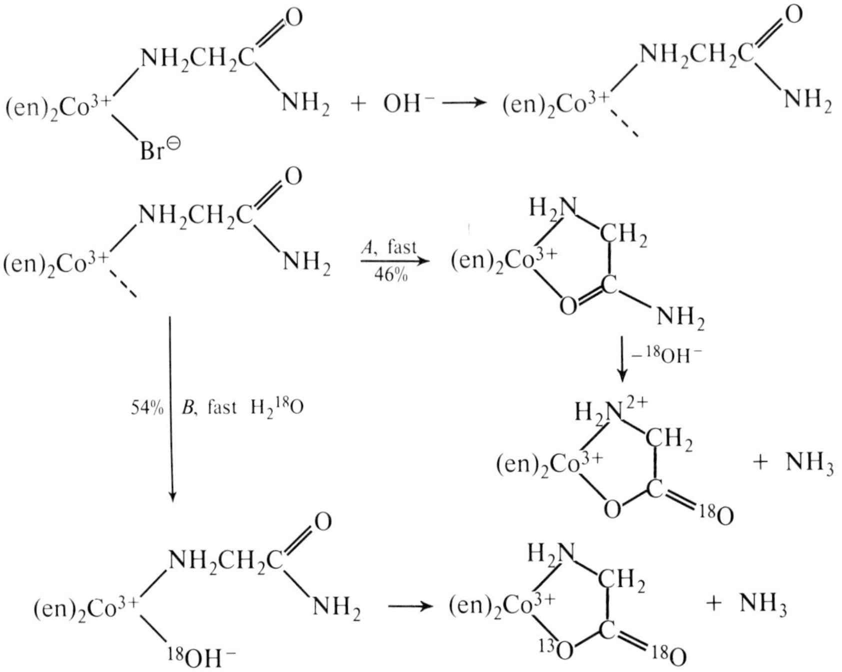

$(en)_2Co^{3+}(NH_2CH_2C(=O)NH_2)(Br^{\ominus}) + OH^- \longrightarrow (en)_2Co^{3+}(NH_2CH_2C(=O)NH_2)$

$(en)_2Co^{3+}(NH_2CH_2C(=O)NH_2) \xrightarrow[46\%]{A,\ fast} (en)_2Co^{3+}(H_2N\text{—}CH_2\text{—}C(NH_2)=O) \xrightarrow{-{}^{18}OH^-} (en)_2Co^{3+}(H_2N^{2+}\text{—}CH_2\text{—}C(=^{18}O)\text{—}O) + NH_3$

54% B, fast $H_2{}^{18}O$

$(en)_2Co^{3+}(NH_2CH_2C(=O)NH_2)({}^{18}OH^-) \longrightarrow (en)_2Co^{3+}(H_2N\text{—}CH_2\text{—}C(=^{18}O)\text{—}{}^{13}O) + NH_3$

nucleophile. The remarkable aspect of the system is in the apparent velocity of path *B*. Since the OH^--containing intermediate cannot be detected, this hydrolysis process must be faster than the rate of loss of Br^- from the starting material. This amounts to an acceleration of at least 10^7 and probably more. In fact it has been suggested that path *B* is comparable in metal ion-accelerating effect with those produced by related metalloenzymes.[653]

Various observations have emphasized the utility of this hydrolyzibility of the peptide linkage joined to a metal ion through the carbonyl linkage and the

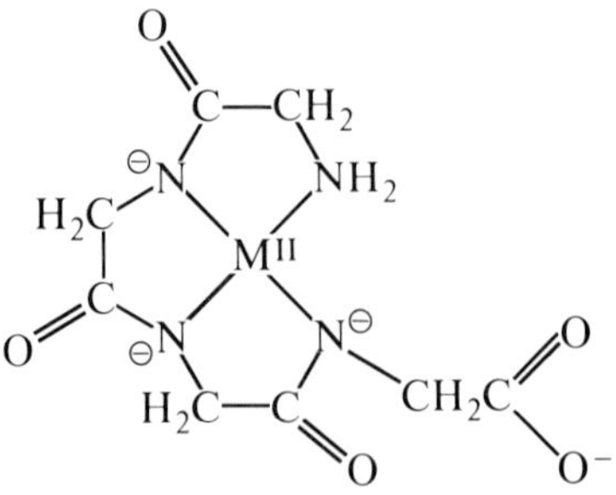

O, C—CH_2, $^{\ominus}$N, NH_2, H_2C, M^{II}, C, O, $^{\ominus}$N, $N^{\ominus}$, H_2C—C, CH_2C, O, O, O^-

113

stabilization, after deprotonation, of the N-bonded peptide linkage, especially when the latter is part of a fused chelate ring. Jones, Cook, and Brammer[653] found that only the dissociated peptide in the Cu(II)–glycylglycine system hydrolyzes. Andreatta, Freeman, Robertson, and Sinclair[654] found that hexaglycine and pentaglycine are selectively cleaved to tetraglycine in the presence of Ni(II) and Cu(II). The products, presumably of structure **113,** are ideally structured to chelate to these metal ions. According to Nakahara et al.,[655] the Schiff base of salicylaldehyde and tetraglycine is hydrolyzed at moderate pH to give the tetradentate complex of structure **114;** the third amino acid residue is not lost at higher pH.

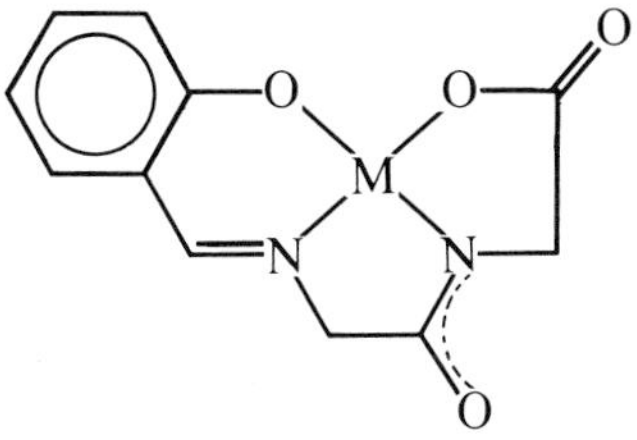

114

A three-site reagent similar to the Co(III) complexes with trien and with two moles of en has been designed to study the effects of the interactions of amino acid esters with inert three-site metal ion promoters.[656] In these reagents, the tridentate ligand diethylenetriamine blocks three sites on inert Co(III). Such species should be capable of adding two amino acid residues sequentially to the N-terminus of a peptide by functioning in accord with mechanistic schemes detailed above. The demonstrated route to tetrapeptide synthesis using this class of reagent is given in Scheme 2-27. Step (a) involves the usual N—O chelation of an amino acid ester followed by nucleophilic attack of a second mole of ester at the coordinated carbonyl function. Step (b) involves base-promoted isomerization of the O-bonded peptide linkage to N-bonded species. In (c) base-promoted dissociation of the remaining Cl^- occurs, producing a five-coordinated intermediate. The carbonyl carbon of the terminal ester group coordinates in (d) and is attacked by the dipeptide amino group in step (e).

Girgis and Legg found that N-terminal trifunctional amino acid esters (α-L-aspartyl-glycine) are not hydrolyzed by such two-site species as $[Co(trien)X_2]^+$ but that the three-site complex $Co(dien)X_3$ does promote the reaction with the formation of $[Co(dien)(L\text{-}Asp)]^+$ wherein the amino acidate is coordinated in a tridentate manner.[657]

A second example of the very rapid nucleophilic attack of a coordinated nucleophile on an adjacent electrophilic center has been reported recently.[658] Coordinated aminoacetonitrile is very quickly changed to oxygen-coordinated glycine amide in $[cis\text{-}Co(en)_2(NH_2CH_2CN)Br]^{2+}$ upon Hg^{2+} promoted removal of the Br^- in aqueous solution. A similar process yields a chelated tri-

Scheme 2-27

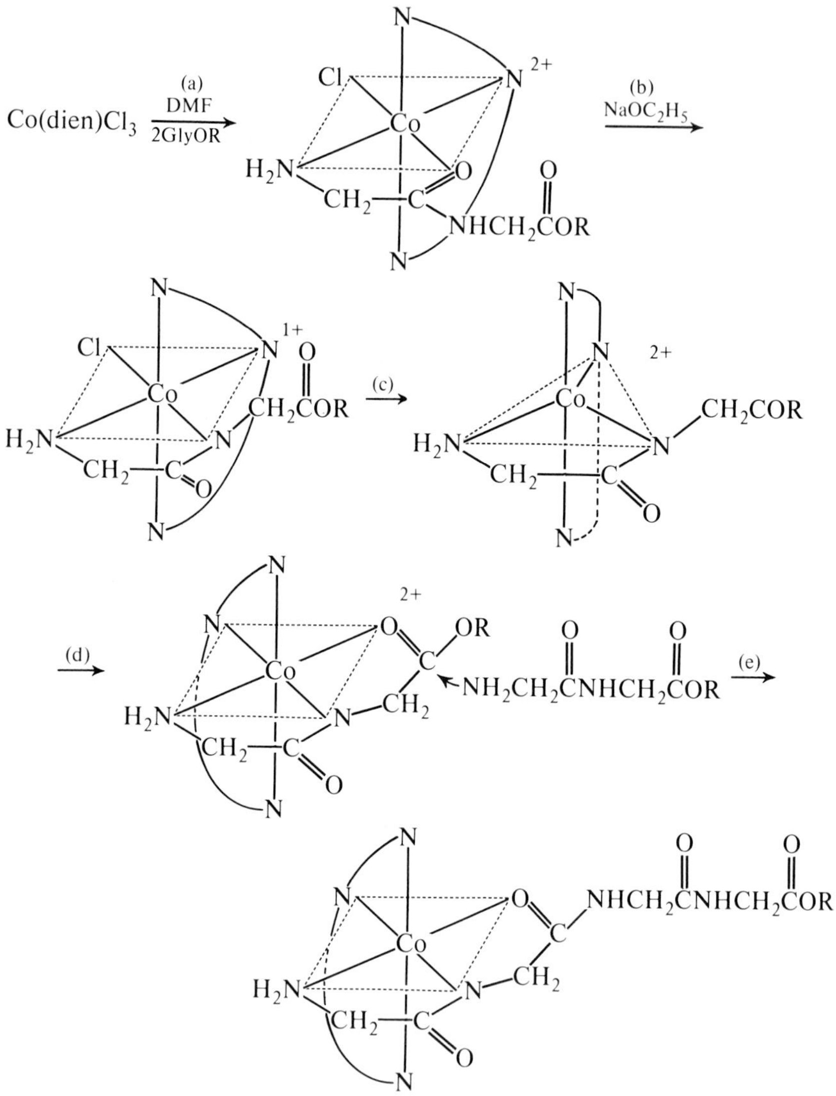

$Co(dien)Cl_3$ (a) DMF, 2GlyOR → (b) $NaOC_2H_5$ → (c) → (d) $NH_2CH_2CNHCH_2COR$ → (e)

dentate ligand in the base hydrolysis of [*cis*-$Co(en)_2(NH_2CH_2CN)Cl]^{2+}$. The product is $[Co(en)(NH_2CH_2C(NH_2){=}NCH_2CH_2NH_2)Cl]^{2+}$.[659]

IX. OXIDATION OF COORDINATED LIGANDS

Among the earliest reactions known arising from the interaction of compounds of transition metals with organic materials is ligand oxidation. This represents the normal expectation in combining positively charged species with compounds having readily removable hydrogen atoms.

The process of ligand oxidation involves electron transfer from the substrate (ligand) to the metal ion together with either hydrogen elimination or attack of electrophiles (oxidants) on the ligand (either before or after initial oxidation), as well as cleavage of carbon–carbon bonds. If the reaction is stoichiometric, the metal ion reduction products are often not readily converted to the active oxidizing species. If the reaction is catalytic, it can involve homolytic cleavage of C—X bonds (with formation of intermediate radicals) or heterolytic C—X bond rupture producing a positively charged species. In metal ion-catalyzed radical intermediate formation, two competing product-forming pathways can occur, with the character of the reaction determined by the relative rates of the two processes. In the presence of air, such a radical intermediate can react directly with molecular oxygen forming peroxy radicals (autoxidation), or it can react with a further equivalent of the metal ion oxidant, resulting in ligand disproportionation, proton elimination, or nucleophilic attack on the oxidized species.

We consider four types of organic ligand oxidations by compounds of transition metals: stoichiometric ligand oxidation (which can involve either homolytic or heterolytic C—X bond cleavage); autoxidation; heterolytic cleavage of C—H bonds in olefins subsequent to nucleophilic attack (the basis of the Wacker process for olefin oxidation); and, finally, oxidative dehydrogenation of coordinated ligands. The latter reaction represents a special case of transition metal ion-catalyzed, homolytic nitrogen–hydrogen bond cleavage in the ligand, which because of the character of the transition metal complex often occurs with retention of the metal–ligand bond.

A. Oxidation of Organic Substrates by Metal Compounds

Historically, much of the early attention focused on the interaction of transition metal ions with organic substances lay in study of the stoichiometric oxidation of these substrates by metal ions in high formal oxidation states [Ce(IV), Mn(III,VII), Co(III), Cr(VI), Pb(IV)].[660] Since our interest lies in the relation of the reaction mechanisms to recognizable chemical features associated with a given oxidant, we briefly examine the common oxidation mechanisms and then consider illustrative examples which provide insight into the mechanistic details for the individual oxidants.

The first consideration in this discussion must be for the number of electrons which can be transferred to each molecule of the oxidant which, of course, determines the nature of the chemical process. Thus, one-electron oxidants generate radical species [Eq. (367)] which are further oxidized to yield the observed products or undergo typical reactions of radicals (coupling, polymerization of unsaturated hydrocarbons, rearrangements, disproportionations, etc.).

$$M^{n+} + S \rightleftharpoons M^{(n-1)+} + R\cdot \tag{367}$$

Two-electron oxidants, however, commonly lead directly to the observed product [Eq. (368)].

$$M^{n+} + S \rightarrow M^{(n-2)+} + P \qquad (368)$$

Finally, several cases of one-step, three-electron oxidations of two substrates have been reported [Eq. (369)].

$$M^{n+} + S + S' \rightarrow M^{(n-3)+} + P + R\cdot \qquad (369)$$

Ce(IV) oxidations, which are almost certainly limited to one-electron transfers generating Ce(III) and a radical species [Eq. (367)], have been examined for many substrates including alcohols and diols.[661] In the former case, oxidations of primary and secondary alcohols (tertiary alcohols are much less reactive) have been determined to occur via a complexed intermediate.[662] The rate-determining step involves cleavage of a C—H bond on the OH^--substituted carbon as shown by the primary isotopic effect found for 1-*d*-cyclohexanol ($kH/kD = 1.9$) [Eq. (370)].[663]

$$C_6H_{10}(OH)D + Ce^{IV} \xrightarrow{\text{slow}} C_6H_{10}^{\cdot}(OH) + Ce^{IV} \xrightarrow{\text{fast}} C_6H_{10}{=}O \qquad (370)$$

In the case of diol oxidation, a different rate-limiting step has been observed. Again, complex formation is indicated, with the slow step apparently being not C—H bond homolysis but rather C—C bond rupture as shown by the facile oxidation of pinacol[664] [Eq. (371)] and the observed absence of a primary isotope effect in the oxidation of 2,3-dideutero-2,3-butanediol [Eq. (372)].

$$CH_3C(OH)(CH_3)—C(OH)(CH_3)—CH_3 + 2Ce^{IV} \longrightarrow 2CH_3C(=O)CH_3 + 2Ce^{III} \qquad (371)$$

$$CH_3C(OH)(D)—C(OH)(D)CH_3 \longrightarrow 2CH_3C(=O)D \qquad (372)$$

Evidence in support of radical intermediates in diol oxidations comes from the observed increase in rate for these oxidations with increasing substitution on the alcoholic carbon atoms,[661] which is in accord with the formation of more stable radicals. The formation of a chelate diol intermediate is not necessary, however, since both ethylene glycol ($HOCH_2CH_2OH$) and 2-methoxyethanol ($CH_3OCH_2CH_2OH$) are oxidized by cerium sulfate via complex formation at about the same rate. In both alcohol and diol oxidations by Ce(IV) there appears to be no O—H bond cleavage in the formation of the intermediate complex since a large number of kinetic reports indicate either no effect on the reaction rate with decreasing pH or a rate acceleration.[661]

Lead tetraacetate,[161,665] $Pb(IV)(OAc)_4$, is a clear example of a two-electron oxidant [Eq. (368)] which in oxidation of alcohols and diols gives evidence of intermediate complex formation. $Pb(OAc)_4$ is known to be rapidly solvolyzed by organic compounds having acidic hydrogens, and alcoholysis products having Pb(IV)—OR bonds have been isolated[665] [Eq. (373)].

$$Pb(OAc)_4 + n\,ROH \rightleftharpoons Pb(OAc)_{4-n}(OR)_n \tag{373}$$

Solutions of $Pb(OAc)_4$ in primary and secondary alcohols are unstable and produce aldehydes and ketones [Eq. (374)] with addition of bases allowing the reaction to occur under milder conditions. The mechanism proposed for these oxidations involves heterolytic cleavage of the O—Pb bond, forming the observed product [Eq. (375)].

$$Pb(OAc)_4 + R'\text{—}CH(R)OH \longrightarrow Pb(OAc)_2 + 2HOAc + R'\text{—}C(R){=}O \tag{374}$$

$$R_2CHOPb^{IV}(OAc)_3 \longrightarrow Pb^{II}(OAc)_2 + R_2C{=}O + HOAc \tag{375}$$

Thus, in the oxidation[666] of 1-alkylcyclohexanol, cyclohexanone is formed in 80% yield [Eq. (376)], and oxidation of diacetone alcohol leads to the formation of acetone and acetoxyacetone [Eq. (377)].[667]

$$\text{1-allylcyclohexanol } (C_6H_{10}(OH)CH_2CH{=}CH_2) + Pb(OAc)_4 \longrightarrow \text{cyclohexanone} \tag{376}$$

$$CH_3COCH_2C(CH_3)_2\text{—}OH \longrightarrow CH_3COCH_3 + CH_3COCH_2OCOCH_3 \tag{377}$$

A mechanism compatible with these results has been suggested[665] which involves nucleophilic attack of acetate ion on the positively charged carbon resulting from rearrangement of the organic cation subsequent to Pb—OR bond heterolysis [Eqs. (378) and (379)].

$$\text{C}_6\text{H}_{10}(\text{O—Pb(OAc)}_4)(\text{CH}_2\text{CH=CH}_2) \rightarrow \text{C}_6\text{H}_{10}(\text{O}^+)(\text{CH}_2\text{CH=CH}_2) \rightarrow$$

$$\text{C}_6\text{H}_{10}\text{=O} + \text{CH}_2\,\text{CHCH}_2{}^+ \xrightarrow{+\text{OAc}^-} \text{C}_6\text{H}_{10}\text{=O} + \text{CH}_2\text{=CHCH}_2\text{OAc} \quad (378)$$

$$\text{CH}_3\text{C(=O)CH}_2\text{C(CH}_3)_2\text{—O—Pb(OAc)}_3 \rightarrow \text{CH}_3\text{C(=O)CH}_2\text{C(CH}_3)_2\text{—O}^+ + \text{OAc}^- \rightarrow$$

$$\text{CH}_3\text{C(=O)CH}_3 + \text{CH}_3\text{C(=O)CH}_2\text{OC(=O)CH}_3 \quad (379)$$

In the case of diol oxidation (cleavage) by $Pb(OAc)_4$[161] [Eq. (380)], a large body of evidence indicates much more rapid oxidation of diols for which chelate ring formation with $Pb(OAc)_4$ by alcoholysis of two Pb—OAc bonds is sterically possible [Eq. (381)].

$$\text{R}_2\text{C(OH)—C(OH)R}_2 + \text{Pb(OAc)}_4 \rightarrow 2\text{R}_2\text{C=O} + \text{Pb(OAc)}_2 + 2\text{HOAc} \quad (380)$$

$$\text{R}_2\text{C(OH)—C(OH)R}_2 + \text{Pb(OAc)}_4 \rightarrow \text{R}_2\text{C(O—)—C(O—)R}_2{>}\text{Pb(OAc)}_2 + 2\text{HOAc} \quad (381)$$

Thus, simple cis diols always have a faster rate of oxidation than the corresponding trans isomers (k cis/k trans is ca. 20 for cyclohexane-1,2-diols, ca. 200 for 1,2-dimethylcyclopentane-1,2-diols, and ca. 3000 for cyclopentane-1,2-diols).[161] For oxidation of trans 1,2-diols, a mechanism similar to Eq. (375) would presumably be operative with deprotonation of the second OH function by external base or by an intramolecular reaction.

One-electron oxidants which have been extensively studied include Co(III), Mn(III), and V(V).[103] The previously mentioned primary and secondary alcohol oxidation by VO_2^+ in acidic aqueous solution [Eqs. (63) and (64)] serves as a model for these reactions. The rate law observed [Eq. (382)] suggests complex formation [Eq. (383)] followed by slow radical formation [Eq. (384)]. In support of this, a primary deuterium isotope effect has been reported for the oxidation of 1-*d*-cyclohexanol.[668] Tertiary alcohols are generally inert to VO_2^+, and al-

$$\frac{-d[\mathrm{V(V)}]}{dt} = k[\mathrm{ROH}][\mathrm{VO_2}^+][\mathrm{H}^+] \qquad (382)$$

$$\mathrm{R_2CHOH} + \mathrm{VO_2}^+ + \mathrm{H_3O}^+ \rightleftharpoons \left[\mathrm{R_2CH{-}\underset{\displaystyle H}{\underset{|}{O}}{-}V(OH)_3}\right]^{2+} \qquad (383)$$

$$\left[\mathrm{R_2CH\underset{\displaystyle H}{\underset{|}{O}}{-}V(OH)_3}\right]^{2+} \longrightarrow \mathrm{R_2\dot{C}OH} + \mathrm{V^{IV}} \qquad (384)$$

cohols which can form relatively stable radical species by hydrogen abstraction from the OH^--substituted carbon atom react at a faster rate. Allyl and crotyl alcohols ($CH_2{=}CHCH_2OH$ and $CH_3CH{=}CHCH_2OH$) react significantly faster than saturated primary and secondary alcohols.[669] Alcohols which can form stable radicals by C—C bond homolysis have been reported to yield anomalous products. Thus, oxidation of β-phenylethanol with V(V) leads not to phenylacetic acid but to benzaldehyde [Eqs. (385) and (386)].[670]

$$\mathrm{C_6H_5CH_2CH_2OH} + \mathrm{V^V} \not\longrightarrow \mathrm{C_6H_5CH_2\overset{\displaystyle O}{\overset{\|}{C}}H} \longrightarrow \mathrm{C_6H_5CH_2\overset{\displaystyle O}{\overset{\|}{C}}{-}OH} \qquad (385)$$

$$\mathrm{C_6H_5CH_2CH_2OH} + \mathrm{V^V} \longrightarrow \mathrm{C_6H_5CH_2}\cdot + \mathrm{CH_2O} \longrightarrow \mathrm{C_6H_5CH_2OH} \longrightarrow \mathrm{C_6H_5\overset{\displaystyle O}{\overset{\|}{C}}H} \qquad (386)$$

Co(III), a much more powerful oxidant, readily oxidizes tertiary alcohols to ketones, presumably via intermediate homolytic C—C bond breakage [Eq. (387)].

$$\mathrm{R_1R_2R_3COH} + \mathrm{Co^{III}} \rightarrow \mathrm{R_1}\cdot + \mathrm{R_2R_3{\cdot}COH} \qquad (387)$$

Permanganate (MnO_4^-) is a well-known two-electron oxidant which has been widely used in the oxidation of alkenes.[663] (MnO_4^- oxidation of unsaturated hydrocarbons constitutes the Baeyar test of analytical chemistry.) In basic solution the products obtained from olefin oxidation include 1,2-diols from two-electron oxidation and ketols and cleavage reaction products resulting from more extensive oxidation [Eq. (388)], the yield of the products depending upon the pH of the solution.

$$\rangle C{=}C\langle + MnO_4^- \longrightarrow \rangle C(OH){-}C(OH)\langle + \rangle C(OH){-}C({=}O){-} + RC({=}O)H \text{ or } RC({=}O)OH + R_2C{=}O \quad (388)$$

Cyclic manganese esters are implicated, and the direction of addition is cis as shown by the following observations. In the oxidation of olefins using ^{18}O-labelled permanganate-oxygen transfer to the substrate was observed [Eq. (389)].[671]

$$\rangle C{=}C\langle + Mn^{18}O_4^- \longrightarrow \text{cyclic } \rangle C{-}O{-}MnO_2^-{-}O{-}C\langle \longrightarrow \rangle C{-}^{18}OH,\ \rangle C{-}^{18}OH \quad (389)$$

Also, oxidation of maleic and fumaric acids leads to *meso*- and *racemic*-tartaric acid, respectively [Eqs. (390) and (391)].[672]

$$(HOOC)HC{=}CH(COOH)\ (cis) + MnO_4^- \longrightarrow meso\text{-}HOOC{-}CH(OH){-}CH(OH){-}COOH \quad (390)$$

$$(HOOC)HC{=}CH(COOH)\ (trans) + MnO_4^- \longrightarrow racemic\text{-}HOOC{-}CH(OH){-}CH(OH){-}COOH \quad (391)$$

Recently, direct observation of cyclic intermediates in the MnO_4^- oxidation of α-β unsaturated acids in aqueous solution has been reported for *trans*-cinnamic [Eq. (392)][673] and crotonic acids.[674] In the former case, use of *trans*-α-d- and *trans*-β-d-cinnamic acid in the oxidations demonstrated an inverse kinetic isotope effect in accord with formation of a nearly symmetric cyclic intermediate.

Chromic acid oxidations of primary and secondary alcohols have long been studied,[675] and general agreement has been reached on the mechanism of alcohol

$$MnO_4^- + C_6H_5CH{=}CHCOOH \longrightarrow C_6H_5CH(O)\text{—}CH(O)COOH\ (\text{cyclic ester with } MnO_2) \longrightarrow Mn^{III} + C_6H_5CHO + HCOOH \quad (392)$$

oxidation by Cr(VI) species. This is suggested to involve initial ester formation with chromic acid [Eq. (393)] followed by rate determining cleavage of the OH-substituted carbon–hydrogen bond, producing a carbonyl compound and Cr(IV) [Eq. (394)].

$$R_2CHOH + H_2CrO_4 + H^+ \rightarrow R_2CHOCrO_3H_2^+ + H_2O \quad (393)$$

$$R_2CHOCrO_3H_2^+ \rightarrow R_2C{=}O + Cr^{IV} \quad (394)$$

A major difficulty with understanding the mechanism of these oxidations is that, in contrast to MnO_4^- oxidations (where the Mn(V) species disproportionates to give Mn(VII) and MnO_2),[676] the fate of the Cr(IV) product has been a matter of contention.

The products of oxidation of cyclobutanol have been demonstrated recently to give strong evidence for the electron transfer number of the oxidizing species.[677,678] Thus, Ce(IV), V(V), Mn(III), and Cu(II) produced either γ-hydroxybuteraldehyde or products derived from reaction of the intermediate radical species formed via initial C—C bond cleavage [Eq. (395)], and Cr(VI) oxidation of cyclobutanol produced cyclobutanone through C—H bond rupture [Eq. (396)].

$$\text{cyclobutanol} + M^{n+} \longrightarrow M^{(n-1)+} + \cdot CH_2CH_2CH_2CHO \xrightarrow[-e]{H_2O} HOCH_2CH_2CH_2CHO \quad (395)$$

$$\text{cyclobutanol} + M^{n+} \longrightarrow M^{(n-2)+} + \text{cyclobutanone} + 2H^+ \quad (396)$$

Other information concerning the details of oxidation of primary and secondary alcohols by solutions of Cr(VI) suggests the mechanism following in Eqs. (397) (398), (399), and (400).

$$R_2CHOH + Cr^{VI} \rightarrow R_2C{=}O + Cr^{IV} \quad (397)$$

$$R_2CHOH + Cr^{IV} \rightarrow R_2\dot{C}OH + Cr^{III} \quad (398)$$

$$R_2\dot{C}OH + Cr^{VI} \rightarrow R_2C{=}O + Cr^{V} \quad (399)$$

$$R_2CHOH + Cr^{V} \rightarrow R_2C{=}O + Cr^{III} \quad (400)$$

Oxidation of alcohols by Cr(IV) was determined to occur via a radical intermediate [Eq. (398)] by conducting the reactions in the presence of radical trapping species, and a primary isotope effect for the Cr(IV) oxidation of 2-*d*-2-propanol indicated rate-determining cleavage of the C—D bond.[679] In addition, 2-propanol oxidation to acetone by Cr(VI) in the presence of acrylonitrile or acrylamide leads to the polymerization of these species, suggesting the presence of radicals in solution with concommitant decrease in the yield of acetone.[680] Finally, in 2-propanol oxidation by Cr(VI) in acetic acid–water, the detection of Cr(V) in the reaction solution was possible, and studies of acetone production as a function of Cr(V) decay supported the mechanism given in Eqs. (397), (398), (399), and (400).[681]

Both oxalic acid[682] and oxalic acid–2-propanol mixtures[683] have been reported to be oxidized by Cr(VI) in a three-electron, one-step process in accord with the following mechanism [Eqs. (401) and (402)].

$$HCrO_4^- + H_2C_2O_4 + H^+ \rightleftharpoons (C_2O_4)CrO_2 + 2H_2O \quad (401)$$

(structure: cyclic chelate O=C–O–Cr(=O)(=O)–O–C=O)

$$(C_2O_4)CrO_2 + S \rightleftharpoons [(C_2O_4)CrO_2\,(S)]$$

$$[(C_2O_4)CrO_2\,(S)] \rightarrow Cr^{III} + CH_3\overset{O}{\overset{\|}{C}}CH_3 + CO_2 + \cdot CO_2^- + 2H^+$$

$$[(C_2O_4)CrO_2\,(S)] \rightarrow Cr^{III} + 3CO_2 + \cdot CO_2^- \quad (402)$$

Subsequent oxidation of the $\cdot CO_2^-(\cdot CO_2H)$ radical by Cr(VI) generates Cr(V) and CO_2. The Cr(V) is observed to be reduced by 2-propanol or oxalic acid to Cr(III), yielding acetone or carbon dioxide, respectively.[684] The evidence in favor of a one-step, three-electron oxidation in the 2-propanol–oxalic acid case is: (a) the kinetics indicate one molecule of both oxalic acid and 2-propanol in the reactive intermediate; (b) both acetone and carbon dioxide are formed at a rate much faster than the rate for either substrate by itself; (c) polymer formation is found in the presence of acrylamide while acrylonitrile traps the $\cdot CO_2^-$-$(\cdot CO_2H)$ radical leading to exactly one CO_2 and one acetone formed with each Cr(VI) reduced to Cr(III); (d) a kinetic isotope effect indicating —COH-D bond cleavage in the rate-limiting step; and (e) isopropyl methyl ether is unreactive. Similarly, in oxalic acid oxidation, the reactive intermediate is a 2:1 oxalic acid:Cr(VI) species, and three molecules of CO_2 are formed in the presence of radical trapping agents for each Cr(VI) reduced to Cr(III). Further, in the case of 2-propanol–oxalic acid oxidation in the absence of radical trapping agents, the CO_2:acetone ratio is found to be within the limits for this mechanism (1:1 to 4:1) [Eqs. (403) and (404)].

$$\cdot CO_2^- + Cr^{VI} \longrightarrow CO_2 + Cr^{V} \tag{403}$$

$$Cr^{V} \xrightarrow{+CH_3CH(OH)CH_3} CH_3C(=O)CH_3 + Cr^{III}$$

$$Cr^{V} \xrightarrow{+(CO_2H)_2} 2CO_2 + Cr^{III} \tag{404}$$

Finally, in oxidation of aromatic ketones with Mn^{3+} in the presence of olefins, four products were obtained, all of which were consistent with initial hydrogen abstraction from the α-carbon of the ketone, followed by radical addition to the olefin, and oxidation by another Mn^{3+} ion completing the reaction [Eqs. (405), (406), (407), and (408)].[674]

$$C_6H_5C(=O)CH_3 + Mn^{3+} \longrightarrow C_6H_5C(=O)CH_2\cdot + RCH{=}CH_2 \longrightarrow RCHCH_2CH_2C(=O)C_6H_5 \tag{405}$$

$$\underset{\bullet}{\mathrm{R}}\mathrm{CHCH_2CH_2}\overset{\mathrm{O}}{\overset{\|}{\mathrm{C}}}\mathrm{C_6H_5} + \mathrm{H}\cdot \longrightarrow \mathrm{RCH_2CH_2CH_2}\overset{\mathrm{O}}{\overset{\|}{\mathrm{C}}}\mathrm{C_6H_5} \tag{406}$$

$$\underset{\bullet}{\mathrm{R}}\mathrm{CHCH_2CH_2}\overset{\mathrm{O}}{\overset{\|}{\mathrm{C}}}\mathrm{C_6H_5} + \mathrm{Mn^{3+}} \longrightarrow \mathrm{RC^{+}HCH_2CH_2}\overset{\mathrm{O}}{\overset{\|}{\mathrm{C}}}\mathrm{C_6H_5} \begin{cases} \xrightarrow{+\mathrm{OAc}} \mathrm{R}\overset{\mathrm{OAc}}{\overset{|}{\mathrm{C}}}\mathrm{HCH_2CH_2}\overset{\mathrm{O}}{\overset{\|}{\mathrm{C}}}\mathrm{C_6H_5} \\ \xrightarrow{-\mathrm{H^+}} \mathrm{RCH{=}CHCH_2}\overset{\mathrm{O}}{\overset{\|}{\mathrm{C}}}\mathrm{C_6H_5} \end{cases} \tag{407}$$

$$\underset{\bullet}{\mathrm{R}}\mathrm{CHCH_2CH_2}\overset{\mathrm{O}}{\overset{\|}{\mathrm{C}}}\mathrm{C_6H_5} \longrightarrow [\text{cyclized radical, H, R}] + \mathrm{Mn^{3+}} \longrightarrow [\text{4-R-1-tetralone}] \tag{408}$$

The roles of organometallic intermediates have been studied in oxidation-reduction processes of free radicals.[685] Such reactions are necessarily limited to one-equivalent electron changes, and their study is based on unambiguous methods for producing transient alkyl radicals. The reaction of an alkyl radical with a metal complex may lead to the formation of an alkyl–metal bond [Eq. (409)], ligand transfer [Eq. (410)], electron transfer (Eq. (411)], or homolytic substitution [Eq. (412)]. As written, Eqs. (410) and (411) involve oxidation by

$$\mathrm{R}\cdot + \mathrm{M}^{n}\mathrm{X}_{m} \rightarrow \mathrm{R\text{-}M}^{n+1}\mathrm{X}_{m} \tag{409}$$

$$\mathrm{R}\cdot + \mathrm{M}^{n}\mathrm{X}_{m} \rightarrow \mathrm{R\text{-}X} + \mathrm{M}^{n-1}\mathrm{X}_{n-1} \tag{410}$$

$$\mathrm{R}\cdot + \mathrm{M}^{n}\mathrm{X}_{m} \rightarrow \mathrm{R}^{+} + \mathrm{M}^{n-1}\mathrm{X}_{n} \tag{411}$$

$$\mathrm{R}\cdot + \mathrm{M}^{n}\mathrm{X}_{m} \rightarrow \mathrm{R\text{-}M}^{n}\mathrm{X}_{n-1} + \mathrm{X}\cdot \tag{412}$$

the metal atom, (409) constitutes reduction, and (412) does not involve oxidation or reduction by the metal atom itself (its oxidation number is unchanged). Alkylation of the metal by radicals is well known and is important as an intermediate step in oxidation by the metal atom. This is illustrated by Eq. (413) for the case of $Cu(OAc)_2$.[686] In such reactions the elimination of the reduced metal atom yields the oxidized product. Olefin formation depends on the availability of β-hydrogens, and *n*-butyl radicals undergo this oxidative elimination exclusively, while neopentyl radicals can only undergo oxidative substitution [Eq. (414)].

$$R\cdot + Cu^{II}(OAc)_2 \rightarrow [R\text{-}Cu^{III}(OAc)_2] \rightarrow \text{olefin} + Cu^{I}OAc + HOAc \quad (413)$$

$$\rangle C{-}C\langle\cdot + HB \longrightarrow \rangle C{-}\overset{|}{\underset{|}{C}}{-}B + e^- + H^+ \quad (414)$$

The overall reaction of Eq. (410) might, on the basis of earlier discussions in this chapter, be most simply conceived as involving addition of R· to the metal and reductive elimination of RX. It has been pointed out,[687] however, that such reactions may occur by direct displacement [Eq. (415)], in analogy to the bridge-activated or inner-sphere mechanisms for electron transfer in purely inorganic systems—e.g., Cr^{II} + $[ClCo^{III}\text{-}(NH_3)_5]^{2+}$.

$$R\cdot + Cu^{II}Br_2 \rightarrow [R \cdots Br \cdots Cu^{II}Br] \rightarrow RBr + Cu^{I}Br \quad (415)$$

The outer-sphere process of Eq. (411) may be involved in the oxidation of radicals capable of producing relatively stable ionic entities such as trityl, tropyl, or pyryl, but more commonly such carbocations follow the heterolysis of an alkyl–metal intermediate [Eq. (416)].[688]

$$PhCH_2CH_2\cdot + Cu^{II} \longrightarrow [C_6H_5CH_2CH_2Cu^{III}] \longrightarrow (\text{spiro-}CH_2{-}CH_2\text{ cyclohexadienyl cation}) + Cu^{I} \quad (416)$$

The oxidation of tetraalkylead compounds by $Ir^{IV}Cl_6^{2-}$ provides an example of ligand transfer (at lead) which appears to involve intermediate radical formation. The rate behavior is not consistent with an alternative electrophilic attack of iridium on a Pb—C bond. Also, there is a linear correlation of rates (log k) of $PbMe_nEt_{4-n}$ with one-electron oxidation potentials and a striking relationship with the vertical ionization potentials of the lead compounds as determined by photoelectron spectroscopy. The proposed mechanism is given in Scheme 2-28. Thus, this two-electron process can be described as occurring via

Scheme 2-28

$$R_4Pb + Ir^{IV}Cl_6^{2-} \rightarrow R_4Pb^+ + Ir^{III}Cl_6^{3-}$$

$$R_4Pb^+ \xrightarrow{\text{fast}} R\cdot + R_3Pb^+$$

$$R\cdot + Ir^{IV}Cl_6^{2-} \xrightarrow{\text{fast}} RCl + Ir^{III}Cl_5^{2-}$$

one-electron steps. Analogously, the oxidation of tetraethyl lead by $Cu^{II}X_2$ (X = acetate, triflate, chloride) involves the ethyl radical as the prime intermediate [Eq. (417)].[690]

$$Et_4Pb + 2Cu^{II}X_2 \rightarrow Et_3PbX + 2Cu^{I}X + CH_2CH_2 + CH_3CH_2OAc + CH_3CH_2Cl \quad (417)$$

In this case, however, selectivity studies strongly suggest the intermediacy of an alkyl–Cu(II) species [Eqs. (418) and (419)].

$$Et_4Pb + Cu^{II}X_2 \rightarrow Et_3PbX + EtCu^{II}X \quad (418)$$

$$EtCu^{II}X \xrightarrow{\text{fast}} Et\cdot + Cu^{I}X \quad (419)$$

The processes described immediately above indicate the oxidative disproportionation and substitution products to be expected in oxidations of this kind involving radical intermediates although, depending on the system, the radical coupling product may also be expected. Reductive elimination by such species as $CH_3CH_2Au(CH_3)_2L$ and certain platinum(IV) and nickel(II) analogs can be attributed to mechanisms not involving free radicals since only the reductive coupling product ($CH_3CH_2CH_3$) and no alkene or products derived from reaction with solvent are found. Di-*n*-butylbis(triphenylphosphine)platinum(II) undergoes thermolysis, forming the alkyl disproportionation products [Eq. (420)]

$$(n\text{-}C_4H_9)_2Pt^{II}L_2 \rightarrow n\text{-}C_4H_{10} + 1\text{-}C_4H_3 + Pt^{0}L_2 \quad (420)$$

and no coupling product.[691,692] In this case, the decomposition has been explained in terms of elimination of Pt—H from one of the bound butyl groups with concommitant binding of the olefin. This process requires prior dissociation of a phosphine ligand. Reductive elimination of butane and olefin dissociation then accounts for the products [equations (421), (422), and (423)].

$$L_2Pt(CH_2CH_2CH_2CH_3)_2 \longrightarrow L{-}Pt(CH_2CH_2CH_2CH_3)_2 + L \quad (421)$$

$$L{-}Pt(CH_2CH_2CH_2CH_3)_2 \longrightarrow L(H)Pt(\eta^2\text{-}H_2C{=}CH{-}CH_2CH_3)(CH_2CH_2CH_2CH_3) \quad (422)$$

$$L(H)Pt(\eta^2\text{-}H_2C{=}CH{-}CH_2CH_3)(CH_2CH_2CH_2CH_3) \longrightarrow C_4H_{10} + C_4H_8 + Pt^0 + L \quad (423)$$

Such reductive eliminations are not unexpected when the metal atom can undergo a two-electron reduction to form a product of substantial stability. Similar reactions, however, have also been observed for alkyl derivatives of metal ions such as Au(I) and Cu(I), that are not expected to undergo two-equivalent changes.[687] For example, $CH_3Au[P(C_6H_5)_3]$ decomposes in Decalen solution to give almost quantitative yields of ethane.

X. AUTOXIDATIONS CATALYZED BY TRANSITION METAL COMPOUNDS

The oxidation reactions of organic substrates with molecular oxygen are classified as autoxidations. Many of these processes have the characteristics of radical chain processes, occurring under mild conditions and involving autocatalysis by oxidation intermediates or products.[693–697] Autoxidations occurring in biological systems also constitute a vast realm, and certain strikingly different features have been observed among them.[698–703] Because of the magnitude of the subject and the extent to which it has been reviewed, it is appropriate here only to present and to illustrate the more general aspects.

The simplest possible view one might take of the subject relates to the general differences in the bonds present in reactants and products. If the element oxygen is incorporated in the product, then the O_2 and C—H (or C—C) bonds are broken, and C—O and O—H bonds are formed. In view of the values for these bond energies, autoxidation processes are generally expected to be rather strongly exothermic. For purposes of general orientation one might ask naively "Why is catalysis required at all?" rather than the question that does confront us, "How do transition element compounds catalyze autoxidations?" The "surprising" fact that organic compounds exist in an oxygen atmosphere is traceable to the structure of the oxygen molecule and the various reduction processes it undergoes. Most superficially, the fact that free O_2 has a triplet ground state suggests that its reaction with singlet organic molecules may be spin forbidden,[703] but it is more enlightening to consider the different couples in which the O_2 molecule may be implicated in redox processes. Scheme 2-29 shows these and the corresponding electrode potentials for 1 atm O_2 and 1*M* H^+ and those for 1 atm O_2 at pH 7. Under both sets of conditions the first one-electron process is highly unfavorable—i.e., the O_2 molecule is not a strong oxidizing agent if the product is superoxide ion, O_2^-, or hydrogen superoxide, HO_2. Indeed the reaction between an oxygen molecule and a hydrocarbon to form HO_2 and a radical [Eq. (424)] is strongly endothermic (ca. +46 kcal) and can occur at a measurable rate only at high temperatures.[702,703] The corresponding electron abstraction reaction is similarly unfavorable [Eq. (425)].

$$RH + O_2 \rightarrow R\cdot + HO_2\cdot \tag{424}$$

$$R\text{-}H + O_2 \rightarrow R\text{-}H^+ + O_2\cdot^- \tag{425}$$

These processes are observed only when unusually stable radicals can be formed. It is particularly striking that in all obvious modes of oxidation-reduction other than this initial one-electron step, oxygen is a relatively vigorous oxidizing agent. This is apparent from Scheme 2-29, but the matter deserves examination in more detail.

Scheme 2-29

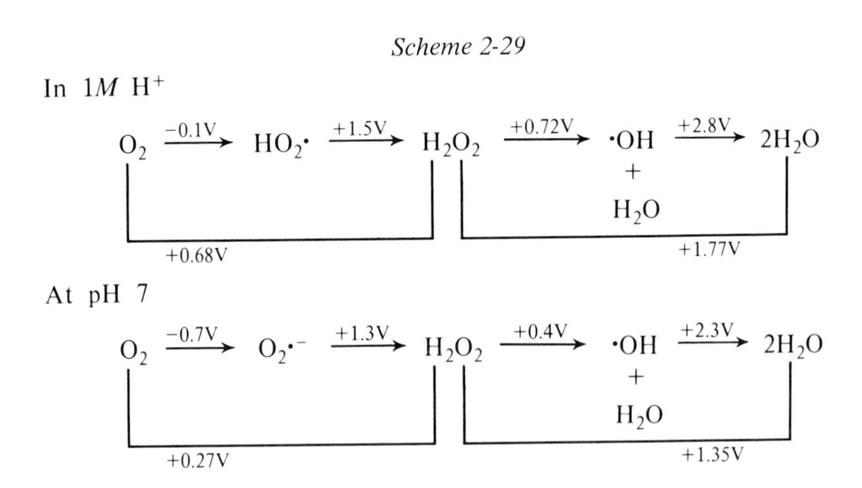

Although the HO_2/H_2O_2 couple enjoys a large positive standard potential, the hydrogen abstraction reactions are usually endothermic (5–20 kcal), again, unless R· is resonance stabilized. Alkanes and unactivated aromatics do not react at low temperatures. $HO_2\cdot$ is, however, believed to be the source of O_2 toxicity because of its oxidizing power.[702] The restrictions on the reactivity of $O_2\cdot^-$ are similar. The potential for the two-electron process (O_2/H_2O_2) is less positive both under standard conditions and at neutral pH (Scheme 2-29), and it is not surprising that this couple is not strongly implicated in the mechanisms of organic autoxidations. Despite the lethargy of the oxygen molecule in oxidation reactions under mild conditions, transition metal compounds are often very effective at catalyzing these reactions by a variety of more or less well-understood mechanisms.

A. Radical Chain Processes

As stated earlier, transition metal-catalyzed reactions may have the characteristics of radical chain reactions. The propagation and termination steps usually assumed[696,697] are shown in Eqs. (426) and (427) (propagation) and Eqs. (428), (429), and (430) (termination).

$$R\cdot + O_2 \rightarrow RO_2\cdot \tag{426}$$

$$RO_2\cdot + RH \rightarrow ROOH + R\cdot \tag{427}$$

$$2R\cdot \rightarrow R_2 \tag{428}$$

$$RO_2\cdot + R\cdot \rightarrow ROOR \tag{429}$$

$$2RO_2\cdot \rightarrow ROH + RCOR + O_2 \tag{430}$$

Eq. (430), along with the decomposition of the organic peroxides, represents the product-forming reactions [e.g., Eq. (431)].

$$RCH_2OOH \longrightarrow RC(=O)H + H_2O \tag{431}$$

These reactions commonly involve induction periods that depend on the initiation process. During this induction period, no oxygen uptake is observed, and the addition of a metal salt or complex may dramatically affect the induction period, perhaps even eliminating it. The role of the transition metal catalyst in initiation may involve a number of reactions. The most commonly reported function of the catalyst is the decomposition of peroxide species into radicals which then initiate[696] the oxidation process [Eqs. (432) and (433)]. In the first reaction [Eq. (432)], the metal ion abstracts an OH radical and reduces it to hydroxide, thereby being oxidized by one equivalent. The OH^- probably remains coordinated in some cases. This reaction is the primary initiating step in the Haber-Weiss mechanism for the decomposition of hydrogen peroxide by aqueous ferrous ion (Fenton's reagent).[705,706] The second mode of destruction of organic peroxides [Eq. (433)] represents the reduction of a metal ion and was suggested in a modification to the Haber-Weiss mechanism.[707] The overall effect of the reactions of Eqs. (432) and (433) is to generate peroxyl and superoxyl radicals while cycling the metal ion between two oxidation states (e.g., Fe^{2+} and Fe^{3+}).

$$ROOH + M^{n+} \rightarrow RO\cdot + OH^- + M^{n+1} \qquad (432)$$

$$ROOH + M^{n+1} \rightarrow RO_2\cdot + H^+ + M^{n+} \qquad (433)$$

Initiation by oxygen activation has also been suggested[696] and is to be expected since numerous metal complexes with O_2 are known. This is illustrated by the catalysis of cumene autoxidation to cumene hydroperoxide by copper phthalocyanine[708] [Eq. (434)].

$$C_6H_5CH(CH_3)_2 + O_2 \rightarrow C_6H_5C(CH_3)_2O_2H \qquad (434)$$

Autoxidation catalyzed by electron transfer from the substrate to the catalyst species has been suggested for a number of low temperature (ca. 100°C) Co(III)-catalyzed alkylbenzene autoxidations that have found commercial use.[709] These include the Teijin process for conversion of *p*-xylene to terephthalic acid [Eq. (435)].[710]

$$CH_3C_6H_4CH_3 \xrightarrow{O_2} HOOCC_6H_4COOH \qquad (435)$$

This mechanism, which is favored under mild conditions and in the presence of high levels of catalyst, involves initial oxidation of Co(II) to Co(III) (presumably by reduction of peroxy species), electron transfer from *p*-xylene to Co(III) [Eq. (436)], peroxy radical formation [Eq. (437)], and reduction of the peroxy species to an aldehyde [Eq. (438)]. Further oxidation of the aldehyde to a carboxyl group is believed to involve participation by cobalt in the series of steps given by Eqs. (439)–(443). The main characteristics of this process which bear on the mechanism are: (a) use of ca. 0.5 mole of Co(II) for each mole of *p*-xylene to maximize the yield of terephthalic acid; (b) addition of co-oxida-

$$Co^{III} + CH_3\text{-}C_6H_4\text{-}CH_3 \longrightarrow Co^{II} + \dot{C}H_2\text{-}C_6H_4\text{-}CH_3 + H^+ \quad (436)$$

$$\dot{C}H_2\text{-}C_6H_4\text{-}CH_3 + O_2 \longrightarrow CH_2OO\cdot\text{-}C_6H_4\text{-}CH_3 \quad (437)$$

$$CH_2OO\cdot\text{-}C_6H_4\text{-}CH_3 + Co^{II} \longrightarrow Co^{III} + H(C{=}O)\text{-}C_6H_4\text{-}CH_3 + OH^- \quad (438)$$

tion/reagent (ketones, ozone, halides) had no effect on the reaction; and (c) analysis of the reaction products with elapsed time gave good agreement with expressions derived for the mechanism proposed. Use of high levels of Co^{3+} in the oxidation process favors short propagation chain lengths since the Co(II) could intercept the peroxy radical formed in Eq. (437) via Eq. (438), effectively competing with hydrogen abstraction from RH groups by the peroxy radical. Also, the absence of a detectable difference in rate, or product selectivity, in the presence of additives argues for an electron transfer mechanism since, as is shown

$$H(C{=}O)\text{-}C_6H_4\text{-}CH_3 + Co^{III} \longrightarrow \dot{C}O\text{-}C_6H_4\text{-}CH_3 + Co^{II} + H^+ \quad (439)$$

$$\mathrm{CH_3C_6H_4\dot{C}O} + \mathrm{O_2} \longrightarrow \mathrm{CH_3C_6H_4C(=O)OO\cdot} \qquad (440)$$

$$\mathrm{CH_3C_6H_4C(=O){-}OO\cdot} + \mathrm{RH} \longrightarrow \mathrm{CH_3C_6H_4C(=O)OOH} + \mathrm{R\cdot} \qquad (441)$$

RH = aldehyde, alkyl group

$$\mathrm{CH_3C_6H_4C(=O)OOH} + \mathrm{Co^{II}} \longrightarrow \mathrm{CH_3C_6H_4C(=O)O\cdot} + \mathrm{OH^-} + \mathrm{Co^{III}} \qquad (442)$$

$$\mathrm{CH_3C_6H_4C(=O)O\cdot} + \mathrm{RH} \longrightarrow \mathrm{CH_3C_6H_4C(=O)OH} + \mathrm{R\cdot} \qquad (443)$$

later, the selectivity of radical propagation for different C—H bonds resulting from hydrogen abstraction by peroxy species differs from that found for electron transfer reactions. Studies of the mechanism of oxidation of toluene to benzoic acid [Eq. (444)][711] and of *n*-butane to acetic acid [Eq. (445)][712] with Co(III) lead to similar results. In the former case for the use of Co(III) salts in the oxi-

$$C_6H_5CH_3 \xrightarrow{O_2} C_6H_5COOH \qquad (444)$$

$$CH_3CH_2CH_2CH_3 \xrightarrow{O_2} 2CH_3\overset{\overset{O}{\|}}{C}OH \qquad (445)$$

dation of toluene (initially to benzaldehyde), kinetic evidence for oxygen consumption and Co(III) reduction was in accord with rate-limiting electron transfer producing a radical which then reacts with oxygen. In both the toluene oxidation and the previously mentioned xylene oxidation, evidence was found for a dimeric cobalt species as the active catalyst.

The alkane oxidation in Eq. (445) is of some interest. Observations in favor of an electron transfer reaction included: (a) increasing rate with increasing Co(III) concentration; (b) selectivity patterns in oxidation products and reaction rates which argue against radical initiation; and (c) the rates were not appreciably retarded by species which would effectively trap long-lived radicals, thus slowing the rate of hydrogen abstraction by radical species. This study also investigated the rates of oxidation of substrates as a function of alkyl group substitution and found that the least substituted material was more rapidly oxidized. Thus, toluene was oxidized more quickly than cumene, cyclohexane faster than methylcyclohexane, *n*-butane faster than isobutane, and *n*-undecane was unreactive. All these reactivity sequences are inconsistent with a normal free radical pathway which predicts increasing reactivity with increasing substitution on the reactive carbon. Also, Co(III) oxidation of toluene and cyclohexane occur at similar rates in aqueous acetonitrile, suggesting that in both cases the electron removed is from an aliphatic C—H bond.[713]

Another mechanism for autoxidation catalyzed by transition metals involves radical initiation by homolysis of a transition metal–ligand bond [Eq. (446)]. The resulting radical can then abstract hydrogen from the substrate, leading to the initiation step [Eq. (447)].

$$L_mM^{n+}\text{—}X \rightarrow L_mM^{(n-1)+} + X\cdot \qquad (446)$$

$$X\cdot + RH \rightarrow HX + R\cdot \qquad (447)$$

Thus, in the cobalt acetate-catalyzed oxidation of *p*-xylene, at lower cobalt concentrations, the addition of Br^- leads to a six-fold increase in rate. This is consistent with initiation by Br· formed by reduction of Co^{3+}.[714] Other evidence which suggests this type of mechanism under appropriate conditions is found in the Co^{3+}-catalyzed autoxidation of substituted alkylbenzenes in the presence of halides.[715] Nonselective radical attack by Cl· on *p-tert*-butyltoluene leads

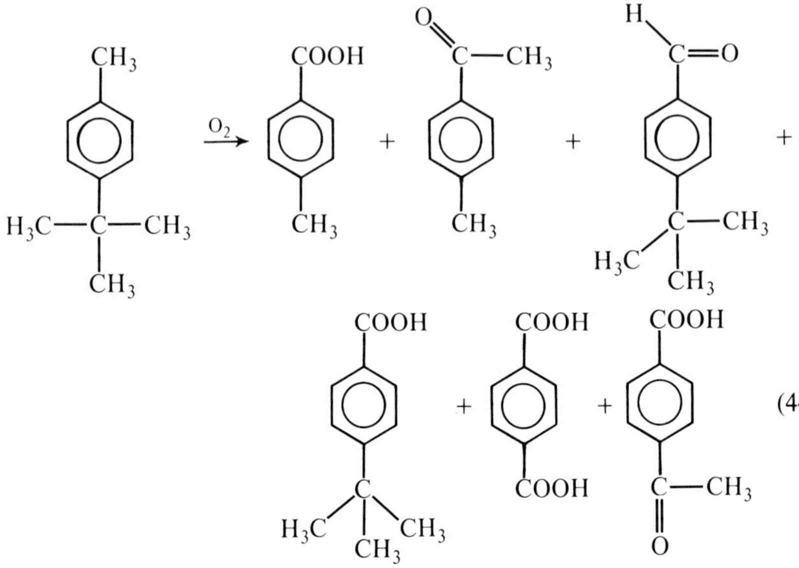

(448)

to a variety of products that are inconsistent with an electron transfer mechanism [Eq. (448)].

Another type of catalyst which has been suggested to function by M—L bond homolysis is involved in the Co(III)–carboxylic acid system.[716] The rate of oxidation of toluene by Co^{3+} in aqueous acetonitrile was found to be increased upon addition of carboxylic acids to the solution; this is in accord with the following reactions [Eqs. (449) and (450)].

$$Co^{3+} + RC(=O)O^- \longrightarrow Co^{III}\text{—}OC(=O)R \longrightarrow R\cdot + CO_2 + Co^{II} \quad (449)$$

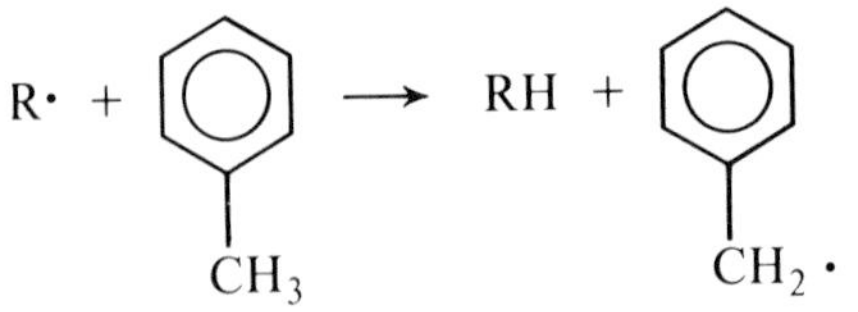

$$R\cdot + C_6H_5CH_3 \longrightarrow RH + C_6H_5CH_2\cdot \quad (450)$$

The mechanisms of autoxidation initiation by transition metal catalysts are thus variable depending upon the catalyst species, its concentration, other substances in solution, the reaction conditions, and the reaction medium. The functions of the catalysts in branching [Eqs. (432) and (433)] and termination steps are more complicated and beyond the scope of this brief review. It is noteworthy, however, that the predominant species in the redox couple found in Eqs. (432) and (433) are variable depending on the nature of the peroxy radicals and the solvent.[697] Thus, in cobalt(II)-/(III)-catalyzed peroxide decomposition in polar solvents, the cobalt is almost all in the divalent state; in nonpolar solvents the trivalent state predominates. Also, for Mn(II)-/(III)-catalyzed autoxidation, the Mn(II) oxidation step analogous to Eq. (433) is much faster than the peroxide oxidation step [Eq. (432)], and, for both Mn(II)-/(III)- and Co(II)-/(III)-catalyzed peracid decomposition, reaction (451) is much faster than reaction (452), so that the catalyst is predominantly in the plus-three state.

$$RC(=O)OOH + M^{2+} \longrightarrow M^{3+}\text{—}OH + RCO_2^{\bullet} \qquad (451)$$

$$RC(=O)OOH + M^{3+} \longrightarrow M^{2+} + RCO_3^{\bullet} + H^+ \qquad (452)$$

B. Oxidation of Organic Substrates Catalyzed by Palladium(II)

An outstanding example of a commercially successful homogeneous transition metal-catalyzed reaction is found in the oxidation of ethylene to acetaldehyde by Pd(II) in the presence of Cu(II) and oxygen [Eq. 453)].[583,717,718,719] The original observation that in aqueous solution palladium chloride was reduced by ethylene to metal with the production of acetaldehyde was reported in 1894 [Eq. (454)].[720]

$$C_2H_4 + H_2O \xrightarrow{Pd^{II} + Cu^{II} + O_2} CH_3C(=O)H + 2H^+ \qquad (453)$$

$$C_2H_4 + PdCl_2 + H_2O \longrightarrow CH_3C(=O)H + 2H^+ + Pd^0 \qquad (454)$$

The work of Smidt et al., utilizing oxygen and $CuCl_2$ for the oxidation of the resultant palladium metal, led to the catalytic ethylene oxidation reaction designated as the Wacker process [Eq. (453)].[580,583,721] The key feature in this catalytic cycle involves the oxidation of Pd^0 by $CuCl_2$ [Eq. (455)] followed by air oxidation of the resultant Cu(I) [Eq. (456)].

$$Pd^0 + 2Cu^{II} \rightarrow Pd^{II} + 2Cu^I \qquad (455)$$

$$2Cu^I + 1/2O_2 + 2H^+ \rightarrow 2Cu^{II} + H_2O \qquad (456)$$

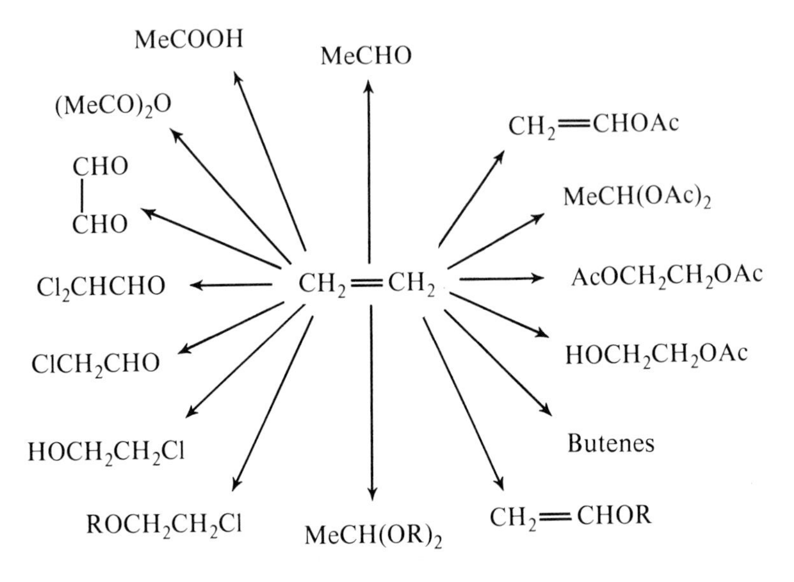

Figure 2-10 Products formed from ethylene in reactions involving palladium catalysts.

Under normal commercial catalytic conditions the rapid air oxidation of Cu(I) to Cu(II) is rate determining.[722] The use of different solvents and reaction conditions has led to the products from the oxidation of ethylene shown in Figure 2-10. The generally accepted reaction mechanism for the Wacker process involves:

(1) Reversible formation of $Pd(C_2H_4)Cl_3^-$ from C_2H_4 and $PdCl_4^{2-}$ [Eq. (457)].

$$PdCl_4^{2-} + C_2H_4 \rightleftharpoons [Pd(C_2H_4)Cl_3]^- + Cl^- \tag{457}$$

(2) Reversible substitution of H_2O for Cl^- to yield $[Pd(H_2O)(C_2H_4)Cl]$ [Eq. (458)].

$$Pd(C_2H_4)Cl_3^- + H_2O \rightleftharpoons Pd(C_2H_4)(H_2O)Cl_2 + Cl^- \tag{458}$$

(3) Reversible dissociation of H^+ to give $[Pd(C_2H_4)(OH)Cl_2]^-$ [Eq. (459)].

$$Pd(C_2H_4)(H_2O)Cl_2 \rightleftharpoons [Pd(C_2H_4)(OH)Cl_2]^- + H^+ \tag{459}$$

(4) Insertion of C_2HR into the Pd—OH bond to yield the σ complex $[Pd(L)Cl_2(CH_2CH_2OH)]^-$ [Eq. (460)].

$$[Pd(C_2H_4)(OH)Cl_2]^- + L \rightarrow [Pd(L)Cl_2(CH_2CH_2OH)]^- \tag{460}$$

(5) A very fast hydride shift in the σ complex to generate acetaldehyde [Eq. (461)].

$$[Pd(L)Cl_2(CH_2CH_2OH)] \longrightarrow Pd^0 + CH_3\overset{\overset{\displaystyle O}{\|}}{C}H \qquad (461)$$

Evidence in support of these mechanistic steps includes the determination of olefin complex formation by changes in olefin solubility in aqueous solution on addition of Pd(II) salts.[723,724,725] In addition, ethanol and vinyl alcohol have been eliminated as intermediates.[726,727] Ethanol is oxidized much more slowly than ethylene and hence would accumulate in detectable quantities if formed; experiments with ethylene oxidation in D_2O result in no incorporation of deuterium into the product aldehyde, a result that is inconsistent with vinyl alcohol's being an intermediate. It was initially suggested that the hydroxy group is coordinated prior to reaction with the Pd-bound olefin [Eq. (459)] since the rate observed for ethylene oxidation would require a rate 10^4 times faster than a diffusion-controlled process for the attack of free OH^- on the coordinated ethylene.[725] This is also supported by deuterium studies which revealed a deuterium isotope effect that is consistent with dissociation of non-coordinated $H_2O(D_2O)$ prior to nucleophilic attack on the bound olefins.[728]

It was also established that the rate-determining step in the ethylene oxidation sequence does not involve a 1,2-hydride shift since the isotope effect for C_2H_4 and C_2D_4 under identical conditions is only 1.07.[725] Thus, the slow step is suggested to involve nucleophilic attack of a coordinated hydroxy group on the Pd(II)-bound olefin to yield the β-hydroxyalkyl species [Eq. (462)].

$$\left[Cl_2Pd(H_2C{=}CH_2)(OH) \right] \xrightarrow{\text{slow}} [Pd(L)Cl_2(CH_2CH_2OH)]^- \xrightarrow{\text{fast}} \text{products} \qquad (462)$$

In support of this, alkylation of $PdCl_2$ with 2-hydroxyethylmercuric chloride yielded acetaldehyde via a β-hydride elimination [Eq. (463)].[729] Similar products were observed for $ROCH_2CH_2HgCl$ reactions [Eq. (464)].

$$PdCl_2 + HgCl(CH_2CH_2OH) \xrightarrow{H_2O} Pd^0 + CH_3\overset{\overset{\displaystyle O}{\|}}{C}H \qquad (463)$$

$$PdCl_2 + HgCl(CH_2CH_2OR) \longrightarrow ROCH{=}CH_2 + Pd^0 \qquad (464)$$

The rate law[726,730,731] determined for this reaction [Eq. (465)] is in accord with the principal features of the proposed mechanisms, being inhibited by H^+ and X^- ions,[721,726] and is now reported to be valid for both high and low Pd(II) concentrations.[732] The requirement for dissociation of two chloride ligands [steps (1) and (2) of the proposed mechanism] and subsequent deprotonation of the Pd(II)-bound H_2O [step (3)] rationalize the observed chloride and acid dependence in the rate law.

$$\frac{-d[C_2H_4]}{dt} = \frac{k[PdCl_4^{2-}][C_2H_4]}{[H^+][Cl^-]^2} \qquad (465)$$

The oxidation of higher α-olefins and cycloolefins to ketones[721] was also reported. The oxidation of propylene to acetone and *cis*- and *trans*-2-butene and 1-butene to methyl ethyl ketone in aqueous solution was reported to occur in accord with the rate law determined for ethylene oxidation [Eq. (465)].[725] Less than an order of magnitude separated the rates observed for those oxidations (in contrast to the results reported for Tl^{3+} catalyst oxidations),[718,733] leading Henry to suggest that in the rate-determining step, transfer of Pd(II)-coordinated OH^- to coordinated olefin, little carbonium ion character was involved.[718] Dienes are converted to vinyl aldehydes[721] and a variety of X-substituted olefins (X = halide, $-CO_2H$, $-C(=O)NH_2$) are converted to ketones with concommitant removal of the X groups[721,734,735] [Eq. (466)].

$$RCH{=}CHX \text{ or } RC(X){=}CH_2 \longrightarrow RC(=O)CH_3 \qquad (466)$$

Since the major products from the oxidation of α-olefins are ketones, the addition of the hydroxy group to the Pd(II)-coordinated olefin is presumed to be largely Markovnikov with attack at the β-carbon although some aldehydes are produced and steric effects are important. Electron-withdrawing groups (X = NO_2, CN) on the ethylene promote attack on the β-carbon [Eq. (467)]. A study of phenyl-substituted styrenes has shown that electron-withdrawing groups on the phenyl ring lead to increasing yields of aldehyde, and electron-releasing substituents promote the formation of ketones.[736]

$$XCH{=}CH_2 \longrightarrow XCH_2C(=O)H \qquad (467)$$

An interesting analogous reaction involves formation of propiophenone from phenylcyclopropane to the presence of $PdCl_2$ in aqueous solution [Eq. (468)].[737]

$$\text{C}_6\text{H}_5\text{-cyclopropane-H(D)} \xrightarrow[\text{H}_2\text{O-glyme}]{\text{PdCl}_2} \text{C}_6\text{H}_5\text{C(=O)-CH(H(D))CH}_3 \quad (468)$$

Parallel experiments with phenylcyclopropane-1-*d* indicate that the reaction occurs by a mechanism similar to olefin oxidation with two 1,2-hydride shifts necessary to account for the presence of the deuterium on the 2-carbon in the product.

C. Olefin Oxidation in Other Solvents

Other solvents which yield nucleophiles on deprotonation, including acetic acid and alcohols, have been used in the catalytic oxidation of olefins leading to the formation of a variety of products [Eqs. (469) and (470)]. A large body of data

$$\begin{aligned} C_2H_4 + HOAc \longrightarrow\ & CH_2{=}CHOAc + CH_3\overset{\overset{\displaystyle O}{\|}}{C}H + CH_3CH(OAc)_2 + \\ & + AcOCH_2CH_2OAc + AcOCH_2CH_2OH + AcOCH_2CH_2Cl \end{aligned} \quad (469)$$

$$C_2H_4 + ROH \longrightarrow CH_2{=}CHOR + CH_3CH(OR)_2 + CH_3\overset{\overset{\displaystyle O}{\|}}{C}H \quad (470)$$

exists concerning the variation in product distribution with change in the catalyst system for the reactions in Eq. (469), but the mechanisms are not unambiguously determined. Several significant observations have been reported, however. The reaction in Eq. (469) requires the presence of base (OAc^-) for optimal reaction rates.[738] Also, in the absence of Cu(II), the oxidation of ethylene leads to vinyl acetate while the other products shown in Eq. (469) are found when Cu(II) is present as a co-catalyst.[739] Use of CH_3COOD as the solvent leads to the incorporation of no deuterium in the reaction products.[740] A mechanism consistent with these observations is suggested to involve conversion of an olefinic π-complex into a β-acetoxy-substituted ethyl ligand via[739] either insertion of ethylene into a Pd—OAc bond or attack of free OAc^- [Eq. (471)].

$$[Pd(C_2H_4)(OAc)_3]^- + L \rightarrow [Pd(L)(OAc)_2(CH_2CH_2OAc)]^- \quad (471)$$

Vinyl acetate would then be formed by β-hydride elimination to produce an unstable Pd—H species [Eq. (472)]; the saturated products are suggested to arise from direct transfer of electrons from $[Pd(L)X_2(CH_2CH_2OAc)]^-$ or its

$$[Pd(L)(OAc)_2(CH_2CH_2OAc)]^- \longrightarrow CH_2{=}CHOAc + [Pd(L)(OAc)_2H] \quad (472)$$

$$[Pd(L)X_2(CH_2CH_2OAc)]^- + 2Cu^{II} \longrightarrow Pd^{II} + XCH_2CH_2OAc + 2Cu^{I} \quad (473)$$

$$[Pd(L)X_2(CH(CH_3){-}OAc)]^- + 2Cu^{II} \longrightarrow Pd^{II} + CH_3CH(OAc)_2 + 2Cu^{I} \quad (474)$$

isomer $[Pd(L)X_2(CH_2(CH_3)\text{-}OAc)]^-$ to Cu(II) with concommitant transfer of X^- to the substituted alkyl groups [Eqs. (473) and (474)]. Analogous reactions were observed for higher olefins, including some production of 1,3- and 1,4-disubstituted alkanes which presumably arose from isomerization of the substituted alkyl groups prior to electron transfer to the Cu(II). This reaction sequence is supported by a recent study on the oxidation of cyclohexene and deuterated cyclohexene in acetic acid catalyzed by Pd(II) and Cu(II) co-catalysts.[741] The products obtained are rationalized on the basis of initial formation of a β-acetoxy-substituted Pd(II) alkyl complex [Eq. (475)].

$$\text{cyclohexene} + [Pd^{II}L_mX_n]^{(2-n)+} \rightleftharpoons \left[C_6H_{10}(OAc)(Pd^{II}L_mX_{n-1}) \right]^{(2-n)+} \quad (475)$$

The product distribution is then suggested to be determined by a competition between Pd—H elimination and direct electron transfer to the $CuCl_2$ present, leading to cycloalkene acetates and disubstituted cycloalkanes, respectively. An additional complication arises from the isomerization of the initial Pd(II) alkyl groups via a series of Pd—H additions and eliminations (to Pd(II)-bound olefin) prior to the decomposition step. These mechanistic suggestions are also supported[729] by the decomposition of $Pd(II)CH_2CH_2OAc$ species in ether and acetic acid/acetate [Eq. (476)]. The formation of the diacetate product is again rationalized on the basis of isomerization of the alkyl group prior to attack by nucleophile (OAc).

$$HgCl(CH_2CH_2OAc) + PdCl_2 \begin{cases} \xrightarrow{\text{ether}} CH_2{=}CHOAc \\ \xrightarrow[\text{OAc}]{\text{HOAc}} CH_2{=}CHOAc + CH_3CH(OAc)_2 \end{cases} \quad (476)$$

Alcohols represent another solvent system for the oxidation of olefins by Pd(II) catalysts [Eq. (470)], resulting in vinyl ethers and acetals.[738,742] Again, reaction in CH_3OD resulted in no incorporation of deuterium into the acetal,[740] and the reaction intermediate is presumably a β-alkoxy-substituted alkyl group since the familiar exchange reaction [Eq. (477)] leads to a vinyl ether.[729]

$$HgCl(CH_2CH_2OC_2H_5) + Pd^{II} \xrightarrow{\text{ether}} CH_2{=}CHOC_2H_5 + Pd^{II}{-}H \quad (477)$$

A plausible mechanism for this reaction would then, by analogy with the acetic acid system, involve a β-alkoxy-substituted Pd(II) alkyl complex which could then decompose by a β-hydride elimination to give the vinyl ether and a Pd(II)—H species— or isomerize, followed by alkoxide attack to give the acetal [Eq. (478)].

$$C_2H_4 + ROH + Pd^{II} \xrightarrow{\text{base}} [Pd^{II}(CH_2CH_2OR)]^- \longrightarrow CH_2{=}CHOR + Pd^{II}{-}H$$

$$[Pd^{II}(CH_3{-}CH{-}OR)]^- \xrightarrow{OR^-} CH_3CH(OR)_2 + Pd^0 \quad (478)$$

Reaction of olefins with alcohols and diols have also been reported to yield ketones and acetals [Eq. (479)].

$$\text{cyclohexene} + PdCl_2 + CuCl_2 \xrightarrow{C_2H_5OH} \text{cyclohexanone}$$

$$\text{cyclohexene} + PdCl_2 + CuCl_2 \xrightarrow{(CH_2OH)_2} \text{cyclohexanone ethylene ketal} \quad (479)$$

Surprisingly, oxidation of ^{14}C-labeled ethylene[744] with $PdCl_2/CuCl_2$ in ethanol produced acetaldehyde which was largely unlabeled along with labeled acetal [eq. (480)].

$$^{14}C_2H_4 + C_2H_5OH \xrightarrow[CuCl_2]{PdCl_2} CH_3\overset{O}{\overset{\|}{C}}H + {}^{14}CH_3{}^{14}CH(OC_2H_5)_2 \quad (480)$$

This indicated that for ethanol and higher alcohols, ethylene was displaced, leading to coordinated alcohol which was then oxidized or to a rapid exchange of H_2O between the alcohol and ethylene.

Alcohols are oxidized to aldehydes, ketones, and acetals by Pd(II).[727,744,745] This process has been suggested[582] to occur by β-hydride elimination from coordinated alkoxide to give unstable Pd(II)—H species [Eq. (481)].

$$\mathrm{RCH_2OH + PdCl_2 \rightleftharpoons Pd(L)Cl_2(\underset{\displaystyle H}{\underset{|}{O}}CH_2R) \rightleftharpoons}$$

$$\mathrm{Pd(L)_2Cl(OCH_2R) \longrightarrow R\overset{\displaystyle O}{\overset{\|}{C}}H + [Pd^{II}—H]} \qquad (481)$$

The subsequent decomposition of these Pd(II) hydrides would then generate palladium metal. In the absence of chloride ion and in acetic acid solution, this oxidation has been found to proceed to formation of carboxylic acids from primary alcohols because of the higher oxidation potential of Pd(II) under these conditions.[746]

Other substrates catalytically oxidized by Pd(II) in the presence of suitable reoxidants include carbon monoxide, formic acid, aromatic ring systems, and alkyl groups on such rings.[582] The addition of organometallic species to olefins catalyzed by Pd(II) in the presence of oxidants also presumably occurs by a mechanism similar to that of olefin oxidation [Eqs. (482) and (483)].

$$\mathrm{MR + Pd^{II} \rightleftharpoons Pd^{II}—R + M^+} \qquad (482)$$

$$\mathrm{R'CH{=}CH_2 + Pd^{II}—R \longrightarrow R'CH{=}CHR + R'\underset{\displaystyle R}{\underset{|}{C}}{=}CH_2 + Pd^0} \qquad (483)$$

Eqs. (482) and (483) lead to the overall reaction in Eq. (484).

$$\mathrm{MR + R'CH{=}CH_2 \xrightarrow[{[O]}]{Pd^{II}} R'CH{=}CHR + RR'C{=}CH_2} \qquad (484)$$

In view of the many reactions known involving alkyl groups coordinated to Pd(II), insertion of coordinated olefin into the Pd(II)—R bond followed by reductive elimination of Pd^0 generating the alkylated olefin [Eq. (484)] represents a reasonable mechanism.

In conclusion, it is appropriate to state that this catalytic oxidation of organic substrates by Pd(II) in the presence of oxidants represents a manifestation of a general property of group VIII transition metal complexes. Thus, a similar catalytic air oxidation of ethylene by $RhCl_3$ in aqueous solution in the presence of Fe(III) has been reported [Eq. (485)].[747]

$$C_2H_4 + H_2O \xrightarrow{RhCl_3 + Fe^{3+}} CH_3\overset{O}{\overset{\|}{C}}H + 2H^+ \qquad (485)$$

D. Oxidative Dehydrogenation of Ligands Coordinated to Transition Metals

As indicated in section V, B., coordination to transition metal sometimes results in stabilization of ligand oxidation products which otherwise would be very fragile. This is particularly evident in the oxidation of coordinated amine ligands to form imine linkages (Eq. (486)].

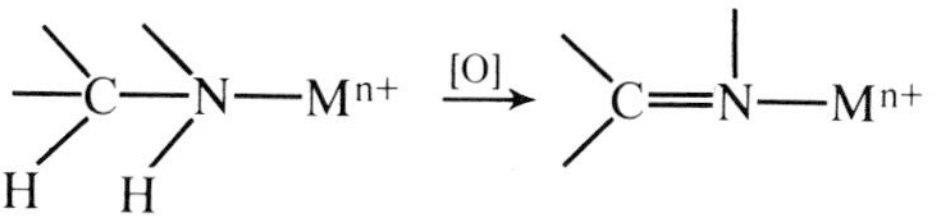

$$>\underset{H}{C}-\underset{H}{N}-M^{n+} \xrightarrow{[O]} >C{=}N-M^{n+} \qquad (486)$$

In this case the site of oxidation on the amine changes. Aliphatic amines are commonly oxidized to yield various nitrogen oxide species. Also, the stability of the product imine is modified, being, in general, insensitive to acid hydrolysis and more susceptible to hydrolysis by nucleophilic attack (e.g., OH^-) at the azomethine carbon but much more readily isolated and characterized. The generality of this reaction is shown by the oxidation of Pt(IV)–ethylenediamine complexes [Eq. (487a)] with chlorine and the formation of a charge delocalized chelate ring by the facile oxidation of the diamine complex illustrated in Eq. (487b).[749]

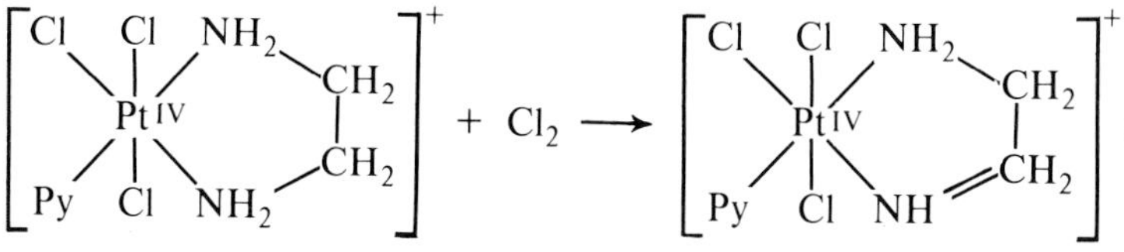

$$[Pt^{IV}Cl_3(Py)(NH_2CH_2CH_2NH_2)]^+ + Cl_2 \longrightarrow [Pt^{IV}Cl_3(Py)(NH_2CH_2CH{=}NH)]^+ \qquad (487a)$$

$$Ni^{II}\text{ complex (N–H, N–H)} \xrightarrow[\text{or } IO_3^-]{O_2} Ni^{II}\text{ complex (delocalized } \ominus \text{ chelate ring)} \qquad (487b)$$

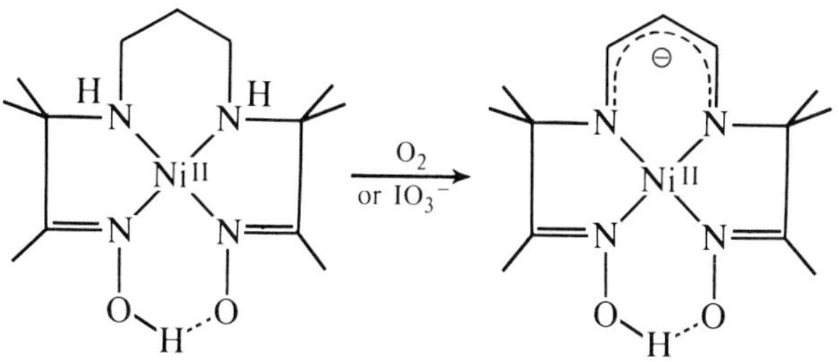

The oxidative dehydrogenation reaction has been extensively utilized in the synthesis of transition metal complexes having increased π-electron density on the ligand structure. Much of this work has been done on metal complexes with macrocyclic ligands because of the inherent stability of such complexes toward stepwise dissociation and because of the interest in generating synthetic analogs of such naturally occurring macrocyclic ligand structures as the corrin ring of vitamin B_{12} and the porphyrin ring found in iron heme complexes.[94]

The first example of this process reported for macrocyclic ligand complexes involves the nitric acid oxidation of **115** leading to **116** [Eq. (488)].[750] Since then many macrocyclic nickel complexes having secondary amine linkages with hy-

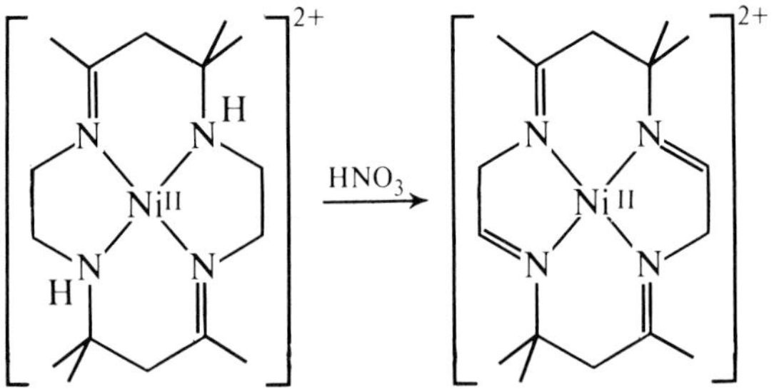

$$\mathbf{115} \xrightarrow{HNO_3} \mathbf{116} \qquad (488)$$

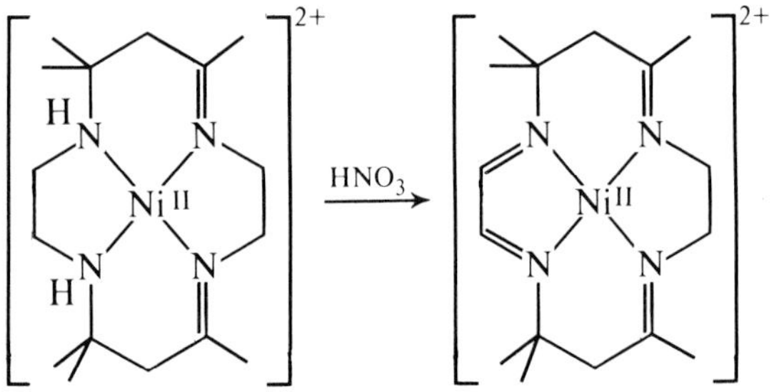

$$\mathbf{117} \xrightarrow{HNO_3} \qquad (489)$$

$$\mathbf{118} \xrightarrow{O_2} \qquad (490)$$

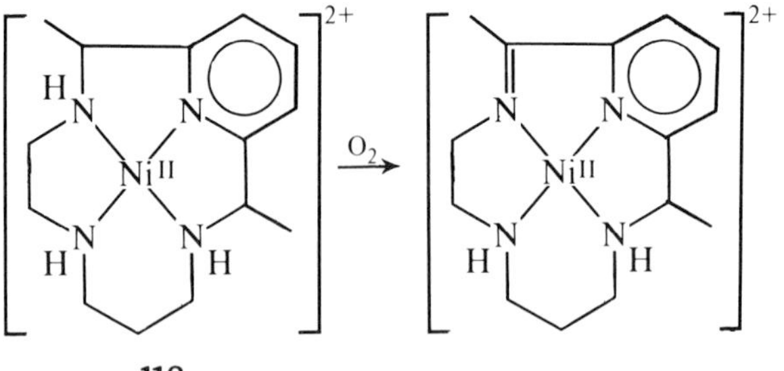

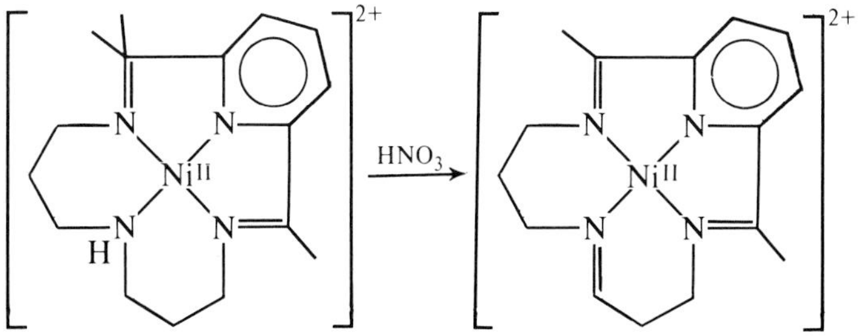

(491)

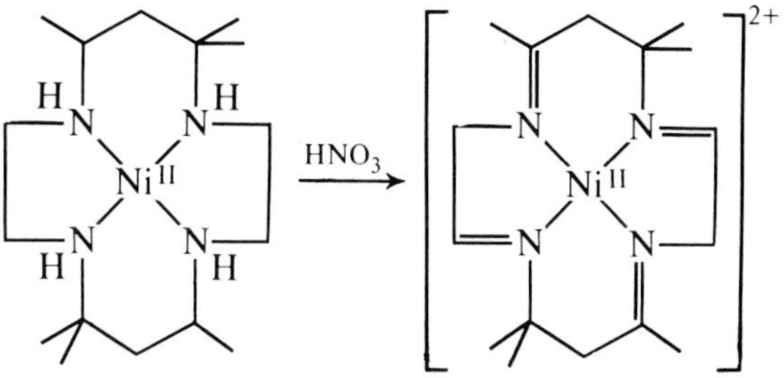

119

(492)

120

drogen atoms on the α-carbon have been oxidized to yield similar results,[245,228] including structures **117** (the methyl group positional isomer of **115**) [Eq. (489)],[751,752] **118** [Eq. (490)],[753] **119** [Eq. (491)],[753] and **120** [Eq. (492)].[751]

An oxidation mechanism consistent with the known chemistry of these systems could involve initial conversion of the Ni(II) complexes in their Ni(III) analogs with subsequent electron abstraction from an N—H bond yielding a Ni(II) species with a ligand radical, structure **121**, and H^+ [Eqs. (493) and (494)].

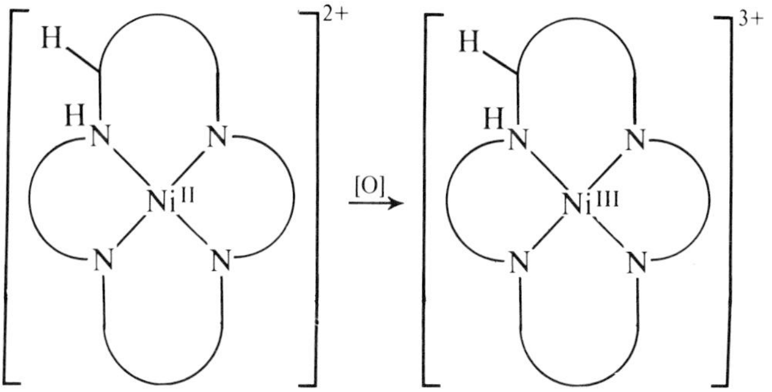

(493)

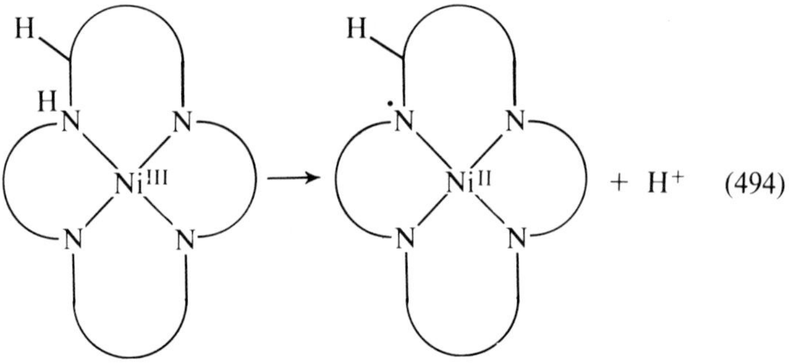

$$\text{[Ni}^{\text{III}}\text{ macrocycle with N–H and C–H]} \longrightarrow \text{[Ni}^{\text{II}}\text{ radical macrocycle]} + H^+ \quad (494)$$

121

Oxidation of the radical species **121** would then yield the observed product [Eq. (495)]. Early support for this mechanism was found in the synthesis of Ni(III) complexes with macrocyclic ligands having secondary amine donor atoms which are stable in the solid state but decompose in solution.[754,755,756]

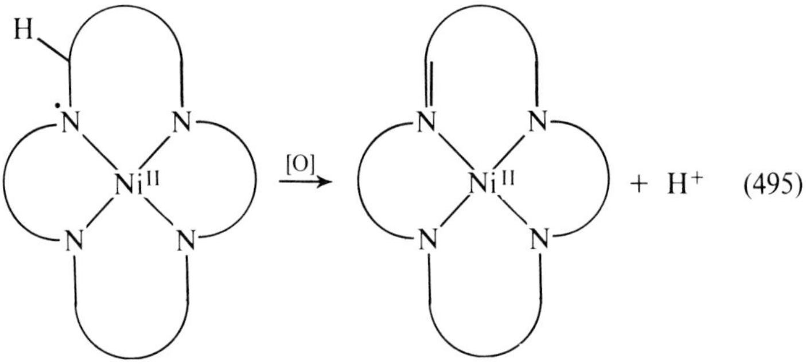

$$\text{[Ni}^{\text{II}}\text{ radical macrocycle]} \xrightarrow{[O]} \text{[Ni}^{\text{II}}\text{ monoimine macrocycle]} + H^+ \quad (495)$$

The detection of the radical intermediate of structure **121** has recently been reported upon base-induced decomposition of solutions of the Ni(III) complex; this strongly supports the proposed mechanism.[757] In the nickel(II) product formed by decomposition of the radical species, both the corresponding Ni(II) tetraamine complex and a monoimine species were present. Thus, in this reaction sequence, a nickel(III) complex converts to a nickel(II) radical species, from which is generated a Ni(II) monoimine product in accord with the proposed mechanism.

This type of reaction has also been reported for a Cu(III) tetraamine macrocyclic complex,[758] in which a Cu(III) intermediate was generated electrochemically. Examination of the Cu(II) product isolated showed a mixture of Cu(II) tetraamine and an oxidized Cu(II) macrocyclic species with evidence for the introduction of imine linkages.

In view of the relatively forceful conditions necessary to generate Ni(III) and Cu(III), oxidative dehydrogenation should occur much more readily in systems where a $M^{n+}/M^{(n+1)+}$ redox couple is available at lower potentials. This has been observed. A number of low spin, six-coordinate iron(II) complexes with macrocyclic ligands having secondary amine donor atoms are oxidatively dehydrogenated to form imine linkages by molecular oxygen.[186,759,760] In Eqs. (496), (497), and (498) axial acetonitrile ligands on the six-coordinate, low-spin iron have been omitted for clarity.

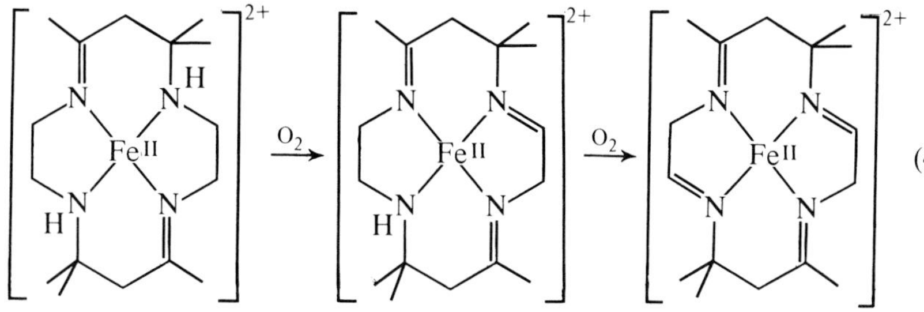

(496)

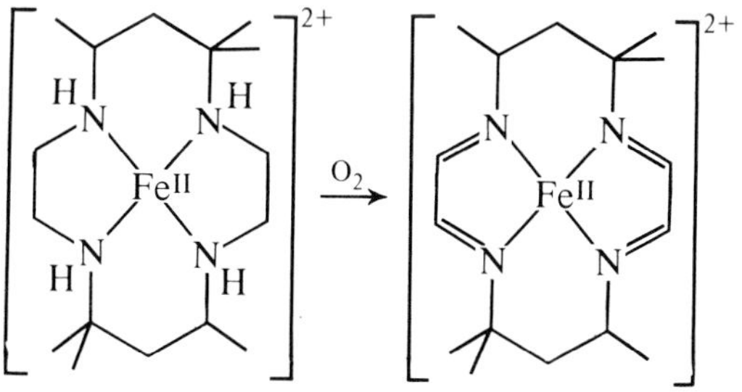

(497)

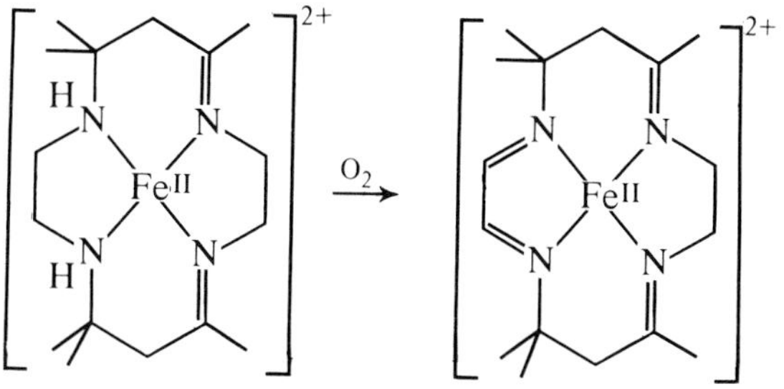

(498)

A mechanism analogous to that for the Ni complexes has been suggested. In accord with this mechanism, an Fe(III) intermediate complex was isolated, which on water-induced decomposition in acetonitrile, led to Fe(II) derivatives of both the starting material and the oxidized ligand. A particularly significant difference in the behaviors of the Ni and Fe complexes is evident upon comparing Eqs. (492) and (497). The iron atom directs the dehydrogenation reaction so as to produce conjugated α-diimine moieties, presumably because of strong metal–ligand π-bonding in the product.[760]

This remarkable process has also been reported for bidentate secondary amine complexes of low-spin Fe(II),[185] leading to coordinated α-diimine species [Eq. (499)] and for monodentate[761] and bidentate[762] Ru(II) amine complexes, forming coordinated CN^- [Eq. (500)] and α-diimine ligands [Eq. (501)], respectively, as well as a Ru(III) bidentate amine complex[762] [yielding the same Ru^{2+} species as shown in Eq. (501)]. Another type of product[763] which can be generated via oxidative dehydrogenation is shown in Eq. (502). Here again, an Fe(III) → Fe(II)—ligand radical intermediate is presumably involved, which then rearranges before oxidation and reacts with the superoxide ion generated in the Fe(III) oxidation. This same class of product has been reported for a cobalt complex.[764]

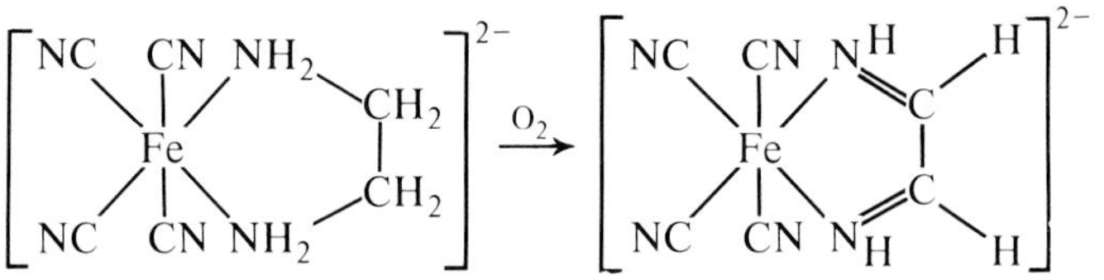

$$[Fe(CN)_4(NH_2CH_2CH_2NH_2)]^{2-} \xrightarrow{O_2} [Fe(CN)_4(HN{=}CH{-}CH{=}NH)]^{2-} \qquad (499)$$

$$Ru(NH_2CH_3)_6^{2+} \xrightarrow{O_2} \text{“}Ru(CN)_3 \cdot 3H_2O\text{”} \qquad (500)$$

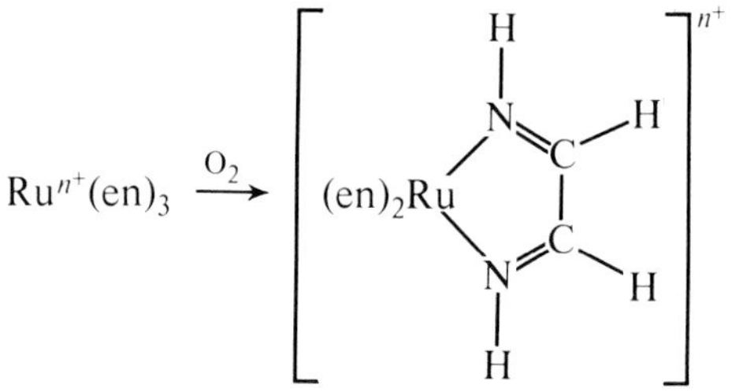

$$Ru^{n+}(en)_3 \xrightarrow{O_2} [(en)_2Ru(HN{=}CH{-}CH{=}NH)]^{n+} \qquad (501)$$

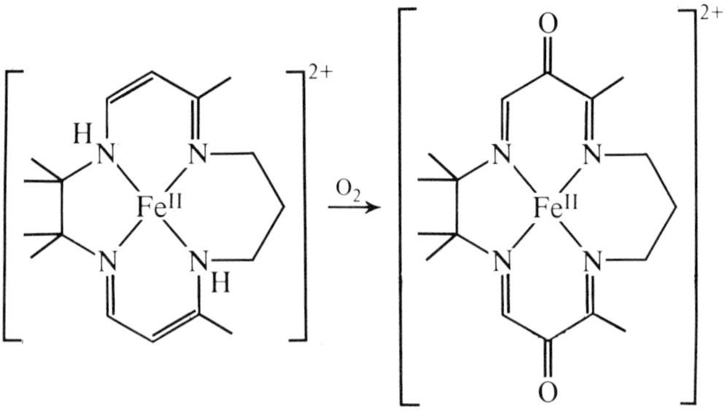

$$[Fe^{II}]^{2+} \xrightarrow{O_2} [Fe^{II}]^{2+} \qquad (502)$$

A plausible explanation for the facile nature of oxidative dehydrogenation in the presence of low-spin Fe(II) and Ru(II) can be found in the properties of the low-spin d^5, oxidized metal intermediate. Taube[97] reports an approximate 10^5 increase in rate of reduction of

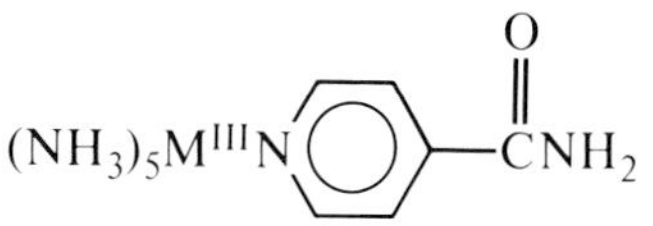

$$(NH_3)_5M^{III}N\bigcirc\!\!-\overset{O}{\overset{\|}{C}}NH_2$$

by Cr^{2+} on going from Co(III) to Ru(III) (which have nearly the same reduction potential) and attributes the much faster reduction in the case of the Ru(III) species to the vacancy in the t_{2g} orbitals which can mix with ligand orbitals of π-symmetry, allowing direct transfer of the electron from Cr^{2+} to Ru^{3+} via the π-electron cloud on the ligand. Similarly, in oxidative dehydrogenation with Fe^{3+} and Ru^{3+} intermediates, direct electron transfer from an N—H bond to the low-spin d^5 metal ion would be expected to be much more facile than in the cases of Ni^{3+} (d^5) and Cu^{3+} (d^8) where no properly oriented empty metal orbital is found in the ground state.

Oxidation of charge-delocalized rings has been observed,[765] under certain conditions, to introduce unsaturation into neighboring chelate rings [Eq. (503)].

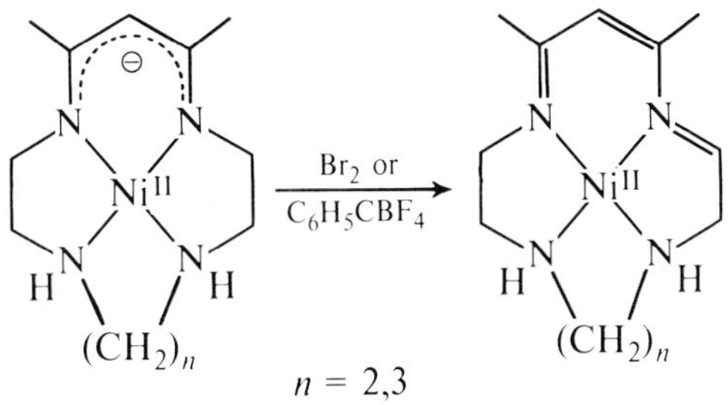

$$\text{Ni}^{II}\text{ complex (delocalized }\ominus\text{ ring, } (CH_2)_n) \xrightarrow[C_6H_5CBF_4]{Br_2\text{ or}} \text{Ni}^{II}\text{ complex (C=C and C=N unsaturation, } (CH_2)_n) \qquad n = 2,3 \tag{503}$$

As shown in Eq. (487), oxidative dehydrogenation in which two imine linkages are introduced into a six-membered chelate ring leads to ionized products [also Eq. (504)].

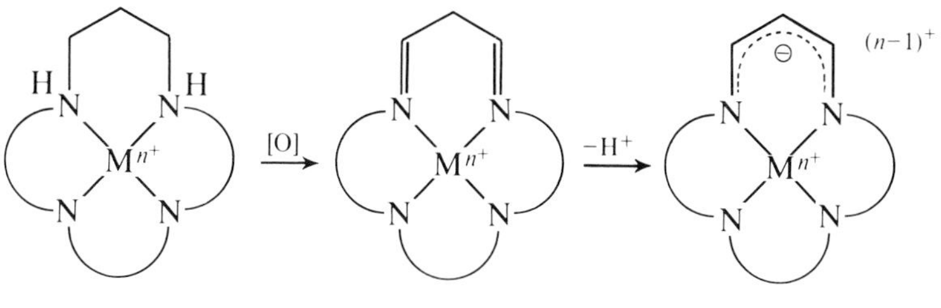

$$M^{n+}\text{ (HN, NH)} \xrightarrow{[O]} M^{n+}\text{ (two C=N)} \xrightarrow{-H^+} [M^{n+}\text{ (delocalized }\ominus\text{ ring)}]^{(n-1)+} \tag{504}$$

Thus, application of the reactions given in Eq. (503) to a complex containing two charge-delocalized, six-membered chelate rings, with an intervening saturated six-membered ring, was expected[94] to produce analogs of the corrin ring of vitamin B_{12} [Eq. (505)].

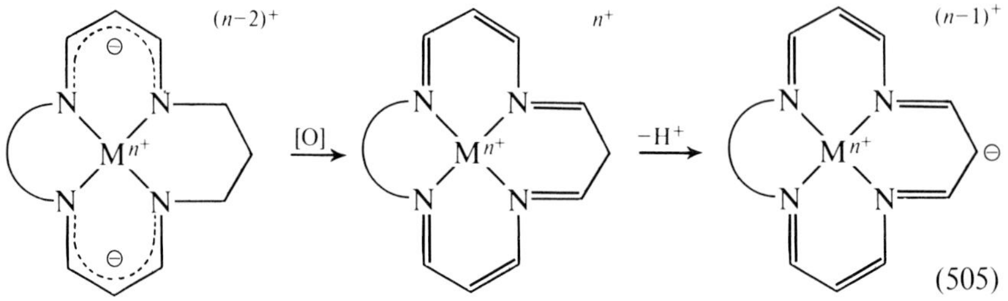

$(n-2)^+$ $\ominus$ N N M^{n+} N N $\ominus$ $\xrightarrow{[O]}$ n^+ N N M^{n+} N N $\xrightarrow{-H^+}$ $(n-1)^+$ N N M^{n+} N N $\ominus$

(505)

The first report of such a synthesis involved the oxidative dehydrogenation of nickel(II) complexes **122** with dichlorodicyanoquinone to produce **123** [Eq. (506)].[766,767]

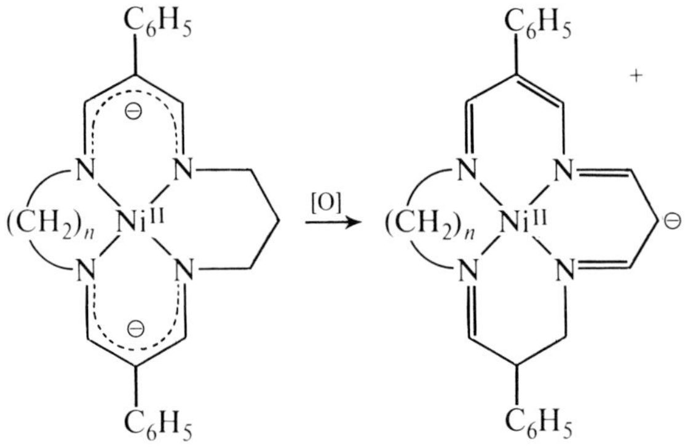

C_6H_5 $\ominus$ N N $(CH_2)_n$ Ni^{II} N N $\ominus$ C_6H_5 $\xrightarrow{[O]}$ C_6H_5 + N N $(CH_2)_n$ Ni^{II} N N $\ominus$ C_6H_5

(506)

122 $n = 2,3$ **123** $n = 2,3$

A related dehydrogenation reaction has led to interesting unsaturated 14-membered ring derivatives [Eq. (507)].[768,769]

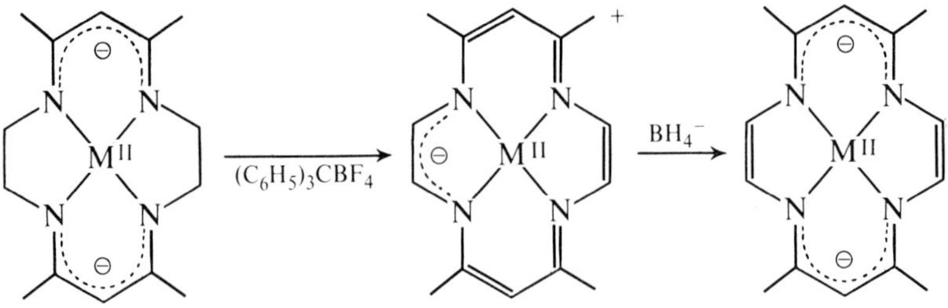

$\ominus$ N N M^{II} N N $\ominus$ $\xrightarrow[(C_6H_5)_3CBF_4]{}$ + N N $\ominus$ M^{II} N N $\xrightarrow{BH_4^-}$ $\ominus$ N N M^{II} N N $\ominus$

(507)

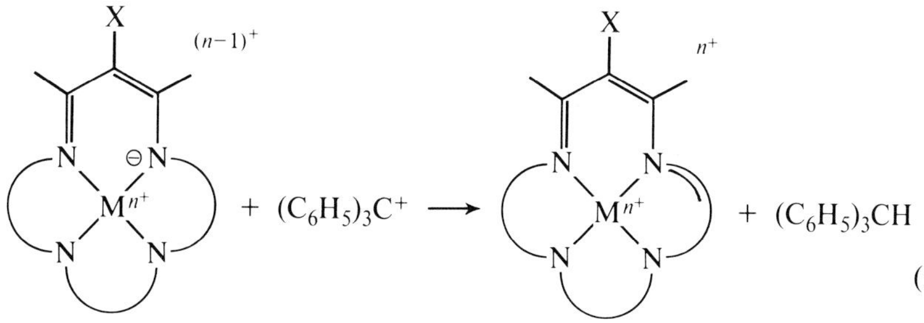

$$+ (C_6H_5)_3C^+ \longrightarrow \quad + (C_6H_5)_3CH \qquad (508)$$

The mechanisms of these reactions [Eqs. (505), (506), and (507)] are not clear however. Trityl fluoroborate $[(C_6H_5)_3CBF_4]$ is a well known hydride extracting agent, suggesting a reaction sequence such as that shown in Eq. (508), but the oxidation in Eq. (503) is also achieved by Br_2 or by electrochemical oxidation of the starting material in Ni(III) which then rearranges to give the observed product.

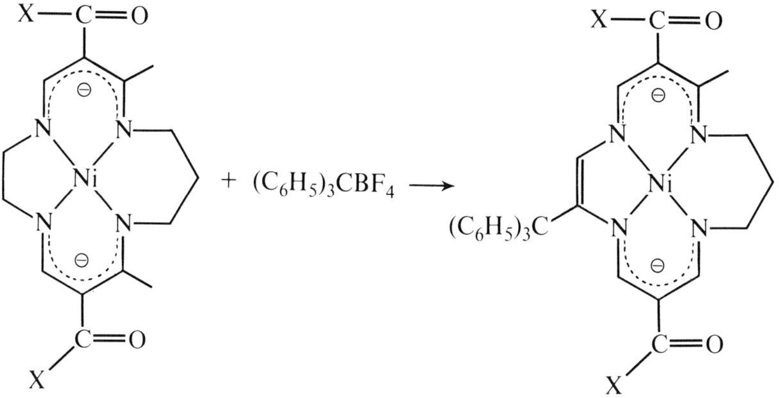

$$+ (C_6H_5)_3CBF_4 \longrightarrow \qquad (509)$$

X = CH_3, OEt

Also, the use of $(C_6H_5)_3CBF_4$ with the 15-membered macrocycles shown in Eq. (509) leads to oxidation of the five-membered chelate ring,[763] as does I_2 oxidation [Eq. (510)], while treatment with Br_2 results in electrophilic addition. Finally, dichlorodicyanoquinone oxidation of several nickel(II) macrocycles analogous to **122** but without the phenyl substituents leads to quinone addition products.[763]

The existence of a radical pathway in such oxidations is strongly indicated by the formation of dimeric oxdiation products in systems of this kind. Thus, Cunningham and Sievers[770] have isolated the highly oxidized dimer of structure **124**, and Dabrowiak has characterized conformational isomers of a related compound.[771]

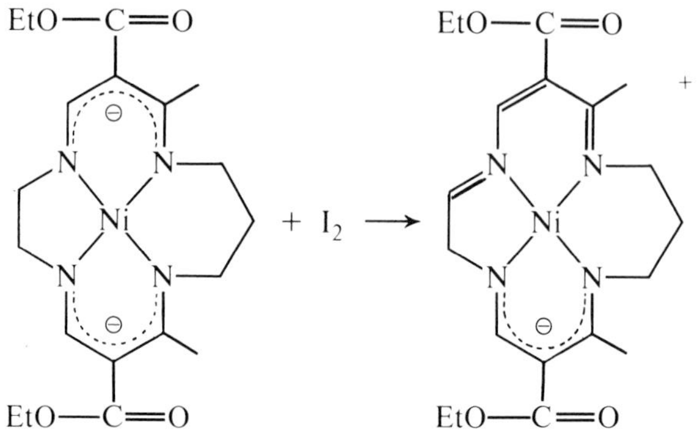

$$\text{(Ni macrocycle complex)} + I_2 \longrightarrow \text{(oxidized Ni macrocycle complex)}^+ \quad (510)$$

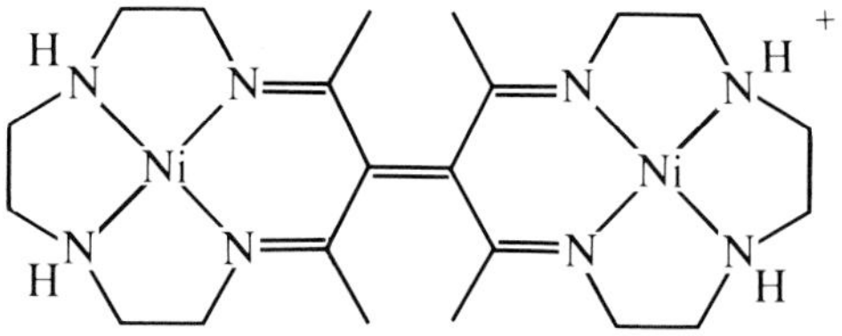

124

E. Oxygenases

Oxidation reactions involving molecular oxygen in biological systems are either identified with the energy production processes—that is, ATP synthesis—or associated with so-called free oxidations.[698–703] The latter is not a part of the electron transport chain. Instead, free oxidations involve an array of enzymes that use O_2 directly to oxidize organic molecules in the cell. These enzymes are roughly classified into two major categories: oxidases, which cause dehydrogenation of the substrate and reduce O_2 to H_2O, and oxygenases, which introduce oxygen atoms into substrate molecules. Many of the oxygenases contain metal atoms at their active sites, and since the oxidation reactions they produce are distinctive, brief consideration is given to them here.

The oxygenases are further classified as monooxygenases and dioxygenases. Dioxygenases introduce both oxygen atoms of O_2 into the substrate molecule [Eq. (511)]; the monooxygenases require both a substrate and a cofactor.

$$SH_2 + O_2 \xrightarrow{\text{dioxygenase}} S(OH)_2 \quad (511)$$

The monooxygenases introduce one oxygen atom into the substrate and simultaneously produce a single water molecule by reaction with the cofactor [Eq. (512)].

$$SH + AH_2 + O_2 \xrightarrow{\text{monooxygenase}} SOH + A + H_2O \qquad (512)$$

These enzymes were originally called mixed-function oxygenases by Mason, who is credited with having discovered them,[772,773] because they both insert oxygen atoms into and dehydrogenate organic molecules.

The monooxygenases occur widely in microorganisms and in higher organisms, both plants and animals. Examples are listed in Table 2-9. The most thoroughly studied is camphor 5-monooxygenase from the bacterium *Pseudomonas putida*,[774] and much of the summary to follow relates to that species. The microsomes in the liver contain monooxygenases that cause hydroxylation of many foreign compounds and several natural metabolites. These substances, however, are not restricted in occurrence to the liver but are also found in the lungs, gastrointestinal tract, skin, kidneys, and adrenal mitochondria. Several microorganisms use hydrocarbons as nutrients, and the associated enzymatic oxidations have been investigated as a route to foodstuffs from petroleum hydrocarbons.[775]

TABLE 2-9
EXAMPLES OF MONOOXYGENASES[a]

Enzyme	Electron Donor	Occurrence
Squalene 2, 3-monooxygenase EC 1.12.1.3.	NADPH	liver
Cholesterol 20-monooxygenase EC 1.14.1.9	NADPH	adrenal mitochondria
Dopamine β-monooxygenase EC 1.14.2.1	ascorbate	adrenal glands
Phenylalanine 4-monooxygenase EC 1.14.3.1	tetrahydrobiopterin	liver cytoplasm
Aliphatic acid ω-oxygenase	NADPH	liver endoplasmic reticulum
Arene *p*-monooxygenase	NADPH	liver endoplasmic reticulum
Camphor 5-monooxygenase	NADPH	*Pseudomonas putida*

[a] Adapted from Ref. 702.

The oxidation reactions catalyzed by monooxygenases are broad-ranging, including the hydroxylation of aliphatic compounds [Eq. (513)], hydroxylation of arenes [Eq. (514)], epoxidation of olefins [Eq. (515)], lactone formation [Eq. (516)], amine oxide formation [Eq. (517)], sulfoxide formation [Eq. (518)], and oxidative dialkylation of heteroatoms [Eq. (519)].

$$\begin{array}{c}|\\ -\text{C}-\text{H}\\ |\end{array} \xrightarrow[O_2]{E} \begin{array}{c}|\\ -\text{C}-\text{OH}\\ |\end{array} \tag{513}$$

$$\text{ArH} \xrightarrow[O_2]{E} \text{ArOH} \tag{514}$$

$$\gt\text{C}=\text{C}\lt \xrightarrow[O_2]{E} \gt\text{C}\overset{O}{\frown}\text{C}\lt \tag{515}$$

$$\text{HRC}(=O)\text{CH}_3 \longrightarrow \text{R}-\text{C}(=O)-\text{O}\ (\text{four-membered ring}) + \text{H}_2\text{CO} \tag{516}$$

$$\text{R}^1\text{R}^2\text{R}^3\text{N} \xrightarrow[O_2]{E} \text{R}^1\text{R}^2\text{R}^3\text{N}-\text{O} \tag{517}$$

$$\text{R}^1\text{R}^2\text{S} \xrightarrow[O_2]{E} \text{R}^1\text{R}^2\text{S}=\text{O} \tag{518}$$

$$C_6H_5OCH_3 \xrightarrow[O_2]{E} C_6H_5OH + CH_2O \tag{519}$$

These hydroxylating enzymes are generally multicomponent systems, consisting of the oxygen and substrate-interactive protein and additional proteins that serve as electron carriers; attention focuses naturally on the former. All such enzymes can be classified into two categories—those utilizing metal-containing active sites and those having no metallic elements at their active sites. Only the former are considered here. These are more numerous, and the metals involved are iron and/or copper. From among these, the largest group contains an iron-heme prosthetic group and are named cytochrome P-450. The name derives from the position of a visible spectral absorption band of the carbon monoxide adduct. This prosthetic group is not present in all iron-containing monooxygenases. The cytochrome P-450 systems are best understood and serve generally as the basis for the discussion that follows.

The bacterial monooxygenases are illustrated by camphor 5-monooxygenase.[774] The enzyme system consists of three proteins, the pure cytochrome P-450$_{cam}$ [P-450–Fe^{II}], a flavoprotein (P–FAD), and an iron–sulfur protein [2P–(FeS)$_2$]. Referring to the general equation for the reaction, the ultimate electron donor AH_2 is reduced diphosphopyridine nucleotide (DPNH). The system given in Eq. (520) constitutes an electron-transport chain with the cytochrome P-450 interacting directly with O_2 and receiving electrons from the iron–sulfur protein 2P–(FeS)$_2$. The catalytic cycle by which the cytochrome oxidizes the substrate is shown in rudimentary form in Scheme 2-30, and its relationship to the total system is displayed in Scheme 2-31.

$$\text{DPNH} \longrightarrow \text{P-FAD} \longrightarrow \text{2P-(FeS)}_2 \longrightarrow \text{P-450-Fe}^{\text{III}} \longrightarrow \text{O}_2 \qquad (520)$$

Scheme 2-30

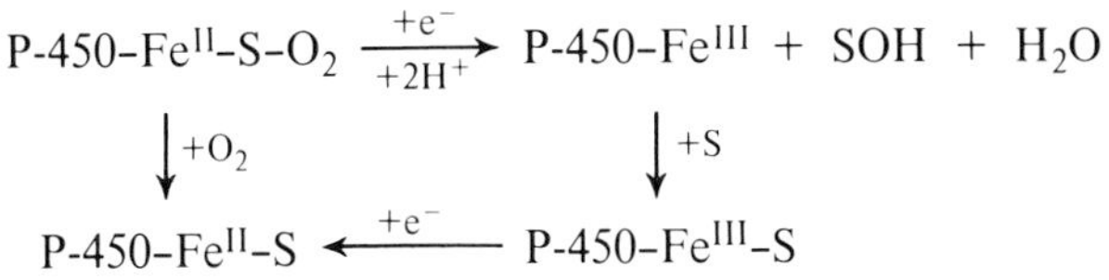

$$\text{P-450-Fe}^{\text{II}}\text{-S-O}_2 \xrightarrow[+2\text{H}^+]{+e^-} \text{P-450-Fe}^{\text{III}} + \text{SOH} + \text{H}_2\text{O}$$

$$\uparrow{+\text{O}_2} \text{ (from P-450-Fe}^{\text{II}}\text{-S)} \qquad \downarrow{+\text{S}} \text{ (from P-450-Fe}^{\text{III}}\text{)}$$

$$\text{P-450-Fe}^{\text{II}}\text{-S} \xleftarrow{+e^-} \text{P-450-Fe}^{\text{III}}\text{-S}$$

In Scheme 2-30 the resting enzyme appears at the upper right. This trivalent iron species complexes with substrate and then receives on electron from the electron transport system. The resulting Fe(II)–enzyme–substrate complex combines with dioxygen, after the fashion of myoglobin or hemoglobin, to form the ternary complex P-450–Fe^{II}–S–O_2. This species is fairly stable, but it slowly decomposes into a high-spin Fe(III) derivative.[776] The transfer of a second electron to the ternary complex is followed by product formation. Scheme 2-31 shows the coupling processes in the oxidation cycle. Although the total system is much more involved, attention is most appropriately focused here on the nature of the product-forming chemical reaction that involves the substrate and molecular oxygen.

Scheme 2-31

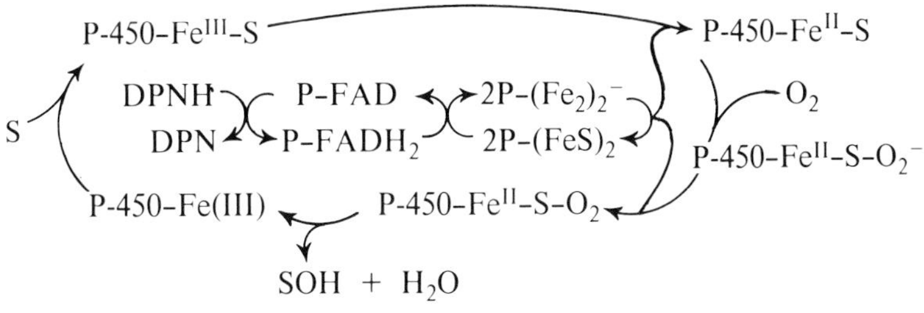

P-450–Fe^{III}–S → P-450–Fe^{II}–S

S; DPNH, DPN; P-FAD, P-FADH$_2$; 2P-(Fe$_2$)$_2^-$, 2P-(FeS)$_2$; O$_2$

P-450–Fe^{II}-S-O$_2^-$

P-450–Fe(III) ← P-450–Fe^{II}-S-O$_2$

SOH + H$_2$O

Studies on monooxygenases have shown that an oxygen atom is transferred to the substrate after the ternary complex has acquired an additional electron. This represents two-electron pre-reduction of the system composed of the resting enzyme, dioxygen, and the substrate. This prior addition of two electrons is general to both the metal-free and metal-containing monooxygenases. Referring to a more detailed analysis of the reactivity of normal triplet O_2 and its one-electron and two-electron reduction products than that presented earlier in this section, Hamilton has suggested boundary conditions for the reactive species that could be involved in such monooxygenases as cytochrome P-450.[703] For those cases where the products appear to be formed by reaction of the substrate with a highly electron-deficient oxygen species, the reactive oxygen species could be merely a metal ion complexed to O_2: this could overcome the spin-forbidden nature of direct substrate–triplet O_2 reaction. Alternatively, it could involve a chemically altered peroxide (free peroxide could act as a powerful nucleophile but not as a strong electrophile) or possibly the hydroxyl radical.

A particularly appealing suggestion has been based on the strong analogy between monooxygenase reactions and the reactions of carbenes.[703,777] Carbenes insert into the C—H bonds of alkyl compounds [Eq. (521)].

$$\text{—}\overset{|}{\underset{|}{C}}\text{—H} + CR_2 \longrightarrow \text{—}\overset{|}{\underset{|}{C}}\text{—}CR_2H \qquad (521)$$

The analogy found in the monooxygenase reaction given in Eq. (513) is obvious. Similarly, the enzyme-catalyzed formation of epoxides from olefins and dioxygen [Eq. (515)] is analogous to cyclopropane formation from carbenes and olefins [Eq. (522)].

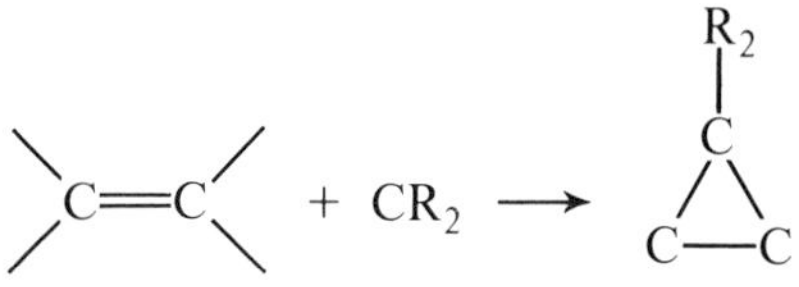

$$>C{=}C< + CR_2 \longrightarrow \text{(cyclopropane: } C(R_2)\text{ bonded to } C\text{—}C) \qquad (522)$$

The reaction with aromatic rings, when viewed in substantial detail, is still more significant. Carbenes combine with aromatic compounds to give alkylbenzenes and norcaradienes [Eq. (523)], which are in rapid equilibrium with cycloheptatrienes [Eq. (524)].

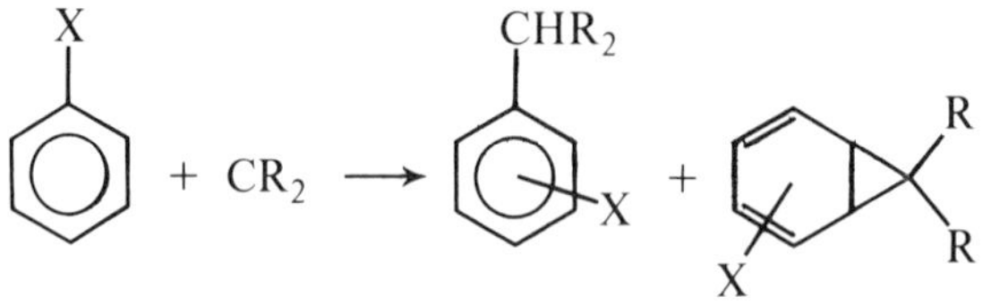

$$C_6H_5X + CR_2 \longrightarrow X\text{-}C_6H_4\text{-}CHR_2 + \text{X-norcaradiene}(R_2) \qquad (523)$$

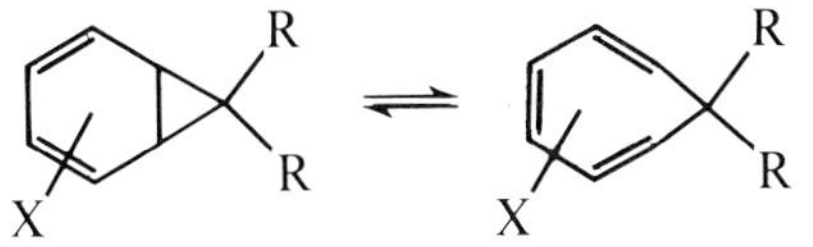

(524)

Aromatic compounds are converted to phenols and arene oxides [Eq. (525)], which are in equilibrium with oxepines [Eq. (526)] by a variety of oxygenases. Totally analogous reactions can be written for nitrenes. It has, therefore, been suggested that the monooxygenase reactions involve an "oxene" intermediate and that a singlet oxygen atom is transferred to the substrate in the enzymatic processes.

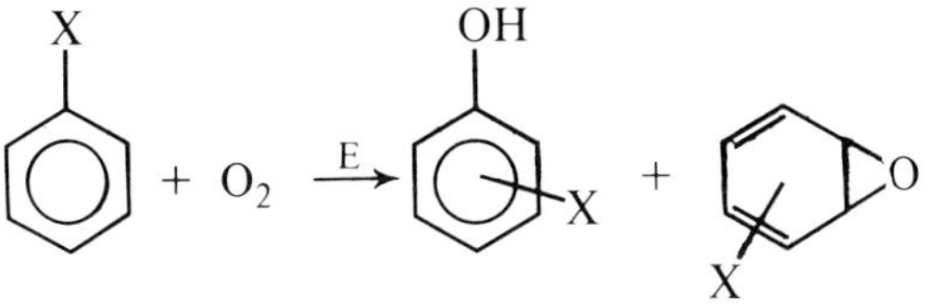

$$C_6H_5X + O_2 \xrightarrow{E} HOC_6H_4X + \text{arene oxide} \qquad (525)$$

$$\text{arene oxide} \rightleftharpoons \text{oxepine} \qquad (526)$$

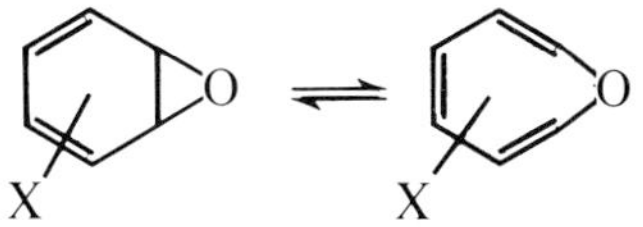

The formation of the arene oxide as an intermediate has been well established and is perhaps best demonstrated in the so-called NIH shift. This unique reaction was discovered in 1966 during a study of the action of the enzyme phenylalanine hydroxylase on phenylalanine-t_4.[778] It was found that >95% of the tritium migrated to the 3-position and was retained in the tyrosine product [Eq. (527)].

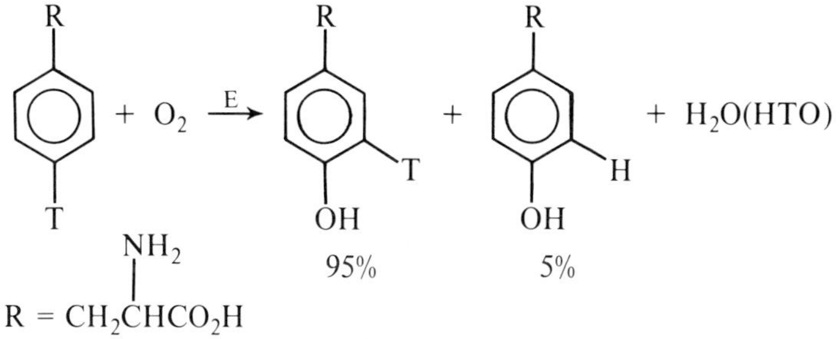

$$R{-}C_6H_4{-}T + O_2 \xrightarrow{E} \underset{95\%}{R{-}C_6H_3(T){-}OH} + \underset{5\%}{R{-}C_6H_3(H){-}OH} + H_2O(HTO) \qquad (527)$$

$R = CH_2CH(NH_2)CO_2H$

Intramolecular rearrangements of this class are so characteristic of monooxygenase enzymes[779,780,781] that they constitute criteria for identifying these species. Of more significance, the NIH shift provides important evidence for the nature of the active species in monooxygenase oxidations of substrates. The mechanism of the NIH shift requires formation of an arene oxide as an intermediate. The migration occurs to form the enol isomer of the phenol, and the position adopted by the OH in the phenol is determined by the relative stabilities of the alternate carbonium ions. This is shown in Eq. (528). The retention of high percentages of heavy hydrogen arises from an isotope effect.

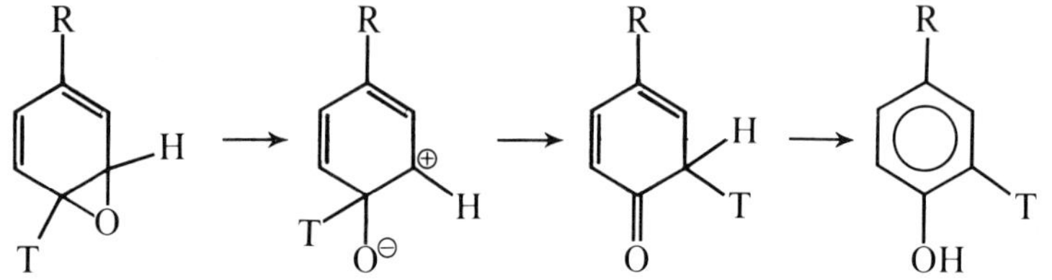

(528)

An additional useful criterion for monooxygenase behavior is better understood on the basis of arene oxide formation. Because of the nature of the process, such substrates as toluene give only the ortho- and para-hydroxylated products with no meta isomer being produced. This is associated with the cationoid intermediates which direct the course of arene oxide rearrangement to phenol.

A free carbene is not involved in many reactions that show the features of carbene reactions; such reactions are described as carbenoid reactions or carbene transfers. Similarly, a free oxygen atom is not presumed to be involved in the oxene reactions, and they are better described as "oxenoid" processes.[703] Perhaps the greatest lesson to be learned from the study of the reactions of the monooxygenases is that oxygen is capable of undergoing reactions by way of intermediates that are vastly different from those described in the preceding sections on autoxidations in this summary. The potential is great for synthetic applications of the distinctly different reaction mode for molecular oxygen. Recognition of these distinctive qualities in the monooxygenase reactions has added further impetus to investigations designed to mimic the behavior of oxygenase systems.

Three types of model systems have been studied (Scheme 2-32).[781] Type I consists of the apparent essential ingredients—molecular oxygen, an organic reducing agent, and a catalytic metal ion. This is illustrated by Udenfriend's system and related systems. Type II models are based on recognition of the possibility that O_2 might have suffered two-equivalent reduction prior to the

Scheme 2-32

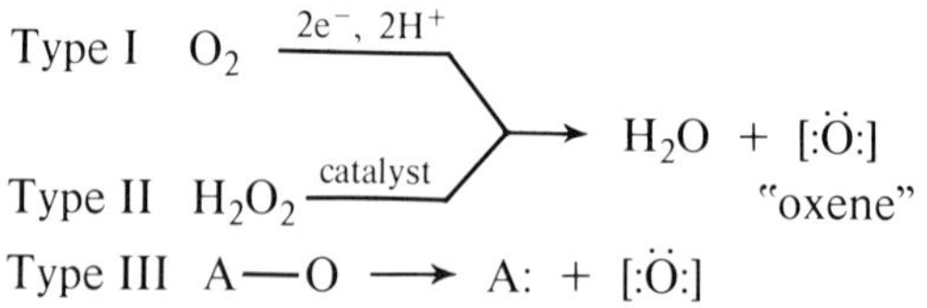

Type I O_2 $\xrightarrow{2e^-,\ 2H^+}$ H_2O + [:Ö:] "oxene"

Type II H_2O_2 $\xrightarrow{\text{catalyst}}$ H_2O + [:Ö:] "oxene"

Type III A—O ⟶ A: + [:Ö:]

generation of its active form and the concept that a peroxide, which is a two-equivalent reduction product of O_2, might be activated by a suitable chemical system. In this sense, such organic peroxides as peroxytrifluoroacetic acid are monooxygenase models. Type III models are oxygen atom donors. Chromylacetate is used as an example below.

It was discovered in 1954 that molecular oxygen will hydroxylate aromatic rings in the presence of Fe(II), EDTA, and ascorbic acid at neutral pH.[782] Subsequent studies have elaborated the system extensively.[703,781] The system will also hydroxylate saturated hydrocarbons, forming alcohols, and epoxidize olefins. The hydroxylation of aromatic rings can be promoted by a variety of other reducing agents, including pyrimidines, tetra-hydropteridines, reduced nicotinamide coenzymes, cysteine, and thiosalicylic acid. While those systems seem to emulate monooxygenase behavior to a certain extent, they suffer from very serious limitations. The yields of the hydroxylation reactions are low, and the occurrence of a variety of ill-defined reactions is indicated. Equally serious, the hydroxylation of arenes does *not* involve the NIH shift so that arene oxides are not implicated as intermediates. It follows that the activation of O_2 in these systems is not identical to that by monooxygenases. It is also generally agreed, however, that O_2 is not reduced primarily to H_2O_2 by the Udenfriend systems and that the hydroxylation is not mediated by $\cdot OH$ or $HO_2\cdot$. Perhaps it is most appropriate to assume that these synthetic hydroxylating systems partake of some of the qualities of the monooxygenases for reasons derived from a basic similarity in the mode of O_2 activation. Simultaneously, they may differ in the important ways that have been observed because of relationships between structure and mechanism. The critical intermediate may have both ionic and radical characteristics and may be able to react by either of the related classes of mechanisms. In such a situation, structural variations may alter the mechanism greatly. These complex systems have been discussed in detail by Norman and Lindsay Smith[698] and by Hamilton.[703]

It has been found that such peroxyacids as peroxytrifluoroacetic acid react with saturated optically active centers in alkanes to produce alcohols with retention of configuration.[703] This is the behavior expected for a carbenoid analog—an oxenoid mechanism is suggested. Such species also exhibit the NIH shift in the hydroxylation of aromatics, further emphasizing the analogy between their mechanisms and those occurring with the monooxygenases. Models used to explain these facts presume that peroxide is activated such that it is a powerful electrophilic reagent [Eq. (529)].[699,781] The promotion of the hydroxylation of aromatic rings by flavins[783] has been rationalized by the assumption that flavin hydroperoxides are formed as the reactive intermediates. Hamilton[699] suggested the alternate tautomeric structure (**125**) which represents the transferring oxygen atom as highly electrophilic. This bears a strong analogy to the carbonyl oxide formed by fluorene carbene, which promotes hydroxylation by O_2 accompanied by an NIH shift. The formation of the carbonyl oxide is shown in Eq. (530). Such species could well serve the oxenoid function.

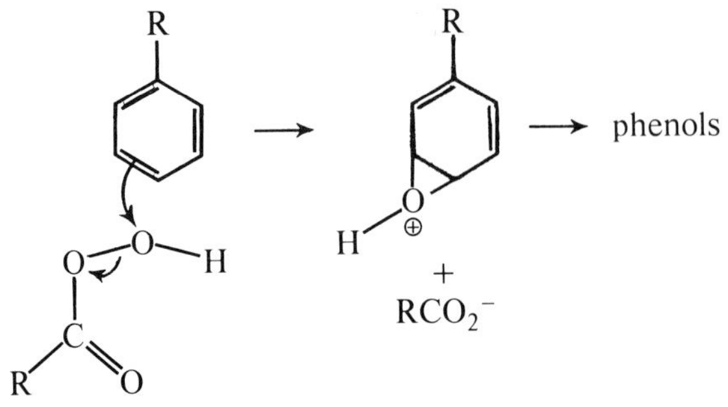

(529)

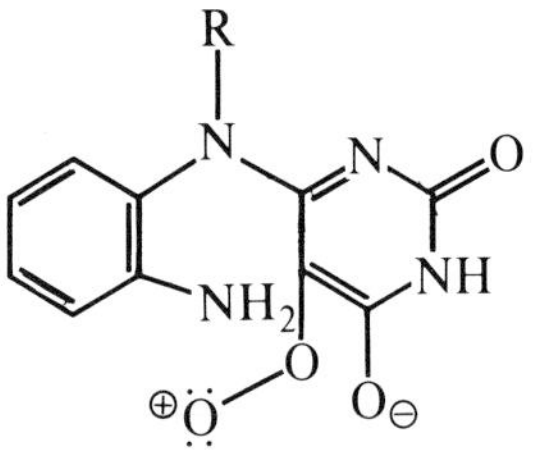

125

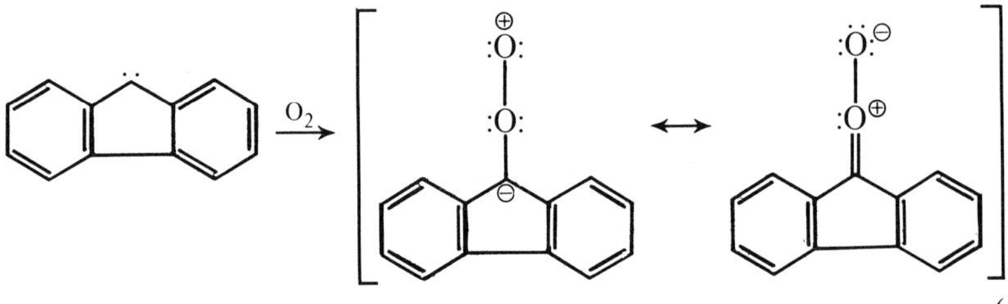

(530)

Chromyl derivatives CrO_2X_2 have long been known to hydroxylate hydrocarbons with at least partial retention of configuration,[660] and chromyl acetate epoxidizes olefins with geometric selectivity.[783] Sharpless and Flood[784] showed that chromyl acetate and chromyl chloride hydroxylate arenes and that the reaction produces the NIH shift. Thus, the structural unit Cr=O appears capable of introducing the oxygen atom into the substrate structure in a manner similar to that typical of monooxygenases. It has been shown that photoactivated heterocyclic amine oxides can produce the full range of oxenoid reactions as well.[703,781] Very recently, Graves and McClusky[785] have observed that *trans*-2-deuterocyclohexanol undergoes hydroxylation by ferrous perchlorate and *m*-chloroperoxybenzoic acid with retention of configuration, and that the added oxygen derives from the peroxy acid. They invoke an FeO unit as the oxygen transfer species.

The ultimate consequences of research in the area of ligand reactions and metal atom promotion of chemical reactions are so broad and far reaching as to inspire even a dull imagination. The potential exists for controlling via homogeneous catalysis many classes of reactions that now either are very difficult to perform or are responsive only to severe conditions. It is tempting to predict enzyme specificity and accelerations in reaction processes not yet conceived. It is appropriate for the coordination chemist to take a jealous view of nature and strive to do chemistry in equally subtle and intricate ways, using his ever expanding field as the touchstone.

REFERENCES

1. Taqui Khan, M. M., and A. E. Martell. *Homogeneous Catalysis by Metal Complexes, Vol. 1,* Academic Press, New York, 1974.
2. Busch, D. H. *Adv. Chem. Ser.* **37,** 1 (1963).
3. Jones, M. M., and W. A. Connor. *Ind. Eng. Chem.* **55,** (9) 15 (1963).
4. Nyholm, R. S. *Proc. Third Int. Congr. Catal., Vol. 1,* W. M. H. Sachtler, G. C. A. Schuit and P. Zwietering, eds. Interscience Publishers, New York, N.Y., 1965, p. 25.
5. Fernando, Q. *Advances in Inorganic Chemistry and Radiochemistry, Vol. 7,* H. J. Emeléus and A. G. Sharpe, eds. Academic Press, New York, 1965, p. 185.
6. Collman, J. P. *Transition Metal Chemistry, Vol. 2,* R. L. Carlin, ed. Marcel Dekker, Inc., New York, 1966, p. 2.
7. Basolo, F., and R. G. Pearson. *Mechanisms of Inorganic Reactions—A Study of Metal Complexes in Solution, 2nd Ed.,* John Wiley, New York, 1967, Chaps. 7–8.
8. Jones, M. M. *Ligand Reactivity and Catalysis,* Academic Press, New York, 1968.
9. Busch, D. H. *Science,* **171,** 241 (1971).
10. Martell, A. E. *Pure Appl. Chem.* **17,** (1) 129 (1968).
11. Orchin, M., and W. Rupilius. *Catal. Rev.* **6,** (1) 85 (1972).
12. Heck, R. F. *Adv. Chem. Ser.* **49,** 181 (1965).
13. Parshall, G. W., and J. J. Mrowca. *Adv. Organomet. Chem.* **7,** 157 (1968).
14. Wiberg, K. B., H. Maltz and M. Okano. *Inorg. Chem.* **7,** 830 (1968).
15. Dickerson, R. E. *Sci. Am.* **226,** (4) 58 (1972).
16. Olah, G. A. *Friedel-Crafts and Related Reactions.* Interscience Publishers, New York, 1965.
17. Chevrier, B., J. LeCarpentier and R. Weiss. *J. Am. Chem. Soc.* **94,** 5718 (1972), and references cited therein.
18. Kroll, H. *J. Am. Chem. Soc.* **74,** 2036 (1952).
19. Kwiatek, J., and J. K. Seyler. *Adv. Chem. Ser.* **70,** 207 (1968).
20. Halpern, J. *Adv. Chem. Ser.* **70,** 1 (1968).
21. Coffey, R. S. *Aspects of Homogeneous Catalysis, Vol. 1,* R. Ugo, Ed., and C. Manfredi, Editore, Milan, Italy, 1970.
22. Reference 8, p. 192.
23. Halpern, J. *Chem. Eng. News.* **44,** p. 68 (Oct. 31, 1966).
24. Schrauzer, G. N. *Acc. Chem. Res.* **1,** 97 (1968).

25. Vaska, L. *Science.* **140,** 809 (1963).
26. Vaska, L. *Acc. Chem. Res.* **1,** 335 (1968).
27. Collman, J. P. *Acc. Chem. Res.* **1,** 136 (1968).
28. Collman, J. P., and W. R. Roper. *Adv. Organomet. Chem.* **7,** 53 (1968).
29. Halpern, J. *Acc. Chem. Res.* **3,** 386 (1970).
30. Chatt, J., and J. M. Davidson. *J. Chem. Soc.* 843 (1965).
31. Cope, A. C., and R. W. Siekman. *J. Am. Chem. Soc.* **87,** 3272 (1965).
32. Parshall, G. W. *Acc. Chem. Res.* **3,** 139 (1970); **8,** 113 (1975).
33. Ugo, R. *Coord. Chem. Rev.* **3,** 319 (1968).
34. Skell, P. S., and J. E. Girard. *J. Am. Chem. Soc.* **94,** 5518 (1972).
35. Vaska, L., and J. W. DiLuzio. *J. Am. Chem. Soc.* **83,** 2784 (1961).
36. Vaska, L., and J. W. DiLuzio. *J. Am. Chem. Soc.* **83,** 679 (1962).
37. Chock, P. B., and J. Halpern. *J. Am. Chem. Soc.* **88,** 3511 (1966).
38. LaPlaca, S. J., and J. A. Ibers. *J. Am. Chem. Soc.* **87,** 2581 (1965).
39. McGinnety, J. A., and J. A. Ibers. *Chem. Commun.* 235 (1968).
40. McGinnety, J. A., R. J. Doedens and J. A. Ibers. *Inorg. Chem.* **6,** 2243 (1967).
41. Mason, R. *Nature.* **217,** 543 (1968).
42. Chatt, J., and L. A. Duncanson. *J. Chem. Soc.* 2939 (1953).
43. Dewar, M. J. S. *Bull. Soc. Chim., Fr.* **18,** C71 (1951).
44. Reference 11, p. 97.
45. Pearson, R. G. *Acc. Chem. Res.* **4,** 152 (1971).
46. Pearson, R. G., and W. R. Muir. *J. Am. Chem. Soc.* **92,** 5519 (1970).
47. Bradley, J. S., D. E. Connor, D. Dolphin, J. A. Labinger and J. A. Osborn. *J. Am. Chem. Soc.* **94,** 4043 (1972).
48. Birk, J. P., J. Halpern and A. L. Pickard. *J. Am. Chem. Soc.* **90,** 4491 (1968); J. P. Birk, J. Halpern and A. L. Pickard, *Inorg. Chem.* **7,** 2672 (1968).
49. Pearson, R. G., and J. Rajaram. *Inorg. Chem.* **13,** 246 (1974).
50. Lau, K. S., R. W. Fries and J. K. Stille. *J. Am. Chem. Soc.* **96,** 4983 (1974); P. K. Wong, K. S. Y. Lau and J. K. Stille. *J. Am. Chem. Soc.* **96,** 5956 (1974).
51. Kramer, A. V., J. A. Labinger, J. S. Bradley and J. A. Osborn. *J. Am. Chem. Soc.* **96,** 7145 (1974); A. V. Kramer and J. A. Osborn. *J. Am. Chem. Soc.* **96,** 7832 (1974).
52. Jones, M. M. *Adv. Chem. Ser.* **62,** 229 (1967).
53. Reference 8, p. 53.
54. Buckingham, D. A., and A. M. Sargeson. *Chelating Agents and Metal Chelates,* F. P. Dwyer and D. P. Mellor, eds., Academic Press, New York, 1964, Chap. 6.
55. Waters, W. A., and J. S. Littler. *Oxidation in Organic Chemistry, Vol. 5A,* K. B. Wiberg, ed., Academic Press, New York, 1965.
56. Towle, P. H., and R. H. Baldwin. *Hydrocarbon Process.* **43,** (11) 149 (1964); T. Yoshimura. *Chem. Eng.* **76,** 78 (May 5, 1969).
57. Ford, P., D. F. P. Rudd, R. Gaunder and H. Taube. *J. Am. Chem. Soc.* **90,** 1187 (1968).
58. Coates, G. E. *Organometallic Compounds,* 2nd Ed., Methuen, London, 1960, p 273.
59. Eichhorn, G. L., and J. C. Bailar, Jr. *J. Am. Chem. Soc.* **75,** 2905 (1953).
60. Eichhorn, G. L., and I. M. Trachtenberg. *J. Am. Chem. Soc.* **76,** 5183 (1954).
61. Eichhorn, G. L., and N. D. Marchand. *J. Am. Chem. Soc.* **78,** 2688 (1956).

62. Benfey, O. T. *J. Am. Chem. Soc.* **70,** 2163 (1948); S. Oae and C. A. Vander Werf. *J. Am. Chem. Soc.* **75,** 2724 (1953).
63. Martell, A. E. *Adv. Chem. Ser.* **37,** 161 (1963).
64. Gustafson, R. L., S. Chaberek and A. E. Martell. *J. Am. Chem. Soc.* **85,** 598 (1963).
65. Lynen, F., E. Reichert and L. Rueff. *Ann.* **574,** 1 (1951); F. Lynen. *Ann.,* **574,** 33 (1951).
66. Bender, M. L. *Chem. Rev.* **60,** 53 (1960).
67. Bender, M. L. *Adv. Chem. Ser.* **37,** 19 (1963).
68. Jensen, K. A. *Z. Anorg. Chem.* **252,** 227 (1943-44).
69. Lindoy, L. F. *Coord. Chem. Rev.* **4,** 41 (1969).
70. Lindoy, L. F., S. E. Livingstone and T. N. Lockyer. *Inorg. Chem.* **6,** 652 (1967).
71. Blomstrand, C. W. *J. Prakt. Chem.* **27,** 161 (1883); **38,** 523 (1888).
72. Bjerrum, J., G. Schwarzenbach and L. G. Sillén. *Stability Constants,* Chemical Society, London, 1957; L. G. Sillén and A. E. Martell. *Stability Constants,* Chemicai Society, London, 1964; R. M. Smith and A. E. Martell. *Critical Stability Constants,* Plenum Publishing Corp., New York, Vol. 1, 1974; Vol. 2, 1975; Vol. 4, 1976.
73. Koltun, W. L., M. Fried and F. R. N. Gurd. *J. Am. Chem. Soc.* **82,** 233 (1960).
74. Watt, G. W., and D. G. Upchurch. *J. Am. Chem. Soc.* **89,** 177 (1967).
75. Jicha, D. C., and D. H. Busch. *Inorg. Chem.* **1,** 872, 878 (1962).
76. Root, C. A., and D. H. Busch. *Inorg. Chem.* **7,** 789 (1968).
77. Pedersen, K. *Acta Chem. Scand.* **2,** 252, 385 (1948).
78. Schutzenberger, P. *Bull. Soc. Chim., Fr.* **14,** 97 (1870).
79. Schutzenberger, P., and M. Fontaine. *Bull. Soc. Chim. Fr.* **17,** 386, 482 (1872); **18,** 101 (1872).
80. Arbuzov, A. E., and V. M. Lovostrova. *Dokl. Akad. Nauk SSSR.* **84,** 503 (1952).
81. Geisenheimer, G. *Ann. Chim. Phys., s6.* **23,** 231 (1891); W. Strecker and M. Schurigin. *Chem. Ber.,* **42,** 1767 (1910).
82. Wilkinson, G. *J. Am. Chem. Soc.* **73,** 5501 (1951).
83. Zanella, A., and H. Taube. *J. Am. Chem. Soc.* **93,** 7166 (1971).
84. Gleu, K., and W. Breuel. *Z. Anorg. Allg. Chem.* **235,** 211 (1937-38).
85. Collman, J. P., M. Kubota and J. Hosking. *J. Am. Chem. Soc.* **89,** 4809 (1967).
86. Boos, R. N., J. E. Carr and T. B. Coun. *Science.* **117,** 603 (1953).
87. Müller, O., and G. Müller. *Biochem. Z.* **336,** 299 (1962).
88. Schrauzer, G. N. *Acc. Chem. Res.* **1,** 97 (1968).
89. Schrauzer, G. N. Adv. Chem. Ser. **100,** 1 (1971) and references cited therein.
90. Ochiai, E., K. M. Long, C. R. Sperati and D. H. Busch. *J. Am. Chem. Soc.* **91,** 3201 (1969).
91. Wolberg, A., and J. Manassen. *Inorg. Chem.* **9,** 2365 (1970).
92. Hush, N. S., and I. S. Woolsey. *J. Am. Chem. Soc.* **94,** 4107 (1972).
93. Tokel, N. E., V. Katovic, K. Farmery, L. B. Anderson and D. H. Busch. *J. Am. Chem. Soc.* **92,** 400 (1970).
94. Busch, D. H., K. Farmery, V. Goedken, V. Katovic, A. C. Melnyk, C. R. Sperati and N. Tokel. *Adv. Chem. Ser.* **100,** 44 (1971).

95. Takvoryan, N., K. Farmery, V. Katovic, F. V. Lovecchio, L. B. Anderson, E. S. Gore and D. H. Busch. *J. Am. Chem. Soc.,* **96,** 731 (1974).
96. Hawkinson, S. W., and E. B. Fleischer. *Inorg. Chem.* **8,** 2402 (1969).
97. Taube, H. *Electron Transfer Reactions of Complex Ions in Solution.* Academic Press, New York, 1970, Chap. 4.
98. French, J. E., and H. Taube. *J. Am. Chem. Soc.* **91,** 6951 (1969).
99. Hager, L. P., D. L. Doubek, R. M. Silverstein, J. H. Hargis and J. C. Martin. *J. Am. Chem. Soc.* **94,** 4364 (1972).
100. Cockle, S. A., H. A. O. Hill and R. J. P. Williams. *J. Am. Chem. Soc.* **94,** 275 (1972).
101. Babior, B. M. *Acc. Chem. Res.* **8,** 376 (1975).
102. Chaffee, E., and J. O. Edwards. *Progress in Inorganic Chemistry, Vol. 13,* J. O. Edwards, ed., Interscience, New York, 1970, p. 234. C. Walling, *Acc. Chem. Res.* **8,** 125 (1975).
103. Waters, W. A., and J. S. Littler. *Oxdiation in Organic Chemistry, Vol. 5A,* K. B. Wiberg, ed., Academic Press, New York, 1965, p. 192.
104. Buckingham, D. A., D. M. Foster and A. M. Sargeson. *J. Am. Chem. Soc.* **90,** 6032 (1968).
105. Ibid. **91,** 3451, 4102 (1969).
106. Ibid. **92,** 6151 (1970).
107. Andrade, C., and H. Taube. *J. Am. Chem. Soc.* **86,** 1328 (1964).
108. Bunton, C. A., and D. R. Llewellyn. *J. Chem. Soc.* 1692 (1953).
109. Jordan, R. B., and H. Taube. *J. Am. Chem. Soc.* **86,** 3890 (1964).
110. Bai, K. S., and D. L. Leussing. *J. Am. Chem. Soc.* **89,** 6126 (1967).
111. Hopgood, D., and D. L. Leussing. *J. Am. Chem. Soc.* **91,** 3740 (1969).
112. Leach, B. E., and D. L. Leussing. *J. Am. Chem. Soc.* **93,** 3377 (1971).
113. Wojcicki, A. *Adv. Organomet. Chem.* **11,** 87 (1973); **12,** 32 (1974).
114. Heck, R. F. *Acc. Chem. Res.* **2,** 10 (1969).
115. Coffield, T. H., J. Kozikowski and R. D. Closson. *J. Org. Chem.* **22,** 598 (1957).
116. Coffield, T. H., J. Koziowski and R. D. Closson. *Chem. Soc. Spec. Publ. n*13, 126 (1959).
117. Calderazzo, F., and F. A. Cotton. *Inorg. Chem.* **1,** 30 (1962).
118. Mawby, R. J., F. Basolo and R. G. Pearson. *J. Am. Chem. Soc.* **86,** 3994 (1964).
119. Mawby, R. J., F. Basolo and R. G. Pearson. *J. Am. Chem. Soc.* **86,** 5043 (1964).
120. Noack, K., and F. Calderazzo. *J. Organomet. Chem.* **10,** 101 (1967).
121. Hitch, R. R., S. K. Gondal and C. T. Sears. *Chem. Commun.* 777 (1971).
122. Noack, K., M. Ruch and F. Calderazzo. *Inorg. Chem.* **7,** 345 (1968).
123. Kraihanzel, C. S., and P. K. Maples. *J. Am. Chem. Soc.* **87,** 5267 (1965).
124. Whitesides, G. M., and D. J. Boschetto. *J. Am. Chem. Soc.* **91,** 4313 (1969).
125. Yamamoto, Y., H. Yamazaki and N. Hagihara. *J. Organomet. Chem.* **18,** 189 (1969).
126. Nesmeyanov, A. N., K. N. Anisimov, N. E. Kolabova and F. S. Denisov, *Izv. Akad. Nauk SSSR, Ser. Khim.* 133 (1968).
127. Su, S. R., J. A. Hanna and A. Wojcicki. *J. Organomet. Chem.* **21,** 21 (1970).
128. Giannotti, C., B. Septe and D. Benlian. *J. Organomet. Chem.* **39,** C5 (1972).

129. Evans, J. A., M. J. Hacker, R. D. W. Kemmitt, D. R. Russell and J. Stocks. *Chem. Commun.* 72 (1972).
130. Giannotti, C., C. Fontaine, B. Septe and D. Doue. *J. Organomet. Chem.* **39,** C74 (1972).
131. Kitching, W., and C. W. Fong. *Organomet. Chem. Rev., Sect. A.* **5,** 281 (1970).
132. Wojcicki, A. *Acc. Chem. Res.* **4,** 344 (1971).
133. Bibler, J. P., and A. Wojcicki. *J. Am. Chem. Soc.* **88,** 4862 (1966); **86,** 5051 (1964).
134. Churchill, M. R., and J. Wormald. *Inorg. Chem.* **10,** 572 (1971).
135. Farrone, F., L. Silvestra, E. Sergi and R. Pretropaolo. *J. Organomet. Chem.* **34,** C55 (1972).
136. Carey, N. A. D., and H. C. Clark. *Can. J. Chem.* **46,** 649 (1968).
137. Pollick, P. J., J. P. Bibler and A. Wojcicki. *J. Organomet. Chem.* **16,** 201 (1969).
138. Duboc, J., P. Mazerolles, M. Joly, W. Kitching, W. Fong and W. H. Atwell. *J. Organomet. Chem.* **34,** 17 (1972).
139. Strohmeier, W., and J. F. Guttenberger. *Chem. Ber.* **97,** 1871 (1964).
140. Jacobson, S. E., and A. Wojcicki. *J. Am. Chem. Soc.* **93,** 2535 (1971).
141. Graziani, M., and A. Wojcikci. *Inorg. Chim. Acta.* **4,** 347 (1970).
142. Kubota, M., and B. M. Loeffler. *Inorg. Chem.* **11,** 469 (1972).
143. Alexander, J. J., and A. Wojcicki. *J. Organomet. Chem.* **15,** P23 (1968).
144. Martell, A. E. *Adv. Chem. Ser.* **62,** 272 (1967) and references cited therein.
145. Kurtz, A. C. *Am. J. Med. Sci.* **194,** 875 (1937).
146. Kurtz, A. C. *J. Biol. Chem.* **122,** 477 (1937–38).
147. Kurtz, A. C. *J. Biol. Chem.* **140,** 705 (1941); **180,** 1253 (1949).
148. Turba, F., and K. H. Schuster. *Z. Physiol. Chem.* **283,** 27 (1948).
149. Brubaker, G. R., and D. H. Busch. *Inorg. Chem.* **5,** 2110 (1966).
150. Neuberger, A., and F. Sanger. *Biochem. J.* **37,** 515 (1943).
151. Wolf, D. E., and J. Valiant, R. L. Peck and K. Folkers. *J. Am. Chem. Soc.* **74,** 2002 (1952).
152. Hanby, W. E., S. G. Waley and J. Watson. *J. Chem. Soc.* 3239 (1950).
153. Krause, R. A., and S. D. Goldby. *Adv. Chem. Ser.* **37,** 143 (1963).
154. Drinkard, W. C., H. F. Bauer and J. C. Bailar, Jr. *J. Am. Chem. Soc.* **82,** 2992 (1960).
155. Collman, J. P., reference 6, p. 40.
156. Hipp, C. J., and D. H. Busch, unpublished results.
157. Alexander, M. D., and D. H. Busch. *J. Am. Chem. Soc.* **88,** 1130 (1966); *Chem. Eng. News.* **43,** 58 (March 8, 1965).
158. Hoppé, J. I., and J. E. Prue. *J. Chem. Soc.* 1775 (1957).
159. Johnson, G. L., and R. J. Angelici. *J. Am. Chem. Soc.* **93,** 1106 (1971).
160. Buckingham, D. A., L. G. Marzilli and A. M. Sargeson. *J. Am. Chem. Soc.* **89,** 4539 (1967).
161. Bunton, C. A. *Oxidation in Organic Chemistry,* Vol. 5A. K. B. Wiberg, ed., Academic Press, New York, 1965, p. 398.
162. Criegee, R. *Angew. Chem.* **70,** 173 (1958).
163. Roćek, J., and F. H. Westheimer. *J. Am. Chem. Soc.* **84,** 2241 (1962).
164. Gelles, E., and R. W. Hay. *J. Chem. Soc.* 3673 (1958).
165. Gelles, E., and A. Salama. *J. Chem. Soc.* 3683, 3689 (1958).

166. Hay, R. W. *J. Chem. Educ.* **42,** 413 (1965).
167. Steinberger, R., and F. H. Westheimer. *J. Am. Chem. Soc.* **73,** 429 (1951).
168. Steinberger, R., and F. H. Westheimer. *J. Am. Chem. Soc.* **71,** 4158 (1949).
169. Rund, J. V., and R. A. Plane. *J. Am. Chem. Soc.* **86,** 367 (1964).
170. Seltzer, S., G. A. Hamilton and F. H. Westheimer. *J. Am. Chem. Soc.* **81,** 4018 (1959).
171. Reference 8, p. 55.
172. Beatty, I. M., and D. I. Magrath. *J. Am. Chem. Soc.* **82,** 4983 (1960).
173. Breslow, R., R. Fairweather, and J. Keana. *J. Am. Chem. Soc.* **89,** 2135 (1967).
174. Longuet-Higgins, H. C., and L. E. Orgel. *J. Chem. Soc.* 1969 (1956).
175. Criegee, R., and G. Schröder. *Angew, Chem.* **71,** 70 (1959).
176. Hübel, W., and E. H. Braye. *J. Inorg. Nucl. Chem.* **10,** 250 (1959).
177. Nunez, L. J., and G. L. Eichhorn. *J. Am. Chem. Soc.* **84,** 901 (1962).
178. Busch, D. H., and J. C. Bailar, Jr. *J. Am. Chem. Soc.* **78,** 1137 (1956).
179. Taylor, L. T., F. L.Urbach and D. H. Busch. *J. Am. Chem. Soc.* **91,** 1072 (1969).
180. Katovic, V., L. T. Taylor and D. H. Busch. *J. Am. Chem. Soc.* **91,** 2122 (1969).
181. Black, D. St. C. *Chem. Commun.* 311 (1967).
182. Black, D. St. C., and R. C. Srivastava. *Aust. J. Chem.* **23,** 2067 (1970).
183. Black, D. St. C., and R. C. Srivastava. *Aust. J. Chem.* **22,** 1439 (1969).
184. Krumholtz, P. *J. Am. Chem. Soc.* **75,** 2163 (1953).
185. Goedken, V. L. *Chem. Commun.* 207 (1972).
186. Dabrowiak, J. C., F. V. Lovecchio, V. L. Goedken and D. H. Busch. *J. Am. Chem. Soc.* **94,** 5502 (1972).
187. Collman, J. P. *Adv. Chem. Ser.* **37,** 78 (1963).
188. Bayer, E. *Angew. Chem.* **76,** 76 (1964).
189. Ibid. **73,** 533 (1961).
190. Lovecchio, F. V., E. S. Gore and D. H. Busch. *J. Am. Chem. Soc.* **96,** 3109 (1974).
191. Richards, A. F., J. H. Ridd and M. L. Tobe. *Chem. Ind.* 1727 (1963).
192. Asperger, R. G., and C. F. Liu. *Inorg. Chem.* **6,** 796 (1967); R. C. Job and T. C. Bruice. *J. Am. Chem. Soc.* **96,** 809 (1974).
193. Glusker, J. P., H. L. Carrell, R. Job and T. C. Bruice. *J. Am. Chem. Soc.* **96,** 5741 (1974).
194. Murakami, M., and K. Takahashi. *Bull. Chem. Soc. Jpn.* **32,** 308 (1959).
195. Williams, D. H., and D. H. Busch. *J. Am. Chem. Soc.* **87,** 4644 (1965).
196. Gillard, R. D., P. R. Mitchell and N. C. Payne.. *Chem. Commun.* 1150 (1968).
197. Thompson, M. C., and D. H. Busch. *J. Am. Chem. Soc.* **84,** 1762 (1962); *Chem. Eng. News.* **40,** 57 (Sept. 17, 1962).
198. Thompson, M. C., and D. H. Busch. *J. Am. Chem. Soc.* **86,** 213 (1964).
199. Thompson, M. C., and D. H. Busch. *J. Am. Chem. Soc.* **86,** 3651 (1964).
200. Busch, D. H. *Rec. Chem. Prog.* **25,** 107 (1964); L. F. Lindoy and D. H. Busch. *Preparative Inorganic Reactions, Vol. 6,* W. L. Jolly, ed., John Wiley and Sons, New York, 1971, p. 1.
201. Root, C. A. Ph.D. thesis. The Ohio State University, 1965, and references cited therein.

202. Höhne, R., and D. H. Busch, unpublished results; D. H. Busch, 145th Meeting of the American Chemical Society, Los Angeles, Calif., 1963.
203. Jicha, D. C., and D. H. Busch. *Inorg. Chem.* **1,** 872, 878, 884 (1962).
204. Wei, C. H., and L. F. Dahl. *Inorg. Chem.* **9,** 1878 (1970).
205. Jadamus, H., Q. Fernando and H. Freiser. *Inorg. Chem.* **3,** 928 (1964).
206. Fernando, Q., and P. J. Wheatly. *Inorg. Chem.* **4,** 1726 (1965).
207. Busch, D. H., D. C. Jicha, M. C. Thompson, J. W. Wrathall and E. Blinn. *J. Am. Chem. Soc.* **86,** 3642 (1964).
208. Brubaker, G. R., and D. H. Busch. *Inorg. Chem.* **5,** 2114 (1966).
209. Barefield, E. K., S. M. Nelson and D. H. Busch. *Quart. Revs. Chem. Soc.* **22,** 457 (1968).
210. Busch, D. H. *Adv. Chem. Ser.* **62,** 616 (1967).
211. Busch, D. H. *Helv. Chim. Acta.* Fasciculus extraordinarius Alfred Werner, 174 (1967).
212. Egen, N. B., and R. A. Krause. *J. Inorg. Nucl. Chem.* **31,** 127 (1969).
213. Thornton, P., M. S. Elder, G. M. Prinz and D. H. Busch. *Inorg. Chem.* **7,** 2426 (1968).
214. Schrauzer, G. N., R. K. Y. Ho and R. P. Murillo. *J. Am. Chem. Soc.* **92,** 3508 (1970).
215. Blinn, E. L., and D. H. Busch. *Inorg. Chem.* **7,** 820 (1968).
216. Wilkins, R. G. *Acc. Chem. Res.* **3,** 408 (1970).
217. Umland, F., and D. Thiereg. *Angew. Chem.* **74,** 388 (1962).
218. Schrauzer, G. N. *Chem. Ber.* **95,** 1438 (1962).
219. Johnson, A. W., and I. T. Kay. *Proc. Chem. Soc. London,* 168 (1961).
220. Johnson, A. W., I. T. Kay and R. Rodrigo. *J. Chem. Soc.* 2336 (1963).
221. Eschenmoser, A. *Pure Appl. Chem.* **7,** 297 (1963); E. Bertele, H. Boos, J. D. Dunitz, F. Elsinger, A. Eschenmoser, I. Felner, H. P. Gribi, H. Gschwend, E. F. Meyer, M. Pesaro and R. Scheffold. *Angew. Chem. Int. Ed. Engl.* **3,** 490 (1964).
222. Yamada, Y., D. Miljkovic, P. Wehrli, B. Golding, P. Loliger, R. Keese, K. Muller and A. Eschenmoser. *Angew. Chem. Int. Ed. Engl.* **8,** 343 (1969).
223. Dolphin, D., R. L. N. Harris, A. W. Johnson, and I. T. Kay. *Proc. Chem. Soc., London.* 359 (1964).
224. Dolphin, D., R. L. N. Harris, J. L. Huppatz, A. W. Johnson and I. T. Kay. *J. Chem. Soc.* (*C*). 30 (1966).
225. Clarke, D. A., R. Grigg, R. L. N. Harris, A. W. Johnson, I. T. Kay and K. W. Skelton. *J. Chem. Soc.* (*C*). 1648 (1967).
226. Harris, R. L. N., A. W. Johnson and I. T. Kay. *Quart. Rev. Chem. Soc.* **20,** 211 (1966).
227. Falk, J. E. *Porphyrins and Metalloporphyrins,* Elsevier, Amsterdam, 1964.
228. Lindoy, L. F., and D. H. Busch. *Preparative Inorganic Reactions, Vol. 6,* W. L. Jolly, ed., John Wiley and Sons, Inc., New York, 1971, p. 1.
229. Johnson, A. W., and I. T. Kay. *J. Chem. Soc.* 2418 (1961).
230. Hurley, T. J., M. A. Robinson and S. I. Trotz. *Inorg. Chem.* **6,** 389 (1967).
231. Bayer, E., and G. Schenk. *Chem. Ber.* **93,** 1184 (1960).
232. Bayer, E. *Angew. Chem.* **73,** 659 (1961).
233. Jadamus, H., Q. Fernando and H. Freiser. *J. Am. Chem. Soc.* **86,** 3056 (1964).
234. Hesse, G., and G. Ludwig. *Ann.* **632,** 158 (1960).
235. Schubert, M. P. *J. Biol. Chem.* **114,** 341 (1936).

236. Taylor, L. T., S. C. Vergez and D. H. Busch. *J. Am. Chem. Soc.* **88,** 3170 (1966).
237. McGeachin, S. G., *Can. J. Chem.* **44,** 2323 (1966).
238. Melson, G. A., and D. H. Busch. *Proc. Chem. Soc., London.* 223 (1963).
239. Melson, G. A., and D. H. Busch. *J. Am. Chem. Soc.* **86,** 4834 (1964).
240. Ibid. **87,** 1706 (1965).
241. Cummings, S. C., and D. H. Busch. *J. Am. Chem. Soc.* **92,** 1924 (1970).
242. Cummings, S. C., and D. H. Busch. *Inorg. Chem.* **10,** 1220 (1971).
243. Curtis, N. F. *Coord. Chem. Rev.* **3,** 3 (1968).
244. Curtis, N. F., and R. W. Hay. *Chem. Commun.* 524 (1966).
245. Curtis, N. F. *J. Chem. Soc.* 4409 (1960).
246. Taylor, L. T., N. J. Rose and D. H. Busch. *Inorg. Chem.* **7,** 785 (1968).
247. Baldwin, D. A., and N. J. Rose. 157th Meeting of the American Chemical Society, Minneapolis, 1969.
248. Jäger, E. G. *Z. Chem.* **4,** 437 (1964).
249. Ibid. **8,** 30, 392 (1968).
250. Jäger, E. G. *Z. Anorg. Allg. Chem.* **364,** 178 (1969).
251. Curry, J. D., and D. H. Busch. *J. Am. Chem. Soc.* **86,** 592 (1964).
252. Karn, J. L., and D. H. Busch. *Nature.* **211,** 160 (1966).
253. Karn, J. L., and D. H. Busch. *Inorg. Chem.* **8,** 1149 (1969).
254. Rich, R. L., and G. L. Stucky. *Inorg. Nucl. Chem. Lett.* **1,** 61 (1965).
255. Fleischer, E. B., and S. W. Hawkinson. *Inorg. Chem.* **7,** 2312 (1968).
256. Green, M., and P. A. Tasker. *Chem. Commun.* 518 (1968).
257. Lindoy, L. F., and D. H. Busch. *Chem. Commun.* 1589 (1968).
258. Lindoy, L. F., and D. H. Busch. *J. Am. Chem. Soc.* **91,** 4690 (1969).
259. Lindoy, L. F., and D. H. Busch. *Inorg. Nucl. Chem. Lett.* **5,** 525 (1969).
260. Black, D. St. C., and A. J. Hartshorn. *Chem. Commun.* 706 (1972).
261. Uhlemann, E., and M. Plath. *Z. Chem.* **9,** 234 (1969).
262. Nelson, S. M., P. Bryan and D. H. Busch. *Chem. Commun.* 641 (1966).
263. Nelson, S. M., and D. H. Busch. *Inorg. Chem.* **8,** 1859 (1969).
264. Fleischer, E., and S. Hawkinson. *J. Am. Chem. Soc.* **89,** 720 (1967).
265. Heuvelen, A. V., M. D. Lundeen, H. G. Hamilton and M. D. Alexander. *J. Chem. Phys.* **50,** 489 (1969).
266. Lindoy, L. F., and D. H. Busch. *Inorg. Chem.* **13,** 2494 (1974).
267. Fleischer, E. B., and P. A. Tasker. *Inorg. Nucl. Chem. Lett.* **6,** 349 (1970); P. A. Tasker and E. B. Fleischer. *J. Am. Chem. Soc.* **92,** 7072 (1970).
268. Bosnich, B., C. K. Poon and M. L. Tobe. *Inorg. Chem.* **4,** 1102 (1965).
269. Poon, C. K., and M. L. Tobe. *J. Chem. Soc.* (*A*), 2069 (1967); 1549 (1968).
270. Stetter, H., and K. H. Mayer. *Chem. Ber.* **94,** 1410 (1961).
271. Barefield, E. K. *Inorg. Chem.* **11,** 2273 (1972).
272. Katovic, V., L. T. Taylor and D. H. Busch. *Inorg. Chem.* **10,** 458 (1971).
273. Boston, D. R., and N. J. Rose. *J. Am. Chem. Soc.* **90,** 6859 (1968).
274. Parks, J. E., B. E. Wagner and R. H. Holm. *J. Am. Chem. Soc.* **92,** 3500 (1970); *Inorg. Chem.* **10,** 2472 (1971).
275. Orchin, M. *Adv. Catal.* **16,** 1 (1966).
276. Bird, C. W. *Transition Metal Intermediates in Organic Syntheses,* Logos Press, London, 1967.
277. Cramer, R., and R. V. Lindsey, Jr. *J. Am. Chem. Soc.* **88,** 3534 (1966).

278. Cramer, R. *J. Am. Chem. Soc.* **88,** 2272 (1966).
279. Hudson, B., D. E. Webster and P. B. Wells. *J. Chem. Soc., Dalton Trans.* 1204 (1972).
280. Johnson, M. *J. Chem. Soc.* 4859 (1963).
281. Goldfarb, I. J., and M. Orchin. *Adv. Catal.* **9,** 609 (1957).
282. Heck, R. F., and D. S. Breslow. *J. Am. Chem. Soc.* **83,** 4023 (1961).
283. Taylor, P., and M. Orchin. *J. Am. Chem. Soc.* **93,** 6504 (1971).
284. Taylor, P., and M. Orchin. *J. Organomet. Chem.* **26,** 389 (1961).
285. Karapinka, G. L., and M. Orchin. *J. Org. Chem.* **26,** 4187 (1961).
286. Takegami, Y., C. Yokohawa, Y. Watanabe, H. Masada and Y. Okuda. *Bull. Chem. Soc., Jpn.* **37,** 1190 (1964).
287. Taqui Khan, M. M., and A. E. Martell, reference 1.
288. Manuel, T. A. *J. Org. Chem.* **27,** 3941 (1962).
289. Manuel, T. A. *Trans N. Y. Acad. Sci.* **26,** 442 (1964).
290. Emerson, G. F., and R. Pettit. *J. Am. Chem. Soc.* **84,** 4591 (1962).
291. Hendrix, W. T., F. G. Cowherd and J. L. von Rosenberg. *Chem. Commun.* 97 (1968).
292. Cowherd, F. G., and J. L. von Rosenberg. *J. Am. Chem. Soc.* **91,** 2157 (1969).
293. Murdoch, H. D., and E. Weiss. *Helv. Chim. Acta.* **45,** 1927 (1962).
294. Pettit, R., G. Emerson and J. Mahler. *J. Chem. Educ.* **40,** 175 (1963).
295. Tayim, H. A., A. Bouldoukian and M. Kharboush. *Inorg. Nucl. Chem. Lett.* **8,** 231 (1972).
296. Arnet, J. E., and R. Pettit. *J. Am. Chem. Soc.* **83,** 2954 (1961).
297. Emerson, G. F., J. E. Mahler, R. Kochhar and R. Pettit. *J. Org. Chem.* **29,** 3620 (1964).
298. Pettit, R., and G. F. Emerson. *Adv. Organomet. Chem.* **1,** 1 (1964).
299. Pettit, R. *Ann. N. Y. Acad. Sci.* **125,** 89 (1965).
300. King, R. B., T. A. Manuel and F. G. A. Stone. *J. Inorg. Nucl. Chem.* **16,** 233 (1961).
301. Alper, H., P. G. LePort and S. Wolfe. *J. Am. Chem. Soc.* **91,** 7553 (1969).
302. Nelson, S. M., and M. Sloan. *Chem Commun.* 745 (1972).
303. Harrod, J. F., and A. J. Chalk. *J. Am. Chem. Soc.* **88,** 3491 (1966).
304. Schunn, R. A. *Inorg. Chem.* **9,** 2567 (1970).
305. Hirabayashi, K., S. Saito, and I. Yasumori. *J. Chem. Soc., Faraday 1,* 978 (1972).
306. Calvin, M. *Trans. Faraday Soc.* **34,** 1181 (1938).
307. Andretta, A., F. Conti, and G. F. Ferrari. *Aspects of Homogeneous Catalysis, Vol. 1*, R. Ugo, ed., Manfredi, Milan, 1970, p. 203.
308. Kwiatek, J. *Transition Metals in Homogeneous Catalysis,* G. N. Schrauzer, ed., Marcel Dekker, New York, 1971, pp. 13–51.
309. James, B. R. *Homogeneous Hydrogenation,* Wiley, New York, 1973.
310. Halpern, J. *Annu. Rev. Phys. Chem.* **16,** 103 (1965).
311. Billig, E., C. B. Strow and R. L. Pruett. *Chem. Commun.* 1307 (1968).
312. Cramer, R. *Acc. Chem. Res.* **1,** 186 (1968).
313. Kwiatek, J., *Catal. Rev.* **1,** 37 (1967).
314. Kwiatek, J., I. L. Mador and J. K. Seyler. *Adv. Chem. Ser.* **37,** 201 (1963).
315. Jackman, L. M., J. A. Hamilton and J. M. Lawlor. *J. Am. Chem. Soc.* **90,** 1914 (1968).

316. Halpern, J., and L. Y. Wong. *J. Am. Chem. Soc.* **90,** 6665 (1968).
317. Simándi, L., and F. Nagy. *Acta Chim. Acad. Sci. Hung.* **46,** 137 (1965).
318. Strohmeier, W., and N. Iglauer. *Z. Phys. Chem.* (*Frankfurt am Main*). **51,** 50 (1966).
319. Simándi, L. I., F. Nagy and E. Budó. *Acta Chim. Acad. Sci. Hung.* **58,** 39 (1968).
320. Burnett, M. G., P. J. Connolly and C. Kemball. *J. Chem. Soc.* (*A*). 991 (1968).
321. Piringer, O., and A. Farcas. *Nature.* **206,** 1040 (1965); A. Farcas, U. Luca, N. Morar and O. Piringer. *Z. Phys. Chem.,* (*Frankfurt am Main*). **58,** 87 (1968); R. Ripan, A. Farcas and O. Piringer. *Z. Anorg. Allg. Chem.* **346,** 211 (1966).
322. Wymore, C. E. *Chem. Eng. News.* **46,** 52 (April 8, 1968).
323. Ohgo, Y., S. Takeuchi and J. Yoshimura. *Bull. Chem. Soc., Jpn.* **43,** 505 (1970).
324. Ohgo, Y., K. Kobayashi, S. Takeuchi and J. Yoshimura. *Bull. Chem. Soc., Jpn.* **45,** 933 (1972).
325. Halpern, J. *Adv. Catal.* **11,** 301 (1959).
326. Chalk, A. J., and J. F. Harrod. *Adv. Organomet. Chem.* **6,** 119 (1968).
327. Feder, H. M., and J. Halpern. *J. Am. Chem. Soc.* **97,** 7186 (1975).
328. Young, J. F., J. A. Osborn, F. H. Jardine and G. Wilkinson. *Chem. Commun.* 131 (1965).
329. Jardine, F. H., J. A. Osborn, G. Wilkinson and J. F. Young. *Chem. Ind.* (London). 560 (1965).
330. Osborn, J. A., F. H. Jardine, J. F. Young and G. Wilkinson. *J. Chem. Soc.* (*A*). 1711 (1966).
331. Jardine, F. H., J. A. Osborn and G. Wilkinson. *J. Chem. Soc.* (*A*). 1574 (1967).
332. Montelatici, S., A. van der Ent, J. A. Osborn and G. Wilkinson. *J. Chem. Soc.* (*A*). 1054 (1968).
333. Tolman, C. A., P. Z. Meakin, D. L. Lindner and J. P. Jesson. *J. Am. Chem. Soc.* **96,** 2762 (1974).
334. Brown, T. H., and P. J. Green. *J. Am. Chem. Soc.* **92,** 2359 (1970).
335. Lehman, D. D., D. F. Shriver and I. Wharf. *Chem. Commun.* 1486 (1970).
336. Eaton, D. R., and S. R. Stuart. *J. Am. Chem. Soc.* **90,** 4170 (1968).
337. Arai, H., and J. Halpern. *Chem. Commun.* 1571 (1971).
338. Sacco, A., R. Ugo, and A. Moles. *J. Chem. Soc.* (*A*). 1670 (1966).
339. Siegel, S., and D. Ohrt. *Inorg. Nucl. Chem. Lett.* **8,** 15 (1972).
340. Strohmeier, W., and R. Endres. *Z. Naturforsch.* **B26,** 362 (1971).
341. Odell, A. L., J. B. Richardson and W. R. Roper. *J. Catal.* **8,** 393 (1967).
342. Hussey, A. S., and Y. Takeuchi. *J. Am. Chem. Soc.* **91,** 672 (1969).
343. Bond, G. C., and R. A. Hillyard. *Discuss. Faraday Soc.* **46,** 20 (1968).
344. Biellmann, J. F., and M. J. Jung. *J. Am. Chem. Soc.* **90,** 1673 (1968).
345. Lawson, D. N., J. A. Osborn and G. Wilkinson. *J. Chem. Soc.* (*A*). 1733 (1966).
346. Baird, M. C., J. T. Mague, J. A. Osborn and G. Wilkinson. *J. Chem. Soc.* (*A*). 1347 (1967).
347. Augustine, R. L., and J. F. van Peppen. *Chem. Commun.* 495, 497 (1970).
348. Ibid. 571 (1970).
349. Bennett, M. A., and P. A. Longstaff. *Chem. Ind.* (London). 846 (1965).

350. Harmon, R. E., J. L. Parson, D. W. Cooke, S. K. Gupta and J. Schoolenberg, *J. Org. Chem.* **34,** 3684 (1969).
351. Birch, A. J., and K. A. M. Walker. *J. Chem. Soc.* (*C*). 1894 (1966).
352. Morandi, J. R., and H. B. Jensen. *J. Org. Chem.* **34,** 1889 (1969).
353. Horner, L., H. Siegel and H. Bûthe. *Angew. Chem. Int. Ed.* **7,** 942 (1968).
354. Knowles, W. S., and M. J. Sabacky. *Chem. Commun.* 1445 (1968).
355. Morrison, J. D., R. E. Burnett, A. M. Aguiar, C. J. Morrow, and C. Phillips. *J. Am. Chem. Soc.* **93,** 1301 (1971).
356. Grubber, R. H., and L. C. Kroll. *J. Am. Chem. Soc.* **93,** 3062 (1971).
357. Bath, S. S., and L. Vaska. *J. Am. Chem. Soc.* **85,** 3500 (1963).
358. Evans, D., G. Yagupsky and G. Wilkinson. *J. Chem. Soc.* (*A*). 2660 (1968).
359. LaPlaca, S. J., and J. A. Ibers. *Acta Crystallogr.* **18,** 511 (1965).
360. O'Connor, C., and G. Wilkinson. *J. Chem. Soc.* (*A*). 2665 (1968).
361. Evans, D., J. A. Osborn and G. Wilkinson. *J. Chem. Soc.* (*A*). 3133 (1968).
362. Yagupsky, M., C. K. Brown, G. Yagupsky and G. Wilkinson. *J. Chem. Soc.* (*A*). 937 (1970).
363. Dewhirst, K. C., W. Keim and C. A. Reilly. *Inorg. Chem.* **7,** 546 (1968).
364. Shapley, J. R., R. R. Schrock and J. A. Osborn. *J. Am. Chem. Soc.* **91,** 2816 (1969).
365. Schrock, R. R., and J. A. Osborn. *J. Am. Chem. Soc.* **93,** 2397 (1971).
366. Ibid. **98,** 2134 (1976).
367. Ibid., 2143.
368. Schrock, R. R., and J. A. Osborn. *Chem. Commun.* 567 (1970).
369. Knowles, W. S., M. J. Sabacky, B. D. Vineyard, and D. J. Weinkauff. *J. Am. Chem. Soc.* **97,** 2567 (1975).
370. Eberhardt, G. G., and L. Vaska. *J. Catal.* **8,** 183 (1967).
371. James, B. R., and N. A. Memon. *Can. J. Chem.* **46,** 217 (1968).
372. James, B. R., and G. L. Rempel. *Discuss. Faraday Soc.* **46,** 48 (1968).
373. Halpern, J., J. F. Harrod and B. R. James. *J. Am. Chem. Soc.* **88,** 5150 (1966).
374. Muetterties, E. L., and F. J. Hirsekorn. *J. Am. Chem. Soc.* **96,** 4063 (1974).
375. Hirsekorn, F. J., M. C. Rakowski and E. L. Muetterties. *J. Am. Chem. Soc.* **97,** 237 (1975).
376. Falbe, J. *Carbon Monoxide in Organic Synthesis,* Springer-Verlag, New York, 1970.
377. Ungvary, F., and L. Marko. *J. Organomet. Chem.* **20,** 205 (1969).
378. Piacenti, F., S. Pucci, M. Bianchi, R. Lazzaroni and P. Pino. *J. Am. Chem. Soc.* **90,** 6847 (1968).
379. Casey, C. P., and C. R. Cyr. *J. Am. Chem. Soc.* **93,** 1280 (1971).
380. Nagy-Magor, Z., G. Bor and L. Markó. *J. Organomet. Chem.* **14,** 205 (1968).
381. Heck, R. F., and D. S. Breslow. *2nd Cong. Int. Catal., Paris, 1960.* Paper 27; Edition Techn., Paris, 1961.
382. Piacenti, F., P. Pino, R. Lazzaroni and M. Bianchi. *J. Chem. Soc.* (*C*). 488 (1966).
383. Pregaglia, G. F., A. Andresta, G. Gregorio, G. Montrasi and G. F. Ferrari. *Chem. Ind.* (*Milan*). **54,** 405 (1972).
384. Rupilius, W., and M. Orchin. *J. Org. Chem.* **37,** 936 (1972).
385. Wakamatsu, H., J. Uda and N. Yamakami. *Chem. Commun.* 1540 (1971).

386. Woodward, R. B., and R. Hoffmann. *The Conservation of Orbital Symmetry,* Academic Press, 1970.
387. Paquette, L. A. *Acc. Chem. Res.* **4,** 280 (1971).
388. Gassman, P. G., and T. J. Atkins. *J. Am. Chem. Soc.* **93,** 4597 (1971).
389. Ibid. 1042.
390. Sakai, M., H. Yamaguchi, H. H. Westberg and S. Masamune. *J. Am. Chem. Soc.* **93,** 1043 (1971).
391. Paquette, L. A., G. R. Allen, Jr. and R. P. Henzel. *J. Am. Chem. Soc.* **92,** 7002 (1970).
392. Sakai, M., and S. Masamune. *J. Am. Chem. Soc.* **93,** 4610 (1971).
393. Sakai, M., H. Yamaguchi and S. Masamune. *Chem. Commun.* 486 (1971).
394. Gassman, P. G., T. J. Atkins and F. J. Williams. *J. Am. Chem. Soc.* **93,** 1812 (1971).
395. Paquette, L. A., R. P. Henzel and S. E. Wilson. *J. Am. Chem. Soc.* **93,** 2335 (1971).
396. Paquette, L. A., and G. Zon. *J. Am. Chem. Soc.* **96,** 203 (1974).
397. Zon, G., and L. A. Paquette. *J. Am. Chem. Soc.* **96,** 215 (1974).
398. Paquette, L. A. and G. Zon, *J. Am. Chem. Soc.* **96,** 224 (1974).
399. Sakai, M., H. H. Westberg, H. Yamaguchi and S. Masamune. *J. Am. Chem. Soc.* **93,** 4611 (1971).
400. Gassman, P. G., and F. J. Williams. *J. Am. Chem. Soc.* **92,** 7631 (1970).
401. Gassman, P. G., and F. J. Williams. *Chem. Commun.* 80 (1972).
402. Gassman, P. G., and F. J. Williams. *Tetrahedron Lett.* 1409 (1971).
403. Gassman, P. G., G. R. Meyer and F. J. Williams. *Chem. Commun.* 842 (1971).
404. Kirmse, W., and K. Horn. *Chem. Ber.* **100,** 2698 (1967); W. R. Moser. *J. Am. Chem. Soc.* **91,** 1135 (1969).
405. Gassman, P. G., and T. Nakai. *J. Am. Chem. Soc.* **93,** 5897 (1971).
406. Ibid. **94,** 2877 (1972).
407. Masamune, S., H. Cuts and M. G. Hogben. *Tetrahedron Lett.* 1017 (1966).
408. Dauben, W. G., and D. L. Whalen. *Tetrahedron Lett.* 3743 (1966).
409. Furstoss, R., and J. M. Lehn. *Bull. Soc. Chim. Fr.* 2497 (1966).
410. Dauben, W. G., C. H. Schallhorn and D. L. Whalen. *J. Am. Chem. Soc.* **93,** 1446 (1971).
411. Paquette, L. A., and J. C. Stowell, *J. Am. Chem. Soc.* **92,** 2584 (1970).
412. Ibid. **93,** 2459 (1971).
413. Dauben, W. G., M. G. Buzzolini, C. H. Schallhorn and D. W. Whalen and K. J. Palmer. *Tetrahedron Lett.* 787 (1970).
414. Cassar, L., P. E. Eaton and J. Halpern. *J. Am. Chem. Soc.* **92,** 6366 (1970).
415. Westberg, H. H., and H. Ona. *Chem. Commun.* 248 (1971).
416. Byrd, J. B., L. Cassar, P. E. Eaton and J. Halpern. *Chem. Commun.* 40 (1971).
417. Cassar, L., P. E. Eaton and J. Halpern. *J. Am. Chem. Soc.* **92,** 3515 (1970).
418. Pettit, R., H. Sugahara, J. Wristers and W. Merk. *Discuss. Faraday Soc.* **47,** 71 (1969).
419. Dauben, W. G., and A. J. Kielbania, Jr. *J. Am. Chem. Soc.* **93,** 7345 (1971).
420. Hogeveen, H., and H. C. Volger. *Chem. Commun.* 1133 (1967).
421. Kaiser, K. L., R. F. Childs and P. M. Maitlis. *J. Am. Chem. Soc.* **93,** 1270 (1971).
422. Menon, B. C., and R. E. Pincock. *Can. J. Chem.* **47,** 3327 (1969).

423. Hogeveen, H., and H. C. Volger. *J. Am. Chem. Soc.* **89,** 2486 (1967).
424. Cassar, L., and J. Halpern. *Chem. Commun.* 1082 (1970).
425. Bailey, N. A., R. D. Gillard, M. Keeton, R. Mason and D. R. Russell. *Chem. Commun.* 396 (1966).
426. Roundhill, D. M., D. N. Lawson and G. Wilkinson. *J. Chem. Soc.* (*A*). 845 (1968).
427. Powell, K. G., and F. J. McQuillin. *Chem. Commun.* 931 (1971).
428. Wolberg, A., and J. Manassen. *J. Am. Chem. Soc.* **92,** 2987 (1970).
429. Merk, W., and R. Pettit. *J. Am. Chem. Soc.* **89,** 4788 (1967).
430. Wristers, J., L. Brener and R. Pettit. *J. Am. Chem. Soc.* **92,** 7499 (1970).
431. Calderon, N., *Acc. Chem. Res.* **5,** 127 (1972); R. J. Haines and G. J. Leigh. *Chem. Soc. Rev.,* **4,** 155 (1975).
432. Eleuterio, H. S. U.S. patent 3,074,918 (1963).
433. Banks, R. L., and G. C. Bailey. *Ind. Eng. Chem. Prod. Res. Dev.* **3,** 170 (1964).
434. Natta, G., G. Dall'Asta, G. Mazzanti and G. Motroni. *Makromol. Chem.* **69,** 163 (1963).
435. Natta, G., G. Dall'Asta, I. W. Bassi and G. Carella. *Makromol. Chem.* **91,** 87 (1966).
436. Natta, G., G. Dall'Asta and G. Mazzanti. *Angew. Chem.* **76,** 765 (1964).
437. Calderon, N., H. Y. Chen and K. W. Scott. *Tetrahedron Lett.* 3327 (1967).
438. Calderon, N., E. A. Ofstead, J. P. Ward, W. A. Judy and K. W. Scott. *J. Am. Chem. Soc.* **90,** 4133 (1968).
439. Wang, J. L., and H. R. Menapace. *J. Org. Chem.* **33,** 3794 (1968).
440. Zuech, E. A. *Chem. Commun.* 1182 (1968).
441. Hughes, W. B. *Chem. Commun.* 431 (1969).
442. Hughes, W. B. *J. Am. Chem. Soc.* **92,** 532 (1970).
443. Mol, J. C., J. A. Moulijn and C. Boelhouwer. *Chem. Commun.* 633 (1968).
444. Clark, A., and C. Cook. *J. Catal.* **15,** 420 (1969).
445. Scott, K. W., N. Calderon, E. A. Ofstead, W. A. Judy and J. P. Ward. *Adv. Chem. Ser.* **91,** 399 (1969).
446. Bradshaw, C. P. C., E. J. Howman and L. Turner. *J. Catal.* **7,** 269 (1967).
447. Bailey, G. C. *Catal. Rev.* **3,** 37 (1969).
448. Lewandos, G. S., and R. Pettit. *J. Am. Chem. Soc.* **93,** 7087 (1971).
449. Mango, F. D., and J. H. Schachtschneider. *J. Am. Chem. Soc.* **93,** 1123 (1971); **89,** 2484 (1967); F. D. Mango. *Adv. Catal.* **20,** 291 (1969).
450. Lewandos, G. S., and R. Pettit. *Tetrahedron Lett.* 789 (1971).
451. Pennella, F., R. L. Banks and G. C. Bailey. *Chem. Commun.* 1548 (1968).
452. Basset, J. M., J. L. Bilhou, R. Mutin and A. Theolier. *J. Am. Chem. Soc.* **97,** 7376 (1975).
453. Hérisson, J. L., and Y. Chauvin. *Makromol. Chem.* **141,** 161 (1970); J. P. Soufflet, D. Commereuc and Y. Chauvin. *C. R. Acad. Sci. Ser. C.* **276,** 169 (1973).
454. Casey, C. P., and T. J. Burkhardt. *J. Am. Chem. Soc.* **96,** 7808 (1974).
455. Katz, T. J., and J. McGinnis. *J. Am. Chem. Soc.* **97,** 1592 (1975).
456. Grubbs, R. H., P. L. Burk and D. D. Carr. *J. Am. Chem. Soc.* **97,** 3265 (1975).
457. Schrock, R. R. *J. Am. Chem. Soc.* **96,** 6796 (1974).
458. Sanders, A., L. Cohen, W. P. Giering, D. Kennedy and C. V. Magatti. *J. Am. Chem. Soc.* **95,** 5430 (1973).
459. Muetterties, E. L. *Inorg. Chem.* **14,** 951 (1975).

460. Farona, M. F., and W. S. Greenlee, *Chem. Commun.* 759 (1975).
461. Kluiber, R. W., and G. Sasso. *Inorg. Chim. Acta.* **4,** 226 (1970); D. L. Johnston and W. De W. Horrocks, Jr. *Inorg. Chem.* **10,** 687 (1971).
462. Ewens, R. V. G., and C. S. Gibson. *J. Chem. Soc.* 431 (1949).
463. Thompson, M. C. Ph.D. thesis, The Ohio State University, 1963.
464. Busch, D. H., J. A. Burke, Jr., D. C. Jicha, M. C. Thompson and M. L. Morris. *Adv. Chem. Ser.* **37,** 125 (1963).
465. Blomstrand, C. W. *J. Prakt. Chem.* **38,** 497 (1888).
466. Ibid. **27,** 161 (1883).
467. Hofmann, K. A., and W. O. Rabe, *Z. Anorg. Chem.* **14,** 293 (1897).
468. Loir, A. *Ann.* **107,** 234 (1858).
469. Smiles, S. *J. Chem. Soc.* **77,** 160 (1900).
470. Hilditch, T. P., and S. Smiles. *J. Chem. Soc.* **91,** 1394 (1907).
471. Rây, P. C. *J. Chem. Soc.* **109,** 131, 603 (1916).
472. Ibid. **111,** 101 (1917).
473. Rây, P. C., and P. C. Guha. *J. Chem. Soc.* **115,** 261, 541, 1148 (1919); R. C. Rây. *J. Chem. Soc.* **115,** 548 (1919).
474. Adams, R., W. Reifschneider and M. D. Nair. *Croat. Chem. Acta.* **29,** 277 (1957); cf. *Chem. Abstr.* **53,** 16145d (1959).
475. Adams, R., and A. Ferretti. *J. Am. Chem. Soc.* **81,** 4939 (1959).
476. Spinelli, D., and A. Salvemini. *Ann. Chim.* (*Rome*). **51,** 1296 (1961); cf. *Chem. Abstr.* **56,** 15530d (1962).
477. Blinn, E. L., and D. H. Busch. *J. Am. Chem. Soc.* **90,** 4280 (1968).
478. Blinn, E. L. Ph.D. thesis, The Ohio State University, 1967.
479. Rose, N. J., C. A. Root and D. H. Busch. *Inorg. Chem.* **6,** 1431 (1967).
480. Lindoy, L. F., and S. E. Livingston. *Inorg. Chem.* **7,** 1149 (1968).
481. *Chem. Eng. News.* **40,** 57, (Sept. 17, 1962).
482. Elder, M. S., G. M. Prinz, P. Thornton and D. H. Busch. *Inorg. Chem.* **7,** 2426 (1968).
483. Schrauzer, G. N., and H. N. Rabinowitz. *J. Am. Chem. Soc.* **90,** 4297 (1968).
484. Kothari, V. Ph.D. thesis, The Ohio State University, 1966.
485. Girling, R. L., and E. L. Amma. *Inorg. Chem.* **6,** 2009 (1967).
486. Tsutsui, M. *Trans. N.Y. Acad. Sci.* **30,** 658 (1968).
487. Imai, H. *Kagaku* (*Kyoto*). **25,** 430, 519, 628 (1970).
488. Maitlis, P. M. *Adv. Organomet. Chem.* **4,** 95 (1966); R. Pettit, *Pure Appl. Chem.* **17,** 253 (1968); L. Watts and R. Pettit. *Adv. Chem. Ser.* **62,** 549 (1967).
489. Collman, J. P. *Angew. Chem. Int. Ed.* **4,** 132 (1965).
490. Kluiber, R. W. *J. Am. Chem. Soc.* **82,** 4839 (1960).
491. Collman, J. P., R. P. Blair, R. L. Marshall and L. Slade. *Inorg. Chem.* **2,** 576 (1963).
492. Collman, J. P., and M. Yamada. *J. Org. Chem.* **28,** 3017 (1963).
493. Collman, J. P., R. L. Marshall, W. L. Young III and C. T. Sears, Jr. *J. Org. Chem.* **28,** 1449 (1963).
494. Barker, R. H., J. P. Collman and R. L. Marshall. *J. Org. Chem.* **29,** 3216 (1964).
495. Collman, J. P., R. L. Marshall, W. L. Young III and S. D. Goldby. *Inorg. Chem.* **1,** 704 (1962).
496. Bose, K. S., and C. C. Patel. *J. Inorg. Nucl. Chem.* **33,** 2947 (1971).

497. Martin, J. G., R. M. C. Wei and S. C. Cummings. *Inorg. Chem.* **11,** 475 (1972).
498. Hipp, C. J., and D. H. Busch. *Chem. Commun.* 737 (1972).
499. Hipp, C. J., and D. H. Busch, *Inorg. Chem.* **12,** 894 (1973).
500. Dolphin, D. and R. H. Felton. *Acc. Chem. Res.* **7,** 26 (1974).
501. Hager, L. P., D. R. Morris, F. S. Brown and H. Eberwein, *J. Biol. Chem.* **241,** 1769 (1966).
502. Corfield, P. W. R., J. D. Mokren, C. J. Hipp and D. H. Busch. *J. Am. Chem. Soc.* **95,** 4465 (1973).
503. Pillsbury, D. G., and D. H. Busch, submitted for publication.
504. Vitzthum, G., and F. Lindner. *Angew. Chem. Int. Ed.* **10,** 315 (1971).
505. Jacobson, S. E., P. Reich-Rohrwig and A. Wojcicki. *Inorg. Chem.* **12,** 717 (1973).
506. Graziani, M., J. P. Bibler, R. M. Montesano and A. Wojcicki. *J. Organomet. Chem.* **16,** 507 (1969).
507. Alexander, J. J., and A. Wojcicki. *J. Organomet. Chem.* **5,** 655 (1971).
508. Whitesides, G. M., and D. J. Boschetto. *J. Am. Chem. Soc.* **93,** 1529 (1971).
509. Cutler, A., R. W. Fish, W. P. Giering and M. Rosenblum. *J. Am. Chem. Soc.* **94,** 4354 (1972).
510. Giering, W. P., and M. Rosenblum. *J. Am. Chem. Soc.* **93,** 5299 (1971).
511. Bannister, W. D., B. L. Booth, R. N. Haszeldine and P. L. Loader. *J. Chem. Soc.* (*A*). 930 (1971).
512. Thomasson, J. E., P. W. Robinson, D. A. Ross and A. Wojcicki. *Inorg. Chem.* **10,** 2130 (1971).
513. Churchill, M. R., and J. Wormald. *J. Am. Chem. Soc.* **93,** 354 (1971).
514. Lindner, E., and R. Grimmer. *Chem. Ber.* **104,** 544 (1971).
515. Lichtenberg, D. W., and A. Wojcicki. *J. Organomet. Chem.* **33,** C77 (1971).
516. Angelici, R. J. *Acc. Chem. Res.* **5,** 335 (1972).
517. Cardin, D. J., B. Cetinkaya and M. F. Lappert. Chem. Rev. **72,** 545 (1972).
518. Cotton, F. A., and C. M. Lukehart. *Progress in Inorganic Chemistry, Vol. 16,* S. J. Lippard, ed., Interscience, 1972, p. 487.
519. Jetz, W., and R. J. Angelici. *J. Am. Chem. Soc.* **94,** 3799 (1972).
520. Behrens, H., H. Krohberger, R. J. Lampe, J. Langer, D. Maertens and P. Passler. *Proc. Int. Conf. Coord. Chem., XIII.* Cracou-Zakopane, Poland, Sept., 1970, Vol. 12, p. 339.
521. Behrens, H., E. Lindner and P. Pässler. *Z. Anorg. Allg. Chem.* **365,** 137 (1969).
522. Edgell, W. F., M. T. Yang, B. J. Bulkin, R. Bayer and N. Koizumi. *J. Am. Chem. Soc.* **87,** 3080 (1965); W. F. Edgell and B. J. Bulkin. *J. Am. Chem. Soc.* **88,** 4839 (1966).
523. Kruse, A. E., and R. J. Angelici. *J. Organomet. Chem.* **24,** 231 (1970).
524. Busetto, L., and R. J. Angelici. *Inorg. Chim. Acta.* **2,** 391 (1968).
525. Busetto, L., M. Graziani and U. Belluco. *Inorg. Chem.* **10,** 781 (1970).
526. Palagyi, J., and L. Markó. *J. Organomet. Chem.* **17,** 453 (1969).
527. Angelici, R. J., and L. J. Blacik. *Inorg. Chem.* **11,** 175 (1972).
528. Malatesta, L., M. Angoletta and G. Caglio. *J. Chem. Soc.* (*A*). 1836 (1970).

529. Cherwinski, W. J., and H. C. Clark. *Inorg. Chem.* **10,** 2263 (1971); *Can. J. Chem.* **47,** 2665 (1969); H. C. Clark, K. R. Dixon and W. J. Jacobs. *J. Am. Chem. Soc.* **91,** 1346 (1969).
530. Deeming, A. J., and B. L. Shaw. *J. Chem. Soc.* (*A*). 443 (1969).
531. Beck, W., M. Bauder, G. LaMonnica, S. Cenini and R. Ugo. *J. Chem. Soc.* (*A*). 113 (1971).
532. Friedrich, W., and M. Moskophidis. *Z. Naturforsch.* **B26,** 879 (1971).
533. Mays, M. J., and F. P. Stefanini. *J. Chem. Soc.* (*A*). 2747 (1971).
534. Brunner, H. *Angew. Chem. Int. Ed.* **10,** 249 (1971).
535. Knebel, W. J., and P. M. Treichel. *Chem. Commun.* 516 (1971).
536. Jetz, W., and R. J. Angelici. *J. Organomet. Chem.* **35,** C37 (1972).
537. King, R. B., M. B. Bisnette and A. Fronzaglia. *J. Organomet. Chem.* **5,** 341 (1966).
538. Clark, H. C., and W. J. Jacobs. *Inorg. Chem.* **9,** 1229 (1970).
539. Angelici, R. J., and L. Busetto. *J. Am. Chem. Soc.* **91,** 3197 (1969).
540. Behrens, H., R. Lindner, D. Maertens, P. Wild and R. Lampe. *J. Organomet. Chem.* **34,** 367 (1972).
541. Fischer, E. O., and A. Massböl. *Angew. Chem. Int. Ed.* **3,** 580 (1964).
542. Fischer, E. O. *Pure Appl. Chem.* **24,** 407 (1970).
543. Aumann, R., and E. O. Fischer. *Chem. Ber.* **101,** 954 (1968); M. Y. Darensbourg and D. J. Darensbourg. *Inorg. Chem.* **9,** 32 (1970).
544. Moser, E., and E. O. Fischer. *J. Organomet. Chem.* **12,** P1 (1968).
545. Connor, J. A., and E. M. Jones. *Chem. Commun.* 570 (1971).
546. Moser, G. A., E. O. Fischer and M. D. Rausch. *J. Organomet. Chem.* **27,** 379 (1971); J. A. Connor, E. M. Jones and J. P. Lloyd. *J. Organomet. Chem.* **24,** C20 (1970).
547. Fischer, E. O., and H. J. Kollmeier. *Angew. Chem. Int. Ed.* **9,** 309 (1970).
548. Darensbourg, D. J., and M. Y. Darensbourg. *Inorg. Chim. Acta.* **5,** 247 (1971).
549. Green, M. L. H., and C. R. Hurley. *J. Organomet. Chem.* **10,** 188 (1967); M. L. H. Green, L. C. Mitchard and M. G. Swanwick. *J. Chem. Soc.* (*A*). 794 (1971).
550. Schöllkopf, U., and F. Gerhart. *Angew. Chem. Int. Ed.* **6,** 560 (1967).
551. Cotton, F. A., and C. M. Lukehart. *J. Am. Chem. Soc.* **93,** 2672 (1971).
552. Badley, E. M., J. Chatt and R. L. Richards. *J. Chem. Soc.* (*A*). 21 (1971); E. M. Badley, J. Chatt, R. L. Richards and G. A. Sim. *Chem. Commun.* 1322 (1969).
553. Badley, E. M., B. J. L. Kilby and R. L. Richards. *J. Organomet. Chem.* **27,** C37 (1971).
554. Clark, H. C., and L. E. Manzer. *J. Organomet. Chem.* **30,** C89 (1971).
555. Miller, J., A. L. Balch and J. H. Enemark. *J. Am. Chem. Soc.* **93,** 4613 (1971).
556. Burke, A., A. L. Balch and J. H. Enemark. *J. Am. Chem. Soc.* **92,** 2555 (1970); W. M. Butler and J. H. Enemark. *Inorg. Chem.* **10,** 2416 (1971); G. Rouschias and B. L. Shaw, *J. Chem. Soc.* (*A*). 2097 (1971).
557. Balch, A. L., and J. Miller. *J. Am. Chem. Soc.* **94,** 417 (1972).
558. Treichel, P. M., J. P. Stenson and J. J. Benedict. *Inorg. Chem.* **10,** 1183 (1971).
559. Chisholm, M. H., and H. C. Clark. *Chem. Commun.* 763 (1970); *Inorg. Chem.* **10,** 1711 (1971).
560. Chisholm, M. H., and H. C. Clark. *J. Am. Chem. Soc.* **94,** 1532 (1972); M. H. Chisholm, H. C. Clark and D. H. Hunter. *Chem. Commun.* 809 (1971).

561. Nozaki, H., H. Takaya, S. Moriuti and R. Noyori. *Tetrahedron.* **24,** 3655 (1968).
562. Mango, F. D., and I. Dvoretzky. *J. Am. Chem. Soc.* **88,** 1654 (1966).
563. Jolly, P. W., and R. Pettit. *J. Am. Chem. Soc.* **88,** 5044 (1966).
564. Green, M. L. H., M. Ishaq and R. N. Whiteley. *J. Chem. Soc.* (*A*), 1508 (1967).
565. Collier, M. R., B. M. Kingston and M. F. Lappert. *Chem. Commun.* 1498 (1970).
566. Crociani, B., and T. Boschi. *J. Organomet. Chem.* **24,** C1 (1970).
567. Casey, C. P., R. A. Boggs and R. L. Anderson. *J. Am. Chem. Soc.* **94,** 8947 (1972).
568. Moser, E., and E. O. Fischer. *J. Organomet. Chem.* **15,** 147 (1968).
569. Connor, J. A., and E. O. Fischer. *Chem. Commun.* 1024 (1967).
570. Klabunde, U., and E. O. Fischer. *J. Am. Chem. Soc.* 89, 7141 (1971).
571. Connor, J. A., and E. O. Fischer. *J. Chem. Soc.* (*A*). 578 (1969).
572. Fischer, E. O., B. Heckl and H. Werner. *J. Organomet. Chem.* **28,** 359 (1971).
573. Kreiter, C. G. *Angew. Chem. Int. Ed.* **7,** 390 (1968).
574. Knaus, L., and E. O. Fischer. *Chem. Ber.* **103,** 1262, 3744 (1970).
575. Heckl, B., H. Werner and E. O. Fischer. *Angew. Chem. Int. Ed.* **7,** 817 (1968); H. Werner, E. O. Fischer and C. G. Kreiter. *J. Organomet. Chem.* **28,** 367 (1971).
576. Green, M. L. H., L. C. Mitchard and M. G. Swanwick, *J. Chem. Soc.* (*A*). 794 (1971).
577. Fischer, E. O., and R. Aumann. *Chem. Ber.* **101,** 963 (1968).
578. Casey, C. P., T. J. Burkhardt, *J. Am. Chem. Soc.* **94,** 6543 (1972).
579. Fischer, E. O., and K. H. Dotz. *Chem. Ber.* **103,** 1273 (1970).
580. Smidt, J., W. Hafner, R. Jira, R. Sieber, T. Sedlmeier and A. Sabel. *Angew. Chem. Int. Ed.* **1,** 80 (1962).
581. Henry, P. M. *Acc. Chem. Res.* **6,** 16 (1973).
582. Maitlis, P. M. *The Organic Chemistry of Palladium, Vol. 2,* Academic Press, New York, 1971, p. 108.
583. Henry, P. M. *Adv. Organomet. Chem.* **13,** 363 (1975).
584. Sabel, A., J. Smidt, R. Jira and H. Prigge. *Chem. Ber.* **102,** 2939 (1969).
585. Volger, H. C. *Rec. Trav. Chim. Pays-Bas.* **87,** 481 (1968).
586. McKeon, J. E., P. Fitton and A. A. Griswold. *Tetrahedron.* **28,** 227 (1972).
587. Henry, P. M. *J. Am. Chem. Soc.* **93,** 3853 (1971).
588. Ibid. *94,* 7316 (1972).
589. Ibid. 7311.
590. Henry, P. M. *J. Org. Chem.* **37,** 2443 (1972).
591. Henry, P. M., and G. A. Ward, *J. Am. Chem. Soc.* **93,** 1494 (1971).
592. Ibid. **94,** 673 (1972).
593. Ibid. **94,** 1527 (1972).
594. Ibid. **94,** 5200 (1972).
595. Henry, P. M. *Inorg. Chem.* **11,** 1876 (1972).
596. Henry, P. M., and R. N. Pandey. *Adv. Chem. Ser.* **132,** 33 (1974).
597. Henry, P. M. *J. Org. Chem.* **38,** 2766 (1973).
598. Winkhaus, G., L. Pratt and G. Wilkinson. *J. Chem. Soc.* 3807 (1961).
599. Jones, D., L. Pratt and G. Wilkinson. *J. Chem. Soc.* 4458 (1962).

600. Fischer, E. O., and M. W. Schmidt. *Chem. Ber.* **100,** 3782 (1967).
601. Jones, D., and G. Wilkinson. *J. Chem. Soc.* 2479 (1964).
602. Khand, I. U., P. L. Pauson and W. E. Watts. *J. Chem. Soc.* (*C*). 2257 (1968).
603. Bird, P. H., and M. R. Churchill. *Chem. Commun.* 777 (1967).
604. Fischer, E. O., and R. D. Fischer. *Angew. Chem.* **72,** 919 (1960).
605. Efraty, A., and P. M. Maitlis. *J. Am. Chem. Soc.* **89,** 3744 (1967).
606. Helling, J. F., and D. M. Braitsch. *J. Am. Chem. Soc.* **92,** 7207 (1970).
607. Zelonka, R. A., and M. C. Baird. *J. Organomet. Chem.* **35,** C43 (1972).
608. Walker, P. J. C., and R. J. Mawby. *Inorg. Chem.* **10,** 404 (1971); *Chem. Commun.* 330 (1972); *J. Chem. Soc., Dalton Trans.* 622 (1973).
609. Kroll, H. *J. Am. Chem. Soc.* **74,** 2036 (1952).
610. Meriwether, L., and F. H. Westheimer. *J. Am. Chem. Soc.* **78,** 5119 (1956).
611. Bamann, E., J. G. Haas and H. Trapmann. *Arch. Pharm.* **294,** 569 (1961).
612. Bender, M. L., and B. W. Turnquest. *J. Am. Chem. Soc.* **79,** 1889 (1957).
613. Westheimer, F. H., and M. W. Shookhoff. *J. Am. Chem. Soc.* **62,** 269 (1940).
614. Hay, R. W., M. L. Jansen and P. L. Cropp. *Chem. Commun.* 621 (1967).
615. Bender, M. L., and B. W. Turnquest. *J. Am. Chem. Soc.* **79,** 1652 (1957).
616. Conley, Jr., H. L., and R. B. Martin. *J. Phys. Chem.* **69,** 2914 (1965).
617. Westheimer, F. W. *Trans. N.Y. Acad. Sci.* **18,** 15 (1955).
618. Conley, Jr., H. L., and R. B. Martin. *J. Phys. Chem.* **69,** 2923 (1965).
619. Hay, R. W., and P. J. Morris. *Chem. Commun.* 23 (1967).
620. Connor, W. A., M. M. Jones and D. L. Tuleen. *Inorg. Chem.* **4,** 1129 (1965).
621. Hix, Jr., J. E., and M. M. Jones. *Inorg. Chem.* **5,** 1863 (1966).
622. Angelici, R. J., and B. E. Leach. *J. Am. Chem. Soc.* **89,** 4605 (1967).
623. Leach, B. E., and R. J. Angelici. *J. Am. Chem. Soc.* **90,** 2504 (1968).
624. Angelici, R. J., and B. E. Leach. *J. Am. Chem. Soc.* **90,** 2499 (1968).
625. Hopgood, D., and R. J. Angelici. *J. Am. Chem. Soc.* **90,** 2508 (1968).
626. Angelici, R. J., and D. Hopgood. *J. Am. Chem. Soc.* **90,** 2514 (1968).
627. Leach, B. E., and R. J. Angelici. *Inorg. Chem.* **8,** 907 (1969).
628. Angelici, R. J., and J. W. Allison. *Inorg. Chem.* **10,** 2238 (1971).
629. Alexander, M. D., and D. H. Busch. *Inorg. Chem.* **5,** 602 (1966).
630. Springer, M. P., and C. Curran. *Inorg. Chem.* **2,** 1270 (1963).
631. Bolin, I. *Z. Anorg. Allg. Chem.* **143,** 201 (1925).
632. Buckingham, D. A., D. M. Foster and A. M. Sargeson. *J. Am. Chem. Soc.* **92,** 5701 (1970).
633. Buckingham, D. A., D. M. Foster and A. M. Sargeson. *J. Am. Chem. Soc* **94,** 4032 (1972).
634. Buckingham, D. A., D. M. Foster, L. G. Marzilli and A. M. Sargeson. *Inorg. Chem.* **9,** 11 (1970).
635. Hay, R. W., R. Bennett and D. J. Barnes. *J. Chem. Soc. Dalton Trans.* 1524 (1972).
636. Wu, Y., and D. H. Busch. *J. Am. Chem. Soc.* **92,** 3326 (1970).
637. Buckingham, D. A., D. M. Foster and A. M. Sargeson. *Aust. J. Chem.* **22,** 2479 (1969).
638. Murakami, M., H. Itatani, K. Takahashi, J. Kang and K. Suzuki. *Mem. Inst. Sci. Ind. Res., Osaka Univ.* **20,** 95 (1963).
639. Hix, Jr., J. E., and M. M. Jones. *J. Am. Chem. Soc.* **90,** 1723 (1968).
640. Bennett, W. E. *J. Am. Chem. Soc.* **81,** 246 (1959).

641. Leach, B. E., and R. J. Angelici. *J. Am. Chem. Soc.* **91,** 6296 (1969).
642. McDonald, C. C., and W. D. Phillips. *J. Am. Chem. Soc.* **85,** 3736 (1963).
643. Brill, A. S., R. B. Martin and R. J. P. Williams. *Electronic Aspects of Biochemistry,* B. Pullman, ed., Academic Press, New York, 1964, p. 540.
644. Collman, J. P., and D. A. Buckingham. *J. Am. Chem. Soc.* **85,** 3039 (1963).
645. Buckingham, D. A., J. P. Collman, D. A. R. Happer and L. G. Marzilli. *J. Am. Chem. Soc.* **89,** 1082 (1967).
646. Buckingham, D. A., and J. P. Collman. *Inorg. Chem.* **6,** 1803 (1967).
647. Collman, J. P., and E. Kimura. *J. Am. Chem. Soc.* **89,** 6096 (1967).
648. Buckingham, D. A., L. G. Marzilli and A. M. Sargeson. *J. Am. Chem. Soc.* **89,** 2772 (1967).
649. Buckingham, D. A., P. A. Marzilli, I. E. Maxwell, A. M. Sargeson, M. Fehlmann and H. Freeman. *Chem. Commun.* 488 (1968).
650. Kim, M. K., and A. E. Martell. *Biochemistry.* **3,** 1169 (1964).
651. Buckingham, D. A., C. E. Davis, D. M. Foster and A. M. Sargeson. *J. Am. Chem. Soc.* **92,** 5571 (1970).
652. Volshtein, L. M., and L. S. Anokhova. *Russ. J. Inorg. Chem. Engl. Transl.* **4,** 142, 782 (1959).
653. Jones, M. M., T. J. Cook and S. Brammer. *J. Inorg. Nucl. Chem.* **28,** 1265 (1966).
654. Andreatta, R. H., H. C. Freeman, A. V. Robertson and R. L. Sinclair. *Chem. Commun.* 203 (1967).
655. Nakahara, A., K. Hamada, Y. Nakao and T. Higashiyama. *Coord. Chem. Rev.* **3,** 207 (1968).
656. Wu, Y., and D. H. Busch. *J. Am. Chem. Soc.* **94,** 4115 (1972).
657. Girgis, A. Y., and J. I. Legg. *J. Am. Chem. Soc.* **94,** 8420 (1972).
658. Buckingham, D. A., A. M. Sargeson and A. Zanella. *J. Am. Chem. Soc.* **94,** 8246 (1972).
659. Buckingham, D. A., B. M. Foxman, A. M. Sargeson and A. Zanella. *J. Am. Chem. Soc.* **94,** 1007 (1972).
660. Wiberg, K. B., ed., *Oxidation in Organic Chemistry, Vol. 5A.* Academic Press, New York, 1965.
661. Richardson, W. H., reference 660, p. 244.
662. Littler, J. S., and W. A. Waters. *J. Chem. Soc.* 2767 (1960).
663. Littler, J. S. *J. Chem. Soc.* 4135 (1959).
664. Huston, R. C., G. L. Goerner, and H. H. György. *J. Am. Chem. Soc.* **70,** 389 (1948); R. C. Huston and G. L. Goerner. *J. Am. Chem. Soc.* **68,** 2504 (1946).
665. Criegee, R. *Oxidation in Organic Chemistry, Vol. 5A,* K. B. Wiberg, ed., Academic Press, New York, 1965, p. 278.
666. Braude, E. A., and O. H. Wheeler. *J. Chem. Soc.* 320 (1955).
667. Baker Castor Oil Co., British patent 759,416 (1956); cf. *Chem. Zentr.,* 1009 (1961).
668. Littler, J. S., and W. A. Waters. *J. Chem. Soc.* 4046 (1959).
669. Jones, J. R., and W. A. Waters. *J. Chem. Soc.* 2068 (1962).
670. Ibid. 2772 (1960).
671. Wiberg, K. B., and K. A. Saegebarth. *J. Am. Chem. Soc.* **79,** 2822 (1957).
672. Böeseken, J., and M. C. de Graaff. *Rec. Trav. Chim. Pays-Bas.* **41,** 199 (1922); J. Böeseken. *Rec. Trav. Chim. Pays-Bas.* **47,** 683 (1928).

673. Lee, D. G., and J. Brownridge. *J. Am. Chem. Soc.* **95,** 3033 (1973).
674. Wiberg, K. B., C. J. Deutsch and J. Roćek. *J. Am. Chem. Soc.* **95,** 3034 (1973).
675. Wiberg, K. B., reference 660, p. 69.
676. Stewart, R., reference 660, p. 1.
677. Roćek, J., and K. Meyer. *J. Am. Chem. Soc.* **94,** 1209 (1972).
678. Roćek, J., and A. E. Radkowski. *J. Org. Chem.* **38,** 89 (1973).
679. Rahman, M., and J. Roćek. *J. Am. Chem. Soc.* **93,** 5455 (1971).
680. Ibid. 5462.
681. Wiberg, K. B., and S. K. Mukherjee. *J. Am. Chem. Soc.* **93,** 2543 (1971).
682. Hason, F., and J. Roćek. *J. Am. Chem. Soc.* **94,** 9073 (1972).
683. Ibid. 3181.
684. Ibid. 8946.
685. Kochi, J. K. *Free Radicals,* Wiley-Interscience, New York, 1973, Chap. 11.
686. Kochi, J. K., A. Bemis and C. L. Jenkins. *J. Am. Chem. Soc.* **90,** 4616 (1968).
687. Kochi, J. K. *Acc. Chem. Res.* **7,** 351 (1974).
688. Kochi, J. K. *Pure Appl. Chem.* **4,** 377 (1971).
689. Gardner, H. C., and J. K. Kochi. *J. Am. Chem. Soc.* **96,** 1982 (1974).
690. Clinton, N. A., and J. K. Kochi. *J. Organomet. Chem.* **56,** 243 (1973).
691. Whitesides, G. M., J. F. Gaasch and E. R. Stedronsky. *J. Am. Chem. Soc.* **94,** 5258 (1972).
692. McDermott, J. X., J. F. White and G. M. Whitesides. *J. Am. Chem. Soc.* **95,** 4451 (1973).
693. Waters, W. A. *Mechanisms of Oxidation of Organic Compounds,* Methuen, London, 1964.
694. Turney, T. A. *Oxidation Mechanisms,* Butterworths, London, 1965.
695. Reich, L. and S. S. Stivala. *Autoxidation of Hydrocarbons and Polyolefins, Kinetics and Mechanisms,* Marcel-Dekker, New York, 1969.
696. Mayo, F. R., ed. *Oxidation of Organic Compounds, Adv. Chem. Ser.* **75, 76** and **77,** 1968.
697. Sheldon, R. A., and J. K. Kochi. "Mechanisms of Metal-Catalyzed Oxidations of Organic Compounds in the Liquid Phase," *Oxidation and Combustion Reviews.* **5,** 135 (1973).
698. Norman, R. O. C., and J. R. Lindsay Smith. *Oxidases and Related Redox Systems,* T. E. King H. S. Mason and M. Morrison, eds., John Wiley, New York, 1964.
699. Hamilton, G. A. *Progress in Bioorganic Chemistry, Vol. 1,* E. T. Kaiser and F. J. Kézdy, eds., Wiley-Interscience, New York, 1971.
700. Hamilton, G. A. *Adv. Enzymol.* **32,** 55 (1969).
701. Hayaishi, O., ed. *Molecular Mechanisms of Oxygen Activation,* Academic Press, New York, 1974.
702. Akhrem, A. A., D. I. Metelitsa and M. E. Skurko. *Russ. Chem. Rev., Engl. Transl.* **44,** (5) 398 (1975).
703. Hamilton, G. A. *Molecular Mechanisms of Oxygen Activation,* O. Hayaishi, ed., Academic Press, New York, 1974.
704. Fridovich, I. *Acc. Chem. Res.* **5,** 321 (1972).
705. Haber, F., and J. Weiss. *Naturwissenschaften.* **20,** 948 (1932).
706. Walling, C. *Acc. Chem. Res.* **8,** 125 (1975).
707. Barb, W. G., J. H. Baxendale, P. George and K. R. Hargrave. *Trans. Faraday Soc.* **47,** 462 (1951).

708. Kropf, H. Ann. **637,** 73 (1960).
709. Heiba, E. I., R. M. Dessau and W. J. Koehl, Jr. *J. Am. Chem. Soc.* **91,** 6830 (1969).
710. Ichikawa, Y., G. Yamashita, M. Tokashiki and T. Yamaji. *Ind. Eng. Chem.* **62,** (4) 38 (1970).
711. Kamiya, Y., and M. Kashima. *J. Catal.* **25,** 326 (1972).
712. Onopchenko, A., and J. G. D. Schulz. *J. Org. Chem.* **38,** 909 (1973).
713. Cooper, T. A., and W. A. Waters. *J. Chem. Soc.* (*B*). 687 (1967).
714. Fields, E. K., and S. Meyerson. *Adv. Chem. Ser.* **76,** 395 (1968).
715. Holtz, H. D. *J. Org. Chem.* **37,** 2069 (1972).
716. Cooper, T. A., A. A. Clifford, D. J. Mills and W. A. Waters. *J. Chem. Soc.* (*B*). 793 (1966).
717. Aguiló, A. *Adv. Organomet. Chem.* **5,** 321 (1967).
718. Henry, P. M. *Adv. Chem. Ser.* **70,** 126 (1968).
719. Stern, E. W. *Catal. Rev.* **1,** 73 (1967).
720. Phillips, F. C. *Am. Chem. J.* **16,** 255 (1894).
721. Smidt, J., W. Hafner, R. Jira, J. Sedlmeier, R. Sieber, R. Rüttinger and H. Kojer. *Angew. Chem.* **71,** 176 (1959).
722. Teramoto, K., T. Oga, S. Kikuchi and M. Ito. *Yuki Gosei Kagaku Kyokai Shi.* **21,** 298 (1963); cf., *Chem. Abstr.* **59,** 7339g (1963).
723. Moiseev, I. I., M. I. Vargaftik and Ya. K. Syrkin. *Dokl. Akad. Nauk SSSR Engl. Transl. Phys. Chem. Sect.* **152,** 773 (1963).
724. Pestrikov, S. V., I. I. Moiseev and L. M. Sverzh. *Russ. J. Inorg. Chem. Engl. Transl.* **11,** 1113 (1966).
725. Henry, P. M. *J. Am. Chem. Soc.* **86,** 3246 (1964); **88,** 1595 (1968).
726. Hafner, W., R. Jira, J. Sedlmeier and J. Smidt. *Chem. Ber.* **95,** 1575 (1962).
727. Nikiforova, A. V., I. I. Moiseev and Ya. K. Syrkin. *Zh. Obshch Khim.* **33,** 3239 (1963).
728. Moiseev, I. I., M. N. Vargaftik and Ya. K. Syrkin. *Izv. Akad. Nauk SSSR, Ser. Khim., Engl. Transl.* 1050 (1963).
729. Moiseev, I. I., and M. N. Vargaftik. *Dokl. Akad. Nauk SSSR Engl. Transl. Chem. Sect.* **166,** 80 (1966).
730. Jira, R., J. Sedlmeier and J. Smidt. *Ann.* **693,** 99 (1966).
731. Moiseev, I. I. *Am. Chem. Soc. Div. Pet. Chem. Prepr.* **14,** B49 (1969).
732. Henry, P. M. *J. Am. Chem. Soc.* **94,** 4437 (1972).
733. Ibid. **87,** 990, 4423 (1965); **88,** 1597 (1966).
734. Smidt, J., and R. Sieber. *Angew. Chem.* **71,** 626 (1959).
735. Smidt, J., R. Sieber, W. Hafner and R. Jira. German patent 1,176,141 (1964); cf. *Chem. Abstr.* **62,** 447e (1965).
736. Okada, H., and H. Hashimoto. *Kogyo Kagaku Zasshi.* **69,** 2137 (1966); H. Okada, T. Noma, Y. Katsuyama and H. Hashimoto. *Bull. Chem. Soc. Jpn.* **41,** 1395 (1968).
737. Ouellette, R. J., and C. Levin. *J. Am. Chem. Soc.* **93,** 471 (1971); **90,** 6889 (1968).
738. Moiseev, I. I., M. N. Vargaftik and Ya. K. Syrkin. *Dokl. Akad. Nauk SSSR Engl. Transl. Phys. Chem. Sect.* **130,** 115 (1960).
739. Henry, P. M. *J. Org. Chem.* **32,** 2575 (1967).

740. Moiseev, I. I., and M. N. Vorgaftik. *Izv. Akad. Nauk SSSR Engl. Transl.* 744 (1965).
741. Henry, P. M., *J. Am. Chem. Soc.* **94,** 7305 (1972).
742. Stern, E. W., and M. L. Spector. *Proc. Chem. Soc., London.* 370 (1961).
743. Lloyd, W. G., and B. J. Luberoff. *J. Org. Chem.* **34,** 3949 (1969).
744. Ketley, A. D., and L. P. Fisher. *J. Organomet. Chem.* **13,** 243 (1968).
745. Lloyd, W. G. *J. Org. Chem.* **32,** 2816 (1967).
746. Brown, R. G., J. M. Davidson and C. Triggs. *Am. Chem. Soc. Div. Pet. Chem. Prepr.* **14,** B23 (1969).
747. James, B. R., and M. Kastner. *Can. J. Chem.* **50,** 1698, 1708 (1972).
748. Kukushkin, Yu N., and Yu. S. Vorshavskii. *Russ. J. Inorg. Chem. Engl. Transl.* **11,** 193 (1966).
749. Vassian, E. G., and R. K. Murmann. *Inorg. Chem.* **6,** 2043 (1967).
750. Curtis, N. F. *Chem. Commun.* 882 (1966).
751. Curtis, N. F. *J. Chem. Soc.* (*A*). 2834 (1971).
752. Barefield, E. K., and D. H. Busch. *Inorg. Chem.* **10,** 108 (1971).
753. Barefield, E. K., F. V. Lovecchio, N. E. Tokel, E. Ochiai and D. H. Busch. *Inorg. Chem.* **11,** 283 (1972).
754. Gore, E. S., and D. H. Busch. *Inorg. Chem.* **12,** 1 (1973).
755. Curtis, N. F., and D. F. Cook. *Chem. Commun.* 962 (1967).
756. Barefield, E. K., and D. H Busch. *Chem. Commun.* 522 (1970).
757. Barefield, E. K., and M. T. Mocelle. *J. Am. Chem. Soc.* **97,** 4238 (1975).
758. Olson, D. C., and J. Vasilevskis. *Inorg. Chem.* **10,** 463 (1971).
759. Goedken, V. L., and D. H. Busch. *J. Am. Chem. Soc.* **94,** 7355 (1972).
760. Dabrowiak, J. C., and D. H. Busch. *Inorg. Chem.* **14,** 1881 (1975).
761. McWhinnie, W. R., J. D. Miller, J. B. Watts and D. Y. Waddan. *Chem. Commun.* 629 (1971).
762. Lane, B. C., J. E. Lester and F. Basolo. *Chem. Commun.* 1618 (1971); S. E. Diamond, G. M. Tom and H. Taube. *J. Am. Chem. Soc.* **97,** 2661 (1975).
763. Riley, D. P. Ph.D. thesis, The Ohio State University, 1975.
764. Weiss, M. C., and V. L. Goedken. *J. Am. Chem. Soc.* **98,** 3389 (1976).
765. Hipp, C. J., L. F. Lindoy and D. H. Busch. *Inorg. Chem.* **11,** 1988 (1972).
766. Tang, S. C., G. N. Weinstein and R. H. Holm. *J. Am. Chem. Soc.* **95,** 613 (1973).
767. Tang, S. C., and R. H. Holm. *J. Am. Chem. Soc.* **97,** 3359 (1975).
768. Truex, T. J., and R. H. Holm. *J. Am. Chem. Soc.* **94,** 4529 (1972).
769. Millar, M., and R. H. Holm. *J. Am. Chem. Soc.* **97,** 6052 (1975).
770. Cunningham, J. A., and R. E. Sievers. *J. Am. Chem. Soc.* **95,** 7183 (1973).
771. Dabrowiak, J. C., private communication.
772. Mason, H. S. *Adv. Enzymol.* **19,** 128 (1957).
773. Mason, H. S. *Ann. Rev. Biochem.* **34,** 595 (1965).
774. Gunsalus, I. C., J. R. Meeks, J. D. Lipscomb, P. Debrunner and E. Münck. *Molecular Mechanisms of Oxygen Activation,* O. Hayaishi, ed., Academic Press, New York, 1974, Chap. 14.
775. Johnson, M. J. *Science.* **155,** 1515 (1967).
776. Sharrock, M., E. Münck, P. G. Debrunner, V. Marshall, J. D. Lipscomb and I. C. Gunsalus, *Biochemistry.* **12,** 258 (1973).
777. Hamilton, G. A. *J. Am. Chem. Soc.* **86,** 3391 (1964).

778. Guroff, G., C. A. Reifsnyder and J. Daly. *Biochem. Biophys. Res. Commun.* **24,** 720 (1966).
779. Guroff, G., J. W. Daly, D. M. Jerina, J. Renson, B. Witkop and S. Udenfriend. *Science.* **157,** 1524 (1967).
780. Daly, J. W., D. M. Jerina and B. Witkop. *Experimentia.* **28,** 1129 (1972).
781. Jerina, D. M. *Chem. Technol.* 120 (1973).
782. Udenfriend, S., C. T. Clark, J. Axelrod and B. B. Brodie. *J. Biol. Chem.* **208,** 731 (1954).
783. Kruse, W. *Chem. Commun.* 1610 (1968).
784. Sharpless, K. B., and T. C. Flood. *J. Am. Chem. Soc.* **93,** 2316 (1971).
785. Groves, J. T., and G. A. McClusky. *J. Am. Chem. Soc.* **98,** 859 (1976).

3

Oxidation-Reduction Reactions of Coordination Complexes

David E. Pennington
Department of Chemistry
Baylor University
Waco, Texas 76703

Foreword

The field of electron transfer has been most active since its beginnings in 1954. During the past several years new avenues of inquiry in the areas of intramolecular electron transfer, intervalence transfer, excited-state electron transfer, radical ion electron transfer, and bioinorganic electron transfer have continued to challenge some of the best minds in chemistry. In addition to treating the more traditional areas this review attempts to distill some of the essence from the more recent areas mentioned above, except that of bioinorganic electron transfer. I refer the reader to the excellent review by Larry Bennett in Vol. 18 of "Progress in Inorganic Chemistry" for a comprehensive coverage of that topic. Coverage of the primary literature is fairly complete through June 1976.

I. THE OUTER-SPHERE MECHANISM

A. Theory

The theory of electron transfer processes has received widespread attention.[1-22] Marcus' work[5-10,16] has perhaps been the most extensively tested against experiment, followed closely by that of Hush[14] and Levich.[12,13,17] These theories are compared in the review of Marcus[16] and in a monograph by Reynolds and Lumry.[24] Unlike the Marcus–Hush adiabatic treatments where the transition probabilities are assumed to be near unity, Levich[12,13,17] has chosen to calculate the transition probabilities for various electron transfer processes and has re-

0-8412-0292-3/78/36-174-476$30.00/1

ported values of κ ranging over about six orders of magnitude (10^{-6}–3.5). Very recently, Kestner, Logan, and Jortner[21] presented a critical review of the quantum mechanical theory of electron transfer. Their results for nonadiabatic electron transfer provide an extension of the classical Marcus–Hush approach and the Levich–Dogonadze quantum mechanical approach[12,13] to include inner-sphere configurational changes. In each of the theoretical treatments normally a model is proposed for the electron transfer process and then the relevant contributions to the overall free energy of activation are evaluated.

The Franck–Condon principle,[24,25,26] which requires short lifetimes of electronic transitions compared with those of nuclear motion, underlies each of these treatments. Despite the apparent simplicity of this principle, it has far-reaching consequences, at least for exchange processes and heteronuclear reactions where the standard free energy changes are near zero. In these systems it is necessary for each reactant to undergo a distortion (generally involving the inner coordination sphere as well as polarization of the surrounding medium) to some nonequilibrium configuration prior to the event of electron transfer to conserve energy. This requirement is best seen through a negative example where, for simplicity, only the distortions involving the inner coordination sphere of the two reactants are considered. (The attendant changes in polarization of the medium are implicit.)

Consider the consequences of no distortion of the iron coordination spheres prior to electron transfer in the iron(II)–iron(III) exchange reaction.

$$Fe(H_2O)_6{}^{2+} + {}^*Fe(H_2O)_6{}^{3+} \rightleftarrows Fe(H_2O)_6{}^{3+} + {}^*Fe(H_2O)_6{}^{2+}$$

Both product species would be produced in excited-state configurations—i.e., Fe(II) would have Fe(III)—O bond lengths and Fe(III) would have Fe(II)—O bond lengths. These excited-state species would then decay to their respective ground states, liberating heat, but the standard enthalpy change for electron exchange reactions is *zero*. Hence, the above mechanism violates the first law of thermodynamics. However, if the Fe—O bonds are first reorganized into some nonequilibrium configuration (say by shortening the Fe(II)—O bond lengths and extending the Fe(III)—O bond lengths), the problem is not encountered. Instead, the energy required to distort the reactants into the nonequilibrium configuration is exactly counterbalanced when the product ions reestablish their equilibrium configurations, thus satisfying the requirement for zero enthalpy change. Note that this argument applies rigorously only for exchange processes and, to a good approximation, to other reactions with small standard free-energy changes. For those reactions having large standard free-energy changes, it is likely that excited-state species form in the sequence of reaction steps.

Since the Franck–Condon principle has been widely utilized in discussions involving the mechanism of electron transfer, it is pertinent to inquire about the magnitude of the barrier which this principle poses for the electron transfer process. Marcus[27] estimated $\Delta F^{\ddagger}$ to be about 8 kcal/mol for the exchange reactions

$$M(H_2O)_6^{2+} \rightarrow M(H_2O)_6^{3+} \quad (M = Fe \text{ or } Co)$$

The observed sluggish exchange rate between $Co(NH_3)_6^{2+}$ and $Co(NH_3)_6^{3+}$ was previously thought to arise from large reorganizational energies associated with the matching of energy levels in the two reactants.[28] Since the actual difference in bond lengths for the cobalt complexes is substantially less[29] than previously supposed,[29] Stynes and Ibers[30] have recalculated the contribution of inner-sphere reorganizational energies to the energy of activation using Stranks' force constant-bond distance model.[28] Their estimate of 6.8 kcal is considerably less than Stranks' estimate and insufficient to account for the 10^{15} difference between the $Co(NH_3)_6^{2+,3+}$ and $Ru(NH_3)_6^{2+,3+}$ self-exchange rates. In the ruthenium system the bond lengths differ only by 0.04 Å, and the Franck–Condon barrier is ≈ 0. While outer-sphere effects have either been neglected or assumed equal for the two systems, similar calculations based on structural studies of low-spin, macrocyclic, tetradentate amine complexes of cobalt(III) by Glick, Schmonsees, and Endicott[31] and Glick, Kuszaj, and Endicott[32] yielded Franck–Condon barriers of 7 ± 1 and ~ 21 kcal, respectively, for the *trans*-Co[14]tetraene-$N_4(H_2O)_2^{2+,3+}$ and *trans*-Co-[14]-diene-$N_4(H_2O)_2^{2+,3+}$ self-exchange reactions. Interestingly, the difference in Franck–Condon barriers *is* sufficient to account for the $>10^6$ difference in self-exchange rates between these two macrocyclic systems. (We return to the $Co(NH_3)_6^{2+,3+}$ self-exchange problem later in discussing the spin multiplicity effects.) Still another estimate of the Franck–Condon barrier is available from the literature of intervalence transfer.

Allen and Hush[33] and Hush[34] have reviewed the subject of intervalence-transfer absorption, the former giving qualitative evidence for the phenomenon in solution and in the solid state and the latter considering the theory and spectroscopic data. Additionally, Robin and Day[35] and more recently Cowan et al.[36] have reviewed the fascinating area of mixed-valency compounds. Robin and Day's class I, II, and III compounds are those with no interaction between metal sites, those with weak interaction and integral valence states, and those with strong interaction and nonintegral valence states, respectively. Optical intervalence transfer is an optical transition (usually in the visible region of the spectrum) which induces electron transfer between two metal centers each of which possesses more than one accessible oxidation state. Hush[34] has shown that the intervalence-transfer absorption is related to the energy barrier of thermal electron transfer between the same two species. For our purposes Hush's simplified model (coupling of the electron with two independent oscillators of identical frequency) is assumed. Here $E^*_{th} = E^2_{op}/4(E_{op} - E_0)$. When E_0, the difference in energies of final and initial states at equilibrium, approaches zero, say, for an exchange reaction, $E_{op} = 4E^*_{th}$. Thus for the simplest approach, the energy for intervalence-transfer absorption is about four times the Franck–Condon barrier for thermal electron transfer. In cases where E_0 is not zero, E_{op} and E^*_{th} can very well be quite similar.

Support for this model in the area of electron transfer reactions came originally from Creutz and Taube[37] who reported the observation of near-infrared bands in a series of pyrazine-bridged binuclear ruthenium(II, III) ammine complexes. The bands were assigned as intervalence-transfer absorptions since they did not occur in the corresponding ruthenium(II, II) or ruthenium(III, III) binuclear complexes. From the simplified Hush model $\Delta F^{\ddagger}_{th}$ is estimated to be 4.7 kcal/mol, corresponding to $k = 3 \times 10^9$ sec^{-1} for the self-exchange rate constant within the binuclear complex. The original report combined with significant new developments in the synthesis of bridged binuclear complexes of ruthenium has generated a recent flurry of activity in this area. Tom, Creutz, and Taube[38] have observed intervalence-transfer bands for a series of 4,4′-bipyridine-bridged binuclear ruthenium(II, III) complexes (in contrast to earlier work[39,40]) and from the solvent dependence of the symmetric pentaamine system concluded from Hush's theory that $\Delta F^{\ddagger}_{op} = 10$ kcal, which implies $\Delta F^{\ddagger}_{th} = 2.5$ kcal. Similar studies by Callahan, Brown, and Meyer[41] extend the above observations and indicate that both bridging and nonbridging ligands affect the energies of these transitions. Why neither the band widths nor the solvent dependencies of the intervalence transitions for $[(NH_3)_5RuLRu(NH_3)_5]^{5+}$ ions (L = pyrazine or cyanogen)[42,43] fits the Hush theory is perplexing. Mayoh and Day[44] have discussed the trapping of valence states in the pyrazine complex, and their recent calculations[45] indicate only ~1% valence delocalization. Finally, intervalence bands have also been observed for the ion $[(NC)_5FeCNFe(CN)_5]^{6-}$,[46] for Cu(II) and Fe(III) complexes with $Fe(CN)_5L^{3-}$ (L = NH_3, py, $pyzCO_2$, dmso, tmso, and CO),[47] for $[L_2V(OH)]_2^{3+}$ (L = 1,10-phenthroline or 2,2-bipyridine),[48] and for $[Cl(NH_3)_4OsN]_2^{5+}$ [49] It has been noted[37] that the value computed for $\Delta F^{\ddagger}_{th}$ should lie below that (8 kcal) estimated by Marcus since changes occur only in the t_{2g} energy level for the ruthenium(II, III) systems.

A general mechanism for the outer-sphere reaction can be envisioned as follows: step (*1*) diffusion of the two reactants together forming a precursor complex; step (*2*) reorganization of the precursor complex to a reactants' activated complex; step (*3*) formation of the products' activated complex by electron transfer; step (*4*) reorganization of the products' activated complex to yield a successor complex; and step (*5*) diffusion of the product species away from the successor complex. In principle any of steps (*2*)–(*4*) could be rate determining.[5] According to the Marcus' theory, the dominant contributions to the overall free energy barrier of reaction are made by the reorganizations of the outer-sphere solvent molecules and the inner-sphere ligands from their equilibrium configurations in the precursor complex to some nonequilibrium configuration in the reactants' activated complex. A Coulombic barrier and a multiplicity barrier, as well as the work required to bring the separated reactants together, can also contribute to the overall free energy but only in a minor way.

Before proceeding to the calculation of an overall rate constant, it is instructive to consider the electron transfer process in terms of potential energy surfaces. (A more detailed account is given by Reynolds and Lumry.[23] In his discussion

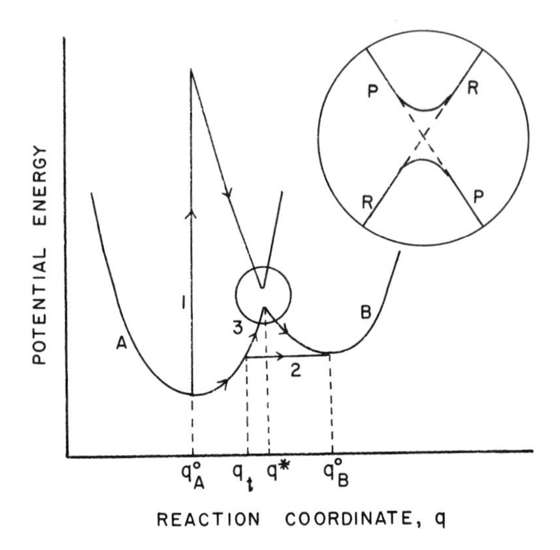

Figure 3-1 First-order potential energy surfaces for (1) intervalence transfer, (2) tunneling, and (3) classical transfer

of potential energy surfaces Marcus[10] noted that the potential energy of the system is a function of the translational, rotational, and vibrational coordinates of the reacting species *and* of the molecules of the surrounding medium. Figure 3-1 presents a profile of a first-order potential energy surface and an enlarged view of the interaction region. "The abcissa represents any concerted (translational, rotational, and/or vibrational) motion(s) leading from any spatial configuration (of all atoms) that is suited to the electronic structure of the reactants to one suited to that of the products."[10] The above configurational changes can be accomplished by at least three mechanisms: (a) nonradiative "classical" transfer; (b) nonradiative tunnel transition *of the system;* and (c) radiative or intervalence transfer.

The enlarged view of the interaction region exhibits the typical quantum mechanical splitting between energy surfaces when the distance between reacting species is sufficiently small. Two situations may be encountered here. If the electronic interaction is sufficiently large, the reactants (R) will pass over to products (P) along the lower surface. When this situation is attained with near unit probability, the process is called adiabatic. Alternatively, when the electronic interaction is weak, the reactants may readily pass over the intersection region without being converted to products. Nevertheless, a finite probability remains that the reactants will be converted to products. When this situation occurs, the process is called nonadiabatic. For nonadiabatic reactions a transition probability must be evaluated in addition to the reaction parameters. Marcus' theory assumes that most electron transfer reactions are of the adiabatic type; however, Levich[17] and Ruff[18] have criticized this assumption.

In the Marcus treatment of electron transfer the rate constant for the overall binuclear reaction is given by

$$k_{bi} = \kappa \rho Z e^{-\Delta F^*/RT} \tag{1}$$

where κ, the probability of electron transfer per passage through the intersection regions of two potential energy surfaces, and ρ, a preexponential term, are taken to be near unity for adiabatic reactions. The quantity Z is taken to be the collision frequency ($\sim 10^{11}$ sec^{-1}) between two uncharged reactant molecules in solution. Also, ΔF^* is not the measured free energy of activation but is related to it by Eq. (2):

$$\Delta F^{\ddagger} = \Delta F^* + RT\ln(kT/Zh) \tag{2}$$

where k and h are Boltzmann's and Planck's constants, respectively. Moreover, ΔF^* is related to a Coulombic term and various reorganizational terms by Eqs. (3)–(8):

$$\Delta F^* = w + m\lambda^2 - T\Delta S_e \tag{3}$$

At infinite dilution

$$w = e_1 e_2 / D_s r \tag{4}$$

and w is usually small enough to be neglected at the high ionic strength used in most reactions.[23] The static dielectric constant is D_s, and r is the internuclear distance in the collision complex. The constant m is defined by the relation

$$m = -\left(\frac{1}{2} + \frac{F^\circ + w^p - w}{2\lambda}\right) \tag{5}$$

where w^p is the work required to form products from the product-activated complex. The λ term arises from Franck–Condon restrictions on the electron transfer process, and for clarity, in the detailed mechanism the reorganizational energy is partitioned between the inner and outer coordination spheres of the reactants. Thus

$$\lambda = \lambda_i + \lambda_0 \tag{6}$$

$$\lambda_i = \sum_j \frac{k_j k_j^p}{k_j k_j^p} (\Delta q_j^0)^2 \tag{7}$$

where the k_j and k_j^p denote the force constants of the jth vibrational coordinate in a species involved in the reaction when that species is a reactant and product, respectively, and Δq_j^0 is the difference in equilibrium bond coordinates between products and reactants for the jth vibrational level. Also,

$$\lambda_0 = (\Delta e)^2 \left(\frac{1}{2a_1} + \frac{1}{2a_2} - \frac{1}{r}\right)\left(\frac{1}{n^2} - \frac{k}{D_s}\right) \tag{8}$$

TABLE 3-1
RATE CONSTANTS CALCULATED FROM THE MARCUS CROSS RELATION

Reaction	k_{12}^{obsd} (M^{-1} sec^{-1})	k_{12}^{calc} (M^{-1} sec^{-1})	k_{11}^{a} (M^{-1} sec^{-1})	k_{22}^{a} (M^{-1} sec^{-1})	ΔE, V(K_{12})	f	Ref. (k^{obsd})
1. $W(CN)_8^{4-}$—$Mo(CN)_8^{3-}$	5.0×10^6	$>3 \times 10^6$	$>4 \times 10^4$	3×10^4	$0.26(2.3 \times 10^4)$	4.0×10^{-1}	52
2. $W(CN)_8^{4-}$—$Fe(CN)_6^{3-}$	4.3×10^4	$>3 \times 10^4$	$>4 \times 10^4$	7.2×10^2	$0.23(7.2 \times 10^3)$	5.5×10^{-1}	52
3. $W(CN)_8^{4-}$—$IrCl_6^{2-}$	6.1×10^7	$>6 \times 10^7$	$>4 \times 10^4$	2.3×10^5	$0.39(3.5 \times 10^6)$	1.3×10^{-1}	52
4. $W(CN)_8^{4-}$—Ce(IV)	$>10^8$	3×10^8	$>4 \times 10^4$	4.4	$0.90(1.3 \times 10^{15})$	4×10^{-4}	52
5. $Mo(CN)_8^{4-}$—Ce(IV)	1.4×10^7	1×10^7	3×10^4	4.4	$0.64(5.6 \times 10^{10})$	1.6×10^{-2}	52
6. $Mo(CN)_8^{4-}$—$IrCl_6^{2-}$	1.9×10^6	8×10^5	3×10^4	2.3×10^5	$0.13(1.5 \times 10^2)$	8.0×10^{-1}	52
7. $Mo(CN)_8^{4-}$—MnO_4^-	2.67×10^2	6×10^1	3×10^4	3×10^3	$-0.24(8.6 \times 10^{-5})$	5.1×10^{-1}	53
8. $Mo(CN)_8^{4-}$—$HMnO_4^-$	1.87×10^7	2×10^7	3×10^4	1.10×10^3	$0.50(3.0 \times 10^8)$	5.8×10^{-2}	53
9. $Fe(CN)_6^{4-}$—$IrCl_6^{2-}$	3.8×10^5	1×10^6	7.4×10^2	2.3×10^5	$0.24(1.2 \times 10^4)$	5.1×10^{-1}	52
10. $Fe(CN)_6^{4-}$—$Fe(phen)_3^{3+}$	$\geqslant 10^8$	$>1 \times 10^8$	7.4×10^2	$>3 \times 10^7$	$0.39(3.5 \times 10^6)$	1.2×10^{-1}	54
11. $Fe(CN)_6^{4-}$—$Mo(CN)_8^{3-}$	3.0×10^4	4×10^4	7.4×10^2	3×10^4	$0.12(1.0 \times 10^2)$	8.5×10^{-1}	52
12. $Fe(CN)_6^{4-}$—MnO_4^-	1.7×10^5	6×10^4	7.4×10^2	3×10^3	$0.20(2.5 \times 10^3)$	6.6×10^{-1}	55
13. $Fe(CN)_6^{4-}$—Ce(IV)	1.9×10^6	6×10^6	7.4×10^2	4.4	$0.76(5.8 \times 10^{12})$	4.7×10^{-3}	52

14. $Fe(phen)_3^{2+}$—Ce(IV)	1.4×10^5	$>7 \times 10^6$	$>3 \times 10^7$	4.4	$0.36(1.1 \times 10^6)$	2.1×10^{-1}	56
15. $Fe(phen)_3^{2+}$—Mn(III)	1.8×10^3	$>2 \times 10^5$	$>3 \times 10^7$	3×10^{-4}	$0.44(2.5 \times 10^{12})$	1.8×10^{-1}	57
16. $Fe(phen)_3^{2+}$—MnO_4^-	6.1×10^3	>3	$>3 \times 10^7$	3×10^3	$-0.50(3.4 \times 10^{-9})$	2.4×10^{-2}	58
17. $Fe(phen)_3^{2+}$—$HMnO_4$	2.4×10^8	$>1 \times 10^7$	$>3 \times 10^7$	1.10×10^3	$0.24(1.2 \times 10^4)$	4.4×10^{-1}	58
18. $Fe(EDTA)^{2-}$–$Mn(CyDTA)^-$	$\sim 4 \times 10^4$	1×10^7	(3×10^4)	1.2	$0.69(4 \times 10^{11})$	1.2×10^{-2}	59
19. Fe^{2+}—$Fe(phen)_3^{3+}$	3.7×10^4	$>5 \times 10^6$	4.0	$>3 \times 10^7$	$0.35(7.6 \times 10^5)$	2.4×10^{-1}	59
20. Fe^{2+}—Ce(IV)	1.3×10^6	5×10^5	4.0	4.4	$0.71(8.3 \times 10^{11})$	2.0×10^{-2}	56
21. Fe^{2+}—$IrCl_6^{2-}$	3.0×10^6	2×10^4	4.0	2.3×10^5	$0.16(5.0 \times 10^2)$	8.9×10^{-1}	60
22. $Ru(NH_3)_6^{2+}$—Fe(III)	3.4×10^5	2×10^6	4.0×10^3	4.0	$0.67(1.7 \times 10^{11})$	1.7×10^{-2}	61
23. $Ru(en)_3^{2+}$—Fe(III)	8.4×10^4	4×10^5	2.0×10^2	4.0	$0.56(2.5 \times 10^9)$	5.7×10^{-2}	61
24. $Cr(EDTA)^{2-}$–$Fe(EDTA)^-$	$\geqslant 10^6$	1×10^{10}	(3×10^3)	(3×10^4)	$1.12(1.4 \times 10^{19})$	9.6×10^{-8}	59
25. Cr^{2+}—Fe^{3+}	2.3×10^3	$<1 \times 10^6$	$\leqslant 2 \times 10^{-5}$	4.0	$1.18(6.6 \times 10^{19})$	1.7×10^{-4}	62
26. Cr^{2+}—$Ru(NH_3)_6^{3+}$	2×10^2	$<3 \times 10^3$	$\leqslant 2 \times 10^5$	4×10^3	$0.51(3.7 \times 10^8)$	1.6×10^{-1}	63
27. Cr^{2+}—$IrCl_6^{2-}$	(10^6)	$<5 \times 10^8$	$\leqslant 2 \times 10^{-5}$	2.3×10^5	$1.34(3.2 \times 10^{22})$	1.2×10^{-6}	64
28. V^{2+}—$Ru(NH_3)_6^{3+}$	1.3×10^3	2×10^3	1.2×10^{-2}	4×10^3	$0.31(1.6 \times 10^5)$	4.7×10^{-1}	65
29. V^{2+}—$IrCl_6^{2-}$	4×10^6	7×10^8	1.2×10^{-2}	2.3×10^5	$1.11(1.6 \times 10^{19})$	1.1×10^{-5}	60

[a] k_{11} and k_{22} are the rate constants for self-exchange for reductant and oxidant, respectively, for the given redox pair.

where Δe is the number of electrons transferred, a_1 and a_2 are the respective reactant radii, and n is the refractive index of the solvent. The terms r and D_s are defined above. The latter term requires that the reactants be spherical and is based on what Marcus[8] terms an equivalent equilibrium distribution—i.e., on ". . . a system having the charges of the reactants a distance r apart but a distribution of orientations of solvent molecules and of positions of ions in the ionic atmosphere which would be in equilibrium with the hypothetical charges, $e_n + m(e_n{}^p)$, $n = 1{,}2$." [8] For example, in an exchange reaction where ($m = -\frac{1}{2}$) between, say, iron(II) and iron(III), the hypothetical charge on each of the two central ions is $[2 - \frac{1}{2}(2 - 3)]$ or $2\frac{1}{2}$. Therefore, the solvent molecules are oriented about the ion cluster as if each ion had the given hypothetical charge. Finally, the free energy contribution resulting from any change in multiplicity accompanying the reaction is embodied in the term $T\Delta S_e$, where $\Delta S_e = R\ln \Omega/\Omega^*$ and Ω^* and Ω are the products of ionic multiplicities in the reactants' and products' activated complexes, respectively. Eqs. (1)–(8) were first developed by treating the solvent as a dielectric continuum and then were confirmed in a more general statistical mechanical treatment.[10] To implement the computation of k_{bi}, the equations are used now in reverse order.

Marcus[9] has derived a remarkably simple expression for the calculation of rate constants for so-called cross reactions. A cross reaction is one in which the two reactants differ in some respect other than oxidation state or isotopic composition. They can be two different metal ion complexes or different complexes of the same metal (in different oxidation states). The cross relation is

$$k_{12} = (k_{11}k_{22}K_{12}f)^{1/2} \tag{9}$$

where k_{11} and k_{22} are the respective self-exchange reactions, K_{12} is the equilibrium constant for the reaction, and f is a correction term for the difference in free energies of the two reacting species:

$$\log f = (\log K_{12})^2/[4 \log(k_{11}k_{22}/Z^2)] \tag{10}$$

The parameter Z is the collision frequency for two uncharged molecules, and it is generally assumed to have a value of 10^{11} sec^{-1}. Newton[50] has given an alternate derivation of the cross relation on a more elementary level than the Marcus treatment. More recently, Rosseinsky[51] modified the cross relation to accommodate the fraction of electron transfer in the transition state. In the following paragraphs the Marcus cross relationship is analyzed, tested against experimentally determined cross-reaction rate constants, and utilized to estimate several self-exchange rate constants which are otherwise difficult of access.

B. Comparison of Marcus' Cross Relationship with Experimental Data

Table 3-1 summarizes a variety of cross reactions for which calculations have been performed to test the Marcus' cross relation. Cross reactions involving cobalt couples are discussed separately. An examination of columns 2 and 3 reveals a remarkable consistency between calculated and observed rate constants for the most part. About two-thirds of the values agree to within an order of magnitude despite variations in K_{12} (column 6) exceeding 30 orders of magnitude and despite self-exchange rate constants (columns 4 and 5) ranging over at least 14 orders of magnitude. There are, however, some notable exceptions (cf. Table 3-1, entries 14–16, 18–19, 21, 25, and 29) to the theory.

Note that each of the systems (except entry 29) not conforming to the theory involves at least one d^6 reactant. However, this is not a sufficient condition for exceptional behavior since other iron(II) and cobalt(III) systems do conform to the theory. Other reasons for the lack of agreement between observed and calculated constants may reside in the uncertainty of mechanism for the self-exchanges—e.g., $Fe^{2+,3+}$—and cross reactions or in the availability of only lower limits for some of the self-exchange constants. However, higher values would not in most cases improve the agreement, especially among the iron–polypyridine complexes.

To simplify the following discussion of redox systems involving macrocyclic ligands the following structures, their names, and the abbreviations to be used are given.

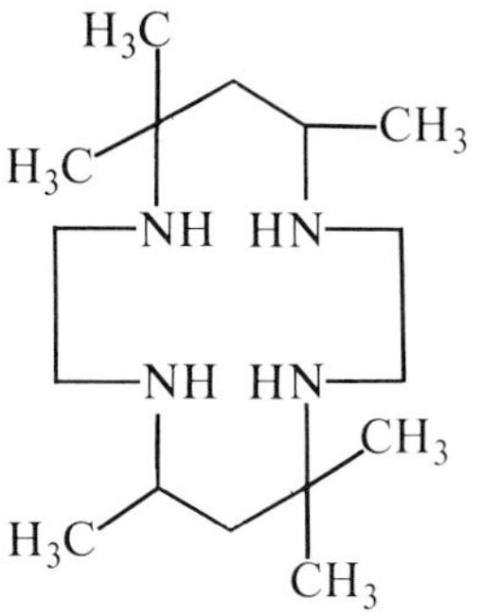

H3C
H3C
CH3
NH HN
NH HN
CH3
H3C
CH3

teta

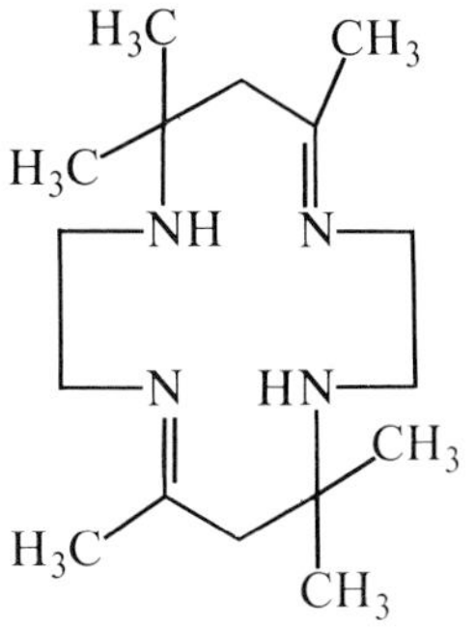

H3C
CH3
H3C
NH N
N HN
CH3
H3C
CH3

trans-[14]diene
(hmtad)

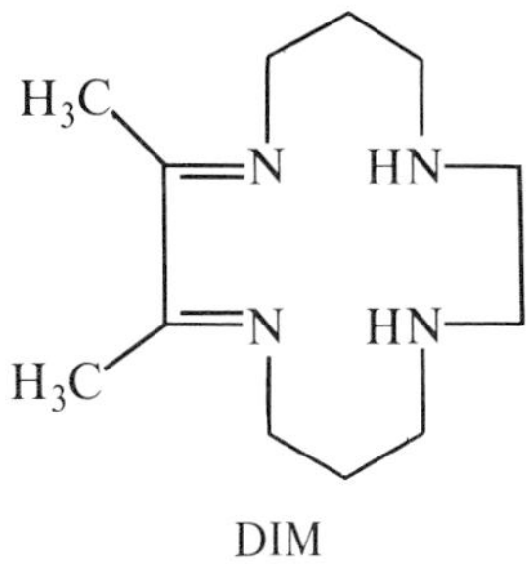

H3C
N HN
N HN
H3C

DIM

H3C
CH3
N N
N N
H3C
CH3

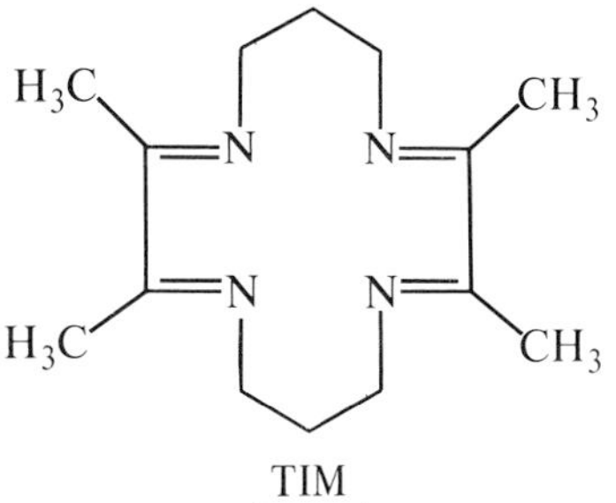

TIM
(tmtat)

teta = 5,7,7,12,14,14-hexamethyl-1,4,8,11-tetraazacyclotetradecane

trans-[14]diene = 5,7,7,12,14,14-hexamethyl-1,4,8,11-tetraazacyclotetradeca-4,11-diene

DIM = 2,3-dimethyl-1,4,8,11-tetraazacyclotetradeca-1,3-diene

TIM = 2,3,9,10-tetramethyl-1,4,8,11-tetraazacyclotetradeca-1,3,8,10-tetraene

Table 3-2 summarizes comparisons of observed vs. calculated rate constants involving cobalt as both an oxidant and a reductant. (Additional data involving macrocyclic complexes for which the Marcus theory is inadequate can be found in Table 3-4 (entries 1–6) under nonbridging ligand effects for outer-sphere reactions.) Agreement of k_{12}^{obsd} and k_{12}^{calc} within a factor of 10 is assumed satisfactory. On this basis there are at least 17 entries which agree and many more which do not agree. For entries 1–4 in Table 3-2 Przystas and Sutin[66] report better agreement (within 10X) than that in Table 3-2 by using a higher estimate ($10^{-9}M^{-1}sec^{-1}$) for $Co(NH_3)_6^{2+,3+}$ self-exchange. No justification was given for this revised estimate, however. Remember that E^0 is not accurately known and that the rate constant for exchange is extrapolated over 40° using an assumed energy of activation. For entries 17–30, Rillema and Endicott[70] used a linear free energy relationship of the form $\Delta G^{\ddagger}_{21} = a_{21} + b\Delta G^0_{21}$ for various oxidations of $CoTIM^{2+}$ and $Co(trans[14]diene)^{2+}$. The values of b were about 0.33 ± 0.05 instead of 0.50 as predicted by the Marcus equation. By contrast, the reductions of the same type of cobalt(III) macrocyclic complexes exhibit the correct dependence on the standard free energy but differ in the free energy independent term a.[70,74,75,76] In Marcus' theory this term is ascribed to the reorganizational barriers, $\lambda_{1,2} = \frac{1}{2}(\lambda_{Co} + \lambda_{red})$. For cobalt(III) complexes the dependence of $\Delta G^{\ddagger}_{12}$ on λ_{Co} is less than predicted by the theory, and the discrepancies become larger as the difference between λ_{Co} and λ_{red} becomes larger.[70] The reason(s) for the discrepancy, whether tunneling of the Franck–Condon barrier, failure of the theoretical model, or some unsuspected contribution, is not known. An additional confusing factor is that some of the macrocyclic ligand complexes—e.g., entries 17, 22, 25–27—as well as the all-cobalt cross reactions (entries 31–35) agree acceptably with the theory.

Finally, Hyde, Davies, and Sykes[73] combined literature data (entries 38–42) with their data on the chromium(II) and vanadium(II) acid independent reductions of aquo cobalt(III) and noted marked deviations from Marcus' theory for plots of $\Delta G^{\ddagger}_{12}$ vs. ΔG^0_{12}. However, they also observed the hint of a minimum in the plot for a very large (55–60 kcal/mol) ΔG^0_{12} as predicted by Marcus[77]—$\alpha \cong (1 + \Delta G^0_{12}/4\Delta G^{\ddagger}_0)/2$ where $\alpha = d(\Delta G^{\ddagger}_{12})/d(\Delta G^0_{12})$ and $\Delta G^{\ddagger}_0$ = free energy of activation for the $Co^{2+,3+}$ exchange (16.7 kcal/mol). In these and other systems the work terms are neglected, and it remains to be determined from theory whether such terms could account for some of the observed discrepancies.

TABLE 3-2

COMPARISON OF RATE CONSTANTS OBSERVED FOR CROSS REACTIONS INVOLVING COBALT WITH THOSE CALCULATED BY THE MARCUS EQUATION

Reaction	k_{12}obsd	k_{12}calc	k_{11}	k_{22}	ΔE, V(K_{12})	f	Ref. k_{12}obsd
1. Cr^{2+}—$Co(NH_3)_6^{3+}$	1.0×10^{-3}	$<8 \times 10^{-5}$	$<2 \times 10^{-5}$	$<3 \times 10^{-12}$	$0.51(44 \times 10^{8})$	3.2×10^{-1}	66
2. $Ru(NH_3)_6^{2+}$—$Co(NH_3)_6^{3+}$	1.1×10^{-2}	$<1 \times 10^{-4}$	4×10^{3}	$<3 \times 10^{-12}$	$\sim 0(1)$	~ 1	63
3. V^{2+}—$Co(NH_3)_6^{3+}$	1.0×10^{-2}	$<1.3 \times 10^{-4}$	1.2×10^{-2}	$<3 \times 10^{-12}$	$0.36(1.3 \times 10^{6})$	5.5×10^{-1}	66
4. Eu^{2+}—$Co(NH_3)_6^{3+}$	2×10^{-2}	$<3 \times 10^{-4}$	$<10^{-4}$	$<3 \times 10^{-12}$	$0.53(9.5 \times 10^{8})$	3.0×10^{-1}	67
5. Cr^{2+}—$Co(en)_3^{3+}$	3.4×10^{-4}	$<5 \times 10^{-4}$	$<2 \times 10^{-5}$	2×10^{-5}	$0.17(7.6 \times 10^{2})$	8.5×10^{-1}	66
6. V^{2+}—$Co(en)_3^{3+}$	7.2×10^{-4}	6×10^{-4}	1.2×10^{-2}	2×10^{-5}	$0.015(1.8)$	1	66
7. $CrEDTA^{2-}$—$CoEDTA^{-}$	3×10^{5}	3×10^{7}	(3×10^{3})	(2×10^{-7})	$1.4(5.4 \times 10^{23})$	2.6×10^{-6}	59
8. $FePDTA^{2-}$—$CoEDTA^{-}$	1.3×10^{1}	1×10^{1}	(3×10^{4})	(2×10^{-7})	$0.28(5.6 \times 10^{4})$	5.9×10^{-1}	59
9. $FePDTA^{2-}$—$CoDyDTA^{-}$	1.2×10^{1}	1×10^{1}	(3×10^{4})	(2×10^{-7})	$0.25(1.7 \times 10^{4})$	8.3×10^{-1}	59
10. Cr^{2+}—$Co(phen)_3^{3+}$	3.0×10^{1}	$<1 \times 10^{4}$	$<2 \times 10^{-5}$	4.8	$0.83(1.2 \times 10^{14})$	1.3×10^{-2}	66
11. V^{2+}—$Co(phen)_3^{3+}$	3.8×10^{3}	2.7×10^{4}	1.2×10^{-2}	4.8	$0.68(3.4 \times 10^{11})$	3.7×10^{-2}	66
12. V^{2+}—$Co(bipy)_3^{3+}$	1.1×10^{3}	1.7×10^{4}	1.2×10^{-2}	9.0	$0.63(4.8 \times 10^{10})$	5.8×10^{-2}	66
13. $CoEDTA^{2-}$—$MnCyDTA^{-}$	9×10^{-1}	8×10^{-1}	2×10^{-7}	(1.2)	$0.41(8.9 \times 10^{6})$	3.8×10^{-1}	59
14. $CoEDTA^{2-}$—$Fe(phen)_3^{3+}$	9.1×10^{4}	$>1 \times 10^{6}$	2×10^{-7}	$>3 \times 10^{7}$	$0.70(7.2 \times 10^{11})$	2.2×10^{-1}	68
15. $CoEDTA^{2-}$—$IrCl_6^{2-}$	4×10^{3}	4×10^{3}	2×10^{-7}	2.3×10^{5}	$0.56(3.1 \times 10^{9})$	1.1×10^{-1}	69
16. $Co(phen)_3^{2+}$—Fe^{3+}	5.3×10^{2}	1.5×10^{3}	4.8	4.0	$0.32(2.6 \times 10^{5})$	4.8×10^{-1}	66
17. $Co(TIM)^{2+}$—Ce(IV)	1.8×10^{6}	4×10^{6}	9×10^{-1}	4.6	$0.90(1.8 \times 10^{15})$	1.9×10^{-3}	70
18. $Co(TIM)^{2+}$—$Ru(bipy)_3^{3+}$	1.0×10^{7}	6×10^{8}	9×10^{-1}	10^{8}	$0.72(1.6 \times 10^{12})$	2.2×10^{-3}	70
19. $Co(TIM)^{2+}$—$Fe(bipy)_3^{3+}$	6.5×10^{5}	$>5 \times 10^{7}$	9×10^{-1}	$>3 \times 10^{7}$	$0.56(3.1 \times 10^{9})$	2.8×10^{-2}	70
20. $Co(TIM)^{2+}$—$Fe(phen)_3^{3+}$	3.5×10^{5}	$>9 \times 10^{7}$	9×10^{-1}	$>3 \times 10^{7}$	$0.52(6.5 \times 10^{8})$	4.7×10^{-2}	70
21. $Co(TIM)^{2+}$—$Fe(Me_2\text{-}phen)_3^{3+}$	9.2×10^{4}	$>2 \times 10^{6}$	9×10^{-1}	$>3 \times 10^{7}$	$0.34(5.8 \times 10^{5})$	2.7×10^{-1}	70

TABLE 3-2 CONTINUED
COMPARISON OF RATE CONSTANTS OBSERVED FOR CROSS REACTIONS INVOLVING COBALT WITH THOSE CALCULATED BY THE MARCUS EQUATION

Reaction	k_{12}^{obsd}	k_{12}^{calc}	k_{11}	k_{22}	ΔE, V(K_{12})	f	Ref. k_{12}^{obsd}
22. $Co(TIM)^{2+}$–$Fe(Me_4$-$phen)_3^{3+}$	4.1×10^4	3×10^5	9×10^{-1}	1.7×10^7	$0.31(1.8 \times 10^5)$	3.4×10^{-1}	70
23. $Co(TIM)^{2+}$–Fe^{3+}	1.0×10^3	2×10^1	9×10^{-1}	4.0	$0.23(7.9 \times 10^3)$	6.6×10^{-1}	70
24. Co([14] diene)$^{2+}$–Ce(IV)	7.9×10^5	$<3 \times 10^3$	$<2 \times 10^{-7}$ (70°C)	4.6	$0.88(8.3 \times 10^{14})$	1.0×10^{-2}	70
25. Co([14] diene)$^{2+}$–Ru-$(bipy)_3^{3+}$	1.7×10^6	$<5 \times 10^5$	$<2 \times 10^{-7}$	10^8	$0.70(7.2 \times 10^{11})$	2.0×10^{-2}	70
26. Co([14] diene)$^{2+}$–Fe-$(bipy)_3^{3+}$	6.9×10^4	$<3 \times 10^4$	$<2 \times 10^{-7}$	$>3 \times 10^7$	$0.54(1.4 \times 10^9)$	1.0×10^{-1}	70
27. Co([14] diene)$^{2+}$–Fe-$(phen)_3^{3+}$	1.1×10^5	$<1.6 \times 10^4$	$<2 \times 10^7$	$>3 \times 10^7$	$0.50(3.0 \times 10^8)$	1.4×10^{-1}	70
28. Co([14] diene)$^{2+}$–Fe-$(Me_2phen)_3^{3+}$	1.8×10^4	$<8 \times 10^2$	$<2 \times 10^{-7}$	$>3 \times 10^7$	$0.32(2.6 \times 10^5)$	4.6×10^{-1}	70
29. Co([14] diene)$^{2+}$–Fe-$(Me_4phen)_3^{3+}$	3.2×10^3	$<4 \times 10^2$	$<2 \times 10^{-7}$	1.7×10^7	$0.29(8.3 \times 10^4)$	5.2×10^{-1}	70
30. Co([14] diene)$^{2+}$–Fe^{3+}	1.5×10^2	$<5 \times 10^{-2}$	$<2 \times 10^{-7}$	4.0	$0.21(3.6 \times 10^3)$	7.8×10^{-1}	70
31. $Co(bipy)_3^{2+}$–$Co(terpy)_2^{3+}$	2.7×10^1	4×10^1	9.0	5×10^1	$-0.03(3.1 \times 10^{-1})$	1.01	71
32. $Co(terpy)_2^{2+}$–$Co(bipy)_3^{3+}$	6.4×10^1	4×10^1	4.8×10^1	9.0	0.03(3.2)	9.9×10^{-1}	71
33. $Co(terpy)_2^{2+}$–$Co(bipy)^{3+}$	6.8×10^2	2×10^4	4.8×10^1	4×10^{-2}	$0.53(9.5 \times 10^8)$	1.2×10^{-1}	71
34. $Co(terpy)_2^{2+}$–Co-$(phen)_3^{3+}$	2.8×10^2	8×10^1	4.8×10^1	4.8	$0.09(3.4 \times 10^1)$	9.3×10^{-1}	71
35. $Co(terpy)_2^{2+}$–$Co(phen)^{3+}$	1.4×10^3	2×10^4	4.8×10^1	4×10^{-2}	$0.53(9.5 \times 10^8)$	1.2×10^{-1}	71
36. $Co(terpy)_2^{2+}$–Co(III)	7.4×10^4	3×10^9	4.8×10^1	6×10^{-1}	$1.5(2.6 \times 10^{25})$	1.4×10^{-8}	71

37. $Co(phen)^{2+}$–$Co(bipy)^{3+}$	6×10^{-2}	4×10^{-2}	4×10^{-2}	4×10^{-2}	0(1)	1	71
38. $Fe(phen)_3{}^{2+}$—Co^{3+}	1.4×10^{4}	$>9 \times 10^{8}$	$>3 \times 10^{7}$	3.3	$0.76(7.6 \times 10^{12})$	1.1×10^{-3}	52
39. $Fe(5\text{-Me–phen})_3{}^{2+} + Co^{3+}$	1.5×10^{4}	$>1 \times 10^{9}$	$>3 \times 10^{7}$	3.3	$0.80(3.6 \times 10^{13})$	5.1×10^{-4}	52
40. $Fe(5\text{-Cl–phen})_3{}^{2+} + Co^{3+}$	5.0×10^{3}	$>5 \times 10^{8}$	$>3 \times 10^{7}$	3.3	$0.70(7.2 \times 10^{11})$	3.0×10^{-3}	52
41. $Fe(5\text{-NO}_2\text{–phen})_3{}^{2+} + Co^{3+}$	1.5×10^{3}	$>1 \times 10^{8}$	$>3 \times 10^{7}$	3.3	$0.57(4.6 \times 10^{9})$	2.1×10^{-2}	52
42. $Fe^{2+} + Co^{3+}$	2.5×10^{2}	5.8×10^{7}	4.0	3.3	$1.09(3.0 \times 10^{18})$	8.5×10^{-5}	72
43. $V^{2+} + Co^{3+}$	8.8×10^{5}	2.4×10^{10}	1.2×10^{-2}	3.3	$2.12(8.5 \times 10^{35})$	1.7×10^{-14}	73
44. $Cr^{2+} + Co^{3+}$	1.2×10^{4}	$<3 \times 10^{11}$	$<2 \times 10^{-5}$	3.3	$2.27(3.0 \times 10^{38})$	4.7×10^{-12}	73

[a] k_{11} and k_{22} are the rate constants for self-exchange for reductant and oxidant, respectively, for the given redox pair.

On two previous occasions the sluggish $Co(NH_3)_6^{2+,3+}$ self-exchange reaction has been noted. In the absence of a substantial Franck–Condon barrier, Stynes and Ibers[30] concluded that spin multiplicity restrictions must account for the very slow rate observed. Assuming that a low-spin cobalt(II) species is the reactive one, they calculated a difference of 24.6 kcal between the $^4T_{1g}$ ground state and the first excited doublet state $^2E_g(t_{2g})^6(e_g)^1$ and noted that the value was consonant with the very slow self-exchange rate. Their mechanism is, therefore:

$$^4T_{1g}Co(NH_3)_6^{2+} \overset{K}{\rightleftarrows} {}^2E_gCo(NH_3)_6^{2+}$$

$$^2E_gCo(NH_3)_6^{2+} + {}^1A_{1g}Co(NH_3)_6^{3+} \xrightarrow{k_{rate}} {}^1A_{1g}Co(NH_3)_6^{3+} + {}^2E_gCo(NH_3)_6^{2+}$$

$$^2E_gCo(NH_3)_6^{2+} \rightarrow {}^4T_{1g}Co(NH_3)_6^{2+}$$

There are several more recent developments, however, which tend to obscure rather than clarify the issue at hand. Studies on the rates of intersystem crossover by Beattie et al.[78,79] have shown that changes in spin multiplicity can be quite rapid—e.g., $t_{1/2}$ = 15–30 nsec. *If* these measurements are typical of other systems, spin multiplicity changes cannot be rate-determining steps in electron transfer reactions. In addition, it has been noted[32] that the self-exchange rates of the $Co(NH_3)_6^{2+,3+}$ and *trans*-Co[14]diene-$N_4(H_2O)_2^{2+,3+}$ systems are comparable in spite of the fact that the latter is a low-spin complex. Moreover Bodek and Davies[80] have cited unpublished calculations by Wherland and Gray indicating an activation energy of only 13.4 kcal for the "nonvertical" transition between the ground state $^1A_{1g}$ and the $^5T_{2g}$ state in $Co(NH_3)_6^{3+}$. The 13.4-kcal value is about one-fifth of the value for the corresponding "vertical" transition energy of 62.5 kcal and about one-half of that for the $^4T_{1g} \rightarrow {}^2E_g$ transition in $Co(NH_3)_6^{2+}$ calculated by Stynes and Ibers. In summary, it would appear that the spin crossover rates and the slow self-exchange rate for the low-spin macrocyclic complex are consistent with the Marcus theory which relegates only a minor role to spin multiplicity effects.

C. Discussion of Additional Mechanistic Details: Stereospecificity, Self-Exchange Calculations, and Negative $\Delta H^‡$

Sutter and Hunt[81] have reported a most unusual outer-sphere reaction between optically active $Co(phen)_3^{3+}$ and racemic $Cr(phen)_3^{2+}$ to produce optically active $Cr(phen)_3^{3+}$. A nonspherically symmetric spatial distribution for the reducing electron was suggested to account for the asymmetric induction. In an elegant series of NMR studies of such outer-sphere chromium(II) reductants La Mar and Van Hecke[82] confirmed that there are, in fact, significant M → L π-contact shifts at the $C_{4,7}$ positions on various substituted phenanthroline ligands. These results suggest strongly a possible mechanism and site for stereospecific electron transfer in accord with the suggestion of Sutter and Hunt. La Mar and Van Hecke also noted that molecular models suggest optimum contact between the $C_{4,7}$ P_z orbitals of the reactants along the C_3 axis when the two reactants are of the same chirality rather than mirror images. However, because of their additional observation that strong solvent interaction exists at the same site on the reductant, they could not definitely rule out a solvent-mediated mechanism. While the above results are novel, a reexamination of the earlier system[81] by Kane-Maguire, Tollinson, and Richards[83] along with an expanded survey of various analogous cobalt(III)–chromium(II) systems failed to reveal any stereospecific effects for these outer-sphere processes. A precedent for this conclusion was set in an earlier study by Grossman and Wilkins[85] for a series of all-cobalt redox processes.

In connection with the above noted contact shifts, another study of such an effect is noteworthy. Goedken and Peng[84] recently interpreted the anomalous NMR spectra of some organocobalt(III) macrocyclic ligand complexes of the type

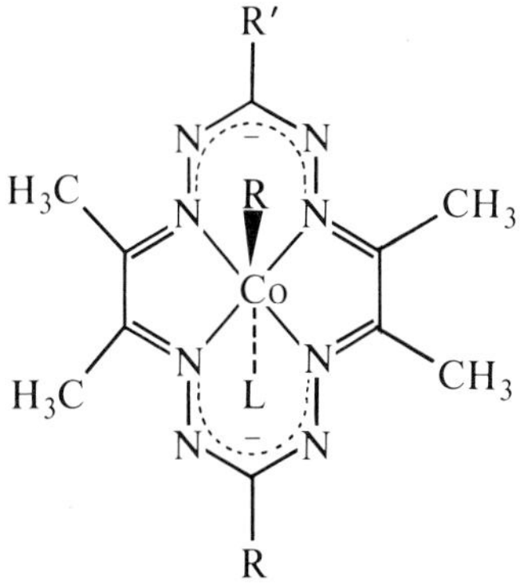

R = Me, Et, Ph
R′ = H, Me, Ph
L = py, CN^-, MeCN, $MeNHNH_2$

to contact shifts arising from a *thermally* populated triplet state of cobalt(III). Except for the fact that there is a totally delocalized π system in the macrocyclic ligand above, there is a striking similarity to $Co(TIM)^{3+}$ complexes reported by Endicott et al. discussed earlier. In view of the above results it is tempting to suggest that there may exist a direct correlation between the self-exchange rates and the thermal accessibility of a paramagnetic triplet state in the CoL_4^{3+} complexes, being greatest for L_4 = TIM > DIM $\approx$ *trans*-[14]diene $\gg$ teta. Larger contact shifts would be expected for the CoL_4^{2+} paramagnetic ground state complexes.

Sykes and co-workers have carried on an active program on the redox chemistry of binuclear cobalt(III) complexes. Earlier work has been reviewed by Sykes and Weil[86] and by Hyde, Scott, and Sykes.[87] The reactions normally proceed in two steps to produce two moles of cobalt(II) per mole of binuclear complex. Scott and Sykes[88] have reported that both chromium(II) and vanadium(II) react with complexes of the type $(NH_3)_4Co(NH_2,OCRO)Co(NH_3)_4{}^{4+}$ where R = H or CH_3 via outer-sphere mechanisms. Weighardt and Sykes[89] also report that $(NH_3)_3Co(OH,OH,OCRO)Co(NH_3)_3{}^{3+}$ is reduced by these ions analogously. Scott and Sykes[90] likewise report that similar reductions of oxalato bridged, tetranuclear complexes of cobalt(III) occur. In all three systems ring opening is competitive with the chromium(II) reductions, and the rate ratio method of Toppen and Linck[91] is used in each case to infer the outer-sphere mechanism for vanadium(II)—i.e., $k_{Cr}/k_V \leq 0.025$.

Several interesting, miscellaneous outer-sphere reactions have been reported. Two studies dealing with reductions by chromium(II)–cyanide species indicate that $Cr(CN)_6{}^{3-}$ is reduced by $Cr(CN)_5OH^{4-}$ [92] while $Co(CN)_5Br^{3-}$ is reduced by $Cr(CN)_6{}^{4-}$.[93] Actually, the secondary reaction in the latter study is more novel than the primary one. The $Cr(CN)_6{}^{4-}$ is utilized in a follow-up reaction to produce $Co(CN)_5H^{3-}$, an active hydrogenating agent for α,β-unsaturated compounds, from $Co(CN)_5{}^{3-}$ and solvent water. Three additional but unrelated studies are the chromium(VI) oxidation of ferrocene,[94] the second-order approach to equilibrium for the reaction $Np^{III} + Ru^{III} = Np^{IV} + Ru^{II}$ with $Ru^{III} = Ru(NH_3)_6{}^{3+}$ or $Ru(NH_3)_5H_2O^{3+}$,[95] and the novel $Fe(Me_2dtc)_3{}^{III,IV}$ exchange reaction where Me_2dtc = dimethyldithiocarbamate.[96]

Application of the Marcus cross relation to the calculation of self-exchange rate constants has met with limited success. Thus, Pladziewicz and Espenson[97] recast the cross relation into the form $k_{ii} + k_{jj} = A_{ij}$ for a series of reactions among eight variously substituted ferrocenes and obtained the "best fit" for some 22 cross reaction constants ranging from 4.2–150 × $10^6 M^{-1}$ sec^{-1}. A plot of $k_{ij}{}^{obsd}$ vs. $k_{ij}{}^{calc}$ using the derived self-exchange values was linear, having a slope of 0.55 in excellent agreement with theoretical expectation. The ferrocene–ferrocenium exchange constant which was calculated to be 5.7 × $10^6 M^{-1}$ sec^{-1} (at 25°C) has been more recently confirmed by Yang, Chan, and Wahl using a line-broadening NMR technique.[98] An earlier value of 1.7 × $10^6 M^{-1}$ sec^{-1} (at −70°C) had been reported by Stranks. These results disagree substantially with recent estimates based on intervalence transfer data[99] and a diffusion coefficient measurement.[100] Use of the cross relation to obtain the $Cu^{+,2+}$ self-exchange constant yields values of 9 × 10^2 and 6 × 10^3 from the $Ru(NH_3)_6{}^{2+}$—Cu^{2+} and V^{2+}—Cu^{2+} reactions, respectively.[63,67] Stasiw and Wilkins[54] have examined several substituted tetra- and pentacyanoferrate(II,III) systems.

Recently, Braddock and Meyer[101] and Cramer and Meyer[102] found that a number of outer-sphere oxidations of iron(II) by polypyridine complexes of ruthenium(III) and iron(III) exhibit negative enthalpies of activation. Application of the Marcus equation to calculate the respective ruthenium and iron polypyridine self-exchange rates led to values which were about three orders of magnitude below reasonable estimates. These results prompted the suggestion[101,102] that a "non-Marcus" contribution was involved in the various cross reactions with iron(II). However, Marcus and Sutin[103] have subsequently shown that the Marcus theory can accommodate negative activation enthalpies on the basis of the following equation derived by them.

$$\Delta H^{\ddagger}_{12} + \tfrac{1}{2}RT = \Delta H^{*}_{12} = [\Delta H^{*}_{11}/2 + \Delta H^{*}_{22}/2](1 - 4\alpha^2) + (\Delta H^{0}_{12}/2)(1 + 2\alpha)$$

$$\alpha = \Delta G^{0}_{12}/4(\Delta G^{*}_{11} + \Delta G^{*}_{22})$$

They noted that ΔH^* (and hence $\Delta H^{\ddagger}$) can become negative when ΔH^{0}_{12} is sufficiently negative. Marcus and Sutin also discussed a molecular interpretation of the negative activation enthalpies, noting that the large negative ΔS^{0}_{12} values imply wide separation of the quantum states of products as compared with reactants. The activated complexes and their respective energies at the intersections of the reactant and product potential energy surfaces are quantum-state dependent. The Boltzmann weighted average of these energies for very negative ΔH^{0}_{12} can therefore lie below the average energies of reactions. In solution, the given energies are essentially equal to the thermodynamic heat contents of the system, and hence negative enthalpies of activation are possible.

Interestingly, Cobble[104,105] had earlier predicted that retrograde kinetics can be observed for reactions in aqueous media where the activation enthalpies approach zero. Thus

$$\Delta H^*(T_2) = \Delta H^*(T_1) + \Delta C^*_p(T_2 - T_1)$$

where ΔC^*_p is the heat capacity of activation. Values of ΔC^*_p have been estimated to range between 100 and −70 cal/mol·deg.[104] For small temperature changes the ΔC^*_p term has been assumed negligible; however, if the above ΔC^*_p values are typical, the maximum contribution from this term would be −1.4 kcal/mol for $\Delta T = 20°$.

Although a majority of the reactions involving chromium(II) and vanadium(II) are of the inner-sphere type, some oxidants do not permit the formation of bridged binuclear intermediates, thus necessitating an outer-sphere mechanism. Among the known outer-sphere reactions for chromium(II) are the reductions of $Co(NH_3)_6^{3+}$,[106] $Ru(NH_3)_6^{3+}$,[63] $Ru(NH_3)_5py^{3+}$, and $Ru(NH_3)_5$-(methyl nicotinate)$^{3+}$ [107] and various RoX^{3+} complexes where Ro = $Co(NH_3)_5^{3+}$ and X = urea, cyanamide, *N*-cyanoguanidine,[108] imidazole, pryazole, pyridine, DMF, or *N,N*-dimethylnicotinamide[109] (3- and 4-methylpyridine).[110] Chen and Gould[109] and more recently Fan and Gould[111] have found excellent correlations of the type $\log k_{V^{2+}} = 1.10 \log k_{Cr^{2+}} + 1.85$ and $\log k_{Ru(NH_3)_6^{2+}} = 1.05 \log k_{M^{2+}} + b$, where M = Cr^{2+}, Eu^{2+}, or V^{2+} and b is constant. The utility of such relationships in calculating outer-sphere components for predominantly inner-sphere reactions will be explored in section II, D. Similar outer-sphere reductions of aromatic nitrile complexes of pentaaminecobalt(III) by chromium(II) have been reported by Balahura, Wright, and Jordan.[112] Similarly, the reactions of vanadium(II) with the last five RoX^{3+} complexes above, $Ru(NH_3)_6^{3+}$ [65] and $Co(NH_3)_6^{3+}$,[106] as well as reductions of the latter by Cu^+,[108] Zn^+, Cd^+, Ni^+,[113] and Eu^{2+} [67] are outer-sphere because no bridging ligand is accessible for these oxidants in acidic solution. In addition to these systems there are a growing number of reductions by vanadium(II) in which the rate constants exceed that for substitution of water on the vanadium(II) ion (90 sec^{-1} or, dividing by the concentration of water, $1.7M^{-1}$ sec^{-1}) as measured by ^{17}O NMR.[114] For example, the reductions of $Fe(H_2O)_5NCS^{2+}$ and $Fe(H_2O)_5N_3^{2+}$,[115] $Ru(H_2O)_5Cl^{2+}$,[116] $Ru(NH_3)_5Cl^{2+}$, and *cis*-$Ru(NH_3)_4Cl_2^+$,[117] $Ru(NH_3)_5X^{2+}$ ($X = Br^-, CH_3CO_2^-, HCO_2^-$),[118] the hydrated form of $Ru(NH_3)_5$(4-formylpyridine)$^{3+}$,[119] $PtCl_6^{2-}$,[120] and numerous cobalt(III) macrocyclic ligand complexes (see Table 3-2) all have rate constants substantially in excess of $1.7M^{-1}$ sec^{-1}. Green, Schug, and Taube[121] have cautioned (footnote 16, Ref. 121) that the maximum rate for electron transfer is set by the water exchange results only if the mechanism is S_N1. Halpern[122] has suggested that the absence of significant isotopic fraction factors (for the nitrogen atoms of ammonia) in the chromium(II) reductions of various *cis*- and *trans*-$Coen_2NH_3X^{2+}$ complexes may indicate limiting substitution on chromium(II). More recently, Sutin[123] suggested that rate-limiting substitution on chromium(II) may account for the lack of discrimination which this reductant exhibits toward $Fe(H_2O)_5NCS^{2+}$ and $Fe(H_2O)_5N_3^{2+}$.[124] Overall, the evidence is less than compelling.

For most outer-sphere reactions it has not been possible to separate the detailed steps of precursor complex formation and electron transfer. However, Gaswick and Haim[125] have reported the direct measurement of an elementary, outer-sphere electron transfer within the ion pair, $[Co(NH_3)_5H_2O^{3+}]\cdot[Fe(CN)_6^{3-}]$, ($Q = 1.5 \times 10^3 M^{-1}$), with $k_{e.t.} = 1.0 \times 10^{-1}$ sec^{-1}. More recent attempts using $Co(phen)_3^{3+}$ as the oxidant were unsuccessful.[126] This result parallels an earlier direct measurement of electron transfer within an inner-sphere precursor by Cannon and Gardiner.[127]

D. Excited State Electron Transfer

One of the highly interesting developments in electron transfer chemistry and closely allied to the question of spin multiplicity is the observation of transient intermediates in the reductions of various bipyridyl and terpyridyl metal complexes by the hydrated electron.[128,129,130] Although the possibility of excited state cobalt(II) had previously been considered,[131] no such species was observed. Waltz and Pearson[128] first observed such a transient in the $Co(bipy)_3^{3+} - e^-_{aq}$ reaction, noting that: (1) the primary reaction is diffusion controlled; (2) the final product is "normal" $Co(bipy)_3^{2+}$; (3) N_2O, an efficient scavenger for e^-_{aq}, reduces the appearance of the intermediate; (4) decay of the intermediate is faster than that of $Co(bipy)_3^{2+}$, is catalyzed by $Co(bipy)_3^{3+}$, and involves substitution; (5) the intermediate is scavenged by known organic triplet scavengers; and (6) the molar absorptivity of the intermediate is higher than the normal $Co(bipy)_3^{2+}$ as is that of the low-spin $Co(terpy)_2^{2+}$. These observations were interpreted on the basis of an excited state, low-spin cobalt(II) complex. The estimated first-order rate constant for the change in spin multiplicity based on this interpretation was calculated to be 5×10^3 sec^{-1}, a value which is substantially below that observed in a subsequent study[79] but in another system.

More recently, Baxendale and Fiti[129] confirmed most of the earlier observations of Waltz and Pearson[128] and extended the original observations to include $Ru(bipy)_3^{3+}$ and $Co(terpy)_2^{3+}$ and gamma radiolysis studies. Similar intermediates for the latter two complex ions were observed in spite of the fact that the terpy complex is low spin. This observation, combined with the observation that $G(Co^{II})$ values obtained in the presence of triplet scavengers are identical to those calculated on the basis of competition between $Co(bipy)_3^{3+}$ and the scavengers (behaving as oxidants) for the intermediate, strongly suggest that the intermediates in these systems are bipy ligand radical complexes rather than low-spin cobalt(II) species. Moreover, the oxidation potential calculated[128] for E^0 [(intermediate)/$Co(bipy)_3^{3+}$] is more positive than that required to oxidize the O_2 and $Cr(CN)_6^{3-}$ scavengers.

Hoffman and Simic[130] reported still another intermediate in the $Co(bipy)_3^{3+} - e^-_{aq}$ system which decays via a first order process with k = 3.5 sec^{-1}, and the same decay constant is observed for the transient in the $Cr(bipy)_3^{3+}$–e^-_{aq} system. The detailed nature of these intermediates is unclear.

Since these early reports appeared, developments in the field of transition metal photochemistry[132] have spawned intensive activity in the field of excited state electron transfer. No effort will be made here to review photoreduction processes per se but rather the infant field of electron transfer reactions involving excited state reactants. Gafney and Adamson[133] first proposed that (^{3}CT)-$Ru(bipy)_3^{2+}$ reacted with $Co(NH_3)_5Br^{2+}$ to produce Co^{2+} and $Ru(bipy)_3^{3+}$. Natarajan and Endicott,[134] however, preferred a quenching mechanism involving triplet-to-triplet energy transfer followed by intramolecular reduction. Navon and Sutin[135] have recently set forth convincing evidence for the electron transfer mechanism. Thus, the rate constants k_q for quenching of (^{3}CT)$Ru(bipy)_3^{2+}$ phosphorescence by $Co(NH_3)_5X^{n+}$ complexes increase in the order $NH_3 < H_2O < Cl^- < Br^-$ as do also the rate constants k_2 for the photoinduced oxidants of $Ru(bipy)_3^{2+}$ by the same oxidants. The order is that also observed for typical outer-sphere reactions. Moreover, $k_2/k_q = 0.86 \pm 0.15$ for the chloro complex, and similar values for the other complexes imply that the Co(III)-quenched reactions all produce $Ru(bipy)_3^{3+}$ in near quantitative yields. The energy transfer mechanism[134] was criticized[135] on three counts: (1) the energies of (^{3}CT)-$Ru(bipy)_3^{2+}$ and (^{3}CT)$Co(NH_3)_5Br^{2+}$ do not overlap; (2) radicals are required to be produced with unit efficiency and then react with $Ru(bipy)_3^{2+}$ quantitatively, whereas $\Phi[Co^{2+}] = 0.3$ in the redox decomposition of (^{1}CT)Co-$(NH_3)_5X^{2+}$; and (3) the main evidence for bromine radicals in the $Co(NH_3)_5$-Br^{2+} quenching is obtained in 50% 2-propanol where the yield of $Ru(bipy)_3^{3+}$ is half that in water. Navon and Sutin also noted, however, that the Stern-Volmer constant K_{SV} is decreased by nearly the same factor, making it unnecessary to invoke competitive radical reactions to account for the decreased yields of Ru-$(bipy)_3^{3+}$ in the mixed solvent.

Navon and Sutin[135] also studied the $(^3CT)Ru(bipy)_3^{2+}$ quenching by $Ru(NH_3)_6^{3+}$ and $Ru(NH_3)_5Cl^{2+}$ and invoked the electron transfer mechanism. Subsequently, Bock, Meyer, and Whitten[136] have confirmed the mechanism

$$(^3CT)Ru(bipy)_3^{2+} + Ru(NH_3)_6^{3+} \rightarrow Ru(bipy)_3^{3+} + Ru(NH_3)_6^{2+}$$

$$Ru(bipy)_3^{3+} + Ru(NH_3)_6^{2+} \rightarrow Ru(bipy)_3^{2+} + Ru(NH_3)_6^{3+}$$

in a flash photolysis study of the $(^3CT)Ru(bipy)_3^{2+}$ luminescence and as well measured the electron transfer quenching constants for iron(III) and several organic species, including paraquat. More recently, they extended the range of organic quenchers to obtain the potential $E[Ru(bipy)_3^{3+}/(^3CT)Ru(bipy)_3^{2+}] = -0.81 \pm 0.02$ V,[137] 2.10 V more negative than the E^0 value for the 3+/2+ couple of +1.29 V. Bolletta, Maestri, Moggi, and Balzani[138] have shown that not only does $Cr(bipy)_3^{3+}$ quench the $(^3CTRu(bipy)_3^{2+}$ luminescence but also

TABLE 3-3
SUMMARY OF ELECTRON TRANSFER QUENCHING CONSTANTS FOR $(^3CT)Ru(bipy)_3^{2+}$

Oxidant	k_q $(M^{-1}\ sec^{-1})$	Ref.	Reductant	k_q $(M^{-1}\ sec^{-1})$	Ref.
$Co(NH_3)_6^{3+}$	$\sim 1 \times 10^7$	135[a]	$Ru(NH_3)_6^{2+}$	2.4×10^9	139[g]
$Co(NH_3)_5H_2O^{3+}$	1.5×10^8	135[a]	$S_2O_4^{2}$	2.4×10^9	139[h]
$Co(NH_3)_5Cl^{2+}$	9.3×10^8	135[a]	$Fe(CN)_6^{4-}$	3.5×10^9	139[h]
$Co(NH_3)_5Br^{2+}$	2.5×10^9	135[a]	$Mo(CN)_8^{4-}$	3.4×10^9	140[c]
$Ru(NH_3)_6^{3+}$	2.7×10^9	142[c]	$Os(CN)_6^{4-}$	1.2×10^9	140[c]
	2.1×10^9	142[a]			
$Ru(NH_3)_5Cl^{2+}$	2.7×10^9	135[a]			
Fe^{3+}	2.3×10^9	142[c]			
	2.9×10^9	141[d]			
Tl^{3+}	1.1×10^8	143[c]			
$Cr(bipy)_3^{3+}$	3.3×10^9	138[e]			
$Co(phen)_3^{3+}$	2.3×10^9	142[c]			
$Os(bipy)_3^{3+}$	3.8×10^9	142[c]			
$Fe(CN)_6^{3-}$	7.3×10^9	142[a]			
O_2	$\leq 3.3 \times 10^9$	142[f]			

[a] 0.5*M* H_2SO_4.
[b] 1.0*M* CF_3CO_2H.
[c] $\mu = 0.5M$.
[d] 1*M* $HClO_4$.
[e] $\mu = 0.2M$.
[f] H_2O.
[g] 0.2*M* acetate buffer.
[h] 0.02*M* acetate buffer.

that $Ru(bipy)_3^{2+}$ quenches the $(^2MC)Cr(bipy)_3^{2+}$ excited state via an electron transfer mechanism with $k_q = 4.0 \times 10^8 M^{-1}\ sec^{-1}$.

Creutz and Sutin[139] have calculated the potential of the $(^3CT)Ru(bipy)_3^{2+} + e^- = Ru(bipy)_3^+$ couple to be +0.84 V (vs. NHE), indicating that (^3CT)-$Ru(bipy)_3^{2+}$ is a moderately strong oxidant. From a comparison of the quenching constants for $Fe(CN)_6^{4-}$, $S_2O_4^{2-}$, and $Ru(NH_3)_6^{2+}$ with the rate constants for outer-sphere oxidations of these species by $Os(bipy)_3^{3+}$ and ferricytochrome *c*, they infer that electron transfer quenching is dominant. Shortly thereafter, Juris, Gandolfi, Manfrin, and Balzani[140] obtained direct chemical evidence for the existence of oxidation products in the quenching of $(^3CT)Ru(bipy)_3^{2+}$ by $Mo(CN)_8^{4-}$ and $Os(CN)_6^{4-}$. Finally, excited state $Eu(phen)_3^{3+}$ is reductively quenched by various organic substrates with direct observation of the one-electron oxidized products.[141] Presently available electron transfer quenching constants for $(^3CT)Ru(bipy)_3^{2+}$ are listed in Table 3-3. Other polypyridine systems of ruthenium[141] and osmium[142] have been reported.

In addition to the $Ru(bipy)_3^{2+}$ work, a most interesting photochemically induced, intramolecular electron transfer mechanism has been advanced by Farr, Hulett, Lane, and Hurst.[144] They have observed that $M \rightarrow L(\pi^*)$ excitation of various πbonded copper(I) complexes of alkenoic acido and primary aminoalkene complexes of pentaamminecobalt(III) leads to reduction of cobalt(III) by copper(I) ($\Phi[Co^{2+}] = 0.24 - 0.65$). For the conjugated organic ligands the transfer of an electron may follow the conventional thermal mechanism; however, these reductions also proceed with up to two methylene groups between the alkene–Cu(I) site and the ligand adjacent to cobalt(III). The proposed activated complex is given as:

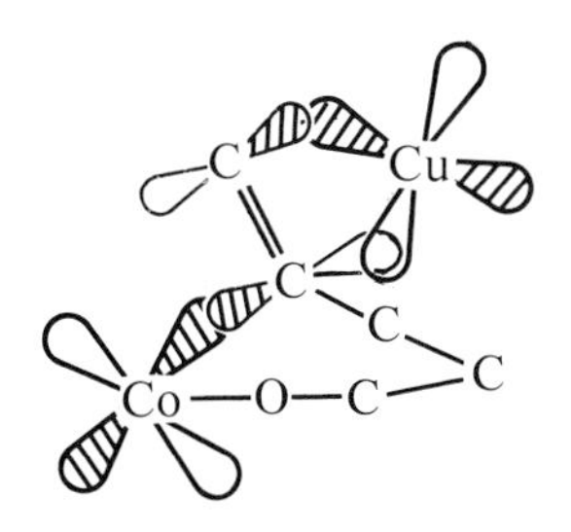

Moreover, ligand isomerization of the fumarato complex accompanies the photoreduction. Further work on such outer-sphere, intramolecular mechanisms is in progress in at least one other laboratory.[145] In a related study Durante and Ford[146] have also found that $(H_3N)_5Ru^{III}pyzCu^{II\,5+}$ undergoes a metal to ligand charge transfer upon flash photolysis yielding $(H_3N)_5Ru^{III}pyzCu^{I}$($\Phi > 0.2$). The latter species apparently reverts to the original reactants via intramolecular electron transfer ($k = 7.8 \times 10^3\ sec^{-1}$), although a bimolecular mechanism could not be fully rejected.

Very recently there has been a revival of interest in the mechanism of luminescence quenching for the uranyl ion, Burrows, Formosinko, Miguel, and

Coehlo[174] have demonstrated an electron transfer mechanism for the quenching by Mn^{2+} (by observing Mn^{3+}). Moreover, they found a linear correlation between log k_q and ionization potentials for a variety of metal ions, suggesting further examples of electron transfer quenching. As is noted below, however, this correlation alone is inadequate for specifying the quenching mechanism. Kato and Fukutomi[148] have proposed that the rather complex light-accelerated exchange between U(VI) and U(IV) proceeds via electron transfer quenching of excited state UO_2^{2+} by labelled U^{4+}, followed by exchange of the U(V) produced. Yokoyoma, Mariyasu, and Ikeda[149] present the latest in a long line of arguments for the halide ion quenching mechanism of the uranyl luminescence. On the basis of thermodynamic considerations they conclude that the higher of two uranyl luminescing states is capable of oxidizing all halide ions except fluoride. Furthermore, the order of k_q values (I > Br > Cl) parallels that expected from the free energies of reaction. However, Laurence and Balzani[143] have cautioned that the mere observation of increased Stern-Volmer quenching constants ($F^- > Cl^- > Br^-$) is not sufficient evidence for an electron transfer quenching mechanism since the same order of ligand efficiencies is observed[150] for the $(^3CT)Ru(bipy)_3^{2+}$–$Cr(en)_2X_2^+$ reactions where an energy transfer mechanism is operative. Rate constants for the reductive quenching of excited states of uranyl ion by CH_3OH and CD_3OH have been measured in flash photolysis experiments, and they exhibit the same k_H/k_D (= 2.76) as the fluorescence quenching constants, thus indicating electron transfer for the latter quenching mechanism.[151]

In summarizing their conclusions about excited state quenching mechanisms, Navon and Sutin[135] concluded that at least four distinguishable mechanisms were possible, two involving energy transfer with or without net chemical change and two involving net chemical change. Clear evidence of the ability of an excited state to act as an electron donor or acceptor can be obtained only with a quenching substrate to which energy transfer cannot occur.[143] Furthermore, the products of the quenching step must exhibit lifetimes which either permit their direct detection or allow their reaction with a scavenger. The stimulus provided by these fundamental studies regarding mechanisms has rekindled interest in the use of excited-state redox systems for conversion of solar energy into electrical and chemical energy.[142,152,153]

II. THE INNER-SPHERE MECHANISM

A. General Considerations

The inner-sphere mechanism for electron transfer reactions was first demonstrated by Taube, Myers, and Rich[154] in the now classic reaction:

$$Co(NH_3)_5Cl^{2+} + Cr^{2+}_{(aq)} \leftrightharpoons [(NH_3)_5CoClCr^{4+}]^{\ddagger} \xrightarrow{H^+} Cr(H_2O)_5Cl^{2+} + Co^{2+}_{(aq)} + 5NH_4^+$$

Several general comments about the mechanism are in order. The formation of a bridged activated complex requires that one of the reagents be substitutionally labile and that there be a suitable group(s) for bridging the two metal ions. Although ligand transfer is often observed for inner-sphere reactions, Seewald, Sutin, and Watkins[116] have demonstrated conclusively that the bridging group is not required to transfer by observing an Ru^{II}—Cl bonded species among the primary products of the $Ru(NH_3)_5Cl^{2+}$—Cr^{2+} reaction. Further examples have been cited by Grossman and Haim[155] for the reactions: $Fe(CN)_6^{3-}$-$Co(CN)_5^{3-}$,[156] $Fe(CN)_6^{3-}$-$CoEDTA^{2-}$,[157,158] and $IrCl_6^{2-}$-$Co(CN)_5^{3-}$ [155] where the inner-sphere mechanism obtains without ligand transfer. While the oxidant usually supplies the bridging group for most inner-sphere reactions, Birk[159] and Birk and Weaver[160] have shown that this is not a necessary feature of the mechanism since Cr(III)-Fe(III) and V(IV)-Fe(III) intermediates are formed in the reduction of Cr(VI) and V(V), respectively, by the substitutionally inert iron(II) complexes $Fe(CN)_6^{4-}$, $Fe(bipy)(CN)_4^{2-}$, and $Fe(bipy)_2(CN)_2$.

Identification of either a binuclear intermediate or a primary reaction product bearing the transferred bridging ligand has been used to establish that some reductions by $Cr(H_2O)_6^{2+}$, $Fe(H_2O)_6^{2+}$, $Co(CN)_5^{3-}$, $V(H_2O)_6^{2+}$, and $Ru(NH_3)_5$ proceed via inner-sphere mechanisms.[115,161-165] Taube[166] and Sutin[161] have cautioned, however, that formation of a ligand-substituted product can arise by substitution subsequent to the redox process. Therefore, proof of the mechanism requires the ligand-substituted product to be in excess of its equilibrium concentration or to be inert to substitution. More recently, Linck[167] has noted the possibility of a postcursor complex decomposing by bond rupture at both metal sites before observing either the binuclear complex or its fragments. For example, he noted that where both substituted and unsubstituted products were observed, dual interpretations of the mechanism are possible: (1) parallel inner-sphere and outer-sphere paths or (2) spontaneous and metal-catalyzed bond rupture. The reaction of $IrCl_6^{2-}$ with chromium(II) has been interpreted on the basis of (1) above,[32] but the ambiguity of the mechanistic interpretation (1) or (2) remains.

A final point of general inquiry concerns the requirement(s) imposed on a bridging ligand. Haim[168] has recently examined this question in detail, concluding that an available, basic pair of electrons on the adjacent atom is both a necessary and sufficient condition for attack on that atom but that it is not a sufficient condition for attack at a site remote from the adjacent atom. The sufficient condition may be ligand reducibility, but this point has not been conclusively demonstrated.[168] An interesting conclusion also is that coordinated water cannot serve as a bridging ligand, presumably because of its very low basicity.[169] Recently, however, three reports of alkyl transfer in redox processes would appear to reopen the question of bridging ligand requirements. Thus, van den Bergen and West[170] report that methyl and ethyl groups are transferred between cobalt ions in the redox reaction below for both forward and reverse reactions (acen and tfen = tetradentate Schiff bases).

$$t\text{-}CH_3CO^{III}(H_2O)(acen) + Co^{II}(tfen) \rightleftarrows Co^{II}(acen) + t\text{-}CH_3Co^{III}(H_2O)(tfen)$$

Moreover, Espenson and Shveima[171] and Espenson and Sellers[172] have demonstrated that alkyl transfer obtains in the chromium(II) reductions of cobaloximes and cobalamines, respectively. Two subtly different and as yet indistinguishable mechanisms, redox bimolecular electrophilic substitution (redox S_E2) and bimolecular homolytic substitution (S_H2), have been postulated for the alkyl transfer reactions.[171]

B. Precursors, Intramolecular Electron Transfer, and Successors

In recent years there has been considerable interest in the detailed mechanism of the inner-sphere process with the focus on precursor and successor (or postcursor) complexes and most recently on the intramolecular electron transfer within the precursor complex.[123,167,168,173,174] A five-step detailed mechanism can be depicted as follows:

step 1 $M^{II}A_6^{+} + {*M^{III}L_5X^{n+}} \leftrightarrows M^{II}A_6^{2+} \cdot X{*M^{III}L_5^{n+}}$ $\qquad k_d/k_{-d}$

step 2 $M^{II}A_6^{2+} \cdot X{*M^{III}L_5^{n+}} \rightleftarrows A_5M^{II}X{*M^{III}L_5^{n+2}} + L$ $\qquad k_s/k_{-s}$

step 3 $A_5M^{II}X{*M^{III}L_5^{n+2}} \rightleftarrows A_5M^{III}X{*M^{II}L_5^{n+2}}$ $\qquad k_e/k_{-e}$

step 4 $A_5M^{III}X{*M^{II}L_5^{n+2}} + L \rightleftarrows A_5M^{III}X^{n+} \cdot {*ML_6^{2+}}$ $\qquad k'_s/k'_{-s}$

step 5 $A_5M^{III}X^{n+} \cdot {*M^{II}L_6^{2+}} \rightleftarrows M^{III}A_5X^{n+} + {*ML_6^{2+}}$ $\qquad k'_d/k'_{-d}$

Steps *1* and *5* involve the diffusion-controlled encounter complex formation; steps *2* and *4* are substitution-controlled reactions involving precursor formation and successor (or postcursor) dissociation; and step *3* involves intramolecular electron transfer between precursor and successor complexes. For simplicity, steps *1* and *2*, as well as *4* and *5*, are usually combined to give a three-step mechanism. In the above model, steps *2*, *3*, or *4* can be rate determining, and Linck[167] has depicted energy profiles covering each possibility: schemes II, I, and III, respectively, in his notations.

When precursor formation is rate-determining, the activation energy required should be on the same order of magnitude as that required for substituting water on the labile species (the reductant in the above model). For divalent metal ions of the first transition series, the activation energies for water exchange range between 5.6 and 12.1 kcal/mol.[175] Thus it would appear that a minimum activation energy of about 5 kcal/mol is to be expected for either precursor formation or successor dissociation. The lower limit given is for water exchange on Cu^{2+}, a Jahn-Teller distorted ion, and might be lowered in electrostatically favorable cases. The activation free energy for intramolecular electron transfer has been estimated at 8 kcal/mol.[176] Although estimates purportedly based on a direct measurement of the Franck-Condon barrier tend to confirm this estimate,[37] Cannon and Gardiner[177] have measured the activation parameters for electron transfer within the precursor $(NH_3)_5Co^{III}NTAFe^{II}$ to be $\Delta H^{\ddagger} = 18.7 \pm 1.3$ and $\Delta S^{\ddagger} = 0 \pm 4$ eu. Moreover, calculations based on the absolute rate theory ($\kappa = 1$) indicate that $\Delta G^{\ddagger}_{et} \approx 17 \pm 5$ kcal on the basis of rate constants for intramolecular electron transfer summarized in Table 3-4. These results appear to disagree substantially with the earlier estimates.[37,176]

An examination of the results in Table 3-4 indicates a remarkable range of nearly 10^6 in the intramolecular electron transfer rate constants. Conjugation between the metal centers markedly enhances the rate (cf entry 2 with 3, 4, or 5 and entry 7 with 8). It is remarkable that entry 5 has a higher constant than entry 3, and it has been suggested that when Ru(II) reduces Co(III), the electron transfer mechanism may actually be outer-sphere within the binuclear intermediate. The copper(I) adduct (entry 9) is the only one of a number of copper(I) precursor complexes, mostly alkene cobalt(III) species, observed to have a measurable electron transfer rate in the absence of photochemical excitation. A common feature of the available data in these systems is the highly favorable equilibrium quotient for precursor formation. Taube[174] has argued that the rate being measured for the Ru(II)–Co(III) processes is that for the irreversible trapping of an electron on Co(III) which originated on Ru(II) and not the rate of electron transfer between Ru(II) and Co(III).

On the basis of some recently determined association constants between $Co(NH_3)_5C_2O_4^+$ and various divalent metal ions,[178] it is also possible to estimate (within a factor of ~50) the electron transfer rate constants for chromium(II) and vanadium(II) reductions of that complex. The measured equilibrium quotients for divalent ions ranged from 15 to $500M^{-1}$. Taking $Q_p \approx 10^2$, $k_{Cr^{2+}} = 4.6 \times 10^4 M^{-1}$ sec^{-1},[164] and $k_{V^{2+}} = 4.5 \times 10^1 M^{-1}$ sec^{-1},[164] one calculates $k_{et}^{Cr^{2+}} \approx 5 \times 10^2$ sec^{-1} and $k_{et}^{V^{2+}} \approx 5 \times 10^{-1}$ sec^{-1} in reasonable agreement with values in Table 3-4.

Several indirect criteria for precursor formation have been advanced, including negative enthalpies of activation[184] and a complex hydrogen ion dependence characteristic of acid–base chemistry involving binuclear species.[185] Negative activation enthalpies can arise when electron transfer within the precursor is rate determining. Since $k_{obsd} = K_{1,2}k_3$, where $K_{1,2}$ is the overall equilibrium constant for *steps 1* and *2* in the detailed mechanism, $\Delta H^{\ddagger}_{obsd} = \Delta H^0_{12} + \Delta H^{\ddagger}_3$, which can be negative if the negative contribution by ΔH^0_{12} is substantial. Unfortunately, no ΔH^0_{12} data are available in the literature at present. With regard to the complex acid dependence argument, the evidence for precursor formation is not compelling since Haim[186] has discussed the form of rate law observed in terms of the net activation process[187] without regard to the oxidation states of the individual metal ions. Therefore, the observed form of the rate law is consistent with either precursor or successor complex formation.

TABLE 3-4
INTRAMOLECULAR ELECTRON TRANSFER REACTIONS[a]

Precursor	Q_P (M^{-1})	$k_{e.t.}$ (sec^{-1})	Ref
1. $(NH_3)_5Co^{III}NTAFe^{II}(H_2O)_n{}^{2+}$[b]	1.1×10^6	1.15×10^{-1}	177
2. $(NH_3)_5Co^{III}O_2C-\langle\bigcirc\rangle NRu^{II}(NH_3)_4SO_4{}^{2+}$	—	$\sim 1 \times 10^2$	179
3. $(NH_3)_5Co^{III}O_2CCH_2-\langle\bigcirc\rangle NRu^{II}(NH_3)_4(H_2O)_2{}^{4+}$	—	1.6×10^{-3}	179
4. $(NH_3)_5Co^{III}O_2C-\langle\bigcirc\rangle N-Ru^{II}(NH_3)_4(H_2O)_2{}^{4+}$	—	1.6×10^{-3}	179
5. $(NH_3)_5Co^{III}O_2CCH_2-\langle\bigcirc\rangle N-Ru^{II}(NH_3)_4(H_2O)_2{}^{4+}$	—	5.5×10^{-3}	179

6. 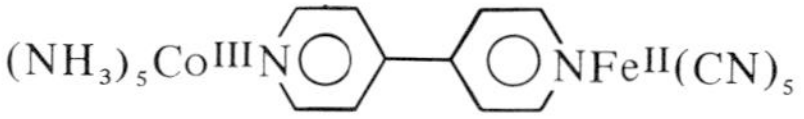 $(NH_3)_5Co^{III}N\langle\bigcirc\rangle-\langle\bigcirc\rangle NFe^{II}(CN)_5$	1.2×10^6	2.6×10^{-3}	180
7. 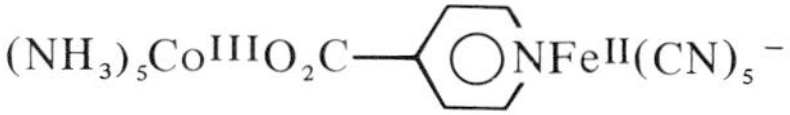 $(NH_3)_5Co^{III}O_2C-\langle\bigcirc\rangle NFe^{II}(CN)_5^-$	6.0×10^5	1.8×10^{-4}	181
8. 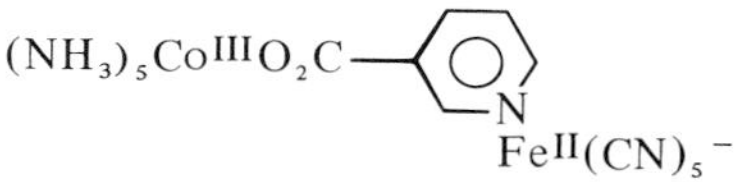 $(NH_3)_5Co^{III}O_2C-\langle\bigcirc\rangle$ (N) $Fe^{II}(CN)_5^-$	7.9×10^5	$\leqslant 3 \times 10^{-5}$	181
9. 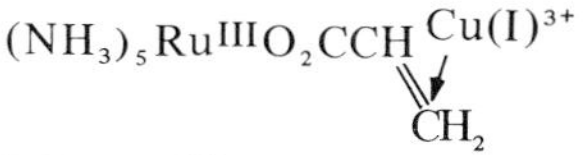 $(NH_3)_5Ru^{III}O_2CCH(=CH_2) \rightarrow Cu(I)^{3+}$	—	$\sim 8 \times 10^{-4}$	182
10. $(NH_3)_5Co^{III}NTACr^{II}(H_2O)_n{}^{2+}$	3×10^7	6×10^2	183

[a] At 25°C and $\mu = 1.0M$.
[b] NTA = nitrilotriacetate.

Most binuclear intermediates observed to date in electron transfer reactions are of the successor type. As early as 1954 Meyers and Taube[188] observed the formation of $M[(NC)_5Fe^{II}CNCr^{III}]$ in the reduction of $Fe(CN)_6{}^{3-}$ by chromium(II). A reinvestigation of this system in highly dilute solutions revealed a rate constant $>6 \times 10^6 M^{-1}\ sec^{-1}$, with the formation of at least two intermediates.[189] Haim and Wilmarth[156] reported that reducing $Fe(CN)_6{}^{3-}$ with $Co(CN)_5{}^{3-}$ yields $[(NC)_5Fe^{II}CNCo^{III}(CN)_5]^{6-}$ in which the iron center is subject to further reversible redox processes. Numerous studies of the Fe(C-N)$_6{}^{3-}$–$CoEDTA^{2-}$ reaction have been reported[190-193] where a cyanide bridged successor complex is observed, and the following mechanism was proposed by Adamson and Gonick:[190]

$$Fe(CN)_6{}^{3-} + CoEDTA^{2-} \underset{k_{-1}}{\overset{k_1}{\rightleftharpoons}} [(NC)_5Fe^{II}CNCo^{III}EDTA] \xrightarrow{k_2} CoEDTA^- + Fe(CN)_6{}^{2-}$$

However, recent measurements on this system at pH10 (where ring closure of $CoEDTAH_2O^-$ is slow) give $CoEDTA^-$ as the sole primary product. Moreover, direct measurement of k_{-2}, as well as similar values for $CoEDTAH_2O^-$ and $CoEDTACl^{2-}$, demonstrates that the back reaction proceeds substantially faster (at least $\times 10^2$) than does substitution on the $CoEDTA^-$ species, requiring an outer-sphere mechanism. Thus, on the basis of microscopic reversibility considerations the forward rate must also proceed via an outer-sphere mechanism, ruling out dissociation of the successor. The alternate mechanism then involves parallel inner- and outer-sphere pathways.[193] These arguments should be considered for the related CoCyDTA[192] and CoDTPA[194] reactions with $Fe(CN)_6{}^{3-}$

Table 3-5 summarizes available data on the various chromium(III) successor complexes, including the rate constants for their formation, aquations, and in a few instances their metal-catalyzed aquations. An examination of column 2 (k_f) reveals a wide range ($\sim 10^5$) of formation rates, but all are greater than 400. For entries 17–22, steric crowding at the lead-in point decreases k_f and increases k_{aq}. That successor complexes are observed in these instances is readily understood on the basis of their relatively slower spontaneous (k_{aq}) and metal-catalyzed ($k_{cat\ aq}$) aquation rates (columns 3 and 4, respectively). It is perhaps surprising that the metal-catalyzed aquations for entries 1, possibly 2, and 3 are substantially more rapid (at, say, $10^{-3}M$ catalyst concentration) than the spontaneous aquations, whereas for entry 6 the analogous processes are competitive. The former group displays no acid dependence and can therefore be outer-sphere for the metal-catalyzed path, while the latter is very likely inner sphere. For entry 1, it is known that the chromium(II)-catalyzed reaction is inner sphere. According to Gordon,[195] tracer experiments using ^{18}O-labelled uranyl ion indicate a quantitative transfer of uranyl oxygens to chromium(III) upon decomposition of the trinuclear species.

TABLE 3-5
RATE COMPILATION OF CHROMIUM(III) SUCCESSOR COMPLEXES

Successor	$k_f{}^a (M^{-1}\ sec^{-1})$	$k_{aq}\ (sec^{-1})$	$k_{cat\ aq}\ (M^{-1}\ sec^{-1})\ (M^{n+})$	Ref.
1. $[OU^{V}OCr^{III}]^{4+}$	1.29×10^4 (14.5°C)	8.3×10^{-2}	$4.16 \times 10^2\ (Cr^{2+})$	196, 197
2. $[OPu^{V}OCr^{III}]^{4+}$	$>10^5$	—	$1.2 \times 10^2\ (Pu^{3+})$ $3.6 \times 10^3\ (Fe^{2+})^b$	196, 198, 199
3. $[ONp^{V}OCr^{III}]^{4+}$	—	2.3×10^{-6}	$1.6 \times 10^3\ (Np^{3+})$ $3.6 \times 10^2\ (Co^{3+})$	200, 201, 202
4. $[U^{IV}(OCr^{III})_2]^{6+}$	4.16×10^2	—	—	197
5. $[V^{III}OCr^{III}]^{4+}$	$>10^4$	$\frac{k_0 Q + k_1[H^+]}{Q + [H^+]}$	—	196, 203
6. $[(NH_3)_5Ru^{II}N\langle pyridine\text{-}4\text{-}yl\rangle\text{—}C(NH_2){=}OCr^{III}]^{5+}$	3.9×10^5	$3.1 \times 10^{-6} +$ $2.8 \times 10^{-7}/[H^+]$	$4.7 \times 10^{-4}\ [Cr^{2+}]/[H^+]$	107, 204
7. $[(NH_3)_5Ru^{II}N\langle pyridine\text{-}3\text{-}yl\rangle\text{—}C(NH_2){=}OCr^{III}]^{5+}$	1.4×10^4 (4°C)	$\sim 10^7 +$ $2.0 \times 10^{-7}/[H^+]$	$4.0 \times 10^{-5}\ [Cr^{2+}]/[H^+]$	107, 204
8. $[(NH_3)_5Ru^{II}N\langle pyridine\text{-}4\text{-}yl\rangle\text{—}C(OCH_3){=}OCr^{III}]^{5+}$	4.1×10^4	$3.6 \times 10^{-2} +$ $2.2 \times 10^{-3}/[H^+]$	—	107

TABLE 3-5 CONTINUED
RATE COMPILATION OF CHROMIUM(III) SUCCESSOR COMPLEXES

Successor	k_f (M^{-1} sec^{-1})	k_{aq} (sec^{-1})	$k_{cat\ aq}$ (M^{-1} sec^{-1}) (M^{n+})	Ref.
9. $[c\text{-}(NH_3)_4Ru^{II}(NC_5H_4\text{—}C(NH_2){=}O)_2Cr^{III}]^{5+}$	7.0×10^6	—	—	107
10. $[c\text{-}(NH_3)_4Ru^{II}(NC_5H_4\text{—}C(OCH_3){=}O)_2Cr^{III}]^{5+}$	1.1×10^6	—	—	107
11. $[(H_2O)_5Ru^{II}ClCr^{III}]^{4+}$	$\geqslant 2 \times 10^4$	$1.20 + 0.36/[H^+]$	—	116
12. $[(NH_3)_5Ru^{II}ClCr^{III}]^{4+}$	—	4.6×10^2	—	117
13. $[c\text{-}(NH_3)_4(H_2O)Ru^{II}ClCr^{III}]^{4+}$	—	1.17×10^2	—	117
14. $[c\text{-}(NH_3)_4ClRu^{II}ClCr^{III}]^{3+}$	—	1.58×10^2	—	117
15. $[c\text{-}(bipy)_2ClRu^{II}ClCr^{III}]^{3+}$	—	6.2	—	117
16. $[(NH_3)_5Ru^{II}OHCr^{III}]^{4+}$	3.5×10^6	$10.5 + 4.5 \times 10^3[H^+]$	—	118
17. $[(NH_3)_5Ru^{II}O\text{—}C(H){=}OCr^{III}]^{4+}$	1.7×10^5	2.4	—	118
18. $[(NH_3)_5Ru^{II}O\text{—}C(CH_3){=}OCr^{III}]^{4+}$	2.6×10^4	2.55×10^1	—	118
19. $[(NH_3)_5Ru^{II}O\text{—}C(CF_3){=}OCr^{III}]^{4+}$	1.4×10^3	$>10^2$	—	118

20. $[(NH_3)_5Ru^{II}O-C(C_6H_5)=OCr^{III}]^{4+}$	5.8×10^3	—	—	118
21. $[(NH_3)_5Ru^{II}O-C(C_6H_4OH)=OCr^{III}]^{4+}$	4.0×10^3	2.9×10^1	—	118
22. $[(NH_3)_5Ru^{II}O-C(C_6H_4CO_2C_2H_5)=OCr^{III}]^{4+}$	6.6×10^3	—	—	118
23. $[(NH_3)_5Ru^{II}NC_5H_4-C(NH_2)=SCr^{III}]^{5+}$	—	$3.9 \times 10^{-4} + 9.0 \times 10^{-5}/[H^+]$	$0.27\ (Cr^{2+})$	205
24. $[(NH_3)_5Co^{III}O-C(=O)-C(=O)-OCr^{III}]^{4+}$	*c*	—	$1.88\ [Cr^{2+}]$	206
25. $[(NH_3)(en)_2Co^{III}-OO-Cr^{III}]^{4+}$	2.1×10^3	—	—	207

[a] k_f is the second-order rate constant for one-electron reduction of the parent complexes by Cr^{2+} to the observed successor.
[b] At $0.2M$ $[H^+]$; k is $f(L/[H^+])$.
[c] $(k_aK[H^+] + k_b)/1 + K[H^+]$, where $k_a = 1.45M^{-1}sec^{-1}$, $k_b = 35.2M^{-1}sec^{-1}$, and $K = 18M^{-1}$.

A remarkable diversity of chemistry exists within these systems among the closely related ruthenium series (entries 6–22). On one hand, quantitative formation of the successor very probably obtains for all entries except 7, 11, and 12–15, while on the other, operation of parallel inner- and outer-sphere reactions are suggested (entries 7 and 11, although the earlier ambiguity of successor dissociation should be recalled). Additionally, for entries 12–15, reversible, equilibrium formation of successor complexes is invoked. Movius and Linck[208] have developed an interesting test for precursor vs. successor formation on the basis of equilibrium quotient comparisons. Quotients in the range 10^{-1}–$10^{1} M^{-1}$ imply precursor formation (Q_P), while substantially larger values imply successor formation (Q_S). The precursor range is based on Cannon and Earley's estimated equilibrium constants for reactions of the type:[209]

$$M^{n+} + MOH^{n-1} = \overset{\displaystyle H}{MOM^{(2n-1)}} \qquad (n = 2 \text{ or } 3)$$

Incidentally, the values of Q_S obtained for the various $Fe(CN)_6{}^{3}$–$Co^{II}L$ (L = EDTA, CyDTA, DPTA) reactions lie in the range 112–831M^{-1}.

Turning now from considerations of precursors, postcursors, and intramolecular electron transfer, we focus on the detailed nature of the bridging groups for e ectron transfer.

C. Formation and Reactions of Linkage Isomers

When an ambidentate ligand such as cyanide, thiocyanate, or nitrite ion functions as a bridging group for an inner-sphere redox process, the possibility exists that linkage isomers may be produced among the primary reaction products depending on the position of attack by the reducing agent. Frequently, however, only one of the two possible isomers is obtained in a given reaction, and, interestingly, it is not always the thermodynamically stable isomer which is formed. For example, according to Candlin, Halpern, and Nakamura,[210] the reaction of $Co(CN)_5{}^{3-}$ with $Co(NH_3)_5NCS^{2+}$ produces the thermodynamically stable $Co(CN)_5SCN^{3-}$ isomer, whereas Halpern and Nakamura reported[211] that the reaction of $Co(CN)_5{}^{3-}$ with $Co(NH_3)_5NO_2{}^{2+}$ produces the metastable $Co(CN)_5ONO^{3-}$ isomer, which spontaneously reverts to the *N*-bonded species. Shea and Haim[212] recently reported that $Co(NH_3)_5SCN^{2+}$ is reduced by $Co(CN)_5{}^{3-}$ exclusively via adjacent attack. Moreover, Haim and Sutin[213] found that the reaction of either $Fe(H_2O)_5NCS^{2+}$ or *trans*-$Co(en)_2(H_2O)NCS^{2+}$ with chromium(II) produces the thermodynamically unstable isomer $Cr(H_2O)_5SCN^{2+}$ in 75 and 100% yields, respectively.[214] Although some $Cr(H_2O)_5NCS^{2+}$ is formed when the iron(III) species is a reactant, it has not been possible to establish that it is formed by adjacent attack at nitrogen bound to cobalt(III). (The *N*-bonded species can arise by an hydroxide bridged reaction between $Fe(H_2O)_5NCS^{2+}$ and $Cr(H_2O)_5NCS^{+}$.) Recently, however, Shea and

Haim[215] presented direct evidence for parallel adjacent and remote attack involving a bridging thiocyanate ion. Thus, $Co(NH_3)_5SCN^{2+}$ reacts with chromium(II) according to the following scheme:

$$Co(NH_3)_5SCN^{2+} + Cr^{2+} \xrightarrow{H^+} \begin{cases} CrSCN^{2+} \text{ (29\% adjacent attack)} \\ \qquad + Co^{2+} + 5NH_4^+ \\ CrNCS^{2+} \text{ (71\% remote attack)} \end{cases}$$

In the above study it was also shown that the reaction of $Co(NH_3)_5NCS^{2+}$ with chromium(II) produces $Cr(H_2O)_5SCN^{2+}$ as the exclusive chromium(III) product by utilizing a parent–daughter decay scheme[216] to calculate the time of maximum buildup on $Cr(H_2O)_5SCN^{2+}$ concentration. (Such a procedure was necessary because $Cr(H_2O)_5SCN^{2+}$ reacts with chromium(II) about a factor of three faster than it does with the parent cobalt(III) species.) Brown and Pennington[217] reported the preparation of a novel complex ion containing both an *N*- and *S*-bonded complex, $Cr(H_2O)_4(SCN)(NCS)^+$.

As early as 1958 Ball and King[218] considered adjacent attack highly unlikely for the exchange reaction between $Cr(H_2O)_5NCS^{2+}$ and $*Cr(H_2O)_6{}^{2+}$ for steric reasons. The reductant would probably have to lose at least two water molecules to achieve the necessary configuration for the activated complex. In this context note that the formation of double-bridged activated complexes also requires the loss of two water molecules on a given reductant. One of the common reasons forwarded for the slow rate of the exchange is that the "wrong-bonded" isomer is produced and is therefore thermodynamically unfavorable. If the latter argument were accepted, then the activation free energies of the $CrSCN^{2+}$–Cr^{2+} reaction and the $CrNCS^{2+}$–Cr^{2+} exchange reaction should differ only by the difference in stabilities of the two linkage isomers. The latter difference had been estimated to be 7.5 kcal/mol,[213] but unfortunately there are no activation parameters available for the exchange reaction. The fact that only remote attack is featured in the $Co(NH_3)_5NCS^{2+}$–Cr^{2+} reaction makes it all the more likely that the $Cr(H_2O)_5NCS^{2+}$–$*Cr^{2+}$ exchange proceeds by an analogous mechanism.

Several features of the reactions involving ambidentate bridging ligands are noteworthy. First, the very fact that unstable linkage isomers are produced indicates that the inner-sphere mechanism is operative. Second, the fact that an electron can be transferred via a two- or three-atom bridge is a significant result in itself. Since the production of linkage isomers continues to provide a novel and stimulating area of study in redox chemistry, a summary of such reactions is given in Table 3-6.

A striking feature of the entries in Table 3-6 is the small variation of oxidant metal atoms investigated. Nevertheless, observed chemistries vary quite a bit. For example, in entries 2 and 27 where $Co(CN)_5{}^{3-}$ is the reductant, there are actually two reaction paths, the inner-sphere path leading to the unstable linkage isomer $Co(CN)_5NC^{3-}$ and the other yielding $Co(CN)_6^{3-}$.[211] The rate law has the form:

$$\text{Rate} = (k_i + k_o[CN^-])[Co(NH_3)_5X^{2+}][Co(CN)_5^{3-}]$$

where X = NO_2^- or CN^- and where k_i and k_o are interpreted as rate constants for the inner- and outer-sphere reductions of $Co(NH_3)_5X^{2+}$, respectively. The values of these constants are such that variations in the cyanide ion concentration can control the amount of reaction which proceeds via the inner- or outer-sphere process. The above systems represent two of approximately five systems[60,211,213,219] where there is reasonably good evidence of competition between inner- and outer-sphere processes. Since the ultimate product of the cyanide bridged reaction is $Co(CN)_6^{3-}$ (the same as the product of the outer-sphere process), detection of the metastable $Co(CN)_5NC^{3-}$ intermediate provided a key to interpreting the reaction mechanism.

For entries 3, 4, 8, 10, 14, 15, and 25 the rate constants are smaller than the rate constants for substitution on the respective reductants so that no metastable intermediates have been observed. Thus, it is not even certain that the inner-sphere mechanism is operative in these systems. Interestingly, entries 9, 11, and 26 (the *S*-bonded counterparts of entries 8, 10, and 25) all yield detectable intermediates with the thiocyanate ion *N*-bonded.

The various redox studies involving linkage isomerization of the cyanide ion are given by entries 27–33 in Table 3-6. As noted earlier, $Co(CN)_5NC^{3-}$ is a product of the reaction between $Co(NH_3)_5CN^{2+}$ and $Co(CN)_5^{3-}$ (entry 27), and it is noteworthy that the subsequent isomerization of this product is presumably not catalyzed by $Co(CN)_5^{3-}$ [211] For entries 28–30 Birk and Espenson reported[226] that a common intermediate is produced, presumably $Cr(H_2O)_5NC^{2+}$, which undergoes both spontaneous and chromium(II)-catalyzed (entry 32) linkage isomerizations at comparable rates. Moreover, it has been suggested[227] that the cyanide bridged inner-sphere reaction for entries 31, 33, and 34 proceed via similar mechanisms, although no evidence for the *N*-bonded intermediate is available. This lack of mechanistic evidence results from the fact that the rate constants for both the spontaneous and the chromium(II)-catalyzed isomerizations are substantially larger than those found for entries 31, 33, and 34. Finally, the results of the vanadium(II) reductions (entries 35 and 36) can be compared with the corresponding reductions by chromium(II) in entries 28 and 29. There is a rather striking reversal in relative reactivities for these systems which may reflect a difference in reaction mechanism for the two reductants.

A brief comparison between the cyanide and thiocyanate ion systems is in order. It would appear that the various *C*-bonded cyanide and *N*-bonded thiocyanate ligand systems react predominantly, if not exclusively, via remote attack mechanisms. Despite a rather thorough research, Birk and Espenson[227] were unable to uncover any evidence for adjacent attack in any of the cyanide ion bridged systems. As noted previously, in only one instance have both adjacent and remote attack mechanisms been reported to occur simultaneously,[215] and

TABLE 3-6
REDOX REACTIONS INVOLVING AMBIDENTATE LIGANDS

Oxidant[a]	Reductant	Mode[b]	k (M^{-1} sec^{-1})	Ref.
1. Ro*O*NO^{2+}	$Co(CN)_5^{3-}$	N	—	211
2. Ro*N*O$_2^{2+}$	$Co(CN)_5^{3-}$	R[c]	3.4×10^4	211
3. Ro*O*NO^{2+}	Eu(II)	?	7.5×10^1	220
4. Ro*N*O$_2^{2+}$	Eu(II)	?	1.0×10^2	220
5. Rr′*O*NO^{2+}	Cr(II)	?	2.75×10^2	221
6. Ro*N*CS^{2+}	$Co(CN)_5^{3-}$	R	1.1×10^6	210, 211, 212
7. Ro*S*CN^{2+}	$Co(CN)_5^{3-}$	A	$>10, <2 \times 10^9$	212
8. Ro*N*CS^{2+}	V(II)	?	3×10^{-1}	222
9. Ro*S*CN^{2+}	V(II)	R	3×10^1	223
10. Ro*N*CS^{2+}	Fe(II)	?	$<3 \times 10^{-6}$	224
11. Ro*S*CN^{2+}	Fe(II)	R	1.2×10^{-1}	223
12. Ro*N*CS^{2+}	Cr(II)	R	1.9×10^1	67, 215
13. Ro*S*CN^{2+}	Cr(II)	A	8×10^4	215
		R	1.9×10^5	
14. Ro*N*CS^{2+}	Eu(II)	?	5.0×10^{-2}	225
15. Ro*S*CN^{2+}	Eu(II)	?	3.1×10^3	225
16. *c*-Coen$_2$(H$_2$O)*N*CS^{2+}	Cr(II)	R	4.5×10^1	213
17. *t*-Coen$_2$(H$_2$O)*N*CS^{2+}	Cr(II)	R	1.4×10^3	213
18. *c*-Coen$_2$(NH$_3$)*N*CS^{2+}	Cr(II)	R	3.1	213
19. *t*-Coen$_2$(NH$_3$)*N*CS^{2+}	Cr(II)	R	3.8	213
20. *c*-Coen$_2$(*N*CS)$_2^+$	Cr(II)	R	3.0×10^1	213
21. *t*-Coen$_2$(*N*CS)$_2^+$	Cr(II)	R	2.8×10^1	213
22. Fe(H$_2$O)$_5$*N*CS^{2+}	Cr(II)	R (?)	$\sim 3 \times 10^7$	124, 213
23. Rr*N*CS^{2+}	*Cr(II)	R (?)	1.4×10^{-4}	218

TABLE 3-6 CONTINUED
REDOX REACTIONS INVOLVING AMBIDENTATE LIGANDS

Oxidant[a]	Reductant	Mode[b]	k (M^{-1} sec^{-1})	Ref.
24. Rr*S*CN^{2+}	Cr(II)	R	4.2×10^{1}	213
25. Rr*N*CS^{2+}	V(II)	?	4.41×10^{-5}	115b
26. Rr*S*CN^{2+}	V(II)	R	8.0	115b
27. Ro*C*N^{2+}	$Co(CN)_5{}^{3-}$	R	2.9×10^{2}	211
28. Ro*C*N^{2+}	Cr(II)	R	3.58×10^{1}	226
29. *t*-$Co(NH_3)_4(H_2O)$*C*N^{2+}	Cr(II)	R	1.45×10^{3}	226
30. *t*-$Coen_2(H_2O)$*C*N^{2+}	Cr(II)	R	1.12×10^{3}	226
31. Rr*C*N^{2+}	*Cr(II)	R	7.7×10^{-2}	227
32. Rr*N*C^{2+}	Cr(II)	R	1.60	226
33. *c*-$Cr(H_2O)_4(CN)_2{}^{2+}$	Cr(II)	R	4.19	227
34. 1,2,3-$Cr(H_2O)_3(CN)_3$	Cr(II)	R	4.56	227
35. Ro*C*N^{2+}	V(II)	?	1.1×10^{1}	228
36. $Co(NH_3)_4(H_2O)$*C*N^{2+d}	V(II)	?	9.4	228
37. Ro*N*H_2CHO^{3+}	Cr(II)	O	1.74	229
38. Ro*O*$CHNH_2{}^{3+}$	Cr(II)	?	8.5×10^{-3}	229
39. Ro*S*$SO_3{}^{+}$	Cr(II)	?	1.8×10^{-1}	230
40. Ro*O*$S(O_2)S^{+}$	Cr(II)	?	1.33×10^{1}	230
41. RrN(C$_5$H$_4$)–C(=O)NH$_2$ $^{3+}$	Cr(II)	O	8.1×10^{-1}	219
42. Rr*O*=C(NH$_2$)–(C$_5$H$_4$)NH^{4+}	Cr(II)	pyN	1.06×10^{-1} $[H^{+}]^{-1}$	219

43. Ro*N*CO^{2+}	Cr(II)	R	5.1×10^{-1}	231, 232
44. *t*-$Coen_2(H_2O)NO^{2+}$[d]	Cr(II)	R	4.4×10^{3}	233
45. *t*-$Co(dmgH)_2$(*t*-Bupy)*N*CS	$Co(dmgH)_2$-*t*-Bupy	R	—	234
46. *t*-L$Co(dmgH)_2$*N*CS[d]	$Co(dmgH)_2L$	S	—	235
47. RoN⟨C₅H₄⟩—CN^{3+}	Cr(II)	R	1.2×10^{7}	236
48. RoNC—⟨C₅H₄⟩N^{3+}	Cr(II)	R	6.4×10^{3}	236
49. RrNC—⟨C₅H₄⟩N^{3+}	Cr(II)	R	7.1×10^{3}	236

[a] Ro = $Co(NH_3)_5{}^{3+}$, Rr′ = $Cr(NH_3)_5{}^{3+}$, Rr = $Cr(H_2O)_5{}^{3+}$.
[b] Mode of attack: A = adjacent, R = remote (or atom attached if more than one site possible).
[c] Inner-sphere path only.
[d] Mixture of isomers.

there the thiocyanate ion was *S*-bonded. In view of the latter observation it is pertinent to inquire whether or not similar mechanisms obtain in the other *S*-bonded thiocyanate ligand systems (entries 9, 11, 24, and 26) although it would be difficult to establish this point experimentally because of the labilities of the iron(III) and vanadium(III) products or because of the efficient chromium(II)-catalyzed linkage isomerization of $Cr(H_2O)_5SCN^{2+}$. This efficient process is to be contrasted to the comparatively sluggish chromium(II)-catalyzed linkage isomerization of the $Cr(H_2O)_5NC^{2+}$ complex.

Of the remaining systems in Table 3-6 only entries 38, 41, 42, and 43–48 require further comment. Unlike the *N*-bonded formamide complex (entry 37) the *O*-bonded isomer (entry 38) reacts with chromium(II) via an outer-sphere mechanism. It is remarkable that the *O*-bonded formamide does not function as a bridging ligand since both O and N possess at least one unshared electron pair. Adjacent attack at the carbonyl would be expected to impose severe steric constraints on an inner-sphere mechanism. As for attack at N, either the inner-sphere successor was too short-lived to be detected or complete protonation blocked the site under the experimental conditions used. The systems represented by entries 41 and 42 can be expressed in terms of the following equilibrium:

$$Cr^{2+} + (H_2O)_5CrNC_5H_4C(=O)NH_2^{3+} + H^+ \underset{k_{42}}{\overset{k_{41}}{\rightleftharpoons}} (H_2O)_5Cr{-}O{=}C(NH_2){-}C_5H_4NH^{4+} + Cr^{2+}$$

The kinetically derived equilibrium quotient $k_{41}/k_{42} = 7.7M^{-1}$ compares well with $5.9M^{-1}$, determined spectrophotometrically.[219] The above isonicotinamide reactions currently represent the longest bridging system for which there is definitive evidence of electron transfer. Entry 44 is quite novel (producing an O-bonded nitroso linkage isomer of chromium(III) which reverts to the more stable N-bonded isomer) in view of the reductions of $Cr(H_2O)_5NO^{2+}$ [237] and $Cr(NH_3)_5NO^{2+}$ [238] by chromium(II) where reduction of the nitrosyl ligand dominates.

Finally, for entry 45 remote attack is implicated by the formation of a sulfur-bonded product. Interestingly, the sole product from the mixture of linkage isomers in entry 46 is sulfur bonded, indicating remote attack for the N-bonded oxidant and adjacent attack for the S-bonded oxidant. Addition of Co(dmgH)$_2$$(C_6H_5)_3P$ to the product solution causes the system to revert to a mixture of linkage isomers. Note the facility with which the nitrile lead-in group affects electron transfer compared with pyridinyl nitrogen (cf. entries 47 and 48). Astonishingly, the 4-cyanopyridine complex of pentaamminechromium(III) reacts *faster* than its pentaamminecobalt(III) counterpart (cf. entries 48 and 49). In the following section the exciting arena of organic bridging groups is explored in considerable detail.

D. Bridging through Organic Structural Units

Electron transfer through organic structural units is not basically different from that for simple inorganic ligands except that considerable versatility can be "built in" for probing the function of the bridging ligand. Traditionally, these functions are divided into three categories: (1) "adjacent attack" with the attendant steric and inductive effects; (2) "adjacent attack" with chelation of the reducing agent in the activated complex; and (3) "remote attack" involving conjugated bond systems. The bulk of reported data[239] has dealt with the reductions of a variety of carboxylate penta- and tetraammine complexes of cobalt(III) and more recently[118] the pentaammine complexes of ruthenium(III), by such reducing agents as chromium(II), vanadium(II), iron(II), and europium(II). In the cobalt(III) systems chromium(II), iron(II), and vanadium(II) appear to be predominantly inner-sphere reductants. Of the ruthenium(III) reductions investigated, however, only those by chromium(II) are definitely of the inner-sphere type. Taube[240] and Gould and Taube[241] have reviewed the role of organic bridging groups in electron transfer reactions. One of the important questions of electron transfer reactions deals with the mechanism(s) of the actual transfer of the electron—namely does the process occur by a resonance or a chemical (radical ion) mechanism? Such a question is perhaps amenable to direct testing using suitably chosen organic bridging groups. The functions of organic bridging ligands are now considered in detail.

First, "adjacent attack" implies attack by the reducing agent on either the carboxylate oxygen bound to the metal center or the carbonyl oxygen of the bound carboxylate ligand. The different rates of oxygen exchange between solvent and carboxylatopentaamminecobalt(III) complexes in acidic solution establish that the two oxygens of the carboxylate group are distinguishable when the ligands are formate, trifluoroacetate, and oxalate.[240,242] In addition, it has since been shown[243] by x-ray crystallography that the solid acetato complex is bound through only one oxygen and that the two carbon–oxygen distances are different (though possibly not by more than 0.01–0.03 Å in view of the large standard deviations reported). It appears likely that they are distinguishable in other systems as well. On the other hand, kinetic and tracer studies in basic solution indicate that the carboxylate oxygens are not distinguishable.[244] If it is accepted that the carboxylate oxygens are distinguishable in acidic solution, several questions naturally arise. Which oxygen is attacked by the reducing agent? Is there competition in some cases for the position of attack? While indirect arguments favor attack on the carbonyl oxygen, no unequivocal evidence has been forthcoming despite several claims.[245,246]

Nevertheless, several indirect arguments bear on this question. Scott and Sykes[245] and Wieghardt and Sykes[246] observed no inner-sphere products in the Cr(II) reductions of various μ-carboxylato dimers of Co(III) where adjacent attack of coordinated oxygen would be required. Observations by Stritar and Taube[118] indicate that the relative rates for the chromium(II) reductions of carboxylatopentaammine complexes of cobalt(III) and ruthenium(III) are essentially the same. Since both systems exhibit the same relative sensitivities to changes in the organic moieties, it was suggested[118] that the influence of the R groups is limited to the formation constants of the binuclear $Ru^{III}—Cr^{II}$ and $Co^{III}—Cr^{II}$ ions and that the structures of the ions are the same. It was further reasoned that if the structures of the μ-hydroxo and μ-formato intermediates, $(NH_3)_5RuORCr(H_2O)_5{}^{4+}$ (R = H, CHO), were the same, than their rates of hydrolyses should parallel the basicities of OR^- to some extent. In fact, they did not, and it was then suggested that the chromium is bound to a different type of oxygen in the two systems. Such an argument then requires that the chromium be bound at the carbonyl oxygen in the various carboxylate complexes. Thus, the carboxyl groups function as three-atom bridging groups. From the observations of Ball and King[218] on the exchange reaction between $Cr(H_2O)_5NCS^{2+}$ and $*Cr(H_2O)_6{}^{2+}$, attack by chromium(II) at a nitrogen adjacent to chromium(III) was rendered highly unlikely. Analogously, on the basis of steric considerations, attack at the bound carboxylate oxygen would not be favored. A final indirect criterion for adjacent attack at carbonyl oxygen is provided by the retardation in rate of the $Co(NH_3)_5(CH_3CO_2)^{2+}$-$Cr^{2+}$ reaction at high acidities, an observation interpreted[247,248] as competition between H^+ and Cr^{2+} for the carbonyl oxygen.

Table 3-7 lists some of the more recent values of rate constants for adjacent attack by chromium(II) on carboxylatopentaamminecobalt(III) complexes. For the simple acetato complexes Linck[249] demonstrated a linear free energy relationship with Taft's σ parameter[250] which reflects electronic effects only. Similarly, Barrett, Swinehart, and Taube[247] reported a correlation for entries 1–7 in Table 3-7 with Taft's σ^* parameter (except for the formato and pivalato complexes, where steric factors are presumed dominant), indicating only a modest dependence of the rates on inductive effects ($\rho^* = -0.37$). Finally, Chen and Gould[251] have reported a correlation with Taft's steric substituent parameter. Aside from these correlations, one is struck by the near constancy of the rate constants, even for the tripositive amino acid complexes. Increased *N*-methyl substitution on the amine nitrogen has a slight rate-retarding effect.

Where adjacent attack with chelation of the reducing agent occurs in the activated complex, general increase in rate occurs over that for reactions proceeding with only adjacent attack.[240] Moreover, $k_{obsd} = k_o + k_{-1}[H^+]^{-1}$, where the latter term arises when proton loss is required for chelation. A variety of Co(III)—Cr(II) reactions in which enhanced reactivities can be attributed to chelation, at least in part, are summarized in Table 3-8. Additional examples are available.[161,255,256,258–261] A survey of the entries reveals that α-hydroxy-

TABLE 3-7
REACTIONS OF CARBOXYLATOPENTAAMMINECOBALT(III) COMPLEXES WITH Cr^{2+} INVOLVING ADJACENT ATTACK AT 25°C AND $\mu = 1.0M$

Carboxylate	k (M^{-1} sec^{-1})	$\Delta H^\ddagger$ (kcal mol^{-1})	$\Delta S^\ddagger$ (eu)	Ref.
1. Formate	7.2	8.3	−27	247
2. Acetate	0.35	8.2	−33	247
3. Trifluoroacetate	0.0170[a] 0.052[b]	9.3	−35	247
4. Chloroacetate	0.12	8.9	−33	247
5. Dichloroacetate	0.075	8.1	−36	247
6. Benzoate	0.15	9.0	−32	247
7. Pivalate	0.0070	11.1	−31	247
8. Glycine[c]	0.064	7.7	−38	252
9. *N*-Methylglycine	0.044	8.0	−38	252
10. *N,N*-Dimethylglycine	0.038	7.5	−40	252
11. *N,N,N*-Trimethylglycine	0.016	7.7	−41	252
12. Alanine[d]	0.049	8.0	−37	253
13. Valine	0.042	8.2	−38	253
14. Isoleucine	0.039	8.6	−40	253
15. Leucine	0.055	7.6	−37	253
16. Serine	0.060	7.6	−37	253
17. Threonine	0.033	9.2	−43	253
18. L-Arginine	0.115	6.7	−33	253
19. 3-Carboxamidopyridine	0.017	10.3	−32	254
20. 4-Carboxamidopyridine	0.078	8.6	−35	254

[a] $\mu = 0.20M$.
[b] $\mu = 3.0M$.
[c] This and the remaining amino acid complexes are tripositive.
[d] For entries 12–17 the amino acids are racemic.

carboxylic acids, α-alkoxy (or aryloxy) carboxylic acids, four- and five-carbon dicarboxylic acids and their half-esters, α-carbonyl carboxylic acids, and several miscellaneous carboxylic acids participate in the chelate mechanism. In addition to the kinetic evidence, chelated chromium(III) products have been identified in most of the cases cited. Implicit in the stoichiometric argument is the assumption that chelate ring closure of any monodentate complex would be slow. Although largely untested, this expectation is fulfilled in at least one case. Olson and Behnke[262] have shown that the monodentate malonate complex undergoes ring closure only slowly. Unfortunately, no data are available for monodentate oxalate ring closure. The identification of chelated products is based upon their relatively higher molar absorptivities in the visible absorption region and on the basis of product charge. These criteria are not always entirely compelling.

Several features of the mechanism involving adjacent attack with chelation are noteworthy. First, the observed rate constants do not exhibit a first-order

TABLE 3-8
REACTIONS OF $Co(NH_3)_5Org^{2+}$ WITH Cr^{+2} INVOLVING ADJACENT ATTACK WITH CHELATION AT 25°C AND $\mu = 1M$

Oxidant[a]	k_0 (M^{-1} sec^{-1})	k_{-1} (sec^{-1})	Ref.
1. $RoOC(O)CH_2OH^{2+}$	3.06	—	255
2. $RoOC(O)CH(OH)CH_3{}^{2+}$	6.65	—	255
3. $RoOC(O)C(OH)(C_6H_5)_2{}^{2+}$	1.93	—	255
4. $RoOC(O)CH(OH)CH_2CO_2H^{2+}$	2.7	3×10^{-1}	255
5. $RoOC(O)CH(OH)CH(OH)CO_2H^{2+}$	1.49	6×10^{-2}	255
6. $RoOC(O)CH_2CH(OH)CO_2H^{2+}$	3.6×10^{-1}	3×10^{-2}	255
7. $RoOC(O)(o\text{-}C_6H_5OH)^{2+}$ [b]	1.5×10^{-1}	2×10^{-8}	256
8. $RoOC(O)CH_2(OCH_3)^{2+}$	4.2×10^{-1}	—	255
9. $RoOC(O)CH_2(OC_2H_5)^{2+}$	4.1×10^{-1}	—	255
10. $RoOC(O)CH_2(OC_6H_5)^{2+}$	1.48×10^{-1}	—	255
11. $RoOC(O)CH(OCH_3)CH_3{}^{2+}$	1.18×10^{1}	—	255
12. $RoOC(O)CH_2CO_2H^{2+}$	1.24	1.00	255
13. $RoOC(O)CH(C_2H_5)CO_2H^{2+}$	1.8×10^{1}	2.0×10^{-1}	255
14. $RoOC(O)C(CH_3)_2CO_2H^{2+}$	7.0×10^{-2}	5.0×10^{-2}	255
15. $RoOC(O)CH(CH_2C_6H_5)CO_2H^{2+}$	9.9×10^{-1}	4.6×10^{-1}	255
16. $RoOC(O)CH_2CO_2CH_3{}^{2+}$	5.00×10^{-1}	—	257
17. $RoOC(O)CH_2CO_2C_2H_5{}^{2+}$	6.70×10^{-1}	—	257
18. $RoOC(O)(o\text{-}C_6H_5CO_2H)^{2+}$	7.5	1.70	258
19. $RoOC(O)C(O)NH_2{}^{2+}$	1.90×10^{2}	—	164
20. $RoOC(O)C(O)OH^{2+}$	1.0×10^{2}	4.6×10^{4}	164
21. $RrOC(O)C(O)OH^{2+}$	1.79×10^{-2}	1.50×10^{-3}	259

[a] $Ro = Co(NH_3)_5{}^{3+}$; $Rr = Cr(NH_3)_5{}^{3+}$
[b] $\mu = 3.0M$.

term in the hydrogen ion concentration. Second, the observed rates are sensitive to steric and inductive effects. For example, it is seen (cf entries 1–3 in Table 3-8) that the lactate complex reacts more rapidly than does the glycolate because of the inductive effect of the methyl substituent, whereas the benzilate complex reacts more slowly, probably because of a combined steric and electron withdrawing effect of the two phenyl substituents. The inductive effect of a single methyl group is all the more remarkable when the rate constants for methoxy acetate and methyl lactate (entries 8 and 11) are compared. The effect is further illustrated in entries 9–10 for changes in the α-alkoxy (or aryloxy) groups. It is also interesting that the assignment of the faster rate constant for α-malate was made on the basis that a five-membered chelate ring would provide a more stable product than a six-membered one.[240,255] However, there is no necessary correlation between kinetic reactivity and thermodynamic stability.

Early studies[263,264] on the malonate system indicated a term first order in hydrogen ion, but more recent work in $LiClO_4/HClO_4$ media has shown that term to be the result of a medium effect.[265] In this context, note that significant ester hydrolysis accompanies electron transfer in the chromium(II) reductions of ethyl and methyl half-esters of malonatopentaamminecobalt(III) (entries 16–17 in Table 3-8). Equal amounts of free alcohol and chelated malonatochromium(III) are formed for the methyl (50%) and ethyl (65%) species. The remaining chromium product in each case is the monodentate half-ester complex of chromium(III). These results indicate competition either between adjacent attack with and without chelation[264] or between carbon–oxygen and chromium–oxygen bond fission.[240] Assuming that adjacent attack dominates, the former alternative accounts for the lack of hydrolysis in the vanadium(II) and europium(II) reductions, as well as for the lack of hydrolysis in the reductions of the methyl succinate half-ester complex by chromium(II), vanadium(II), or europium(II). (Formation of a seven-membered chelate ring involving succinate can be unfavorable.) On the other hand, the bond fission alternative is equally attractive (since both vanadium(III) and europium(III) are labile) and requires only that no chelate with the succinate half-ester be formed. The latter requirement is imposed on both alternatives but for different reasons.

The isotopic exchanges between $Cr(H_2O)_4C_2O_4^+$ and $^*Cr^{2+}$ ($k = 0.13M^{-1}sec^{-1}$) and chromium(II)-catalyzed aquation of *cis*-$Cr(H_2O)_2(C_2O_4)_2^-$ to $Cr(H_2O)_4C_2O_4^+$ and $HC_2O_4^-$ ($R = k[Cr^{2+}][Cr^{III}][H^+]$, $k = 0.04M^{-2}sec^{-1}$) are also likely to proceed via chelated activated complexes[266] where the bridging function is:

$$\left[Cr \begin{matrix} O - C - O \\ O - C - O \end{matrix} Cr \right]$$

An entirely analogous formulation has been proposed by Scott, Green, and Sykes[267] as a possible second stage final product in the reduction of

$$(NH_3)_4Co\overset{NH_2}{\frown}Co(NH_3)_4^{4+}, \text{ both Co bridged by } O{-}C(O){-}, \; C(=O)OH$$

by chromium(II). Replacing one of the chromium atoms by another metal atom gives the activated complex for the chromium(II) reduction of $Co(NH_3)_4C_2O_4^+$.[268] An analogous structure likely obtains in the inner-sphere $Co(C_2O_4)_3^{3-}$—Fe^{2+} reaction[162] as well.

Gould[259] has reported a most unusual reaction between 2-carboxylatopyrazinepentaamminecobalt(III) and Cr^{2+} in which at least two observable intermediates are formed. The intermediates postulated were an N-,O-chelated precursor complex and a radical ion intermediate. The precursor is formed virtually quantitatively in excess Cr^{2+} and decays either directly to the chelated Cr(III) product or by ligand reduction to the second intermediate ($k = 2.63 \times 10^2\ sec^{-1}$) which subsequently reduces the cobalt(III) center ($k = 2.4\ sec^{-1}$) to give the postcursor complex. In the analogous reduction of the 2-carboxamidopyridine complex reported by Balahura,[254] no such intermediates were detected, but the rate was markedly enhanced relative to the 2 and 3 isomers, presumably because of adjacent attack with chelation.

Several possible explanations have been forwarded to account for the enhanced reactivities observed for reactions proceeding via the chelate mechanism. First, chelation in the activated complex stabilizes the chromium(II). It has been noted[255] that stability of the chelated chromium(III) product is not a necessary condition for this mechanism. For instance, the glycolate chelate is metastable and reverts to a species identical to glycolatopentaaquochromium(III). It is not unreasonable to expect that chelate formation will assist in raising the energy of the d_{z^2} orbital on chromium(II) toward the nonequilibrium configuration required for electron transfer. In support of this idea is the observation that chelated chromium(III) products have higher $10Dq$ values than do the corresponding monodentate forms, at least for the lactate[255] and malonate[262,264] ligands where both forms are known. Furthermore, it is likely that the same trend obtains for chromium(II). For complexes containing the glycolate ion, however, $10Dq$ is apparently the same for both forms.[255]

Price and Taube have suggested[164] that a factor which may be responsible for the large differences in rates observed for various chromium(II) reductions is the accessibility to the reducing electron of a low-lying orbital on the ligand. Moreover, such an orbital would be needed whether electron transfer takes place by intermediate formation of a radical ion or by a resonance (or superexchange) mechanism. It appears likely that the observed trends reflect the interplay of these factors and perhaps others yet unrecognized.

Early in the history of chromium(II) reductions of pentaaminecobalt(III) complexes containing organic bridging functions, it was recognized[166,269,270] that attack by the reducing agent could occur at a site on the organic ligand remote from its point of attachment to cobalt(III). Direct evidence for this phenomenon is, of course, desirable. As noted earlier (see section II C.) remote attack for simple inorganic ligands such as N_3^-, NCS^-, and NO_2^- has been demonstrated by forming products containing metastable linkage isomers. Furthermore, Nordmeyer and Taube have advanced[219] the first unequivocal evidence that electron transfer can occur over extended π systems in the reductions of nicotinamide- and isonicotinamidepentaamminecobalt(III) ions by chromium(II). In each case reduction of the original cobalt(III) complex by chromium(II) produces a metastable chromium(III) intermediate formulated below for the nicotinamide and isonicotinamide complexes, respectively.

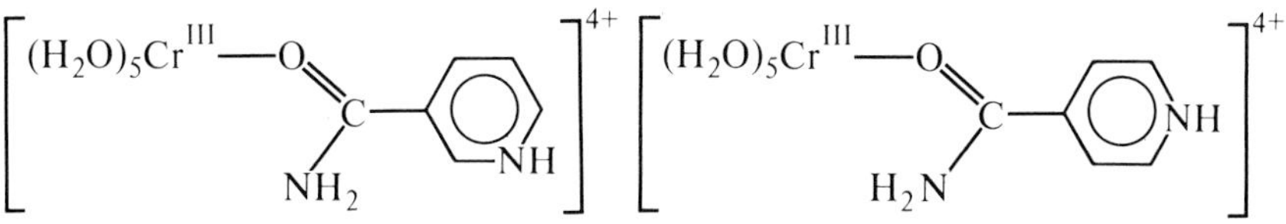

$$\left[(H_2O)_5Cr^{III}{-}O{=}C(NH_2){-}C_5H_4NH\right]^{4+} \qquad \left[(H_2O)_5Cr^{III}{-}O{=}C(H_2N){-}C_5H_4NH\right]^{4+}$$

Thus, the dominant mechanisms for reduction of the nicotinamide- and isonicotinamidepentaamminecobalt(III) ions involve "remote attack" by chromium(II) at the amide carbonyl oxygens. Note that an outer-sphere mechanism is competitive with remote attack in the case of the nicotinamide (but not the isonicotinamide) complex and accounts for some of 29% of the total reduction. Furthermore, the isonicotinamidepentaaquochromium(III) product reacts further with chromium(II) by remote attack to establish a linkage isomer equilibrium in which the isonicotinamide ligand is bound to chromium(III) via either the pyridinyl nitrogen or the carbonyl oxygen. In a similar study of the nicotinamide- and isonicotinamidepentaamminecobalt(III) complex ions Norris and Nordmeyer[271] have concluded that the vanadium(II) reductions, too, are inner-sphere on the basis that the reactions are faster than those involving the stronger reductant chromium(II). Itzkowitz and Nordmeyer[418] have advanced further evidence (developed more fully below) for the chemical or radical ion mechanism based on the insensitivity of k_H/k_D ratios for reduction of $(NH_3)_5Co(i\text{-nic})^{3+}$ vs. $(ND_3)_5Co(i\text{-nic})^{3+}$ as compared with other outer-sphere and resonance transfer inner-sphere reactions.

Two additional observations of postcursor complexes in the remote attack of chromium(II) at the *p*-formyl group in both *p*-formylbenzoato-[272] and *p*-formylcinnamatopentaamminecobalt(III)[260] further exemplify this mechanism. Both reactions exhibit terms first order in $[H^+]$. An unresolved feature of the mechanism for the latter pathway is the site(s) of protonation. Several interpretations have been advanced with the proton located on the adjacent carbonyl: (1) where it can (presumably) improve conjugation along the unsaturated π-way leading to the cobalt(III) center in a resonance mechanism[273] or (2) where it can affect the energy of the lowest empty ligand acceptor orbital in a radical ion mechanism. In the context of first-order hydrogen ion dependencies, note that medium effects may give rise to ambiguities in the form of an apparent dependence of this type. For example, the rates of chromium(II) reductions of various alkene carboxylatopentaaminecobalt(III) complexes appear to increase with increasing acidity, and the protonation of the unsaturated functional group is implicated.[274] Subsequent studies on the reductions of these complexes by other metal ions revealed only slight dependencies on $[H^+]$,[275] suggesting that medium effects may have accounted for results in the chromium(II) study. In contrast to the *p*-formyl systems Balahura and Purcell[276] have reported that the rate of reduction of *p*-acetylbenzonitrilepentaaminecobalt(III) by chromium(II), where the postcursor complex was detected, is independent of the acidity. Balahura and Lewis[277] have proposed that the 3-formylacetylacetonato-bis(ethylenediamine)cobalt(III) complex reacts with chromium(II) via the remote formyl group. Although no postcursor complex was observed, the rate enhancement of ~$10^3\times$ over that of the corresponding, unreactive acetylacetonato complex ($k \leq 10^{-4} M^{-1}\text{sec}^{-1}$) is consistent with the proposal.

Another series of chromium(II) reductions of recent interest and involving "remote attack" are the reductions of carboxylato-bridged, binuclear cobalt(III) complexes of the type:

$$\left[(H_3N)_3Co(\mu\text{-OH})_2(\mu\text{-O}_2CR)Co(NH_3)_3\right]^{3+}$$

where

R = *o*- and *p*- C_6H_4–CHO, HO–C(=O)–CH=CH– (maleato and fumarato), and H–C(–C(=O)–OH)=C(–H)–

When there is no lead-in group remote from the carboxylate bridge, the chromium(II) reductions produce chromic ion and are likely outer sphere.[245,246] However, mixed metal Co(III)—Cr(III) binuclear intermediates were observed in the *o*- and *p*-formylbenzoato systems by Hyde, Wieghardt, and Sykes[278] and in the maleato and fumarato systems by Hyde, Scott, Wieghardt, and Sykes.[279] The μ-amido-μ-maleatobis[tetraamminecobalt(III)]$^{4+}$ ion also reacts via remote attack.[279] More recently, Spiecker and Wieghardt[392] have extended the above tribridged series to include the 2-, 3- and 4-pyridinecarboxylates, where the 2- and 4-isomers proceed via remote attack.

Table 3-9 summarizes the kinetic results for the above reactions as well as for other chromium(II) reductions of organopentaamminecobalt(III) complex ions believed to occur via remote attack. Two significant features of the ligands

TABLE 3-9
REDUCTIONS BY CHROMIUM(II) INVOLVING REMOTE ATTACK

Oxidant[a]	μ	k_o (M^{-1} sec^{-1})	Ref.
1. RoN(pyridine, 3-C(=O)NH$_2$) 3+	1.0	0.033	219
2. RoN(pyridine)–C(=O)–NH$_2$ 3+	1.0	17.4	219
3. Rr′N(pyridine)–C(=O)–NH$_2$ 3+	1.0	1.8	219
4. Rr′O=C(NH$_2$)–(pyridine)NH 4+, c	1.0	0.3[H$^+$]$^{-1}$	219
5. RoN(pyridine)–C(=O)–(phenyl) 3+	0.01 0.003	230(0°C) 220(0°C)	281
6. RoN(pyridine, 3-C(=O)–phenyl) 3+	1.2 0.013	250 63	281
7. RoN(pyridine, 3-C(=O)–CH$_3$) 3+	1.2 0.013	130 25	281
8. RoO–C(=O)–(pyridine)N 2+	3.0	1500 + 1.3[H$^+$]	256
9. RoO–C(=O)–(phenyl)–C(=O)H 2+	1.0	53 + 380[H$^+$]	272
10. RoO–C(=O)–(quinoline)N 2+	1.0	>1100	272

TABLE 3-9 CONTINUED
REDUCTIONS BY CHROMIUM(II) INVOLVING REMOTE ATTACK

	Oxidant[a]	μ	k_o (M^{-1} sec^{-1})	Ref.
11.	$Ro{-}O{-}C(=O){-}CH{=}CH{-}CH{=}CH{-}CO_2H^{2+}$	1.0	2.2 + 15[H^+]	272
12.	$Ro{-}O{-}C(=O){-}CH{=}CH{-}C_6H_4{-}CHO^{2+}$	1.2	260 + 1800[H^+]	260
13.	$RoNC{-}C_6H_4{-}C(=O){-}CH_3$	1.0	6×10^3	276
14.	$RoN C_5H_4{-}CN^{3+}$	1.0	124	236
15.	$RoNC{-}C_5H_4N^{3+}$	1.0	6370	236
16.	$RoNC{-}C_5H_4N^{3+}$	1.0	2.67	236
17.	$RoNC{-}C_5H_4N^{3+}$	1.0	>16000	236
18.	$RrNC{-}C_5H_4N^{3+}$	1.0	7100	236
19.	$Coen_2[O{-}C(CH_3){-}C(CHO){-}C(CH_3){-}O]$	1.0	0.0863	277

[a] Ro = $Co(NH_3)_5^{3+}$; Rr′ = $Cr(H_2O)_5^{3+}$.
[b] Inverse depends on [H^+].

in these complexes are noteworthy: (1) they are easily reduced by chromium(II) and (2) they have conjugated bond systems between the remote functional group and their point of attachment to cobalt(III). Although direct evidence is lacking for the majority of reactions thought to proceed by this mechanism, the kinetic evidence is, nevertheless, strongly suggestive. For example, the ketopyridine complexes (entries 5–7 of Table 3-9) have rate constants five to six orders of magnitude greater than that reported for the analogous reduction of pyridinepentaamminecobalt(III) ion which is known to proceed via an outer-sphere mechanism. The rate enhancement results specifically from the keto function since there can be no question of adjacent attack with chelation. Unfortunately, the chromium(III)–keto complexes have not been detected in these reactions. The remaining complexes listed in the table have organic ligands coordinated to cobalt(III) through a carboxylate group, and most of the rate laws for reduction by chromium(II) exhibit a first-order term in the hydrogen ion concentration. Note here that Taube has suggested[240] the possibility that the proton may be situated on the remote carboxyl or elsewhere apart from the adjacent carboxyl group and that it facilitates electron transfer to the ligand via adjacent attack by virtue of its positive charge.

Additional evidence for remote attack has been sought in the form of ester hydrolysis upon reducing various (half-ester) pentaamminecobalt(III) ions with chromium(II) (entries 16–17 in Table 3-9), vanadium(II), and europium(II). No appreciable ester hydrolysis occurs in the methyl fumarato[280] or methyl succinato[257] systems. Even though no significant amount of ester hydrolysis occurs during these reductions, Hurst and Taube[280] have noted that remote attack is not precluded by such observations, which only indicate that a newly formed chromium–oxygen bond, if formed, is more labile than the carbon–oxygen bond broken upon hydrolysis. As noted previously ester hydrolysis does occur in the chromium(II)–malonatopentaamminecobalt(III) reaction; however, chelation, not remote attack, is the mechanism for that system.

A most interesting indirect approach to inference of remote attack has been reported by Fan and Gould.[282] Casting the rate ratio method of Linck[167,173] into a linear free energy relationship, $\log k_{Ru(NH_3)_6^{2+}} = 1.05 \log k_M + 6$ (M = Cr, V, Eu and b = 2.30, 0.96, 0.48, respectively), for a series of organopentaamminecobalt(III) oxidants, they computed the outer-sphere contributions to the rates. For chromium(II) reductions (known to be predominantly inner sphere) the results indicated that from 10^{-4}% to 3% of the rate were outer sphere. For vanadium(II) and europium(II) the results were 10^{-4}% to 4% and 0.3% to 33%, respectively. Since the outer-sphere contributions to reductions of the benzoylpyridine complex by chromium(II) and europium(II) were $<10^{-4}$%, an inner sphere mechanism via remote attack is implied.

Obviously, considerable thought and experimental ingenuity have been focused on the remote attack mechanism, resulting at least in part from its important role in biological energy transport. Since ligand reducibility is suspected[256,258,283] as a necessary condition for remote attack, a direct examination of the electron transfer mechanism itself may be possible.

The direct observation of radicals bound to metal ions is now widely accepted. In 1970 Hoffman and Simic[284] reported forming such a species in the reaction of hydroxyl radicals with benzoatopentaamminecobalt(III). No oxidation of the metal center was observed, however. Subsequently they reported[285] that reduction of the analogous *p*-nitrobenzoato complex by e^-(aq), CO_2^-, and

$(CH_3)_2COH$ produced a common radical intermediate which does decay with the production of some cobalt(II). Gould[259] has also reported the direct observation of two intermediates in reducing 2-carboxylatopyrazinepentaammine-cobalt(III) with chromium(II). The intermediates were interpreted to be an N-, O-chelated precursor complex and a radical ion intermediate. The precursor is formed virtually quantitatively with excess chromium(II) and decays either directly to a chelated chromium(III) product or by ligand reduction to the second radical intermediate ($k = 2.63 \times 10^2$ sec^{-1}). The radical ion intermediate then reduces the cobalt(III) center ($k = 2.4$ sec^{-1}) to give the postcursor complex. In the analogous reduction of the 2-carboxamidopyridine complex Balahura[254] did not detect any intermediate although the rate was markedly enhanced relative to the 2- and 3-isomers. More recently, Hoffman and Kimmel[286] have observed yet another intermediate radical in the reaction of hydroxyl radical with pyridinepentaamminecobalt(III). The first-order decay ($k = 6.0$ sec^{-1}) is attributed to intramolecular electron transfer to the cobalt(III) center. The interpretation is, however, surprising in view of the previous work on the benzoato complex[284] and requires that the pyridine be oxidized by a net two-electron reaction. Finally, Rowan, Hoffman, and Milburn[287] report observing a coordinated oxalate radical in the ligand-to-metal charge transfer photochemistry of $Co(C_2O_4)_3^{3-}$

Further evidence that coordinated radicals can exist in these systems comes from stoichiometric observations in the reductions of various organopentaamminecobalt(III) complexes by chromium(II).[256] Two extremes in behavior have been observed. For the *p*-nitrobenzoate complex, the nitro group and none of the cobalt(III) is reduced, while for the *p*-formylbenzoato complex the reduction produces cobalt(II) quantitatively. A cobalt(II) yield substantially less than 100% signifies that some entity is competing with cobalt(III) for the reducing agent chromium(II). This supposition is consistent with the decreased yields of cobalt(II) when the hydrogen ion concentration is increased since the competing reduction by an organic ligand to a more saturated state almost certainly requires protons. Gould has extended these observations and has noted[283] that reducing *o*-iodosobenzoatopentaamminecobalt(III) with chromium(II) produces only 11% cobalt(II). Moreover, the reduction of this complex with excess chromium(II) proceeds at a rate identical to that for the *o*-iodobenzoato complex,[283,288] strongly indicating that the initial step is merely a fast step (which produces the *o*-iodo derivative) preceding the rate-determining reduction of cobalt(III) in the *o*-iodobenzoato complex.

Candlin and Halpern have advanced arguments for a radical ion intermediate $[(NH_3)_5CoCO_2]^{2+}$ in the permanganate oxidation of the formatopentaamminecobalt(III) ion.[289] However, Taube has noted[290] that a low-spin cobalt(II) intermediate of the formula $Co(NH_3)_5^{2+}$ would explain their data equally well. Moreover, he cited additional unpublished tracer data[291] which demonstrates that up to one third of the oxygen in the aquopentaamminecobalt(III) ion product derives from the permanganate oxidant when the reaction is run in acetate buffers at pH 5.2. This result is more readily accounted for on the basis of the low-spin $Co(NH_3)_5^{2+}$ intermediate which can be reoxidized to $Co(NH_3)_5H_2O^{3+}$, deriving its oxygen from the permanganate oxidant.[291]

For those systems in which the radical intermediate is not detected, a criterion to distinguish the chemical (radical ion) and resonance transfer mechanisms is needed. In the radical ion mechanism the metal ion is not required to undergo reorganization as a part of the activation process. Thus, Nordmeyer and Taube[219] have suggested that changing the central metal ion and/or the ligands should not influence the stability of the activated complex, particularly if the metal ions are of the same charge. The rates of reduction by a common reducing agent should be insensitive to such changes if the radical ion mechanism is operative. However, the overall rate may remain somewhat sensitive to the nature of the acceptor ion since the radical ion may decay back to reactants as well as forward to the products.[219,280] Sufficient data are now available to test these ideas directly.

Table 3-10 summarizes the results pertinent to this examination. The heading k_M/k_{Cr} is defined as the ratio of rate constants for reduction by chromium(II) of the given metal complex to that for the corresponding aquochromium(III) complex. For entries 1, 2, and 8, 13–15 and 18, 20 and 21, and 23 and 24, it would appear that within a factor of 50 k_M/k_{Cr} is relatively insensitive to changes in metal ions and/or nonbridging ligands, indicating the possibility for a radical ion mechanism. Recall that for simple bridging ligands such as acetate or chloride ion, k_M/k_{Cr} ranges between $>10^4$ and 10^7; consequently, the range of 50 (being "relatively insensitive") does not appear too liberal.

TABLE 3-10
RELATIVE RATE COMPARISONS FOR DISTINGUISHING BETWEEN RADICAL ION AND RESONANCE TRANSFER MECHANISMS

Oxidant	k_o[a] (M^{-1} sec^{-1})	k_M/k_{Cr}[b]	Ref.
1. $(NH_3)_5CoNC_5H_4C(=O)NH_2^{3+}$	1.74×10^1	9.7	219
2. $(bipy)_2Co(NC_5H_4C(=O)NH_2)_2^{3+}$	1.4×10^1	7.8	292
3. $(NH_3)_5RuNC_5H_4C(=O)NH_2^{3+}$	3.92×10^5	2.2×10^5	107
4. $(NH_3)_5RuNC_5H_4C(=O)OCH_3^{3+}$	4.08×10^4	2.3×10^4	107

TABLE 3-10 CONTINUED
RELATIVE RATE COMPARISONS FOR DISTINGUISHING BETWEEN RADICAL ION AND RESONANCE TRANSFER MECHANISMS

Oxidant	k_O[a] (M^{-1} sec^{-1})	k_M/k_{Cr}	Ref.
5. $(NH_3)_5RuN$(pyridine-3-C(=O)NH$_2$)	1.35×10^4 [c]	$>7.5 \times 10^3$	107
6. *cis*-$(NH_3)_4Ru(N$(C$_6$H$_4$)–C(=O)–NH$_2)_2^{3+}$	7.0×10^6 [c]	3.9×10^6	107
7. *cis*-$(NH_3)_4Ru(N$(C$_6$H$_4$)–C(=O)–OCH$_3)_2^{3+}$	1.10×10^6	6.1×10^5	107
8. $(H_2O)_5CrN$(C$_6$H$_4$)–C(=O)NH$_2$ $^{3+}$	1.8	1	219
9. $(NH_3)_5CoN_3^{2+}$	$\sim 3 \times 10^5$	5×10^4	67
10. $(NH_3)_5CrN_3^{2+}$	1.8×10^{-2}	2.9×10^{-3}	293
11. $(H_2O)_5FeN_3^{2+}$	2.9×10^7	4.7×10^6	124
12. $(H_2O)_5CrN_3^{2+}$	6.2	1	294
13. *cis*-$(NH_3)_4Co(H_2O)fumH^{2+}$ [d]	6.07×10^1	$\sim 2 \times 10^1$	273
14. *cis*-$(en)_2Co(H_2O)fumH^{2+}$ [d]	9.10	3	273
15. $(NH_3)_5CofumH^{2+}$ [d]	1.32	$\sim 4 \times 10^{-1}$	280
16. $(NH_3)_5RhfumH^{2+}$ [d]	4.4×10^{-2} [e]	1.5×10^{-2}	280
17. $(H_2O)_5CrfumH^{2+}$ [d]	~ 3	1	273
18. *cis*-$(NH_3)_4Co(H_2O)(CH_3CO_2)$	4.7×10^1	$>4.7 \times 10^5$	295
19. $(H_2O)_5Cr(CH_3CO_2)^{2+}$	$<10^{-4}$	1	296
20. $(NH_3)_5CoC_2O_4H^{2+}$	1.6×10^2	4×10^1	164
21. $(H_2O)_5CrC_2O_4H^{2+}$	4	1	297
22. $(NH_3)_4CoC_2O_4^+$	$\sim 2 \times 10^5$	2×10^6	268
23. $(NH_3)_4CrC_2O_4^+$	6.4×10^{-3}	4.9×10^{-2}	259
24. $(H_2O)_4CrC_2O_4^+$	1.3×10^{-1}	1	266
25. $(NH_3)_5ComalH^{2+}$ [d]	2×10^2	5×10^1	298
26. $(NH_3)_5CrmalH^{2+}$ [d]	1.8×10^{-2}	4.5×10^{-3}	259
27. $(H_2O)_5CrmalH^{2+}$ [d]	4.0	1	298

[a] 25°C.
[b] See text for definition.
[c] 4°C, inner-sphere path.
[d] fumH = bifumarate and malH = bimaleate.
[e] 35°C (results of a single experiment at $1M[H^+]$).

Remote attack is a feature of the first 11 entries in Table 3-10, and except for the first three entries, the resonance transfer mechanism would be indicated by the k_M/k_{Cr} ratios. Nevertheless, Gaunder and Taube[107] have noted for the ruthenium(III) systems that the π-symmetry matching between the acceptor orbitals on the metal ion and the nicotinamide carrier orbitals could lead to the higher k_M/k_{Cr} ratios even for a radical ion mechanism. No such symmetry matching is possible between the nicotinamide ligand π orbitals and the cobalt(III), chromium(III), or rhodium(III) acceptor orbitals. Entries 9–12 represent examples of the resonance transfer mechanism, and it is noteworthy that the acceptor orbital for iron(III) has π symmetry to match that of the azide ion mediator. If one accepts the provisional conclusion of Gauder and Taube,[107] then the radical ion mechanism can be distinguished from resonance transfer only when the k_M/k_{Cr} ratio is insensitive to the nature of the complex ions.

In addition to the isonicotinamide ligands, the monohydrogen fumarate ligands are also believed to proceed via the radical ion intermediate. However, they may differ from the isonicotinamide reactions in one important respect. Indirect evidence indicates that reduction of the fumarato complexes by chromium(II) involves adjacent rather than remote attack. Thus, Hurst and Taube[280] have found that the values of k_0 for fumOH$^-$, fumOCH$_3$$^-$, and fumNH$_2$$^-$, as ligands in pentaamminecobalt(III) complexes, do not depend upon the nature of the substituents on the remote end of the fumarato group. In contrast to these results the various isonicotinamide complexes of ruthenium(III) which are reduced by chromium(II) via remote attack are sensitive to the remote substituents on the isonicotinamide group (cf. entries 3 and 4 or 6 and 7). Again considering the radical ion mechanism, one notes that entries 15 and 17 further support the arguments of Nordmeyer and Taube[219] although it is somewhat surprising that the cobalt(III) species is less reactive than that of chromium(III). (A similar inversion in reactivity was noted earlier for the nitrile bound 4-cyanopyridine complexes, an observation adding weight to the radical ion proposal.) The corresponding rhodium(III) species is even less reactive (entry 16). To include entry 16 in the radical ion class would require extending (by 20×) the k_M/k_{Cr} range. However, the quoted value is from an incidental observation based on a single experiment.

If the mechanism for the fumarato ligands does involve adjacent attack, then Diaz and Taube[273] argue that a more nearly correct comparison for k_M/k_{Cr} involves the *cis*-aquocobalt(III) complex constant as opposed to the pentaamminecobalt(III) complex if solvation effects at the adjacent site are considered. While such an argument is entirely plausible, the comparisons involving various fumarato complexes (except for the rhodium complex) imply the operation of a radical ion mechanism without resort to these special effects. The slower rate observed for the aquoethylenediamine complex is expected on the basis of the chelate effect, but it is also consistent with an alteration of solvation effects at the adjacent carbonyl resulting from the hydrophobic nature of the amine ligands. The gravity of solvation effects in redox reactions may have been treated too lightly in view of recent findings of Bennett et al.[299] Although the article deals with enhanced acidities of protic chelate complexes, electron transfer reactions involving conjugate base species are clearly implicated. In summary, they conclude "... that enhanced acidity arises not only from solvation changes which occur on substitution at the acidic site but also from a modification of the influence of the amine (or other polar) function(s) on the acidic group in the chelate complex."[299] Presumably these effects are transmitted via inductive effects passed along by polarization of adjacent bonds and/or via electric field effects operating in the region of the polar and reactive groups, the field being dominant.

Entries 18 and 19 of Table 3-10 both involve resonance transfer via adjacent attack and are included to emphasize the sensitivity of the resonance transfer mechanism to a change in the oxidizing center. For entries 20–27 the mechanism probably involves adjacent attack with chelation since the chromium(III) product in each case contains a chelated oxalate group. In view of the above discussion on solvation effects, a direct comparison of entries 20 and 21 may be questionable. Moreover, it was noted earlier that the radical ion mechanism is probably not operative for the ligands oxalate and maleate. As noted earlier, Itzkowicz and Nordmeyer[411] have developed an independent criterion for distinguishing the radical ion from the resonance mechanism. Deuteration of the nonbridging ammonia molecules in a variety of monosubstituted pentaamminecobalt(III) species gives rise to an isotope effect as measured by k_H/k_D—i.e., when the reactivity of the undeuterated complex is compared with that of the fully deuterated species. Two distinct groups of k_H/k_D values were observed, one group lying in the range of 1.45 ± 0.15 and the other in the range 1.10 ± 0.10. The former was exemplary of both inner- and outer-sphere mechanisms for resonance transfer (sixth ligand = py, nicotinamide, OH^-, CN^-, NCS^-, OAc^-), whereas the latter is interpreted as exemplary of the radical ion mechanism (sixth ligand = isonicotinamide, 4-pyridinecarboxylate, and fumarate). Thus, as noted above, the rate of a reaction involving the radical ion mechanism is expected to be relatively insensitive to the chemical nature of the oxidant. This expectation is realized in the results of the present study. By contrast, the reaction rates involving resonance transfer involve substantial dislocations on the oxidizing center. A substantial isotope effect in the precursor formation step was ruled out. Taken as a whole, the available data tend to bear out the original proposals of Nordmeyer and Taube[219] and thus firmly establish the radical ion mechanism.

Finally, Table 3-11 summarizes data for the chromium(II) reductions of chelated tetraammine- and bis(ethylenediamine)cobalt(III) complexes. The malonato complex (entry 2) yields both mono- and bidentate chromium products. For entries 3–6 the products are $(H_2O)_5CrOC(O)\text{-}CHRNH_3^{3+}$, indicating adjacent attack. Interestingly, the rate constants for these reactions are virtually identical with those observed for adjacent attack in the simple carboxylatopentaamine class. The constraints of the five-membered ring may lead to some

TABLE 3-11
REDUCTIONS OF CHELATE TETRAAMMINE- AND BIS(ETHYLENEDIAMINE)COBALT(III) SPECIES BY CHROMIUM(II) AT 25°C AND $\mu = 1.0$

Oxidant	k (M^{-1} sec^{-1})	$\Delta H^‡$ (kcal/mol)	$\Delta S^‡$ (eu)	Ref.
1. $Co(NH_3)_4(C_2O_4)^+$	2×10^5	—	—	
2. $Co(NH_3)_4(mal)^+$	26.4	7.2 ± 3	−27.8 ± 0.9	300
3. $Coen_2(Gly)^{2+}$	2.2 (Li^+)	8.8	−27	303
	1.65 (Na^+)	9.7	−25	301
4. $Coen_2$(D,L-Ala)$^{2+}$	0.367	10.9	−24	301
5. $Coen_2$(D,L-Leu)$^{2+}$	0.358	11.5	−22	301
6. $Coen_2$(D,L-phenylAla)$^{2+}$	0.529	11.2	−22	301
7. $Coen_2(CH_3SCH_2CO_2)^{2+}$	274	8.5	−19	303
8. $Coen_2(HOCH_2CO_2)^{2+}$	$38 + 0.99/[H^+]$	—	—	302
9. $Coen_2(HSCH_2CO_2)^{2+}$	$>2 \times 10^6$	—	—	302
10. $Coen_2(CH_3SCH_2CH_2NH_2)^{3+}$	0.38	5.4	−42	303

rate acceleration, but it is likely attenuated by the chelate effect of ethylenediamine. The cis effect of sulfur can be seen by comparing entries 7 and 9 with entry 8. That electron transfer is not occurring at sulfur can be seen in entry 10 where the reaction is slow and outer sphere. The cis-rate enhancing effect of sulfur, however, does not carry over into the analogous chromium(III)–chromium(II) system.[335]

E. Double Bridging

Following the discovery of the bridged activated complex[154] in which a single chlorine atom was transferred quantitatively from a cobalt(III) center to a chromium(III) center, Taube and Myers[188] surveyed the reaction of *cis*-$Coen_2Cl_2^+$ with chromium(II) with the expressed intent of learning whether more than one chlorine was transferred by an oxidant with chlorine atoms in cis positions. Only one atom, however, was found associated with the chromium(III) per cobalt(III) reduced. In addition Kruse and Taube[304] examined the reactions of chromium(II) with *cis*-$Co(NH_3)_4(OH_2)_2^{3+}$ and *cis*-$Coen_2(OH_2)_2^{3+}$ where the aquo compounds were enriched in ^{18}O-labelled water. In each of these systems only one oxygen was transferred per cobalt(III) reduced, thus ruling out the possibility of double bridging in these systems. Chia and King[305] sought evidence for double bridging in the exchange reaction between *cis*-$Cr(OH_2)_4F_2^+$ and $^{51}Cr^{2+}$. However, no evidence for the exchange was found. Chromium(II) was found to catalyze the aquation of *cis*-$Cr(OH_2)_4F_2^+$ to $Cr(OH_2)_5F^{2+}$

Finally, Snellgrove and King[306] obtained unequivocal evidence for a double-bridged activated complex in a study of the exchange between *cis*-$Cr(OH_2)_4(N_3)_2^+$ and $^{51}Cr^{2+}$. The activated complex for the exchange was thus formulated:

$$[(H_2O)_4Cr\begin{smallmatrix}NNN\\ \\NNN\end{smallmatrix}Cr(H_2O)_4{}^{3+}]$$

Subsequently, Haim[307] reported that chromium(II) also catalyzes the aquation of *cis*-$Cr(OH_2)_4(N_3)_2^+$ to $Cr(OH_2)_5N_3^{2+}$ but at a rate 1/30 that of the double-bridged exchange process. This result, combined with the exchange results above, indicates that the reaction between *cis*-$Cr(OH_2)_4(N_3)_2^+$ and chromium(II) proceeds via parallel single- and double-bridged activated complexes:

$$cis\text{-}Cr(H_2O)_4(N_3)_2^+ \xrightarrow{*Cr^{2+}} \begin{cases} \xrightarrow{k_d} cis\text{-}{*Cr}(H_2O)_4(N_3)_2^+ \\ \xrightarrow[H_3O^+]{k_s} {*Cr}(H_2O)_5N_3^{2+} + HN_3 \end{cases}$$

In the latter study *cis*-$Cr(OH_2)_4(N_3)_2^+$ was prepared by the reaction of an excess of *cis*-$Co(NH_3)_4(N_3)_2^+$ or *cis*-$Coen_2(N_3)_2^+$ with chromium(II). These preparations also yield substantial amounts of $Cr(OH_2)_5N_3^{2+}$, the relative yields of the two azido complexes being different for the two oxidants. These results clearly demonstrate the operation of parallel single- and double-bridged activated complexes.

Additional evidence for this type of kinetic scheme was obtained from a report by Fraser[308] that >69% of the chromium(III) produced in the reduction of *cis*-$Co(NH_3)_4(CH_3CO_2)_2^+$ by chromium(II) has a unipositive charge (presumably *cis*-$Cr(OH_2)_4(CH_3CO_2)_2^+$. While the kinetics (and likely the yield of double-bridged product) are reported in error,[309] the double-bridged feature of the reaction has not been set aside. More recently, Ward and Haim[309] have reported a thorough kinetic and stoichiometric study of the chromium(II) reduction of *cis*-$Coen_2(HCO_2)_2^+$ in which single and double bridging involving the formate ion is observed. Moreover, Wood and Higginston[310,311] have reported that at least some transfer of two carboxylate groups accompanies the chromium(II) reduction of $Co^{III}EDTA^-$ and $Co(C_2O_4)_3^{3-}$. Although it was observed that three carboxylate groups appeared in the chromium(III) products under certain conditions, the authors concluded that there was insufficient evidence to support the argument for triple bridging. Similarly, Huchital[312] has interpreted the chromium(II)-catalyzed aquation of $Cr(C_2O_4)_3^{3-}$ to *cis*-$Cr(H_2O)_2(C_2O_4)_2^-$, as well as the unusual chromium(II)-catalyzed trans-to-cis isomerization of the $Cr(H_2O)_2(C_2O_4)_2^-$ ion in terms of double-bridged mechanisms. The intermediate was interpreted to be $Cr(H_2O)_4(C_2O_4H)_2^+$, resulting from the attack by chromium(II) on two adjacent oxygens. The final product is then catalytically converted to the thermodynamically stable cis isomer via a second double-bridged process ($k = 5.0M^{-1}sec^{-1}$). An entirely analogous mechanism is likely responsible for the *cis*-$Cr(H_2O)_2(acac)_2^+$ product in the $Coen(acac)_2^+$–chromium(II) reaction.[313]

All of the above examples involving double bridging exhibit a common feature namely, that two ambidentate ligands serve to bridge the oxidant and reductant. There are several reactions involving a different mode of double bridging where both bridges are supplied by one ligand. These include the chromium(II) reductions of $Co(NH_3)_4(C_2O_4)^+$,[268] $Co(C_2O_4)_3^{3-}$ (at $[H^+] \geq 0.2M$) $Co(EDTA)^-$,[311] and $Cr(H_2O)_4(C_2O_4)^+$.[266] Recall from the previous section that Butler and Taube[255] observed that the formation of a chelated product via a redox reaction does not require chelation in the activated complex since ring closure can conceivably occur as a rapid, independent step subsequent to the electron transfer. Both Wood and Higginson[311] and Huchital[312] have reported slow ring closure for monodentate oxalato complexes. However, the process is chromium(II) catalyzed. Moreover, since the monodentate malonatochromium(III) ion also exhibits slow ring closure,[262] it appears likely that the postulated[308] chelation in the activated complex is justified, though certainly not proved.

Besides the direct evidence for double bridging cited above, several indirect lines of evidence have been used to suggest the possibility of this phenomenon. Thus, Kopple and Miller[314] have suggested that the chromium(II) reduction of *cis*-$Co(NH_3)_4(CH_3CO_2)H_2O^{2+}$ may proceed via a double-bridged activated complex. Chromium(II) reacts 280 times faster with *cis*-$Co(NH_3)_4(CH_3CO_2)$-$(H_2O)^{2+}$ than with $Co(NH_3)_5(CH_3CO_2)^{2+}$ at $1M$ acidity. Moreover, at hydrogen ion concentrations of 0.05 and $1.0M$ >80% of the chromium(III) was recovered as $Cr(OH_2)_5(CH_3CO_2)^{2+}$, and the observed rate constant was of the form $k_{obsd} = 47 + 2.8[H^+]^{-1}$. Since rate terms inverse in hydrogen ion concentration involve transfer of hydroxide ion for chromium(II) reductions of certain aquocobalt(III) complexes[304] and since acetate ion is transferred to an extent >80% in the acetatocobalt(III) complex above, double bridging is certainly a possibility (particularly at the lower acidities). The same argument has been advanced for the analogous kinetic term in the chromium(II) reductions of *cis*-$Co(NH_3)_4(SO_4)(H_2O)^+$,[315] *cis*-$Co(NH_3)_4$(bifumarato)$(H_2O)^{2+}$,[273] and *cis*-$Coen_2$(bifumarato)$(H_2O)^{2+}$.[273]

In addition to the above results, the exchanges between chromium(II) and $Cr(H_2O)_5(H_2PO_2)^{2+}$,[316] $Cr(H_2O)_5(CH_3CO_2)^{2+}$,[296] and $Cr(H_2O)_5$-$(CO_2CH_2CH_2CO_2H)^{2+}$,[273] as well as the chromium-catalyzed linkage isomerization of $Cr(H_2O)_5SCN^{2+}$,[213] have kinetic terms inverse first order in hydrogen ion concentration. For these systems the double-bridging mechanism has also been invoked to interpret the acid-dependent kinetic term. Provided that this interpretation is correct, the combined acid-dependent and independent kinetic terms observed in the above systems correspond to the operation of parallel double- and single-bridged mechanisms discussed earlier.[307]

Although the above interpretation has been widely accepted, it is pertinent to recall that only one ^{18}O atom was found to transfer in the chromium(II) reduction of *cis*-$Coen_2(H_2O)_2^{3+}$ [304] Moreover, rate laws with inverse acid dependencies have been observed for the spontaneous aquations of acidopentaaquochromium(III) ions, as well as for the spontaneous[213] and mercury(II)-catalyzed[214] linkage isomerizations of $Cr(H_2O)_5SCN^{2+}$. Since double bridging cannot be a factor in these systems, it is entirely possible that double bridging is not a factor in the redox systems either. It would appear that the above argument for double bridging is amenable to a direct test by using ^{18}O-labelled water cis to the bridging ligand to indicate whether or not an oxygen atom, as well as the bridging ligand, is transferred. Such an experiment would require that the acidopentaaquochromium(III) product exchange with chromium(II) at a rate slower than that for the initial cobalt(III)–chromium(II) redox process. According to the recent data of Deutsch and Taube,[296] this requirement is met for the acetato exchange reaction. Interestingly, an experiment of this type was considered by Miller and Frazer during the oral presentation (which the author attended) of their results on the sulfato system above.[315] However, to the author's knowledge the results of such an experiment have not been published.

A relative rate comparison has also been used by Haim[317] to infer double bridging by azide ions in the reduction of *cis*-$Co(NH_3)_4(N_3)_2^+$ by iron(II). Reduction of the cis isomer is 20-fold faster than that of $Co(NH_3)_5N_3^{2+}$ and a factor of two faster than that of *trans*-$Co(NH_3)_4(N_3)_2^+$ at 25°C.

Table 3-12 presents a summary of the reactions known to exhibit double bridging. In every case, except perhaps for the oxalato complexes, the reactions proceed via parallel single- and double-bridged activated complexes. The ratio of double to single bridging k_d/k_s is seen to vary over a factor of about 150. This variation is seen to be a function of the metal ion and the nonbridging ligands as well as the bridging groups.

An interesting sidelight to the double-bridging mechanism is the inverse hydrogen ion dependence for the diformato complex which has been interpreted to mean that the protonated cis complex is unreactive. An internally hydrogen-bonded structure with the proton attached to the carbonyl oxygen of both formates was suggested[309] to account for this lack of reactivity. The contention that the protonated form is unreactive is, at first glance, surprising since blocking of the double-bridged pathway does not necessarily preclude the possibility of single-bridging. If a single-bridging mechanism were operative, the rate would be diminished as observed since k_s is one sixth of k_d. However, the latter interpretation would require that the ratio k_d/k_s be sensitive to the hydrogen ion concentration, a condition not substantiated by the data.[309] However, for the stoichiometric experiments used to evaluate k_d/k_s, the range of acidities was experimentally restricted to values $0.01M < [H^+] < 0.1M$ and did not appreciably overlap with the range of acidities used in the kinetic studies.

TABLE 3-12
RATE CONSTANTS (AT μ = 1.0M) FOR REACTIONS OF CHROMIUM(II) INVOLVING DOUBLE BRIDGING

Oxidant	T (°C)	k_{obsd}	k_d	k_s[a]	k_d/k_s[a]	Ref.
1. *Cis*-$Cr(OH_2)_4(N_3)_2^+$	0	60	60	0.95[b]	62	306, 307
2. *Cis*-$Co(NH_3)_4(N_3)_2^+$	0	—	$>10^3$	$>10^3$	1.2	307
3. *Cis*-$Coen_2(N_3)_2^+$	0	—	$>10^3$	$>10^3$	0.4	307
4. *Cis*-$Coen_2(HCO_2)_2^+$	25	$\frac{434}{1 + 0.44\,[H^+]}$	334	50	3.3	309
5. *Cis*-$Co(NH_3)_4(CH_3CO_2)_2^+$	25	c	—	—	(>4.4)	308
6. *Cis*-$Cr(H_2O)_4(C_2O_4H_2)_2^+$	25	5.0[d]	5.0	—	(⩾20)[e]	312
7. $Cr(C_2O_4)_3^{3-}$	25	0.129	0.129	—	(⩾20)[e]	312
8. *Trans*-$Cr(H_2O)_2(C_2O_4)_2^-$	25	0.108	0.108	—	(⩾20)[e]	312
9. $Coen(acac)_2^+$	25	6.4×10^{-3}	$\geqslant 2.0 \times 10^{-3}$	1.2×10^{-3}	1.7	313
10. $Cr(H_2O)_3(N_3)_3$	4.8	112	112	—	>20	411

[a] All k_s values are corrected for symmetry number factors; units of k's are M^{-1} sec^{-1}
[b] Extrapolated from results at 15° and 25°C.
[c] No values are given since it has been noted [309] that acid-catalyzed aquation of the cobalt(III) complex is comparable with the rate of reduction by chromium(II); observation not taken into account in the original study.
[d] Preliminary value.
[e] Estimated assuming 4% single-bridged product may have been undetected.

In view of the results presented for double bridging, it is pertinent to inquire about those requirements necessary for the reactants to achieve the double-bridged activated complex. For double bridging to be observed, it is obvious that stabilization of the activated complex for this mechanism must be approximately comparable with that for the single-bridged mechanism. However, thermodynamic stability of the chromium(III) product cannot be a significant factor, at least for the azido and formato systems, since the disubstituted species are less stable than the corresponding monosubstituted ones. It has been suggested[306] that the electronic structure of the bridging group and steric considerations are important. Thus, azide ions can provide an extended π electronic system between the two metal ion centers. Moreover, if adjacent attack (by chromium(II)) at the carbonyl oxygen(s) is assumed for the various carboxylato complexes, there is at least some π character associated with the three-atom bridge separating the metal ions in most of these systems. Unfortunately, the chromium(II) reduction of *cis*-$Coen_2(CN)_2^+$ [226] has not been extensively investigated because of the complexity of the post-redox reactions which occur. This system is of interest because the cyanide ions could provide a good π ligand route of two atoms for the electron transfer.

Finally, steric constraints have been proposed[308] to play an important role in double bridging. Thus, in the case of the inflexible oxalato chelate ring (with appropriately oriented chelating atoms) double bridging is virtually the only path observed, whereas the more flexible monodentate ligands require considerably more ordering in the activation process, leading to variable k_d/k_s ratios. It is indeed remarkable that double bridging is observed at all for the last three entries in Table 3-12 since adjacent attack (at the oxygen bonded to the metal center) appears mandatory for entry 9 and quite likely for entries 7 and 8. From the above considerations one can at best conclude that further innovative thought and work will be required to unravel the detailed bases of the double-bridging phenomenon.

III. NONBRIDGING LIGAND EFFECTS

Historically, the potential for observing nonbridging ligand effects was recognized[318] very early in the literature of electron transfer chemistry, and it was first dealt with theoretically within the framework of ligand field theory.[319] Although there has been a review published on the subject by Earley,[320] recent developments, especially of anion effects on outer-sphere reactions, have provided some new insights on the mechanisms of nonbridging ligand effects. These effects here are divided into two categories: (1) effects on the oxidant, the more conventional approach, and (2) effects on the reductant, where mechanistic interpretations are ofter rendered difficult as a result of the lability of the metal center. Rate laws of the general form rate = $(\Sigma k_n[X]^n)[Ox][Red]$ (n = 0,1,2) are observed for these effects. It is now possible to address the question of mechanisms for both inner- and outer-sphere processes with supporting data.

A. Inner-Sphere Oxidant Effects

For the inner-sphere reactions much of the early work on nonbridging effects was prompted by Orgel's theory[319] and the ready availability of numerous geometric isomers of cobalt(III) and chromium(III) ammine and amine complexes. For inner-sphere reactions in which the d_{z^2} orbital accepts the reducing electron, the theory predicts that the rate of electron transfer is inversely proportional to the ligand field strength of the group trans to the bridging ligand. Ogard and Taube[321] extended these observations to include stretching of the trans metal–ligand bond to achieve energy matching between donor and acceptor orbitals. The first systematic test of these ideas was reported by Benson and Haim[322] who found general agreement between their results and Orgel's theory ($k_{trans}/k_{cis} > 1$ and $k_{trans} \sim$ ligand field strength for $Coen_2XCl^{n+}$–Fe^{2+} reactions). They further invoked the bond stretching criterion to account for the especially high reactivity of *trans*-$Coen_2(H_2O)Cl^{2+}$ with water trans to the bridging group. The early successes[317,322] stimulated intense activity[51,121,209,213,305,309,323–330] and ignited a controversy over the M—O stretching from *trans*-H_2O, most of which is summarized in Table 3-13. The latter controversy has been resolved in favor of significant bond stretching for the slower redox reactions based on isotopic fractionation studies by Dechant and Hunt[325] along with results for entries 1, 7, 12, 14, 22, 25, 27, and 30. However, for the faster redox reactions another isotopic fractionation study,[121] the results of entry 18, and the more recent[331] reductions of *cis*- and *trans*-$Coen_2(H_2O)OH^{2+}$ by titanium(III) argue against the importance of this effect. More recently, Williams and Garner[330] have found evidence for linear correlations between E_a and $(\Delta Dq_L)^2$ for a series of chloroaquoamminechromium(III) reductions (entries 25 and 29–31) by chromium(II) for an inner-sphere Marcus model and a bond stretching model. There are exceptions to the Orgel-Taube theory as may be noted in entries 4, 13, 20, 23, and 29 with respect to relative reactivities of trans vs. cis isomers.

One of the major shortcomings of Orgel's theory is that it does not accommodate the effects of varying ligands cis to the bridging group. Bifano and Linck[329] have justly criticized this deficiency. Moreover, they proposed that the σ-bonding strength of the variable ligand would be a better measure of the relative stabilization of the d_{z^2} orbital than ligand field strength. The model is equally applicable to cis or trans isomers. Except for the latter point, the model is a special case of the Orgel model since, in principle, ligand field strengths can be broken down into σ and π components.[332] The application of this model based on pK_a (for the variable ligand) as the σ-bond strength criterion is limited in view of the fact that at least four different scales (pK_a, nucleophilicity, electrode potentials, and polarizability) correlate the rates of nucleophilic displacement reactions. More recently, Linck[167] has again invoked this model to interpret (with success) the trends in rate constants for chromium(II) reductions of various chloroaquoamminechromium(III) complexes.[330]

TABLE 3-13
NONBRIDGING LIGAND EFFECTS ON THE RATES OF SOME Fe(II), V(II), AND Cr(II) REDOX REACTIONS

Oxidant[a] Redn. by Fe^{2+} (16 & 17 V^{2+})	k (M^{-1} sec^{-1}) trans	k (M^{-1} sec^{-1}) cis	Ref.	Oxidant[a] Redn. by Cr^{2+}	k (M^{-1} sec^{-1}) trans	k (M^{-1} sec^{-1}) cis	Ref.
1. $R(H_2O)\underline{Cl}^{2+}$ [b]	2.4×10^{-1}	4.5×10^{-4}	322	18. $R(H_2O)\underline{OH}^{2+}$ [c]	1.3×10^{6}	4.0×10^{5}	209
2. $R(Br)\underline{Cl}^{+}$	3.6×10^{-2}	—	322	19. $R(NH_3)\underline{OH}^{2+}$	2.2×10^{5}	2.0×10^{5}	209
3. $R(Cl)\underline{Cl}^{+}$ [c]	1.6×10^{-2}	8.0×10^{-4}	322	20. $R(HCO_2)\underline{HCO_2}^{+}$ [c]	5.1	5.0×10[d]	309
4. $R(NCS)\underline{Cl}^{2+}$	1.3×10^{-4}	1.7×10^{-4}	322	21. $R(NCS)\underline{Cl}^{+}$	2.5×10^{6}	1.8×10^{6}	213
5. $R(NH_3)\underline{Cl}^{2+}$	6.6×10^{-5}	1.8×10^{-5}	324	22. $R(H_2O)NC\underline{S}^{2+}$	1.4×10^{3}	4.5×10^{1}	213
6. $R(Br)\underline{Br}$	1.8×10^{-2}	—	324	23. $R(NCS)NC\underline{S}^{+}$ [c]	1.4×10^{1}	1.5×10^{1}	213
7. $R(H_2O)\underline{Br}$	9.4×10^{-2}	2.8×10^{-4}	329	24. $R(NH_3)NC\underline{S}^{2+}$	3.8	3.1	213
8. $R(3,5\text{-}(CH_3)_2py)\underline{Cl}^{2+}$	—	4.6×10^{-4}	329	25. $R_2(H_2O)\underline{Cl}^{2+}$	4.3×10^{-1}	1.2×10^{-2}	327
9. $R(3\text{-}(CH_3)py)\underline{Cl}^{2+}$	—	5.8×10^{-4}	329	26. $R_2(Cl)\underline{Cl}^{+}$ [c]	5.1×10^{-1}	2.0×10^{-2}	327
10. $R(py)\underline{Cl}^{2+}$	—	7.9×10^{-4}	329	27. $R_3(H_2O)\underline{Cl}^{2+}$	1.2	1.1×10^{-1}	325
11. $R(3\text{-}Clpy)\underline{Cl}^{2+}$	—	2.0×10^{-3}	317	28. $R_4(Cl)\underline{Cl}^{+}$ [c]	2.0×10^{2}	1.4×10^{2}	326
12. $R_1(H_2O)\underline{N_3}^{2+}$	2.4×10^{1}	3.6×10^{-1}	317	29. $R_4(F)\underline{F}^{+}$ [c]	7.5×10^{-4}	5.0×10^{-3}	305
13. $R_1(N_3)\underline{N_3}^{+}$ [c]	3.7×10^{-2}	9.2×10^{-2} [e]	323	30. $R_3(H_2O)\underline{I}^{2+}$	1.6×10^{2}	1.5×10^{1}	382
14. $R_1(H_2O)\underline{Cl}^{2+}$	$\sim 10^{1}$	3.5×10^{-2}	323	31. $Cr(NH_3)_3(H_2O)_2\underline{Cl}^{2+}$ [f]	—	—	330
15. $R_1(Cl)\underline{Cl}^{+}$ [c]	1.1	—	323	32. $Cr(NH_3)_2(H_2O)_3\underline{Cl}^{2+}$ [f]	6.9	—	330
16. $R(\underline{N_3})_2^{+}$	2.6×10^{1}	3.3×10^{1}	163, 165	33. $Cr(NH_3)(H_2O)_4\underline{Cl}^{2+}$ [f]	1.9×10^{1}	—	330
17. $R(H_2O)\underline{N_3}^{+}$	1.8×10^{1}	1.7×10^{1} [g]	165				

[a] All rate constants given to two significant figures for (H^+)-independent terms only, 25°C.
[b] Bridging group is underlined; R = $Coen_2^{3+}$; R_1 = $Co(NH_3)_4^{3+}$; R_2 = $Cren_2^{3+}$; R_3 = $Cr(NH_3)_4^{3+}$; and R_4 = $Cr(H_2O)_4^{3+}$
[c] Corrected for symmetry number factor.
[d] Singly bridged cis.
[e] Singly and doubly bridged path for cis.
[f] Geometry uncertain but H_2O *believed* to be trans to Cl.
[g] Oxidant is *cis*-$Co(NH_3)_4(H_2O)N_3^{2+}$

One last type of nonbridging ligand effect on the oxidant is to be noted. Chelation of the oxidant with bi- or multidentate ligands generally slows down the rate of a redox process relative to the unchelated form up to 40-fold for the slower Co^{III}—Fe^{II} reactions, whereas the effect is less than three-fold for the faster Co^{III}—Cr^{II} reactions.[323] More recently, Linck[173] has discussed observations of these effects for tri-, tetra-, and pentadentately ligated oxidants. One of the more interesting of these effects is that of nonbridging sulfur in the iron(II) reductions of *s-cis*-dichloro- and oxalato-(1,8-diamino-3,6-dithiaoctane)cobalt(III) where rates are ~10^3-fold greater than the corresponding *cis*-(ligand)$Coen_2^+$ reductions.[333,334] However, the interpretation of a mechanism for this effect is rendered difficult by the fact that the sulfurs are both cis and trans to the bridging ligand. Rate enhancement by a cis nonbridging sulfur ligand has been noted.[302,303,335]

A similar difficulty to that noted for nonbridging sulfur above arises in the interpretation of the chromium(II)-catalyzed aquations of various chromium(III) complexes. Although the relative order of nonbridging ligand efficiency has been established[336-341] as $I^- > Br^- > Cl^- > NH_3 > H_2O > F^- > N_3^-$, the effect could arise from a position either cis or trans to the bridging hydroxide ion in these systems.

Although the reactions of platinum(II,IV) complexes are formally distinguished as two-electron transfer systems, marked parallels exist with the cobalt(III)- and chromium(III)-chromium(II) reactions. The change in coordination number accompanying electron transfer creates considerable bond stretching in the activated complex. Consequently, groups trans to the bridge on the oxidant have a profound influence on the rate,[342-346] whereas cis effects are relatively small.[347]

B. Outer-Sphere Oxidant Effects

Although early attention to these effects was restricted to inner-sphere reactions, Patel and Endicott[348] and Linck et al.[323,349] provided convincing evidence that these effects are manifested in the outer-sphere reactions as well. For example, plots of log $k_{M^{2+}}$ vs. log $k_{Fe^{2+}}$ ($M^{2+} = Ru(NH_3)_6^{2+}$ or V^{2+}) are linear for reductions of chlorocobalt(III) complexes. A slope of 0.83 for $Ru(NH_3)_6^{2+}$ indicates nearly identical sensitivities to nonbridging ligand effects as the presumed inner-sphere reductant Fe^{2+}.[349] Table 3-14 summarizes some of the more pertinent comparisons. The vanadium(II) reductions are classified[349] as outer-sphere on the basis of an observation that a plot of log $k_{V^{2+}}$ vs. log $k_{Fe^{2+}}$ exhibits no "break point" as $k_{V^{2+}}$ passes beyond the substitution limit for vanadium(II). The argument is strengthened by the observation of Grossman and Haim[350] that such a "break point" does occur in an analogous plot for reductions of oxalatocobalt(III) complexes by iron(II) and vanadium(II). While the σ-bond strength model has been invoked[349] to account for these observations, the argument is substantially weakened by the results of Patel and Endicott[348] which indicate

TABLE 3-14
NONBRIDGING LIGAND EFFECTS FOR OUTER-SPHERE REDUCTIONS

Oxidant[a]	$k_{Ru(NH_3)_6^{2+}}$ [b] trans	cis	Ref.	$k_{V^{2+}}$ [b] trans	cis	Ref.
1. RCl_2^+	9×10^3	$8. \times 10^2$	63	6.4×10^1	1.2×10^1	349
2. $R(H_2O)Cl^{2+}$	$>10^5$	2.3×10^2	348	2.6×10^2	1.0×10^1	349
3. $R(NH_3)Cl^{2+}$	—	1.2×10^1	348	—	1.9	349
4. $R(CH_3CO_2)_2^+$	$1 + 12[H^+]$	$0.2 + 4[H^+]$	63	—	—	
5. $R_1(H_2O)Cl^{2+}$	—	—	—	4.6×10^2	5.8×10^1	323
6. $RpyCl^{2+}$	—	6.6×10^2	348	—	3.2	349
7. $R[NH_2CH_2C_6H_5]Cl^{2+}$	—	3.7×10^1	348	—	2.8	349

[a] $R = Coen_2^{3+}$; $R_1 = Co(NH_3)_4^{3+}$.
[b] At 25°C; units of k are M^{-1} sec^{-1}.

little or no correlation between reactivity and pK_a for an entire series of *cis*-$Coen_2[NH_2R]Cl^{2+}$–$Ru(NH_3)_6^{2+}$ reactions (not included in the table).

From a mechanistic point of view it has been argued[348] that in the activated complex cobalt(III) must be similarly distorted for both the inner- and outer-sphere reactions. One is therefore strongly tempted to suggest that orientation

TABLE 3-15
NONBRIDGING LIGAND EFFECTS FOR MACROCYCLIC COMPLEXES

trans-Oxidant	$k_{Ru(NH_3)_6^{2+}}$	$k_{V^{2+}}$	$k_{Cr^{2+}}$	Ref.
1. Co(*trans*[14]diene)$(H_2O)_2^{3+}$	8×10^2	2×10^2	$<1.5 \times 10^1$	74
2. Co(teta)$(H_2O)_2^{3+}$	3.0×10^3	2×10^3	$<10^2$	74
3. Co(TIM)$(H_2O)_2^{3+}$	1.9×10^4	1.0×10^3	—	75
4. Co(*trans*[14]diene)$(NH_3)_2^{3+}$	2.9	2.2×10^{-1}	4.0×10^{-3}	76
5. Co(teta)$(NH_3)_2^{3+}$	8.4	1.7×10^{-1}	4.8×10^{-3}	76
6. Co(TIM)$(NH_3)_2^{3+}$	3.7	4.1×10^{-1}	1.2×10^{-2}	76
7. Co(DIM)$(NH_3)_2^{3+}$	6	2.2×10^{-1}	—	76
8. Co(*trans*[14]diene)$(H_2O)NCS^{2+}$	1.1×10^1	—	—	75
9. Co(*trans*[14]diene)$(H_2O)Cl^{2+}$	3.8×10^2	1.7×10^3	—	75
10. Co(*trans*[14]diene)$(H_2O)Br^{2+}$	7×10^3	1.2×10^4	—	75
11. Co(*trans*[14]diene)$(H_2O)OH^{2+}$	1.0×10^2	8×10^2	3.6×10^6	74
12. Co(teta)$(H_2O)OH^{2+}$	5×10^2	8×10^3	7×10^6	74
13. Co(*trans*[14]diene)$(NO_2)_2^+$	5.7×10^{-1}	—	—	75
14. Co(*trans*[14]diene)Cl_2^+	2.0×10^3	6×10^3	—	75
15. Co(teta)Cl_2^+	7.5×10^4	1.6×10^3	—	75
16. Co(*trans*[14]diene)Br_2^+	5×10^5	2.3×10^4	—	75
17. Co(teta)Br_2^+	1.0×10^6	3.4×10^4	—	75

of the chloride ion on the oxidant toward the reductant ion obtains for the outer-sphere mechanism. However, it is noteworthy that the rate constants for the $Ru(NH_3)_6^{2+}$ reductions of *cis*-$Coen_2(NH_2R)Cl^{2+}$ are not very sensitive to increased steric hindrance by R.[348] From R = H to $-CH_2CH_3$ to $-C_6H_{11}$, the rate constants actually increase from 12 to 32 to $45M^{-1}sec^{-1}$—a very puzzling trend. Conceivably the increasing hydrophobic character of R leads to its orientation away from the more hydrophilic solvation site near the adjacent chloride ion. For additional effects see Linck.[173]

Table 3-15 summarizes the various macrocyclic ligand systems reported by Liteplo and Endicott[74] and Rillema et al.[75,76] for which nonbridging ligands can be appraised. A general trend (with exceptions for V^{2+} in entries 3 and 15) in relative rates is that Co(TIM) > Co(teta) > Co(trans[14]diene), indicating a possible connection with the number of imine nitrogen donor atoms. A puzzling feature among the entries is that the $Ru(NH_3)_6^{2+}$ reductions are all faster than the V^{2+} reductions for symmetrically disubstituted species (entries 1–7 and 13–17), whereas the unsymmetrically disubstituted ones (entries 8–12) exhibit the opposite reactivity order. All the vanadium(II) rate constants exceed the substitution rates of that ion. Comparisons between Table 3-14 and Table 3-15 permit further evaluation of cis nonbridging ligand effects. Entries 1, 2, and 5 from the former with entries 14, 15, and 9 from the latter indicate opposite trends for the macrocylic ligand effects for $Ru(NH_3)_6^{2+}$ and V^{2+}.

C. Inner-Sphere Reductant Effects

Studies of nonbridging ligand effects on reductant ions are largely centered on anion effects for both inner- and outer-sphere reactions. Perhaps the earliest report of an anion effect for an inner-sphere reaction was that for the chromium(II)-catalyzed exchange of chloride ion between $Cr(H_2O)_5Cl^{2+}$ and the free ion reported by Taube and King,[351] where the following mechanism was proposed.

$$Cr(H_2O)_5Cl^{2+} + Cr^{2+} + {}^*Cl^- \rightleftarrows Cr(H_2O)_4Cl{}^*Cl^+ + Cr^{2+} + H_2O$$

Exchange is immeasurably slow in the absence of the catalyst. Subsequently, Taube[318] reported that both $P_2O_7H_n{}^{n\text{-}4}$ and $SO_4{}^{2-}$, but not Cl^-, are efficiently trapped by chromium(III) in the $Co(NH_3)_5H_2O^{3+}$–Cr^{2+} reaction. Similar observations were made for $P_2O_7H_n{}^{n\text{-}4}$ in the $Co(NH_3)_5Cl^{2+}$–Cr^{2+} reaction,[318,352] for various anions in the Cr^{3+}–Cr^{2+} reaction,[353] and for Cl^- in the $Co(NH_3)_5fum^{2+}$–Cr^{2+} [354] and $Co(NH_3)_5Cl^{2+}$–Cr^{2+} [355] reactions. In the latter study about 6% $Cr(H_2O)_4Cl_2{}^+$ was observed as a mixture of cis and trans isomers, indicating that the added chloride ion can exert its nonbridging ligand effect from a position either cis or trans to the bridging group.

In Table 3-16 a summary of anion effects on the kinetics of a variety of inner- and outer-sphere reactions is presented. The values of k_X in columns 5–7 represent the reaction pathway in which the added anion does not penetrate the first coordination sphere of the oxidant. An examination of columns 5–7 reveals one

TABLE 3-16
HALIDE ION EFFECTS ON THE RATES OF ELECTRON TRANSFER REACTIONS BASED UPON THE RATE LAW: RATE = $(k_O + k_X[X^-])$ [OX] [RED] AT 25°C

Reaction[a]	μ (M)	Mech.	k_O (M^{-1} sec^{-1})	k_F(k_{NCS}) (M^{-1} sec^{-1})	k_{Cl} (M^{-1} sec^{-1})	k_{Br} (M^{-1} sec^{-1})	k_{Cl}/k_O	Ref.
1. $RoNH_3^{3+} + Cr^{2+}$	1.0	O.S.	1.0×10^{-3}	(7.3×10^{-1})	1.1×10^{-2}	—	1.1×10^{1}	356
2. $RoNH_3^{3+} + Cr^{2+}$	2.60	O.S.	7.2×10^{-3}	—	6.0×10^{-1}	3.1×10^{-1}	8.3×10^{1}	106, 354
3. $Coen_3^{3+} + Cr^{2+}$	1.0	O.S.	3.4×10^{-4}	(3.2×10^{-1})	1.5×10^{-3}	—	2.3×10^{1}	356
4. $Ru(NH_3)_6^{3+} + Cr^{2+}$	0.055	O.S.	8.4×10^{1}	—	1.2×10^{3}	—	1.4×10^{1}	357
5. $Co(phen)_3^{3+} + Cr^{2+}$	1.0	O.S.	3.0×10^{1}	(2.2×10^{5})	1.7×10^{2}	—	5.7	356
6. $RoNH_3^{3+} + V^{2+}$	0.40	O.S.	4.4×10^{-3}	9.2×10^{1}	3.5×10^{-2}	—	7.9	358
7. $RoNH_3^{3+} + V^{2+}$	1.0	O.S.	1.0×10^{-2}	(2.4×10^{1})	2.4×10^{-2}	—	2.4	356
8. $Coen_3^{3+} + V^{2+}$	1.0	O.S.	7.2×10^{-4}	(2.4)	1.9×10^{-3}	—	2.6	356
9. $Co(phen)_3^{3+} + V^{2+}$	1.0	O.S.	3.8×10^{3}	(6.3×10^{7})	6.0×10^{3}	—	1.6	356
10. $Fe(bipy)_3^{3+} + Fe^{2+}$	1.0	O.S.	1.88×10^{4}	(8.9×10^{8})	2.9×10^{5}	—	1.5×10^{1}	359
11. $Fe(dimbipy)_3^{3+} + Fe^{2+}$	1.0	O.S.	3.2×10^{2}	(3.8×10^{7})	7.4×10^{3}	—	2.3×10^{1}	359
12. $Fe(phen)_3^{3+} + Fe^{2+}$	1.0	O.S.	3.40×10^{4}	(2.0×10^{9})	4.9×10^{5}	3.8×10^{5}	1.4×10^{1}	359
13. $Co(phen)_3^{3+} + Fe^{2+}$	1.0	O.S.	5.3×10^{2}	(2.0×10^{7})	2.0×10^{4}	—	3.8×10^{1}	356
14. $RoOAc^{2+} + Cr^{2+}$	1.65	I.S.	3.5×10^{-1}	—	2.0×10^{-1}	6.4×10^{-2}	5.9×10^{-1}	354
15. $RofumH^{2+} + Cr^{2+}$	3.4	I.S.	1.00	—	6.9×10^{-1}	3.0×10^{-1}	6.9×10^{-1}	354
16. $RrCl^{2+} + Cr^{2+}$	1.15	I.S.	5.83×10^{-2}	$<3 \times 10^{1}$	$<10^{-2}$	$<10^{-2}$	$<6 \times 10^{-1}$	321
17. $Rr'Cl^{2+} + Cr^{2+}$	1.0	I.S.	9[b]	—	1.0	—	1.1×10^{-1}	351
		O.S.	$<10^{-5}$	—	5.0×10^{-1}	—	$>5 \times 10^{4}$	

18. $Rr'I^{2+} + Cr^{2+}$	2.0	I.S.	$\sim 10^3$	5.3×10^4	2.74×10^2	6.3×10^1	$\sim 3 \times 10^{-1}$	336
		O.S.	10^{-4}	—	—	—	$>3 \times 10^6$	
19. $RoH_2O^{3+} + V^{2+}$	1.0	?	5.3×10^{-1}	4.5×10^3	2.5	—	4.7	358
20. $RoF^{2+} + Fe^{2+}$	1.7	?	7.6×10^{-3}	1.3×10^1	2.1×10^{-2}	3.5×10^{-3}	2.8	360
21. $RoBr^{2+} + Fe^{2+}$	1.7	?	9.2×10^{-4}	4	$<10^{-3}$	—	~ 1	360
22. $Fe^{3+} + Fe^{2+}$	1.0	?	3.0	(2.4×10^3)	9.8×10^1 [c]	—	3.3×10^1	361
23. $Fe^{3+} + Eu^{2+}$	1.0	?	3.38×10^3	—	7.0×10^3	3.8×10^3 [d]	2.1	339, 362
24. $Fe^{3+} + Cr^{2+}$	1.0	?	2.3×10^3	—	2×10^4	3.9×10^3 [d]	9	339, 363
25. $FeTpyP(H_2O)_2{}^{3+} + Cr^{2+}$	0.5	?	2.7×10^2	(2.3×10^7)	9.1×10^5	4.2×10^5	3.0×10^3	364, 365
26. $CoTMpyP(H_2O)_2{}^{3+} + Cr^{2+}$	0.5	O.S.	1.6×10^1	(2.7×10^4)	4.9×10^2	—	3.0×10^1	365
27. $Co^{3+} + Cr^{2+}$	3.0	?	3.4×10^3	—	2.4×10^5	—	7.0×10^1	73
28. $CoTPPS(H_2O)_2{}^{-} + Cr^{2+}$	0.25	I.S.[e]	—	(1.3×10^6)	2.9×10^4	—	—	366
29. Ferricyt. *c* + Cr^{2+}	0.10	O.S.	10	(1×10^6)	3×10^2	—	30	367, 368
30. $CoTMpyP(H_2O)_2{}^{3+}$– $Ru(NH_3)_6{}^{2+}$		O.S.	1.0×10^5	(6.6×10^5)	3.3×10^5	—	2.8	369

[a] Ro = $Co(NH_3)_5{}^{3+}$, Rr = $Cr(NH_3)_5{}^{3+}$, Rr′ = $Cr(H_2O)_5{}^{3+}$.
[b] 0°C.
[c] Contributions from inner-sphere (I.S.) and outer-sphere (O.S.) paths an upper limit.
[d] 1.6°C.
[e] Attack by chromium(II) at a remote sulfonate oxygen on the periphery of the porphine ring is proposed.

trend immediately, that the order of catalytic efficiencies for the halides is F^- $\approx NCS^- > Cl^- > Br^-$. Moreover, it can be seen (cf. column 8) that the ratio $k_{Cl}/k_O > 1$ for those reactions which proceed via an outer-sphere mechanism and $k_{Cl}/k_O < 1$ for reactions of the inner-sphere type.

In the formally analogous Pt^{IV}–Pt^{II}–X^- systems, where a recent view by Mason has appeared,[370] the evidence is overwhelming that the anion effects are primarily localized in a position trans to the bridging group.

One of the most novel nonbridging ligand effects on the reductant iron(II) was recently proposed by Cannon and Stillman,[371] where an anionic metal complex possibly serves as a nonbridging ligand. The previously reported[162] inner-sphere reaction, $Co(ox)_3^{3-}$–Fe^{2+}, was found to exhibit a second-order dependence on Co(III) when the oxidant was in excess. The proposed activated complex is

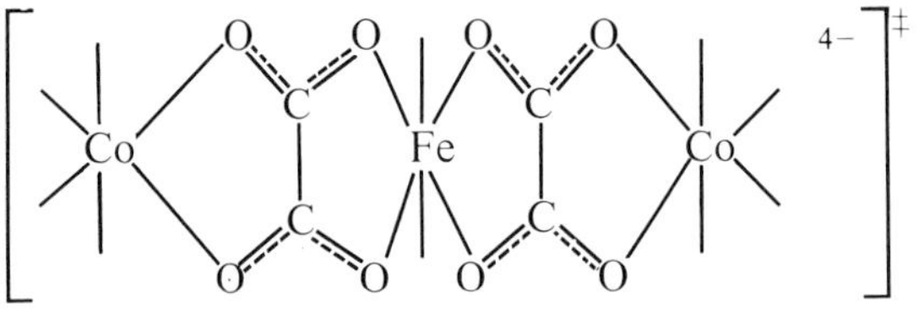

Although it has long been known that platinum(II) catalyzes the substitution of ligands on platinum(IV) by an efficient redox mechanism, parallel systems not involving a change in coordination number are not so widely known. Pennington and Haim[336] found that chromium(II) catalyzes the substitution of I^- in $Cr(H_2O)_5I^{2+}$ by F^-, Cl^-, and Br^- with 100% efficiency and with rates decreasing in the listed order. (Without the catalyst only about 15% substitution has been observed.[372,373]) The preferred mechanism ascribes a nonbridging ligand role to the added anion:

$$Cr(H_2O)_5I^{2+} + Cr^{2+} + X^- \rightleftarrows Cr(H_2O)_4IX^+ + Cr^{2+} + H_2O$$

$$Cr(H_2O)_4IX^+ + Cr^{2+} \rightarrow Cr(H_2O)_5X^{2+} + Cr^{2+} + I^-$$

The observed efficiency of substitution is understandable since the intermediate can decay only to products or back to reactants.

In addition to the effects of simple anions on the reductant, other type ligand substitutions have been observed. For example, Williams and Garner[374] reported a study on the effect of replacing coordinated water on chromium(II) by ethylenediamine. Relative rate comparisons based on the $Co(NH_3)_5Cl^{2+}$–Cr^{2+} reaction and product analysis indicated that the rate increases from six to eight times for each added amine up to two amines. Kochi and Powers[375] have utilized RX–$Cren_2^{2+}$ reactions to prepare a wide range of alkyl chromium(III) species of the type $RCren_2DMF^{2+}$ (and $Cren_2(DMF)Cl^{2+}$), and they report as well that similar reactions with Br_2 and I_2 yield $Cren_2(DMF)X^{2+}$. Additional reductions of cobalt(III) and chromium(III) complexes by variously ligated species of chromium(II) have been reported. For example, Ogino and Tanaka[376] proposed an inner-sphere mechanism for the $Cr(NH_3)_5X^{n+}$–$Cr(H_2O)HEDTA^-$ reactions based on a slope of one observed for a log $k_{Cr^{2+}}$ vs. log $k_{Cr(H_2O)HEDTA^-}$ plot. Cannon and Gardiner[183] have studied the effect of *N*-methyliminodiacetate (MIDA) on the $Co(NH_3)_5OAc^{2+}$–Cr^{2+} reaction. The $Cr(H_2O)_3MIDA$ reduction proceeds 60 times faster than that for $Cr(aq)^{2+}$. Similar rate enhancements were observed for the $Co(NH_3)_5OAc^{2+}$–Fe(MIDA)[377] and $CrOH^{2+}$–Cr(MIDA)[378] reactions. Davies and Earley[379] report that a variety of $Cr^{II}L$ complexes (L = EDDaDp, EDTA, HEDTA, MIDA) reduce $Cr(NH_3)_5Cl^{2+}$ with rates correlated by the linear free energy relationship $\Delta \log k = 0.38\ \Delta E^0$. In a glycine buffer system the $Co(NH_3)_5OAc^{2+}$–Cr^{2+} exhibits a glycine dependence, $k_{obsd} = k_0 + k_1[Gly^-]$. Recently, Cannon[380] has proposed a way of estimating the stability constants for some $Cr^{II}L$ systems, including glycine. Mason[370] has noted that steric bulk of nonbridging ligands on Pt(II) can be important in determining its reactivity toward Pt(IV) species. Peloso and Basato[381] have noted that there is very little difference in reactivities of *cis*- or *trans*-$Ptam_2X_2$ toward iron(II) nor is there much effect as *am* is varied from NH_3 to $(n\text{-}C_3H_7)NH_2$ for either cis or trans isomers. Peloso[382] has recently reviewed the mechanisms of complementary and noncomplementary redox reactions of the Pt^{IV}–Pt^{II} system. Numerous examples of substitutions on the reductant species were covered earlier in Table 3-2 for outer-sphere reactions. Further emphasis is explored below.

D. Outer-Sphere Reductant Effects

Until relatively recently there has been little systematic effort directed to the effects of anions on the rates of outer-sphere reactions although scattered observations have been made. Zwickel and Taube[106] and later Manning and Jarnagin[354] examined the effect of chloride ion on the $Co(NH_3)_6^{3+}$–Cr^{2+} reaction. In the latter work the effect of bromide ion was also examined, and on the basis that chloride ion was more efficient in mediating the transfer it was concluded that an activated complex of the type $[(NH_3)_5CoNH_3\cdots X\cdots Cr(OH_2)_5]^{\ddagger}$ was inconsistent with the observed trend. Anion catalysis of the rates for the vanadium(II) reductions of $Co(NH_3)_5L^{3+}$ (L = NH_3, H_2O) is marked ($F^- > SO_4^{2-} \gg Cl^- > I^-$). Dodel and Taube[358] ascribe the observed order of efficiency to stabilization of the activated complex toward the final vanadium(III)–X product. From the close parallels observed in relative rate comparisons between anion effects in the $[Co(NH_3)_5]_2NH_2^{5+}$–V^{2+} reaction ($F^- > SO_4^{2-} \gg Cl^-$) and the $Co(NH_3)_5NH_3^{3+}$–V^{2+} reaction above, Doyle and Sykes[383] suggested that the anion is brought into the activated complex by vanadium(II) but on the side remote from the oxidant. Otherwise, relative rates should have been sensitive to the charge difference between the ions.

By contrast, in a recent extensive investigation of anion effects Przystas and Sutin[356] argue strongly for a mechanism involving orientation of anions between the cationic metal centers. Clearly, electrostatic considerations favor such an orientation. More importantly, they noted that the ratio k_{NCS}/k_{Cl} varies substantially as the oxidant is varied, whereas it would be expected to remain constant both on thermodynamic grounds and on the basis of the small anion effects observed for inner-sphere reactions (where the anion is *not* interposed between the metal centers). The conclusion here would appear to be somewhat more firmly based than the earlier ones.[354,383] Of further interest is the observation by Przystas and Sutin that thiocyanate ion is virtually without effect in the $Co(phen)_3^{3+}$–V^{2+} reaction unless it first complexes with the vanadium(II). Przystas and Sutin[356] have offered the interesting suggestion that the trends in rate ratios for halide effects, viz. k_X/k_O, may be rationalized on the basis of a symmetry argument analogous to that of Stritar and Taube[118] for inner-sphere reactions. Thus for π-bonded anions the faster reactions should be observed for π-acceptor oxidants and/or π-donor reductants, whereas for σ-type anions the analogous σ-acceptor and donor systems should exhibit the faster reactions. Examining the data for outer-sphere reactions in Table 3-16 reveals that $VNCS^+/V^{2+}$ and $FeNCS^+/Fe^{2+}$ values are indeed greater than the corresponding $CrNCS^+/Cr^{2+}$ values, while the $CrCl^+/Cr^{2+}$ values are in turn greater than those for VCl^+/V^{2+}. However, there is a breakdown in the argument for the $FeCl^+/Fe^{2+}$ system where k_{Cl}/k_O exceeds the corresponding value for $CrCl^+/Cr^{2+}$. Przystas and Sutin[356] noted that $FeCl^{2+}$ is substantially more stable than either $CrCl^{2+}$ or VCl^{2+}, a fact which may account for the reversal in expected reactivities. A similar argument has been used to account for the greater reactivity of $Pt(am)_2Cl_2^{2+}$ toward Fe^{2+} than $Pt(am)_2Br_2^{2+}$.[381] While halide ions are classically referred to as σ-type ligands, there are also nonbonding electron pairs of π symmetry potentially available to facilitate electron transfer. If a mechanism involving these orbitals were operative, the so-called σ-symmetry designation for halides would be ambiguous.

In addition to the above modes of anion behavior Sutin and Forman[359] have postulated a new mechanism for certain anion catalyses in the outer-sphere reductions of $Fe(phen)_3^{3+}$, $Fe(bipy)_3^{3+}$, and $Fe(dimbipy)_3^{3+}$. While chloride and bromide ions catalyze these reactions to about the same extent as they do the Fe^{3+}–Fe^{2+} reaction, thiocyanate, azide, and iodide ions exhibit dramatically higher (by 10^5) effects. Nucleophilic attack of the anion on an electrophilic carbon in the phenanthroline or bipyridine ring system or, alternatively, the addition of the anion to the π system of the ligand have been advanced (as possible initial mechanistic steps) to account for the unusually effective catalysis. For thiocyanate ion Fe*N*CS^{2+} is formed quantitatively. Sutin and Forman pointed out that the π-type addition mechanism can better accommodate the observation that replacement of the hydrogen atoms in the 4 and 4′ positions of bipyridine by methyl groups does not significantly alter the ratio k_X/k_O. Strikingly similar effects in the ferricytochrome *c*–Cr^{2+} reaction at pH 6–7 were observed by Yandell, Fay, and Sutin[367] provided that the anions were not preequilibrated with the protein; otherwise, substitution on the iron center leads to prior binding by the anion and adjacent attack by the reductant, chromium(II).

Pasternack and Sutin[365] have examined the effects of chloride and thiocyanate ions on the chromium(II) reductions of diaquotetrakis(4-*N*-methylpyridyl)-porphinecobalt(III) and iron(III). The former is an outer-sphere process and the latter an inner-sphere process involving substitution of water by the anion prior to electron transfer. The rate ratios $k_{Cl}:k_{O} = 30$ and $k_{NCS}:k_{Cl} = 55$ on the CoTMpyP–Cr(II) reaction are analogous to those in the $Co(NH_3)_6^{3+}$–Cr^{2+} reaction (R = 11 and 61) rather than those for the $Co(phen)_3^{3+}$–Cr^{2+} reaction (R = 6 and 1300), implying an activated complex, $[Co—H_2O—X—Cr]^{\ddagger}$, without electron transfer through the porphyrin ring system. By contrast, Przystas and Sutin[356] have shown that the anion effects (Cl^- and NCS^-) for the ferricytochrome *c*–Cr^{2+} reaction at low pH yield $k_{NCS}:k_{Cl} = 3 \times 10^3$, indicating behavior analogous to the $Co(phen)_3^{3+}$–Cr^{2+} system and possibly electron transfer via the heme ring. The observation of a Cr—S linkage strongly indicates some sort of binding of thiocyanate ion to the heme or some inner-sphere contribution. While the above arguments appear to be consistent, Fleisher, Krishnamurthy, and Cheung[366] report that chromium(II) attacks a sulfonato oxygen on the periphery of the porphine in the diaquotetrakis[*p*-sulfonatophenyl]porphinatocobalt(III)–Cr^{2+} reaction. Their data yield $k_{NCS}:k_{Cl} = 45$ for the anion effects, indicating a nonbridging ligand role for the thiocyanate ion even when electron transfer proceeds via the porphine ring. If this interpretation is correct, the previous relative rate arguments are substantially weakened. The role of the anion must be linked intimately to the reductant, however, for the anion effects on the outer-sphere $CoTMpyP^{3+}$–$Ru(NH_3)_6^{2+}$ reaction are quite modest.[269]

The anion effects reported by Espenson and Webb[384] for the outer-sphere chromium(II) reduction of the $Ta_6Br_{12}^{3+}$ cluster exhibit an inverse order (Cl $<$ Br $<$ I $<$ NCS) from that usually observed; however, the anions are likely to be coordinated to the tantalum cluster since each tantalum atom has an available coordination site. Moreover, $CrSCN^{2+}$ was produced as well as the usually expected $CrNCS^{2+}$ product. Parallel inner- and outer-sphere mechanisms are apparently operative since both Cr^{3+} and CrX^{2+} are observed products.

Ulstrup[385] has proposed an unusual radical ion mechanism in the reaction of $Co(NH_3)_5X^{n+}$ species by $Cr(2,2'\text{-bipy})_2(4,4'\text{-bipy})(H_2O)^{2+}$, but the 4,4′-bipyridyl ligand was not found attached to the final chromium(III) product. A theory was also developed to accommodate such a mechanism involving nonbridging ligand effects for outer-sphere reactions.[386] However, in view of the substitution lability of the reductant and the likely presence of free 4,4′-bipyridine in these solutions, a ligand-catalyzed reduction of the cobalt(III) oxidants such as that discussed in section VII may better account for the experimental observations. Nevertheless, the idea is a fascinating one and should be further explored using a nonlabile reductant—e.g., $Ru(NH_3)_5 4,4'\text{-bipy}^{2+}$.

Further, Gillard, Heaton, and Vaughan[387] have shown that there is an induction period in the substitution of Y^- into complexes of the type *trans*-$Rhpy_4X_2^+$ and that the subsequent rapid substitution rate is independent of the Y^- concentration. These observations indicate that a catalyst (demonstrated to be Rh(I)) is generated in the initial induction period and that, unlike the Pt^{IV}–Pt^{II}–Y^- systems, Y^- is involved subsequent to the rate-determining step:

$$trans\text{-}Rhpy_4X_2^+ + Rhpy_4^+ \xrightarrow{\text{slow}} py_4RhXRhpy_4X^{2+}$$

$$py_4RhXRhpy_4X^{2+} + Y^- \xrightarrow{\text{fast}} py_4YRhXRhpy_4X^+$$

$$py_4YRhXRhpy_4X^+ \xrightarrow{\text{fast}} trans\text{-}Rhpy_4XY^+ + X^- + Rhpy_4^+$$

Such a mechanism may account for the earlier observations of Bott, Bounsall, and Poe[388] on the substitution of iodide ion in *trans*-$Rhen_2IX^+$ by Cl^- and Br^-.

Recently Kallen and Earley[389] examined the kinetics of the equilibrium of aquoruthenium(II) and (III) complexes with halide ions (see Chapter 1, section II,B.) Both the rate law, rate $= k[Ru^{2+}(aq)][X^-][Ru^{3+}(aq)]^0$, and the high (~20 kcal/mol) activation energy support a mechanism which is substitution controlled. It is quite likely that a similar mechanism applies to the rapid equilibration of halide ions with $Ru(NH_3)_5H_2O^{3+}$ in the presence of ruthenium(II)–ammine solutions.[357,390] James, Murray, and Higginson[391] more recently reported that $Fe^{II}(CN)_5X^{3-}$ catalyzes the substitution of Y for X in $Fe^{III}(CN)_5X^{2-}$ ($X = H_2O$, NH_3 and $Y = N_3^-$, NCS^-, OH^-, and $Co(CN)_5^{3-}$). Again the rate is limited by the substitution rate, here of Y for X on iron(II).

IV. SPECIFIC ION EFFECTS

The effects which specific ions exert on the rates of electron transfer reactions usually fall into one of two general categories: (1) medium effects—e.g., replacement of one ion, say H^+, by another, say Li^+ or Na^+, and (2) the effects of ions whose charges are opposite those of the reactants. In the previous section the effects of anions on reactions involving cationic reactants were classified as nonbridging ligand effects. Apart from these, however, are the effects of cations on reactions between negatively charged reactants. Although extensive data are lacking, these effects are challenging from the standpoint of mechanisms in their own right.

A. Medium Effects

Medium effects are usually encountered in those systems where the dependence of the rate on the hydrogen ion concentration is tested. Pethybridge and Prue[393] have recently reviewed kinetic salt effects and the specific influence of ions on rate constants; however, while the subject is well done overall, there is not a great deal on medium effects in redox systems. Newton and Baker[187] have given a concise discussion of these effects which is summarized below. For those reactions run at constant ionic strength or constant total anion concentration, hydrogen ion is usually replaced by lithium ion or sodium ion, thus changing the medium. It is, therefore, pertinent to inquire about possible changes in activity coefficients under these conditions. According to the Brönsted principle of specific interaction of ions,[394,395] the activity coefficients of ions, if positively charged, will not be influenced by the concentrations of other ions at constant ionic strength. It is then anticipated that such medium effects will be small. Experimentally it is generally observed that replacement of hydrogen ions by lithium or sodium ions gives rise to rates which are reduced from 0–35%, smaller effects being observed with lithium ion than for sodium ion.

However, many oxidation–reduction reactions involve rate laws with hydrogen ion-dependent terms. If only one such term is observed, either the medium effect is small or there is a fortuitous cancellation of medium effects by an actual path. In the other cases it is convenient to express the medium effect on an individual rate constant k_i by the equation

$$k_i = k_i^0 (1 + a_i[H^+])$$

which is derived from Harned's rule[395]

$$\ln(k_i) = \ln(k_i^0) + a_i[H^+]$$

for small values of a_i. At 25°C the values of a_i usually lie in the vicinity of $\leq 0.1M^{-1}$ for lithium ion and $0.25M^{-1}$ for sodium ion. A Harned-type correction of the type

$$\ln(k_i) = \ln(k_i^0) - a_2[H^+]$$

was recently utilized by Toppen and Linck[91] to suggest that the acid independent term given in the rate law for the $Co(NH_3)_5H_2O^{3+}$–Cr^{2+} reaction may very well be a medium effect instead. In an $HClO_4$–$LiClO_4$ medium the acid-independent term obtained from the usual treatment, $k_{obsd} = a + b[H^+]^{-1}$, is $-0.72 \pm 0.14M^{-1}sec^{-1}$, a number which is physically meaningless. Application of the above Harned-type correction yields a value of $k_i^0 = 3.12\ sec^{-1}$ and a value of $a_i = 0.25 \pm 0.05M^{-1}$; the latter value is somewhat high for an $HClO_4$–$LiClO_4$ medium. If the medium-effect interpretation is accepted, it may very well follow that acid independent kinetic terms reported in this and other systems correspond to an outer-sphere mechanism rather than to a water-bridged inner-sphere mechanism as previously interpreted. A probable far-ranging consequence of this observation is that water cannot function as a bridging group in electron transfer reactions because of its low basicity.[168]

Thamburaj and Gould[275] have reinterpreted an early study[274] of the acid dependence in the chromium(II) reduction of various α-,β-unsaturated carboxylatopentaamminecobalt(III) complexes as a medium effect. As noted in an earlier section the acid-catalyzed path observed[255,263] in the chromium(II) reduction of malonatopentaamminecobalt(III) has been shown[265] to arise from a medium effect. Lavallee and Newton[199] have reported substantial medium effects (changes in activity coefficient quotients exceeding 37%) in the plutonium(III)-catalyzed aquation of the Cr^{III}–Pu^{V} dimer. Thus the value k_{Li}/k_{Na} = 1.60 for $\Delta Z^2 = 24$ in the above system may be compared with similar values of 1.37 and 1.29 for U^{3+}–$Co(NH_3)_5H_2O^{3+}$ [396] and V^{2+}–Np^{4+} [198] systems where $\Delta Z^2 = 18$ and 1.20 and 1.25 for the Np^{3+}–UO_2^{2+} [397] and Pu^{3+}–NpO_2^{2+} [398] systems where $\Delta Z^2 = 12$. Substantial medium effects in the reduction of various oxalato complexes—the latter even in $HClO_4/LiClO_4$—have been noted.[334,350] The take-home lesson is that Li^+ is the preferred replacement ion for H^+ in redox studies, especially when apparent first-order hydrogen ion dependencies are encountered. Even then the medium effect does not always vanish.[336,350]

Ionic strength effects in the range 0.01–4M are usually handled by the extended Debye-Hückel equation

$$\log k' = \log k^0 + (A\Delta Z^2\mu^{1/2})/(1 + Ba^0\mu^{1/2}) + C\mu$$

with k^0, a^0, and C as adjustable parameters (See, e.g., Ref. 199). Recently Stalnaker and Wahl[399] reported that specific ion effects such as ion association were more important than ion atmosphere effects on activity coefficients, at least in the $Os(4,4'\text{-}Me_2bipy)_3^{2+}$–$Fe(4,4'\text{-}Me_2bipy)_3^{3+}$ reaction in acetonitrile. Koren[400] has given an extensive report on the variation of ln k with ionic strength in various alcohol–water mixtures. The results are interpreted on the basis of making or breaking water structure.

B. Cation Effects

Relatively few systems in coordination chemistry have been thoroughly investigated for cationic effects on reactions between negatively charged reactants. These are the $MnO_4^{1-,2-}$ [401,402] and the $Fe(CN)_6^{3-,4-}$ [403,404,405] exchange reactions investigated primarily by Wahl and co-workers and the oxidations of iodide ion by $Fe(CN)_6^{3-}$ [406] and $W(CN)_8^{4-}$.[407] The $MnO_4^{1-,2-}$ exchange rate is enhanced by the following cations in order of increasing efficiency: $Li^+ = Na^+ < K^+ < Cs^+$. (At constant ionic strength the rate is a linear function of the cesium ion concentration.[402]) Cation effects on the $Fe(CN)_6^{3-,4-}$ exchange rate have been more widely investigated. The rate enhancement follows the order: $H^+ < Li^+ < Na^+ < K^+ < NH_4^+ < Rb^+$ and $Mg^{2+} < Ca^{2+} < Sr^{2+}$;[404] $K^+ < Ba^{2+} < Ca^{2+} < H^+$ and $[(n\text{-}C_5H_{11})_4N]^+ < [(n\text{-}C_4H_9)_4N]^+ < [(C_6H_5)_4N]^+ < [(n\text{-}C_3H_7)_4N]^+ < K^+ < [(C_2H_5)_4N]^+ < [Co(C_5H_5)_2]^+ < [(CH_3)_4N]^+$.[405] Campion, Deck, King, and Wahl[405] have suggested that the differences in catalytic activity of H^+ and Ca^{2+} in the two studies lie in leveling effects for the concentrated solutions required for Shporer, Ron, Loewenstein, and Navon's NMR work,[404] which were not observed in the isotopic exchange studies made in more dilute solutions. They also noted that the electrolyte effects on the exchange rates are principally associated with the nature and concentrations of the cations, not the anions or the ionic strength.

Bok, Leipoldt, and Basson[407] have recently presented initial rate data which indicate a cation efficiency order ($Cs^+ > Rb^+ > K^+ > Na^+$) for the oxidation of iodide ion by $W(CN)_8{}^{3-}$. The Cs^+ ion is eight times more efficient than Na^+. Remarkably similar effects were also observed for the oxidation of iodide ion by $Fe(CN)_6{}^{3-}$ by Majid and Howlett.[406] The rates decreased ninefold from Cs^+ to Li^+ with $NH_4{}^+$ between Cs^+ and K^+.

Table 3-17 summarizes representative data for the effects of cations on the $Fe(CN)_6{}^{3-,4-}$ exchange reaction along with several properties of the cations or the system. Several trends are noteworthy in these comparisons. The rate constants observed in concentrated solutions by NMR methods vary with the tabulated properties as follows: (1) directly as the E values [vs. N.H.E.] for the $Fe(CN)_6{}^{3-,4-}$ couple in the presence of the respective cation except for H^+ and possibly Mg^{2+}, Ca^{2+}, and Sr^{2+}; (2) inversely with the approximate Stokes' solution radii (or directly with the metal ion polarizabilities). The results for the tetraalkylammonium ions obtained by the isotopic labeling technique parallel

TABLE 3-17
INFLUENCE OF CATIONS ON THE $Fe(CN)_6{}^{3-,4-}$ EXCHANGE REACTION AND VARIOUS PHYSICAL PROPERTIES OF THE CATION REDOX SYSTEM

Cation	$10^{-4} \cdot k$[a] (M^{-1} sec^{-1})	E[c] (V)	r_s[d] (Å)	μ (M)
1. H^+	0.4	(0.573)	—	1.75
2. Li^+	3	0.448	2.4	1.75
3. Na^+	5.8	0.448	1.8	1.75
4. K^+	7.9	0.458	1.3	1.75
	(0.0355)[b]	—	—	(0.01)
5. $NH_4{}^+$	8.4	0.459	1.3	1.75
6. Rb^+	13.1	0.469	1.2(1.5)[e]	1.75
7. Cs^+	13.7	(0.48)	1.2(1.7)[e]	0.166
8. Mg^{2+}	6.7	0.476	3.5	0.875
9. Ca^{2+}	6.7	0.470	3.1	0.875
10. Sr^{2+}	6.7	0.468	3.1	0.875
11. $[(CH_3)_4N]^+$	0.12_6[b]	—	2.0	0.01
12. $[(C_2H_5)_4N]^+$	0.025[b]	—	2.8	0.01
13. $[(n\text{-}C_3H_7)_4N]^+$	0.0040_6[b]	—	3.9	0.01
14. $[(n\text{-}C_4H_9)_4N]^+$	0.0023_1[b]	—	4.7	0.01
15. $[(n\text{-}C_5H_{11})_4N]^+$	0.0015_6[b]	—	5.2	0.01

[a] Rate constants for exchange at 32°C.
[b] At 0.1°C.
[c] At 25°C vs. N.H.E. (Ref. 408).
[d] Stokes' radii for hydrated ions.
[e] Atomic radius.

those obtained by the NMR technique. It is not unexpected that there should be some correlation of the exchange rate with E which is directly related to free energy. In fact, it is somewhat surprising that there is not a better correlation, a fact which either suggests that factors other than free energy changes are important or that the difference in media for exchange and potential measurements is physically significant.

At least two mechanisms have been proposed to account for the observed trend in exchange rates with size.[404] First, ion pairing between the cation and either (or both) of the negatively charged reactants would serve to facilitate the approach of the reactant ions by reducing the Coulombic barrier and perhaps to lower the requirements for the outer-sphere solvent reorganizations. As Campion, Deck, King, and Wahl have noted,[405] there is abundant evidence for ion pairing in ferro- and ferricyanide solutions. For tetraalkylammonium cations they noted that these constants are in the range of $\sim 20 M^{-1}$ for ferrocyanide solutions and $\sim 5 M^{-1}$ for ferricyanide. For H^+ the former constant is $1.8 \times 10^4 M^{-1}$,[408] a value which undoubtedly reflects the basicity of the coordinated cyanide.

A second mechanism is actually a special case of the first, but it requires a special orientation of the cation with respect to the reactants. Thus, the cation may serve as an electron mediator by bridging the two reactants in much the same way that anions or neutral molecules serve as bridges for inner-sphere reactions. For the exchange reactions, at least, the motion of the cation would be in the direction of the electron transfer, although such a requirement is not necessarily applicable in cross-reactions, just as atom transfer is not required of the inner-sphere process. In support of the second mechanism, direct evidence that electron transfer is accompanied by sodium ion transfer was recently obtained from ESR measurements of sodium benzophenone ketyl upon the addition of benzophenone.[409] The fine structure resulting from the nuclear spin (3/2) of sodium is not changed upon electron transfer, whereas that resulting from the proton is. These observations demonstrate that the unpaired electron does not change its sodium ion environment upon electron transfer.

The latter mechanism also receives support from the trend in cation polarizabilities as noted by Shporer, Ron, Loewenstein, and Navon.[404] They further suggested the possibility of a two-step transfer process in which the metal ion was first reduced for a finite time although they concluded that a one-step, resonance-type electron transfer was more likely. An interesting test of the two-step process may be possible in a study of the NMR Knight shift for the metal ions (or ^{14}N and/or ^{1}H in the case of the tetraalkylammonium cations) provided that the unpaired electron spends some time at the indicated nucleus. Even the observation of a Knight shift for an ion pair would lend an air of credence to the mechanism. A ^{14}N Knight shift study[410] of alkali metal solutions in liquid ammonia, for example, has demonstrated that the electron spends a considerable amount of time at the nitrogen nucleus rather than on the hydrogens. An ESR detection of the intermediate, analogous to those being sought in the reductions of organopentaamine–cobalt(III) systems, would also provide evidence for the second mechanism. It has been correctly noted[405] that there is no direct evidence that the electron is transferred through the cationic group. However, the argument that such a mechanism is unlikely for the tetraalkyl-ammonium cations because they are difficult to reduce is considerably weakened by the observations of Taube[240] that electron transfer through organic structural units does take place even when the reducing agent will *not* reduce the bridging ligand in its free state. Until further evidence becomes available, there appears to be no way to distinguish between the two mechanisms.

V. REACTIONS OF UNCERTAIN CLASSIFICATION

For those reactions which proceed via uncertain mechanisms, inner- vs. outer-sphere or resonance vs. radical ion, indirect criteria must be applied to support the assignment. Considerable effort has been expended to develop such criteria. In a few cases the use of rapid kinetic techniques has allowed the detection of unstable intermediates as primary products of the redox process—e.g. the iron(II) reductions of $Co(C_2O_4)_3^{3-}$ and *trans*-$Co(en)_2H_2OCl^{2+}$ [162] and the vanadium(II) reductions of $Co(NH_3)_5N_3^{2+}$,[165] CrN_3^{2+},[411] and *cis*-$Co(en)_2$-$(N_3)_2^{+}$.[163] Otherwise, relative rate comparisons (including isotopic fractionation factors), activation parameters ($\Delta V^{\ddagger}$, $\Delta H^{\ddagger}$, and $\Delta S^{\ddagger}$), and relative stabilities of activated complexes have proven useful though subject to certain limitations. No reliable method, however, has been advanced to distinguish between outer-sphere and hydrogen atom transfer mechanisms.

Recent relative rate comparisons involving reductions of carboxylatopentaamminecobalt(III) ions by copper(I),[110] titanium(III),[412] and europium(II)-[111,282] with those of chromium(II) have been summarized by Fan and Gould[111] and are presented here in Table 3-18. After the hydrolysis of Ti^{3+} to $TiOH^{2+}$ is taken into account, there are sufficient similarities in the reactivity patterns for each of these reductants and chromium(II) to warrant their provisional assignment as inner-sphere reductants, though supporting evidence is probably weakest for Cu^{+}. The reactions of Cu^{+} and $TiOH^{2+}$ are consistently lower and those of Eu^{2+} slightly higher than for Cr^{2+} for typical adjacent attack reagents. The acid dependencies for Cu^{+}, $TiOH^{2+}$, and Eu^{2+} reductions of binoxalato and 2-pyridinecarboxylato complexes follow the Cr^{2+} pattern. In the Cu^{+} and $TiOH^{2+}$ series there are deviations from the Cr^{2+} trends for those oxidants where radical ion mechanisms may obtain for Cr^{2+} but not for Cu^{+} or $TiOH^{2+}$ because of unfavorable reduction potentials. It is perhaps of interest that Birk[413] has recently obtained evidence for an inner-sphere mechanism in the Cr-$(NH_3)_5N_3^{2+}$–Ti(III) reaction. Orhanovic and Earley[414] also reported a kinetic study of three chlorocobalt(III)–Ti(III) reductions. Thompson and Sykes[415] have reported linear log–log plots (of slope 0.48 ± 0.08) in comparing reactivities of $TiOH^{2+}$ against $Ru(NH_3)_6^{2+}$ and V^{2+} for a common series of pentaamminecobalt(III) oxidants. Since the acetato, sulfato, and μ-superoxo complexes

TABLE 3–18
RATE COMPARISONS OF CARBOXYLATOPENTAAMMINECOBALT(III) COMPLEXES WITH VARIOUS REDUCTANTS AT 25°C AND μ = 1M

Ligand	k_{Cr}[a]	k_{Cu}[b]	$k_{Ti}/[H^+]$[c]	k_{Eu}[a]	k_V[d]
1. Formate	7.2	0.034	0.0070	16	3.5
2. Acetate	0.35	<0.004	.0049	1.97	1.25
3. Trifluoroacetate	0.344	—	—	1.3	0.86
4. Triethylacetate	0.0022	—	—	.068	0.127
5. Trimethylacetate	0.0070	—	—	.18	0.22
6. Benzoate	0.15	<0.002	—	0.84	0.59
7. Glycolate	3.1	<0.006	0.31	88.	8.2
8. Lactate	6.7	<0.005	1.35	91	—
9. Binoxalate	$100 + 40/[H^+]$	$0.058 + 0.03/[H^+]$	$1.20/[H^+]$	5×10^2	$12.5 + 0.4[H^+]$
10. 2-Pyridinecarboxylate	$36/[H^+]$	$0.12/[H^+]$	$0.0018 + 0.0044/[H^+]$	2.8	0.94
11. Fumarate	$1.6 + 4.0[H^+]$ [c]	—	0.0055	—	—
12. *p*-Formylbenzoato	$53 + 380[H^+]$	<0.01	—	$1.08 + 0.31[H^+]$	$0.88 + 0.24[H^+]$

[a] Ref. 275 and 282.
[b] Ref. 110.
[c] Ref. 412.
[d] Ref. 251.

did not fit either of these correlations, which included known outer-sphere reactions, they were assigned to an inner-sphere category in keeping with Fan and Gould's observations. These results were rationalized in terms of HSAB theory.[421]

In earlier work Gunther and Linck[349] found a linear, log–log correlation in comparing the Fe^{2+} and V^{2+} reductions of various chlorocobalt(III) complexes. They suggested an outer-sphere mechanism for the V^{2+} reactions subject to finding a system which displayed a change in sensitivity to nonbridging ligands as the mechanism changed from inner- to outer-sphere. Subsequently, Grossman and Haim[350] demonstrated such a break-point in a log–log plot for reductions of various oxalatocobalt(III) complexes. The vanadium(II) rates were substitution controlled (therefore constant) for the mono- and bis(oxalato) species, but the rate for the $Co(C_2O_4)_3{}^{3-}$–V^{2+} reaction was markedly higher for the outer-sphere reduction. Very recently Prince and Segal[416] have reported another break-point in the reduction of various $Co(Hdmg)_2L_2{}^+$ complexes by V^{2+} as the base strength of the axial groups is changed.

Sutin has noted[123] that rate constants much in excess of about 40 $M^{-1}sec^{-1}$ for vanadium(II) reductions are very probably outer sphere. Moreover, he noted that the enthalpies of activation for the substitution-controlled process usually exceed 10 kcal/mol, whereas values much below that are representative of outer-sphere reactions. Examples conforming to this generalization are recorded in Refs. 123, 167, 173, and 411. Entropies of activation have been disappointing to the extent that they are not generally diagnostic of mechanism. Only one study has appeared[417] reporting volumes of activation and that was for a series of cobalt(III) complexes with iron(II) as the reductant. Positive volumes of activation were obtained as expected for an inner-sphere mechanism, but no model inner- or outer-sphere reactions were examined. Sutin,[161] however, has cautioned that both entropies and volumes of activation are influenced by electrostatic factors and reorganizations in the coordination spheres of the reactants.

Relative rates of azide vs. isothiocyanate complexes have frequently been cited[123,161,168,223] in attempts to distinguish inner- vs. outer-sphere mechanisms. When k_{N_3}/k_{NCS} exceeds $\sim 10^3$, the mechanism is likely inner-sphere, and when it is in the range of 1–10^2, the outer-sphere mechanism is presumed to be operable. A notable exception to this criterion is given in Ref. 223. Similar rate ratios involving aquo vs. hydroxo arguments must now be viewed with reservation.[168,169]

As noted in the previous discussion of the radical ion mechanism (section IID.) a nonbridging isotopic fractionation study was found to be adequate in distinguishing between a resonance and a radical ion mechanism.[418] In the context of isotopic fractionation factors, it is noteworthy that Diebler, Dodel, and Taube[258] have suggested that oxygen fractionation factors in the reductions of aquopentaamminecobalt(III) by various reducing agents might be useful in distinguishing inner- and outer-sphere mechanisms. The suggestion is based upon the rather large difference in the fractionation factors ($d \ln {}^{16}O/d \ln {}^{18}O$)

obtained for reduction of $Co(NH_3)_5OH^{2+}$ by chromium(II) and hexaammine-ruthenium(II), 1.047 and 1.017, respectively. Unfortunately, the reductions of the aquo complex by vanadium(II) and europium(II) do not involve the hydrolyzed form and the fractionation factors for these ions cannot be compared directly with those cited above. Nevertheless, they are virtually identical to that obtained for hexaammineruthenium(II), 1.021, under conditions where the aquopentaamminecobalt(III) ion is the reactant. Since the ruthenium(II) reductant is known to react largely via an outer-sphere mechanism, the suggestion from the above reactions is that vanadium(II) and europium(II) also react similarly.

Haim has suggested that a comparison of the relative stabilities of the transition states, rather than the reactivity order for the series (X = F, Cl, Br, I), could provide a useful indirect criterion for distinguishing between inner- and outer-sphere mechanisms. Such comparisons can be made on the basis of formal equilibrium quotients involving activated complexes—e.g.,

$$[(NH_3)_5CoFM^{n+}]^{\ddagger} + I^- \overset{Q_5}{\rightleftarrows} [(NH_3)_5CoIM^{n+}]^{\ddagger} + F^-$$

Assuming that the value of the transmission coefficient κ in the transition state theory expression $k = \kappa(RT/Nh)K^{\ddagger}$ is unity for the reactions considered, values of Q_5 can be obtained from the appropriate combination of the constants for the following reactions, namely k_2Q_4/k_1Q_3.

$$Co(NH_3)_5F^{2+} + M(II) \rightleftarrows [(NH_3)_5CoFM^{n+}]^{\ddagger} \qquad k_1 \qquad (1)$$

$$Co(NH_3)_5I^{2+} + M(II) \rightleftarrows [(NH_3)_5CoIM^{n+}]^{\ddagger} \qquad k_2 \qquad (2)$$

$$Co(NH_3)_5H_2O^{3+} + F^- \rightleftarrows Co(NH_3)_5F^{2+} + H_2O \qquad Q_3 \qquad (3)$$

$$Co(NH_3)_5H_2O^{3+} + I^- \rightleftarrows Co(NH_3)_5I^{2+} + H_2O \qquad Q_4 \qquad (4)$$

The results of these calculations are presented in Table 3-19.

Two extremes in the values of Q_5 are observed, those for which $Q_5 \leq 1$ and those for which $Q_5 \geq 10^2$. The lower range of values is associated with an inner-sphere mechanism, while the upper range is characterized by an outer-sphere mechanism with but one exception (last entry). In rationalizing the exceptional behavior of the pentacyanocobaltate(II) reduction, Haim noted[420] that this reductant is classified as a class B or "soft" complex ion, whereas the other inner-sphere reductants examined are class A or "hard" complexes.[421] Thus, it was proposed that "soft" complex ion reductants, in contrast to their "hard" counterparts, have Q_5 ($Q_{I,F}$ below) >1. While the above method of comparing redox data is a useful one, Taube[167] has noted that the division of reductants into the "hard" and "soft" classes for the inner-sphere mechanism does not appear to be internally consistent when generalized. For example, if a "hard" metal ion reductant reacts with a complex ion which is reducible to a "soft" metal center, and if the reaction is cast in the form

$$M^{3+} + X^- + R^{2+}{}_{hard} \rightleftarrows M^{2+}{}_{soft} + X^- + R^{3+}$$

then the equilibrium constant is independent of the concentration and chemical nature of X^-. Thus, if fluoride ion enhances the rate of the forward reaction over that observed for iodide ion, then it must also enhance the rate of the reverse reaction. However, the "soft" reductant classification would require $Q_{I,F} > 1$ (i.e., a greater enhancement by iodide ion) for the reverse reaction contrary to the above equilibrium requirement. Nevertheless, this method of data analysis is instructive.

Among the many actinide redox systems, Newton and Rabideau[427] noted an approximate correlation between the entropy of the net activated complex and its formal charge. [$S^*_{complex} = \Delta S^* + \Sigma S°$ (react.) $- \Sigma S°$ (other prod.)] Newton and Baker[428] extended the earlier observations to include activated

TABLE 3-19
FORMAL EQUILIBRIUM QUOTIENTS INVOLVING ACTIVATED COMPLEXES

Equilibrium			Mech.	Q_5
$[RoFCr^{4+}]^‡ + I^-$	$\rightleftarrows$	$[RoICr^{4+}]^‡ + F^-$	I.S.	6.4×10^{-2} [a]
$[RoFCr^{4+}]^‡ + Br^-$	$\rightleftarrows$	$[RoBrCr^{4+}]^‡ + F^-$	I.S.	7.8×10^{-2} [b]
$[RoFCr^{4+}]^‡ + Cl^-$	$\rightleftarrows$	$[RoICr^{4+}]^‡ + F^-$	I.S.	1×10^{-1} [b]
$[RoClCr^{4+}]^‡ + I^-$	$\rightleftarrows$	$[RoICr^{4+}]^‡ + Cl^-$	I.S.	6×10^{-1} [b]
$[RoClCr^{4+}]^‡ + Br^-$	$\rightleftarrows$	$[RoBrCr^{4+}]^‡ + Cl^-$	I.S.	5×10^{-1} [b]
$[RoBrCr^{4+}]^‡ + I^-$	$\rightleftarrows$	$[RoBrCr^{4+}]^‡ + Br^-$	I.S.	7×10^{-1} [b]
$[RoFV^{4+}]^‡ + I^-$	$\rightleftarrows$	$[RoIV^{4+}]^‡ + Br^-$	?	2.2×10^{-1} [b]
$[RoFV^{4+}]^‡ + Br^-$	$\rightleftarrows$	$[RoBrV^{4+}]^‡ + F^-$	?	7.3×10^{-2} [a]
$[RoFFe^{4+}]^‡ Br^-$	$\rightleftarrows$	$[RoBrFe^{4+}]^‡ + F^-$	?	1.6×10^{-3} [a]
$[RoFEu^{4+}]^‡ + I^-$	$\rightleftarrows$	$[RoIEu^{4+}]^‡ + F^-$	?	2.2×10^{-5} [a]
$[RoFEu^{4+}]^‡ + Br^-$	$\rightleftarrows$	$[RoBrEu^{4+}]^‡ + F^-$	?	4.6×10^{-5} [b]
$[RoFTi^{5+}]^‡ + I^-$	$\rightleftarrows$	$[RoITi^{5+}]^‡ + F^-$	?	9.4×10^{-5} [c]
$[RoFTi^{5+}]^‡ + Br^-$	$\rightleftarrows$	$[RoBrTi^{5+}]^‡ + F^-$	?	1.0×10^{-4} [c]
$[RoFU^{5+}]^‡ + Br^-$	$\rightleftarrows$	$[RoBrU^{5+}]^‡ + F^-$	?	5.3×10^{-4} [d]
$[RoClRu(NH_3)^{4+}{}_6]^‡ + I^-$	$\rightleftarrows$	$[RoIRu(NH_3)_6{}^{4+}]^‡ + Cl^-$	O.S.	2.5×10^{2} [a]
$[RoFCr(bipy)_3{}^{4+}]^‡ + Br^-$	$\rightleftarrows$	$[RoICr(bipy)_3{}^{4+}]^‡ + F^-$	O.S.	3.9×10^{2} [a]
$[RoFCu^{3+}]^‡ + Br^-$	$\rightleftarrows$	$[RoBrCu^{3+}]^‡ + F^-$	?	5.6×10^{3} [e]
$[RoFH^{2+}]^‡ + I^-$	$\rightleftarrows$	$[RoIH^{2+}]^‡ + F^-$	?	8×10^{2} [f]
$[RoF(e_{aq})^+]^‡ + Br^-$	$\rightleftarrows$	$[RoBr(e_{aq})^+]^‡ + F^-$	?	4.2×10^{-2} [f]
$[RoFNi^{3+}]^‡ + Br^-$	$\rightleftarrows$	$[RoBrNi^{3+}]^‡ + F^-$	?	1 [g]
$[RoFZn^{3+}]^‡ + Br^-$	$\rightleftarrows$	$[RoBrZn^{3+}]^‡ + F^-$	?	1.5×10^{-2} [g]
$[RoFCd^{3+}]^‡ + Br^-$	$\rightleftarrows$	$[RoBrCd^{3+}]^‡ + F^-$	?	2.2×10^{-2} [g]
$[RoFCo(CN)_5{}^-]^‡ + Br^-$	$\rightleftarrows$	$[RoBrCo(CN)_5]^‡ + F$	I.S.	1.6×10^{4} [a]

[a] Ref. 420. Remaining values of Q_5 were calculated from references given in the footnotes of Table I, Ref. 420, and from rate constants given by footnotes b–g here.
[b] Ref. 175, p. 481
[c] Ref. 422.
[d] Ref. 423.
[e] Ref. 424.
[f] Ref. 425.
[g] Ref. 426.

complexes of charge 0–6+ with $S^*_{complex}$ varying from +28 to −127 cal/deg·mol. More recently Hinderberger and Thompson[201] have further extended the range to +7 ($S^* = -144$ cal/deg·mol). Substantial discrepancies (30–60 cal/deg·mol) have been noted between two net activated complexes of the same charge $[NpO_2Hg_2{}^{5+}]^*$ and $[NpO_2VO^{5+}]^*$ [429] and in the Cr^{III}–Cl^{III} reaction.[486] The reason(s) for failure of this remarkable correlation is not immediately obvious. Ekstrom, McLaren, and Smythe[430] have most recently added a series of U^{3+} reductions to the S^*/Z^* correlation for $Z^* = 5$ and 6+. These results indicate that the charge is the dominant, though not the only, factor in determining S^*. In addition to this work Ekstrom[431] has reinvestigated the kinetics of uranium(V) disproportionation as well as the uranium(IV, VI) exchange kinetics.[432]

Interest has continued in the search for the new reducing agents among the lanthanide series ions. Thus, Faraggi and Feder[433] have reported the reductions of a broad series of pentaamminecobalt(III) and ruthenium(III) ions by samarium(II), ytterbium(II), and europium(II), as well as several conventional reductants, at neutral pH. The reactivity order for any given oxidant was found to be Sm > Yb > Eu. The Yb^{2+} ion exhibits a curious behavior, yielding an inverted reactivity order for Co(III) and a normal order for Ru(III). Christensen, Espenson, and Butcher[434] have expanded the Yb^{2+} series to include halogenopentaammine- and pentaaquochromium(III) ions. The rate constants go through a minimum at chloride ion.

Finally, two studies of higher valent manganese are examined. The stoichiometry of the MnO_4^-–Mn^{2+} reaction is $MnO_4^- + 4Mn^{2+} + 8H^+ = 5Mn^{3+} + 4H_2O$, and the rate law has the form $R = k_{obsd}[MnO_4^-][Mn^{2+}]^2$, where $k_{obsd} = k_0 + k_1[H^+] + k_2[H^+]^2$ with k_0 dominant. While the authors favor a termolecular mechanism, they could not rule out the intermediate formation of $Mn_2O_4^+$.[435] The disproportionation of manganate ion is wavelength dependent: at 610 nm the rate is first-order in [Mn(VI)] and $[H^+]$; at 510 nm the rate is second-order in [Mn(VI)] and first-order in $[H^+]$. A mechanism involving formation of MnO_3 for the first-order path and its reaction with $HMnO_4^-$ in the second-order path has been postulated.[436]

VI. TWO-ELECTRON TRANSFERS

Using time-dependent perturbation theory, Gurnee and Magee[437] have calculated the reaction cross section for the two-electron, gas phase exchange between Ne and Ne^{2+} to be a factor of two lower than that for the Ne–Ne^+ exchange. Experimentally, the values obtained by Wolf[438] suggest a factor closer to four. These results have important implications for they demonstrate not only that simultaneous two-electron transfers are possible but also that they can be competitive with the corresponding one-electron processes. Westheimer [439] has noted, however, that whether a reaction is fundamentally a series of one-electron transfer processes or of two-electron processes is a conclusion from the experimental data and not a direct observation. A distinction between these two processes may not be possible in some cases. Nevertheless, mechanisms considering two-electron transfers for transition metal ions have long been of interest.[440] Recently, evidence of three-electron transfers is accumulating.[473]

Perhaps the "classic" two-electron reaction is the exchange between Pt^{II}–Pt^{IV} [441,442] and Cl^- and $*Cl^-$ [443] in the reaction

$$trans\text{-}Pten_2Cl_2^{2+} + *Pten_2^{2+} + *Cl^- \rightarrow trans\text{-}*Pten_2Cl*Cl^{2+} + Pten_2^{2+} + Cl^-$$

where the activated complex is depicted[443] as

$$\left[\begin{array}{c} \text{en} \quad\;\; \text{en} \\ \text{ClPt*Cl*PtCl} \\ \text{en} \quad\;\; \text{en} \end{array}\right]^{3+}$$

Peloso[444] has given a brief review of both complementary and noncomplementary redox reactions involving Pt^{II} and Pt^{IV}. Several reports by Beattie et al.[445,446] indicate that the reduction of platinum(IV) species by chromium(II) occurs via two-electron transfers. For example, the reduction of *cis*-$Pt(pn)_2Cl_2^{2+}$ by Cr^{2+} is postulated to occur by the following mechanism[446] on the basis of stoichiometric observations.

$$cis\text{-}Pt^{IV}(pn)_2Cl_2^{2+} + Cr^{II} \xrightarrow{H^+} Pt^{II}pn(pnH)Cl^{2+} + Cr^{IV}Cl$$

$$Cr^{IV}Cl + Cr^{2+} \xrightarrow{\text{fast}} Cr^{III}Cl^{2+} + Cr^{III}$$

Further indirect support for this mechanism is derived from the successive one-electron reductions by Eu^{2+} [446] which produce two ionic chlorides and no monodentate propylenediamineplatinum(II).

During 1974 and 1975 one of the long-standing controversies of one- vs. two-electron transfers, the thallium(I, III) exchange,[447–452] has been resolved in favor of the two-electron mechanism. Schwarz, Comstock, Yandell, and Dodson[453] have given an excellent summary of the various indirect arguments on this problem, along with a convincing argument against the participation of Tl^{2+} in the exchange process. Working independently, Falcinella, Felgate, and Laurence[454] arrived at the same conclusion. Both groups utilized the same approach, studying the reactions

$$Tl^{2+} + Fe^{2+} \rightarrow Tl^+ + Fe^{3+} \qquad (k_2)$$

$$2Tl^{2+} \rightleftarrows Tl^+ + Tl^{3+} \qquad (k_5)$$

in combination with

$$Tl^{3+} + 2Fe^{2+} \rightarrow Tl^+ + 2Fe^{3+}$$

to obtain E^0 values for the $Tl^{3+,2+}$ and $Tl^{2+,1+}$ couples followed by evaluation of K_5 and finally k_{-5} from consideration of microscopic reversibility. Thus, the Tl^{3+}–Fe^{2+} reaction has long been thought[455] to occur by the mechanism:

$$Tl^{3+} + Fe^{2+} \rightleftarrows Tl^{2+} + Fe^{3+} \qquad (k_1, k_{-1})$$

$$Tl^{2+} + Fe^{2+} \rightarrow Tl^{+} + Fe^{3+} \qquad (k_2)$$

In 1M $HClO_4$, $k_1 = 1.39 \times 10^{-2}$ $M^{-1}sec^{-1}$ and $k_{-1}/k_2 = 5.1 \times 10^{-2}$.[453] Combining k_2 (6.7×10^6 $M^{-1}sec^{-1}$) with the latter ratio yields k_{-1} (3.4×10^5 $M^{-1}sec^{-1}$), and from the microscopic reversibility principle $K_1 = k_1/k_{-1} = 4.1 \times 10^{-8}$. Now the latter value and the known potential for the $Fe^{3+,2+}$ couple can be used to calculate E^0 ($Tl^{3+,2+}$) = 0.30V which in turn leads to E^0 ($Tl^{2+,1+}$) = 2.22V from E^0($Tl^{3+,1+}$). If Tl^{2+} were involved in the exchange, then the calculated value of $k_{-5}^{calc} = k_5/K_5 = 7.4 \times 10^{-24}$ $M^{-1}sec^{-1}$ compared with $k_5^{obsd} = 7 \times 10^{-5}$ $M^{-1}sec^{-1}$. Clearly, the exchange between thallium(I) and thallium(III) cannot proceed via thallium(II) as an intermediate, and the mechanism is necessarily a two-electron process. It is of interest that the photochemically induced $Tl^{+,3+}$ exchange does appear to occur via a Tl^{2+} intermediate.[456]

Two-electron transfer processes can often be identified by virtue of product analyses. From evidence given in Table 3-20 it can be seen (cf. columns 3 and 4) that the percentages of one- vs. two-electron transfer range from zero to 100 for each path and that the nature of the oxidant appears to govern whether one- or two-electron transfer dominates. (Compare, for example, entries 2, 5, and 10; 1 and 9; 6 and 17; 7 and 16.) It is interesting that while the chromium(III)--chlorine(III) reaction proceeds via parallel one- and two-electron transfers (entry 15), the analogous iron(II) reaction proceeds via the one-electron mechanism only.[457] Overall rate constants given in column 2 can be divided into the respective one- and two-electron transfer contributions using the stoichiometric results in columns 3 and 4. Interestingly, the evidence for two-electron transfer in the various iron(II) and chromium(II) reactions is based upon "dimer" formation—i.e., the M^{IV} is not directly observed—whereas the product in the oxidation of vanadium(III) is VO^{2+}, which reacts only slowly (k = 0.97 $M^{-1}sec^{-1}$) to form VOV^{4+}.

Several additional examples have appeared where two-electron transfers are inferred from stoichiometric measurements. In the V^V–Sn^{II} reaction where both vanadium(III) and vanadium(IV) are formed in chloride media,[458] vanadium(III) is the dominant product. Since the distribution of reaction products is virtually unaffected by adding tris(oxalato)cobalt(III), a known scavenger for tin(III), it was concluded that the dominant step in the reaction mechanism is a two equivalent change. The small consumption (<2%) of the scavenger was attributed to the formation of tin(III) by a minor one-equivalent step which produces vanadium(IV). The latter is also produced, but more slowly, by the subsequent reaction of vanadium(III) with vanadium(V). Cornelius and Gordon[474] also report that in the presence of excess chlorine, vanadium(III) is oxidized via competitive one- and two-electron transfers yielding 20–30% V(V).

A comparison of induction factors (I = moles of substrate oxidized or reduced/moles of inducing agent added) in various reactions can often be utilized to distinguish reactive intermediates or possibly to indicate the absence of in-

TABLE 3-20
RATE CONSTANTS AND PERCENT YIELDS OF ONE- AND TWO-ELECTRON TRANSFER PROCESSES

Reaction	k^a (M^{-1} sec^{-1})	% Yields 1e$^-$	% Yields 2e$^-$	Ref.
1. $V^{II} + O_2$	$>10^2$	~40	~60	459
2. $V^{II} + H_2O_2$	$>10^2$	~70	~30	459
3. $V^{II} + Tl^{III}$	6.8×10^1	—	6	460
4. $V^{II} + Hg^{II}$	1.0^c, 8.5^d	~12	~88	461, 462
5. $Fe^{II} + H_2O_2$	5.8×10^1	>99	<1	463, 464
6. $Fe^{II} + Cl_2$	8.0×10^1	>95	<5	463
7. $Fe^{II} + HOCl$	3.2×10^3	~85	~15	463
8. $Fe^{II} + O_3$	1.7×10^5	~60	~40	463
9. $Cr^{II} + O_2$	1.6×10^8	~0	~100	465, 466 467
10. $Cr^{II} + H_2O_2$	2.1×10^3 (2.8×10^4)	~86	~14	465, 468, 469
11. $Cr^{II} + Cr^{VI}$	—	50	50	465
12. $Cr^{II} + Tl^{III}$	—	~0	~100	465
13. $Cr^{II} + HClO_3$	5.5×10^1	(~33)37	(~67)63	465, 470
14. $Cr^{II} + ClO_2$	3.5×10^5–4.8×10^4	49–65	51–35	465, 470
15. $Cr^{II} + HClO_2$	8.1×10^4	65	35	465, 470
16. $Cr^{II} + HOCl$	2.1×10^4	(69)61	(31)39	465, 470
17. $Cr^{II} + Cl_2$	9.7×10^3	(~100)90	(~0)9.9	465, 470
18. $Cr^{II} + HN_3$	1.8×10^1	90–92	8–10	471
19. $U^{4+} + Cr^{VI}$	$5.1 \times 10^4 [H^+]^{-1}$	~0	~100	472

[a] The rate constants are not corrected for stoichiometric factors from the following rate expressions: entries 1–4, $d[V^{IV}]/dt$; entries 5–10, $d[M^{III}]/2dt$.
[b] Dominant reaction path.
[c] One-electron path.
[d] $k/[H^+]$ (two-electron path) at $1M$ $[H^+]$.

termediates when $I = 0$. Baker, Brewer, and Newton[460] have presented such evidence for the two-electron oxidation of vanadium(II) by thallium(III). Their arguments follow. Since vanadium(III) induces the normally slow Fe^{II}–Tl^{III} reaction[455] with an induction factor ranging from 0.63[475] to 0.93[460] and the U^{IV}–Tl^{III} reaction with $I = 0.2$–0.5[460] it is likely that both systems involve the same intermediate, presumably thallium(II). In contrast, however, the vanadium(II) induced U^{IV}–Tl^{III} reaction has $I = 0$.[460] (Now, differences in induction factors may be rationalized in either of two ways: (1) by different reactivities between the proposed intermediate and two substrates or (2) by assuming that *different* intermediates are generated in the reactions compared.) Baker, Brewer, and Newton concluded that either thallium(II) reacts very much faster with vanadium(II) than with vandium(III) or, since $I = 0$, that no thallium(II) is produced in the oxidation of vanadium(II) by thallium(III). They ruled out the former possibility by demonstrating that the V^{II}–Tl^{III} reaction is slower than that between V^{III} and Tl^{III}. Therefore, the V^{II}–Tl^{III} reaction must proceed via a two-electron process, leaving open the possibility that thallium(II) is the intermediate in the V^{III}-induced reactions cited above.

On the basis of patterns in rate laws Davies, Kipling, and Sykes[476] have suggested that two-electron reductions by $(Hg^{I})_2$ give rise to a term inverse first-order in $[Hg^{2+}]$ whereas one-electron reductions are independent of $[Hg^{2+}]$. At present there are no exceptions to this suggestion, but only two two-electron reagents, Tl^{III} [477] and BrO_3^-,[476] have been utilized to test the suggestion. Alternative mechanisms for the Tl^{III}–$(Hg^{I})_2$ reaction can be formulated:

Mechanism A	Mechanism B
$(Hg^{I})_2 \rightleftarrows Hg^0 + Hg^{II}$	$(Hg^{I})_2 + Tl^{III} \rightleftarrows Tl^{II} + Hg^{I} + Hg^{II}$
$Hg^0 + Tl^{III} \rightarrow Hg^{II} + Tl^{I}$	$Hg^{I} + Tl^{II} \rightarrow Hg^{II} + Tl^{I}$

[Flash photolysis studies of $(Hg^{I})_2$ in solution with potential oxidants would prove most interesting.] Similar mechanistic ambiguities are encountered in the thallium(III) oxidation of uranium(IV)[478] and the molybdenum(VI) oxidation of tin(II).[479] In the latter study, molybdenum(IV), the intermediate whose existence was sought, arises in either the one- or two-electron transfer mechanisms for concentrated(9–12*M*) hydrochloric acid media. (For an informative discussion of the ambiguity in mechanistic interpretations of rate laws, see Ref. 186.)

Actually, the distinction between a two-electron transfer and two successive one-electron transfer processes becomes difficult, if not impossible, when the intermediate oxidation state is very short-lived. According to Westheimer,[439] to be considered, an intermediate must survive long enough ($t_{1/2} > 10^{-11}$ sec) to break out of the "solvent cage" in which it is formed so that it may react with some species other than those surrounding it at the moment of its creation. Rabideau and Masters[480] have used such an argument to support a two-electron transfer mechanism for the reaction:

$$Pu^{VI} + Sn^{II} \rightarrow Pu^{IV} + Sn^{IV}$$

Three alternative one-electron mechanisms are possible.

Mechanism A	Mechanism B
$Pu^{VI} + Sn^{II} \rightarrow Pu^{V} + Sn^{III}$	$Pu^{VI} + Sn^{II} \rightarrow Pu^{V} + Sn^{III}$
$Pu^{VI} + Sn^{III} \rightarrow Pu^{V} + Sn^{IV}$	$Pu^{VI} + Sn^{III} \rightarrow Pu^{V} + Sn^{IV}$
$2Pu^{V} \rightarrow Pu^{IV} + Pu^{VI}$	$Pu^{V} + Sn^{II} \rightarrow Pu^{IV} + Sn^{III}$
	$Pu^{V} + Sn^{III} \rightarrow Pu^{IV} + Sn^{IV}$

Mechanism C

$$Pu^{VI} + Sn^{II} \rightarrow Pu^{V} + Sn^{III}$$
$$Pu^{V} + Sn^{III} \rightarrow Pu^{IV} + Sn^{IV}$$

Mechanism A was ruled out because the rate of disproportionation of plutonium(V) was found to be slow compared with that for the Pu^{VI}–Sn^{II} reaction, and mechanism B was eliminated because the Pu^{V}–Sn^{II} reaction is also slower than that for the Pu^{VI}–Sn^{II} reaction. Mechanism C was considered unlikely because it requires that plutonium(V) and tin(III) react in a separate step where each is in a low, steady-state concentration, whereas they could have reacted (at a higher level of probability) within the solvent cage in which they were originally produced—i.e., effectively, in a two-electron transfer process. If this appealing argument were accepted, then those reactions cited in the previous paragraph could very well involve two-electron transfer processes. However, it should be noted that some very reactive intermediates can escape their original "solvent cages" to react with other substrates. For example, the reaction $Tl^{III} + 2Fe^{II} = Tl^{I} + 2Fe^{III}$ almost certainly involves thallium(II) as an intermediate,[455] and presumably it escapes its original "solvent cage" to react with another iron(III) ion.

From the above considerations it would appear that a distinction between the "simultaneous" two-electron transfer and two successive one-electron transfer processes may be made only when the rates for either of the two successive reactions are slower than the rates for diffusion of the intermediate ions from their "solvent cages." Such a requirement is necessary before chemical scavengers can be expected to react with these intermediates with any reasonable degree of probability.

In the reductions of chromium(VI) by arsenic(III)[481] and by sulfur(IV)[482] it is possible that mechanisms involving two-equivalent changes are operative. An interesting tracer study in the reaction between radiolabeled chromium(VI) and chromium(II) was reported by Hegedus and Haim.[483] Although the mechanism consisting of three successive one-equivalent steps (originally proposed by Ardon and Plane[465] is basically correct, the tracer results suggest that there is a small but significant contribution from a two-equivalent change in the mechanistic scheme. According to the mechanism originally proposed, all of the $Cr(H_2O)_6{}^{3+}$ product originates in the chromium(II) reactant, whereas the dimer $[Cr(H_2O)_4OH]_2{}^{4+}$ derives one chromium atom each from chromium(II) and chromium(VI). However, in the tracer study ~10% of the $Cr(H_2O)_6{}^{3+}$ derives from the radiolabeled chromium(VI). A mechanism consistent with these observations[483] is given as follows:

$$*Cr^{VI} + Cr^{II} \longrightarrow *Cr^{V} + Cr^{3+}$$

$$*Cr^{V} + Cr^{II} \longrightarrow \begin{cases} 80\%\ Cr^{IV} + Cr^{3+} \\ 20\%\ *Cr^{3+} + Cr^{IV} \end{cases}$$

$$*Cr^{IV} + Cr^{II} \longrightarrow *Cr(OH)_2Cr^{4+}$$

$$Cr^{IV} + Cr^{II} \longrightarrow Cr(OH)_2Cr^{4+}$$

Since the second step produces only 50% of the $Cr(H_2O)_6^{3+}$, an ~20% contribution by a two-equivalent reaction between $*Cr^V$ and Cr^{II} is required to account for the ~10% $*Cr(H_2O)_6^{3+}$ produced overall.

Kenny and Carlyle[484] have extended the observations in the Cr^{VI}–Cr^{II} system to chloride media where the elementary reaction(s) yielding $CrCl^{2+}$ and $(CrOH)_2^{4+}$ compete more effectively with elementary steps yielding Cr^{3+}. A chloro dimer (Cr_2Cl^{3+}) of chromium(II) was invoked to account for the increased yields of $CrCl^{2+}$. Interestingly, it will be recalled that a chloro dimer of chromium(III) was proposed in the Pt^{IV}–Cr^{II} reactions,[445,446] discussed earlier in this section, and it presumably reacted to yield 50% Cr^{3+} and 50% $CrCl^{2+}$—yields not inconsistent with the reaction $Cr_2Cl^{3+} + Cr^{VI} \rightarrow Cr^{IV} + Cr^{3+} + CrCl^{2+}$ proposed by Kenny and Carlyle[484] in which Cr_2Cl^{3+} is oxidized to Cr_2Cl^{5+} via a two-electron process.

Several ^{18}O-tracer studies have been reported in which the transfer of an oxygen atom(s) from the oxidant was interpreted in terms of a two equivalent oxidation. Thus, Gordon and Taube[485] have suggested that the ozone oxidation of $U^{4+}(aq)$ involves oxygen-atom transfer, where one of the uranyl oxygens is derived from ozone. A most striking result is obtained for the corresponding permanganate ion oxidation. Under certain conditions all of the contained "yl" oxygen is derived from the permanganate ion. Since the net change in oxidation state for the oxidizing agent does not match that for U^{4+}($5U^{4+} + 2MnO_4^- + 2H_2O = 5UO_2^{2+} + 2Mn^{2+} + 4H^+$), the reaction must occur in stages involving at least intermediate oxidation states of the oxidant: either $Mn^{VII} \rightarrow Mn^V \rightarrow Mn^{III}$ or $Mn^{VIII} \rightarrow Mn^{VI}$ followed by $Mn^{VI} \rightarrow Mn^{IV} \rightarrow Mn^{II}$ with efficient oxygen atom *and* oxide ion transfer in each two-equivalent step. For chromate ion the transfer is less efficient. Buchacek and Gordon[486] have found that ^{18}O is efficiently transferred from chlorine(III) to uranium(IV) in the reaction $U^{IV} + Cl^{III} + C_6H_5OH \rightarrow U^{VI} + H + ClC_6H_4OH$. The heterogeneous oxidations of U^{4+} with PbO_2 and MnO_2 also give rise to oxygen atom transfers in excess of one per uranyl ion. Additional heterogeneous oxidations of chromium(II) have been performed using various transition metal oxides in ^{18}O enriched water. Only in the reactions of Pb_2PbO_4, Ca_2PbO_4, and $PbO_{1.75}$ was the dimer of chromium(III) formed although some oxygen atom transfer was noted for MnO_2, Tl_2O_3, and Mn_2O_3. Each of the latter reactions produces the one-electron oxidation product of chromium(II). These results serve to demonstrate that oxygen atom transfer is not necessarily evidence for a two-electron process. Zabin and Taube[487] made the interesting observation that while oxidation of chromium(II) by $Tl^{3+}(aq)$ produces "dimer",[465] its oxidation by $Tl_2O_3(s)$ produces only $Cr^{3+}(aq)$. Espenson and Taube[488] have also shown that ozonization of $Mn^{2+}(aq)$ and $Tl^+(aq)$ produces MnO_2 and a thallium(III) oxide, respectively, in which at least one oxygen atom in the metal oxide product derives from the ozone.

VII. REACTIONS WITH NONMETALLIC SPECIES

The reactions of transition metal complexes with nonmetallic species cover a vast expanse of literature. Therefore, the coverage here is quite selective, largely encompassing the generation and subsequent reactions of organic and inorganic free radicals. Coverage of the free radical reactions is extended over both inner- and outer-sphere mechanistic types.

The reactions of metal ions with nonmetallic species have long been of interest, particularly as reagents in qualitative and quantitative analysis. As early as 1949 Westheimer[440] published a "classic" review of chromium(VI) oxidations. Edwards[489,490,491] has reviewed the reactions of oxyanions and edited a book[492] on peroxide reaction mechanisms. The reaction of iron(II) with peroxide was reviewed by Uri[493] in 1952. Presumably, the reaction generates iron(III), hydroxyl radicals, and hydroxide ions. In the absence of other oxidizable species, the hydroxyl radical can either oxidize another iron(II) or react with hydrogen peroxide to form hydroperoxo radical, involved in the chain decomposition of hydrogen peroxide. The nature of the intermediate produced in the reaction and the analogous reaction with the peroxydisulfate ion has been investigated by means of chemical competition studies by Woods, Kolthoff, and Meehan.[494,495] These studies have established that the intermediates in the two systems are different and that sulfate radical ion is the intermediate in the latter. Either hydroxyl radicals or iron(IV) could be intermediates in the former. However, it has since been shown[463] that iron(IV) is not produced in the iron(II)-hydrogen peroxide reaction. Therefore, it is very likely that hydroxyl radical is the intermediate.

The dramatic decrease in substitution rate which occurs when chromium(II) is oxidized to chromium(III) has been exploited to obtain mechanistic information about reductions by chromium(II). In the vast majority of such reductions where the oxidant contains an atom that can potentially act as a donor, the reduced form of the oxidant, or at least a part of it, is trapped in the coordination sphere of the primary chromium(III) product. This feature applies to ordinary, long-lived oxidants as well as to many unstable, short-lived species such as bromine and iodine atoms,[188] di-, and trichloromethyl radicals as studied by Anet and Leblanc[496] and Anet[497] and benzyl radicals reported by Kochi and Davis.[498] Hanson and Premuzic[499] have recently reviewed the use of chromium(II) as a reducing agent in organic chemistry. Although aqueous chromium(II) apparently does not reduce alkyl halides, its ethanolamine and ethylenediamine complexes do according to Kochi and Mocadlo[500] and Kochi and Powers.[501] Schmidt, Swinehart, and Taube[502] have produced an interesting series of σ-bonded organochromium(III) complexes by using hydroxyl radicals (generated in situ by the Cr^{2+}–H_2O_2 reaction) to abstract hydrogen atoms from organic molecules and then by capturing the resulting organic radicals with chromium(II). In this manner C—Cr bonds have been formed where methyl, ethyl, *n*-propyl, and iso-propyl alcohols, as well as diethyl ether, were used. Even a methyl σ-bonded species is formed via the chromium(II)–*tert*-butyl hydroperoxide reaction.[503,504]

Nohr and Spreer have examined the reactions of $CrCHI_2^{2+}$[505] and $CrCH_2I^{2+}$[506] with chromium(II) with quite interesting results. While CrI^{2+}, Cr^{3+}, and $CrCH_2I^{2+}$ are products of the reaction, labeling experiments require the formation of an intermediate binuclear intermediate

$$\mathrm{Cr}\overset{\mathrm{H}}{\underset{\mathrm{I}}{\mathrm{C}}}\mathrm{Cr}^{4+}$$

to account for the appearance of the chromium label in all products. The binuclear intermediate is not a feature of Castro and Kray's mechanism[507] for chromium(II) reduction of geminal dihalides. In a followup to the above reaction evidence was found for a long-lived organochromium(III) product.[506] In a very recent study Sevcik[508] found that the chromium(II) reduction of trichloroacetic acid produces $CrCl^{2+}$ and a dichloroacetate complex of chromium(III) *without* (or <20%) a C—Cr bond. Finally, Asher and Deutsch[509] have reported the first reduction of a disulfide to produce a Cr—S bonded species.

Since such a variety of substrates is efficiently trapped by chromium(II) upon its oxidation, the question arises, "Is this a completely general phenomenon of chromium(II) reductions?" Two recent studies have clearly demonstrated that this feature of chromium(II) reductions is not completely general. Thus, Pennington and Haim[510] found that the SO_4^- radical ion reacts with chromium(II) without being incorporated into the coordination sphere of the primary chromium(III) product. In an independent study Schmidt, Swinehart, and Taube[504] reported entirely analogous results for the nitrogen-containing radicals NH_3^+, $(CH_3)_3N^+$, and $(C_6H_5)(CH_3)_2N^+$. In principle, these reactions can occur either by an outer-sphere mechanism or by abstraction of a hydrogen atom from the hydration sphere of chromium(II). It is likely that these radical ions react with chromium(II) at rates which exceed the substitution rate on this extremely labile reductant. Indirect support for this argument is as follows: since reductions by chromium(II) are, in general, faster than corresponding reductions by iron(II) and since the absolute rate constant for the reduction of sulfate radical ion by iron(II) has a value of $0.9 \times 10^8 M^{-1}sec^{-1}$,[511] it follows that the reaction of sulfate radical ion with chromium(II) approaches diffusion control. Interestingly, the substrates used to generate the above radical ions do react with chromium(II) in such a way that a part of the oxidant is transferred to the primary chromium(III) product. Thus, $Cr(H_2O)_5SO_4^+$ is produced along with sulfate radical ions in the rate-determining step when peroxydisulfate is reduced by chromium(II).[510] Moreover, the rate-determining step of the reaction of ^{18}O-labeled hydroxylamine with chromium(II) yields ^{18}O in the chromium(III) product, as well as free NH_3^+ radicals. Presumably, dimethylaniline and trimethylamine *N*-oxides react similarly to the hydroxylamine.[504] The rate constant for the hydroxylamine–chromium(II) reaction and the observed inertness of hydrazine toward chromium(II) are at variance with an earlier communication by Wells and Salam[512] on these reactions. It is quite probable that this earlier work[512] was in error since a more detailed account of the experimental results is given in the later work.[504]

The reductions of nonmetallic species by the pentacyanocobaltate(II) ion parallel those of chromium(II) in that the oxidant (or a portion thereof) is frequently retained in the first coordination sphere of the oxidized cobalt(III) product. Two limiting situations are usually observed: (1) Either a binuclear intermediate is formed involving the reduced form of the oxidant as a bridge or (2) two substituted pentacyanocobaltate(III) products are formed in a two-step, free radical process. For example, oxygen[156] and acetylene[513] react with pentacyanocobaltate(III) in a 1:2 mole ratio to form binuclear intermediates. These reactions may involve an initial two-electron transfer followed by attack of another reductant ion on the cobalt(IV) intermediate. On the other hand, hydrogen[514–518] and water,[515] bromine and iodine,[518] hydrogen peroxide,[156,519] hydroxylamine and cyanogen iodide,[519] and numerous organic halides[520–525] appear to react via a common mechanism conforming to the stoichiometry:

$$2\,Co(CN)_5^{3-} + X{-}Y \rightarrow Co(CN)_5X^{3-} + Co(CN)_5Y^{3-}$$

The general mechanism proposed for this type reaction[525] is patterned after that for the reaction with the halogens[526] and involves the production of free radicals in the initial, rate-determining step:

$$Co(CN)_5^{3-} + X{-}Y \xrightarrow{k_{rate}} Co(CN)_5X^{3-} + Y$$

$$Co(CN)_5^{3-} + Y \xrightarrow{fast} Co(CN)_5Y^{3-}$$

This mechanism is entirely analogous to that proposed for reactions of organic halides with chromium(II).[527,528]

The question of where the reductant attacks the unsymmetrical oxidants in the initial step of the above mechanism must be dealt with at this point. For the organic halides and dihalides, Halpern and Maher[520,521] have proposed that the initial step involves halogen atom abstraction followed by the formation of a σ-bonded organopentacyanocobaltate(III) species in the fast step. Alternatively, Kwiatek and Seyler[523] have proposed that the fast step may, in some cases, involve hydrogen atom abstraction. Presumably, cyanogen iodide also involves halogen atom abstraction in the rate-determining step.[519] For water and hydroxylamine the initial attack of the reductant is most probably at oxygen. In the case of hydroxylamine the absence of a free radical scavenger effect by iodide ion supports this mechanistic contention.[519] (Iodide ion reacts virtually quantitatively with hydroxyl radicals produced in the reaction of hydrogen peroxide with pentacyanocobaltate(II).)

It is surprising that pentacyanocobaltate efficiently captures the NH_2 radical produced in the hydroxylamine reaction, whereas the reductant chromium(II) does not. However, it is noteworthy, and perhaps significant, that $Co(CN)_5^{3-}$, in contrast to $Cr(H_2O)_6^{2+}$, has no hydrogen atoms available for abstraction.

Schneider, Phelan, and Halpern[529] and Halpern and Phelan[530] have reported that a group of low-spin cobalt(II) complexes of the type $Co(dmgH)_2B_n$ (n = 1,2, dmg = various substituted glyoximes, and B = organic base) react with aryl halides via a mechanism analogous to that proposed for the pentacyanocobaltate(II) reductions.[521] The inner-sphere mechanism has been confirmed in the $Co(dmgH)_2$ reductions of $Co(NH_3)_5X^{2+}$.[531] The rate constants for these reductions are relatively insensitive to the substituents on the glyoxime, and they increase modestly with increasing basicity of the axial ligand B. An interesting feature of these reactions is that the complexes with B = 1 and B = 2 react at the same rates, suggesting that when B = 2, one of the two basic groups is dissociated prior to the rate-determining step. More recently, pyridine solutions of $Co(dmgH)_2$ have been utilized for hydrogenations.[532,533] Schiff-base type complexes,[534,535] iron(II) porphyrins,[536] a cobalt(II) macrocyclic complex,[537] and $V(py)_4Cl_2$[538] react with organic halides or pseudohalides similarly.

The reaction of halogen species with transition metal ions has received considerable attention recently, especially the dihalide radical ion species. Malone and Endicott[539] provide a good summary bibliography for such species as well as estimates for the various halogen reduction potentials in their report of the near-diffusion controlled (k = 1–7 × 10^9 $M^{-1}sec^{-1}$) rates for the oxidations of Co(*trans*[14]diene)$^{2+}$ by Cl_2^- and Br_2^-, of $Co(NH_3)_5Br^{2+}$ by Br_2^-, and of $Co(NH_3)_6^{3+}\cdot I^-$ by I_2^-. $Ru(bipy)_3^{2+}$ [540] and Tl^I [541] also react with Br_2^- and Cl_2^-, respectively, within the above range. A summary of the available potential data from Malone and Endicott and the more recent data of Woodruff and Margerum[542] is given below, all species being aqueous. Interestingly, the halogen atoms are all 0.3–0.4V better oxidants and reductants than the radical species.

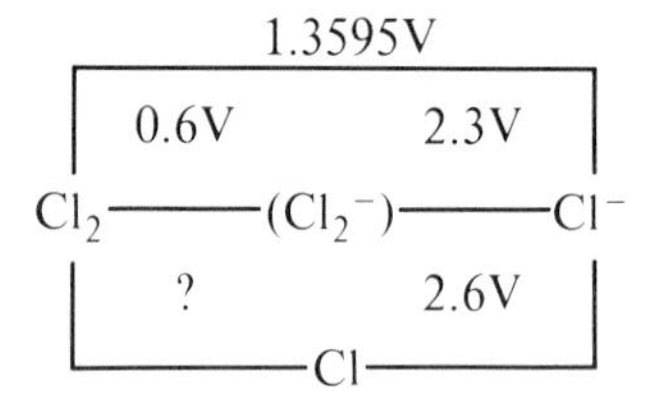

1.087V
0.51V 1.67V
Br_2 — (Br_2^-) — Br^-
0.11V 2.06V
Br

0.6197V
0.11V 1.13V
I_2 — (I_2^-) — I^-
−0.18V 1.42V
I

Thornton and Laurence[543,544,545] have examined the kinetics for the oxidations of Fe^{2+} and Mn^{2+} by Cl_2^- and Br_2^- and of Co^{2+} by Cl_2^-. An inner-sphere mechanism for the Fe^{2+}–Br_2^- reaction, as well as the fraction of the Fe^{2+}–Cl_2^- reaction, is indicated by rate constants nearly identical to that for substitution on iron(II) and by the direct observation of about 30% $FeCl^{2+}$ as a primary product for the latter. The authors noted that this result carries the implication of a 100% inner-sphere mechanism for the one-electron Fe^{2+}–Cl_2 oxidation[463] where >70% $FeCl^{2+}$ was observed. The major fraction of the Fe^{2+}–Cl_2^- reaction must be outer-sphere since the rate constant ($k = 1.0 \times 10^7 M^{-1}sec^{-1}$) exceeds the substitution rate by nearly one order of magnitude. In the Mn^{2+}–Br_2^-(Cl_2^-) reactions a most unanticipated result was the observation of rate saturation with respect to [Mn^{2+}].[546,547] Moreover, in spite of a large difference in the standard free energies, the derived association quotients and the rate constants are nearly identical. These results were interpreted on the basis of an inner-sphere, electron transfer controlled mechanism, but the interpretation is not unequivocal. The corresponding Co^{2+}–Cl_2^- reaction is substitution controlled, inner-sphere as judged by $k = 1.4 \times 10^6 M^{-1}sec^{-1}$, and $CoCl^{2+}$ is observed as the primary product. For oxidations of V^{2+} by I_2^-, Br_2^-, and Cl_2^- the rates increase in the order given and all are outer-sphere.[546] For Cr^{2+} the same order is observed and the mechanism is inner-sphere except for Cl_2^- which appears to be about 50:50 inner- and outer-sphere.[546]

The oxidations of halide and pseudohalide ions by cobalt(III) and iron(III) have received recent attention. Thus, Davies and Watkins[548] reported the cobalt(III) oxidations of Br^-, I^-, and NCS^- (along with numerous other substrates) and interpretated their mixed second-order kinetic results in terms of a common dissociative interchange mechanism except for I^-, which reacted much faster. A similar conclusion was reported independently for the Co^{III}–Br^- system.[549] Both authors subscribed to a mechanism involving radical formation. However, it is noteworthy that the redox potentials diagramed above suggest that Co^{3+} ($E^0 = 1.82V$) is incapable of oxidizing Br^- to Br though the one-electron oxidation to Br_2^- is accessible. The potential is favorable for oxidation of I^- to I· (and even more favorable for I_2^-), which could account for its greater rate. For the corresponding oxidations by $CoOH^{2+}$ similar restrictions formally apply, but ΔE^0 for the production of I· is only $-0.08V$, assuming $K_h = 2 \times 10^{-3}$.[548] For the more complicated oxidation of Cl^- [463] the potentials indicate that neither Co^{3+} nor $CoOH^{2+}$ can serve as one-electron oxidants. The two-electron oxidation of Cl^- to HOCl *may* be possible with the formation of cobalt(I).

Also related to the above arguments is the oxidation of I^- by Fe^{3+} which was recently reinvestigated by Laurence and Ellis.[550] In spite of a thorough search, no spectral evidence was obtained for FeI^{2+}, an intermediate proposed in an earlier study.[551] The authors noted that an ion triplet would account for the observed second-order dependence on the iodide ion. Again, however, the potentials are unfavorable for the one-electron oxidation of I^- to either I· or I_2^-, suggesting that a two-electron process is operative. Such a process has been proposed for a similar kinetic term in the VO_2^+–I^- reaction.[552,553] Even the first-order terms in iodide ion are interpreted in terms of HOI formation. Lawani and Sutter[554] claim a similar rate-determining formation of HOBr in the MnO_4^-–Br^- reaction in acidic media. The species *cis*-$[Co(NH_3)_2(H_2O)_4]^{3+}$ has recently found utility as an oxidant for Br^-.[555] Most recently, Mureinik[556] has proposed that the oxidation of halide ions by $Co(phen)_3^{3+}$ also leads to racemization in the order of efficiency $I^- > Br^- > Cl^-$.

The oxidation of various substrates by halogen species continues to attract attention. Woodruff and Margerum[557] have proposed a general inner-sphere mechanism for the oxidation of polydentately ligated metal ions:

$$\text{I. } M^{II}L^{2+} + \begin{bmatrix} X_2 + X^- \\ \rightleftharpoons \\ X_3^- \end{bmatrix} \overset{K_{eq}}{\rightleftharpoons} M^{II}LX_2^{2-} \xrightarrow{k_{ox}} M^{III}LX^{2-} + X^{\bullet} \quad (\text{or } M^{III}L^{2-} + X_2^-)$$

II. Rapid post-electron transfer process

III. Elimination of coordinated halide ion

where M = Fe, Co, or Mn; X = I or Br; and L = EDTA or CyDTA. In all cases the rate constants stand in the order Fe > Co > Mn; Br > I; $X_2 > X_3^-$; and EDTA > CyDTA.[557,558] Earlier work had established that $k_{obsd} > k_{Br_2}$ and that $k_{OBr^-} > k_{HOBr}$ for both EDTA and CyDTA complexes of Co(II).[559]

Malin and Swinehart[560] have reported the halogen (Br_2, I_2, I_3^-) oxidations of V^{2+} to be outer-sphere, exceeding the rate of substitution on the metal center. By contrast the oxidations of Fe^{2+} by Cl_2,[463] Br_2,[561] and I_2[551] are slow, decreasing in the given order, and the first is inner-sphere. Adegite and Ford-Smith[562] have reported the correspondingly sluggish oxidations of U^{4+} where the rate law is $R = k[U^{4+}][X_2]/[H^+]^2$. There is a reported discrepancy in the experimental hydrogen ion dependence in the VO^{2+}–Cl_2 reaction.[563,564] The Ti^{3+}–Cl_2 reaction[565] offers an interesting challenge on mechanistic assignments, being consistent with either $TiCl^{2+}$ + HOCl or $TiOH^{2+}$ + Cl_2 or both for combinations in the rate-determining step. Finally, there is a need for revision in the improperly balanced mechanism cited for the V^{3+}–Br_2 reaction.[566]

While the previously cited reactions of transition metal complexes with organic and inorganic free radicals involve an inner-sphere mechanism, there are now new lines of evidence for radical-catalyzed reactions of the outer-sphere variety. Thus, Norris and Nordmeyer[567] have reported that the reaction

$$Co(NH_3)_5N\langle\bigcirc\rangle\!-\!\underset{\underset{O}{\|}}{C}NH_2{}^{3+}\text{–}Eu^{2+}$$

is autocatalytic, an observation which they interpreted by the following mechanism (L = isonicotinamide):

$$(NH_3)_5CoL^{3+} + Eu^{2+} \overset{H^+}{\rightleftarrows} Co^{2+} + Eu^{3+} + HL^+ + 5NH_4^+$$

$$HL^+ + Eu^{2+} \rightarrow HL + Eu^{3+}$$

$$HL + (NH_3)_5CoL^{3+} \rightarrow Co^{2+} + HL^+ + 5NH_4^+$$

Once the protonated isonicotinamide cation is produced in the initial reaction, it is rapidly converted to its radical form which then reacts via a radical transfer, outer-sphere mechanism. The interpretation is considerably strengthened by independent observations of Barber and Gould[568] that the slow outer-sphere reductions of $(NH_3)_5CoX^{3+}$ (X = NH_3, pyridine, or dimethylformamide) by chromium(II) and europium(II) are dramatically catalyzed by 4-pyridinecarboxylic acid (and its carboxyl bound aquochromium(III) complex) and 1,2-bis(4-pyridyl)-ethylene. Entirely analogous results have been observed in the corresponding vanadium (III), but not chromium(II), reductions of the 2,4- and 2,5-pyridinedicarboxylic acid complexes by Chen and Gould.[109] In neither case do the free ligands react with the cobalt(III) complexes in the absence of the reducing cations. Autocatalysis is also a feature of the europium(II) reductions of *N*-bonded 4-carbethoxypyridine- and the 1,2-bis(4-pyridyl)ethylenepentaamminecobalt(III) complexes as well as the chromium(II) reduction of the latter.[572] In the systems investigated to date only those organic ligands with substituents γ, but not α or β, to the pyridine nitrogen were found to be catalytically active. More recently, Ulstrup[573] has observed catalysis of the Ce(IV) oxidations of Fe^{2+} and $Fe(phen)_3{}^{2+}$ by pyrazine and several pyridine derivatives. In connection with radical reductions for outer-sphere processes Cohen and Meyerstein[569] have reported that the *p*-nitrobenzoate radical reduces various substituted pentaamminecobalt(III) complexes, and Olson and Hoffman[570] have reported that $\cdot CH_2OH$ reduces $Co(NH_3)_6{}^{3+}$.

At least two additional outer-sphere reactions are thought to involve radical reactions. First, Davies, Sutin, and Watkins[571] have reported that the $Cr(CN)_6{}^{4-}$–H_2O_2 reaction yields $Cr(CN)_6{}^{3-}$, OH, and OH^-. Subsequent reaction of the OH radical with the reductant also yields $Cr(CN)_6{}^{3-}$. Both stoichiometric and kinetic results support, but do not prove, the one-electron oxidation of Cr(II).

ADDENDUM

During the last six months the field of electron transfer has continued to focus on excited-state chemistry, intervalence transfer, bridging through organic structural groups (especially in the area of intramolecular electron transfer), and reactions with nonmetallic species. An update on the emerging field of bioinorganic electron transfer by Wherland and Gray is forthcoming.[574]

Earlier (Sect. I-D) it was proposed that $Ru(bipy)_3^{2+}$ quenches the metal-centered doublet of $Cr(bipy)_3^{3+}$ by an electron transfer mechanism. More recently this mechanism has been confirmed in flash photolysis studies in which $Cr(bipy)_3^{2+}$ was identified as a transient product in the quenched solution.[575] Remarkably, the same reduced chromium species is also formed when $Cr(bipy)_3^{3+}$ quenches $(^3CT)Ru(bipy)_3^{2+}$ or Fe^{2+}. Creutz and Sutin[576] have provided another example of the oxidizing properties of $(^3CT)Ru(bipy)_3^{2+}$ in the electron transfer quenching of the excited-state species by Eu^{2+} ($k_q = 2.8 \times 10^7\ M^{-1}\ sec^{-1}$), with the thermal back reaction occurring with $k = 2.7 \times 10^7\ M^{-1}\ sec^{-1}$. Lin and co-workers[577] have noted, as have others[143,578], however, that energy transfer can lead to the same products as electron transfer quenching and *vice versa,* except[575] when there is no overlap between the energies of donor and acceptor species. In order to test the relative importance of k_{energy} vs. $k_{e.t.}$, Lin *et al.*,[577] have studied the Fe^{3+}, Eu^{3+}, and Cr^{3+} quenching of (^3CT)Ru-(polypyridine)$_3^{2+}$ complexes which have virtually identical absorption and emission spectra but variable redox potentials. The nearly diffusion-controlled quenching constants for Fe^{3+} correlate only weakly with the reduction potentials of the polypyridine complexes, whereas there is a strong correlation for Eu^3 and none at all for Cr^{3+}. Continuous and flash photolysis results were used to identify the Eu^{2+} product. Thus, electron transfer is indicated for the Fe^{3+} and Eu^{3+} quenching and energy transfer for Cr^{3+} quenching. For Eu^{3+} quenching a plot of log k_q vs. log K_q is linear with a slope of 0.49 and an intercept of 1.55, yielding the first Marcus-type correlation involving excited-state species and an estimated lower limit, self-exchange rate constant of $2 \times 10^6\ M^{-1}\ sec^{-1}$ for the $*RuL_3^{2+}$–RuL_3^{3+} couples. Finally, Creutz and Sutin[579] have utilized several electron transfer quenching processes to probe the so-called inverted region, $-\Delta G_{12}^0 > 2(\Delta G_{11}^* + \Delta G_{22}^*)$, in the Marcus theory. Thus the luminescence quenching constants of $(^3CT)Ru(bipy)_3^{2+}$ and (^3CT)Ru(4,4′-$(CH_3)_2$bipy)$_3^{2+}$ first increase with increasing standard potentials with $Cr(bipy)_3^{3+}$ and $Cs(bipy)_3^{3+}$ as the quenchers, then decrease rapidly for Ru(4,4′-$(CH_3)_2$bipy)$_3^{3+}$ and $Ru(bipy)_3^{3+}$ as quenchers as the standard potential further increases.

Closely related to the Marcus theory is the observation of Pelizzeti and Mentasti[605] that the oxidations of quinols by cobalt(III) obey the relationship $\Delta G^* = a + b\Delta G_{12}^0$ ($a = 17.5 \pm 2.8$ kcal/mol and $b = 0.35 \pm 0.06$) rather than the Marcus theory. Rillema and Endicott[580] have reported that Co(TIM)-$(H_2O)_2^{2+}$ catalyzes the outer-sphere chromium(II) reduction of Co(*trans*-14-diene)$(NH_3)_2^{3+}$ *via* a mechanism involving Co(TIM)$(H_2O)_2^+$. Such a mechanism circumvents the high Franck–Condon barrier of the uncatalyzed reaction.

Recent temperature-dependent, intramolecular electron transfer results of Fischer, Tom, and Taube[581] for a related series of 4,4′-bipyridyl-bridged Co(III)–Ru(II) complexes gave remarkably constant enthalpies of activation (20.0 ± 0.5 kcal/mol) and very low entropies of activation (0 ± 2 eu). Thus the

Franck–Condon barrier to electron transfer appears to be higher, at least for these and two other related systems,[32,177] than previously noted (*see* Sect. I-A and II-B.). In addition, the low entropies of activation observed for intramolecular electron transfer imply that the normally observed entropies of activation (−30 eu) for 2+/3+ interactions are due to the concentration of charges in the dielectric medium, and not to the electron transfer event. Piering and Malin[582] have reported an intramolecular electron transfer rate constant of 0.012 sec^{-1} for the complex

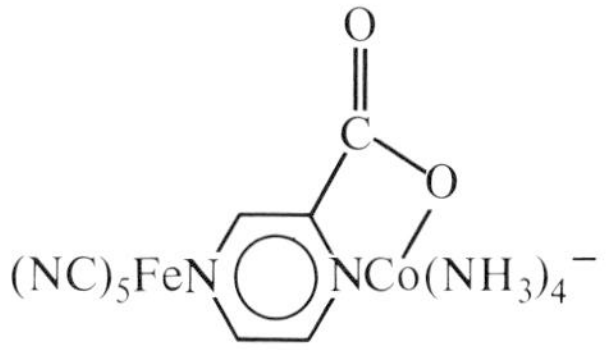

O
C
O
$(NC)_5FeN$ ⟨ring⟩ $NCo(NH_3)_4^-$

formed from the parent complex of cobalt(III) and $Fe(CN)_5H_2O^{3-}$. Hurst[583] has utilized the affinity of copper(I) for olefins to study the competitive inter- and intramolecular electron transfer rates between copper(I) and $Ru(NH_3)_5(4\text{-vinylpyridine})^{3+}$. Rate saturation in Ru(III) was observed. If the second-order constant for the 4-ethyl derivative is taken as representative of the intermolecular rate, then the ratio of inter- to intramolecular processes is 0.7% for the 4-vinylpyridine system. Finally, the $V(O)(EDTA)^{2-}$–$V(EDTA)^{2-}$ and $V(O)(HEDTA)^-$–$V(HEDTA)^-$ reactions lead to precursor formation followed by intramolecular electron transfer ($k_{e.t.}{}^{EDTA}$ = 14.3 sec^{-1}).[584] These results contrast with the observation of a successor complex in the related $V(O)^{2+}$–V^{2+} system.[585]

In the related area of intervalence transfer there have been three studies on the solvent dependencies of IVT bands for various Ru(II)L–Ru(III) complexes. For the complexes $(NH_3)_5Ru(pyrazine)Ru(bipy)_2Cl^{4+}$ [586] and $[Ru(bipy)_2Cl]_2L^{3+}$ (L = pyrazine, 4,4′-bipyridine, and *trans*-1,2-bis(4-pyridyl)-ethylene)[587] the solvent effect is entirely in accord with the predictions of Hush's theory. However, for the ion $(NH_3)_5Ru$–NC–CH(*t*-butyl)–$CNRu(NH_3)_5{}^{5+}$ the solvent dependence does not correlate with the theory.[588] As yet there is no hypothesis to account for these divergent observations.

The greatest concentration of new work has focused on organic bridging groups. The bis-μ-hydroxo-μ-carboxylatobis(triammine)cobalt(III) complexes have been most widely exploited. Hery and Wieghardt[589] have investigated the chromium(II) and vanadium(II) reductions of several μ-acetylenedicarboxylic acid derivatives. Where remote attack *via* a conjugated bond system is accessible, the reductions by Cr^{2+} are inner-sphere; otherwise they are outer-sphere. By contrast the corresponding μ-terephthalate complexes are not reduced by chromium(II) *via* an inner-sphere, remote attack mechanism even though a remote site and conjugated bond system would appear accessible.[590] The latter ligands are not themselves reduced by Cr^{2+}. Likewise the bis-μ-hydroxo-μ-malonato and dimethylmalonato complexes, as well as the μ-amido-μ-malonato and dimethylmalonato analogs, appear to undergo outer-sphere reduction for the acid-independent paths ($k_{Cr}/k_V \leqslant 0.036$). The interesting interpretation of the inverse acid-dependent term—that the reductant is attached to the remote carboxylate although electron transfer occurs *via* an outer-sphere mechanism, i.e. not over the saturated bond system—bears consideration.[591] (*See also* Sect. I-D and Ref. 581.)

Several papers treating adjacent attack with chelation (*see* Sect. II-D) have recently appeared. Martin, Liang, and Gould[592] have reported a comparative study of various *ortho*-hydroxy and aminobenzoato-pentaamminecobalt(III) reductions by chromium(II), where $k_{obs} = k_0 + k_{-1}/[H^+]$. The rates are insensitive to other substituents on the aryl ring. For the amino series of reductions the aminium group must be deprotonated in order to form the chelated chromium(III) product. Values of $k_b(k_{-1}/K_A)$ can therefore be calculated without the ambiguity associated with the site of proton loss in the *ortho*-hydroxy series, i.e. loss of H^+ from the $-OH$ or from the reductant. In a related study[593] the titanium(III) reductions of substituted salicylatopentaamminecobalt(III) complexes are interpreted on the basis of precursor complex formation, although there is a kinetic ambiguity associated with the nature of the rate-determining electron transfer process.

Gould, Johnson, and Morland[594] have investigated the Cr^{2+}, V^{2+}, and Eu^{2+} reductions of 3- and 4-(acetyl, benzoyl, and butyryl)pyridine complexes of pentaamminecobalt(III). By relative rate comparisons the reactions all appear to be inner-sphere, with $\log k_{Eu} = 0.61(\log k_{Cr}) + 0.36$ being the first linear free energy correlation observed where structural variation occurs in the path of electron transfer. A chromium(III)-bound radical is apparently observed in the chromium(II) reduction of the related 4-pyridinecarboxaldehyde derivative. Both here and in the 4-formylbenzonitrile complex[595] hydration of the aldehyde group serves to inactivate the lead-in group. Interestingly, Ogino, Tsukahara, and Tanaka[596] have shown that the rates of adjacent attack for amino acid complexes of pentaamminecobalt(III) are enhanced as the aminium group is moved away from the carboxylate lead-in group. Deprotonation has been shown to be important in the reduction of 1-aminoethylphosphonic acidopentaamminecobalt(III) by chromium(II).[597]

In the area of nonbridging ligand effects Peloso[598] has reported a linear correlation between the sigma donor strengths of various diamines and the iodide-dependent rate constant for the $Pt(diam)Cl_4$–I^- reaction. Welch and Pennington[599] have interpreted the appearance of chromium(II) rate saturation in the $Cr(NCS)F^+$–Cr^{2+} reaction in terms of a dead-end, thiocyanate-bridged intermediate. Tanaka *et al.*[600] report vastly accelerated rates of reduction of $Co(NH_3)_5OAc^{2+}$ by $Cr(HEDTRA)H_2O^-$ and $Cr(EDTA)H_2O^-$ compared with Cr^{2+} (*see* Sect. III-C).

Finally, there have been some interesting observations made in the reactions of coordination complexes with nonmetallic species. (*See* Sect. VI.) Hyde and

Espenson[601] have reported stoichiometries and kinetics (ranging over 10^{12}) for the chromium(II) reductions of organic peroxides, hydroperoxides, and peracids. The product distributions were taken to imply rate-determining, inner-sphere attack by the reductant at the least sterically hindered oxygen. Organochromium products were formed in a subsequent fast reaction with a second mole of the reductant. Verton, Zipp, and Ragsdale[602] and Zipp and Ragsdale[603] have examined the reductions of pyridine–N–oxides by titanium(III) and chromium(II), respectively. Espenson and Leslie[604] have investigated the kinetics of halogen atom abstraction of $CrCHCl_2^{2+}$ by Cr^{2+} in which $CrCl^{2+}$, Cr^{3+}, and $CrCH_2Cl^{2+}$ are products. Labelling experiments with $^*Cr^{2+}$ require the formation of an intermediate, Cr_2CHCl^{4+}, to account for the distribution of the label in both Cr^{3+} and $CrCH_2Cl^{2+}$ fractions.

REFERENCES

1. Libby, W. F. *J. Phys. Chem.* **56,** 863 (1952).
2. Marcus, R. J., B. J. Zwolinski, and H. Eyring. *J. Phys. Chem.* **58,** 432 (1954).
3. Zwolinski, B. J., R. J. Marcus, and H. Eyring. *Chem. Rev.* **55,** 157 (1955).
4. Weiss, J. *Proc. R. Soc. London.* **A222,** 128 (1954).
5. Marcus, R. A. *J. Chem. Phys.* **24,** 966 (1956).
6. Marcus, R. A. *J. Chem. Phys.* **26,** 867 (1957).
7. Ibid., 872.
8. Marcus, R. A. *Discuss. Faraday Soc.* **29,** 21 (1960).
9. Marcus, R. A. *J. Phys. Chem.* **67,** 853 (1963).
10. Marcus, R. A. *J. Chem. Phys.* **43,** 679 (1965).
11. Laidler, K. J. *Can. J. Chem.* **37,** 138 (1959).
12. Levich, V. G., and R. R. Dogonadze. *Proc. Acad. Sci. USSR, Phys. Chem. Sect.* **133,** 591 (1960).
13. Levich, V. G., and R. R. Dogonadze. *Collect. Czech. Chem. Commun.* **26,** 193 (1961).
14. Hush, N. S. *Trans. Faraday Soc.* **57,** 557 (1961).
15. Sacher, E., and K. J. Laidler. *Trans. Faraday Soc.* **59,** 396 (1963).
16. Marcus, R. A. *Annu. Rev. Phys. Chem.* **15,** 155 (1964).
17. Levich, V. G. *Adv. Electrochem. Electrochem. Eng.* **4,** 249 (1966).
18. Ruff, I. *Q. Rev.* **22,** 199 (1968).
19. Dogonadze, R. R., J. Ulstrup, and Yu. I. Kharkats. *J. Chem. Soc. Faraday Trans. 2.* **68,** 744 (1972).
20. Schmidt, P. P. *Aust. J. Chem.,* **22,** 673 (1969); *J. Chem. Phys.,* **56,** 2775 (1972); **57,** 3749 (1972); *J. Phys. Chem.,* **77** 488 (1973); **78,** 1684 (1974).
21. Kestner, N. R., J. Logan, and J. Jortner. *J. Phys. Chem.* **78,** 2148 (1974).
22. Efrima, S., and M. Bixon. *J. Chem. Phys.* **64,** 3639 (1976).
23. Reynolds, W. L., and R. W. Lumry. *Mechanisms of Electron Transfer.* The Ronald Press Co., New York, 1966.
24. Libby, W. F. Abstracts, Physical and Inorganic Section, 115th National Meeting of the American Chemical Society, San Francisco, CA, March 27–April 1, 1949.

25. Platzman, C. R., and J. Franck. *Z. Phys.* **138,** 411 (1954).
26. Condon, E. U. *Am. J. Phys.* **15,** 365 (1947).
27. Marcus, R. A. *Trans. N. Y. Acad. Sci.* **19,** 423 (1957).
28. Stranks, D. R. *Discuss. Faraday Soc.* **29,** 116 (1960).
29. Barnet, T., B. M. Craven, H. C. Freeman, N. E. Kime, and J. A. Ibers. *Chem. Commun.,* 307 (1966); N. E. Kime and J. A. Ibers, *Acta Crystallogr., Sect. B.* **25,** 168 (1969).
30. Stynes, H. C., and J. A. Ibers. *Inorg. Chem.* **10,** 2304 (1971).
31. Glick, M. D., W. G. Schmonsees, and J. F. Endicott. *J. Am. Chem. Soc.* **96,** 5661 (1974).
32. Glick, M. D., J. M. Kuszaj, and J. F. Endicott. *J. Am. Chem. Soc.* **95,** 5097 (1973).
33. Allen, G. C., and N. S. Hush. *Prog. Inorg. Chem.* **8,** 357 (1967).
34. Hush, N. S. *Prog. Inorg. Chem.* **8,** 391 (1967).
35. Robin, M. B., and P. Day. *Adv. Inorg. Chem. Radiochem.* **10,** 247 (1967).
36. Cowan, D. O., C. LeVanda, J. Park, and F. Kaufman. *Acc. Chem. Res.* **6,** 1 (1973).
37. Creutz, C., and H. Taube. *J. Am. Chem. Soc.* **91,** 3988 (1969).
38. Tom, G. M., C. Creutz, and H. Taube. *J. Am. Chem. Soc.* **96,** 7827 (1974).
39. Adeyemi, S. A., J. N. Braddock, G. M. Brown, J. A. Ferguson, F. J. Miller, and T. J. Meyer. *J. Am. Chem. Soc.* **94,** 300 (1972).
40. Lavallee, D. K., and E. B. Fleischer. *J. Am. Chem. Soc.* **94,** 2583 (1972); E. B. Fleischer and D. K. Lavallee. *J. Am. Chem. Soc.* **94,** 2599 (1972).
41. Callahan, R. W., G. M. Brown, and T. J. Meyer. *J. Am. Chem. Soc.* **96,** 7829 (1974); *Inorg. Chem.* **14,** 1443 (1975).
42. Creutz, C., and H. Taube. *J. Am. Chem. Soc.* **95,** 1086 (1973).
43. Tom, G. M., and H. Taube. *J. Am. Chem. Soc.* **97,** 5310 (1975).
44. Mayoh, B., and P. Day. *J. Am. Chem. Soc.* **94,** 2885 (1972).
45. Mayoh, B., and P. Day. *Inorg. Chem.* **13,** 2273 (1974).
46. Glauser, R., H. Hauser, F. Herren, A. Ludi, P. Roder, E. Schmidt, H. Siegenthaler, and F. Wenk. *J. Am. Chem. Soc.* **95,** 8457 (1973).
47. Toma, H. E. *J. Inorg. Nucl. Chem.* **38,** 431 (1976).
48. Murray, K. S., and R. S. Sheahan. *J. Chem. Soc. Dalton Trans.* 1182 (1973).
49. Magnuson, R. H., and H. Taube. *J. Am. Chem. Soc.* **94,** 7213 (1972).
50. Newton, T. W. *J. Chem. Educ.* **45,** 571 (1968).
51. Rosseinsky, D. R. *Chem. Commun.* 225 (1972).
52. Campion, R. J., N. Purdie, and N. Sutin. *Inorg. Chem.* **3,** 1091 (1964).
53. Thomas, L., and K. W. Hicks. *Inorg. Chem.* **13,** 749 (1974).
54. Stasiw, R., and R. G. Wilkins. *Inorg. Chem.* **8,** 156 (1969).
55. Rawoof, M. A., and J. R. Sutter. *J. Phys. Chem.* **71,** 2767 (1967).
56. Dulz, G., and N. Sutin. *Inorg. Chem.* **2,** 917 (1963).
57. Diebler, H., and N. Sutin. *J. Phys. Chem.* **68,** 174 (1964).
58. Hicks, K. W., and J. R. Sutter. *J. Phys. Chem.* **75,** 1107 (1971).
59. Wilkins, R. G., and R. E. Yelin. *Inorg. Chem.* **7,** 2667 (1968).
60. Thorneley, R. N. F., and A. G. Sykes. *J. Chem. Soc. A.* 1036 (1970).
61. Meyer, T. J., and H. Taube. *Inorg. Chem.* **7,** 2369 (1968).
62. Caryle, D. W., and J. H. Espenson. *J. Am. Chem. Soc.* **91,** 599 (1969).
63. Endicott, J. F., and H. Taube. *J. Am. Chem. Soc.* **86,** 1686 (1964).

64. Sykes, A. G., and R. N. F. Thorneley. *J. Chem. Soc. A.* 232 (1970).
65. Jacks, C. A., and L. E. Bennett. *Inorg. Chem.* **13,** 2035 (1974).
66. Przystas, T. J., and N. Sutin. *J. Am. Chem. Soc.* **95,** 5545 (1973).
67. Candlin, J. P., J. Halpern, and D. L. Trimm. *J. Am. Chem. Soc.* **86,** 1019 (1964).
68. Wilkins, R. G., and R. E. Yelin. *J. Am. Chem. Soc.* **92,** 1191 (1970).
69. Dyke, R., and W. C. E. Higginson. *J. Chem. Soc.* 2802 (1963).
70. Rillema, D. P., and J. F. Endicott. *J. Am. Chem. Soc.* **94,** 8711 (1972).
71. Farina, R., and R. G. Wilkins. *Inorg. Chem.* **7,** 514 (1968).
72. Davies, G. *Inorg. Chem.* **10,** 1155 (1971).
73. Hyde, M. R., R. Davies, and A. G. Sykes. *J. Chem. Soc. Dalton Trans.* 1838 (1972).
74. Liteplo, M. P., and J. F. Endicott. *Inorg. Chem.* **10,** 1420 (1971).
75. Rillema, D. P., J. F. Endicott, and R. C. Patel. *J. Am. Chem. Soc.* **94,** 394 (1972).
76. Rillema, D. P., and J. F. Endicott. *Inorg. Chem.* **11,** 2361; **12,** 1712 (1973).
77. Marcus, R. A. *J. Phys. Chem.* **72,** 891 (1968).
78. Yardley, J. T., and J. K. Beattie. *J. Am. Chem. Soc.* **94,** 8925 (1972).
79. Beattie, J. K., N. Sutin, D. H. Turner, and G. W. Flynn. *J. Am. Chem. Soc.* **95,** 2052 (1973).
80. Bodek, I., and G. Davies. *Coord. Chem. Rev.* **14,** 269 (1974).
81. Sutter, J. H., and J. B. Hunt. *J. Am. Chem. Soc.* **91,** 3107 (1969).
82. La Mar, G. N. *J. Am. Chem. Soc.* **94,** 9042, 9049, 9055 (1972).
83. Kane-Maguire, N. A. P., R. M. Tollison, and D. E. Richardson. *Inorg. Chem.* **15,** 499 (1976).
84. Goedken, V. L., and S. M. Peng. *Chem. Commun.* 258 (1975).
85. Grossman, B., and R. G. Wilkins. *J. Am. Chem. Soc.* **89,** 4230 (1967).
86. Sykes, A. G., and J. A. Weil. *Prog. Inorg. Chem.* **13,** 1 (1970).
87. Hyde, M. R., K. L. Scott, and A. G. Sykes. *Coord. Chem. Rev.* **8,** 121 (1972).
88. Scott, K. L., and A. G. Sykes. *J. Chem. Soc. Dalton Trans.* 1832 (1972).
89. Wieghardt, K., and A. G. Sykes. *J. Chem. Soc. Dalton Trans.* 651 (1974).
90. Scott, K. L., and A. G. Sykes. *J. Chem. Soc. Dalton Trans.* 736 (1973).
91. Toppen, D. L., and R. G. Linck. *Inorg. Chem.* **10,** 2635 (1971).
92. Jeftic, L., and S. W. Feldberg. *J. Phys. Chem.* **75,** 2381 (1971).
93. Espenson, J. H., and D. A. Holm. *J. Am. Chem. Soc.* **94,** 5709 (1972).
94. Pladziewicz, J. R., and J. H. Espenson. *Inorg. Chem.* **11,** 3136 (1972).
95. Lavallee, D. K., C. Lavallee, J. C. Sullivan, and E. Deutsch. *Inorg. Chem.* **12,** 570 (1973).
96. Palazzotto, M. C.. and L. H. Pignolet. *Inorg. Chem.* **13,** 1781 (1974).
97. Pladziewicz, J. R., and J. H. Espenson. *J. Phys. Chem.* **75,** 3381 (1971); J. R. Pladziewicz, and J. H. Espenson. *J. Am. Chem. Soc.* **95,** 56 (1973).
98. Yang, E. S., M. Chan, and A. C. Wahl. *J. Phys. Chem.* **79,** 2049 (1975).
99. Brown, G. M., T. J. Meyer, D. O. Cowan, C. LeVanda, F. Kaufman, P. V. Roling, and M. D. Rausch. *Inorg. Chem.* **14,** 506 (1975).
100. Ruff, I., and I. Korösi-Odor. *Inorg. Chem.* **9,** 186 (1970); I. Ruff, V. J. Friedrich, K. Demeter, and K. Csillag. *J. Phys. Chem.* **75,** 3303 (1971).
101. Braddock, J. N., and T. J. Meyer. *J. Am. Chem. Soc.* **95,** 3158 (1973).
102. Cramer, J. L., and T. J. Meyer. *Inorg. Chem.* **13,** 1250 (1974).

103. Marcus, R. A., and N. Sutin. *Inorg. Chem.* **14,** 213 (1975).
104. Cobble, J. W. *J. Am. Chem. Soc.* **86,** 5394 (1964).
105. Cobble, J. W. *Science.* **152,** 1479 (1966).
106. Zwickel, A., and H. Taube. *J. Am. Chem. Soc.* **83,** 793 (1961).
107. Gaunder, R. G., and H. Taube. *Inorg. Chem.* **9,** 2627 (1970).
108. Balahura, R. J., and R. B. Jordan. *J. Am. Chem. Soc.* **93,** 625 (1971).
109. Chen, J. C., and E. S. Gould. *J. Am. Chem. Soc.* **95,** 5539 (1973).
110. Dockal, E. R., E. T. Everhart, and E. S. Gould. *J. Am. Chem. Soc.* **93,** 5661 (1971).
111. Fan, F. F., and E. S. Gould. *Inorg. Chem.* **13,** 2647 (1974).
112. Balahura, R. J., G. B. Wright, and R. B. Jordan. *J. Am. Chem. Soc.* **95,** 1137 (1973).
113. Meyerstein, D., and W. A. Mulac. *J. Phys. Chem.* **73,** 1091 (1969).
114. Olson, M. V., Y. Kanazawa, and H. Taube. *J. Chem. Phys.* **51,** 289 (1969).
115. Baker, B. R., M. Orhanović, and N. Sutin. *J. Am. Chem. Soc.* **89,** 722 (1967); M. Orhanović, H. N. Po, and N. Sutin. *J. Am. Chem. Soc.* **90,** 7224 (1968).
116. Seewald, D., N. Sutin, and K. O. Watkins. *J. Am. Chem. Soc.* **91,** 7307 (1969); **92,** 3523 (1970).
117. Movius, W. G., and R. G. Linck. *J. Am. Chem. Soc.* **92,** 2677 (1970).
118. Stritar, J. A., and H. Taube. *Inorg. Chem.* **8,** 2281 (1969).
119. Zanella, A., and H. Taube. *J. Am. Chem. Soc.* **93,** 7166 (1971).
120. Bakac, A., T. D. Hand, and A. G. Sykes. *Inorg. Chem.* **14,** 2540 (1975).
121. Green, M., K. Schug, and H. Taube. *Inorg. Chem.* **4,** 1184 (1965).
122. Halpern, J., private communication cited in Ref. *121.*
123. Sutin, N. *Acc. Chem. Res.* **1,** 225 (1968).
124. Orhanović, M., and N. Sutin, *J. Am. Chem. Soc.* **90,** 4286 (1968).
125. Gaswick, D., and A. Haim. *J. Am. Chem. Soc.* **93,** 7347 (1971).
126. Haim, A., and N. Sutin. *Inorg. Chem.* **15,** 476 (1976).
127. Cannon, R. D., and J. Gardiner. *J. Am. Chem. Soc.* **92,** 3800 (1970).
128. Waltz, W. L., and R. G. Pearson. *J. Phys. Chem.* **73,** 1941 (1969).
129. Baxendale, J. H., and M. Fiti. *J. Chem. Soc. Dalton Trans.* 1995 (1972).
130. Hoffman, M. Z., and M. Simic. *Chem. Commun.* 640 (1973).
131. Baxendale, J. H., E. M. Fielden, and J. P. Keene. *Proc. R. Soc. London.* **A286,** 320 (1965).
132. Fleischauer, P. D., A. W. Adamson, and G. Sartori. *Prog. Inorg. Chem.* **17,** 1 (1972).
133. Gafney, H. D., and A. W. Adamson. *J. Am. Chem. Soc.* **94,** 8238 (1972).
134. Natarajan, P., and J. F. Endicott. *J. Phys. Chem.* **77,** 971, 1823 (1973).
135. Navon, G., and N. Sutin. *Inorg. Chem.* **13,** 2159 (1974).
136. Bock, C. R., T. J. Meyer, and D. G. Whitten. *J. Am. Chem. Soc.* **96,** 4710 (1974).
137. Bock, C. R., T. J. Meyer, and D. G. Whitten. *J. Am. Chem. Soc.* **97,** 2909 (1975).
138. Bolletta, F., M. Maestri, L. Moggi, and V. Balzani. *Chem. Commun.* 901 (1975).
139. Creutz, C., N. Sutin. *Inorg. Chem.* **15,** 496 (1976).
140. Juris, A., M. T. Gandolfi, M. F. Manfrin, and V. Balzani. *J. Am. Chem. Soc.* **98,** 1047 (1976).

141. Young, R. C., T. J. Meyer, and D. G. Whitten. *J. Am. Chem. Soc.* **98,** 286 (1976).
142. Lin, C. T., and N. Sutin. *J. Phys. Chem.* **80,** 97 (1976).
143. Laurence, G. S., and V. Balzani. *Inorg. Chem.* **13,** 2976 (1974).
144. Farr, J. K., L. G. Hulett, R. H. Lane, and J. K. Hurst. *J. Am. Chem. Soc.* **97,** 2654 (1975).
145. Haim, A., private communication.
146. Durante, V. A., and P. C. Ford. *J. Am. Chem. Soc.* **97,** 6898 (1975).
147. Burrows, H. D., S. J. Formosinho, M. Da G. Miguel, and F. P. Coelho. *J. Chem. Soc. Faraday Trans. 1* **72,** 163 (1976).
148. Kato, Y., and H. Fukutomi. *J. Inorg. Nucl. Chem.* **38,** 1323 (1976).
149. Yokoyoma, Y., M. Moriyasu, and S. Ikeda. *J. Inorg. Nucl. Chem.* **38,** 1329 (1976).
150. Bolletta, F., M. Maestri, L. Moggi, and V. Balzani. *J. Am. Chem. Soc.* **95,** 7864 (1973).
151. Allen, D. M., H. D. Burrows, A. Cox, R. J. Hill, T. J. Kemp, and T. J. Stone. *Chem. Commun.* 59 (1973).
152. Young, R. C., T. J. Meyer, and D. G. Whitten. *J. Am. Chem. Soc.* **97,** 4781 (1975).
153. Balzani, V., L. Moggi, M. F. Manfrin, F. Boletta, and M. Gleria. *Science.* **189,** 852 (1975).
154. Taube, H., H. Myers, and R. L. Rich. *J. Am. Chem. Soc.* **75,** 4118 (1953).
155. Grossman, B., and A. Haim. *J. Am. Chem. Soc.* **92,** 4835 (1970).
156. Haim, A., and W. K. Wilmarth. *J. Am. Chem. Soc.* **83,** 509 (1961).
157. Adamson, A. W., and E. Gonick. *Inorg. Chem.* **2,** 129 (1963).
158. Huchital, D. H., and R. G. Wilkins. *Inorg. Chem.* **6,** 1022 (1967).
159. Birk, J. P. *J. Am. Chem. Soc.* **91,** 3189 (1969).
160. Birk, J. P., and S. V. Weaver. *Inorg. Chem.* **11,** 95 (1972).
161. Sutin, N. *Annu. Rev. Phys. Chem.* **17,** 119 (1966).
162. Haim, A., and N. Sutin. *J. Am. Chem. Soc.* **88,** 5343 (1966).
163. Espenson, J. H. *J. Am. Chem. Soc.* **89,** 1276 (1967).
164. Price, H. J., and H. Taube. *Inorg. Chem.* **7,** 1 (1968).
165. Hicks, K. W., D. L. Toppen, and R. G. Linck. *Inorg. Chem.* **11,** 310 (1972).
166. Taube, H. *Adv. Inorg. Chem. Radiochem.* **1,** 1 (1959).
167. Linck, R. G. *MTP Int. Rev. Sci. Inorg. Chem. Ser. One.* **9,** 303 (1971) University Park Press, Baltimore.
168. Haim, A. *Acc. Chem. Res.* **8,** 264 (1975).
169. Toppen, D. L., and R. G. Linck. *Inorg. Chem.* **10,** 2635 (1971).
170. van den Bergen, A., and B. O. West. *Chem. Commun.* 52 (1971).
171. Espenson, J. H., and J. S. Shveima. *J. Am. Chem. Soc.* **95,** 4468 (1973).
172. Espenson, J. H., and T. D. Sellers, Jr. *J. Am. Chem. Soc.* **96,** 94 (1974).
173. Linck, R. G. MTP *Int. Rev. Sci., Inorg. Chem. Ser. Two.* **9,** 173 (1974) University Park Press, Baltimore.
174. Taube, H. *Pure Appl. Chem.* **44,** 25 (1975).
175. Basolo, F., and R. G. Pearson. *Mechanisms of Inorganic Reactions.* 2nd ed., John Wiley and Sons, Inc., New York, 1967, p. 152.
176. Patel, R. C., R. E. Ball, J. F. Endicott, and R. G. Hughes. *Inorg. Chem.* **9,** 23 (1970).

177. Cannon, R. D., and J. Gardiner. *Inorg. Chem.* **13,** 390 (1974).
178. Nanda, R. K., and A. C. Dash. *J. Inorg. Nucl. Chem.* **36,** 1595 (1974).
179. Isied, S. S., and H. Taube. *J. Am. Chem. Soc.* **95,** 8198 (1973).
180. Gaswick, D., and A. Haim. *J. Am. Chem. Soc.* **96,** 7845 (1974).
181. Jwo, J. J., and A. Haim. *J. Am. Chem. Soc.* **98,** 1172 (1976).
182. Hurst, J. K., and R. H. Lane. *J. Am. Chem. Soc.* **95,** 1703 (1973).
183. Cannon, R. D., and J. Gardiner. *J. Chem. Soc. Dalton Trans.* 622 (1976).
184. Patel, R. C., R. E. Ball, J. F. Endicott, and R. G. Hughes. *Inorg. Chem.* **9,** 23 (1970).
185. Liteplo, M. P., and J. F. Endicott. *J. Am. Chem. Soc.* **91,** 3982 (1969).
186. Haim, A. *Inorg. Chem.* **5,** 2081 (1966).
187. Newton, T. W., and F. B. Baker. *J. Phys. Chem.* **67,** 1425 (1963).
188. Taube, H., and H. Myers. *J. Am. Chem. Soc.* **76,** 2103 (1954).
189. Thorneley, R. N. F., and A. G. Sykes. *J. Chem. Soc. A.* 862 (1970).
190. Adamson, A. W., and E. Gonick. *Inorg. Chem.* **2,** 129 (1963).
191. Huchital, D. H., and R. G. Wilkins. *Inorg. Chem.* **6,** 1022 (1967).
192. Huchital, D. H., and R. J. Hodges. *Inorg. Chem.* **12,** 998 (1973).
193. Rosenhein, L., D. Speiser, and A. Haim. *Inorg. Chem.* **13,** 1571 (1974).
194. Huchital, D. H., and R. J. Hodges. *Inorg. Chem.* **12,** 1004 (1973).
195. Gordon, G. *Inorg. Chem.* **2,** 1277 (1963).
196. Ekstrom, A., and Y. Farrar. *Inorg. Chem.* **11,** 2610 (1972).
197. Ekstrom, A. *Inorg. Chem.* **12,** 2455 (1973).
198. Newton, T. W., and M. J. Burkhart. *Inorg Chem.* **10,** 2323 (1971).
199. Lavallee, C., and T. W. Newton. *Inorg. Chem.* **11,** 2616 (1972).
200. Sullivan, J. C. *Inorg. Chem.* **3,** 315 (1964).
201. Hinderberger, E. H., and R. C. Thompson. *Inorg. Chem.* **4,** 784 (1975).
202. Sullivan, J. C., and R. C. Thompson. *Inorg. Chem.* **6,** 1795 (1967).
203. Espenson, J. H. *Inorg. Chem.* **4,** 1533 (1965).
204. Craft, R. W., and R. G. Gaunder. *Inorg. Chem.* **13,** 1005 (1974).
205. Craft, R. W., and R. G. Gaunder. *Inorg. Chem.* **14,** 1283 (1975).
206. Scott, K. L., and M. Green, and A. G. Sykes. *J. Chem. Soc. A.* 3651 (1971).
207. Hyde, M. R., and A. G. Sykes. *J. Chem. Soc. Dalton Trans.* 1550 (1974).
208. Movius, W. G., and R. G. Linck. *J. Am. Chem. Soc.* **91,** 5394 (1969).
209. Cannon, R. D., and J. E. Earley. *J. Am. Chem. Soc.* **87,** 5264 (1965); **88,** 1872 (1966).
210. Candlin, J. P., J. Halpern, and S. Nakamura. *J. Am. Chem. Soc.* **85,** 2517 (1963).
211. Halpern, J., and S. Nakamura. *J. Am. Chem. Soc.* **87,** 3002 (1965).
212. Shea, C. J., and A. Haim. *Inorg. Chem.* **12,** 3013 (1973).
213. Haim, A., and N. Sutin. *J. Am. Chem. Soc.* **87,** 4210 (1965); **88,** 434 (1966).
214. Orhanović, M., and N. Sutin. *J. Am. Chem. Soc.* **90,** 538, 4286 (1968).
215. Shea, C., and A. Haim. *J. Am. Chem. Soc.* **93,** 3055 (1971).
216. Friedlander, G., J. W. Kennedy, and J. M. Miller. *Nuclear and Radiochemistry.* 2nd ed. John Wiley and Sons, Inc., New York, 1964, p. 71.
217. Brown, L. D., and D. E. Pennington. *Inorg. Chem.* **10,** 2117 (1971).
218. Ball, D. L., and E. L. King. *J. Am. Chem. Soc.* **80,** 1091 (1958).
219. Nordmeyer, F. R., and H. Taube. *J. Am. Chem. Soc.* **88,** 4295 (1966).

220. Miller, R. G., D. E. Peters, and R. T. M. Fraser. *Exchange Reactions*. IAEA, Vienna, 1965, p. 203.
221. Fraser, R. T. M. *J. Chem. Soc.* 3641 (1965).
222. Orhanović, M., H. N. Po, and N. Sutin. *J. Am. Chem. Soc.* **90,** 7224 (1968).
223. Fay, D. P., and N. Sutin. *Inorg. Chem.* **9,** 1291 (1970).
224. Espenson, J. H. *Inorg. Chem.* **4,** 121 (1965).
225. Adegite, A., and T. A. Kuke. *J. Chem. Soc. Dalton Trans.* 158 (1976).
226. Birk, J. P., and J. H. Espenson. *J. Am. Chem. Soc.* **90,** 1153 (1968).
227. Ibid., 2266.
228. Ibid.
229. Balahura, R. J., and R. B. Jordan. *J. Am. Chem. Soc.* **92,** 1533 (1970).
230. Peters, D. E., and R. T. M. Fraser. *J. Am. Chem. Soc.* **87,** 2758 (1965).
231. Burmeister, J. L., and N. J. DeStefano. *Inorg. Chem.* **9,** 972 (1970).
232. Balahura, R. J., and R. B. Jordan. *Inorg. Chem.* **10,** 198 (1971).
233. Roberts, R. L., D. W. Caryle, and G. L. Blackmer. *Inorg. Chem.* **14,** 2739 (1975).
234. Epps, L. A., and L. G. Marzilli. *Chem. Commun.* 109 (1972).
235. Marzilli, L. G., R. C. Stewart, L. A. Epps, and J. B. Allen. *J. Am. Chem. Soc.,* **95,** 5796 (1973).
236. Balahura, R. J. *J. Am. Chem. Soc.* **98,** 1487 (1976).
237. Armor, J. N., and M. Buchbinder. *Inorg. Chem.* **12,** 1086 (1973).
238. Armor, J. N., M. Buchbinder, and R. Cheney. *Inorg. Chem.* **13,** 2990 (1974).
239. Sykes, A. G. *Adv. Inorg. Chem. Radiochem.* **10,** 153 (1967).
240. Taube, H. *Mechanisms of Inorganic Reactions. Adv. Chem. Ser.* **49,** 107 (1965).
241. Gould, E. S., and H. Taube. *Acc. Chem. Res.* **2,** 321 (1969).
242. Andrade, C., R. B. Jordan, and H. Taube. *Inorg. Chem.* **9,** 711 (1970).
243. Fleischer, E. B., and R. Frost. *J. Am. Chem. Soc.* **87,** 3998 (1965).
244. Jordan, R. B., and H. Taube. *J. Am. Chem. Soc.* **88,** 4406 (1966).
245. Scott, K. L., and A. G. Sykes. *J. Chem. Soc. Dalton Trans.* 2364 (1972).
246. Wieghardt, K., and A. G. Sykes. *J. Chem. Soc. Dalton Trans.* 651 (1974).
247. Barrett, M. B., J. H. Swinehart, and H. Taube. *Inorg. Chem.* **10,** 1983 (1971).
248. Thomas, J. C., J. W. Reed, and E. S. Gould. *Inorg. Chem.* **14,** 1696 (1975).
249. Linck, R. G., *Mechanisms of Inorganic Reactions. Adv. Chem. Ser.* **49,** 124 (1965).
250. Taft, R. W. Jr., *Steric Effects in Organic Chemistry.* John Wiley and Sons, Inc., New York, 1956, pp. 598, 619.
251. Chen, J. C., and E. S. Gould. *J. Am. Chem. Soc.* **95,** 5539 (1973).
252. Holwerda, R., E. Deutsch, and H. Taube. *Inorg. Chem.* **11,** 1965 (1972).
253. Bembi, R., and W. U. Malik. *J. Inorg. Nucl. Chem.* **37,** 570 (1975).
254. Balahura, R. J. *Inorg. Chem.* **13,** 1350 (1974).
255. Butler, R. D., and H. Taube. *J. Am. Chem. Soc.* **87,** 5597 (1965).
256. Gould, E. S., and H. Taube. *J. Am. Chem. Soc.* **86,** 1318 (1964).
257. Huchital, D. H., and H. Taube. *J. Am. Chem. Soc.* **87,** 5371 (1965).
258. Gould, E. S. *J. Am. Chem. Soc.* **88,** 2983 (1966).
259. Gould, E. S. *J. Am. Chem. Soc.* **94,** 4360 (1972).
260. Gould, E. S. *J. Am. Chem. Soc.* **96,** 2373 (1974).

261. Thornton, A. T., K. Wieghardt, and A. G. Sykes. *J. Chem. Soc. Dalton Trans.* 147 (1976).
262. Olson, M. V., and C. E. Behnke. *Inorg. Chem.* **13,** 1329 (1974).
263. Svatos, G., and H. Taube. *J. Am. Chem. Soc.* **83,** 4172 (1961).
264. Huchital, D. H., and H. Taube. *Inorg. Chem.* **4,** 1660 (1965).
265. Lavallee, C., and E. Deutsch. *Inorg. Chem.* **11,** 3133 (1972).
266. Spinner, T., and G. M. Harris. *Inorg. Chem.* 1067 (1972).
267. Scott, K. L., M. Green, and A. G. Sykes. *J. Chem. Soc. A.* 3651 (1971).
268. Hwang, C., and A. Haim. *Inorg. Chem.* **9,** 500 (1970).
269. Sebera, D. K., and H. Taube. *J. Am. Chem. Soc.* **83,** 1785 (1961).
270. Fraser, R. T. M., D. K. Sebera, and H. Taube. *J. Am. Chem. Soc.* **81,** 2906 (1959).
271. Norris, C., and F. R. Nordmeyer. *Inorg. Chem.* **10,** 1235 (1971).
272. Zanella, A., and H. Taube. *J. Am. Chem. Soc.* **94,** 6403 (1972).
273. Diaz, H., and H. Taube. *Inorg. Chem.* **9,** 1304 (1970).
274. Liang, A., and E. S. Gould. *Inorg. Chem.* **12,** 12 (1973).
275. Thamburaj, P. K., and E. S. Gould. *Inorg. Chem.* **14,** 15 (1975).
276. Balahura, R. J., and W. L. Purcell. *Inorg. Chem.* **14,** 1469 (1975).
277. Balahura, R. J., and N. A. Lewis. *Can. J. Chem.* **5,** 1154 (1975).
278. Hyde, M. R., K. Wieghardt, and A. G. Sykes. *J. Chem. Soc. Dalton Trans.* 690 (1976).
279. Hyde, M. R., K. L. Scott, K. Wieghardt, and A. G. Sykes. *J. Chem. Soc. Dalton Trans.* 153 (1976).
280. Hurst, J. K., and H. Taube. *J. Am. Chem. Soc.* **90,** 1178 (1968).
281. Gould, E. S., *J. Am. Chem. Soc.* **89,** 5792 (1967).
282. Fan, F. F., and E. S. Gould. *Inorg. Chem.* **13,** 2639 (1974).
283. Gould, E. S. *J. Am. Chem. Soc.* **87,** 4730 (1965).
284. Hoffman, M. Z., and M. Simic. *J. Am. Chem. Soc.* **92,** 5533 (1970).
285. Hoffman, M. Z., and M. Simic. *J. Am. Chem. Soc.* **94,** 1757 (1972).
286. Hoffman, M. Z., and M. Simic. *Chem. Commun.* 549 (1975).
287. Rowan, N. S., M. Z. Hoffman, and R. M. Milburn. *J. Am. Chem. Soc.* **96,** 6060 (1974).
288. Fraser, R. T. M., *J. Am. Chem. Soc.* **84,** 3436 (1962).
289. Candlin, J P., and J. Halpern. *J. Am. Chem. Soc.* **85,** 2518 (1963).
290. Taube, H. *Electron Transfer Reactions of Complex Ions in Solution.* Academic Press, New York, 1970, pp. 92–94.
291. Ibid.
292. Gaunder, R. G., and H. Taube. *Inorg. Chem.* **9,** 2627 (1970).
293. Davies, R., and R. B. Jordan. *Inorg. Chem.* **10,** 1102 (1971).
294. Snellgrove, R., and E. L. King. *Inorg. Chem.* **3,** 288 (1964).
295. Kopple, K. D., and R. R. Miller. *Inorg. Chem.* **2,** 1204 (1963).
296. Deutsch, E., and H. Taube. *Inorg. Chem.* **7,** 1532 (1968).
297. Olson, M. V., Ph.D. Thesis, Stanford University, Stanford, CA, 1969.
298. Olson, M. V., and H. Taube. *Inorg. Chem.* **9,** 2072 (1970).
299. Bennett, L. E., R. H. Lane, M. Gilroy, F. A. Sedor, and J. P. Bennett, Jr. *Inorg. Chem.* **12,** 1200 (1973).
300. Edwards, J. D., Y. Sulfab, and A. G. Sykes. *Inorg. Chem.* **14,** 1474 (1975).
301. Williams, R. D., and D. E. Pennington, *J. Coord. Chem.,* in press.

302. Lane, R. H. and L. E. Bennett. *J. Am. Chem. Soc.* **92,** 1089 (1970).
303. Gilroy, M., F. A. Sedor, and L. E. Bennett. *Chem. Commun.* 181 (1972).
304. Kruse, W., and H. Taube. *J. Am. Chem. Soc.* **82,** 526 (1960).
305. Chia, Y. T., and E. L. King. *Discuss. Faraday Soc.* **29,** 109 (1960).
306. Snellgrove, R., and E. L. King. *J. Am. Chem. Soc.* **84,** 4609 (1962).
307. Haim, A. *J. Am. Chem. Soc.* **88,** 2324 (1966).
308. Fraser, R. T. M. *J. Am. Chem. Soc.* **85,** 1747 (1963).
309. Ward, J. R., and A. Haim. *J. Am. Chem. Soc.* **92,** 475 (1970).
310. Wood, P. B., and W. C. E. Higginson. *Proc. Chem. Soc. London.* 109 (1964).
311. Wood, P. B., and W. C. E. Higginson. *J. Chem. Soc. A.* 1645 (1966).
312. Huchital, D. H. *Inorg. Chem.* **9,** 486 (1970).
313. Balahura, R. J., and N. A. Lewis. *Chem. Commun.* 268 (1976).
314. Kopple, K. D., and R. R. Miller. *Proc. Chem. Soc. London.* 306 (1962).
315. Miller, R. G., and R. T. M. Fraser, *Abstr. 151st National Meeting, Am. Chem. Soc.,* Pittsburgh, Paper No. **83N,** 1966.
316. Schroeder, K. A., and J. H. Espenson. *J. Am. Chem. Soc.* **89,** 2548 (1967).
317. Haim, A. *J. Am. Chem. Soc.* **85,** 1016 (1963).
318. Taube, H. *J. Am. Chem. Soc.* **77,** 4481 (1955).
319. Orgel, L. E. *Inst. Int. Chim. Solvay, Xe Cons. Chim. Rapp. Discuss.* 289 (1956).
320. Earley, J. E. *Prog. Inorg. Chem.* **13,** 243 (1970).
321. Ograd, A. E., and H. Taube. *J. Am. Chem. Soc.* **80,** 1084 (1958).
322. Benson, P., and A. Haim. *J. Am. Chem. Soc.* **87,** 3826 (1965).
323. Linck, R. G. *Inorg. Chem.* **7,** 2394 (1968).
324. Linck, R. G. *Inorg. Chem.* **9,** 2529 (1970).
325. DeChant, Sr. M. J., and J. B. Hunt. *J. Am. Chem. Soc.* **89,** 5988 (1967); **90,** 3695 (1968).
326. Espenson, J. H., and S. G. Slocum. *Inorg. Chem.* **6,** 906 (1967).
327. Pennington, D. E., and A. Haim. *Inorg. Chem.* **5,** 1887 (1966).
328. Hoppenjans, D. W., G. Gordon, and J. B. Hunt. *Inorg. Chem.* **10,** 754 (1971).
329. Bifano, C., and R. G. Linck. *J. Am. Chem. Soc.* **89,** 3945 (1967).
330. Willians, T. J., and C. S. Garner. *Inorg. Chem.* **9,** 2058 (1970).
331. Earley, J. E., and S. Z. Ali. *Abstr. 167th National Meeting, Am. Chem. Soc.,* Los Angeles, CA, Paper No. **INORG. 160,** 1974.
332. McClure, D. S. *Advances in the Chemistry of Coordination Compounds.* Stanley Kirshner, ed., The Macmillan Co., New York, 1961, p. 498.
333. Worrell, J. H., and T. A. Jackman. *J. Am. Chem. Soc.* **93,** 1044 (1971).
334. Worrell, J. H., R. A. Goodard, E. M. Gupton, Jr., and T. A. Jackman. *Inorg. Chem.* **11,** 2734 (1972).
335. Weschler, C. J., and E. Deutsch. *Inorg. Chem.* **15,** 139 (1976).
336. Pennington, D. E., and A. Haim. *Inorg. Chem.* **6,** 2138 (1967).
337. Pennington, D. E., and A. Haim. *J. Am. Chem Soc.* **88,** 3450 (1966).
338. Adin, A., and A. G. Sykes. *J. Chem. Soc. A.* 1518 (1966).
339. Espenson, J. H., and D. W. Carlyle. *Inorg. Chem.* **5,** 586 (1966).
340. Adin, A., J. Doyle, and A. G. Sykes. *J. Chem. Soc. A.* 1504 (1967).
341. Doyle, J., A. G. Sykes, and A. Adin. *J. Chem. Soc. A.* 1314 (1968).
342. Johnson, R. C., and E. R. Berger. *Inorg. Chem.* **7,** 1656 (1968).
343. Mason, W. R., III, and R. C. Johnson. *Inorg. Chem.* **4,** 1258 (1965).

344. Johnson, R. C., and E. R. Berger. *Inorg. Chem.* **4,** 1262 (1965).
345. Mason, W. R., III, E. R. Berger and R. C. Johnson. *Inorg. Chem.* **6,** 248 (1967).
346. Rettew, R. R., and R. C. Johnson. *Inorg. Chem.* **4,** 1565 (1965).
347. Bailey, S. G., and R. C. Johnson. *Inorg. Chem.* **8,** 2596 (1969).
348. Patel, R. C., and J. F. Endicott. *J. Am. Chem. Soc.* **90,** 6364 (1968).
349. Guenther, P. R., and R. G. Linck. *J. Am. Chem. Soc.* **91,** 3769 (1969).
350. Grossman, B., and A. Haim. *J. Am. Chem. Soc.* **93,** 6490 (1971).
351. Taube, H., and E. L. King. *J. Am. Chem. Soc.* **76,** 4053 (1954).
352. Earley, J. E., and J. H. Gørbitz. *J. Inorg. Nucl. Chem.* **25,** 306 (1963).
353. Hunt, J. B., and J. E. Earley. *J. Am. Chem. Soc.* **82,** 5312 (1960).
354. Manning, P. V., and R. C. Jarnagin. *J. Phys. Chem.* **67,** 2884 (1963).
355. Pennington, D. E., and A. Haim. *Inorg. Chem.* **7,** 1659 (1968).
356. Przystas, T. J., and N. Sutin. *J. Am. Chem. Soc.* **95,** 5545 (1973).
357. Endicott, J. F., and H. Taube. *Inorg. Chem.* **4,** 437 (1965).
358. Dodel, P. H., and H. Taube. *2. Phys. Chem.* (*Frankfurt am Main*). **44,** 92 (1965).
359. Sutin, N., and A. Forman. *J. Am. Chem. Soc.* **93,** 5274 (1971).
360. Diebler, H., and H. Taube. *Inorg. Chem.* **4,** 1029 (1965).
361. Sutin, N., J. K. Rowley. and R. W. Dodson. *J. Phys. Chem.* **65,** 1248 (1961).
362. Carlyle, D. W., and J. H. Espenson. *J. Am. Chem. Soc.* **90,** 2272 (1968).
363. Dulz, G., and N. Sutin. *J. Am. Chem. Soc.* **86,** 829 (1964).
364. Hambright, P., and E. B. Fleischer. *Inorg. Chem.* **4,** 912 (1965).
365. Pasternack, R. F., and N. Sutin. *Inorg. Chem.* **13,** 1956 (1974).
366. Fleischer, E. B., M. Krishnamurthy, and S. K. Cheung. *J. Am. Chem. Soc.* **97,** 3873 (1975).
367. Yandell, J. K., D. P. Fay, and N. Sutin. *J. Am. Chem. Soc.* **95,** 1131 (1973).
368. Przystas, T. J., and N. Sutin. *Inorg. Chem.* **14,** 2103 (1975).
369. Pasternack, R. F. *Inorg. Chem.* **15,** 643 (1976).
370. Mason, W. R. *Coord. Chem. Rev.* **7,** 241 (1972).
371. Cannon, R. D., and J. S. Stillman. *J. Chem. Soc. Dalton Trans.* 428 (1976).
372. Ardon, M. *Inorg. Chem.* **4,** 372 (1965).
373. Carey, L. R., W. E. Jones, and T. W. Swaddle. *Inorg. Chem.* **10,** 1566 (1971).
374. Williams, T. J., and C. S. Garner, *Inorg. Chem.* **10,** 975 (1971).
375. Kochi, J. K., and J. W. Powers. *J. Am. Chem. Soc.* **92,** 137 (1970).
376. Ogino, H., and N. Tanaka. *Bull. Chem. Soc. Jpn* **41,** 2411 (1968).
377. Cannon, R. D., and J. Gardiner. *J. Chem. Soc. Dalton Trans.* 887 (1972).
378. Cannon, R. D., and J. E. Earley. *J. Chem. Soc. A.* 1102 (1968).
379. Davies, K. M., and J. E. Earley. *Inorg. Chem.* **15,** 1074 (1976).
380. Cannon, R. D. *J. Inorg. Nucl. Chem.* **38,** 1222 (1976).
381. Peloso, A., and A. Basato. *J. Chem. Soc. Dalton Trans.* 2040 (1972).
382. Peloso, A., and A. Basato. *Coord. Chem. Rev.* **8,** 111 (1972).
383. Doyle, J., and A. G. Sykes. *J. Chem. Soc. A.* 795 (1967).
384. Espenson, J. H., and T. R. Webb. *Inorg. Chem.* **11,** 1909 (1972).
385. Ulstrup, J. *Trans. Faraday Soc.* **67,** 2645 (1971).
386. Ulstrup, J. *Acta Chem. Scand* **25,** 3397 (1971).
387. Gillard, R. D., B. T. Heaton, and D. H. Vaughan. *J. Chem. Soc. A.* 1840 (1971).

388. Bott, H. L., E. J. Bounsall, and H. J. Poë. *J. Chem. Soc. A.* 1275 (1966).
389. Kallen, T. W., and J. E. Earley. *Inorg. Chem.* **10,** 1149 (1971).
390. Endicott, J. F., and H. Taube. *J. Am. Chem. Soc.* **84,** 4984 (1962).
391. James, A. D., R. S. Murray, and W. C. E. Higginson. *J. Chem. Soc. Dalton Trans.* 1273 (1974).
392. Spiecker, H., and K. Wieghardt. *Inorg. Chem.* **15,** 909 (1976).
393. Pethybridge, A. D., and J. E. Prue. *Prog. Inorg. Chem.* **17,** 327 (1972).
394. Brönsted, J. N. *J. Am. Chem. Soc.* **44,** 877 (1922).
395. Robinson, R. A., and R. H. Stokes. *Electrolyte Solutions,* 2nd ed., Butterworths, London, 1959, pp. 436–438.
396. White, J. D., and T. W. Newton. *J. Phys. Chem.* **75,** 2117 (1971).
397. Newton, T. W. *J. Phys. Chem.* **74,** 1655 (1970).
398. Fulton, R. B., and T. W. Newton. *J. Phys. Chem.* **74,** 1661 (1970).
399. Stalnaker, N. D., and A. C. Wahl. *Abstr. 167th National Meeting, Am. Chem. Soc.* Los Angeles, CA, March 1974, Paper No. **INORG. 98.**
400. Koren, R. *J. Inorg. Nucl. Chem.* **36,** 2341 (1974).
401. Sheppard, J. C., and A. C. Wahl. *J. Am. Chem. Soc.* **79,** 1020 (1957).
402. Gjertsen, L., and A. C. Wahl. *J. Am. Chem. Soc.* **81,** 1572 (1959).
403. Wahl, A. C. *Z. Elektrochem.* **60,** 90 (1960).
404. Shporer, M., G. Ron, A. Loewenstein, and G. Navon. *Inorg. Chem.* **4,** 361 (1965).
405. Campion, R. J., C. F. Deck, P. King, Jr., and A. C. Wahl. *Inorg. Chem.* **6,** 672 (1967).
406. Majid, Y. A., and K. E. Howlett. *J. Chem. Soc. A.* 679 (1968).
407. Bok, L. D. C., J. G. Leipoldt, and S. S. Basson. *J. Inorg. Nucl. Chem.* **37,** 2151 (1975).
408. Kolthoff, I. M., and W. J. Tomsieck. *J. Phys. Chem.* **39,** 945, 955 (1935).
409. Adam, F. C., and S. I. Weissman. *J. Am. Chem. Soc.* **80,** 1518 (1958).
410. McConnell, H. M., and C. H. Holm. *J. Chem. Phys.* **26,** 1517 (1957).
411. Thompson, R. C. *Inorg. Chem.* **15,** 1080 (1976).
412. Martin, A. H., and E. S. Gould. *Inorg. Chem.* **14,** 873 (1976).
413. Birk, J. P. *Inorg. Chem.* **14,** 1724 (1975).
414. Orhanović, M., and J. E. Earley. *Inorg. Chem.* **14,** 1478 (1975).
415. Thompson, G. A. K., and A. G. Sykes. *Inorg. Chem.* **15,** 638 (1976).
416. Prince, R. H., and M. G. Segal. *J. Chem. Soc. Dalton Trans.* 1245 (1975).
417. Candlin, J. P., and J. Halpern. *Inorg. Chem.* **4,** 1086 (1965).
418. Itzkowitz, M. M., and F. R. Nordmeyer. *Inorg. Chem.* **14,** 2124 (1975).
419. Diebler, H., P. Dodel, and H. Taube. *Inorg. Chem.* **5,** 1688 (1966).
420. Haim, A. *Inorg. Chem.* **7,** 1475 (1968).
421. Pearson, R. G. *J. Am. Chem. Soc.* **85,** 3533 (1963).
422. Cope, V. W., R. G. Miller, and R. T. M. Fraser. *J. Chem. Soc. A.* 301 (1967).
423. Espenson, J. H., and R. T. Wang. *Chem. Commun.* 207 (1970).
424. Parker, O. J., and J. H. Espenson. *J. Am. Chem. Soc.* **91,** 1968 (1969).
425. Anbar, M., and P. Neta. *Int. J. Appl. Radiat. Isot.* **18,** 493 (1967).
426. Meyerstein, D., and W. A. Mulac. *J. Phys. Chem.* **73,** 1091 (1969).
427. Newton, T. W., and S. W. Rabideau. *J. Phys. Chem.* **63,** 365 (1959).
428. Newton, T. W., and F. B. Baker. *Adv. Chem. Ser.* No. **67,** 268, 1967.
429. Watkins, K. O., J. C. Sullivan, and E. Deutsch. *Inorg. Chem.* **13,** 1712 (1974).

430. Ekstrom, A., A. B. McLaren, and L. E. Smythe. *Inorg. Chem.* **14,** 1035, 2899 (1975).
431. Ekstrom, A. *Inorg. Chem.* **13,** 2237 (1974).
432. Ekstrom, A., G. E. Batley, T. M. Florence, and Y. J. Farrar. *J. Inorg. Nucl. Chem.* **36,** 2355 (1974).
433. Faraggi, M., and A. Feder. *Inorg. Chem.* **12,** 236 (1973).
434. Christensen, R. J., J. H. Espenson, and A. B. Butcher. *Inorg. Chem.* **12,** 564 (1973).
435. Morrow, J. I., and S. Perlman. *Inorg. Chem.* **12,** 2453 (1973).
436. Sutter, J. H., K. Colquitt, and J. R. Sutter. *Inorg. Chem.* **13,** 1444 (1974).
437. Gurnee, E. F., and J. L. Magee. *J. Chem. Phys.* **26,** 1237 (1957).
438. Wolf, F. *Ann. Phys.* (*Leipzig*) **34** (1938).
439. Westheimer, F. H. *Mechanism of Enzyme Action.* W. D. McElroy and B. Glass, eds., Johns Hopkins Press, Baltimore, 1954, p. 321.
440. Westheimer, F. H. *Chem. Rev.* **45,** 419 (1949).
441. Basolo, F., P. H. Wilkes, R. G. Pearson, and R. G. Wilkins. *J. Inorg. Nucl. Chem.* **6,** 161 (1958).
442. Cox, L. T., S. B. Collins, and D. S. Martin, Jr. *J. Inorg. Nucl. Chem.* **17,** 383 (1961).
443. Basolo, F., M. L. Morris, and R. G. Pearson. *Discuss. Faraday Soc.* **29,** 80 (1960).
444. Peloso, A. *Coord. Chem. Rev.* **10,** 123 (1973).
445. Beattie, J. K., and F. Basolo. *Inorg. Chem.* **10,** 486 (1971).
446. Beattie, J. K., and J. Starink. *Inorg. Chem.* **14,** 996 (1975).
447. Prestwood, R. J., and A. C. Wahl. *J. Am. Chem. Soc.* **71,** 3137 (1949).
448. Harbottle, G., and R. W. Dodson. *J. Am. Chem. Soc.* **73,** 2442 (1951).
449. Dodson, R. W. *J. Am. Chem. Soc.* **75,** 1795 (1953).
450. Rossotti, F. J. C. *J. Inorg. Nucl. Chem.* **1,** 159 (1955).
451. Gilks, S. W., and G. M. Waind. *Discuss. Faraday Soc.* **29,** 102 (1960).
452. Roig, E., and R. W. Dodson. *J. Phys. Chem.* **65,** 2175 (1961).
453. Schwarz, H. A., D. Comstock, J. K. Yandell, and R. W. Dodson. *J. Phys. Chem.* **78,** 488 (1974).
454. Balcinella, B., P. D. Felgate, and G. S. Laurence. *J. Chem. Soc. Dalton Trans.* 1367 (1974); 1 (1975).
455. Ashhurst, K. G., and W. C. E. Higginson. *J. Chem. Soc.* 3044 (1953).
456. Stranks, D. R., and J. K. Yandell. *J. Phys. Chem.* **73,** 840 (1969).
457. Ondrus, M. G., and G. Gordon. *Inorg. Chem.* **11,** 985 (1972).
458. Drye, D. J., W. C. E. Higginson, and P. Knowles. *J. Chem. Soc.* 1137 (1962).
459. Swinehart, J. H. *Inorg. Chem.* **4,** 1069 (1965).
460. Baker, F. B., W. D. Brewer, and T. W. Newton. *Inorg. Chem.* **5,** 1294 (1966).
461. Sykes, A. G. *Chem. Commun.* 241 (1970).
462. Green, M., and A. G. Sykes. *J. Chem. Soc. A.* 3067 (1971).
463. Conocchioli, T. J., E. J. Hamilton, Jr., and N. Sutin. *J. Am. Chem. Soc.* **87,** 926 (1965).
464. Po, H. N., and N. Sutin. *Inorg. Chem.* **7,** 621 (1968).
465. Ardon, M., and R. A. Plane. *J. Am. Chem. Soc.* **81,** 3197 (1959).
466. Kolaczkowski, R. W., and R. A. Plane. *Inorg. Chem.* **3,** 322 (1964).
467. Sellers, R. M., and M. G. Simic. *Chem. Commun.* 401 (1975).

468. Davies, G., N. Sutin and K. O. Watkins. *J. Am. Chem. Soc.* **92,** 1892 (1970).
469. Samuni, A., D. Meisel, and G. Czapski. *J. Chem. Soc. Dalton Trans.* 1273 (1972).
470. Thompson, R. C., and G. Gordon. *Inorg. Chem.* **5,** 557, 562 (1966).
471. Linck, R. G. *Inorg. Chem.* **11,** 61 (1972).
472. Espenson, J. H., and R. T. Wang. *Inorg. Chem.* **11,** 955 (1972).
473. Hasan, F., and J. Rocek. *J. Am. Chem. Soc.* **96,** 6802 (1974).
474. Cornelius, R. D., and G. Gordon. *Abstr. 167th National Meeting, Am. Chem. Soc.* Los Angeles, CA, Paper No. **INORG 134,** 1974.
475. Daugherty, N. A. *J. Am. Chem. Soc.* **87,** 5026 (1965).
476. Davies, R., B. Kipling, and A. G. Sykes. *J. Am. Chem. Soc.* **95,** 7250 (1973).
477. Armstrong, A. M., and J. Halpern. *Can. J. Chem.* **35,** 1020 (1957).
478. Harkness, A. C., and J. Halpern. *J. Am. Chem. Soc.* **81,** 3526 (1959).
479. Bergh, A. A., and G. P. Haight Jr. *Inorg. Chem.* **1,** 688 (1962).
480. Rabideau, S. W., and B. J. Masters. *J. Phys. Chem.* **65,** 1256 (1961).
481. Mason, J. G., and A. D. Kowalak. *Inorg. Chem.* **3,** 1248 (1964).
482. Haight, G. P., Jr., E. Perchonock, F. Emmenegger, and G. Gordon. *J. Am. Chem. Soc.* **87,** 3835 (1965).
483. Hegedus, L. S., and A. Haim. *Inorg. Chem.* **6,** 664 (1967).
484. Kenney, J. C., and D. W. Carlyle. *Inorg. Chem.* **12,** 1952 (1973).
485. Gordon, G., and H. Taube. *Inorg. Chem.* **1,** 69 (1962).
486. Buchacek, R., and G. Gordon. *Inorg. Chem.* **11,** 2154 (1972).
487. Zabin, B. A., and H. Taube. *Inorg. Chem.* **3,** 963 (1964).
488. Espenson, J. H., and H. Taube. *Inorg. Chem.* **4,** 704 (1965).
489. Edwards, J. O. *Chem. Rev.* **50,** 455 (1952).
490. Edwards, J. O. *Inorganic Reaction Mechanisms.* Benjamin, New York, 1965, p. 137.
491. Edwards, J. O. *Coord. Chem. Rev.* **8,** 87 (1972).
492. Edwards, J. O. ed., *Peroxide Reaction Mechanisms.* Interscience, New York, 1962.
493. Uri, N. *Chem. Rev.* **50,** 375 (1952).
494. Woods, R., I. M. Kolthoff, and E. J. Meehan. *J. Am. Chem. Soc.* **85,** 2385 (1963).
495. Woods, R., I. M. Kolthoff, and E. J. Meehan. *J. Am. Chem. Soc.* **86,** 1698 (1964).
496. Anet, F. A. L., and E. Leblanc. *J. Am. Chem. Soc.* **79,** 2649 (1957).
497. Anet, F. A. L. *Can. J. Chem.* **37,** 58 (1959).
498. Kochi, J. K., and D. D. Davis. *J. Am. Chem. Soc.* **86,** 5264 (1964).
499. Hanson, J. R., and E. Premuzic. *Angew Chem. Int. Ed. Engl.* **7,** 247 (1968).
500. Kochi, J. K., and P. E. Mocadlo. *J. Am. Chem. Soc.* **88,** 4094 (1966).
501. Kochi, J. K., and J. W. Powers. *J. Am. Chem. Soc.* **92,** 137 (1970).
502. Schmidt, W., J. H. Swinehart, and H. Taube. *J. Am. Chem. Soc.* **93,** 1117 (1971).
503. Kochi, J. K. *J. Am. Chem. Soc.* **84,** 1193 (1962).
504. Schmidt, W., J. H. Swinehart, and H. Taube. *Inorg. Chem.* **7,** 1984 (1968).
505. Nohr, R. S., and L. O. Spreer. *J. Am. Chem. Soc.* **96,** 2618 (1974).
506. Nohr, R. S., and L. O. Spreer. *Inorg. Chem.* **13,** 1239 (1974).
507. Castro, C. E., and W. C. Kray, Jr. *J. Am. Chem. Soc.* **88,** 4447 (1966).

508. Sevcik, P. *J. Inorg. Nucl. Chem.* **38,** 553 (1976).
509. Asher, L. E., and E. Deutsch. *Inorg. Chem.* **14,** 2799 (1975).
510. Pennington, D. E., and A. Haim. *J. Am. Chem. Soc.* **90,** 3700 (1968).
511. Heckel, E., A. Henglein, and G. Beck. *Ber. Bunsenges. Phys. Chem.* **70,** 149 (1966).
512. Wells, C. F., and M. A. Salam. *Chem. Ind.* (*London*). 2079 (1967).
513. Griffith, W. P., and G. Wilkinson. *J. Chem. Soc.* 1629 (1959).
514. King, N. K., and M. E. Winfield. *J. Am. Chem. Soc.* **83,** 3366 (1961).
515. de Vries, B. *J. Catal.* **1,** 489 (1962).
516. Simandi, L., and F. Nagy. *Acta Chim. Acad. Sci. Hung.* **46,** 101 (1965).
517. Burnett, M. G., P. J. Conolly, and C. Kemball. *J. Chem. Soc. A.* 800 (1967).
518. Banks, R. G. S., and J. M. Pratt. *J. Chem. Soc. A.* 854 (1968).
519. Chock, P. B., R. B. K. Dewar, J. Halpern, and L. Y. Wong. *J. Am. Chem. Soc.* **91,** 82 (1969).
520. Halpern, J., and J. P. Maher. *J. Am. Chem. Soc.* **86,** 2311 (1964).
521. Halpern, J., and J. P. Maher, *J. Am. Chem. Soc.* **87,** 5361 (1965).
522. Kwiatek, J., and J. K. Seyler. *J. Organomet. Chem.* **3,** 421 (1965).
523. Kwiatek, J., and J. K. Seyler. *Adv. Chem. Ser.* **70,** 207 (1968).
524. Kwiatek, J. *Catal. Rev.* **1,** 37 (1967).
525. Chock, P. B., and J. Halpern. *J. Am. Chem. Soc.* **91,** 582 (1969).
526. Adamson, A. *J. Am. Chem. Soc.* **78,** 4260 (1956).
527. Castro, C. E., and W. C. Kray, Jr. *J. Am. Chem. Soc.* **85,** 2768 (1963).
528. Castro, C. E., and W. C. Kray, Jr. *J. Am. Chem. Soc.* **86,** 4603 (1964).
529. Schneider, P. W., P. F. Phelan, and J. Halpern. *J. Am. Chem. Soc.* **91,** 77 (1969).
530. Halpern, J., and P. F. Phelan. *J. Am. Chem. Soc.* **94,** 1881 (1972).
531. Adin, A., and J. H. Espenson. *Inorg. Chem.* **11,** 686 (1972).
532. Simandi, L. I., Z. Szeverenyi, and E. Budó-Zahonyi. *Inorg. Nucl. Chem. Lett.* **11,** 773 (1975).
533. Simadi, L. I., Z. Szeverenyi, and E. Budó-Zahonyi. *Inorg. Nucl. Chem. Lett.* **12,** 237 (1976).
534. Marzilli, L. G., P. A. Marzilli, and J. Halpern. *J. Am. Chem. Soc.* **92,** 5752 (1970).
535. Marzilli, L. G., P. A. Marzilli, and J. Halpern. *J. Am. Chem. Soc.* **93,** 1374 (1971).
536. Wade, R. S., and C. E. Castro. *J. Am. Chem. Soc.* **95,** 226, 231 (1972).
537. Weiss, M. C., and V. L. Goedken. *J. Am. Chem. Soc.* **98,** 3389 (1976).
538. Cooper, T. A. *J. Am. Chem. Soc.* **95,** 4158 (1973).
539. Malone, S. D., and J. F. Endicott. *J. Phys. Chem.* **76,** 2223 (1972).
540. Natarajan, P., and J. F. Endicott. *J. Phys. Chem.* **77,** 971 (1973).
541. Dodson, R. W., and H. A. Schwarz. *J. Phys. Chem.* **78,** 892 (1974).
542. Woodruff, W. H., and D. W. Margerum. *Inorg. Chem.* **12,** 962 (1973).
543. Thornton, A. T., and G. S. Laurence. *J. Chem. Soc. Dalton Trans.* 804 (1973).
544. Laurence, G. S., and A. T. Thornton. *J. Chem. Soc. Dalton Trans.* 1637 (1973).
545. Thornton, A. T., and G. S. Laurence. *J. Chem. Soc. Dalton Trans.* 1632 (1973).
546. Thornton, A. T., and G. S. Laurence. *J. Chem. Soc Dalton Trans.* 1142 (1974).

547. Rossiensky, D. R., and R. J. Hill. *J. Chem. Soc. Dalton Trans.* 715 (1972).
548. Davies, G., and K. O. Watkins. *J. Phys. Chem.* **74,** 3388 (1970).
549. Malik, M. N., J. Hill, and A. McAuley. *J. Chem. Soc. A.* 643 (1970).
550. Laurence, G. S., and K. J. Ellis. *J. Chem. Soc. Dalton Trans.* 2229 (1972).
551. Hershey, A. V., and W. C. Bray. *J. Am. Chem. Soc.* **58,** 1760 (1936).
552. Secco, F., S. Celsi, and C. Grati. *J. Chem. Soc. Dalton Trans.* 1675 (1972).
553. Celsi, S., F. Secco, and M. Venturini. *J. Chem. Soc. Dalton Trans.* 793 (1974).
554. Lawani, S. A., and J. R. Sutter. *J. Phys. Chem.* **77,** 1547 (1973).
555. Bodek, I., G. Davies, and J. H. Ferguson. *Inorg. Chem.* **14,** 1708 (1975).
556. Mareinik, R. J. *Inorg. Nucl. Chem. Lett.* **12,** 319 (1976).
557. Woodruff, W. H., and D. W. Margerum. *Inorg. Chem.* **13,** 2578 (1974).
558. Woodruff, W. H., B. A. Burke, and D. W. Margerum. *Inorg. Chem.* **13,** 2573 (1974).
559. Woodruff, W. H., D. W. Margerum, M. J. Milano, H. L. Pardue, and R. E. Santini. *Inorg. Chem.* **12,** 1490 (1973).
560. Malin, J. N., and J. H. Swinehart. *Inorg. Chem.* **8,** 1407 (1969).
561. Carter, P. R., and N. Davidson. *J. Phys. Chem.* **56,** 877 (1952).
562. Adegite, A., and M. H. Ford-Smith. *J. Chem. Soc. Dalton Trans.* 134, 138 (1973).
563. Dreyer, K., and G. Gordon. *Inorg. Chem.* **11,** 1174 (1972).
564. Adegite, A., and S. Edeogu. *J. Chem. Soc. Dalton Trans.* 1203 (1975).
565. Adegite, A. *J. Chem. Soc. Dalton Trans.* 1199 (1975).
566. Adegite, A., and M. H. Ford-Smith. *J. Chem. Soc. Dalton Trans.* 2113 (1972).
567. Norris, C., and F. Nordmeyer. *J. Am. Chem. Soc.* **93,** 4044 (1971).
568. Barber, J. R., Jr. and E. S. Gould. *J. Am. Chem. Soc.* **93,** 4045 (1971).
569. Cohen, H., and D. Meyerstein. *J. Chem. Soc. Dalton Trans.* 2477 (1975).
570. Olson, K. R., and M. Z. Hoffman. *Chem. Commun.* 938 (1975).
571. Davies, G., N. Sutin, and K. O. Watkins. *J. Am. Chem. Soc.* **92,** 1892 (1970).
572. Dockal, E. R., and E. S. Gould. *J. Am. Chem. Soc.* **94,** 6673 (1972).
573. Ulstrup, J. *J. Chem. Soc. Faraday Trans.* **71,** 435 (1975).

ADDENDUM REFERENCES

574. Wherland, S., and H. B. Gray. "Biological Aspects of Inorganic Chemistry," D. Dolphin, Ed., John Wiley and Sons, New York, in press.
575. Ballardini, R., G. Varani, F. Scandola and V. Balzani. *J. Am. Chem. Soc.* **98,** 7433 (1976).
576. Creutz, C., and N. Sutin. *J. Am. Chem. Soc.* **98,** 6385 (1976).
577. Lin, C-T., W. Bottcher, M. Chou, C. Creutz and N. Sutin. *J. Am. Chem. Soc.* **98,** 6536 (1976).
578. Lin, C-T., and N. Sutin. *J. Am. Chem. Soc.* **97,** 3543 (1975).
579. Creutz, C., and N. Sutin. *J. Am. Chem. Soc.* **99,** 241 (1977).
580. Rillema, D. P., and J. F. Endicott. *Inorg. Chem.* **15,** 1459 (1976).
581. Fischer, H., G. M. Tom and H. Taube. *J. Am. Chem. Soc.* **98,** 5512 (1976).
582. Piering, D. A., and J. A. Malin. *J. Am. Chem. Soc.* **98,** 6045 (1976).
583. Hurst, J. K. *J. Am. Chem. Soc.* **98,** 4001 (1976).
584. Kristine, F. J., D. R. Gard and R. E. Shepherd. *J. Chem. Soc., Chem. Commun.,* 994 (1976).

585. Newton, T. W., and F. B. Baker. *Inorg. Chem.* **3,** 569 (1964).
586. Powers, M. J., R. W. Callahan, D. J. Salmon and T. J. Meyer. *Inorg. Chem.* **15,** 1457 (1976).
587. Powers, M. J., D. J. Salmon, R. W. Callahan and T. J. Meyer. *J. Am. Chem. Soc.* **98,** 6731 (1976).
588. Krentzien, H., and H. Taube. *J. Am. Chem. Soc.* **98,** 6379 (1976).
589. Hery, M., and K. Wieghardt. *J. Chem. Soc., Dalton Trans.* **1976** 1536.
590. Hery, M., and K. Wieghardt. *Inorg. Chem.* **15,** 2315 (1976).
591. Kipling, B., K. Wieghardt, M. Hery, and A. G. Sykes. *J. Chem. Soc., Dalton Trans.,* 2176 (1976).
592. Martin, A. H., A. Liang, and E. S. Gould. *Inorg. Chem.* **15,** 1925 (1976).
593. Martin, A. H., and E. S. Gould. *Inorg. Chem.* **15,** 1934 (1976).
594. Gould, E. S., N. A. Johnson, and R. B. Morland. *Inorg. Chem.* **15,** 1929 (1976).
595. Balahura, R. J., and W. L. Purcell. *J. Am. Chem. Soc.* **98,** 4457 (1976).
596. Ogino, H., K. Tsukahara, and N. Tanaka. *Bull. Chem. Soc. Jpn.* **49,** 2743 (1976).
597. Busch, M. A., and D. E. Pennington. *Inorg. Chem.* **15,** 1940 (1976).
598. Peloso, A., *J. Chem. Soc., Dalton Trans.,* 984 (1976).
599. Welch, F. N., and D. E. Pennington. *Inorg. Chem.* **15,** 1515 (1976).
600. Ogino, H. M. Shimura, T. Watanabe, and N. Tanaka. *Inorg. Nucl. Chem. Lett.* **12,** 911 (1976).
601. Hyde, M. R., and J. H. Espenson. *J. Am. Chem. Soc.* **98,** 4463 (1976).
602. Verton, T. E., A. P. Zipp, and R. O. Ragsdale. *J. Chem. Soc., Dalton Trans.,* 2449 (1976).
603. Zipp, A. P., and R. O. Ragsdale. *J. Chem. Soc., Dalton Trans.,* 2452 (1976).
604. Espenson, J. H., and J. P. Leslie, II. *Inorg. Chem.* **15,** 1886 (1976).
605. Pelizzeti, E., and E. Mentasti. *J. Chem. Soc., Dalton Trans.,* 2222 (1976).

Author Index

AUTHOR INDEX

Subject Index

SUBJECT INDEX